Fractional Differentiation Inequalities

George A. Anastassiou

Fractional Differentiation
Inequalities

 Springer

George Anastassiou
University of Memphis
Department of Mathematical Sciences
Memphis, TN 38152
ganastss@memphis.edu

ISBN 978-1-4419-3106-1   e-ISBN 978-0-387-98128-4
DOI 10.1007/978-0-387-98128-4
Springer Dordrecht Heidelberg London New York

Mathematics Subject Classification (2000): 26A33, 26D10, 26D15

Cover illustration: Created by Sven Geier of the California Institute of Technology. The chaos image is
entitled "Sufficiently Advanced Technology." The fractal image is entitled "Unbalanced."

Printed on acid-free paper

Springer is part of Springer Science+Business Media (www.springer.com)

*To the unknown mathematician struggling for the truth,
and my wife Koula and daughters Angela and Peggy*

Contents

Preface

In 1991, although I had graduated in 1984, my great Ph.D. thesis advisor
J. H. B. Kemperman recommended that I read the very interesting paper of
J. Canavati [101], having to do with fractional derivatives and containing a
fractional Taylor formula. The hope was to use this in the study of positive
linear operators in approximation theory, so I could extend my Ph.D. thesis
results to arbitrary order derivatives. At that time I was not successful regarding that goal. So I put this important paper aside for a few years, but it
was always in my mind to use it somewhere else. This fortunately happened
in 1996 when I saw the proof of the basic Opial-type inequality (see Theorem 1.2). I was able to extend it, via the use of [101], to the fractional level.
That led to my paper [17], for which I received an award in 2001 from the
Academy of Athens, Greece, as the best article of the year for mathematical analysis, the "Pallas Award and Prize", an international competition
for Greeks. All the above encouraged to me to do more research in the area
of fractional inequalities (see [15–66]). So finally a natural consequence
was this research monograph, the first of its kind in the area of fractional
inequalities. Fractional differentiation inequalities are by themselves an important and great mathematical topic for research. Furthermore these have
many applications; the most important ones are in establishing the uniqueness of the solution in fractional differential equations and systems and in
fractional partial differential equations. Also they provide upper bounds to
the solution of the above equations. In this monograph we give many such
applications. Each chapter is self-contained and can be read independently
of the others and several graduate courses and seminars can be designed

based on this monograph. The list of fractional differentiation inequalities we study includes Opial, Poincaré, Sobolev, Hilbert, information theory, and Ostrowski. We give results for the above using three different types of fractional derivatives: Canavati, Riemann – Liouville, and Caputo. We examine the univariate and multivariate cases. The final preparation of this book took place in Memphis, USA, and in Athens, Greece, during 2008.

Fractional calculus has become very useful over the last forty years due to its many applications in almost all the applied sciences. We now see applications in acoustic wave propagation in inhomogeneous porous material, diffusive transport, fluid flow, dynamical processes in self-similar structures, dynamics of earthquakes, optics, geology, viscoelastic materials, biosciences, bioengineering, medicine, economics, probability and statistics, astrophysics, chemical engineering, physics, splines, tomography, fluid mechanics, electromagnetic waves, nonlinear control, signal processing, control of electronic power, converters, chaotic dynamics, polymer science, proteins, polymer physics, electrochemistry, statistical physics, rheology, thermodynamics, and neural networks, among others. By now almost all fields of research in science and engineering use fractional calculus to better describe them.

So as expected, this monograph, as part of fractional calculus, is useful for researchers and graduate students in research, seminars, and advanced graduate courses, in pure and applied mathematics, engineering, and all other applied sciences.

I would like to thank my family for their support and for their tolerance in accepting my continuous mathematical habit. Also I am greatly indebted and thankful to Raluca Pop and Razvan Mezei for the heroic and fantastic technical preparation of the manuscript in a very short time.

George A. Anastassiou
November 1, 2008
Memphis, TN, USA

1
Introduction

This monograph is about fractional differentiation inequalities. These inequalities have applications in fractional differential equations in establishing the uniqueness of the solution of initial value problems, giving upper bounds to their solutions, and these inequalities by themselves are of great interest and deserve to be studied thoroughly.

This monograph is the first of its kind, and is a natural outgrowth of the author's research work (see [13–66]) over the last fifteen years. In Chapters 2–11 and 14–16 we study Opial fractional inequalities, in Chapters 12 and 13 we study multivariate fractional Taylor formulae needed for our multivariate results.

In Chapter 17 we present Poincaré fractional inequalities, in Chapter 18 we present Sobolev fractional inequalities, and in Chapters 19–21 we study Hilbert–Pachpatte fractional inequalities. In Chapters 22–23 we derive results using fractional derivatives in the study of Csiszar's f-divergence in information theory and we finish with Chapters 24–26 where we present results for Ostrowski fractional inequalities.

The writing style in the monograph was chosen to help the reader. Each chapter can be read independently of any other chapter and each contains a complete story in itself. So all tools needed per chapter to describe the main results and applications can be found therein.

Our results are presented for all three main types of fractional derivatives: Riemann–Liouville, Caputo, and Canavati types for the univariate and multivariate cases.

At the end we have an appendix where we present conversion formulae from one type of fractional derivative to another, we justify the treatment

G.A. Anastassiou, *Fractional Differentiation Inequalities*,
DOI 10.1007/978-0-387-98128-4_1, © Springer Science+Business Media, LLC 2009

of fractional inequalities per type separately, and we give a table of basic functions' fractional derivatives.

The monograph concludes with a very rich list of 412 references.

At this point we would like to mention the basic inequalities involving ordinary derivatives. They are presented here at the fractional level and expanded in all possible directions.

Theorem 1.1. (1960, Z. Opial [315]) *Let $f \in C^1([0,h])$ such that*

$$f(0) = f(h) = 0, \ f(x) > 0 \text{ on } (0,h).$$

Then

$$\int_0^h |f(x) f'(x)| \, dx \le \frac{h}{4} \int_0^h f'^2(x) \, dx, \tag{1.1}$$

where $h/4$ is the best constant.

Opial's inequality (1.1) has an important modification which is used in many applications to differential equations (see D. Willett [406]). We emulate the next inequality at the fractional level, regarding our study of Opial fractional inequalities.

Theorem 1.2. ([4, p. 8]) *Let $x(t)$ be absolutely continuous in $[0,a]$, and $x(0) = 0$. Then*

$$\int_0^a |x(t) x'(t)| \, dt \le \frac{a}{2} \int_0^a (x'(t))^2 \, dt, \tag{1.2}$$

the last inequality holds as equality, iff $x(t) = ct$, where c is a constant.

We continue with the Poincaré inequality.

Theorem 1.3. [2] *Let $\Omega \subset \mathbb{R}^n$, $n \in \mathbb{N}$; it holds that*

$$\|u\|_{L^p(\Omega)} \le C \|\nabla u\|_{L^p(\Omega)}, \tag{1.3}$$

for functions u with vanishing mean value over Ω, $1 \le p \le \infty$, for general Ω, where $\|\nabla u\|_{L^p(\Omega)}$ is defined as the L^p-norm of the Euclidean norm of ∇u.

Next we give the Sobolev inequality.

Theorem 1.4. ([159, p. 263], the Gagliardo–Nirenberg–Sobolev inequality) *Let $1 \le p < n$. Then there exists a constant C, depending only on p and n, such that*

$$\|u\|_{L^{p*}(\mathbb{R}^n)} \le C \|Du\|_{L^p(\mathbb{R}^n)}, \tag{1.4}$$

for all $u \in C_c^1 (\mathbb{R}^n)$. Here $p^* := np/n - p$, $p^* > p$, and Du is the gradient of u.

We also mention Hilbert's inequality.

Theorem 1.5. ([191], Theorem 316) *If $p > 1$, $q = p/(p-1)$ and*

$$\int_0^\infty f^p(x) \, dx \le F, \quad \int_0^\infty g^q(y) \, dy \le G,$$

then

$$\int_0^\infty \int_0^\infty \frac{f(x) g(y)}{x + y} dx dy < \frac{\pi}{\sin(\pi/p)} F^{1/p} G^{1/q}, \tag{1.5}$$

where f, g are nonnegative measurable functions, unless $f \equiv 0$ or $g \equiv 0$. The constant $\pi \cos ec(\pi/p)$ is the best possible in (1.5).

See also the ordinary Hilbert–Pachpatte inequality, Theorem 19.2 here, and in B. Pachpatte [320, Theorem 1]. For the Csiszar's f-divergence description please see Preliminaries 22.1.

We finish with the famous Ostrowski inequality.

Theorem 1.6. [318] *Let $f \in C'([a,b])$, $x \in [a,b]$. Then*

$$\left| \frac{1}{b-a} \int_a^b f(t) \, dt - f(x) \right| \le \left(\frac{1}{4} + \frac{\left(x - \frac{a+b}{2} \right)^2}{(b-a)^2} \right) (b-a) \|f'\|_\infty. \tag{1.6}$$

Inequality (1.6) is sharp.

For the last also see Theorem 22.1, p. 498 of [19]; there we prove that the optimal function for the sharpness of (1.6) is

$$f^*(t) = |t - x|^\alpha (b - a), \quad \alpha > 1.$$

The author was the first to study Opial fractional derivative inequalities; in 1998, see [15, 17] and [18], which involve fractional derivatives of the Canavati type. For the last papers the author's motivation was Theorem 1.2. Many years later he discovered the following very important theorem, giving one more reason for writing this monograph.

Theorem 1.7. ([261, Theorem 2, 1985], Opial inequality for fractional integrals, by E. Love) *If $p > 0$, $q > 0$, $p + q = r \ge 1$, $0 \le a < b \le \infty$, $\gamma < r$, $\omega(x)$ is decreasing and positive in (a,b), $f(x)$ is measurable and* *nonnegative on* (a, b)*,* I^α *is the Riemann–Liouville operator of fractional integration defined by*

$$(I^\alpha f)(x) = \int_a^x \frac{(x-t)^{\alpha-1}}{\Gamma(\alpha)} f(t)\, dt \text{ for } \alpha > 0,$$

$$I^0 f(x) = f(x),$$

and $I^\beta f$ *is defined similarly for* $\beta \geq 0$*. Here* $\Gamma(\alpha) = \int_0^\infty e^{-t} t^{\alpha-1} dt$*, the gamma function.*

Then

$$\int_a^b [(I^\alpha f)(x)]^p [(I^\beta f)(x)]^q x^{\gamma-\alpha p - \beta q - 1} \omega(x)\, dx$$

$$\leq C \int_a^b [f(x)]^r x^{\gamma-1} \omega(x)\, dx, \tag{1.7}$$

where

$$C = \left(\frac{\Gamma(1-(\gamma/r))}{\Gamma(\alpha+1-(\gamma/r))} \right)^p \left(\frac{\Gamma(1-(\gamma/r))}{\Gamma(\beta+1-(\gamma/r))} \right)^q.$$

In a 1695 letter to L'Hospital, Leibniz asked, "Can we generalize ordinary derivatives to ones of arbitrary order?" L'Hospital replied to Leibniz with another question, "What if the order is 1/2?" Then Leibniz, in a letter dated September 30, 1695 replied, "It will lead to a paradox, from which one day many useful consequences will be drawn." That was the beginning of fractional calculus. The subject has been ongoing for more than 300 years now. Many famous mathematicians contributed to the topic over the years such as Liouville, Euler, Laplace, Lagrange, Riemann, Weyl, Fourier, Abel, Lacroix, Grunwald, and Letnikov. For more details on the history of the subject see [311] and [295].

The first attempt to give a rigorous definition of fractional derivatives and study the subject in depth was by Liouville during 1832–1855. The concept grew out of his earlier work in electromagnetism.

Fractional derivatives describe solutions of fractional integral equations, many times arising from physics, as done by Abel in 1823 to solve the brachistochrone problem.

Fractional calculus has a long history but, from the applicative point of view, it fell into oblivion for hundreds of years because of lack of applications to other sciences such as physics and engineering. This oblivion was also due to its complexity and lack of physical and geometric connection. Only in 2002 (see [335]) did Igor Podlubny show a serious and convincing geometric and physical interpretation of fractional integration and fractional differentiation. However, fractional calculus has emerged as very useful over the last forty years due to its many applications in almost all applied sciences. We now see applications in acoustic wave propagation in inhomogeneous porous material, diffusive transport, fluid flow,

dynamical processes in self-similar structures, dynamics of earthquakes, optics, geology, viscoelastic materials, biosciences, bioengineering, medicine,
economics, probability and statistics, astrophysics, chemical engineering,
physics, splines, tomography, fluid mechanics, electromagnetic waves, nonlinear control, signal processing, control of electronic power, converters,
chaotic dynamics, polymer science, proteins, polymer physics, electrochemistry, statistical physics, rheology, thermodynamics, neural networks, and
many others. By now almost all fields of research in science and engineering
use fractional calculus to better describe them (see also [197]).

The current mathematical theory of fractional calculus seems to lag be-
hind the needs for mathematical modeling of all the above applications.
Therefore there exists a lot of room for mathematical theoretical expansion on the subject.

Most of the above applications lead to the study of fractional differential
equations; an excellent source on the topic is [333].

The recent development of fractal theory has shown great connections
to fractional calculus. In [248], Y. Liang and W. Su have shown the linear
relationship between the Hausdorff dimension of the graph of a generalized
Weierstrass fractal function and the order of its fractional calculus. And
there exists a large set of literature on the last topic.

In general, fractional calculus currently provides the best descriptions for
fractals and chaotic situations.

We must not overlook the great fractional calculus encyclopedic monograph by S. Samko, A. Kilbas, and O. Marichev [366], and supports which
strongly all the above recent scientific activities; see also [134] by
K. Diethelm. Fractional derivatives might exist more easily than ordinary
ones (see [352]), and thus potentially be more attractive.

Fractional calculus has developed a lot in recent years and especially
since 1974 when the first international conference in the field took place,
organized by B. Ross, in New Haven, Connecticut, USA. Now we frequently
see such conferences.

Overall we observed for fractional calculus a very long but essential in-
fancy of about 250 years, a great rich adolescence of about 50 years, and
we foresee a magnificent mature future for many years to come providing
the best descriptions of the complexity of our modern science.

2

Opial-Type Inequalities for Functions and Their Ordinary and Canavati Fractional Derivatives

Several L_p-form Opial-type inequalities [315] are presented involving functions and their ordinary and generalized fractional derivatives. The above follow a generalization of Taylor's formula for generalized fractional derivatives. The chapter ends with the application of the derived inequalities in proving the uniqueness of solution/upper bound to the solution of some known very general fractional differential equations. This treatment is based on [18].

2.1 Preliminaries

In the sequel we follow [101]. Let $g \in C([0,1])$, $n := [\nu]$, $[\cdot]$ the integral part, $\nu > 0$, and $\alpha := \nu - n$ $(0 < \alpha < 1)$. Define

$$(J_\nu g)(x) := \frac{1}{\Gamma(\nu)} \int_0^x (x-t)^{\nu-1} g(t) dt, \quad 0 \le x \le 1, \qquad (2.1)$$

the *Riemann–Liouville integral*, where Γ is the gamma function; $\Gamma(\nu) = \int_0^\infty e^{-t} t^{\nu-1} dt$. We define the subspace $C^\nu([0,1])$ of $C^n([0,1])$:

$$C^\nu([0,1]) := \{g \in C^n([0,1]) \colon J_{1-\alpha} D^n g \in C^1([0,1])\},$$

where $D := d/dx$. $C^\nu([0,1])$ is a Banach space with norm

$$[[f]]_\nu = \max \left\{ \|f\|_\infty, \|f'\|_\infty, \ldots, \left\|f^{(n-1)}\right\|_\infty, \|D^\nu f\|_\infty \right\}, \quad \text{if } \nu \ge 1,$$

G.A. Anastassiou, *Fractional Differentiation Inequalities*,
DOI 10.1007/978-0-387-98128-4_2, © Springer Science+Business Media, LLC 2009

and

$$[[f]]_\nu = \|D^\nu f\|_\infty \,, \text{ if } 0 < \nu < 1,$$

for more see [101].

So let $g \in C^\nu([0,1])$; we define the Canavati ν-*fractional derivative* of g as

$$D^\nu g := D J_{1-\alpha} D^n g. \tag{2.2}$$

When $\nu \geq 1$ we have the Taylor's formula

$$g(t) = g(0) + g'(0)t + g''(0)\frac{t^2}{2!} + \cdots + g^{(n-1)}(0)\frac{t^{n-1}}{(n-1)!} + (J_\nu D^\nu g)(t), \; \forall t \in [0,1]. \tag{2.3}$$

When $0 < \nu < 1$ we get

$$g(t) = (J_\nu D^\nu g)(t), \quad \forall t \in [0,1]. \tag{2.4}$$

Next we carry the above notions over to arbitrary $[a,b] \subseteq \mathbb{R}$. Let $x, x_0 \in [a,b]$ such that $x \geq x_0$, x_0 is fixed. Let $f \in C([a,b])$ and define

$$(J_\nu^{x_0} f)(x) := \frac{1}{\Gamma(\nu)} \int_{x_0}^x (x-t)^{\nu-1} f(t) dt, \quad x_0 \leq x \leq b, \tag{2.5}$$

the *generalized Riemann–Liouville integral*. We define the subspace $C_{x_0}^\nu([a,b])$ of $C^n([a,b])$:

$$C_{x_0}^\nu([a,b]) := \{ f \in C^n([a,b]) : J_{1-\alpha}^{x_0} D^n f \in C^1([x_0,b]) \}.$$

So let $f \in C_{x_0}^\nu([a,b])$; we define the *generalized ν-fractional derivative of f* over $[x_0, b]$ as

$$D_{x_0}^\nu f := D J_{1-\alpha}^{x_0} f^{(n)} \quad (f^{(n)} := D^n f). \tag{2.6}$$

Notice that

$$(J_{1-\alpha}^{x_0} f^{(n)})(x) = \frac{1}{\Gamma(1-\alpha)} \int_{x_0}^x (x-t)^{-\alpha} f^{(n)}(t) dt$$

exists for $f \in C_{x_0}^\nu([a,b])$.

We present the following generalization of Taylor's formula, the fractional Taylor formula (see also [101]).

Theorem 2.1. Let $f \in C_{x_0}^\nu([a,b])$, $x_0 \in [a,b]$ fixed.
(i) If $\nu \geq 1$ then

$$f(x) = f(x_0) + f'(x_0)(x-x_0) + f''(x_0)\frac{(x-x_0)^2}{2} + \cdots + f^{(n-1)}(x_0)\frac{(x-x_0)^{n-1}}{(n-1)!}$$

$$+ (J_\nu^{x_0} D_{x_0}^\nu f)(x), \quad \text{all } x \in [a,b] : x \geq x_0. \tag{2.7}$$

(ii) *If* $0 < \nu < 1$ *we get*

$$f(x) = (J_\nu^{x_0} D_{x_0}^\nu f)(x), \quad \text{all } x \in [a, b] : x \geq x_0. \tag{2.8}$$

Proof. For $x = x_0$ (2.7) holds trivially. Otherwise assume that $x \neq x_0$ and define

$$w := \varphi(t) := x_0 + t(x - x_0), \quad 0 \leq t \leq 1.$$

Thus $\varphi(0) = x_0$ and $\varphi(1) = x$.

Let $g(t) := f(x_0 + t(x - x_0))$; that is, $g(0) = f(x_0)$ and $g(1) = f(x)$. Here we see that $(\nu \geq 1)$,

$$g^{(i)}(t) = f^{(i)}(x_0 + t(x - x_0))(x - x_0)^i, \quad i = 0, 1, \dots, n - 1.$$

In particular $g^{(i)}(0) = f^{(i)}(x_0)(x - x_0)^i$.

Furthermore $g \in C^\nu([0, 1])$. Consequently from (2.3) we have

$$
\begin{aligned}
f(x) &= g(1) = g(0) + g'(0) + \frac{g''(0)}{2} + \cdots + \frac{g^{(n-1)}(0)}{(n-1)!} + (J_\nu D^\nu g)(1) \\
&= f(x_0) + f'(x_0)(x - x_0) + \frac{f''(x_0)}{2}(x - x_0)^2 \\
&\quad + \cdots + \frac{f^{(n-1)}(x_0)}{(n-1)!}(x - x_0)^{n-1} + (J_\nu D^\nu g)(1), \quad \nu \geq 1.
\end{aligned}
$$

It remains to translate $(J_\nu D^\nu g)(1)$ for any $\nu > 0$.

Notice that

$$t = \frac{w - x_0}{x - x_0}; \quad t = 0 \text{ iff } w = x_0,$$

and

$$dt = \frac{dw}{x - x_0}.$$

Denote

$$z := x_0 + y(x - x_0) = \varphi(y), \quad y \in [0, 1],$$

thus

$$y = \frac{z - x_0}{x - x_0}, \quad \text{and} \quad y - t = \frac{z - w}{x - x_0};$$

in particular $t = y$ iff $w = z$.

We observe that

$$
\begin{aligned}
(J_{1-\alpha} g^{(n)})(y) &= \frac{1}{\Gamma(1-\alpha)} \int_0^y (y - t)^{(1-\alpha)-1} \cdot \\
&\qquad f^{(n)}(x_0 + t(x - x_0))(x - x_0)^n dt \\
&= \frac{(x - x_0)^n}{\Gamma(1-\alpha)} \int_0^y (y - t)^{(1-\alpha)-1} f^{(n)}(x_0 + t(x - x_0)) dt \\
&= (x - x_0)^n \frac{(J_{1-\alpha}^{x_0} f^{(n)})(z)}{(x - x_0)^{(1-\alpha)}}.
\end{aligned}
$$

That is,

$$(J_{1-\alpha}g^{(n)})(y) = (x-x_0)^{n-(1-\alpha)} \cdot (J_{1-\alpha}^{x_0}f^{(n)})(z)$$

or

$$(J_{1-\alpha}g^{(n)})(y) = (x-x_0)^{\nu-1}(J_{1-\alpha}^{x_0}f^{(n)})(x_0 + y(x-x_0)).$$

Then

$$\begin{aligned}(D^\nu(J_{1-\alpha}g^{(n)}))(y) &= (x-x_0)^\nu D(J_{1-\alpha}^{x_0}f^{(n)})(x_0 + y(x-x_0))\\ &= (x-x_0)^\nu (D_{x_0}^\nu f)(z).\end{aligned}$$

That is,

$$(D^\nu g)(y) = (x-x_0)^\nu (D_{x_0}^\nu f)(z). \tag{2.9}$$

In particular

$$(D^\nu g)(1) = (x-x_0)^\nu (D_{x_0}^\nu f)(x).$$

Next by using (2.9) we examine

$$\begin{aligned}(J_\nu D^\nu g)(1) &= \frac{1}{\Gamma(\nu)}\int_0^1 (1-t)^{\nu-1}(D^\nu g)(t)dt\\ &= \frac{(x-x_0)^\nu}{\Gamma(\nu)}\int_0^1 (1-t)^{\nu-1}(D_{x_0}^\nu f)(w)dt\\ &= \frac{(x-x_0)^\nu}{\Gamma(\nu)}\int_0^1 \frac{(x-w)^{\nu-1}}{(x-x_0)^{\nu-1}}(D_{x_0}^\nu f)(w)\frac{dw}{(x-x_0)}\\ &= \frac{1}{\Gamma(\nu)}\int_{x_0}^x (x-w)^{\nu-1}(D_{x_0}^\nu f)(w)dw\\ &= (J_\nu^{x_0}D_{x_0}^\nu f)(x).\end{aligned}$$

That is,

$$(J_\nu D^\nu g)(1) = (J_\nu^{x_0}D_{x_0}^\nu f)(x), \tag{2.10}$$

so that (2.7) becomes clear. Also (2.8) is obvious by (2.4) and (2.10). □

Note. (1) $(D_{x_0}^n f) = f^{(n)}, \quad n \in \mathbb{N}$.
(2) Let $f \in C_{x_0}^\nu([a,b])$, $\nu \geq 1$ and $f^{(i)}(x_0) = 0$, $i = 0,1,\ldots,n-1$, $n := [\nu]$. Then by (2.7),

$$f(x) = (J_\nu^{x_0}D_{x_0}^\nu f)(x).$$

That is,

$$f(x) = \frac{1}{\Gamma(\nu)}\int_{x_0}^x (x-t)^{\nu-1}(D_{x_0}^\nu f)(t)dt, \tag{2.11}$$

all $x \in [a,b]$: $x \geq x_0$.
Notice that (2.11) is true also when $0 < \nu < 1$.

2.2 Main Results

We give our first result on Opial-type inequalities for fractional derivatives.

Theorem 2.2. Let $f \in C_{x_0}^{\nu}([a,b])$, $\nu \geq 1$ and $f^{(i)}(x_0) = 0$, $i = 0, 1, \ldots$, $n-1$, $n := [\nu]$. Here $x, x_0 \in [a,b]$: $x \geq x_0$. Let $p, q > 1$ such that $1/p + 1/q = 1$. Then

$$\int_{x_0}^{x} |f(w)| |(D_{x_0}^{\nu} f)(w)| dw \leq \left(\frac{2^{-1/q}(x-x_0)^{(p\nu-p+2)/p}}{\Gamma(\nu)((p\nu-p+1)(p\nu-p+2))^{1/p}} \right)$$

$$\cdot \left(\int_{x_0}^{x} |(D_{x_0}^{\nu} f)(w)|^q dw \right)^{2/q}. \tag{2.12}$$

Proof. From (2.11) and Hölder's inequality we get ($x \geq x_0$)

$$\begin{aligned}
|f(x)| &\leq \frac{1}{\Gamma(\nu)} \int_{x_0}^{x} (x-t)^{\nu-1} |(D_{x_0}^{\nu} f)(t)| dt \\
&\leq \frac{1}{\Gamma(\nu)} \left(\int_{x_0}^{x} ((x-t)^{\nu-1})^p dt \right)^{1/p} \left(\int_{x_0}^{x} (|D_{x_0}^{\nu} f|(t))^q dt \right)^{1/q} \quad (2.13) \\
&= \frac{(x-x_0)^{(p\nu-p+1)/p}}{\Gamma(\nu)(p\nu-p+1)^{1/p}} \cdot \left(\int_{x_0}^{x} (|D_{x_0}^{\nu} f|(t))^q dt \right)^{1/q}.
\end{aligned}$$

Set

$$z(x) := \int_{x_0}^{x} (|D_{x_0}^{\nu} f|(t))^q dt, \quad (z(x_0) = 0).$$

Then

$$z'(x) = (|D_{x_0}^{\nu} f|(x))^q$$

and

$$|D_{x_0}^{\nu} f|(x) = (z'(x))^{1/q}, \quad \text{all } x_0 \leq x \leq b.$$

Therefore from (2.13) we have

$$|f(w)| |D_{x_0}^{\nu} f|(w) \leq \frac{(w-x_0)^{(p\nu-p+1)/p}}{\Gamma(\nu)(p\nu-p+1)^{1/p}}$$

$$\cdot \left\{ \left(\int_{x_0}^{w} |D_{x_0}^{\nu} f|(t))^q dt \right) \cdot z'(w) \right\}^{1/q}, \quad \text{all } x_0 \leq w \leq x. \tag{2.14}$$

Next we integrate (2.14) over $[x_0, x]$

$$\int_{x_0}^{x} |f(w)|\, |D_{x_0}^{\nu} f|(w) dw$$

$$\leq \frac{1}{\Gamma(\nu)(p\nu - p + 1)^{1/p}} \int_{x_0}^{x} (w - x_0)^{p\nu - p + 1/p} \cdot (z(w)z'(w))^{1/q} dw$$

$$\leq \frac{1}{\Gamma(\nu)(p\nu - p + 1)^{1/p}} \left(\int_{x_0}^{x} (w - x_0)^{p\nu - p + 1} dw \right)^{1/p}$$

$$\cdot \left(\int_{x_0}^{x} z(w)z'(w) dw \right)^{1/q}$$

$$= \frac{(x - x_0)^{(p\nu - p + 2)/p}}{\Gamma(\nu)(p\nu - p + 1)^{1/p}(p\nu - p + 2)^{1/p}} \cdot \frac{(z(x))^{2/q}}{2^{1/q}}$$

$$= \frac{2^{-1/q}(x - x_0)^{(p\nu - p + 2)/p}}{\Gamma(\nu)((p\nu - p + 1)(p\nu - p + 2))^{1/p}} \cdot \left(\int_{x_0}^{x} |D_{x_0}^{\nu} f(w)|^q dw \right)^{2/q},$$

Thus establishing (2.12). □

The counterpart of the previous result follows.

Theorem 2.3. *Let* $f \in C_{x_0}^{\nu}([a, b])$, $\nu > 0$, *and* $f^{(i)}(x_0) = 0$, $i = 0, 1, \ldots, n - 1$, $n := \lceil \nu \rceil$ *when* $\nu \geq 1$. *Here* $x, x_0 \in [a, b]$: $x > x_0$. *Assume that* $(D_{x_0}^{\nu} f) \neq 0$ *over* $[x_0, b]$ *and is of fixed sign. Let* $0 < p < 1$ *and* $q: 1/p + 1/q = 1$. *Then*

$$\int_{x_0}^{x} |f(w)|\, |(D_{x_0}^{\nu} f)(w)| dw \geq \frac{2^{-1/q}(x - x_0)^{(p\nu - p + 2)/p}}{\Gamma(\nu)((p\nu - p + 1)(p\nu - p + 2))^{1/p}}$$

$$\cdot \left(\int_{x_0}^{x} |(D_{x_0}^{\nu} f)(w)|^q dw \right)^{2/q}. \tag{2.15}$$

Proof. From (2.7), (2.8), and (2.11) we have

$$|f(w)| = \frac{1}{\Gamma(\nu)} \int_{x_0}^{w} (w - t)^{\nu - 1} |(D_{x_0}^{\nu} f)(t)| dt, \quad \text{all } x_0 \leq w \leq x.$$

Here $0 < p < 1$ and $q = p/p - 1 < 0$. By Hölder's inequality we get

$$|f(w)| \geq \frac{1}{\Gamma(\nu)} \left(\int_{x_0}^{w} (w - t)^{p(\nu - 1)} dt \right)^{1/p} \cdot \left(\int_{x_0}^{w} |(D_{x_0}^{\nu} f)(t)|^q dt \right)^{1/q}$$

$$= \frac{1}{\Gamma(\nu)} \frac{(w - x_0)^{p\nu - p + 1/p}}{(p\nu - p + 1)^{1/p}} \left(\int_{x_0}^{w} |(D_{x_0}^{\nu} f)(t)|^q dt \right)^{1/q}, \tag{2.16}$$

all $w > x_0$. Set

$$z(w) := \int_{x_0}^{w} (|D_{x_0}^{\nu} f|(t))^q dt, \quad (z(x_0) = 0),$$

so that $z'(w) = (|D_{x_0}^\nu f|(w))^q$ and

$$|(D_{x_0}^\nu f)(w)| = (z'(w))^{1/q}, \quad \text{all } x_0 \le w \le x.$$

Then by (2.16) we obtain

$$|f(w)| \, |(D_{x_0}^\nu f)(w)| \ge \frac{1}{\Gamma(\nu)} \frac{(w - x_0)^{(p\nu - p + 1)/p}}{(p\nu - p + 1)^{1/p}} \cdot (z(w)z'(w))^{1/q},$$

all $x_0 < w \le x$.

Let $x_0 < \theta \le w \le x$ and $\theta \downarrow x_0$; then by integration of the last inequality we find

$$\int_{x_0}^x |f(w)| \, |(D_{x_0}^\nu f)(w)| dw = \lim_{\theta \downarrow x_0} \int_\theta^x |f(w)| \, |(D_{x_0}^\nu f)(w)| dw$$

$$\ge \frac{1}{\Gamma(\nu)(p\nu - p + 1)^{1/p}} \cdot \lim_{\theta \downarrow x_0} \int_\theta^x (w - x_0)^{(p\nu - p + 1)/p} (z(w)z'(w))^{1/q} dw$$

$$(\text{here } z(w)z'(w) > 0)$$

$$\ge \frac{1}{\Gamma(\nu)(p\nu - p + 1)^{1/p}} \cdot \lim_{\theta \downarrow x_0} \left(\int_\theta^x (w - x_0)^{p\nu - p + 1} dw \right)^{1/p}$$

$$\cdot \lim_{\theta \downarrow x_0} \left(\int_\theta^x z(w)z'(w) dw \right)^{1/q}$$

$$= \frac{(x - x_0)^{p\nu - p + 2/p}}{\Gamma(\nu)(p\nu - p + 1)^{1/p}(p\nu - p + 2)^{1/p}} \cdot \lim_{\theta \downarrow x_0} \frac{(z^2(x) - z^2(\theta))^{1/q}}{2^{1/q}}$$

$$= \frac{2^{-1/q}(x - x_0)^{p\nu - p + 2/p}}{\Gamma(\nu)((p\nu - p + 1)(p\nu - p + 2))^{1/p}} (z(x))^{2/q}.$$

That establishes (2.15).   □

An extreme case follows.

Proposition 2.4. *Here* $p = 1$, $q = \infty$. *Let* $f \in C_{x_0}^\nu([a, b])$, $\nu > 0$, *and* $f^{(i)}(x_0) = 0$, $i = 0, 1, \ldots, n - 1$, $n := [\nu]$ *when* $\nu \ge 1$. *Here* $x, x_0 \in [a, b]$: $x \ge x_0$. *Then*

$$\int_{x_0}^x |f(w)| \, |D_{x_0}^\nu f|(w) dw \le \frac{(x - x_0)^{\nu + 1} \|D_{x_0}^\nu f\|_\infty^2}{\Gamma(\nu + 2)}. \tag{2.17}$$

Proof. From (2.7), (2.8), and (2.11) we have for all $x_0 \le w \le x$,

$$|f(w)| \le \frac{1}{\Gamma(\nu)} \int_{x_0}^w (w - t)^{\nu - 1} |(D_{x_0}^\nu f)(t)| dt$$

$$\le \frac{1}{\Gamma(\nu)} \left(\int_{x_0}^w (w - t)^{\nu - 1} dt \right) \|D_{x_0}^\nu f\|_{\infty, [x_0, w]}$$

$$= \frac{\|D_{x_0}^\nu f\|_{\infty, [x_0, w]}}{\Gamma(\nu)} \cdot \frac{(w - x_0)^\nu}{\nu}.$$

That is,

$$|f(w)| \le \frac{\|D_{x_0}^\nu f\|_{\infty,[x_0,w]}}{\Gamma(\nu+1)}(w-x_0)^\nu;$$

that is,

$$|f(w)| \le \frac{\|D_{x_0}^\nu f\|_\infty}{\Gamma(\nu+1)}(w-x_0)^\nu, \quad \text{all } x_0 \le w \le x.$$

Thus

$$|f(w)|\,|(D_{x_0}^\nu f)(w)| \le \frac{\|D_{x_0}^\nu f\|_\infty}{\Gamma(\nu+1)}(w-x_0)^\nu |(D_{x_0}^\nu f)(w)|,$$

and

$$
\begin{aligned}
\int_{x_0}^x |f(w)|\,|(D_{x_0}^\nu f)(w)|dw &\le \frac{\|D_\nu^{x_0} f\|_\infty}{\Gamma(\nu+1)} \int_{x_0}^x (w-x_0)^\nu |(D_{x_0}^\nu f)(w)|dw \\
&\le \frac{(\|D_{x_0}^\nu f\|_\infty)^2}{\Gamma(\nu+1)} \left(\int_{x_0}^x (w-x_0)^\nu dw \right).
\end{aligned}
$$

Therefore

$$\int_{x_0}^x |f(w)|\,|(D_{x_0}^\nu f)(w)|dw \le \frac{(x-x_0)^{\nu+1}}{\nu+1} \frac{(\|D_{x_0}^\nu f\|_\infty)^2}{\Gamma(\nu+1)}. \quad \square$$

Remark 2.5. Let $g \in C([x_0,b])$; then one can easily see that

$$(J_\nu^{x_0} g)^{(\ell)}(x) = (J_{\nu-\ell}^{x_0} g)(x), \quad \text{all } x \in [x_0,b],$$

$\ell \in \{1,\dots,n-1\}$, $n := [\nu]$, $\nu \ge 1$.
    For $f \in C_{x_0}^\nu([a,b])$, we therefore have

$$(J_\nu^{x_0} D_{x_0}^\nu f)^{(\ell)}(x) = (J_{\nu-\ell}^{x_0} D_{x_0}^\nu f)(x),$$

all $x \in [x_0,b]$, $\ell \in \{1,\dots,n-1\}$. Assuming that $f^{(i)}(x_0) = 0$, $i = 0,1,\dots,n-1$, by (2.7) we obtain

$$f(x) = (J_\nu^{x_0} D_{x_0}^\nu f)(x), \quad x \ge x_0,$$

and we get

$$f^{(\ell)}(x) = (J_{\nu-\ell}^{x_0} D_{x_0}^\nu f)(x), \quad x \ge x_0$$

and $\ell \in \{1,\dots,n-1\}$. That is,

$$f^{(\ell)}(x) = \frac{1}{\Gamma(\nu-\ell)} \int_{x_0}^x (x-t)^{\nu-\ell-1}(D_{x_0}^\nu f)(t)dt, \quad (2.18)$$

all $x \in [x_0,b]$, $\ell \in \{1,\dots,n-1\}$, $n := [\nu]$, $\nu \ge 1$.

We present

Theorem 2.6. *Let* $f \in C_{x_0}^{\nu}([a,b])$, $\nu \geq 1$ *and* $f^{(i)}(x_0) = 0$, $i = 0, 1, \ldots,$ $n-1$, $n := [\nu]$. *Here* $x, x_0 \in [a,b]$: $x \geq x_0$, *and* $\ell = 1, \ldots, n-1$. *Let* $p, q > 1$ *such that* $1/p + 1/q = 1$. *Then*

$$\int_{x_0}^{x} |f^{(\ell)}(w)|\,|(D_{x_0}^{\nu}f)(w)|dw \leq \frac{2^{-1/q}(x-x_0)^{(\nu p - \ell p - p + 2)/p}}{\Gamma(\nu-\ell) \cdot ((\nu p - \ell p - p + 1)(\nu p - \ell p - p + 2))^{1/p}}$$

$$\cdot \left(\int_{x_0}^{x} |(D_{x_0}^{\nu}f)(w)|^q dw \right)^{2/q}. \tag{2.19}$$

Proof. From (2.18) and Hölder's inequality we get ($x \geq x_0$)

$$|f^{(\ell)}(x)| \leq \frac{1}{\Gamma(\nu-\ell)} \int_{x_0}^{x} (x-t)^{\nu-\ell-1}|(D_{x_0}^{\nu}f)(t)|dt$$

$$\leq \frac{1}{\Gamma(\nu-\ell)} \left(\int_{x_0}^{x} ((x-t)^{\nu-\ell-1})^p dt \right)^{1/p} \cdot \left(\int_{x_0}^{x} |(D_{x_0}^{\nu}f)(t)|^q dt \right)^{1/q}$$

$$= \frac{(x-x_0)^{(\nu p - \ell p - p + 1)/p}}{\Gamma(\nu-\ell)(\nu p - \ell p - p + 1)^{1/p}} \cdot (z(x))^{1/q}, \tag{2.20}$$

where

$$z(x) := \int_{x_0}^{x} |(D_{x_0}^{\nu}f)(t)|^q dt, \quad (z(x_0) = 0).$$

Here

$$z'(x) = |(D_{x_0}^{\nu}f)(x)|^q,$$

and

$$|(D_{x_0}^{\nu}f)(x)| = (z'(x))^{1/q}, \quad x \in [x_0, b].$$

Hence from (2.20) we have

$$|f^{(\ell)}(w)|\,|(D_{x_0}^{\nu}f)(w)| \leq \frac{(w-x_0)^{(\nu p - \ell p - p + 1)/p}}{\Gamma(\nu-\ell)(\nu p - \ell p - p + 1)^{1/p}} \cdot (z(w)z'(w))^{1/q},$$

all $x_0 \leq w \leq x$.

Integrating the last inequality over $[x_0, x]$ we get

$$\int_{x_0}^{x} |f^{(\ell)}(w)|\,|(D_{x_0}^{\nu} f)(w)|\,dw$$

$$\leq \frac{1}{\Gamma(\nu - \ell)(\nu p - \ell p - p + 1)^{1/p}} \int_{x_0}^{x} (w - x_0)^{(\nu p - \ell p - p + 1)/p}$$

$$\cdot (z(w) z'(w))^{1/q}\,dw \leq \frac{1}{\Gamma(\nu - \ell)(\nu p - \ell p - p + 1)^{1/p}}$$

$$\left(\int_{x_0}^{x} (w - x_0)^{\nu p - \ell p - p + 1}\,dw \right)^{1/p} \left(\int_{x_0}^{x} z(w) z'(w)\,dw \right)^{1/q}$$

$$= \frac{(x - x_0)^{(\nu p - \ell p - p + 2)/p}}{\Gamma(\nu - \ell)(\nu p - \ell p - p + 1)^{1/p}(\nu p - \ell p - p + 2)^{1/p}} \cdot \frac{(z(x))^{2/q}}{2^{1/q}},$$

proving (2.19). □

Next we present the counterpart of the last result.

Theorem 2.7. Let $f \in C_{x_0}^{\nu}([a, b])$, $\nu \geq 1$ and $f^{(i)}(x_0) = 0$, $i = 0, 1, \ldots$, $n - 1$, $n := [\nu]$. Here $x, x_0 \in [a, b]$: $x > x_0$, and $\ell = 1, \ldots, n - 1$. Assume that $(D_{x_0}^{\nu} f) \neq 0$ over $[x_0, b]$ and of fixed sign. Let $0 < p < 1$ and $q: 1/p + 1/q = 1$. Then

$$\int_{x_0}^{x} |f^{(\ell)}(w)|\,|(D_{x_0}^{\nu} f)(w)|\,dw \geq \frac{2^{-1/q}(x - x_0)^{(\nu p - \ell p - p + 2)/p}}{\Gamma(\nu - \ell)((\nu p - \ell p - p + 1)(\nu p - \ell p - p + 2))^{1/p}}$$

$$\cdot \left(\int_{x_0}^{x} |D_{x_0}^{\nu} f(w)|^q\,dw \right)^{2/q}. \tag{2.21}$$

Proof. From (2.18) and the assumption of the theorem we have

$$|f^{(\ell)}(x)| = \frac{1}{\Gamma(\nu - \ell)} \int_{x_0}^{x} (x - t)^{\nu - \ell - 1} |D_{x_0}^{\nu} f(t)|\,dt.$$

Here $q = p/p - 1 < 0$. By Hölder's inequality $(0 < p < 1)$ we obtain

$$|f^{(\ell)}(x)| \geq \frac{1}{\Gamma(\nu - \ell)} \left(\int_{x_0}^{x} (x - t)^{\nu p - \ell p - p}\,dt \right)^{1/p} \cdot \left(\int_{x_0}^{x} (|D_{x_0}^{\nu} f|(t))^q\,dt \right)^{1/q}$$

$$= \frac{(x - x_0)^{(\nu p - \ell p - p + 1)/p}}{\Gamma(\nu - \ell)(\nu p - \ell p - p + 1)^{1/p}} \cdot (z(x))^{1/q}, \quad x > x_0.$$

Call

$$z(x) := \int_{x_0}^{x} (|D_{x_0}^{\nu}f|(t))^q dt, \quad z(x_0) = 0;$$

then

$$z'(x) = |(D_{x_0}^{\nu}f)(x)|^q$$

with

$$|(D_{x_0}^{\nu}f)(x)| = (z'(x))^{1/q}, \quad \text{all } x \geq x_0.$$

Therefore

$$|f^{(\ell)}(w)|\,|(D_{x_0}^{\nu}f)(w)| \geq \frac{(w-x_0)^{(\nu p - \ell p - p + 1)/p}}{\Gamma(\nu-\ell)(\nu p - \ell p - p + 1)^{1/p}} \cdot (z(w)z'(w))^{1/q},$$

all $x_0 < w \leq x$.

Consequently

$$\int_{x_0}^{x} |f^{(\ell)}(w)|\,|(D_{x_0}^{\nu}f)(w)|dw = \lim_{\theta \downarrow x_0} \int_{\theta}^{x} |f^{(\ell)}(w)|\,|(D_{x_0}^{\nu}f)(w)|dw$$

$$\geq \frac{1}{\Gamma(\nu-\ell)(\nu p - \ell p - p + 1)^{1/p}} \cdot \lim_{\theta \downarrow x_0} \left(\int_{\theta}^{x} (w-x_0)^{\nu p - \ell p - p + 1} dw \right)^{1/p}$$

$$\cdot \lim_{\theta \downarrow x_0} \left(\int_{\theta}^{x} z(w)z'(w)dw \right)^{1/q}$$

$$= \frac{(x-x_0)^{(\nu p - \ell p - p + 2)/p}}{\Gamma(\nu-\ell)(\nu p - \ell p - p + 1)^{1/p}(\nu p - \ell p - p + 2)^{1/p}} \cdot \frac{(z(x))^{2/q}}{2^{1/q}}.$$

That proves (2.21). □

An extreme case follows.

Proposition 2.8. *Here $p = 1$, $q = \infty$. Let $f \in C_{x_0}^{\nu}([a,b])$ and $f^{(i)}(x_0) = 0$, $i = 0, 1, \ldots, n-1$, $n := [\nu]$, $\nu \geq 1$. Here $x, x_0 \in [a,b]$: $x \geq x_0$ and $\ell = 1, \ldots, n-1$. Then*

$$\int_{x_0}^{x} |f^{(\ell)}(w)|\,|(D_{x_0}^{\nu}f)(w)|dw \leq \frac{(x-x_0)^{\nu-\ell+1}}{\Gamma(\nu-\ell+2)}(\|D_{x_0}^{\nu}f\|_{\infty})^2. \qquad (2.22)$$

Proof. From (2.18) we get

$$|f^{(\ell)}(w)| \leq \frac{1}{\Gamma(\nu-\ell)} \left(\int_{x_0}^{w} (w-t)^{\nu-\ell-1} dt \right) \|D_{x_0}^{\nu}f\|_{\infty}$$

$$= \frac{1}{\Gamma(\nu-\ell)} \frac{(w-x_0)^{\nu-\ell}}{\nu-\ell} \|D_{x_0}^{\nu}f\|_{\infty}.$$

That is,

$$|f^{(\ell)}(w)| \leq \frac{(w-x_0)^{\nu-\ell}}{\Gamma(\nu-\ell+1)} \|D_{x_0}^{\nu} f\|_{\infty},$$

all $x_0 \leq w \leq x$. Furthermore we see that

$$|f^{(\ell)}(w)|\,|(D_{x_0}^{\nu} f)(w)| \leq \frac{(w-x_0)^{\nu-\ell}}{\Gamma(\nu-\ell+1)} (\|D_{x_0}^{\nu} f\|_{\infty})^2.$$

Finally it holds

$$
\begin{aligned}
\int_{x_0}^{x} |f^{(\ell)}(w)|\,|(D_{x_0}^{\nu} f)(w)| dw \;&\leq\; \left(\int_{x_0}^{x} (w-x_0)^{\nu-\ell} dw \right) \frac{(\|D_{x_0}^{\nu} f\|_{\infty})^2}{\Gamma(\nu-\ell+1)} \\
&=\; \frac{(x-x_0)^{\nu-\ell+1}}{(\nu-\ell+1)} \frac{(\|D_{x_0}^{\nu} f\|_{\infty})^2}{\Gamma(\nu-\ell+1)},
\end{aligned}
$$

proving (2.22). $\square$

2.3 Applications

(i) *Uniqueness of solution* to (see also [3])

$$
\begin{cases}
(D_a^{\nu} f)(t) = F(t, f, f', \ldots, f^{(n-1)}): \\
f^{(i)}(a) = a_i,\ 0 \leq i \leq n-1,\ n := [\nu],\ \nu \geq 1.
\end{cases}
\tag{2.23}
$$

Here $f \in C_a^{\nu}([a,b])$; F is a continuous function on $[a,b] \times \mathbb{R}^n$, $t \in [a,b]$. Also F fulfills the Lipschitz condition:

$$|F(t, z_0, z_1, \ldots, z_{n-1}) - F(t, z_0', z_1', \ldots, z_{n-1}')| \leq \sum_{i=0}^{n-1} q_i(t)|z_i - z_i'|,$$

where $q_i(t) \geq 0$, $0 \leq i \leq n-1$ are continuous functions on $[a,b]$.
    Call

$$\phi(b) := \sum_{i=0}^{n-1} \|q_i\|_{\infty} \frac{(b-a)^{\nu-i}}{2\Gamma(\nu-i)\sqrt{(\nu-i)(2\nu-2i-1)}}.$$

Assume that

$$\phi(b) < 1. \tag{2.24}$$

Next we prove uniqueness to the solution of (2.23).
    Let $f_1, f_2 \in C_a^{\nu}([a,b])$ fulfilling (2.23); that is,

$$(D_a^{\nu} f_j)(t) = F(t, f_j, f_j', \ldots, f_j^{(n-1)}): f_j^{(i)}(a) = a_i,\ 0 \leq i \leq n-1;\ j = 1, 2.$$

Call $y := f_1 - f_2$. Then

$$\begin{cases} (D_a^\nu g)(t) = F(t, f_1, f_1', \ldots, f_1^{(n-1)}) - F(t, f_2, f_2', \ldots, f_2^{(n-1)}), \\ \text{such that } g^{(i)}(a) = 0, \ 0 \le i \le n-1. \end{cases} \quad (2.25)$$

Here

$$|F(t, f_1, f_1', \ldots, f_1^{(n-1)}) - F(t, f_2, f_2', \ldots, f_2^{(n-1)})|$$
$$\le \sum_{i=0}^{n-1} q_i(t)|f_1^{(i)}(t) - f_2^{(i)}(t)| = \sum_{i=0}^{n-1} q_i(t)|g^{(i)}(t)|.$$

From

$$(D_a^\nu g(t))^2 = (D_a^\nu g(t))\{F(t, f_1, f_1', \ldots, f_1^{(n-1)}) - F(t, f_2, f_2', \ldots, f_2^{(n-1)})\},$$

we obtain

$$(D_a^\nu g(t))^2 \le |D_a^\nu g(t)| \left(\sum_{i=0}^{n-1} q_i(t)|g^{(i)}(t)| \right)$$
$$= \sum_{i=0}^{n-1} q_i(t)(|D_a^\nu g(t)|\, |g^{(i)}(t)|).$$

Integrating the last inequality we get

$$\int_a^b (D_a^\nu g(t))^2 dt \le \sum_{i=0}^{n-1} \int_a^b q_i(t)|g^{(i)}(t)|\, |D_a^\nu g(t)| dt$$
$$\le \sum_{i=0}^{n-1} \|q_i\|_\infty \int_a^b |g^{(i)}(t)|\, |D_a^\nu g(t)| dt. \quad (2.26)$$

From inequalities (2.12), (2.19), when $p = q = 2$ and $x_0 = a$, we find that

$$\int_a^b |g^{(i)}(t)|\, |D_a^\nu g(t)| dt$$
$$\le \frac{(b-a)^{\nu-i}}{2\Gamma(\nu-i)\sqrt{(2\nu-2i-1)(\nu-i)}} \int_a^b ((D_a^\nu g)(t))^2 dt, \quad (2.27)$$

for $i = 0, 1, \ldots, n-1$.

Combining (2.26) and (2.27) we derive

$$\int_a^b ((D_a^\nu g)(t))^2 dt \le \phi(b) \int_a^b ((D_a^\nu g(t))^2 dt. \quad (2.28)$$

If $\int_a^b (D_a^\nu g(t))^2 dt \ne 0$, then from (2.28) we get $\phi(b) \ge 1$, a contradiction by (2.24).

Thus
$$\int_a^b (D_a^\nu g(t))^2 dt = 0,$$
and
$$(D_a^\nu g(t))^2 = 0, \quad \text{a.e. in } t \in [a,b];$$
that is,
$$D_a^\nu g(t) = 0, \quad \text{a.e. in } t \in [a,b].$$
Furthermore $(J_\nu^a D_a^\nu g)(t) \equiv 0$, over $[a,b]$. But $g^{(i)}(a) = 0$, $0 \le i \le n-1$ by (2.25).

Finally from (2.7), Theorem 2.1 (fractional Taylor's formula) we find that $g(t) \equiv 0$ on $[a,b]$. That is, $f_1 = f_2$ on $[a,b]$, proving the uniqueness to the solution of initial value problem (2.23).

(ii) *Upper bounds on $D_a^\nu f$, solution f, and so on* (see also [3]).

Let the initial value problem
$$\begin{cases} (D_a^\nu f)'(t) = F(t, f, f', f'', \ldots, f^{(n-1)}, (D_a^\nu f)), \ a \le t \le b \\ a \le t \le b \\ f^{(i)}(a) = 0, \ 0 \le i \le n-1, \ n := [\nu], \ \nu \ge 1 \\ \text{and } D_a^\nu f(a) = A \in \mathbb{R}. \end{cases} \tag{2.29}$$

Here F is continuous on $[a,b] \times \mathbb{R}^{n+1}$ and
$$|F(t, x_0, x_1, \ldots, x_n)| \le \sum_{i=0}^{n-1} q_i(t)|x_i|,$$

$q_i(t) \ge 0$, $0 \le i \le n-1$ continuous on $[a,b]$, also $f \in C_a^\nu([a,b])$.

See that
$$(D_a^\nu f)(t)(D_a^\nu f)'(t) = (D_a^\nu f)(t)F(t, f, f', \ldots, f^{(n-1)}, (D_a^\nu f)),$$
and $(a \le x \le b)$
$$\int_a^x (D_a^\nu f)(t)(D_a^\nu f)'(t)dt$$
$$= \int_a^x (D_a^\nu f)(t)F(t, f, f', \ldots, f^{(n-1)}, (D_a^\nu f))dt$$
$$\le \int_a^x |(D_a^\nu f)(t)| \, |F(t, f, f', \ldots, f^{(n-1)}, (D_a^\nu f))|dt$$
$$\le \int_a^x |(D_a^\nu f)(t)| \left(\sum_{i=0}^{n-1} q_i(t)|f^{(i)}(t)| \right) dt$$
$$= \int_a^x \sum_{i=0}^{n-1} q_i(t)|f^{(i)}(t)| \, |(D_a^\nu f)(t)|dt$$
$$\le \sum_{i=0}^{n-1} \|q_i\|_\infty \int_a^x |f^{(i)}(t)| \, |(D_a^\nu f)(t)|dt.$$

That is,

$$\left.\frac{((D_a^\nu f)(t))^2}{2}\right|_a^x \le \sum_{i=0}^{n-1} \|q_i\|_\infty \cdot \int_a^x |f^{(i)}(t)| \, |(D_a^\nu f)(t)| dt.$$

That is,

$$((D_a^\nu f)(x))^2 \le A^2 + 2\sum_{i=0}^{n-1} \|q_i\|_\infty \int_a^x |f^{(i)}(t)| \, |(D_a^\nu f)(t)| dt. \qquad (2.30)$$

From inequalities (2.12), (2.19), when $p = q = 2$ and $x_0 = a$, we get that

$$\int_a^x |f^{(i)}(t)| \, |(D_a^\nu f)(t)| dt \le \frac{(x-a)^{\nu-i}}{2\Gamma(\nu-i)\sqrt{(\nu-i)(2\nu-2i-1)}}$$

$$\cdot \int_a^x ((D_a^\nu f)(t))^2 dt, \qquad (2.31)$$

for all $i = 0, 1, \ldots, n-1$. Call

$$\theta(x) := ((D_a^\nu f)(x))^2, \qquad (2.32)$$

$$\rho := A^2, \qquad (2.33)$$

and

$$Q(x) := \left(\sum_{i=0}^{n-1} \|q_i\|_\infty \frac{(x-a)^{\nu-i}}{\Gamma(\nu-i)\sqrt{(\nu-i)(2\nu-2i-1)}}\right), \qquad (2.34)$$

all $a \le x \le b$.

From (2.30) and using (2.31) through (2.34) we obtain

$$\theta(x) < \rho + Q(x)\int_a^x \theta(t)dt, \qquad (2.35)$$

all $a \le x \le b$. Here ρ, $Q(x)$, $\theta(x) \ge 0$ and $Q(a) = 0$.

Set

$$w(x) := \int_a^x \theta(t)dt, \quad w(a) = 0; \qquad (2.36)$$

that is, $w'(x) = \theta(x)$.

Therefore (2.35) becomes $w'(x) \le \rho + Q(x)w(x)$, all $a \le x \le b$; that is,

$$w'(t) \le \rho + Q(t)w(t), \quad \text{all } a \le t \le x. \qquad (2.37)$$

Next we see that (see also [76])

$$w'(t)e^{-\int_a^t Q(s)ds} \le \rho e^{-\int_a^t Q(s)ds} + e^{-\int_a^t Q(s)ds}Q(t)w(t),$$

and

$$\left(w(t)e^{-\int_a^t Q(s)ds}\right)' \le \rho e^{-\int_a^t Q(s)ds}, \qquad (2.38)$$

all $a \leq t \leq x$.

Integrating (2.38) over $[a, x]$ we find

$$w(x)e^{-\int_a^x Q(s)ds} \leq \rho \int_a^x \left(e^{-\int_a^t Q(s)ds}\right) dt,$$

and

$$w(x) \leq \rho \left(e^{\int_a^x Q(s)ds}\right) \left(\int_a^x \left(e^{-\int_a^t Q(s)ds}\right) dt\right), \qquad (2.39)$$

all $a \leq x \leq b$.

Using (2.39) into (2.35) we obtain

$$\theta(x) \leq \rho + Q(x)\rho \left(e^{\int_a^x Q(s)ds}\right) \left(\int_a^x \left(e^{-\int_a^t Q(s)ds}\right) dt\right).$$

So we have proved that

$$|(D_a^\nu f)(x)| \leq |A| \left\{ 1 + Q(x) \left(e^{\int_a^x Q(s)ds}\right) \right.$$

$$\left. \cdot \left(\int_a^x \left(e^{-\int_a^t Q(s)ds}\right) dt\right) \right\}^{1/2} =: \gamma(x), \qquad (2.40)$$

all $a \leq x \leq b$.

Given $f^{(i)}(a) = 0$, $i = 0, 1, \ldots, n-1$; $\nu \geq 1$ from (2.11) and (2.18) we have that

$$f^{(i)}(x) = \frac{1}{\Gamma(\nu - i)} \int_a^x (x - t)^{\nu - i - 1}((D_a^\nu f)(t))dt. \qquad (2.41)$$

Using (2.40) into (2.41) we derive the *upper bounds*

$$|f^{(i)}(x)| \leq \frac{1}{\Gamma(\nu - i)} \int_a^x (x - t)^{\nu - i - 1}\gamma(t)dt, \qquad (2.42)$$

all $i = 0, 1, \ldots, n-1$, for all $a \leq x \leq b$.

3

Canavati Fractional Opial-Type Inequalities and Fractional Differential Equations

A series of very general Opial-type inequalities [315] is presented involving fractional derivatives of different orders. These are based on Taylor's formula for fractional derivatives. The results are applied in proving uniqueness to the solutions of very general fractional initial value problems of fractional ordinary differential equations. This treatment is based on [61].

3.1 Introduction

The Canavati fractional Opial inequalities that we deal with have the general form

$$
\int_a^b |D^{\nu_1} f(w)|^\alpha |D^{\nu_2} f(w)|^\beta g(w)\, dw
$$

$$
\leq K \left(\int_a^b |D^{\nu_1} f(w)|^\delta |D^{\nu_2} f(w)|^\varepsilon p(w)\, dw \right)^\zeta
$$

for certain continuous functions f on $[a, b]$. Here q, p are weight functions and $D^\gamma f$ denotes the Canavati fractional derivative of f of order γ. These results are presented in Section 3.3, after Section 3.2 (Preliminaries). In Section 3.4, these results are applied to obtain uniqueness criteria for initial value problems for fractional differential equations of the form

$$
D^\nu f + \sum_{j=1}^k q_j D^{\gamma_j} f + q_{k+1} f = h.
$$

G.A. Anastassiou, *Fractional Differentiation Inequalities*,
DOI 10.1007/978-0-387-98128-4_3, © Springer Science+Business Media, LLC 2009

Other sorts of fractional differential equations are studied in the literature. Section 3.5 contains remarks on this and attempts to put the results of Section 3.4 in some perspective.

3.2   Preliminaries

In the sequel we follow [101]. Let $g \in C([0,1])$. Let ν be a positive number, $n := [\nu]$, and $\alpha := \nu - n$ $(0 < \alpha < 1)$. Define

$$(J_\nu g)(x) := \frac{1}{\Gamma(\nu)} \int_0^x (x-t)^{\nu-1} g(t)\, dt, \quad 0 \le x \le 1, \qquad (3.1)$$

the *Riemann–Liouville integral*, where Γ is the gamma function. We define the subspace $C^\nu([0,1])$ of $C^n([0,1])$ as follows.

$$C^\nu([0,1]) := \{ g \in C^n([0,1]) : J_{1-\alpha} D^n g \in C^1([0,1]) \},$$

where $D := d/dx$. So for $g \in C^\nu([0,1])$, we define the Canavati ν-*fractional derivative* of g as

$$D^\nu g := D J_{1-\alpha} D^n g. \qquad (3.2)$$

When $\nu \ge 1$ we have the fractional Taylor's formula

$$
\begin{aligned}
g(t) \;=\; & g(0) + g'(0)t + g''(0)\frac{t^2}{2!} + \cdots + g^{(n-1)}(0)\frac{t^{n-1}}{(n-1)!} \\
& + (J_\nu D^\nu g)(t), \quad \text{for all } t \in [0,1].
\end{aligned}
\qquad (3.3)
$$

When $0 < \nu < 1$ we find

$$g(t) = (J_\nu D^\nu g)(t), \quad \text{for all } t \in [0,1]. \qquad (3.4)$$

Next we carry the above notions over to arbitrary $[a,b] \subseteq \mathbb{R}$ (see [17]). Let $x, x_0 \in [a,b]$ such that $x \ge x_0$, where x_0 is fixed. Let $f \in C([a,b])$ and define

$$(J_\nu^{x_0} f)(x) := \frac{1}{\Gamma(\nu)} \int_{x_0}^x (x-t)^{\nu-1} f(t)\, dt, \quad x_0 \le x \le b, \qquad (3.5)$$

the *generalized Riemann–Liouville integral*. We define the subspace $C_{x_0}^\nu([a,b])$ of $C^n([a,b])$:

$$C_{x_0}^\nu([a,b]) := \{ f \in C^n([a,b]) : J_{1-\alpha}^{x_0} D^n f \in C^1([x_0, b]) \}.$$

For $f \in C_{x_0}^\nu([a,b])$, we define the *generalized ν-fractional derivative of f* over $[x_0, b]$ as

$$D_{x_0}^\nu f := D J_{1-\alpha}^{x_0} f^{(n)} \quad (f^{(n)} := D^n f). \qquad (3.6)$$

Notice that

$$(J_{1-\alpha}^{x_0} f^{(n)})(x) = \frac{1}{\Gamma(1-\alpha)} \int_{x_0}^{x} (x-t)^{-\alpha} f^{(n)}(t)\, dt$$

exists for $f \in C_{x_0}^{\nu}([a,b])$.

We recall the following fractional generalization of Taylor's formula (see [17, 101]).

Theorem 3.1. Let $f \in C_{x_0}^{\nu}([a,b])$, $x_0 \in [a,b]$ fixed.
(i) If $\nu \geq 1$ then

$$f(x) = f(x_0) + f'(x_0)(x-x_0) + f''(x_0)\frac{(x-x_0)^2}{2} + \cdots +$$

$$f^{(n-1)}(x_0)\frac{(x-x_0)^{n-1}}{(n-1)!} + (J_{\nu}^{x_0} D_{x_0}^{\nu} f)(x), \quad \text{for all } x \in [a,b] : x \geq x_0. \quad (3.7)$$

(ii) If $0 < \nu < 1$ then

$$f(x) = (J_{\nu}^{x_0} D_{x_0}^{\nu} f)(x), \quad \text{for all } x \in [a,b] : x \geq x_0. \quad (3.8)$$

We make

Remark 3.2.
(1) $(D_{x_0}^{n} f) = f^{(n)}$; $n \in \mathbb{N}$.
(2) Let $f \in C_{x_0}^{\nu}([a,b])$, $\nu \geq 1$ and $f^{(i)}(x_0) = 0$, $i = 0, 1, \ldots, n-1$, $n := [\nu]$. Then by (3.7)

$$f(x) = (J_{\nu}^{x_0} D_{x_0}^{\nu} f)(x).$$

That is,

$$f(x) = \frac{1}{\Gamma(\nu)} \int_{x_0}^{x} (x-t)^{\nu-1} (D_{x_0}^{\nu} f)(t)\, dt, \quad (3.9)$$

for all $x \in [a,b]$ with $x \geq x_0$. Notice that (3.9) is true, also when $0 < \nu < 1$.

We need the following lemma from [17].

Lemma 3.3. Let $f \in C([a,b])$, $\mu, \nu > 0$. Then

$$J_{\mu}^{x_0}(J_{\nu}^{x_0} f) = J_{\mu+\nu}^{x_0}(f). \quad (3.10)$$

We also make

Remark 3.4. Let $\nu \geq 1, \gamma \geq 0$, be such that $\nu - \gamma \geq 1$, so that $\gamma < \nu$. Call $n := [\nu]$, $\alpha := \nu - n$; $m := [\gamma]$, $\rho := \gamma - m$. Note that $\nu - m \geq 1$ and $n - m \geq 1$. Let $f \in C_{x_0}^\nu([a,b])$ be such that $f^{(i)}(x_0) = 0$, $i = 0, 1, \ldots, n-1$. Hence by (3.7)

$$f(x) = (J_\nu^{x_0} D_{x_0}^\nu f)(x), \quad \text{for all } x \in [a,b] : x \geq x_0.$$

Therefore by Leibnitz's formula and $\Gamma(p+1) = p\Gamma(p)$, $p > 0$, we find that

$$f^{(m)}(x) = (J_{\nu-m}^{x_0} D_{x_0}^\nu f)(x), \quad \text{for all } x \geq x_0. \tag{3.11}$$

It follows that $f \in C_{x_0}^\gamma([a,b])$ and thus

$$(D_{x_0}^\gamma f)(x) := (D J_{1-\rho}^{x_0} f^{(m)})(x) \quad \text{exists for all } x \geq x_0. \tag{3.12}$$

By the use of (3.11) we have on $[x_0, b]$,

$$
\begin{aligned}
J_{1-\rho}^{x_0}(f^{(m)}) &= J_{1-\rho}^{x_0}(J_{\nu-m}^{x_0} D_{x_0}^\nu f) = (J_{1-\rho}^{x_0} \circ J_{\nu-m}^{x_0})(D_{x_0}^\nu f) \\
&= (J_{\nu-m+1-\rho}^{x_0}(D_{x_0}^\nu f)) = J_{\nu-\gamma+1}^{x_0}(D_{x_0}^\nu f).
\end{aligned}
$$

by (3.10). That is,

$$(J_{1-\rho}^{x_0} f^{(m)})(x) = \frac{1}{\Gamma(\nu-\gamma+1)} \int_{x_0}^x (x-t)^{\nu-\gamma}(D_{x_0}^\nu f)(t)\, dt.$$

Therefore

$$(D_{x_0}^\gamma f)(x) = D((J_{1-\rho}^{x_0} f^{(m)})(x)) = \frac{1}{\Gamma(\nu-\gamma)} \cdot \int_{x_0}^x (x-t)^{(\nu-\gamma)-1}(D_{x_0}^\nu f)(t)\, dt; \tag{3.13}$$

hence

$$(D_{x_0}^\gamma f)(x) = (J_{\nu-\gamma}^{x_0}(D_{x_0}^\nu f))(x) \quad \text{and is continuous in } x \text{ on } [x_0, b]. \tag{3.14}$$

3.3 Main Results

The next theorem is a specialization of Theorem 2 from [15]. It presents a fractional Opial-type inequality.

Theorem 3.5. Let $\gamma_1, \gamma_2 \geq 0$, $\nu \geq 1$ be such that $\nu - \gamma_1$, $\nu - \gamma_2 \geq 1$ and $f \in C_{x_0}^\nu([a,b])$ with $f^{(i)}(x_0) = 0$, $i = 0, 1, \ldots, n-1$, $n := [\nu]$. Here $x, x_0 \in [a,b]$ with $x \geq x_0$. Let q be a nonnegative continuous function on $[a,b]$. Denote

$$Q := \left(\int_{x_0}^x (q(w))^2 dw \right)^{1/2}. \tag{3.15}$$

Then

$$\int_{x_0}^{x} q(w)|D_{x_0}^{\gamma_1}(f)(w)|\,|D_{x_0}^{\gamma_2}(f)(w)|dw$$

$$\leq K(q,\gamma_1,\gamma_2,\nu,x,x_0)\cdot\left(\int_{x_0}^{x}(D_{x_0}^{\nu}f(w))^2dw\right), \qquad (3.16)$$

where

$$K(q,\gamma_1,\gamma_2,\nu,x,x_0):=\left(\frac{Q}{\sqrt[3]{6}}\right)\cdot\frac{1}{\Gamma(\nu-\gamma_1)\cdot\Gamma(\nu-\gamma_2)}\cdot$$

$$\frac{(x-x_0)^{2\nu-\gamma_1-\gamma_2-1/2}}{(\nu-\gamma_1-\frac{5}{6})^{1/6}\cdot(\nu-\gamma_2-\frac{5}{6})^{1/6}\cdot(4\nu-2\gamma_1-2\gamma_2-\frac{7}{3})^{1/2}} \qquad (3.17)$$

Proof. From (3.13) we have, for $x_0\leq w\leq x$, $j=1,2$,

$$(D_{x_0}^{\gamma_j}f)(w)\ =\ \frac{1}{\Gamma(\nu-\gamma_j)}\int_{x_0}^{w}(w-t)^{\nu-\gamma_j-1}(D_{x_0}^{\nu}f)(t)dt$$

$$=\ \frac{1}{\Gamma(\nu-\gamma_j)}\int_{x_0}^{x}(w-t)_{+}^{\nu-\gamma_j-1}(D_{x_0}^{\nu}f)(t)dt. \quad (3.18)$$

Hence

$$\int_{x_0}^{x}q(w)\cdot\prod_{j=1}^{2}|(D_{x_0}^{\gamma_j}(f))(w)|dw\leq Q\left(\int_{x_0}^{x}\prod_{j=1}^{2}((D_{x_0}^{\gamma_j}(f))(w))^2dw\right)^{1/2}$$

by (3.18),

$$\leq Q\left\{\int_{x_0}^{x}\prod_{j=1}^{2}\left(\frac{1}{\Gamma(\nu-\gamma_j)}\int_{x_0}^{x}(w-t)_{+}^{\nu-\gamma_j-1}|(D_{x_0}^{\nu}f)(t)|dt\right)^2dw\right\}^{1/2}$$

$$\leq Q\cdot\left\{\int_{x_0}^{x}\prod_{j=1}^{2}\left(\frac{1}{(\Gamma(\nu-\gamma_j))^2}\cdot(x-x_0)^{2/3}\right)\right.$$

$$\cdot\left(\int_{x_0}^{x}|(D_{x_0}^{\nu}f)(t)|^{3/2}(w-t)_{+}^{(\nu-\gamma_j-1)3/2}dt\right)^{4/3}dw\right\}^{\frac{1}{2}}=Q\cdot(x-x_0)^{2/3}\cdot$$

$$\prod_{j=1}^{2}\frac{1}{\Gamma(\nu-\gamma_j)}\cdot\left\{\int_{x_0}^{x}\prod_{j=1}^{2}\left(\int_{x_0}^{x}|D_{x_0}^{\nu}f(t)|^{3/2}(w-t)_{+}^{(\nu-\gamma_j-1)3/2}dt\right)^{4/3}dw\right\}^{1/2}$$

$$\leq Q\cdot(x-x_0)^{2/3}\cdot\prod_{j=1}^{2}\frac{1}{\Gamma(\nu-\gamma_j)}$$

$$\cdot \left\{ \int_{x_0}^x \prod_{j=1}^2 \left(\int_{x_0}^x ((D_{x_0}^\nu f)(t))^2 dt \right) \cdot \left(\int_{x_0}^x (w-t)_+^{(\nu-\gamma_j-1)\cdot 6} dt \right)^{1/3} \cdot dw \right\}^{1/2}$$

$$= Q \cdot (x-x_0)^{2/3} \cdot \left(\prod_{j=1}^2 \frac{1}{\Gamma(\nu-\gamma_j)} \right) \cdot \left(\int_{x_0}^x ((D_{x_0}^\nu f)(t))^2 dt \right)$$

$$\cdot \left\{ \int_{x_0}^x \prod_{j=1}^2 \left(\int_{x_0}^x (w-t)_+^{(\nu-\gamma_j-1)\cdot 6} dt \right)^{1/3} \cdot dw \right\}^{1/2}$$

$$= Q \cdot (x-x_0)^{2/3} \cdot \left(\prod_{j=1}^2 \frac{1}{\Gamma(\nu-\gamma_j)} \right) \cdot \left(\int_{x_0}^x ((D_{x_0}^\nu f)(w))^2 dw \right)$$

$$\cdot \left\{ \int_{x_0}^x \prod_{j=1}^2 \frac{(w-x_0)^{((\nu-\gamma_j-1)\cdot 6+1)/3}}{((\nu-\gamma_j-1)6+1)^{1/3}} dw \right\}^{1/2}$$

$$= Q \cdot (x-x_0)^{2/3} \cdot \left(\prod_{j=1}^2 \frac{1}{\Gamma(\nu-\gamma_j)} \right) \cdot \left(\int_{x_0}^x ((D_{x_0}^\nu f)(w))^2 dw \right)$$

$$\cdot \left(\prod_{j=1}^2 \frac{1}{((\nu-\gamma_j-1)6+1)^{1/6}} \right) \cdot \left(\int_{x_0}^x (w-x_0)^{(\sum_{j=1}^2 (\nu-\gamma_j-1)2)+2/3} \cdot dw \right)^{1/2}$$

$$= Q \cdot \frac{(x-x_0)^{2\nu-\gamma_1-\gamma_2-1/2}}{\Gamma(\nu-\gamma_1)\Gamma(\nu-\gamma_2)} \cdot \left(\int_{x_0}^x ((D_{x_0}^\nu f)(w))^2 dw \right)$$

$$\cdot \frac{1}{(6\nu-6\gamma_1-5)^{1/6}(6\nu-6\gamma_2-5)^{1/6}(4\nu-2\gamma_1-2\gamma_2-\frac{7}{3})^{1/2}} \cdot$$

We have proved that

$$\int_{x_0}^x q(w) \cdot \prod_{j=1}^2 |(D_{x_0}^{\gamma_j}(f))(w)| dw \le \frac{Q}{\sqrt[3]{6}} \cdot$$

$$\frac{(x-x_0)^{2\nu-\gamma_1-\gamma_2-1/2}}{\{\Gamma(\nu-\gamma_1) \cdot \Gamma(\nu-\gamma_2) \cdot (\nu-\gamma_1-\frac{5}{6})^{1/6}(\nu-\gamma_2-\frac{5}{6})^{1/6} \cdot (4\nu-2\gamma_1-2\gamma_2-\frac{7}{3})^{1/2}\}}$$

$$\cdot \left(\int_{x_0}^x ((D_{x_0}^\nu f)(w))^2 dw \right).$$

We have established (3.16).   □

We next give the following very general fractional Opial-type inequality.

Theorem 3.6. *Let* $\gamma \geq 0$, $\nu \geq 1$, $\nu - \gamma \geq 1$, $\alpha, \beta > 0$, $r > \alpha$, $r > 1$; *let* $p > 0$, $q \geq 0$ *be continuous functions on* $[a,b]$. *Let* $f \in C_{x_0}^{\nu}([a,b])$ *with* $f^{(i)}(x_0) = 0$, $i = 0,1,\ldots,n-1$, $n := [\nu]$. *Let* $x, x_0 \in [a,b]$ *with* $x \geq x_0$. *Then*

$$\int_{x_0}^{x} q(w)|D_{x_0}^{\gamma}f(w)|^{\beta}|D_{x_0}^{\nu}f(w)|^{\alpha}dw \leq K(p,q,\gamma,\nu,\alpha,\beta,r,x,x_0)$$

$$\cdot \left(\int_{x_0}^{x} p(w)|D_{x_0}^{\nu}f(w)|^{r}dw\right)^{(\alpha+\beta/r)}. \tag{3.19}$$

Here

$$K(p,q,\gamma,\nu,\alpha,\beta,r,x,x_0) := \left(\frac{\alpha}{\alpha+\beta}\right)^{\alpha/r} \cdot \frac{1}{(\Gamma(\nu-\gamma))^{\beta}}$$

$$\cdot \left(\int_{x_0}^{x} ((q(w))^{r} \cdot (p(w))^{-\alpha})^{1/(r-\alpha)} \cdot (P_1(w))^{(\beta(r-1)/r-\alpha)} \cdot dw\right)^{r-\alpha/r} \tag{3.20}$$

with

$$P_1(w) := \int_{x_0}^{w} (p(t))^{-1/r-1} \cdot (w-t)^{(\nu-\gamma-1)(r/(r-1))} \cdot dt. \tag{3.21}$$

Proof. Again from (3.13) we have

$$(D_{x_0}^{\gamma}f)(w) = \frac{1}{\Gamma(\nu-\gamma)}\int_{x_0}^{w}(w-t)^{\nu-\gamma-1}(D_{x_0}^{\nu}f)(t)dt, \quad \text{for all } x_0 \leq w \leq x.$$

That is,

$$|D_{x_0}^{\gamma}f(w)| \leq \frac{1}{\Gamma(\nu-\gamma)}\int_{x_0}^{w}(w-t)^{\nu-\gamma-1}|D_{x_0}^{\nu}f(t)|dt$$

$$= \frac{1}{\Gamma(\nu-\gamma)}\int_{x_0}^{w}p(t)^{-1/r}(w-t)^{\nu-\gamma-1}p(t)^{1/r}|(D_{x_0}^{\nu}f(t)|dt$$

$$\leq \frac{1}{\Gamma(\nu-\gamma)}\cdot\left(\int_{x_0}^{w}p(t)^{-1/r-1}(w-t)^{(\nu-\gamma-1)r/(r-1)}dt\right)^{(r-1)/r}$$

$$\cdot\left(\int_{x_0}^{w}p(t)|D_{x_0}^{\nu}f(t)|^{r}dt\right)^{1/r}$$

$$= \frac{1}{\Gamma(\nu-\gamma)}(P_1(w))^{r-1/r}\left(\int_{x_0}^{w}p(t)|D_{x_0}^{\nu}f(t)|^{r}dt\right)^{1/r}.$$

Call

$$z(w) := \int_{x_0}^{w}p(t)|D_{x_0}^{\nu}f(t)|^{r}dt, \quad x_0 \leq w \leq x, \quad (\text{so } z(x_0) = 0).$$

Then
$$z'(w) = p(w)|D_{x_0}^\nu f(w)|^r.$$

For $\alpha > 0$ we get

$$|D_{x_0}^\nu f(w)|^\alpha = p(w)^{-\alpha/r} \cdot (z'(w))^{\alpha/r}.$$

For $\beta > 0$ we find

$$q(w)|D_{x_0}^\gamma f(w)|^\beta |D_{x_0}^\nu f(w)|^\alpha$$
$$\leq q(w)\frac{1}{(\Gamma(\nu - \gamma))^\beta}(P_1(w))^{\beta(r-1)/r}(z(w))^{\beta/r}(p(w))^{-\alpha/r}(z'(w))^{\alpha/r}.$$

Hence

$$\int_{x_0}^x q(w)|D_{x_0}^\gamma f(w)|^\beta |D_{x_0}^\nu f(w)|^\alpha dw$$

$$\leq \int_{x_0}^x \frac{q(w)}{(\Gamma(\nu - \gamma))^\beta}(P_1(w))^{\beta(r-1)/r}p(w)^{-\alpha/r}(z(w))^{\beta/r}(z'(w))^{\alpha/r}dw$$

$$\leq \frac{1}{(\Gamma(\nu - \gamma))^\beta} \cdot \left(\int_{x_0}^x ((q(w))^r (p(w))^{-\alpha})^{1/r-\alpha} \cdot (P_1(w))^{\beta(r-1)/(r-\alpha)} \cdot dw \right)^{(r-\alpha)/r}$$

$$\cdot \left(\int_{x_0}^x (z(w))^{\beta/\alpha} z'(w)dw \right)^{\alpha/r} = (*).$$

Set

$$K_0 \quad : \quad = \frac{1}{(\Gamma(\nu - \gamma))^\beta} \left(\int_{x_0}^x ((q(w))^r (p(w))^{-\alpha})^{1/r-\alpha} \cdot (P_1(w))^{\beta(r-1)/(r-\alpha)} \cdot dw \right)^{(r-\alpha)/r}$$

$$(*) \quad = \quad K_0 \left(\int_{x_0}^x (z(w))^{\beta/\alpha} dz(w) \right)^{\alpha/r}$$

$$= \quad K_0 \cdot \left(\frac{(z(w))^{(\beta/\alpha)+1}}{\frac{\beta}{\alpha} + 1} \Big|_{x_0}^x \right)^{\alpha/r}$$

$$= \quad K_0 \cdot \left(\frac{z(x)^{(\beta+\alpha)/\alpha}}{\frac{\beta+\alpha}{\alpha}} \right)^{\alpha/r}$$

$$= \quad K_0 \cdot \left(\frac{\alpha}{\alpha+\beta} \right)^{\alpha/r} \cdot (z(x))^{(\beta+\alpha)/r}.$$

Clearly,

$$K(p, q, \gamma, \nu, \alpha, \beta, r, x, x_0) = K_0 \left(\frac{\alpha}{\alpha+\beta} \right)^{\alpha/r}.$$

We have proved (3.19). $\square$

Corollary 3.7. *Let* $\gamma \geq 0$, $\nu \geq 1$, $\nu - \gamma \geq 1$, *let* q *be a nonnegative continuous function on* $[a, b]$. *Let* $f \in C_{x_0}^\nu([a, b])$ *with* $f^{(i)}(x_0) = 0$, $i = 0, 1, \ldots, n-1$, $n := [\nu]$. *Let* $x, x_0 \in [a, b]$ *with* $x \geq x_0$. *Then*

$$\int_{x_0}^{x} q(w)|D_{x_0}^{\gamma}f(w)|\,|D_{x_0}^{\nu}f(w)|dw$$

$$\leq K(q,\gamma,\nu,x,x_0)\left(\int_{x_0}^{x}(D_{x_0}^{\nu}f(w))^2 dw\right), \qquad (3.22)$$

where

$$K(q,\gamma,\nu,x,x_0) = \frac{1}{\sqrt{2(2\nu-2\gamma-1)}}\cdot\frac{1}{\Gamma(\nu-\gamma)}\cdot$$

$$\left(\int_{x_0}^{x}(q(w))^2(w-x_0)^{2\nu-2\gamma-1}dw\right)^{1/2}. \qquad (3.23)$$

Proof. Apply Theorem 3.6 with $\alpha = \beta = 1$, $r = 2$, $p = 1$. □

Next we give another basic fractional Opial-type inequality.

Theorem 3.8. *Let* $\nu \geq 1$, $\alpha,\beta > 0$, $r > \alpha$, $r > 1$; *let* $p > 0$, $q \geq 0$ *be continuous functions on* $[a,b]$. *Let* $f \in C_{x_0}^{\nu}([a,b])$ *with* $f^{(i)}(x_0) = 0$, $i = 0,1,\ldots,n-1$, $n := [\nu]$. *Let* $x, x_0 \in [a,b]$ *with* $x \geq x_0$. *Then*

$$\int_{x_0}^{x} q(w)|f(w)|^{\beta}|D_{x_0}^{\nu}f(x)|^{\alpha}dw \leq K^*(p,q,\nu,\alpha,\beta,r,x,x_0)$$

$$\cdot\left(\int_{x_0}^{x}p(w)|D_{x_0}^{\nu}f(w)|^r dw\right)^{((\alpha+\beta)/r)}. \qquad (3.24)$$

Here

$$K^*(p,q,\nu,\alpha,\beta,r,x,x_0) := \left(\frac{\alpha}{\alpha+\beta}\right)^{\alpha/r}\cdot\frac{1}{(\Gamma(\nu))^{\beta}}\cdot$$

$$\left(\int_{x_0}^{x}((q(w))^r\cdot(p(w))^{-\alpha})^{1/(r-\alpha)}\cdot(P_1^*(w))^{\beta(r-1)/(r-\alpha)}\cdot dw\right)^{(r-\alpha)/r}, \qquad (3.25)$$

with

$$P_1^*(w) := \int_{x_0}^{w}(p(t))^{-1/r-1}\cdot(w-t)^{(\nu-1)(r/r-1)}dt. \qquad (3.26)$$

Proof. From (3.9) we have

$$f(w) = \frac{1}{\Gamma(\nu)}\int_{x_0}^{w}(w-t)^{\nu-1}(D_{x_0}^{\nu}f)(t)dt, \quad x_0 \leq w \leq x.$$

That is,

$$\begin{aligned}
|f(w)| &\leq \frac{1}{\Gamma(\nu)} \int_{x_0}^{w} (w - t)^{\nu-1} |D_{x_0}^{\nu} f(t)| dt \\
&= \frac{1}{\Gamma(\nu)} \int_{x_0}^{w} p(t)^{-1/r} (w - t)^{\nu-1} p(t)^{1/r} |D_{x_0}^{\nu} f(t)| dt
\end{aligned}$$

(applying Hölder's inequality with indices $r/(r-1), r$)

$$\leq \frac{1}{\Gamma(\nu)} \left(\int_{x_0}^{w} p(t)^{-1/r-1} (w - t)^{(\nu-1)r/r-1} dt \right)^{(r-1)/r}.$$

$$\left(\int_{x_0}^{w} p(t) |D_{x_0}^{\nu} f(t)|^r dt \right)^{1/r}$$

$$= \frac{1}{\Gamma(\nu)} (P_1^*(w))^{r-1/r} \left(\int_{x_0}^{w} p(t) |D_{x_0}^{\nu} f(t)|^r dt \right)^{1/r}.$$

Call

$$z(w) := \int_{x_0}^{w} p(t) |D_{x_0}^{\nu} f(t)|^r dt, \quad x_0 \leq w \leq x, \ (z(x_0) = 0).$$

Then

$$z'(w) = p(w) |D_{x_0}^{\nu} f(w)|^r.$$

For $\alpha > 0$ we obtain

$$|D_{x_0}^{\nu} f(w)|^{\alpha} = (p(w))^{-\alpha/r} (z'(w))^{\alpha/r}.$$

For $\beta > 0$ we have

$$\begin{aligned}
q(w) |f(w)|^{\beta} |D_{x_0}^{\nu} f(w)|^{\alpha} &\leq q(w) \cdot \frac{1}{(\Gamma(\nu))^{\beta}} \cdot (P_1^*(w))^{\beta(r-1)/r} \\
&\quad \cdot (z(w))^{\beta/r} \cdot (p(w))^{-\alpha/r} (z'(w))^{\alpha/r}.
\end{aligned}$$

Thus

$$\begin{aligned}
&\int_{x_0}^{x} q(w) |f(w)|^{\beta} |D_{x_0}^{\nu} f(w)|^{\alpha} dw \\
&\leq \int_{x_0}^{x} \frac{q(w)}{(\Gamma(\nu))^{\beta}} (P_1^*(w))^{\beta(r-1)/r} (p(w))^{-\alpha/r} (z(w))^{\beta/r} (z'(w))^{\alpha/r} dw \\
&\leq \frac{1}{(\Gamma(\nu))^{\beta}} \left(\int_{x_0}^{x} ((q(w))^r (p(w))^{-\alpha})^{1/(r-\alpha)} (P_1^*(w))^{\beta(r-1)/(r-\alpha)} dw \right)^{(r-\alpha)/r} \\
&\quad \cdot \left(\int_{x_0}^{x} (z(w))^{\beta/\alpha} z'(w) dw \right)^{\alpha/r} = (*)
\end{aligned}$$

call

$$K_0^* : = \frac{1}{(\Gamma(\nu))^\beta} \left(\int_{x_0}^x \left((q(w))^r (p(w))^{-\alpha} \right)^{1/(r-\alpha)} \right.$$

$$\left. (P_1^*(w))^{\beta(r-1)/r-\alpha} dw \right)^{r-\alpha/r} \right)$$

$$(*) = K_0^* \left(\int_{x_0}^x (z(w))^{\beta/\alpha} dz(w) \right)^{\alpha/r}$$

$$= K_0^* \left(\frac{(z(w))^{(\beta/\alpha)+1}}{\frac{\beta}{\alpha}+1} \Big|_{x_0}^x \right)^{\alpha/r}$$

$$= K_0^* \left(\frac{z(x)^{\beta+\alpha/\alpha}}{\frac{\beta+\alpha}{\alpha}} \right)^{\alpha/r} = K_0^* \left(\frac{\alpha}{\alpha+\beta} \right)^{\alpha/r} (z(x))^{(\alpha+\beta)/r}.$$

Clearly

$$K^*(p,q,\nu,\alpha,\beta,r,x,x_0) = K_0^* \left(\frac{\alpha}{\alpha+\beta} \right)^{\alpha/r}.$$

We have established (3.24). □

Corollary 3.9. *Let $\nu \geq 1$, and let $q \geq 0$ be a continuous function on $[a,b]$. Let $f \in C_{x_0}^\nu([a,b])$ with $f^{(i)}(x_0) = 0$, $i = 0,1,\ldots,n-1$, $n := [\nu]$. Let $x, x_0 \in [a,b] : x \geq x_0$. Then*

$$\int_{x_0}^x q(w)|f(w)| \, |D_{x_0}^\nu f(w)| dw$$

$$\leq K^*(q,\nu,x,x_0) \cdot \left(\int_{x_0}^x (D_{x_0}^\nu f(w))^2 dw \right), \qquad (3.27)$$

where

$$K^*(q,\nu,x,x_0) = \frac{1}{\sqrt{2(2\nu-1)}} \cdot \frac{1}{\Gamma(\nu)}$$

$$\cdot \left(\int_{x_0}^x (q(w))^2 (w-x_0)^{2\nu-1} dw \right)^{1/2}. \qquad (3.28)$$

Proof. Apply Theorem 3.8 for $\alpha = \beta = 1$, $r = 2$, $p = 1$. □

Remark 3.10. Notice that $K(q,\gamma_1,\gamma_2,\nu,x,x_0)$ (see (3.17)), $K(q,\gamma,\nu, x,x_0)$ (see (3.23)), and $K^*(q,\nu,x,x_0)$ (see (3.28)), all converge to zero as $x \to x_0$.

Remark 3.11. Let $\nu \geq 1$ and $\gamma := \nu - 1$. Clearly: $\gamma \geq 0$ and $\nu - \gamma = 1$. Let $f \in C_{x_0}^{\nu}([a,b])$ such that $f^{(i)}(x_0) = 0$, $i = 0, 1, \ldots, n-1$, $n := [\nu]$. Then by (3.13) we find

$$(D_{x_0}^{\nu-1}f)(x) = \int_{x_0}^{x} (D_{x_0}^{\nu}f)(t)dt, \quad ((D_{x_0}^{\nu-1}f)(x_0) = 0);$$

that is,

$$(D_{x_0}^{\nu-1}f)'(x) = (D_{x_0}^{\nu}f)(x), \quad \text{all } x \geq x_0.$$

3.4 Applications

(i) We prove locally the uniqueness of the solution of the following fractional initial value problem (IVP).

Theorem 3.12. *Let* $\nu \geq 1$, $\gamma_j \geq 0$; $j = 1, 2, \ldots, k \in \mathbb{N}$ *be such that all* $\nu - \gamma_j \geq 1$ *for all* j. *Also we assume the ordering*

$$\nu > \gamma_1 > \gamma_2 > \gamma_3 > \cdots > \gamma_k.$$

We consider continuous functions $q_1, q_2, \ldots, q_k$, q_{k+1}, h *from* $[a,b]$ *into* $\mathbb{R}$. *Let* $x, x_0 \in [a,b]$ *such that* $x \geq x_0$. *Consider the following fractional differential equation*

$$(D_{x_0}^{\nu}f)(w) + \sum_{j=1}^{k} q_j(w) \cdot (D_{x_0}^{\gamma_j}f)(w) + q_{k+1}(w) \cdot f(w) = h(w), \qquad (3.29)$$

for all $x_0 \leq w \leq x$, *such that*

$$f^{(i)}(x_0) = a_i \in \mathbb{R}, \quad i = 0, 1, \ldots, n-1,$$

where $n := [\nu]$. *Then there is at most one solution* f *to this problem.*

Proof. Let $f_1, f_2 \in C_{x_0}^{\nu}([x_0, x])$ fulfill the initial value problem (3.29). Call $g := f_1 - f_2 \in C_{x_0}^{\nu}([x_0, x])$. Then by (3.22) we obtain

$$(D_{x_0}^{\nu}g)(w) + \sum_{j=1}^{k} q_j(w) \cdot (D_{x_0}^{\gamma_j}g)(w) + q_{k+1}(w) \cdot g(w) = 0,$$

true for all $x_0 \leq w \leq x$, such that

$$g^{(i)}(x_0) = 0, \quad \text{all } i = 0, 1, \ldots, n-1.$$

Thus

$$(D_{x_0}^{\nu}g)(w) = -\sum_{j=1}^{k} q_j(w)(D_{x_0}^{\gamma_j}g)(w) - q_{k+1}(w) \cdot g(w),$$

and

$$((D_{x_0}^\nu g)(w))^2 = -\sum_{j=1}^k q_j(w)\cdot(D_{x_0}^{\gamma_j}g)(w)\cdot(D_{x_0}^\nu g)(w)-q_{k+1}(w)\cdot g(w)\cdot(D_{x_0}^\nu g)(w).$$

That is,

$$((D_{x_0}^\nu g)(w))^2 \le \sum_{j=1}^k |q_j(w)|\cdot|D_{x_0}^{\gamma_j}g(w)|\cdot|D_{x_0}^\nu g(w)|$$
$$+ |q_{k+1}(w)|\cdot|g(w)|\cdot|D_{x_0}^\nu g(w)|,$$

which is true for all $x_0 \le w \le x$.

Consequently we have

$$\delta := \int_{x_0}^x ((D_{x_0}^\nu g)(w))^2\cdot dw$$
$$\le \sum_{j=1}^k \int_{x_0}^x |q_j(w)|\cdot|D_{x_0}^{\gamma_j}g(w)|\cdot|D_{x_0}^\nu g(w)|\cdot dw$$
$$+ \int_{x_0}^x |q_{k+1}(w)|\cdot|g(w)|\cdot|D_{x_0}^\nu g(w)|\cdot dw =: (*).$$

From (3.22) and (3.27) we find

$$(*) \le \sum_{j=1}^k K(|q_j|,\gamma_j,\nu,x,x_0)\cdot\left(\int_{x_0}^x (D_{x_0}^\nu g(w))^2 dw\right)$$
$$+ K^*(|q_{k+1}|,\nu,x,x_0)\cdot\left(\int_{x_0}^x (D_{x_0}^\nu g(w))^2 dw\right)$$
$$= \left\{\left(\sum_{j=1}^k K(|q_j|,\gamma_j,\nu,x,x_0)\right) + K^*(|q_{k+1}|,\nu,x,x_0)\right\}\cdot\delta.$$

Set

$$\overline{K} := \sum_{j=1}^k K(|q_j|,\gamma_j,\nu,x,x_0) + K^*(|q_{k+1}|,\nu,x,x_0).$$

By Remark 3.10, $\overline{K} \to 0$ as $x \to x_0$. So we get

$$\delta \le \overline{K}\cdot\delta;$$

that is,

$$\delta(1 - \overline{K}) \le 0;$$

also notice that $\delta \geq 0$. So for $\varepsilon > 0$ sufficiently small such that $|x - x_0| \leq \varepsilon$ and $\overline{K} \leq 1$, we obtain $\delta \leq 0$. That is, $\delta = 0$ for $x : |x - x_0| \leq \varepsilon$. That is,

$$\int_{x_0}^{x} ((D_{x_0}^{\nu} g)(w))^2 dw = 0 \quad \text{for } x : |x - x_0| \leq \varepsilon.$$

That means $D_{x_0}^{\nu} g(w) = 0$ for $w : |w - x_0| \leq \varepsilon$.

Hence by (3.9) we obtain

$$g(w) = 0 \quad \text{for } w : |w - x_0| \leq \varepsilon.$$

We have proved the local uniqueness for the solution of IVP (3.29); more precisely we have

$$f_1(w) = f_2(w) \quad \text{for all } w \text{ with } |w - x_0| \leq \varepsilon.$$

(ii) Let $\nu \geq 2$, $\gamma_1 := \nu - 1$, $\gamma_j \geq 0$; $j = 2, 3, \ldots, k \in \mathbb{N}$ such that $\nu - \gamma_j \geq 2$; $j = 2, 3, \ldots, k$. Also we assume the order

$$\nu > \gamma_1 > \gamma_2 > \gamma_3 > \cdots > \gamma_k.$$

Let $n := [\nu]$; then $n - 1 = [\nu - 1]$. We consider again IVP (3.29). Now we prove the uniqueness of the solution to (3.29) over the whole interval $[x_0, b]$. Again let $f_1, f_2 \in C_{x_0}^{\nu}([x_0, b])$ fulfill the IVP (3.29). Call $g := f_1 - f_2 \in C_{x_0}^{\nu}([x_0, b])$. Clearly $g \in C_{x_0}^{\nu-1}([x_0, b])$. Then again by (3.29) we obtain

$$(D_{x_0}^{\nu} g)(w) + \sum_{j=1}^{k} q_j(w) \cdot (D_{x_0}^{\gamma_j} g)(w) + q_{k+1}(w) \cdot g(w) = 0,$$

all $x_0 \leq w \leq b$, $g^{(i)}(x_0) = 0$, $i = 0, 1, \ldots, n - 1$. Hence

$$(D_{x_0}^{\nu} g)(w) \cdot (D_{x_0}^{\nu-1} g)(w) \quad + \quad \sum_{j=1}^{k} q_j(w) \cdot (D_{x_0}^{\gamma_j} g)(w) \cdot (D_{x_0}^{\nu-1} g)(w)$$

$$+ \quad q_{k+1}(w) \cdot g(w) \cdot (D_{x_0}^{\nu-1} g)(w) = 0.$$

Next we integrate over $[x_0, \tilde{x}] \subseteq [x_0, b]$,

$$\int_{x_0}^{\tilde{x}} (D_{x_0}^{\nu} g)(w) \cdot (D_{x_0}^{\nu-1} g)(w) \cdot dw + \sum_{j=1}^{k} \int_{x_0}^{\tilde{x}} q_j(w) \cdot (D_{x_0}^{\gamma_j} g)(w) \cdot (D_{x_0}^{\nu-1} g)(w) dw$$

$$+ \int_{x_0}^{\tilde{x}} q_{k+1}(w) \cdot g(w) \cdot (D_{x_0}^{\nu-1} g)(w) dw = 0.$$

By Remark 3.11 we get that

$$((D_{x_0}^{\nu-1} g)(\tilde{x}))^2 = -2 \sum_{j=1}^{k} \int_{x_0}^{\tilde{x}} q_j(w) \cdot (D_{x_0}^{\gamma_j} g)(w) \cdot (D_{x_0}^{\nu-1} g)(w) \cdot dw$$

$$-2 \int_{x_0}^{\tilde{x}} q_{k+1}(w) \cdot g(w) \cdot (D_{x_0}^{\nu-1} g)(w) \cdot dw.$$

Therefore we have

$$((D_{x_0}^{\nu-1}g)(\tilde{x}))^2 \;\leq\; 2\sum_{j=1}^{k}\int_{x_0}^{\tilde{x}}|q_j(w)|\cdot|(D_{x_0}^{\gamma_j}g)(w)|\cdot|(D_{x_0}^{\nu-1}g)(w)|\cdot dw$$

$$+\,2\cdot\int_{x_0}^{\tilde{x}}|q_{k+1}(w)|\cdot|g(w)|\cdot|(D_{x_0}^{\nu-1}g)(w)|\cdot dw =: (*).$$

Next we use Corollaries 3.7 and 3.9 in the above setting.
We have

$$(*) \;\leq\; 2\cdot\sum_{j=1}^{k}K(|q_j|,\gamma_j,\nu-1,\tilde{x},x_0)\cdot\left(\int_{x_0}^{\tilde{x}}(D_{x_0}^{\nu-1}g(w))^2dw\right)$$

$$+\,2\cdot K^*(|q_{k+1}|,\nu-1,\tilde{x},x_0)\cdot\left(\int_{x_0}^{\tilde{x}}(D_{x_0}^{\nu-1}g(w))^2dw\right)$$

$$=\; 2\cdot\left\{\left(\sum_{j=1}^{k}K(|q_j|,\gamma_j,\nu-1,\tilde{x},x_0)\right)+K^*(|q_{k+1}|,\nu-1,\tilde{x},x_0)\right\}$$

$$\cdot\left(\int_{x_0}^{\tilde{x}}(D_{x_0}^{\nu-1}g(w))^2dw\right).$$

Set

$$T(\tilde{x}) := 2\cdot\left\{\left(\sum_{j=1}^{k}K(|q_j|,\gamma_j,\nu-1,\tilde{x},x_0)\right)+K^*(|q_{k+1}|,\nu-1,\tilde{x},x_0)\right\}\geq 0.$$

We have established that

$$((D_{x_0}^{\nu-1}g)(\tilde{x}))^2 \leq T(\tilde{x})\cdot\left(\int_{x_0}^{\tilde{x}}(D_{x_0}^{\nu-1}y(w))^2\cdot dw\right),\quad \text{for all } \tilde{x}\colon x_0\leq\tilde{x}\leq x\leq b.$$

Clearly $T(x) = \max_{x_0\leq\tilde{x}\leq x}T(\tilde{x}) \geq 0$, a constant for each $x \in [x_0,b]$. Thus we have shown that

$$((D_{x_0}^{\nu-1}g)(\tilde{x}))^2 \leq T(x)\cdot\left(\int_{x_0}^{\tilde{x}}((D_{x_0}^{\nu-1}g)(w))^2\cdot dw\right),\quad \text{all } x_0\leq\tilde{x}\leq x.$$

By Gronwall's inequality we find

$$((D_{x_0}^{\nu-1}g)(\tilde{x}))^2 = 0,\quad x_0\leq\tilde{x}\leq x,$$

and

$$(D_{x_0}^{\nu-1}g)(\tilde{x}) = 0,\quad \text{for all } \tilde{x} \text{ with } x_0\leq\tilde{x}\leq x.$$

By (3.9) we now get that

$$g(\tilde{x}) = 0, \quad \text{all } \tilde{x} : x_0 \le \tilde{x} \le x.$$

That is, $f_1 = f_2$ over $[x_0, x]$, proving the uniqueness of the solution to IVP (3.29) over the whole interval $[x_0, x]$.

(iii) Consider again IVP (3.29) as in (ii). Additionally assume that $h \equiv 0$ and $a_i = 0$, all $i = 0, 1, \ldots, n-1$. Then the unique solution exists and it is the trivial solution zero.

3.5   Other Fractional Differential Equations

In this section we review the basic theory of fractional differential equations based on Miller – Ross fractional derivatives. The exposition follows Podlubny [333]. Consider the initial value problem

$$\mathbb{D}^{\sigma_n} u(t) + \sum_{j=1}^{n-1} p_j(t) \mathbb{D}^{\sigma_{n-j}} u(t) + p_n(t) u(t) = h(t), \qquad (3.30)$$

$$\mathbb{D}^{\sigma_k - 1} u(t)\big|_{t=0} = b_k, \quad k = 1, \ldots, n, \qquad (3.31)$$

where

$$\mathbb{D}^{\sigma_k} := D^{\alpha_k} D^{\alpha_{k-1}} \cdots D^{\alpha_1},$$

$$\mathbb{D}^{\sigma_k - 1} := D^{\alpha_k - 1} D^{\alpha_{k-1}} \cdots D^{\alpha_1},$$

$$\sigma_k = \sum_{j=1}^{k} \alpha_j, \quad k = 1, \ldots, n,$$

$$0 < \alpha_j \le 1, \quad j = 1, \ldots, n,$$

and $u \in L^1[0, \tau]$. Note that we have switched from x to τ, and we have replaced x_0 by 0 (and we are calling the unknown function u rather than f).

For an example of a concrete partial differential equation coming from the applications consider

$$(D^\gamma)^2 u + 2a D^\gamma u = \Delta u,$$

where $u = u(t, x)$ for $t \ge 0$ and $x \in \mathbb{R}^N$, $\Delta = \sum_{j=1}^{N} \partial^2 / \partial x_j^2$ is the spatial Laplacian, a is a positive constant and $0 < \gamma < 1$; and D^γ is the usual fractional derivative of order γ with respect to time t. This fractional telegraph equation was introduced in [150]; see [151] for the experimental background. Let $\hat{u}(t, \xi)$ (for $t \ge 0$, $\xi \in \mathbb{R}^N$) be the spatial Fourier transform of $u(t, x)$. Then $\hat{u}$ satisfies

$$(D^\gamma)^2 \hat{u} + 2a D^\gamma \hat{u} = -|\xi|^2 \hat{u},$$

which is a family of fractional differential equations of the form (3.30), one for each $\xi \in \mathbb{R}^N$. This problem arises in the study of suspensions, coming from the fluid dynamical modeling of certain blood flow phenomena (see [150] and [151]).

We return to the initial value problem (3.30), (3.31). According to Theorem 3.2 of [333], if each p_j is in $C[0, \tau]$, the problem (3.30), (3.31) has a unique solution in the class $L^1[0, \tau]$. Thus both existence and uniqueness hold for (3.30), (3.31). The number of required initial conditions is n, whereas the order of the equation is σ_n, and σ_n can be any number in the half-open interval $(0, n]$. For instance, we can have $\sigma_n = 5/2$ and $n = 4$ or even $n = 6$. Thus the number of initial conditions required to specify the solution uniquely is not only a function of the order of the equation; it depends on the decomposition (involving $\mathbb{D}^{\sigma_k}$) which leads to the rigorous interpretation of the equation. In the case of our uniqueness criteria (see Section 3.4), ν is the order of the equation and $n = [\nu]$ is the number of specified initial conditions. (Thus if $\nu = 5/2$, necessarily $n = 2$.)

4
Riemann–Liouville Opial-Type Inequalities for Fractional Derivatives

This chapter provides Opial-type inequalities for generalized Riemann–Liouville fractional derivatives. The inequalities are given for integrable functions with a minimal restriction on the order of the derivatives. This treatment relies on [64].

4.1 Introduction and Preliminaries

The original Opial inequality [315] (see also [297, p. 114]) states the following.

Theorem 4.1. *If* $f \in C^1[0, a]$ *with* $f(0) = f(a) = 0$ *and* $f(x) > 0$ *on* $(0, a)$, *then*

$$\int_0^a |f(x)f'(x)| \, dx \le \frac{a}{4} \int_0^a (f'(x))^2 \, dx.$$

The constant $a/4$ *is the best possible.*

This result for classical derivatives has been generalized in several directions (see, for instance, [3, 4]). This chapter is analogous to [17] by the author. Using the different Riemann–Liouville definition of the fractional derivative, we obtain inequalities for integrable rather than continuous functions, while being able to relax the conditions on the order of fractional derivatives.

G.A. Anastassiou, *Fractional Differentiation Inequalities*,
DOI 10.1007/978-0-387-98128-4_4, © Springer Science+Business Media, LLC 2009

We give a brief survey of some facts about Riemann–Liouville fractional derivatives needed in the sequel; for more details see the monograph [366, Chapter 1].

Let $x > 0$. By $C^m[0, x]$ we denote the space of all functions on $[0, x]$ that have continuous derivatives up to order m, and $AC[0, x]$ is the space of all absolutely continuous functions on $[0, x]$. By $AC^m[0, x]$ we denote the space of all functions $g \in C^m[0, x]$ with $g^{(m-1)} \in AC[0, x]$. For any $\nu \in \mathbb{R}$ we denote by $[\nu]$ the integral part of ν (the integer k satisfying $k \leq \nu < k+1$).

By $L_1(0, x)$ we denote the space of all functions integrable on the interval $(0, x)$, and by $L^\infty(0, x)$ the set of all functions measurable and essentially bounded on $(0, x)$. Clearly, $L^\infty(0, x) \subset L_1(0, x)$. Let $\nu > 0$. For any $f \in L_1(0, x)$ the *Riemann–Liouville fractional integral of f* of order ν is defined by

$$I^\nu f(s) = \frac{1}{\Gamma(\nu)} \int_0^s (s - t)^{\nu-1} f(t)\, dt, \quad s \in [0, x], \tag{4.1}$$

and the *Riemann–Liouville fractional derivative of f* of order ν by

$$D^\nu f(s) = \left(\frac{d}{ds}\right)^m I^{m-\nu} f(s) = \frac{1}{\Gamma(m - \nu)} \left(\frac{d}{ds}\right)^m \int_0^s (s - t)^{m-\nu-1} f(t)\, dt, \tag{4.2}$$

where $m = [\nu] + 1$. In addition, we make the conventions

$$D^0 f := f =: I^0 f, \quad I^{-\nu} f := D^\nu f \text{ if } \nu > 0, \quad D^{-\nu} f := I^\nu f \text{ if } 0 < \nu \leq 1. \tag{4.3}$$

If ν is a positive integer, then $D^\nu f = (d/ds)^\nu f$. Let us remark that a somewhat more general definition of the Riemann–Liouville fractional derivative is used in the literature with an anchor point a other than 0: Let $a \in \mathbb{R}$ be fixed, $s \geq a$, and let $f_a(t) = f(a + t)$ be a translation of f. Then set

$$D_a^\nu f(s) := D^\nu f_a(s - a).$$

All our results stated for the fractional derivative defined by (4.2) have an interpretation for the fractional derivative with a general anchor point.

Let $\nu > 0$ and $m = [\nu] + 1$. We define the space $I^\nu(L_1(0, x))$ as the set of all functions f on $[0, x]$ of the form $f = I^\nu \varphi$ for some $\varphi \in L_1(0, x)$ (see Definition 1.2.3 in [366, p. 43]). According to Theorem 1.2.3 in [366, p. 43], the latter characterization is equivalent to the condition

$$I^{m-\nu} f \in AC^m[0, x], \tag{4.4}$$

$$\left(\frac{d}{ds}\right)^j I^{m-\nu} f(0) = 0, \quad j = 0, 1, \ldots, m - 1. \tag{4.5}$$

A function $f \in L_1(0, x)$ satisfying (4.4) is said to have an *integrable fractional derivative $D^\nu f$* (see Definition 1.2.4 in [366, p. 44]). We express these conditions in terms of fractional derivatives.

Lemma 4.2. *Let $\nu > 0$ and $m = [\nu] + 1$. A function $f \in L_1(0, x)$ has an integrable fractional derivative $D^\nu f$ if and only if*

$$D^{\nu-k}f \in C[0, x], \ k = 1, \ldots, m, \quad \text{and} \quad D^{\nu-1}f \in AC[0, x]. \tag{4.6}$$

Furthermore, $f \in I^\nu(L_1(0, x))$ if and only if f has an integrable fractional derivative $D^\nu f$ and satisfies the conditions

$$D^{\nu-k}f(0) = 0 \text{ for } k = 1, \ldots, m. \tag{4.7}$$

Proof. Note that

$$\left(\frac{d}{ds}\right)^k I^{m-\nu}f = \left(\frac{d}{ds}\right)^k I^{k-(\nu-m+k)}f = D^{\nu-m+k}f$$

in view of the definition of the fractional derivative and the equation $[\nu - m + k] + 1 = k$. Then (4.6) is equivalent to (4.4) and (4.7) is equivalent to (4.5). (For $k = m$ we use the convention $D^{\nu-m}f = I^{m-\nu}f$ in (4.6).) □

We need the following result on the law of indices for fractional integration and differentiation using the unified notation (4.3).

Lemma 4.3. (Theorem 1.2.5 [366, p. 46]) *The law of indices*

$$I^u I^v f = I^{u+v}f \tag{4.8}$$

is valid in the following cases.
  (*i*) $v > 0$, $u + v > 0$, and $f \in L_1(0, x)$.
  (*ii*) $v < 0$, $u > 0$, and $f \in I^{-v}(L_1(0, x))$.
  (*iii*) $u < 0$, $u + v < 0$, and $f \in I^{-u-v}(L_1(0, x))$.

Finally we give an integral representation of the fractional derivative $D^\gamma f$ (see also [187]).

Theorem 4.4. *Let $\nu > \gamma \geq 0$, let $f \in L_1(0, x)$ have an integrable fractional derivative $D^\nu f \in L^\infty(0, x)$, and let $D^{\nu-k}f(0) = 0$ for $k = 1, \ldots, [\nu] + 1$. Then*

$$D^\gamma f(s) = \frac{1}{\Gamma(\nu - \gamma)} \int_0^s (s - t)^{\nu-\gamma-1} D^\nu f(t)\, dt, \quad s \in [0, x]. \tag{4.9}$$

Here $D^\gamma f \in AC[0, x]$ for $\nu - \gamma \geq 1$, and $D^\gamma f \in C[0, x]$ for $\nu - \gamma \in (0, 1)$.

Proof. Set $u = \nu - \gamma > 0$ and $v = -\nu < 0$. According to Lemma 4.2, $f \in I^{-v}(L_1(0, x))$. Then case (ii) of Lemma 4.3 guarantees that the law of indices holds for this choice of u, v; namely

$$I^{\nu-\gamma}D^\nu f = I^u I^v f = I^{u+v}f = I^{-\gamma}f = D^\gamma f;$$

this is (4.9). □

4.2   Main Results

The first result here is an Opial-type inequality involving Riemann–Liouville fractional derivatives. We assume that x, ν, γ are real numbers, $x > 0$, $\nu, \gamma \geq 0$, and that $f \in L_1(0, x)$. The standard assumption on f is that $f \in I^\nu(L_1(0, x))$; we prefer to spell this out in the formulation of each theorem by specifying that f has an integrable fractional derivative $D^\nu f$ satisfying (4.6).

Theorem 4.5. *Let* $1/p + 1/q = 1$ *with* p, $q > 1$, *let* $\gamma \geq 0$, $\nu > \gamma + 1 - 1/p$, *and let* $f \in L_1(0, x)$ *have an integrable fractional derivative* $D^\nu f \in L^\infty(0, x)$ *such that* $D^{\nu-j}f(0) = 0$ *for* $j = 1, \ldots, [\nu] + 1$. *Then*

$$\int_0^x |D^\nu f(s)\, D^\nu f(s)|\, ds \leq \Omega(x) \cdot \left(\int_0^x |D^\nu f(s)|^q\, ds \right)^{2/q}, \qquad (4.10)$$

where

$$\Omega(x) = \frac{x^{(rp+2)/p}}{2^{1/q}\Gamma(r+1)((rp+1)(rp+2))^{1/p}}, \qquad r = \nu - \gamma - 1. \qquad (4.11)$$

Proof. We write $\Phi(t) = |D^\nu f(t)|$ and $r = \nu - \gamma - 1$. Because $1 - 1/p > 0$, we have $\nu > \gamma$, and Theorem 4.4 applies. Furthermore, $r > -1$, and $t \mapsto (s-t)^r \in L_1(0, s)$ for any $s \in [0, x]$. Let $0 < s \leq x$. Applying Hölder's inequality to (4.9), we get

$$|D^\nu f(s)| \leq \frac{1}{\Gamma(r+1)} \frac{s^{(rp+1)/p}}{(rp+1)^{1/p}} \left(\int_0^s \Phi(t)^q\, dt \right)^{1/q}. \qquad (4.12)$$

Write $z(s) = \int_0^s \Phi(t)^q\, dt$. Then $z'(t) = \Phi(t)^q$ almost everywhere in $(0, s)$,

$$|D^\nu f(s)| = (z'(s))^{1/q} \quad \text{a.e. in } (0, x),$$

and

$$|D^\nu f(s)\, D^\nu f(s)| \leq \frac{s^{(rp+1)/p}(z(s)z'(s))^{1/q}}{\Gamma(r+1)(rp+1)^{1/p}} \quad \text{a.e. in } (0, x).$$

The function $s^{rp+1}(z(s)z'(s))^{1/q}$ is integrable over $(0, x)$ as $rp + 1 \geq 0$ and $z(s)z'(s)$ is measurable and essentially bounded on $(0, x)$. Applying Hölder's inequality, we obtain

$$\int_0^x s^{rp+1}(z(s)z'(s))^{1/q}\, ds \leq \left(\int_0^x s^{rp+1}\, ds \right)^{1/p} \left(\int_0^x z(s)z'(s)\, ds \right)^{1/q}$$

$$\leq \frac{x^{(rp+2)/p}}{(rp+2)^{1/p}} \frac{(z(s))^{2/q}}{2^{1/q}}.$$

The result then follows when we observe that $|D^\gamma f(s) D^\nu f(s)|$ is integrable as $D^\gamma f \in AC[0, x]$ for $\nu - \gamma \geq 1$, and $D^\gamma f \in C([0, x])$ for $\nu - \gamma \in (0, 1)$, and $D^\nu f \in L^\infty(0, x)$. $\square$

The following result deals with the extreme case of the preceding theorem when $p = 1$ and $q = \infty$.

Theorem 4.6. *Let $\nu > \gamma \geq 0$, and let $f \in L_1(0, x)$ have an integrable fractional derivative $D^\nu f \in L^\infty(0, x)$ such that $D^{\nu-j} f(0) = 0$ for $j = 1, \ldots, [\nu] + 1$. Then*

$$\int_0^x |D^\gamma f(s) D^\nu f(s)| \, ds \leq \Omega_1(x) \cdot \operatorname*{ess\,sup}_{s \in [0,x]} |D^\nu f(s)|^2, \tag{4.13}$$

where

$$\Omega_1(x) = \frac{x^{r+2}}{\Gamma(r+3)}, \quad r = \nu - \gamma - 1.$$

Proof. A straightforward application of Theorem 4.4. $\square$

Theorem 4.5 has the following counterpart for the case $0 < p < 1$.

Theorem 4.7. *Let $1/p + 1/q = 1$ with $0 < p < 1$, let $\nu > \gamma \geq 0$, and let $f \in L_1(0, x)$ have an integrable fractional derivative $D^\nu f \in L^\infty(0, x)$ that is of the same sign a.e. in $(0, x)$, with $(D^\nu f)^{-1} \in L^\infty(0, x)$ such that $D^{\nu-j} f(0) = 0$ for $j = 1, \ldots, [\nu] + 1$. Then*

$$\int_0^x |D^\gamma f(s) D^\nu f(s)| \, ds \geq \Omega(x) \cdot \left(\int_0^x |D^\nu f(s)|^q \, ds \right)^{2/q}, \tag{4.14}$$

where $\Omega(x)$ is defined by (4.11).

Proof. The proof follows a similar pattern as the proof of Theorem 4.5. Because $0 < p < 1$, we need to apply the reverse Hölder's inequality [297, p. 135]

$$\int_0^x |u(s)v(s)| \, ds \geq \left(\int_0^x |u(s)|^p \right)^{1/p} \left(\int_0^x |v(s)|^q \right)^{1/q}$$

valid for any $u \in L^p(0, x)$ and $v \in L^q(0, x)$. Secondly, the assumption that $(D^\nu f)^{-1} \in L^\infty(0, x)$ is needed because $q < 0$. The details of the proof are omitted. $\square$

Under slightly strengthened hypotheses of Theorem 4.5 we obtain the following inequality involving fractional derivatives of three orders.

Theorem 4.8. *Let $1/p+1/q = 1$ with p, $q > 1$, let $\gamma \geq 0$, $\nu \geq \gamma+2-1/p$, and let $f \in L_1(0, x)$ have an integrable fractional derivative $D^\nu f \in L^\infty(0, x)$ such that $D^{\nu-j}f(0) = 0$ for $j = 1, \ldots, [\nu] + 1$. Then*

$$\int_0^x |D^\gamma f(s) \, D^{\gamma+1} f(s)| \, ds \leq \Omega_2(x) \cdot \left(\int_0^x |D^\nu f(s)|^q \, ds \right)^{2/q}, \qquad (4.15)$$

where

$$\Omega_2(x) = \frac{x^{2(rp+1)/p}}{2(\Gamma(r+1))^2 (rp+1)^{2/p}}, \qquad r = \nu - \gamma - 1. \qquad (4.16)$$

Proof. Write $\Phi(t) = |D^\nu f(t)|$ and $r = \nu - \gamma - 1$. From Theorem 4.4 and from the definition of the fractional integral we obtain

$$|D^\gamma f(x)| \leq U(x) := I^{r+1}\Phi(x), \quad |D^{\gamma+1} f(x)| \leq I^r \Phi(x) = U'(x).$$

Observing that $U'(x) = (I^{r+1}\Phi(x))' = I^r \Phi(x)$ and using Hölder's inequality, we get

$$\int_0^x |D^\gamma f(t) \, D^{\gamma+1} f(t)| \, dt \leq \int_0^x U(t) U'(t) \, dt = \tfrac{1}{2} U^2(x)$$

$$= \frac{1}{2(\Gamma(r+1))^2} \left(\int_0^x (x-t)^r \Phi(t) \, dt \right)^2$$

$$\leq \frac{1}{2(\Gamma(r+1))^2} \left(\int_0^x (x-t)^{rp} \, dt \right)^{2/p} \left(\int_0^x \Phi(t)^q \, dt \right)^{2/q}$$

$$= \frac{1}{2(\Gamma(r+1))^2} \frac{x^{2(rp+1)/p}}{(rp+1)^{2/p}} \left(\int_0^x \Phi(t)^q \, dt \right)^{2/q}. \qquad \square$$

The following result is concerned with the case when $p = 1$ and $q = \infty$ in the preceding theorem. The proof is straightforward, and we skip it.

Theorem 4.9. *Let $\gamma \geq 0$, $\nu > \gamma + 1$, and let $f \in L_1(0, x)$ have an integrable fractional derivative $D^\nu f \in L^\infty(0, x)$ such that $D^{\nu-j}f(0) = 0$ for $j = 1, \ldots, [\nu] + 1$. Then*

$$\int_0^x |D^\gamma f(s) \, D^{\gamma+1} f(s)| \, ds \leq \Omega_3(x) \cdot \operatorname{ess\,sup}_{t \in [0,x]} |D^\nu f(t)|^2, \qquad (4.17)$$

where

$$\Omega_3(x) = \frac{x^{2(\nu-\gamma)}}{2(\Gamma(\nu-\gamma+1))^2}. \qquad (4.18)$$

Remark 4.10. We show that inequality (4.17) is sharp, attained for the function $f(t) = t^\nu$. From the known properties of the gamma function,

$$\int_0^s (s-t)^{u-1} t^{v-1} \, dt = \frac{\Gamma(u)\Gamma(v)}{\Gamma(u+v)} s^{u+v-1}, \quad u, v > 0.$$

Let $0 \le j \le [\nu] + 1$, $m = [\nu] - j + 1$ and $\alpha = \nu - [\nu]$. Then $1 - \alpha > 0$, $\nu + 1 > 0$, and

$$
\begin{aligned}
D^{\nu-j} f(s) &= \frac{1}{\Gamma(1-\alpha)} \left(\frac{d}{ds} \right)^m \int_0^s (s-t)^{(1-\alpha)-1} t^{(\nu+1)-1}\, dt \\
&= \frac{1}{\Gamma(1-\alpha)} \frac{\Gamma(1-\alpha)\Gamma(\nu+1)}{\Gamma(m+j+1)} \left(\frac{d}{ds} \right)^m s^{m+j} \\
&= \frac{\Gamma(\nu+1)}{j!} s^j.
\end{aligned}
$$

Then $D^{\nu-j} f(0) = 0$ for $j = 1, \ldots, [\nu] + 1$, and $D^\nu f(s) = \Gamma(\nu + 1)$. Using Theorem 4.4, we obtain

$$
D^\gamma f(s) = \frac{\Gamma(\nu+1)}{\Gamma(\nu-\gamma+1)} s^{\nu-\gamma}, \quad D^{\gamma+1} f(s) = \frac{\Gamma(\nu+1)}{\Gamma(\nu-\gamma)} s^{\nu-\gamma-1}.
$$

Hence

$$
\begin{aligned}
\int_0^x |D^\gamma f(s)\, D^{\gamma+1} f(s)|\, ds &= \frac{(\Gamma(\nu+1))^2}{\Gamma(\nu-\gamma+1)\Gamma(\nu-\gamma)} \int_0^x s^{2(\nu-\gamma)-1}\, ds \\
&= \frac{1}{2} \left(\frac{\Gamma(\nu+1)}{\Gamma(\nu-\gamma+1)} \right)^2 x^{2(\nu-\gamma)}.
\end{aligned}
$$

On the other hand, with $\Omega_3(x)$ given by (4.18) and

$$
\text{ess sup} |D^\nu f(s)|^2 = (\Gamma(\nu+1))^2,
$$

the left side of (4.17) is equal to the right side of (4.17) for $f(t) = t^\nu$.

We give a counterpart of Theorem 4.8 for the case $0 < p < 1$. The proof again depends on the reverse Hölder's inequality, and is omitted.

Theorem 4.11. *Let* $1/p + 1/q = 1$ *with* $0 < p < 1$, *let* $\gamma \ge 0$, $\nu > \gamma + 1$, *and let* $f \in L_1(0, x)$ *have an integrable fractional derivative* $D^\nu f \in L^\infty(0, x)$ *that is of the same sign a.e. in* $(0, x)$, *with* $(D^\nu f)^{-1} \in L^\infty(0, x)$ *such that* $D^{\nu-j} f(0) = 0$ *for* $j = 1, \ldots, [\nu] + 1$. *Then*

$$
\int_0^x |D^\gamma f(s)\, D^{\gamma+1} f(s)|\, ds \ge \Omega_2(x) \cdot \left(\int_0^x |D^\nu f(s)|^q\, ds \right)^{2/q}, \quad (4.19)
$$

where $\Omega_2(x)$ *is given by (4.16).*

We derive yet another useful variant of an Opial-type inequality.

Theorem 4.12. *Let* $1/p + 1/q = 1$ *with* $p, q > 1$, *let* $\gamma \ge 0$, $\nu > \gamma + 1 - 1/p$, *and let* $f \in L_1(0, x)$ *have an integrable fractional derivative* $D^\nu f \in L^\infty(0, x)$ *such that* $D^{\nu-j} f(0) = 0$ *for* $j = 1, \ldots, [\nu] + 1$. *Then, for any* $m > 0$,

$$
\int_0^x |D^\gamma f(s)|^m\, ds \le \Omega_4(x) \cdot \left(\int_0^x |D^\nu f(s)|^q\, ds \right)^{m/q}, \quad (4.20)
$$

where

$$\Omega_4(x) = \frac{x^{(rm+1+(m/p))}}{(\Gamma(r+1))^m(rm+1+(m/p))(rp+1)^{m/p}}, \quad r = \nu-\gamma-1. \quad (4.21)$$

Proof. Inequality (4.12) holds under the hypotheses of the theorem. Raising both sides of (4.12) to the power of m and integrating from 0 to x, we get the result. $\square$

The extreme case of Theorem 4.12 with $p = 1$ and $q = \infty$ follows. The proof is omitted, as it is again a straightforward application of Theorem 4.4.

Theorem 4.13. *Let $\nu > \gamma \geq 0$, and let $f \in L_1(0,x)$ have an integrable fractional derivative $D^\nu f \in L^\infty(0,x)$ such that $D^{\nu-j}f(0) = 0$ for $j = 1,\ldots,[\nu]+1$. Then, for any $m > 0$,*

$$\int_0^x |D^\gamma f(s)|^m \, ds \leq \Omega_5(x) \cdot \text{ess sup}_{t\in[0,x]}|D^\nu f(t)|^m, \quad (4.22)$$

where

$$\Omega_5(x) = \frac{x^{m(\nu-\gamma)+1}}{(m(\nu-\gamma)+1)(\Gamma(\nu-\gamma+1))^m}.$$

4.3   Applications

(i) **Uniqueness of solution to fractional initial value problem**

$$\begin{cases} \text{Let } \gamma_i \geq 0, \nu > \gamma_i + 1/2, \, i = 1,\ldots,r \in \mathbb{N}. \\ \text{Let } f \in L_1(0,x) \text{ have an integrable fractional derivative } D^\nu f \in L^\infty(0,x) \\ \text{such that } D^{\nu-j}f(0) = \alpha_j \in \mathbb{R}, \, j = 1,\ldots,[\nu]+1. \\ \text{Furthermore, let} \\ D^\nu f(t) = F(t,\{D^{\gamma_i}f(t)\}_{i=1}^r) \text{ for all } t \in [0,x]. \end{cases}$$
$$(4.23)$$

Here $F(t,x_1,\ldots,x_r)$ is continuous for $(x_1,\ldots,x_r) \in \mathbb{R}^r$, bounded for $t \in [0,x]$, and fulfills the Lipschitz condition

$$|F(t,z_1,\ldots,z_r) - F(t,z_1',\ldots,z_r')| \leq \sum_{i=1}^r q_i(t)|z_i - z_i'|, \quad (4.24)$$

where $q_i(t) \geq 0$ are bounded on $[0,x]$, $i = 1,\ldots,r$. For $i = 1,\ldots,r$ and $0 \leq s \leq x$ we define

$$\Delta_i(s) := \frac{s^{\nu-\gamma_i}}{2\Gamma(\nu-\gamma_i)\sqrt{(\nu-\gamma_i)(2\nu-2\gamma_i-1)}}, \quad \psi(s) = \sum_{i=1}^r \|q_i\|_\infty \Delta_i(s).$$
$$(4.25)$$

where $\|q_i\|_\infty = \sup_{t\in[0,x]} |q_i(t)|$. We assume that

$$\psi(x) := \sum_{i=1}^{r} \|q_i\|_\infty \Delta_i(x) < 1. \tag{4.26}$$

Let $g \in L_1(0,x)$ have an integrable fractional derivative $D^\nu g \in L^\infty(0,x)$ such that $D^{\nu-j}g(0) = 0$, $j = 1, \ldots, [\nu]+1$. Then, by Theorem 4.4, we have

$$D^{\gamma_i}g(s) = \frac{1}{\Gamma(\nu - \gamma_i)} \int_0^s (s-t)^{\nu-\gamma_i-1} D^\nu g(t)\, dt, \quad s \in [0,x], \; i = 1, \ldots, r. \tag{4.27}$$

When $p = q = 2$, from (4.10) we get for $i = 1, \ldots, r$,

$$\int_0^x |(D^{\gamma_i}g)(w)|\,|(D^\nu g)(w)|\, dw \le \Delta_i(x) \int_0^x |(D^\nu g)(w)|^2\, dw. \tag{4.28}$$

Let f_1, f_2 solve (4.23); that is, let for $k = 1, 2$,

$$(D^\nu f_k)(t) = F(t, \{(D^{\gamma_i} f_k)(t)\}_{i=1}^r), \quad t \in [0,x],$$

and

$$D^{\nu-j}f_k(0) = \alpha_j \in \mathbb{R}, \quad j = 1, \ldots [\nu]+1.$$

If $g := f_1 - f_2$, then

$$D^\nu f(t) = F(t, \{(D^{\gamma_i} f_1)(t)\}_{i=1}^r) - F(t, \{(D^{\gamma_i} f_2)(t)\}_{i=1}^r), \tag{4.29}$$

and

$$D^{\nu-j}g(0) = 0, \quad j = 1, \ldots, [\nu]+1.$$

By (4.24),

$$|F(t, D^{\gamma_1} f_1(t), \ldots, D^{\gamma_r} f_1(t)) - F(t, D^{\gamma_1} f_2(t), \ldots, D^{\gamma_r} f_2(t))|$$

$$\le \sum_{i=1}^{r} q_i(t)|D^{\gamma_i} f_1(t) - D^{\gamma_i} f_2(t)|$$

$$= \sum_{i=1}^{r} q_i(t)|D^{\gamma_i} g(t)|$$

$$\le \sum_{i=1}^{r} \|q_i\|_\infty |D^{\gamma_i} g(t)|. \tag{4.30}$$

Thus

$$|D^\nu g(t)|^2 = |D^\nu g(t)|\,|F(t, \{D^{\gamma_i} f_1(t)\}_{i=1}^r) - F(t, \{D^{\gamma_i} f_2(t)\}_{i=1}^r)|$$

$$\le |D^\nu g(t)| \sum_{i=1}^{r} \|q_i\|_\infty |D^{\gamma_i} g(t)|$$

$$= \sum_{i=1}^{r} \|q_i\|_\infty |D^{\gamma_i} g(t)|\,|D^\nu g(t)|.$$

Consequently,

$$\int_0^x |D^\nu g(t)|^2 \, dt \le \sum_{i=1}^r \|q_i\|_\infty \int_0^x |D^{\gamma_i} g(t)| \, |D^\nu g(t)| \, dt$$

$$\overset{(4.28)}{\le} \sum_{i=1}^r \|q_i\|_\infty \Delta_i(x) \int_0^x |D^\nu g(t)|^2 \, dt$$

$$\overset{(4.26)}{\le} \psi(x) \int_0^x |D^\nu g(t)|^2 \, dt;$$

that is,

$$\int_0^x |D^\nu g(t)|^2 \, dt \le \psi(x) \int_0^x |D^\nu g(t)|^2 \, dt. \tag{4.31}$$

If $\int_0^x |D^\nu g(t)|^2 \, dt \ne 0$, then by (4.31) we obtain $\psi(x) \ge 1$, a contradiction to the assumption $\psi(x) < 1$. Hence $\int_0^x |D^\nu g(t)|^2 \, dt = 0$; that is, $D^\nu g(t) = 0$ a.e. in $[0, x]$. But $D^{\nu-j} g(0) = 0$, $j = 1, \ldots, [\nu] + 1$. Then from (4.9) for $\gamma = 0$ we find $g(t) \equiv 0$ in $[0, x]$. This implies $f_1 = f_2$ on $[0, x]$, proving the uniqueness of the solution to the initial value problem (4.23).

(ii) **Upper bounds on $D^\nu f$, solution f, and others**

$$
\left\{
\begin{array}{l}
\text{Consider the initial value problem for } 0 \le t \le x: \\[4pt]
(D^\nu f)'(t) = F(t, \{D^{\gamma_i} f(t)\}_{i=1}^r, D^\nu f). \\[4pt]
\gamma_i \ge 0, \ \nu > \gamma_i + 1/2, \ i = 1, \ldots, r \in \mathbb{N}. \\[4pt]
\text{Here } f \in L_1(0, x) \text{ has an integrable fractional derivative} \\[4pt]
D^\nu f \in L^\infty(0, x) \text{ which is absolutely continuous.} \\[4pt]
\text{We assume that } D^{\nu-j} f(0) = 0, \ j = 1, \ldots, [\nu] + 1 \text{ and } D^\nu f(0) = A \in \mathbb{R}.
\end{array}
\right. \tag{4.32}
$$

Here F is Lebesgue measurable on $[0, x] \times \mathbb{R}^{r+1}$, and fulfills the condition

$$|F(t, x_1, \ldots, x_r, x_{r+1})| \le \sum_{i=1}^r q_i(t) |x_i|, \tag{4.33}$$

where $q_i(t) \ge 0$ are bounded on $[0, x]$, $i = 1, \ldots, r$. We see that

$$D^\nu f(t)(D^\nu f)'(t) = D^\nu f(t) F(t, \{D^{\gamma_i} f(t)\}_{i=1}^r, D^\nu f(t)),$$

and for $0 \le s \le x$ we have

$$\int_0^s D^\nu f(t)(D^\nu f)'(t) \, dt = \int_0^s D^\nu f(t) F(t, \{D^{\gamma_i} f(t)\}_{i=1}^r, D^\nu f(t)) \, dt.$$

Hence

$$\frac{1}{2}|D^\nu f(t)|^2\Big|_0^s \leq \int_0^s |D^\nu f(t)| |F(t, \{D^{\gamma_i} f(t)\}_{i=1}^r, D^\nu f(t))| \, dt$$

$$\overset{(4.33)}{\leq} \int_0^s |D^\nu f(t)| \left(\sum_{i=1}^r \|q_i\|_\infty |D^{\gamma_i} f(t)| \right) dt$$

$$= \sum_{i=1}^r \|q_i\|_\infty \left(\int_0^s |D^{\gamma_i} f(t)| |D^\nu f(t)| \, dt \right).$$

Recall the notation (4.25) for $\Delta_i(s)$ and $\psi(s)$. Then

$$|D^\nu f(s)|^2 \leq A^2 + 2 \sum_{i=1}^r \|q_i\|_\infty \left(\int_0^s |D^{\gamma_i} f(t)| |D^\nu f(t)| \, dt \right)$$

$$\overset{(4.10)}{\leq} A^2 + \left(\sum_{i=1}^r \|q_i\|_\infty \Delta_i(s) \right) \left(\int_0^s |D^\nu f(t)|^2 \, dt \right)$$

$$= A^2 + \psi(s) \int_0^s |D^\nu f(t)|^2 \, dt;$$

that is,

$$|D^\nu f(s)|^2 \leq A^2 + \psi(s) \int_0^s |D^\nu f(t)|^2 \, dt. \tag{4.34}$$

Set $\theta(s) := |D^\nu f(s)|^2$ for $0 \leq s \leq x$ and $\rho := A^2$. Then

$$\theta(s) \leq \rho + \psi(s) \int_0^s \theta(t) \, dt,$$

where $\rho \geq 0$, $\psi(s) \geq 0$, $\psi(0) = 0$, $\theta(s) \geq 0$ for all $0 \leq s \leq x$. We can apply the generalized Gronwall lemma [144, Corollary 1.1.2] (with $H(x) = x$) to obtain

$$\theta(s) \leq \rho \left(1 + \psi(s) \exp(\Psi(s)) \int_0^s \exp(-\Psi(t)) \, dt \right), \quad \Psi(t) = \int_0^t \psi(u) \, du. \tag{4.35}$$

We have shown that

$$|D^\nu f(s)| \leq |A| \left(1 + \psi(s) \exp(\Psi(s)) \int_0^s \exp(-\Psi(t)) \, dt \right)^{1/2} =: K(s) \tag{4.36}$$

for all $0 \leq s \leq x$. From (4.9) with $\gamma = 0$ we get

$$|f(s)| \leq \frac{1}{\Gamma(\nu)} \int_0^s (s-t)^{\nu-1} |D^\nu f(t)| \, dt$$

for all $0 \leq s \leq x$. Applying (4.36), we obtain

$$|f(s)| \leq \frac{1}{\Gamma(\nu)} \int_0^s (s-t)^{\nu-1} K(t) \, dt. \tag{4.37}$$

Also from (4.9) we have

$$|D^{\gamma_i} f(s)| \leq \frac{1}{\Gamma(\nu - \gamma_i)} \int_0^s (s-t)^{\nu-\gamma_i-1} |D^{\nu} f(t)| \, dt$$

for all $0 \leq s \leq x$, $i = 1, \ldots, r$. Finally, by (4.36) we find

$$|D^{\gamma_i} f(s)| \leq \frac{1}{\Gamma(\nu - \gamma_i)} \int_0^s (s-t)^{\nu-\gamma_i-1} K(t) \, dt \qquad (4.38)$$

for all $0 \leq s \leq x$ and $i = 1, \ldots, r$.

5
Opial-type L^p-Inequalities for Riemann–Liouville Fractional Derivatives

This chapter presents a class of L^p-type Opial inequalities for generalized Riemann–Liouville fractional derivatives of integrable functions. The novelty of this approach is the use of the index law for fractional derivatives instead of a Taylor's formula, which enables us to relax restrictions on the orders of fractional derivatives. This treatment relies on [65].

5.1 Introduction and Preliminaries

The Opial inequality, which appeared in [315], is of great interest in differential equations and other areas of mathematics, and has attracted a great deal of attention in the recent literature. For classical derivatives it has been generalized in several directions (see, for instance, [3, 4, 323]). Love gave a generalization for fractional integrals [261]. This chapter is inspired by the author's paper [15]. Here we consider Lebesgue integrable functions, whereas [15] dealt with continuous functions using a different definition of the fractional derivative.

Our brief survey of basic facts about fractional derivatives is based on the monograph [366] by Samko et al.; most of the results needed in the sequel are contained in Chapter 1 of [366]. The crucial result is Theorem 5.4, which replaces Taylor's formula in the derivation of various estimates.

Throughout the chapter, x denotes a fixed positive number. By $C^m[0, x]$ we denote the space of all functions on $[0, x]$ that have continuous derivatives up to order m, and $AC[0, x]$ is the space of all absolutely continuous

G.A. Anastassiou, *Fractional Differentiation Inequalities*,
DOI 10.1007/978-0-387-98128-4_5, © Springer Science+Business Media, LLC 2009

functions on $[0, x]$. By $AC^m[0, x]$ we denote the space of all functions
$g \in C^m[0, x]$ with $g^{(m-1)} \in AC[0, x]$. For any $\alpha \in \mathbb{R}$ we denote by $[\alpha]$
the integral part of α (the integer k satisfying $k \leq \alpha < k + 1$). If $p \in \mathbb{R}$,
$p > 0$, by $L^p(0, x)$ we denote the space of all Lebesgue measurable func-
tions f for which $|f|^p$ is Lebesgue integrable on the interval $(0, x)$, and by
$L^\infty(0, x)$ the set of all functions measurable and essentially bounded on
$(0, x)$. For any $f \in L^\infty(0, x)$ we write $\|f\|_\infty = \sup_{t \in [0,x]} |f(t)|$. We observe
that $L^\infty(0, x) \subset L^p(0, x)$ for all $p > 0$.

For any $a \in \mathbb{R}$ we write $a_+ = \max(a, 0)$ and $a_- = (-a)_+$.

For the sake of completeness we give a proof of the following known result
which provides a basis for the existence of fractional integrals and is needed
in another context in the chapter.

Lemma 5.1. *Let $f \in L^1(0, x)$ and let $\alpha > -1$ be a real number. Then*

$$F(s) = \int_0^s (s - t)^\alpha f(t)\, dt$$

exists for almost all $s \in [0, x]$, and $F \in L^1(0, x)$.

Proof. Define $k : \Omega := [0, x] \times [0, x] \to \mathbb{R}$ by $k(s, t) = (s - t)_+^\alpha$; that is,

$$k(s, t) = \begin{cases} (s - t)^\alpha & \text{if } 0 \leq t < s \leq x, \\ 0 & \text{if } 0 \leq s \leq t \leq x. \end{cases}$$

Then k is measurable on Ω, and

$$\int_0^x k(s, t)\, ds = \int_0^t k(s, t)\, ds + \int_t^x k(s, t)\, ds = \int_t^x (s - t)^\alpha\, ds = (\alpha + 1)^{-1}(x - t)^{\alpha + 1}.$$

Because the repeated integral

$$\int_0^x dt \int_0^x k(s, t)|f(t)|\, ds = (\alpha + 1)^{-1} \int_0^x (x - t)^{\alpha + 1}|f(t)|\, dt$$

exists and is finite, the function $(s, t) \mapsto k(s, t)f(t)$ is integrable over Ω by
Tonelli's theorem, and the conclusion follows from Fubini's theorem. $\square$

Let $\alpha > 0$. For any $f \in L^1(0, x)$ the *Riemann–Liouville fractional integral
of f of order α* is defined by

$$I^\alpha f(s) = \frac{1}{\Gamma(\alpha)} \int_0^s (s - t)^{\alpha - 1} f(t)\, dt, \quad s \in [0, x]. \tag{5.1}$$

By Lemma 5.1 the integral on the right side of (5.1) exists for almost all
$s \in [0, x]$, and $I^\alpha f \in L^1(0, x)$. The *Riemann–Liouville fractional derivative
of $f \in L^1(0, x)$ of order α* is defined by

$$D^\alpha f(s) = \left(\frac{d}{ds}\right)^m I^{m-\alpha} f(s) = \frac{1}{\Gamma(m - \alpha)} \left(\frac{d}{ds}\right)^m \int_0^s (s - t)^{m - \alpha - 1} f(t)\, dt, \tag{5.2}$$

where $m = [\alpha] + 1$, provided that the derivative exists. In addition, we set

$$D^0 f := f =: I^0 f, \quad I^{-\alpha} f := D^\alpha f \text{ if } \alpha > 0, \quad D^{-\alpha} f := I^\alpha f \text{ if } 0 < \alpha \le 1. \tag{5.3}$$

If α is a positive integer, then $D^\alpha f = (d/ds)^\alpha f$.

A more general definition of fractional integrals and derivatives uses an anchor point other than 0: let $f \in L^1(a,b)$, where $-\infty < a < b < \infty$. For any $s \in [a,b]$ set

$$I_{a+}^\alpha f(s) := \frac{1}{\Gamma(\alpha)} \int_a^s (s-t)^{\alpha-1} f(t)\, dt, \quad I_{b-}^\alpha f(s) := \frac{1}{\Gamma(\alpha)} \int_s^b (s-t)^{\alpha-1} f(t)\, dt.$$

The two fractional derivatives are then defined by an obvious modification of (5.2). All our results stated for the specialized fractional derivative (5.2) have an interpretation for the fractional derivatives with a general anchor point.

Let $\alpha > 0$ and $m = [\alpha] + 1$. A function $f \in L^1(0,x)$ is said to have an *integrable fractional derivative* $D^\alpha f$ (see Definition 2.4 in [366, p. 44]) if

$$I^{m-\alpha} f \in AC^m[0,x]. \tag{5.4}$$

We define the space $I^\alpha(L^1(0,x))$ as the set of all functions f on $[0,x]$ of the form $f = I^\alpha \varphi$ for some $\varphi \in L^1(0,x)$ (see Definition 2.3 in [366, p. 43]). We express these conditions in terms of fractional derivatives.

Lemma 5.2. *Let* $\alpha > 0$ *and* $m = [\alpha]+1$. *A function* $f \in L^1(0,x)$ *has an integrable fractional derivative* $D^\alpha f$ *if and only if*

$$D^{\alpha-k} f \in C[0,x], \ k = 1,\ldots,m, \quad \text{and} \quad D^{\alpha-1} f \in AC[0,x]. \tag{5.5}$$

Furthermore, $f \in I^\alpha(L^1(0,x))$ *if and only if* f *has an integrable fractional derivative* $D^\alpha f$ *and satisfies the conditions*

$$D^{\alpha-k} f(0) = 0 \ \text{for} \ k = 1,\ldots,m. \tag{5.6}$$

Proof. Notice that

$$\left(\frac{d}{ds}\right)^k I^{m-\alpha} f = \left(\frac{d}{ds}\right)^k I^{k-(\alpha-m+k)} f = D^{\alpha-m+k} f$$

in view of the definition of fractional derivative and the equation $[\alpha - m + k] + 1 = k$. Then (5.5) is equivalent to (5.4) and (5.6) is equivalent to condition (2.56) in [366, p. 43]. (For $k = m$ we use the convention $D^{\alpha-m} f = I^{m-\alpha} f$ in (5.5).) □

We need the following result on the law of indices for fractional integration and differentiation using the unified notation (5.3).

Lemma 5.3. (Theorem 2.5 in [366, p. 46]) *The law of indices*

$$I^\mu I^\nu f = I^{\mu+\nu} f \tag{5.7}$$

is valid in the following cases.:
  (*i*) $\nu > 0$, $\mu + \nu > 0$ *and* $f \in L^1(0, x)$.
  (*ii*) $\nu < 0$, $\mu > 0$ *and* $f \in I^{-\nu}(L^1(0, x))$.
  (*iii*) $\mu < 0$, $\mu + \nu < 0$ *and* $f \in I^{-\mu-\nu}(L^1(0, x))$.

The following theorem is a powerful analogue of Taylor's formula with vanishing fractional derivatives of lower orders. In this chapter it is used as the main tool for deriving inequalities. Observe that we do not require $\alpha \geq \beta + 1$ but merely $\alpha > \beta$.

Theorem 5.4. *Let* $\alpha > \beta \geq 0$, *let* $f \in L^1(0, x)$ *have an integrable fractional derivative* $D^\alpha f$, *and let* $D^{\alpha-k}f(0) = 0$ *for* $k = 1, \ldots, [\alpha] + 1$. *Then*

$$D^\beta f(s) = \frac{1}{\Gamma(\alpha - \beta)} \int_0^s (s - t)^{\alpha-\beta-1} D^\alpha f(t) \, dt, \quad s \in [0, x]. \tag{5.8}$$

Proof. Set $\mu = \alpha - \beta > 0$ and $\nu = -\alpha < 0$. According to Lemma 5.2, $f \in I^{-\nu}(L^1(0, x))$. Then case (ii) of Lemma 5.3 guarantees that the law of indices holds for this choice of μ, ν; namely

$$I^{\alpha-\beta} D^\alpha f = I^\mu I^\nu f = I^{\mu+\nu} f = I^{-\beta} f = D^\beta f;$$

this proves the result. Notice that the existence of the integral on the right in (5.8) is guaranteed by Lemma 5.1. $\square$

5.2   Main Results

We assume here that x, ν are positive real numbers, and that $f \in L^1(0, x)$. The standard assumption on f is that $f \in I^\nu(L^1(0, x))$; this is equivalent to f having an integrable fractional derivative $D^\nu f$ satisfying (5.5). In addition we require that $D^\nu f$ be essentially bounded to guarantee that $D^\nu f \in L^p(0, x)$ for $p > 0$. We tabulate the notation used in this section. The inequalities between ν and μ_i are assumed throughout.

l	A positive integer.
x, ν, r_i	Positive real numbers, $i = 1, \dots, l$.
r	$r = \sum_{i=1}^{l} r_i$.
μ_i	Real numbers satisfying $0 \le \mu_i < \nu$, $i = 1, \dots, l$.
α_i	$\alpha_i = \nu - \mu_i - 1$, $i = 1, \dots, l$.
α	$\alpha = \max \{(\alpha_i)_- : i = 1, \dots, l\}$.
β	$\beta = \max \{(\alpha_i)_+ : i = 1, \dots, l\}$.
ω_1, ω_2	Continuous positive weight functions on $[0, x]$.
ω	Continuous nonnegative weight function on $[0, x]$.
s_k, s_k'	$s_k > 0$ and $1/s_k + 1/s_k' = 1$, $k = 1, 2$.

For brevity we write $\boldsymbol{\mu} = (\mu_1, \dots, \mu_l)$ for a selection of the orders μ_i of fractional derivatives, and $\boldsymbol{r} = (r_1, \dots, r_l)$ for a selection of the constants r_i.

We derive a very general Opial-type inequality involving Riemann–Liouville fractional derivatives of an integrable function f, which is analogous to [323, Theorem 1.3] for ordinary derivatives and to [15, Theorem 2] for fractional derivatives.

Theorem 5.5. *Let $f \in L^1(0, x)$ have an integrable fractional derivative $D^\nu f \in L^\infty(0, x)$ such that $D^{\nu-j}f(0) = 0$ for $j = 1, \dots, [\nu] + 1$. For $k = 1, 2$, let $s_k > 1$, let $p \in \mathbb{R}$ satisfy*

$$\alpha s_2 < 1, \qquad p > \frac{s_2}{1 - \alpha s_2}, \tag{5.9}$$

and let $\sigma = 1/s_2 - 1/p$. Finally let

$$Q_1 = \left(\int_0^x \omega_1(\tau)^{s_1'} d\tau \right)^{1/s_1'} \quad and \quad Q_2 = \left(\int_0^x \omega_2(\tau)^{-s_2'/p} d\tau \right)^{r/s_2'}. \tag{5.10}$$

Then

$$\int_0^x \omega_1(\tau) \prod_{i=1}^{l} |D^{\mu_i}f(\tau)|^{r_i} d\tau \le Q_1 Q_2 \, C_1 x^{\rho + 1/s_1} \left(\int_0^x \omega_2(\tau)|D^\nu f(\tau)|^p d\tau \right)^{r/p}, \tag{5.11}$$

where $\rho := \sum_{i=1}^{l} \alpha_i r_i + \sigma r$ and

$$C_1 = C_1(\nu, \boldsymbol{\mu}, \boldsymbol{r}, p, s_1, s_2) := \frac{\sigma^{r\sigma}}{\prod_{i=1}^{l} \Gamma(\nu - \mu_i)^{r_i}(\alpha_i + \sigma)^{r_i \sigma}(\rho s_1 + 1)^{1/s_1}}. \tag{5.12}$$

Proof. First we show that the conditions on s_2 and p guarantee that, for $i = 1, \dots l$,

$$\text{(a) } p > s_2 > 1, \quad \text{(b) } \alpha_i s_2 > -1, \quad \text{(c) } \alpha_i + \sigma > 0. \tag{5.13}$$

This is clear if $\alpha = 0$. If $\alpha > 0$, then $0 < 1 - \alpha s_2 < 1$ and $p > s_2/(1 - \alpha s_2) > s_2 > 1$. For each $i \in \{1, \ldots, l\}$, $\alpha_i \geq -\alpha$, and $\alpha_i s_2 \geq -\alpha s_2 > -1$; furthermore,

$$\alpha_i + \sigma = \alpha_i + \frac{1}{s_2} - \frac{1}{p} = \frac{1 + \alpha_i s_2}{s_2} - \frac{1}{p} \geq \frac{1 - \alpha s_2}{s_2} - \frac{1}{p} > 0.$$

For brevity we write

$$k_i(\tau, t) = (\tau - t)_+^{\alpha_i}, \ i = 1, \ldots, l, \ \Phi(t) = |D^\nu f(t)|, \ 0 \leq \tau, t \leq x.$$

From (5.13) it follows that

$$k_i(\tau, \cdot) \in L^{s_2}(0, x) \text{ and } k_i(\tau, \cdot) \in L^{1/\sigma}(0, x). \tag{5.14}$$

Let $i \in \{1, \ldots, l\}$ and $\tau \in [0, x]$. We then apply Hölder's inequality twice (with the conjugate indices s'_2, s_2 and p/s_2, $p/(p - s_2)$) taking into account (5.14) and the fact that w_2^{-1}, w_2, and Φ are (essentially) bounded:

$$
\begin{aligned}
\int_0^x k_i(\tau, t) \Phi(t)\, dt &= \int_0^x w_2(t)^{-1/p} w_2(t)^{1/p} \Phi(t) k_i(\tau, t) dt \\
&\leq \left(\int_0^x w_2(t)^{-s'_2/p} dt \right)^{1/s'_2} \\
&\quad \left(\int_0^x w_2(t)^{s_2/p} \Phi(t)^{s_2} k_i(\tau, t)^{s_2} dt \right)^{1/s_2} \\
&\leq Q_2^{1/r} \left(\int_0^x w_2(t) \Phi(t)^p dt \right)^{1/p} \left(\int_0^x k_i(\tau, t)^{1/\sigma} dt \right)^\sigma \\
&= Q_2^{1/r} \left(\int_0^x w_2(t) \Phi(t)^p dt \right)^{1/p} \frac{\sigma^\sigma \tau^{\alpha_i + \sigma}}{(\alpha_i + \sigma)^\sigma}.
\end{aligned}
$$

By Theorem 5.4,

$$\Gamma(\nu - \mu_i)|D^{\mu_i} f(\tau)| \leq \int_0^\tau (\tau - t)^{\alpha_i} \Phi(t)\, dt = \int_0^x k_i(\tau, t) \Phi(t)\, dt. \tag{5.15}$$

Therefore

$$\int_0^x \omega_1(\tau) \prod_{i=1}^l |D^{\mu_i}f(\tau)|^{r_i}\, d\tau$$

$$\le \int_0^x \omega_1(\tau) \prod_{i=1}^l \frac{1}{\Gamma(\nu - \mu_i)^{r_i}} \left(\int_0^x k_i(\tau,t)\Phi(t)\, dt \right)^{r_i} d\tau$$

$$\le \int_0^x \omega_1(\tau) \prod_{i=1}^l \frac{1}{\Gamma(\nu - \mu_i)^{r_i}} Q_2^{r_i/r} \left(\int_0^x \omega_2(t)\Phi(t)^p\, dt \right)^{r_i/p}$$

$$\cdot \frac{\sigma^{r_i\sigma}}{(\alpha_i + \sigma)^{r_i\sigma}} \tau^{(\alpha_i+\sigma)r_i}\, d\tau$$

$$= \frac{\sigma^{r\sigma}}{\prod_{i=1}^l \Gamma(\nu - \mu_i)^{r_i}(\alpha_i + \sigma)^{r_i\sigma}} Q_2 \left(\int_0^x \omega_2(t)\Phi(t)^p\, dt \right)^{r/p}$$

$$\cdot \int_0^x \omega_1(\tau) \left(\prod_{i=1}^l \tau^{(\alpha_i+\sigma)r_i} \right) d\tau$$

$$= \Delta Q_2 \left(\int_0^x \omega_2(t)\Phi(t)^p\, dt \right)^{r/p} \int_0^x \omega_1(\tau)\tau^\rho\, d\tau$$

$$\le \Delta Q_2 \left(\int_0^x \omega_2(t)\Phi(t)^p\, dt \right)^{r/p} \left(\int_0^x \omega_1(\tau)^{s_1'}\, d\tau \right)^{1/s_1'} \left(\int_0^x \tau^{\rho s_1}\, d\tau \right)^{1/s_1}$$

$$= \frac{\Delta}{(\rho s_1 + 1)^{1/s_1}} Q_2 \left(\int_0^x \omega_2(t)\Phi(t)^p\, dt \right)^{r/p} Q_1\, x^{\rho+1/s_1}$$

where $\Delta := \sigma^{r\sigma}/(\prod_{i=1}^l \Gamma(\nu - \mu_i)^{r_i}(\alpha_i + \sigma)^{r_i\sigma})$. This completes the proof. $\square$

Next we consider the extreme case $p = \infty$ in analogy with [15, Proposition 1].

Theorem 5.6. *Let* $f \in L^1(0,x)$ *have an integrable fractional derivative* $D^\nu f \in L^\infty(0,x)$ *such that* $D^{\nu-j}f(0) = 0$ *for* $j = 1, \ldots, [\nu] + 1$. *Then*

$$\int_0^x \omega(\tau) \prod_{i=1}^l |D^{\mu_i}f(\tau)|^{r_i}\, d\tau \le \frac{\|\omega\|_\infty\, x^\rho}{\rho \prod_{i=1}^l \Gamma(\nu - \mu_i + 1)^{r_i}} \|D^\nu f\|_\infty^r, \qquad (5.16)$$

where $\rho = \sum_{i=1}^l (\nu - \mu_i)r_i + 1$.

Proof. By Theorem 5.4,

$$|D^{\mu_i}f(\tau)| \le \frac{1}{\Gamma(\nu - \mu_i)} \int_0^\tau (\tau - t)^{\alpha_i} |D^\nu f(t)|\, dt,$$

which implies

$$|D^{\mu_i}f(\tau)| \leq \frac{\|D^\nu f\|_\infty}{\Gamma(\nu - \mu_i)} \frac{\tau^{\nu - \mu_i}}{\nu - \mu_i} = \frac{\|D^\nu f\|_\infty \tau^{\nu - \mu_i}}{\Gamma(\nu - \mu_i + 1)}. \tag{5.17}$$

The result then follows when we raise (5.17) to the power of r_i, take the product from 1 to l, multiply by $\omega(\tau)$, and integrate with respect to τ from 0 to x. □

We have the following counterpart of Theorem 5.5 with s_1, $s_2 \in (0,1)$ and p negative.

Theorem 5.7. *Let $f \in L^1(0, x)$ have an integrable fractional derivative $D^\nu f \in L^\infty(0, x)$ that is of the same sign a.e. in $(0, x)$ and satisfies $D^{\nu - j}f(0) = 0$, $j = 1, \ldots, [\nu] + 1$. For $k = 1, 2$, let $0 < s_k < 1$, let $p < 0$, and let $\sigma = 1/s_2 - 1/p$. Then*

$$\int_0^x \omega_1(\tau) \prod_{i=1}^l |D^{\mu_i}f(\tau)|^{r_i} \, d\tau \geq Q_1 Q_2 C_1 x^{\rho + 1/s_1} \left(\int_0^x \omega_2(\tau)|D^\nu f(\tau)|^p \, d\tau \right)^{r/p},$$

$$\tag{5.18}$$

where $\rho = \sum_{i=1}^l \alpha_i r_i + \sigma r$, Q_1 and Q_2 are defined by (5.10), and C_1 is defined by (5.12).

Proof. Combining Theorem 5.4 with the hypotheses on $D^\nu f$, we have

$$\Gamma(\nu - \mu_i)|D^{\mu_i}f(\tau)| = \int_0^\tau (\tau - t)^{\alpha_i} \Phi(t) \, dt = \int_0^x k_i(\tau, t)\Phi(t) \, dt, \tag{5.19}$$

where $\Phi(t) = D^\nu f$ or $\Phi(t) = -D^\nu f$ (depending on the sign of $D^\nu f$ in $(0, x)$).

Because $\alpha_i > -1$ and $0 < s_2 < 1$, we have $\alpha_i s_2 > -1$. Furthermore, $\sigma = 1/s_2 - 1/p > 0$. Writing $k_i(\tau, t) = (\tau - t)_+^{\alpha_i}$ $(i = 1, \ldots, l)$, we have

$$k_i(\tau, \cdot) \in L^{s_2}(0, x) \quad \text{and} \quad k_i(\tau, \cdot) \in L^{1/\sigma}(0, x). \tag{5.20}$$

We can now retrace the proof of Theorem 5.5, relying on (5.20) and using the reverse Hölder's inequality in place of Hölder's inequality proper (as $0 < s_k < 1$ for $k = 1, 2$ and $p < 0$). □

A possible choice of p in this theorem is $p = (s_1 s_2^2)/(s_1 s_2 - 1)$. This results in an inequality similar to the one obtained earlier by the author [15, Theorem 3].

We obtain yet another counterpart of Theorem 5.5 if we assume that s_1, s_2, and p lie in the interval $(0, 1)$. In this case the hypotheses on s_1, s_2, and p are of necessity more restrictive.

Theorem 5.8. *Let $f \in L^1(0, x)$ have an integrable fractional derivative $D^\nu f \in L^\infty(0, x)$ that is of the same sign a.e. in $(0, x)$ and satisfies*

$D^{\nu-j}f(0) = 0$, $j = 1, \ldots, [\nu] + 1$. *For* $k = 1, 2$, *let* $0 < s_k < 1$, *let* $r s_1 \leq 1$, $p \in \mathbb{R}$,

$$\frac{s_2}{1 - \alpha s_2 + s_2} < p < \frac{s_2}{1 + \beta s_2}, \tag{5.21}$$

and let $\sigma = 1/s_2 - 1/p$. *Then the inequality* (5.18) *holds where* $\rho = \sum_{i=1}^{l} \alpha_i r_i + \sigma r$, Q_1 *and* Q_2 *are defined by* (5.10), *and* C_1 *is defined by* (5.12).

Proof. We show that condition (5.21) guarantees that, for $i = 1, \ldots l$,

$$\text{(a) } 0 < p < s_2 < 1, \qquad \text{(b) } -1 < \alpha_i + \sigma < 0. \tag{5.22}$$

Because $1 - \alpha s_2 + s_2 > 0$ and $1 + \beta s_2 \geq 1$, inequality (5.22) (a) follows directly from (5.21). Furthermore, we have $\alpha_i + \sigma = (1 + \alpha_i s_2)/s_2 - 1/p$, and

$$-1 < \frac{1 - \alpha s_2}{s_2} - \frac{1}{p} \leq \frac{1 + \alpha_i s_2}{s_2} - \frac{1}{p} < \frac{1 + \beta s_2}{s_2} - \frac{1}{p} < 0.$$

This proves (5.22) (b).

Because $\alpha_i > -1$ and $0 < s_2 < 1$, we have $\alpha_i s_2 > -1$. Furthermore, $\sigma < 0$, and $\alpha_i/\sigma > -1$. Writing $k_i(\tau, t) = (\tau - t)_+^{\alpha_i}$ $(i = 1, \ldots, l)$, we have

$$k_i(\tau, \cdot) \in L^{s_2}(0, x) \quad \text{and} \quad k_i(\tau, \cdot) \in L^{1/\sigma}(0, x). \tag{5.23}$$

As in the proof of the preceding theorem we have

$$\Gamma(\nu - \mu_i)|D^{\mu_i}f(\tau)| = \int_0^\tau (\tau - t)^{\alpha_i} \Phi(t) \, dt = \int_0^x k_i(\tau, t) \Phi(t) \, dt, \tag{5.24}$$

where $\Phi(t) = D^\nu f$ or $\Phi(t) = -D^\nu f$ (depending on the sign of $D^\nu f$ in $(0, x)$).

We can now retrace the proof of Theorem 5.5, relying on (5.23) and using the reverse Hölder's inequality in place of Hölder's inequality proper (as $0 < s_k < 1$ for $k = 1, 2$ and $0 < p < 1$). For the last application of Hölder's inequality we need $\tau^\rho \in L^{s_1}(0, x)$. This follows from

$$\rho s_1 = \sum_{i=1}^{l} (\alpha_i + \sigma) r_i s_1 > -r s_1 \geq -1$$

taking into account the assumption $r s_1 \leq 1$.   □

We present a version of Opial's inequality with $l = 2$ motivated by Pang and Agarwal's extension [323, Theorem 1.1] of an inequality due to Fink [160] for classical derivatives. This was further extended in [15, Theorem 4] to fractional derivatives. Our proof is similar to the one given in [323]. In view of the auxiliary inequalities used, in particular of (5.27), the theorem does not extend easily to $l > 2$.

Theorem 5.9. *Let $f \in L^1(0, x)$ have an integrable fractional derivative $D^\nu f \in L^\infty(0, x)$ such that $D^{\nu-j}f(0) = 0$ for $j = 1, \ldots, [\nu]+1$. Let $\nu > \mu_2 \geq \mu_1 + 1 \geq 1$. If $p, q > 1$ are such that $1/p + 1/q = 1$, then*

$$\int_0^x |D^{\mu_1}f(\tau)||D^{\mu_2}f(\tau)|\, d\tau \leq C_2 x^{2\nu - \mu_1 - \mu_2 - 1 + 2/q} \left(\int_0^x |D^\nu f(\tau)|^p\, d\tau \right)^{2/p},$$
$$(5.25)$$

where $C_2 = C_2(\nu, \mu_1, \mu_2, p)$ is given by

$$C_2 := \frac{(1/2)^{1/p}}{\Gamma(\nu - \mu_1)\Gamma(\nu - \mu_2 + 1)((\nu - \mu_1)q + 1)^{1/q}((2\nu - \mu_1 - \mu_2 - 1)q + 2)^{1/q}}.$$
$$(5.26)$$

Proof. First an auxiliary inequality: write $\alpha_i = \nu - \mu_i - 1$ for $i = 1, 2$; in view of the hypothesis $\mu_2 \geq \mu_1 + 1$ we have $\alpha_1 - \alpha_2 - 1 \geq 0$. Let $0 \leq t \leq s \leq x$. Then

$$\int_0^x \left[(\tau - t)_+^{\alpha_1}(\tau - s)_+^{\alpha_2} + (\tau - s)_+^{\alpha_1}(\tau - t)_+^{\alpha_2} \right] d\tau$$

$$\leq \frac{1}{(\nu - \mu_2)}(x - t)^{\alpha_1}(x - s)^{\alpha_2 + 1}. \qquad (5.27)$$

This is verified by estimating the integrand in (5.27) (with $\tau \geq s \geq t$):

$$(\tau - t)^{\alpha_1}(\tau - s)^{\alpha_2} + (\tau - s)^{\alpha_1}(\tau - t)^{\alpha_2}$$
$$= (\tau - t)^{\alpha_1 - \alpha_2 - 1}(\tau - t)^{\alpha_2 + 1}(\tau - s)^{\alpha_2} + (\tau - s)^{\alpha_1 - \alpha_2 - 1}(\tau - s)^{\alpha_2 + 1}(\tau - t)^{\alpha_2}$$
$$\leq (x - t)^{\alpha_1 - \alpha_2 - 1} \left[(\tau - t)^{\alpha_2 + 1}(\tau - s)^{\alpha_2} + (\tau - s)^{\alpha_2 + 1}(\tau - t)^{\alpha_2} \right]$$

(where the last inequality requires $\alpha_1 - \alpha_2 - 1 \geq 0$); (5.27) follows from

$$\int_0^x \left[(\tau - t)_+^{\alpha_2 + 1}(\tau - s)_+^{\alpha_2} + (\tau - s)_+^{\alpha_2 + 1}(\tau - t)_+^{\alpha_2} \right] d\tau =$$

$$\frac{1}{\alpha_2 + 1}[(x - t)(x - s)]^{\alpha_2 + 1}.$$

In the following calculation we abbreviate

$$c_1 := (\Gamma(\nu - \mu_2)\Gamma(\nu - \mu_1))^{-1}, \qquad c_2 := (\Gamma(\nu - \mu_2 + 1)\Gamma(\nu - \mu_1))^{-1},$$
$$c_3 := (\nu - \mu_2)q + 1, \qquad \epsilon := 2\nu - \mu_1 - \mu_2 - 1 + 1/q.$$

By Theorem 5.4,

$$D^{\mu_i}f(\tau) = \frac{1}{\Gamma(\nu - \mu_i)}\int_0^x (\tau - t)_+^{\alpha_i} D^\nu f(t)\, dt, \quad i = 1, 2.$$

Using this representation, the auxiliary inequality (5.27), and Hölder's inequality, we obtain

$$\int_0^x |D^{\mu_1}f(\tau)||D^{\mu_2}f(\tau)|\,d\tau$$

$$\leq c_1 \int_0^x \left(\int_0^x |D^\nu f(t)|(\tau-t)_+^{\alpha_1}\,dt \right) \left(\int_0^x |D^\nu f(s)|(\tau-s)_+^{\alpha_2}\,ds \right) d\tau$$

$$= c_1 \int_0^x |D^\nu f(t)| \left(\int_t^x |D^\nu f(s)| \right.$$

$$\left. \cdot \left(\int_0^x [(\tau-t)_+^{\alpha_1}(\tau-s)_+^{\alpha_2} + (\tau-s)_+^{\alpha_1}(\tau-t)_+^{\alpha_2}]\,d\tau \right) ds \right) dt$$

$$\leq c_2 \int_0^x |D^\nu f(t)| \left(\int_t^x |D^\nu f(s)|(x-t)^{\alpha_1}(x-s)^{\alpha_2+1}\,ds \right) dt$$

$$= c_2 \int_0^x |D^\nu f(t)|(x-t)^{\alpha_1} \left(\int_t^x |D^\nu f(s)|(x-s)^{\alpha_2+1}\,ds \right) dt$$

$$\leq c_2 \int_0^x |D^\nu f(t)|(x-t)^{\alpha_1} \left(\int_t^x |D^\nu f(s)|^p\,ds \right)^{1/p} \left(\int_t^x (x-s)^{q(\alpha_2+1)}ds \right)^{1/q} dt$$

$$= c_2 c_3^{-1/q} \int_0^x |D^\nu f(t)|(x-t)^{\epsilon q} \left(\int_t^x |D^\nu f(s)|^p\,ds \right)^{1/p} dt$$

$$\leq c_2 c_3^{-1/q} \left(\int_0^x |D^\nu f(t)|^p \left(\int_t^x |D^\nu f(s)|^p\,ds \right) dt \right)^{1/p} \left(\int_0^x (x-t)^{\epsilon q}\,dt \right)^{1/q}$$

$$\leq c_2 c_3^{-1/q}(\epsilon q+1)^{-1/q} x^{(\epsilon q+1)/q} \left(\frac{1}{2} \left(\int_0^x |D^\nu f(t)|^p\,dt \right)^2 \right)^{1/p}.$$

This implies (5.25). $\square$

In the following theorem we address the case when the function $|D^\nu f|$ is monotone.

Theorem 5.10. *Let* $f \in L^1(0,x)$ *have an integrable fractional derivative* $D^\nu f \in L^\infty(0,x)$ *such that* $D^{\nu-j}f(0) = 0$ *for* $j = 1,\ldots,[\nu]+1$ *and that* $|D^\nu f|$ *is decreasing on* $[0,x]$. *Let* $l \geq 2$. *If* $p, q > 1$ *are such that* $1/p + 1/q = 1$ *and* $\sum_{i=1}^l \alpha_i p > -1$, *then*

$$\int_0^x \prod_{i=1}^l |D^{\mu_i}f(\tau)|\,d\tau \leq C_3 x^{(\gamma p+lp+1)/p} \left(\int_0^x |D^\nu f(t)|^{lq}\,dt \right)^{1/q}, \qquad (5.28)$$

where $\gamma := \sum_{i=1}^l \alpha_i$ *and*

$$C_3 = C_3(\nu,\boldsymbol{\mu},p) := \frac{p}{(\gamma p+1)^{1/p}(\gamma p+p+1) \prod_{i=1}^l \Gamma(\nu-\mu_i)}. \qquad (5.29)$$

Proof. By Theorem 5.4,

$$|D^{\mu_i}f(\tau)| \leq \frac{1}{\Gamma(\nu - \mu_i)} \int_0^x (\tau - t)_+^{\alpha_i} |D^\nu f(t)| \, dt.$$

The integrand $t \mapsto (\tau - t)_+^{\alpha_i} |D^\nu f(t)|$ is decreasing (and integrable) on $[0, x]$ for all $\tau \in [0, x]$. By Chebyshev's inequality for the product of integrals [183, p. 1099],

$$
\begin{aligned}
\prod_{i=1}^l |D^{\mu_i}f(\tau)| &\leq \frac{x^{l-1}}{\prod_{i=1}^l \Gamma(\nu - \mu_i)} \int_0^x \prod_{i=1}^l (\tau - t)_+^{\alpha_i} |D^\nu f(t)| \, dt \\
&\leq \frac{x^{l-1}}{\prod_{i=1}^l \Gamma(\nu - \mu_i)} \int_0^x (\tau - t)_+^{\gamma} |D^\nu f(t)|^l \, dt \\
&\leq \frac{x^{l-1}}{\prod_{i=1}^l \Gamma(\nu - \mu_i)} \left(\int_0^\tau (\tau - t)^{\gamma p} \, dt \right)^{1/p} \left(\int_0^x |D^\nu f(t)|^{lq} \, dt \right)^{1/q} \\
&\leq \frac{x^{l-1}}{\prod_{i=1}^l \Gamma(\nu - \mu_i)} \left(\frac{\tau^{\gamma p+1}}{\gamma p + 1} \right)^{1/p} \left(\int_0^x |D^\nu f(t)|^{lq} \, dt \right)^{1/q} \\
&= \frac{x^{l-1} \tau^{(\gamma p+1)/p}}{(\gamma p + 1)^{1/p} \prod_{i=1}^l \Gamma(\nu - \mu_i)} \left(\int_0^x |D^\nu f(t)|^{lq} \, dt \right)^{1/q}.
\end{aligned}
$$

Integrating with respect to τ from 0 to x, we get the result. Condition $\sum_{i=1}^l \alpha_i > -1$ was needed in order to apply Hö lder's inequality to $\int_0^x (\tau - t)_+^{\gamma} |D^\nu f(t)|^l \, dt$. $\square$

The following extreme case of the theorem resembles [15, Proposition 4].

Theorem 5.11. *Let the hypotheses of Theorem 5.10 be satisfied, but let $p = 1$ and $q = \infty$. Then*

$$\int_0^x \prod_{i=1}^l |D^{\mu_i}f(\tau)| \, d\tau \leq C_4 x^{\gamma+l+1} \|D^\nu f\|_\infty^l, \tag{5.30}$$

where $\gamma := \sum_{i=1}^l \alpha_i$ and

$$C_4 = C_4(\nu, \boldsymbol{\mu}) := \frac{1}{(\gamma + 1)(\gamma + l + 1) \prod_{i=1}^l \Gamma(\nu - \mu_i)}. \tag{5.31}$$

Proof. As in the proof of Theorem 5.10 we have

$$
\prod_{i=1}^{l} |D^{\mu_i} f(\tau)| \leq \frac{1}{\prod_{i=1}^{l} \Gamma(\nu - \mu_i)} \prod_{i=1}^{l} \int_0^{\tau} |D^{\nu} f(t)| (\tau - t)^{\alpha_i} \, dt
$$

$$
\leq \frac{\tau^{l-1}}{\prod_{i=1}^{l} \Gamma(\nu - \mu_i)} \|D^{\nu} f\|_{\infty}^{l} \int_0^{\tau} (\tau - t)^{\gamma} \, dt
$$

$$
\leq \frac{\tau^{\gamma+l} \|D^{\nu} f\|_{\infty}^{l}}{(\gamma + 1) \prod_{i=1}^{l} \Gamma(\nu - \mu_i)}.
$$

Integrating over $[0, x]$ with respect to τ we obtain (5.30). $\quad\square$

6

Opial-Type Inequalities Involving Canavati Fractional Derivatives of Two Functions and Applications

A wide variety of very general but basic L_p ($1 \leq p \leq \infty$)-form Opial-type inequalities [315] is established involving generalized Canavati fractional derivatives [17, 101] of two functions in different orders and powers.

The above rely on a generalization of Taylor's formula for generalized Canavati fractional derivatives [17]. Several other concrete results of special interest are derived from the developed results. The sharpness of inequalities is established there. Finally applications of some of these special inequalities are given in establishing the uniqueness of solution and in giving upper bounds to solutions of initial value problems involving a very general system of two fractional differential equations. Also upper bounds to various fractional derivatives of the solutions that are involved in the above systems are presented. This treatment relies on [26].

6.1   Introduction

Opial inequalities appeared for the first time in [315] and since then many authors have dealt with them in different directions and for various cases. For a complete recent account of activity in this field see [4], and there still remains a very active area of research. One of the main attractions to these inequalities is their applications, especially to establishing uniqueness

G.A. Anastassiou, *Fractional Differentiation Inequalities*,
DOI 10.1007/978-0-387-98128-4_6, © Springer Science+Business Media, LLC 2009

and upper bounds of the solution of initial value problems in differential equations. The author was the first to present Opial inequalities involving fractional derivatives of functions [15, 17] with applications to fractional differential equations.

Fractional derivatives come up naturally in a number of fields, especially in physics; see the recent book [197]. Some topics include fractional kinetics of Hamiltonian chaotic systems, polymer physics and rheology, regular variation in thermodynamics, biophysics, fractional time evolution, and fractal time series among others. One also deals there with stochastic fractional-difference equations and fractional diffusion equations. Great applications of these can be found in the study of DNA sequences. Other fractional differential equations arise in the study of suspensions, coming from the fluid dynamical modeling of certain blood flow phenomena. An excellent account of the study of fractional differential equations can be found in the recent book [333]. One can also have applications of fractional calculus to viscoelasticity, Bode's analysis of feedback amplifiers, capacitor theory, electrical circuits, electronanalytical chemistry, biology, control theory, fitting of experimental data, and fractional-order physics. The study of fractional differential equations ranges from the very theoretical topics of existence and uniqueness of solution to finding numerical solutions.

The study of fractional calculus started in 1695 with L'Hospital and Leibniz, and was continued later by Fourier in 1822 and Abel in 1823, and continues to our day in an increased fashion due to its many applications and the necessity of dealing with fractional phenomena and structures. So this field is keeping a lot of people active and interested.

In this chapter the author is greatly motivated and inspired by the very important article [3]. Of course there the authors are dealing with other kinds of derivative. So here the author continues his study of fractional Opial inequalities now involving two different functions and produces a wide variety of corresponding results with important applications to systems of two fractional differential equations. Dealing with two functions makes the study more complicated and involved.

We start in Section 6.2 with preliminaries, we continue in Section 6.3 with the main results and we finish in Section 6.4 with the applications.

To give a flavor to the reader of the kind of inequalities we are dealing with, we briefly mention

$$\int_a^b q(w)[|(D_a^{\gamma_1} f_1)(w)|^{\lambda_\alpha}|(D_a^\nu f_1)(w)|^{\lambda_\nu} + |(D_a^{\gamma_1} f_2)(w)|^{\lambda_\alpha}|(D_a^\nu f_2)(w)|^{\lambda_\nu}]dw$$
$$\leq C(a,b,q(w),\gamma_1,\nu,\lambda_\alpha,\lambda_\nu,p(w),p)$$
$$\cdot \left[\int_a^b p(w)[|(D_a^\nu f_1)(w)|^p + |(D_a^\nu f_2)(w)|^p]dw\right]^{((\lambda_\alpha+\lambda_\nu)/p)},$$

for certain continuous functions f_1, f_2, $p(w)$, $q(w)$ on $[a, b]$, all exponents and orders are fractional, and so on. Furthermore, one system of fractional differential equations we are working on briefly looks like

$$(D_a^\nu f_j)(t) = F_j(t, \{(D_a^{\gamma_i} f_1)(t)\}_{i=1}^r, \{(D_a^{\gamma_i} f_2)(t)\}_{i=1}^r), \quad \text{all } t \in [a, b],$$

for $j = 1, 2$, and with $f_j^{(i)}(a) = a_{ij} \in \mathbb{R}$, $i = 0, 1, \ldots, n - 1$, where $n := [\nu]$, $\nu \geq 2$, and so on.

In the literature there are many different definitions of fractional derivatives, some of them being equivalent; see [197, 333]. In this chapter we use one of the most recent due to J. Canavati [101], generalized in [15] and [17] by the author.

One of the advantages of the Canavati fractional derivative is that in applications to fractional initial value problems we need only n initial conditions, as with the ordinary derivative case, whereas with other definitions of fractional derivatives we need $n + 1$ or more conditions; see [333].

6.2   Preliminaries

In the sequel we follow [101]. Let $g \in C([0, 1])$. Let ν be a positive number, $n := [\nu]$, and $\alpha := \nu - n$ ($0 < \alpha < 1$). Define

$$(J_\nu g)(x) := \frac{1}{\Gamma(\nu)} \int_0^x (x - t)^{\nu-1} g(t) dt, \quad 0 \leq x \leq 1, \qquad (6.1)$$

the *Riemann–Liouville integral*, where Γ is the gamma function. We define the subspace $C^\nu([0, 1])$ of $C^n([0, 1])$ as follows.

$$C^\nu([0, 1]) := \{g \in C^n([0, 1]) \colon J_{1-\alpha} D^n g \in C^1([0, 1])\},$$

where $D := d/dx$. So for $g \in C^\nu([0, 1])$, we define the Canavati ν-*fractional derivative* of g as

$$D^\nu g := D J_{1-\alpha} D^n g. \qquad (6.2)$$

When $\nu \geq 1$ we have the fractional Taylor's formula

$$g(t) = g(0) + g'(0)t + g''(0)\frac{t^2}{2!} + \cdots + g^{(n-1)}(0)\frac{t^{n-1}}{(n - 1)!}$$
$$+ (J_\nu D^\nu g)(t), \quad \text{for all } t \in [0, 1]. \qquad (6.3)$$

When $0 < \nu < 1$ we find

$$g(t) = (J_\nu D^\nu g)(t), \quad \text{for all } t \in [0, 1]. \qquad (6.4)$$

Next we carry the above notions over to arbitrary $[a,b] \subseteq \mathbb{R}$ (see [17]). Let $x, x_0 \in [a,b]$ such that $x \geq x_0$, where x_0 is fixed. Let $f \in C([a,b])$ and define

$$(J_\nu^{x_0} f)(x) := \frac{1}{\Gamma(\nu)} \int_{x_0}^x (x-t)^{\nu-1} f(t)dt, \quad x_0 \leq x \leq b, \tag{6.5}$$

the *generalized Riemann–Liouville integral*. We define the subspace $C_{x_0}^\nu([a,b])$ of $C^n([a,b])$:

$$C_{x_0}^\nu([a,b]) := \{f \in C^n([a,b]): J_{1-\alpha}^{x_0} D^n f \in C^1([x_0,b])\}.$$

For $f \in C_{x_0}^\nu([a,b])$, we define the *generalized ν-fractional derivative of f over $[x_0,b]$* as

$$D_{x_0}^\nu f := D J_{1-\alpha}^{x_0} f^{(n)} \quad (f^{(n)} := D^n f). \tag{6.6}$$

Notice that

$$(J_{1-\alpha}^{x_0} f^{(n)})(x) = \frac{1}{\Gamma(1-\alpha)} \int_{x_0}^x (x-t)^{-\alpha} f^{(n)}(t)dt$$

exists for $f \in C_{x_0}^\nu([a,b])$.

We recall the following fractional generalization of Taylor's formula (see [17, 101]).

Theorem 6.1. *Let* $f \in C_{x_0}^\nu([a,b])$, $x_0 \in [a,b]$ *fixed.*
(i) *If* $\nu \geq 1$ *then*

$$f(x) = f(x_0) + f'(x_0)(x-x_0) + f''(x_0)\frac{(x-x_0)^2}{2} + \cdots +$$

$$f^{(n-1)}(x_0)\frac{(x-x_0)^{n-1}}{(n-1)!} + (J_\nu^{x_0} D_{x_0}^\nu f)(x), \text{ for all } x \in [a,b]: x \geq x_0. \tag{6.7}$$

(ii) *If* $0 < \nu < 1$ *then*

$$f(x) = (J_\nu^{x_0} D_{x_0}^\nu f)(x), \text{ for all } x \in [a,b]: x \geq x_0. \tag{6.8}$$

We make

Remark 6.2.
(1) $(D_{x_0}^n f) = f^{(n)}$, $n \in \mathbb{N}$.
(2) Let $f \in C_{x_0}^\nu([a,b])$, $\nu \geq 1$, and $f^{(i)}(x_0) = 0$, $i = 0,1,\ldots,n-1$, $n := [\nu]$.
Then by (6.7)
$$f(x) = (J_\nu^{x_0} D_{x_0}^\nu f)(x).$$

That is,

$$f(x) = \frac{1}{\Gamma(\nu)} \int_{x_0}^{x} (x-t)^{\nu-1} (D_{x_0}^{\nu} f)(t) dt, \tag{6.9}$$

for all $x \in [a,b]$ with $x \geq x_0$. Notice that (6.9) is true also when $0 < \nu < 1$.

We need from [17]

Lemma 6.3. *Let $f \in C([a,b])$, $\mu, \nu > 0$. Then*

$$J_{\mu}^{x_0}(J_{\nu}^{x_0} f) = J_{\mu+\nu}^{x_0}(f). \tag{6.10}$$

We also make

Remark 6.4. Let $\nu \geq 1$, $\gamma \geq 0$, be such that $\nu - \gamma \geq 1$, so that $\gamma < \nu$. Call $n := [\nu]$, $\alpha := \nu - n$; $m := [\gamma]$, $\rho := \gamma - m$. Note that $\nu - m \geq 1$ and $n - m \geq 1$. Let $f \in C_{x_0}^{\nu}([a,b])$ be such that $f^{(i)}(x_0) = 0$, $i = 0, 1, \ldots, n-1$. Hence by (6.7)

$$f(x) = (J_{\nu}^{x_0} D_{x_0}^{\nu} f)(x), \text{ for all } x \in [a,b] : x \geq x_0.$$

Therefore by Leibnitz's formula and $\Gamma(p+1) = p\Gamma(p)$, $p > 0$, we get that

$$f^{(m)}(x) = (J_{\nu-m}^{x_0} D_{x_0}^{\nu} f)(x), \text{ for all } x \geq x_0. \tag{6.11}$$

It follows that $f \in C_{x_0}^{\gamma}([a,b])$ and thus

$$(D_{x_0}^{\gamma} f)(x) := (D J_{1-\rho}^{x_0} f^{(m)})(x) \text{ exists for all } x \geq x_0. \tag{6.12}$$

By the use of (6.11) we have on $[x_0, b]$,

$$\begin{aligned} J_{1-\rho}^{x_0}(f^{(m)}) &= J_{1-\rho}^{x_0}(J_{\nu-m}^{x_0} D_{x_0}^{\nu} f) = (J_{1-\rho}^{x_0} \circ J_{\nu-m}^{x_0})(D_{x_0}^{\nu} f) \\ &= J_{\nu-m+1-\rho}^{x_0}(D_{x_0}^{\nu} f) = J_{\nu-\gamma+1}^{x_0}(D_{x_0}^{\nu} f), \end{aligned}$$

by (6.10). That is,

$$(J_{1-\rho}^{x_0} f^{(m)})(x) = \frac{1}{\Gamma(\nu-\gamma+1)} \int_{x_0}^{x} (x-t)^{\nu-\gamma} (D_{x_0}^{\nu} f)(t) dt.$$

Therefore

$$(D_{x_0}^{\gamma} f)(x) = D((J_{1-\rho}^{x_0} f^{(m)})(x)) = \frac{1}{\Gamma(\nu-\gamma)} \cdot \int_{x_0}^{x} (x-t)^{(\nu-\gamma)-1} (D_{x_0}^{\nu} f)(t) dt; \tag{6.13}$$

hence $(D_{x_0}^{\gamma} f)(x) = (J_{\nu-\gamma}^{x_0}(D_{x_0}^{\nu} f))(x)$ and is continuous in x on $[x_0, b]$. In particular when $\nu \geq 2$ we have

$$(D_{x_0}^{\nu-1} f)(x) = \int_{x_0}^{x} (D_{x_0}^{\nu} f)(t) dt, \quad x \geq x_0. \tag{6.14}$$

That is,

$$(D_{x_0}^{\nu-1} f)' = D_{x_0}^{\nu} f, \quad (D_{x_0}^{\nu-1} f)(x_0) = 0.$$

6.3   Main Results

We present our first main result here:

Theorem 6.5. *Let $\nu \geq 1$, $\gamma_1, \gamma_2 \geq 0$ be such that $\nu - \gamma_1 \geq 1$, $\nu - \gamma_2 \geq 1$, and $f_1, f_2 \in C_{x_0}^\nu([a,b])$ with $f_1^{(i)}(x_0) = f_2^{(i)}(x_0) = 0$, $i = 0, 1, \ldots, n-1$, $n := [\nu]$. Here $x, x_0 \in [a,b] : x \geq x_0$. Consider also $p(t) > 0$, and $q(t) \geq 0$ continuous functions on $[x_0, b]$. Let $\lambda_\nu > 0$ and $\lambda_\alpha, \lambda_\beta \geq 0$ such that $\lambda_\nu < p$, where $p > 1$. Set*

$$P_k(w) := \int_{x_0}^w (w-t)^{(\nu-\gamma_k-1)p/p-1}(p(t))^{-1/p-1}dt, \ k = 1,2, \ x_0 \leq x \leq b; \tag{6.15}$$

$$A(w) := \frac{q(w) \cdot (P_1(w))^{\lambda_\alpha(p-1/p)} \cdot (P_2(w))^{\lambda_\beta(p-1/p)}(p(w))^{-\lambda_\nu/p}}{(\Gamma(\nu-\gamma_1))^{\lambda_\alpha} \cdot (\Gamma(\nu-\gamma_2))^{\lambda_\beta}}, \tag{6.16}$$

$$A_0(x) := \left(\int_{x_0}^x A(w)^{p/p-\lambda_\nu} dw \right)^{(p-\lambda_\nu)/p}, \tag{6.17}$$

and

$$\delta_1 := \begin{cases} 2^{1-((\lambda_\alpha+\lambda_\nu)/p)}, & \text{if } \lambda_\alpha + \lambda_\nu \leq p, \\ \\ 1, & \text{if } \lambda_\alpha + \lambda_\nu \geq p. \end{cases} \tag{6.18}$$

If $\lambda_\beta = 0$, we obtain that

$$\int_{x_0}^x q(w) \left[|(D_{x_0}^{\gamma_1} f_1)(w)|^{\lambda_\alpha} \cdot |(D_{x_0}^\nu f_1)(w)|^{\lambda_\nu} + \right.$$

$$\left. + |(D_{x_0}^{\gamma_1} f_2)(w)|^{\lambda_\alpha} \cdot |(D_{x_0}^\nu f_2)(w)|^{\lambda_\nu} \right] dw$$

$$\leq (A_0(x)|_{\lambda_\beta=0}) \cdot \left(\frac{\lambda_\nu}{\lambda_\alpha + \lambda_\nu} \right)^{\lambda_\nu/p} \cdot \delta_1$$

$$\cdot \left[\int_{x_0}^x p(w) \left[|(D_{x_0}^\nu f_1)(w)|^p + |(D_{x_0}^\nu f_2)(w)|^p \right] dw \right]^{((\lambda_\alpha+\lambda_\nu)/p)} \tag{6.19}$$

Proof. From (6.13) we have

$$(D_{x_0}^{\gamma_k} f_j)(w) = \frac{1}{\Gamma(\nu-\gamma_k)} \int_{x_0}^w (w-t)^{\nu-\gamma_k-1}(D_{x_0}^\nu f_j)(t)dt,$$

for $k = 1, 2$, $j = 1, 2$, and for all $x_0 \leq w \leq b$.

Next applying Hölder's inequality with indices p, $p/(p-1)$ we get

$$|(D_{x_0}^{\gamma_k} f_j)(w)| \leq \frac{1}{\Gamma(\nu - \gamma_k)} \int_{x_0}^{w} (w-t)^{\nu-\gamma_k-1} (p(t))^{-1/p} (p(t))^{1/p} |(D_{x_0}^{\nu} f_j)(t)| dt$$

$$\leq \frac{1}{\Gamma(\nu - \gamma_k)} \left(\int_{x_0}^{w} \left((w-t)^{\nu-\gamma_k-1} (p(t))^{-1/p} \right)^{p/(p-1)} dt \right)^{(p-1)/p}$$

$$\left(\int_{x_0}^{w} p(t) |(D_{x_0}^{\nu} f_j)(t)|^p dt \right)^{1/p}$$

$$= \frac{1}{\Gamma(\nu - \gamma_k)} (P_k(w))^{p-1/p} \left(\int_{x_0}^{w} p(t) |(D_{x_0}^{\nu} f_j)(t)|^p dt \right)^{1/p}.$$

That is, it holds

$$|(D_{x_0}^{\gamma_k} f_j)(w)| \leq \frac{1}{\Gamma(\nu - \gamma_k)} (P_k(w))^{p-1/p} \left(\int_{x_0}^{w} p(t) |(D_{x_0}^{\nu} f_j)(t)|^p dt \right)^{1/p}. \tag{6.20}$$

Put

$$z_j(w) := \int_{x_0}^{w} p(t) |(D_{x_0}^{\nu} f_j)(t)|^p dt,$$

thus

$$z_j'(w) = p(w) |(D_{x_0}^{\nu} f_j)(w)|^p, \quad z_j(x_0) = 0, \quad j = 1, 2.$$

Hence we have

$$|(D_{x_0}^{\gamma_k} f_j)(w)| \leq \frac{1}{\Gamma(\nu - \gamma_k)} (P_k(w))^{(p-1)/p} (z_j(w))^{1/p},$$

and

$$|(D_{x_0}^{\nu} f_j)(w)|^{\lambda_\nu} = p(w)^{-\lambda_\nu/p} (z_j'(w))^{\lambda_\nu/p}, \quad j = 1, 2.$$

Therefore we obtain

$$q(w) |(D_{x_0}^{\gamma_1} f_1)(w)|^{\lambda_\alpha} |(D_{x_0}^{\gamma_2} f_2)(w)|^{\lambda_\beta} |(D_{x_0}^{\nu} f_1)(w)|^{\lambda_\nu}$$

$$\leq q(w) \frac{1}{(\Gamma(\nu - \gamma_1))^{\lambda_\alpha}} (P_1(w))^{\lambda_\alpha (p-1/p)} (z_1(w))^{\lambda_\alpha/p} \frac{1}{(\Gamma(\nu - \gamma_2))^{\lambda_\beta}}$$

$$(P_2(w))^{\lambda_\beta (p-1/p)} (z_2(w))^{\lambda_\beta/p} (p(w))^{-\lambda_\nu/p} (z_1'(w))^{\lambda_\nu/p}$$

$$= A(w) (z_1(w))^{\lambda_\alpha/p} (z_2(w))^{\lambda_\beta/p} (z_1'(w))^{\lambda_\nu/p}.$$

Consequently, by another Hölder's inequality application, we find (by $p/\lambda_\nu > 1$)

$$\int_{x_0}^{x} q(w) |(D_{x_0}^{\gamma_1} f_1)(w)|^{\lambda_\alpha} |(D_{x_0}^{\gamma_2} f_2)(w)|^{\lambda_\beta} |(D_{x_0}^{\nu} f_1)(w)|^{\lambda_\nu} dw$$

$$\leq A_0(x) \left[\int_{x_0}^{x} (z_1(w))^{\lambda_\alpha/\lambda_\nu} (z_2(w))^{\lambda_\beta/\lambda_\nu} z_1'(w) dw \right]^{\lambda_\nu/p}. \tag{6.21}$$

Similarly one finds

$$\int_{x_0}^x q(w)|(D_{x_0}^{\gamma_2} f_1)(w)|^{\lambda_\beta}|(D_{x_0}^{\gamma_1} f_2)(w)|^{\lambda_\alpha}|(D_{x_0}^{\nu} f_2)(w)|^{\lambda_\nu} dw$$

$$\leq A_0(x) \left[\int_{x_0}^x (z_1(w))^{\lambda_\beta/\lambda_\nu} (z_2(w))^{\lambda_\alpha/\lambda_\nu} z_2'(w) dw \right]^{\lambda_\nu/p}. \quad (6.22)$$

Taking $\lambda_\beta = 0$ and adding (6.21) and (6.22) we obtain

$$\int_{x_0}^x q(w) \left[|(D_{x_0}^{\gamma_1} f_1)(w)|^{\lambda_\alpha}|(D_{x_0}^{\nu} f_1)(w)|^{\lambda_\nu} + |(D_{x_0}^{\gamma_1} f_2)(w)|^{\lambda_\alpha}|(D_{x_0}^{\nu} f_2)(w)|^{\lambda_\nu} \right] dw$$

$$\leq (A_0(x)|_{\lambda_\beta=0}) \left\{ \left[\int_{x_0}^x (z_1(w))^{\lambda_\alpha/\lambda_\nu} z_1'(w) dw \right]^{\lambda_\nu/p} \right.$$

$$\left. + \left[\int_{x_0}^x (z_2(w))^{\lambda_\alpha/\lambda_\nu} z_2'(w) dw \right]^{\lambda_\nu/p} \right\}$$

$$= (A_0(x)|_{\lambda_\beta=0}) \left\{ (z_1(x))^{(\lambda_\alpha+\lambda_\nu)/p} + (z_2(x))^{(\lambda_\alpha+\lambda_\nu)/p} \right\} \left(\frac{\lambda_\nu}{\lambda_\alpha + \lambda_\nu} \right)^{\lambda_\nu/p}$$

$$= (A_0(x)|_{\lambda_\beta=0}) \left(\frac{\lambda_\nu}{\lambda_\alpha + \lambda_\nu} \right)^{\lambda_\nu/p} \left\{ \left(\int_{x_0}^x p(t)|(D_{x_0}^{\nu} f_1)(t)|^p dt \right)^{(\lambda_\alpha+\lambda_\nu)/p} \right.$$

$$\left. + \left(\int_{x_0}^x p(t)|(D_{x_0}^{\nu} f_2)(t)|^p dt \right)^{(\lambda_\alpha+\lambda_\nu)/p} \right\} =: (*).$$

In this chapter we frequently use the following basic inequalities.

$$2^{r-1}(a^r + b^r) \quad \leq \quad (a+b)^r \leq a^r + b^r, \quad a, b \geq 0, \ 0 \leq r \leq 1, \quad (6.23)$$

$$a^r + b^r \quad \leq \quad (a+b)^r \leq 2^{r-1}(a^r + b^r), \quad a, b \geq 0, \ r \geq 1. \quad (6.24)$$

Finally using (6.23), (6.24), and (6.18) we get

$$(*) \quad \leq \quad (A_0(x)|_{\lambda_\beta=0}) \cdot \left(\frac{\lambda_\nu}{\lambda_a + \lambda_\nu} \right)^{\lambda_\nu/p} \cdot \delta_1$$

$$\cdot \left\{ \int_{x_0}^x p(t) \left[|(D_{x_0}^{\nu} f_1)(t)|^p + |(D_{x_0}^{\nu} f_2)(t)|^p \right] dt \right\}^{(\lambda_\alpha+\lambda_\nu)/p}.$$

Inequality (6.19) has been established.   □

It follows the counterpart of the last theorem.

Theorem 6.6. *All here are as in Theorem 6.5. Denote*

$$\delta_3 := \begin{cases} 2^{\lambda_\beta/\lambda_\nu} - 1, & \text{if } \lambda_\beta \geq \lambda_\nu, \\ 1, & \text{if } \lambda_\beta \leq \lambda_\nu. \end{cases} \quad (6.25)$$

If $\lambda_\alpha = 0$, then it holds

$$\int_{x_0}^x q(w) \left[|(D_{x_0}^{\gamma_2} f_2)(w)|^{\lambda_\beta} \cdot |(D_{x_0}^\nu f_1)(w)|^{\lambda_\nu} + |(D_{x_0}^{\gamma_2} f_1)(w)|^{\lambda_\beta} |(D_{x_0}^\nu f_2)(w)|^{\lambda_\nu} \right] dw$$

$$\leq (A_0(x)|_{\lambda_\alpha=0}) \, 2^{(p-\lambda_\nu)/p} \left(\frac{\lambda_\nu}{\lambda_\beta + \lambda_\nu} \right)^{\lambda_\nu/p} \delta_3^{\lambda_\nu/p} \qquad (6.26)$$

$$\cdot \left(\int_{x_0}^x p(w) \left[|(D_{x_0}^\nu f_1)(w)|^p + |(D_{x_0}^\nu f_2)(w)|^p \right] dw \right)^{((\lambda_\nu+\lambda_\beta)/p)}, \quad \text{all } x_0 \leq x \leq b.$$

Proof. When $\lambda_\alpha = 0$ from (6.21) and (6.22) we obtain

$$\int_{x_0}^x q(w) |(D_{x_0}^{\gamma_2} f_2)(w)|^{\lambda_\beta} |(D_{x_0}^\nu f_1)(w)|^{\lambda_\nu} \, dw$$

$$\leq (A_0(x)|_{\lambda_\alpha=0}) \left[\int_{x_0}^x (z_2(w))^{\lambda_\beta/\lambda_\nu} z_1'(w) dw \right]^{\lambda_\nu/p}, \quad (6.27)$$

and

$$\int_{x_0}^x q(w) |(D_{x_0}^{\gamma_2} f_1)(w)|^{\lambda_\beta} |(D_{x_0}^\nu f_2)(w)|^{\lambda_\nu} \, dw$$

$$\leq (A_0(x)|_{\lambda_\alpha=0}) \left[\int_{x_0}^x (z_1(w))^{\lambda_\beta/\lambda_\nu} z_2'(w) dw \right]^{\lambda_\nu/p}, \quad (6.28)$$

all $x_0 \leq x \leq b$. Adding (6.27) and (6.28) we get

$$\int_{x_0}^x q(w) \left[|(D_{x_0}^{\gamma_2} f_2)(w)|^{\lambda_\beta} \cdot |(D_{x_0}^\nu f_1)(w)|^{\lambda_\nu} + |(D_{x_0}^{\gamma_2} f_1)(w)|^{\lambda_\beta} \cdot \right.$$

$$\left. |(D_{x_0}^\nu f_2)(w)|^{\lambda_\nu} \right] dw \leq (A_0(x)|_{\lambda_\alpha=0})$$

$$\left\{ \left[\int_{x_0}^x (z_2(w))^{\lambda_\beta/\lambda_\nu} z_1'(w) dw \right]^{\lambda_\nu/p} + \left[\int_{x_0}^x (z_1(w))^{\lambda_\beta/\lambda_\nu} z_2'(w) dw \right]^{\lambda_\nu/p} \right\}$$

$$\leq (A_0(x)|_{\lambda_\alpha=0}) \cdot 2^{p-\lambda_\nu/p} \cdot (M(x))^{\lambda_\nu/p} =: (*), \qquad (6.29)$$

by $0 < \lambda_\nu/p < 1$ and (6.23), where

$$M(x) := \int_{x_0}^x (z_2(w))^{\lambda_\beta/\lambda_\nu} z_1'(w) + (z_1(w))^{\lambda_\beta/\lambda_\nu} z_2'(w)) dw. \qquad (6.30)$$

Call

$$\delta_2 := \begin{cases} 1, & \text{if } \lambda_\beta \geq \lambda_\nu, \\ 2^{1-(\lambda_\beta/\lambda_\nu)}, & \text{if } \lambda_\beta \leq \lambda_\nu. \end{cases}$$

Next we work on $M(x)$. We have that

$$M(x) = \int_{x_0}^x \left((z_1(w))^{\lambda_\beta/\lambda_\nu} + (z_2(w))^{\lambda_\beta/\lambda_\nu} \right) (z_1'(w) + z_2'(w)) dw$$

$$- \int_{x_0}^x \left[(z_1(w))^{\lambda_\beta/\lambda_\nu} z_1'(w) + (z_2(w))^{\lambda_\beta/\lambda_\nu} z_2'(w) \right] dw$$

$$\overset{\substack{(by\ (6.23),\ (6.24))}}{\leq} \delta_2 \int_{x_0}^x (z_1(w) + z_2(w))^{\lambda_\beta/\lambda_\nu} (z_1(w) + z_2(w))' dw$$

$$- \left(\frac{\lambda_\nu}{\lambda_\beta + \lambda_\nu} \right) \left[(z_1(x))^{((\lambda_\nu + \lambda_\beta)/\lambda_\nu)} + (z_2(w))^{(\lambda_\nu + \lambda_\beta/\lambda_\nu)} \right]$$

$$= \delta_2 \left((z_1(x) + z_2(x))^{((\lambda_\nu + \lambda_\beta)/\lambda_\nu)} \right) \left(\frac{\lambda_\nu}{\lambda_\nu + \lambda_\beta} \right) - \left(\frac{\lambda_\nu}{\lambda_\beta + \lambda_\nu} \right)$$

$$\left[(z_1(x))^{((\lambda_\nu + \lambda_\beta)/\lambda_\nu)} + (z_2(w))^{((\lambda_\nu + \lambda_\beta)/\lambda_\nu)} \right]$$

$$= \left(\frac{\lambda_\nu}{\lambda_\beta + \lambda_\nu} \right) \left[\begin{array}{c} \delta_2(z_1(x) + z_2(x))^{((\lambda_\nu + \lambda_\beta)/\lambda_\nu)} - \\ ((z_1(x))^{((\lambda_\nu + \lambda_\beta)/\lambda_\nu)} + (z_2(x))^{((\lambda_\nu + \lambda_\beta)/\lambda_\nu)}) \end{array} \right]$$

$$\overset{(6.24)}{\leq} \left(\frac{\lambda_\nu}{\lambda_\beta + \lambda_\nu} \right) \left[\delta_2 2^{((\lambda_\nu + \lambda_\beta)/\lambda_\nu) - 1} \left((z_1(x))^{((\lambda_\nu + \lambda_\beta)/\lambda_\nu)} \right. \right.$$

$$\left. \left. + (z_2(x))^{((\lambda_\nu + \lambda_\beta)/\lambda_\nu)} \right) - \left((z_1(x))^{((\lambda_\nu + \lambda_\beta)/\lambda_\nu)} + (z_2(x))^{((\lambda_\nu + \lambda_\beta)/\lambda_\nu)} \right) \right]$$

$$= \left(\frac{\lambda_\nu}{\lambda_\beta + \lambda_\nu} \right) (\delta_2 2^{\lambda_\beta/\lambda_\nu} - 1)$$

$$\left[(z_1(x))^{((\lambda_\nu + \lambda_\beta)/\lambda_\nu)} + (z_2(x))^{((\lambda_\nu + \lambda_\beta)/\lambda_\nu)} \right] \left(\text{notice } \delta_3 = \delta_2 2^{\lambda_\beta/\lambda_\nu} - 1 \right)$$

$$\overset{(6.24)}{\leq} \left(\frac{\lambda_\nu}{\lambda_\beta + \lambda_\nu} \right) \delta_3 (z_1(x) + z_2(x))^{((\lambda_\nu + \lambda_\beta)/\lambda_\nu)}.$$

That is, we present that

$$M(x) \leq \left(\frac{\lambda_\nu}{\lambda_\beta + \lambda_\nu} \right) \delta_3 (z_1(x) + z_2(x))^{((\lambda_\nu + \lambda_\beta)/\lambda_\nu)}. \tag{6.31}$$

Consequently by (6.29) and (6.31) we get

$$(*) \leq (A_0(x)|_{\lambda_\alpha=0}) 2^{p - \lambda_\nu/p} \left(\frac{\lambda_\nu}{\lambda_\beta + \lambda_\nu} \right)^{\lambda_\nu/p} \delta_3^{\lambda_\nu/p} (z_1(x) + z_2(x))^{((\lambda_\nu + \lambda_\beta)/p)}$$

$$= (A_0(x)|_{\lambda_\alpha=0}) 2^{p - \lambda_\nu/p} \left(\frac{\lambda_\nu}{\lambda_\beta + \lambda_\nu} \right)^{\lambda_\nu/p} \delta_3^{\lambda_\nu/p}$$

$$\cdot \left(\int_{x_0}^x p(t) \left[|(D_{x_0}^\nu f_1)(t)|^p + |(D_{x_0}^\nu f_2)(t)|^p \right] dt \right)^{((\lambda_\nu + \lambda_\beta)/p)}.$$

We have established (6.26). $\quad\square$

The full case when $\lambda_\alpha, \lambda_\beta \neq 0$ follows.

Theorem 6.7. *All here are as in Theorem 6.5. Denote*

$$\tilde{\gamma}_1 := \begin{cases} 2^{((\lambda_\alpha + \lambda_\beta)/\lambda_\nu)} - 1, & \text{if } \lambda_\alpha + \lambda_\beta \geq \lambda_\nu, \\ 1, & \text{if } \lambda_\alpha + \lambda_\beta \leq \lambda_\nu, \end{cases} \tag{6.32}$$

and

$$\tilde{\gamma}_2 := \begin{cases} 1, & \text{if } \lambda_\alpha + \lambda_\beta + \lambda_\nu \geq p, \\ 2^{1 - (\lambda_\alpha + \lambda_\beta + \lambda_\nu/p)} & \text{if } \lambda_\alpha + \lambda_\beta + \lambda_\nu \leq p. \end{cases} \tag{6.33}$$

Then

$$\int_{x_0}^x q(w) \left[|(D_{x_0}^{\gamma_1} f_1)(w)|^{\lambda_\alpha} |(D_{x_0}^{\gamma_2} f_2)(w)|^{\lambda_\beta} |(D_{x_0}^\nu f_1)(w)|^{\lambda_\nu} \right.$$

$$\left. + |(D_{x_0}^{\gamma_2} f_1)(w)|^{\lambda_\beta} |(D_{x_0}^{\gamma_1} f_2)(w)|^{\lambda_\alpha} |(D_{x_0}^\nu f_2)(w)|^{\lambda_\nu} \right] dw$$

$$\leq A_0(x) \left(\frac{\lambda_\nu}{(\lambda_\alpha + \lambda_\beta)(\lambda_\alpha + \lambda_\beta + \lambda_\nu)} \right)^{\lambda_\nu/p} \left[\lambda_\alpha^{\lambda_\nu/p} \tilde{\gamma}_2 + 2^{p - \lambda_\nu/p} (\tilde{\gamma}_1 \lambda_\beta)^{\lambda_\nu/p} \right]$$

$$\cdot \left(\int_{x_0}^x p(w)(|(D_{x_0}^\nu f_1)(w)|^p + |(D_{x_0}^\nu f_2)(w)|^p) dw \right)^{((\lambda_\alpha + \lambda_\beta + \lambda_\nu)/p)}, \tag{6.34}$$

all $x_0 \leq x \leq b$.

Proof. Here we use the basic inequality

$$ab \leq \frac{a^p}{p} + \frac{b^q}{q}, \tag{6.35}$$

where $a, b \geq 0$ and $p, q > 1$ such that $1/p + 1/q = 1$. From (6.21) we obtain

$$\int_{x_0}^x q(w) |(D_{x_0}^{\gamma_1} f_1)(w)|^{\lambda_\alpha} |(D_{x_0}^{\gamma_2} f_2)(w)|^{\lambda_\beta} |(D_{x_0}^\nu f_1)(w)|^{\lambda_\nu} dw$$

$$\leq A_0(x) \left[\int_{x_0}^x \left(\left(\frac{\lambda_\alpha}{\lambda_\alpha + \lambda_\beta} \right) (z_1(w))^{((\lambda_\alpha + \lambda_\beta)/\lambda_\nu)} + \right. \right.$$

$$\left. \left(\frac{\lambda_\beta}{\lambda_\alpha + \lambda_\beta} \right) (z_2(w))^{((\lambda_\alpha + \lambda_\beta)/\lambda_\nu)} \right) z_1'(w) dw \right]^{\lambda_\nu/p}$$

$$\overset{(6.23)}{\leq} A_0(x) \left[\left(\left(\frac{\lambda_\alpha \lambda_\nu}{\lambda_\alpha + \lambda_\beta} \right) \frac{(z_1(x))^{((\lambda_\alpha + \lambda_\beta + \lambda_\nu)/\lambda_\nu)}}{(\lambda_\alpha + \lambda_\beta + \lambda_\nu)} \right)^{\lambda_\nu/p} \right.$$

$$\left. + \left(\int_{x_0}^x \left(\frac{\lambda_\beta}{\lambda_\alpha + \lambda_\beta} \right) (z_2(w))^{((\lambda_\alpha + \lambda_\beta)/\lambda_\nu)} z_1'(w) dw \right)^{\lambda_\nu/p} \right].$$

Therefore

$$
\int_{x_0}^{x} q(w)|(D_{x_0}^{\gamma_1} f_1)(w)|^{\lambda_\alpha}|(D_{x_0}^{\gamma_2} f_2)(w)|^{\lambda_\beta}|(D_{x_0}^{\nu} f_1)(w)|^{\lambda_\nu} dw
$$

$$
\leq \ A_0(x) \cdot \left\{ \left(\frac{\lambda_\nu \lambda_\alpha}{(\lambda_\alpha + \lambda_\beta)(\lambda_\alpha + \lambda_\beta + \lambda_\nu)} \right)^{\lambda_\nu/p} (z_1(x))^{((\lambda_\alpha + \lambda_\beta + \lambda_\nu)/p)} \right.
$$

$$
\left. + \left(\frac{\lambda_\beta}{\lambda_\alpha + \lambda_\beta} \right)^{\lambda_\nu/p} \left(\int_{x_0}^{x} (z_2(w))^{((\lambda_\alpha + \lambda_\beta)/\lambda_\nu)} z_1'(w) dw \right)^{\lambda_\nu/p} \right\} . \quad (6.36)
$$

Similarly using (6.22) we find

$$
\int_{x_0}^{x} q(w)|(D_{x_0}^{\gamma_2} f_1)(w)|^{\lambda_\beta}|(D_{x_0}^{\gamma_1} f_2)(w)|^{\lambda_\alpha}|(D_{x_0}^{\nu} f_2)(w)|^{\lambda_\nu} dw
$$

$$
\leq \ A_0(x) \left\{ \left(\frac{\lambda_\nu \lambda_\alpha}{(\lambda_\alpha + \lambda_\beta)(\lambda_\alpha + \lambda_\beta + \lambda_\nu)} \right)^{\lambda_\nu/p} (z_2(x))^{((\lambda_\alpha + \lambda_\beta + \lambda_\nu)/p)} \right.
$$

$$
\left. + \left(\frac{\lambda_\beta}{\lambda_\alpha + \lambda_\beta} \right)^{\lambda_\nu/p} \left(\int_{x_0}^{x} (z_1(w))^{((\lambda_\alpha + \lambda_\beta)/\lambda_\nu)} z_2'(w) dw \right)^{\lambda_\nu/p} \right\} . \quad (6.37)
$$

Next adding (6.36) and (6.37) we observe

$$
\Omega := \int_{x_0}^{x} q(w) \left[|(D_{x_0}^{\gamma_1} f_1)(w)|^{\lambda_\alpha}|(D_{x_0}^{\gamma_2} f_2)(w)|^{\lambda_\beta}|(D_{x_0}^{\nu} f_1)(w)|^{\lambda_\nu} \right.
$$

$$
\left. + |(D_{x_0}^{\gamma_2} f_1)(w)|^{\lambda_\beta}|(D_{x_0}^{\gamma_1} f_2)(w)|^{\lambda_\alpha}|(D_{x_0}^{\nu} f_2)(w)|^{\lambda_\nu} \right] dw
$$

$$
\leq A_0(x) \left\{ \left(\frac{\lambda_\nu \lambda_\alpha}{(\lambda_\alpha + \lambda_\beta)(\lambda_\alpha + \lambda_\beta + \lambda_\nu)} \right)^{\lambda_\nu/p} \right.
$$

$$
\left[(z_1(x))^{((\lambda_\alpha + \lambda_\beta + \lambda_\nu)/p)} + (z_2(x))^{((\lambda_\alpha + \lambda_\beta + \lambda_\nu)/p)} \right]
$$

$$
+ \left(\frac{\lambda_\beta}{\lambda_\alpha + \lambda_\beta} \right)^{\lambda_\nu/p} \left[\left(\int_{x_0}^{x} (z_1(w))^{(\lambda_\alpha + \lambda_\beta/\lambda_\nu)} z_2'(w) dw \right)^{\lambda_\nu/p} \right.
$$

$$
\left. \left. + \left(\int_{x_0}^{x} (z_2(w))^{((\lambda_\alpha + \lambda_\beta)/\lambda_\nu)} z_1'(w) dw \right)^{\lambda_\nu/p} \right] \right\}
$$

$$
\overset{(6.23)}{\leq} A_0(x) \left\{ \left(\frac{\lambda_\nu \lambda_\alpha}{(\lambda_\alpha + \lambda_\beta)(\lambda_\alpha + \lambda_\beta + \lambda_\nu)} \right)^{\lambda_\nu/p} \right.
$$

$$
\left[(z_1(x))^{((\lambda_\alpha + \lambda_\beta + \lambda_\nu)/p)} + (z_2(x))^{((\lambda_\alpha + \lambda_\beta + \lambda_\nu)/p)} \right]
$$

$$
\left. + \left(\frac{\lambda_\beta}{\lambda_\alpha + \lambda_\beta} \right)^{\lambda_\nu/p} \cdot 2^{p - \lambda_\nu/p} \cdot (\overline{\Gamma}(x))^{\lambda_\nu/p} \right\} ,
$$

where

$$\overline{\Gamma}(x) := \int_{x_0}^{x} \left((z_1(w))^{((\lambda_\alpha + \lambda_\beta)/\lambda_\nu)} z_2'(w) + (z_2(w))^{((\lambda_\alpha + \lambda_\beta)/\lambda_\nu)} z_1'(w) \right) dw.$$

Again, see (6.30) and (6.31), we get

$$\overline{\Gamma}(x) \le \left(\frac{\tilde{\gamma}_1 \lambda_\nu}{\lambda_\alpha + \lambda_\beta + \lambda_\nu} \right) (z_1(x) + z_2(x))^{((\lambda_\alpha + \lambda_\beta + \lambda_\nu)/\lambda_\nu)}. \qquad (6.38)$$

Hence by (6.38) we obtain

$$\Omega \le A_0(x) \left\{ \left(\frac{\lambda_\nu \lambda_\alpha}{(\lambda_\alpha + \lambda_\beta)(\lambda_\alpha + \lambda_\beta + \lambda_\nu)} \right)^{\lambda_\nu/p} \right.$$

$$\left[(z_1(x))^{((\lambda_\alpha + \lambda_\beta + \lambda_\nu)/p)} + (z_2(x))^{((\lambda_\alpha + \lambda_\beta + \lambda_\nu)/p)} \right]$$

$$\left. + \left(\frac{\lambda_\beta}{\lambda_\alpha + \lambda_\beta} \right)^{\lambda_\nu/p} 2^{p - \lambda_\nu/p} \left(\frac{\tilde{\gamma}_1 \lambda_\nu}{\lambda_\alpha + \lambda_\beta + \lambda_\nu} \right)^{\lambda_\nu/p} (z_1(x) + z_2(x))^{((\lambda_\alpha + \lambda_\beta + \lambda_\nu)/p)} \right\}$$

$$\le A_0(x) \left\{ \left(\frac{\lambda_\nu \lambda_\alpha}{(\lambda_\alpha + \lambda_\beta)(\lambda_\alpha + \lambda_\beta + \lambda_\nu)} \right)^{\lambda_\nu/p} \tilde{\gamma}_2 \right.$$

$$\left. + 2^{p - \lambda_\nu/p} \left(\frac{\lambda_\beta}{\lambda_\alpha + \lambda_\beta} \right)^{\lambda_\nu/p} \left(\frac{\tilde{\gamma}_1 \lambda_\nu}{\lambda_\alpha + \lambda_\beta + \lambda_\nu} \right)^{\lambda_\nu/p} \right\} (z_1(x) + z_2(x))^{((\lambda_\alpha + \lambda_\beta + \lambda_\nu)/p)}$$

$$= A_0(x) \left(\frac{\lambda_\nu}{(\lambda_\alpha + \lambda_\beta)(\lambda_\alpha + \lambda_\beta + \lambda_\nu)} \right)^{\lambda_\nu/p} \left[\lambda_\alpha^{\lambda_\nu/p} \tilde{\gamma}_2 + 2^{p - \lambda_\nu/p} (\tilde{\gamma}_1 \lambda_\beta)^{\lambda_\nu/p} \right]$$

$$\cdot \left[\int_{x_0}^{x} p(t) \left[|(D_{x_0}^\nu f_1)(t)|^p + |(D_{x_0}^\nu f_2)(t)|^p \right] dt \right]^{((\lambda_\alpha + \lambda_\beta + \lambda_\nu)/p)}.$$

We have proved

$$\Omega \le A_0(x) \left(\frac{\lambda_\nu}{(\lambda_\alpha + \lambda_\beta)(\lambda_\alpha + \lambda_\beta + \lambda_\nu)} \right)^{\lambda_\nu/p} \left[\lambda_\alpha^{\lambda_\nu/p} \tilde{\gamma}_2 + 2^{(p - \lambda_\nu)/p} (\tilde{\gamma}_1 \lambda_\beta)^{\lambda_\nu/p} \right]$$

$$\cdot \left[\int_{x_0}^{x} p(w) \left[|(D_{x_0}^\nu f_1)(w)|^p + |(D_{x_0}^\nu f_2)(w)|^p \right] dw \right]^{((\lambda_\alpha + \lambda_\beta + \lambda_\nu)/p)}.$$

We have established (6.34). □

A special important case follows.

Theorem 6.8. *Let* $\nu \ge 2$ *and* $\gamma_1 \ge 0$ *such that* $\nu - \gamma_1 \ge 2$. *Let* $f_1, f_2 \in C_{x_0}^\nu([a, b])$ *with* $f_1^{(i)}(x_0) = f_2^{(i)}(x_0) = 0$, $i = 0, 1, \ldots, n - 1$, $n := [\nu]$. *Here*

$x, x_0 \in [a, b]$: $x \geq x_0$. *Consider also* $p(t) > 0$, *and* $q(t) \geq 0$ *continuous functions on* $[x_0, b]$. *Let* $\lambda_\alpha \geq 0$, $0 < \lambda_{\alpha+1} < 1$, *and* $p > 1$. *Denote*

$$\theta_3 := \left\{ \begin{array}{ll} 2^{\lambda_\alpha/\lambda_{\alpha+1}} - 1, & \text{if } \lambda_\alpha \geq \lambda_{\alpha+1}, \\ 1, & \text{if } \lambda_\alpha \leq \lambda_{\alpha+1} \end{array} \right\}, \tag{6.39}$$

$$L(x) := \left(2 \int_{x_0}^x (q(w))^{(1/1-\lambda_{\alpha+1})} dw \right)^{(1-\lambda_{\alpha+1})} \left(\frac{\theta_3 \lambda_{\alpha+1}}{\lambda_\alpha + \lambda_{\alpha+1}} \right)^{\lambda_{\alpha+1}}, \tag{6.40}$$

and

$$P_1(x) := \int_{x_0}^x (x - t)^{(\nu - \gamma_1 - 1)p/(p-1)} (p(t))^{-1/(p-1)} dt, \tag{6.41}$$

$$T(x) := L(x) \cdot \left(\frac{P_1(x)^{(p-1/p)}}{\Gamma(\nu - \gamma_1)} \right)^{(\lambda_\alpha + \lambda_{\alpha+1})}, \tag{6.42}$$

and

$$\omega_1 := 2^{((p-1)/p)(\lambda_\alpha + \lambda_{\alpha+1})}, \tag{6.43}$$

$$\Phi(x) := T(x)\omega_1. \tag{6.44}$$

Then

$$\int_{x_0}^x q(w) \left[|(D_{x_0}^{\gamma_1} f_1)(w)|^{\lambda_\alpha} |(D_{x_0}^{\gamma_1+1} f_2)(w)|^{\lambda_{\alpha+1}} \right. \tag{6.45}$$

$$\left. + |(D_{x_0}^{\gamma_1} f_2)(w)|^{\lambda_\alpha} |(D_{x_0}^{\gamma_1+1} f_1)(w)|^{\lambda_{\alpha+1}} \right] dw$$

$$\leq \Phi(x) \left[\int_{x_0}^x p(w) \left(|(D_{x_0}^\nu f_1)(w)|^p + |(D_{x_0}^\nu f_2)(w)|^p \right) dw \right]^{((\lambda_\alpha + \lambda_{\alpha+1})/p)},$$

all $x_0 \leq x \leq b$.

Proof. For convenience we set $\gamma_2 := \gamma_1 + 1$. From (6.13) we obtain

$$|(D_{x_0}^{\gamma_k} f_j)(w)| \leq \frac{1}{\Gamma(\nu - \gamma_k)} \int_{x_0}^w (w - t)^{\nu - \gamma_k - 1} |(D_{x_0}^\nu f_j)(t)| dt =: g_{j,\gamma_k}(w), \tag{6.46}$$

where $j = 1, 2$, $k = 1, 2$, all $x_0 \leq w \leq b$. We observe that

$$((D_{x_0}^{\gamma_1} f_j(x))' = (D_{x_0}^{\gamma_1+1} f_j)(x) = (D_{x_0}^{\gamma_2} f_j)(x), \tag{6.47}$$

all $x_0 \leq x \leq b$. And also

$$(g_{j,\gamma_1}(w))' = g_{j,\gamma_2}(w); \quad g_{j,\gamma_k}(x_0) = 0. \tag{6.48}$$

Notice that if $\nu - \gamma_2 = 1$, then

$$g_{j,\gamma_2}(w) = \int_{x_0}^w |(D_{x_0}^\nu f_j)(t)| dt.$$

Next we apply Hölder's inequality with indices $1/\lambda_{\alpha+1}$, $1/(1-\lambda_{\alpha+1})$; we obtain

$$\int_{x_0}^{x} q(w)|(D_{x_0}^{\gamma_1} f_1)(w)|^{\lambda_\alpha}|(D_{x_0}^{\gamma_1+1} f_2(w)|^{\lambda_{\alpha+1}} dw$$

$$\leq \int_{x_0}^{x} q(w)(g_{1,\gamma_1}(w))^{\lambda_\alpha}((g_{2,\gamma_1}(w))')^{\lambda_{\alpha+1}} dw \qquad (6.49)$$

$$\leq \left(\int_{x_0}^{x} (q(w))^{(1/1-\lambda_{\alpha+1})} dw\right)^{(1-\lambda_{\alpha+1})}$$

$$\left(\int_{x_0}^{x} (g_{1,\gamma_1}(w))^{\lambda_\alpha/\lambda_{\alpha+1}} (g_{2,\gamma_1}(w))' dw\right)^{\lambda_{\alpha+1}}.$$

Similarly we get

$$\int_{x_0}^{x} q(w)|(D_{x_0}^{\gamma_1} f_2)(w)|^{\lambda_\alpha}|(D_{x_0}^{\gamma_1+1} f_1(w)|^{\lambda_{\alpha+1}} dw$$

$$\leq \left(\int_{x_0}^{x} (q(w))^{(1/1-\lambda_{\alpha+1})} dw\right)^{(1-\lambda_{\alpha+1})} \qquad (6.50)$$

$$\left(\int_{x_0}^{x} (g_{2,\gamma_1}(w))^{\lambda_\alpha/\lambda_{\alpha+1}} (g_{1,\gamma_1}(w))' dw\right)^{\lambda_{\alpha+1}}.$$

Adding (6.49) and (6.50) we observe

$$\int_{x_0}^{x} q(w) \left[|(D_{x_0}^{\gamma_1} f_1)(w)|^{\lambda_\alpha}|(D_{x_0}^{\gamma_1+1} f_2)(w)|^{\lambda_{\alpha+1}}\right.$$

$$\left. + |(D_{x_0}^{\gamma_1} f_2)(w)|^{\lambda_\alpha}|(D_{x_0}^{\gamma_1+1} f_1)(w)|^{\lambda_{\alpha+1}}\right] dw$$

$$\leq \left(\int_{x_0}^{x} (q(w))^{(1/1-\lambda_{\alpha+1})} dw\right)^{(1-\lambda_{\alpha+1})}$$

$$\left[\left(\int_{x_0}^{x} (g_{1,\gamma_1}(w))^{\lambda_\alpha/\lambda_{\alpha+1}} (g_{2,\gamma_1}(w))' dw\right)^{\lambda_{\alpha+1}}\right.$$

$$\left. + \left(\int_{x_0}^{x} (g_{2,\gamma_1}(w))^{\lambda_\alpha/\lambda_{\alpha+1}} (g_{1,\gamma_1}(w))' dw\right)^{\lambda_{\alpha+1}}\right]$$

$$\overset{(6.23)}{\leq} \left(2\int_{x_0}^{x} (q(w))^{(1/1-\lambda_{\alpha+1})} dw\right)^{(1-\lambda_{\alpha+1})}$$

$$\cdot \left[\int_{x_0}^{x} [(g_{1,\gamma_1}(w))^{\lambda_\alpha/\lambda_{\alpha+1}} (g_{2,\gamma_1}(w))' + (g_{2,\gamma_1}(w))^{\lambda_\alpha/\lambda_{\alpha+1}} (g_{1,\gamma_1}(w))'] dw\right]^{\lambda_{\alpha+1}}$$

(note (6.30) and the proof of (6.31); accordingly here we have)

$$\leq \left(2\int_{x_0}^{x}(q(w))^{(1/1-\lambda_{\alpha+1})}dw\right)^{(1-\lambda_{\alpha+1})} \cdot \left(\frac{\lambda_{\alpha+1}\theta_3}{\lambda_\alpha + \lambda_{\alpha+1}}\right)^{\lambda_{\alpha+1}}$$

$$\cdot \left(g_{1,\gamma_1}(x) + g_{2,\gamma_1}(x)\right)^{(\lambda_\alpha+\lambda_{\alpha+1})}$$

$$= L(x)\left(g_{1,\gamma_1}(x) + g_{2,\gamma_1}(x)\right)^{(\lambda_\alpha+\lambda_{\alpha+1})}$$

$$= \frac{L(x)}{(\Gamma(\nu-\gamma_1))^{(\lambda_\alpha+\lambda_{\alpha+1})}}\left\{\int_{x_0}^{x}(x-t)^{\nu-\gamma_1-1}(p(t))^{-1/p}(p(t))^{1/p}\left[|(D_{x_0}^\nu f_1)(t)|\right.\right.$$

$$\left.\left. + |(D_{x_0}^\nu f_2)(t)|\right]dt\right\}^{(\lambda_\alpha+\lambda_{\alpha+1})}$$

(applying Hölder's inequality with indices $p/p-1$ and p we find)

$$\leq \frac{L(x)}{(\Gamma(\nu-\gamma_1))^{(\lambda_\alpha+\lambda_{\alpha+1})}}$$

$$\cdot \left(\int_{x_0}^{x}(x-t)^{(\nu-\gamma_1-1)p/p-1}(p(t))^{-1/p-1}dt\right)^{(p-1/p)(\lambda_\alpha+\lambda_{\alpha+1})}$$

$$\cdot \left(\int_{x_0}^{x}p(t)\left[|(D_{x_0}^\nu f_1)(t)| + |(D_{x_0}^\nu f_2)(t)|\right]^p dt\right)^{(\lambda_\alpha+\lambda_{\alpha+1}/p)}$$

$$= T(x)\cdot\left[\int_{x_0}^{x}p(t)(|(D_{x_0}^\nu f_1)(t)| + |(D_{x_0}^\nu f_2)(t)|)^p dt\right]^{(\lambda_\alpha+\lambda_{\alpha+1}/p)}$$

$$\leq \Phi(x)\cdot\left[\int_{x_0}^{x}p(t)(|(D_{x_0}^\nu f_1)(t)|^p + |(D_{x_0}^\nu f_2)(t)|^p)dt\right]^{(\lambda_\alpha+\lambda_{\alpha+1}/p)}.$$

We have proved (6.45).   □

Next we treat the case of exponents $\lambda_\beta = \lambda_\alpha + \lambda_\nu$.

Theorem 6.9. *All here are as in Theorem 6.5. Consider the special case* $\lambda_\beta = \lambda_\alpha + \lambda_\nu$. *Denote*

$$\tilde{T}(x) := A_0(x)\left(\frac{\lambda_\nu}{\lambda_\alpha + \lambda_\nu}\right)^{\lambda_\nu/p} 2^{(p-2\lambda_\alpha-3\lambda_\nu)/p}. \tag{6.51}$$

Then

$$\int_{x_0}^{x}q(w)\left[|(D_{x_0}^{\gamma_1}f_1)(w)|^{\lambda_\alpha}|(D_{x_0}^{\gamma_2}f_2)(w)|^{\lambda_\alpha+\lambda_\nu}|(D_{x_0}^\nu f_1)(w)|^{\lambda_\nu}\right.$$

$$\left. + |(D_{x_0}^{\gamma_2}f_1)(w)|^{\lambda_\alpha+\lambda_\nu}|(D_{x_0}^{\gamma_1}f_2)(w)|^{\lambda_\alpha}|(D_{x_0}^\nu f_2)(w)|^{\lambda_\nu}\right]dw$$

$$\leq \tilde{T}(x) \left(\int_{x_0}^x p(w)(|(D_{x_0}^\nu f_1)(w)|^p + |(D_{x_0}^\nu f_2)(w)|^p)dw \right)^{2((\lambda_\alpha + \lambda_\nu)/p)},$$

$$(6.52)$$

all $x_0 \leq x \leq b$.

Proof. We apply (6.21) and (6.22) for $\lambda_\beta = \lambda_\alpha + \lambda_\nu$ and add to find

$$\int_{x_0}^x q(w) \left[|(D_{x_0}^{\gamma_1} f_1)(w)|^{\lambda_\alpha} |(D_{x_0}^{\gamma_2} f_2)(w)|^{\lambda_\alpha + \lambda_\nu} |(D_{x_0}^\nu f_1)(w)|^{\lambda_\nu} \right.$$

$$\left. + |(D_{x_0}^{\gamma_2} f_1)(w)|^{\lambda_\alpha + \lambda_\nu} |(D_{x_0}^{\gamma_1} f_2)(w)|^{\lambda_\alpha} |(D_{x_0}^\nu f_2)(w)|^{\lambda_\nu} \right] dw$$

$$\leq A_0(x) \left\{ \left[\int_{x_0}^x (z_1(w))^{\lambda_\alpha/\lambda_\nu} (z_2(w))^{(\lambda_\alpha/\lambda_\nu)+1} z_1'(w)dw \right]^{\lambda_\nu/p} \right.$$

$$\left. + \left[\int_{x_0}^x (z_1(w))^{(\lambda_\alpha/\lambda_\nu)+1} (z_2(w))^{\lambda_\alpha/\lambda_\nu} z_2'(w)dw \right]^{\lambda_\nu/p} \right\}$$

$$\overset{(6.23)}{\leq} A_0(x) 2^{1-(\lambda_\nu/p)} \left\{ \int_{x_0}^x \left[(z_1(w))^{\lambda_\alpha/\lambda_\nu} (z_2(w))^{(\lambda_\alpha/\lambda_\nu)+1} z_1'(w) + \right. \right.$$

$$\left. \left. (z_1(w))^{\lambda_\alpha/\lambda_\nu+1} (z_2(w))^{\lambda_\alpha/\lambda_\nu} z_2'(w) \right] dw \right\}^{\lambda_\nu/p}$$

$$= A_0(x) 2^{1-\lambda_\nu/p} \left\{ \int_{x_0}^x (z_1(w)z_2(w))^{\lambda_\alpha/\lambda_\nu} [z_2(w)z_1'(w) + z_1(w)z_2'(w)]dw \right\}^{\lambda_\nu/p}$$

$$= A_0(x) 2^{1-\lambda_\nu/p} \cdot \left\{ \int_{x_0}^x (z_1(w)z_2(w))^{\lambda_\alpha/\lambda_\nu} (z_1(w)z_2(w))' dw \right\}^{\lambda_\nu/p}$$

$$= A_0(x) 2^{1-\lambda_\nu/p} \left(\frac{(z_1(x)z_2(x))^{(\lambda_\alpha/\lambda_\nu)+1}}{\frac{\lambda_\alpha}{\lambda_\nu}+1} \right)^{\lambda_\nu/p}$$

$$= A_0(x) 2^{p-\lambda_\nu/p} \left(\frac{\lambda_\nu}{\lambda_\alpha+\lambda_\nu} \right)^{\lambda_\nu/p} (z_1(x)z_2(x))^{(\lambda_\alpha+\lambda_\nu)/p}$$

$$\leq A_0(x) 2^{p-\lambda_\nu/p} \left(\frac{\lambda_\nu}{\lambda_\alpha+\lambda_\nu} \right)^{\lambda_\nu/p} \left(\frac{(z_1(x)+z_2(x))^2}{4} \right)^{(\lambda_\alpha+\lambda_\nu/p)}$$

$$= \tilde{T}(x)(z_1(x)+z_2(x))^{2(\lambda_\alpha+\lambda_\nu)/p}$$

$$= \tilde{T}(x) \left(\int_{x_0}^x p(w) (|(D_{x_0}^\nu f_1)(w)|^p + |(D_{x_0}^\nu f_2)(w)|^p) dw \right)^{2((\lambda_\alpha+\lambda_\nu)/p)}.$$

We have established (6.52). □

Next follow special cases of the above theorems.

Corollary 6.10. (to Theorem 6.5; $\lambda_\beta = 0$, $p(t) = q(t) = 1$)
It holds

$$\int_{x_0}^{x} \left[|(D_{x_0}^{\gamma_1} f_1)(w)|^{\lambda_\alpha} |(D_{x_0}^{\nu} f_1)(w)|^{\lambda_\nu} + |(D_{x_0}^{\gamma_1} f_2)(w)|^{\lambda_\alpha} |(D_{x_0}^{\nu} f_2)(w)|^{\lambda_\nu} \right] dw$$

$$\leq C_1(x) \cdot \left(\int_{x_0}^{x} (|(D_{x_0}^{\nu} f_1)(w)|^p + |(D_{x_0}^{\nu} f_2)(w)|^p) dw \right)^{((\lambda_\alpha + \lambda_\nu)/p)} \tag{6.53}$$

all $x_0 \leq x \leq b$, where

$$C_1(x) := \left(A_0(x)|_{\lambda_\beta = 0} \right) \cdot \left(\frac{\lambda_\nu}{\lambda_\alpha + \lambda_\nu} \right)^{\lambda_\nu/p} \cdot \delta_1, \tag{6.54}$$

$$\delta_1 := \left\{ \begin{array}{ll} 2^{1-(\lambda_\alpha + \lambda_\nu/p)}, & \text{if } \lambda_\alpha + \lambda_\nu \leq p, \\ 1, & \text{if } \lambda_\alpha + \lambda_\nu \geq p \end{array} \right\}.$$

We find that

$$\left(A_0(x)|_{\lambda_\beta=0} \right) = \left\{ \left(\frac{(p-1)^{((\lambda_\alpha p - \lambda_\alpha)/p)}}{(\Gamma(\nu - \gamma_1))^{\lambda_\alpha} (\nu p - \gamma_1 p - 1)^{((\lambda_\alpha p - \lambda_\alpha)/p)}} \right) \right.$$

$$\left. \cdot \left(\frac{(p - \lambda_\nu)^{(p - \lambda_\nu/p)}}{(\lambda_\alpha \nu p - \lambda_\alpha \gamma_1 p - \lambda_\alpha + p - \lambda_\nu)^{(p-\lambda_\nu/p)}} \right) \right\} \cdot (x - x_0)^{((\lambda_\alpha \nu p - \lambda_\alpha \gamma_1 p - \lambda_\alpha + p - \lambda_\nu)/p)}. \tag{6.55}$$

Proof. Here we need to calculate $A_0(x)|_{\lambda_\beta=0}$. From (6.17) we have

$$A_0(x)|_{\lambda_\beta=0} = \left(\int_{x_0}^{x} \left(A(w)|_{\lambda_\beta=0} \right)^{p/p-\lambda_\nu} dw \right)^{(p-\lambda_\nu)/p},$$

where from (6.16) we find

$$A(w)|_{\lambda_\beta=0} = \frac{(P_1(w))^{\lambda_\alpha((p-1)/p)}}{(\Gamma(\nu - \gamma_1))^{\lambda_\alpha}},$$

and here from (6.15) it is

$$P_1(w) = \int_{x_0}^{w} (w - t)^{((\nu - \gamma_1 - 1)/(p-1))p} dt.$$

Therefore we obtain

$$P_1(w) = \left(\frac{p-1}{\nu p - \gamma_1 p - 1} \right) (w - x_0)^{(\nu p - \gamma_1 p - 1)/(p-1)},$$

and

$$A(w)|_{\lambda_\beta=0} = \left(\frac{p-1}{\nu p - \gamma_1 p - 1} \right)^{(\lambda_\alpha p - \lambda_\alpha/p)} \cdot \frac{1}{(\Gamma(\nu - \gamma_1))^{\lambda_\alpha}}$$

$$\cdot (w - x_0)^{(\lambda_\alpha \nu p - \lambda_\alpha \gamma_1 p - \lambda_\alpha)/p}.$$

That is,

$$
\begin{aligned}
A_0(x)|_{\lambda_\beta=0} &= \left(\frac{(p-1)^{(\lambda_\alpha p-\lambda_\alpha/p)}}{(\nu p-\gamma_1 p-1)^{(\lambda_\alpha p-\lambda_\alpha/p)}(\Gamma(\nu-\gamma_1))^{\lambda_\alpha}}\right. \\
&\quad \left.\cdot\left(\int_{x_0}^x (w-x_0)^{(\lambda_\alpha\nu p-\lambda_\alpha\gamma_1 p-\lambda_\alpha/p-\lambda_\nu)}dw\right)\right)^{((p-\lambda_\nu)/p)} \\
&= \left\{\left(\frac{(p-1)^{(\lambda_\alpha p-\lambda_\alpha/p)}}{(\Gamma(\nu-\gamma_1))^{\lambda_\alpha}(\nu p-\gamma_1 p-1)^{((\lambda_\alpha p-\lambda_\alpha)/p)}}\right)\right. \\
&\quad \left.\cdot\left(\frac{(p-\lambda_\nu)^{(p-\lambda_\nu/p)}}{(\lambda_\alpha\nu p-\lambda_\alpha\gamma_1 p-\lambda_\alpha+p-\lambda_\nu)^{((p-\lambda_\nu)/p)}}\right)\right\} \\
&\quad \cdot(x-x_0)^{(\lambda_\alpha\nu p-\lambda_\alpha\gamma_1 p-\lambda_\alpha+p-\lambda_\nu/p)}. \quad \square
\end{aligned}
$$

Corollary 6.11. (to Theorem 6.5; $\lambda_\beta=0$, $p(t)=q(t)=1$, $\lambda_\alpha=\lambda_\nu=1$, $p=2$). *In detail:*

Let $\nu\geq 1$, $\gamma_1,\gamma_2\geq 0$, *such that* $\nu-\gamma_1\geq 1$, $\nu-\gamma_2\geq 1$ *and* $f_1,f_2\in C^\nu_{x_0}([a,b])$ *with* $f_1^{(i)}(x_0)=f_2^{(i)}(x_0)=0$, $i=0,1,\dots,n-1$, $n:=[\nu]$. *Here* $x,x_0\in[a,b]$: $x\geq x_0$. *Then*

$$
\int_{x_0}^x \left[|(D_{x_0}^{\gamma_1}f_1)(w)|\,|(D_{x_0}^\nu f_1)(w)|+|(D_{x_0}^{\gamma_1}f_2)(w)|\,|(D_{x_0}^\nu f_2)(w)|\right]dw
$$

$$
\leq \left(\frac{(x-x_0)^{(\nu-\gamma_1)}}{2\Gamma(\nu-\gamma_1)\sqrt{\nu-\gamma_1}\sqrt{2\nu-2\gamma_1-1}}\right)
$$
$$
\left(\int_{x_0}^x [((D_{x_0}^\nu f_1)(w))^2+((D_{x_0}^\nu f_2)(w))^2]dw\right), \tag{6.56}
$$

all $x_0\leq x\leq b$.

Proof. We apply Corollary 6.10. Here $\delta_1=1$ and

$$
\left(\frac{\lambda_\nu}{\lambda_\alpha+\lambda_\nu}\right)^{\lambda_\nu/p}=\frac{1}{\sqrt 2}.
$$

Furthermore we have

$$
(A_0(x)|_{\lambda_\beta=0}) = \left(\frac{1}{\Gamma(\nu-\gamma_1)\sqrt{2\nu-2\gamma_1-1}}\right)
$$
$$
\cdot\left(\frac{1}{\sqrt 2\sqrt{\nu-\gamma_1}}\right)\cdot(x-x_0)^{(\nu-\gamma_1)}.
$$

Finally we get

$$
C_1(x)=\frac{(x-x_0)^{(\nu-\gamma_1)}}{2\Gamma(\nu-\gamma_1)\sqrt{\nu-\gamma_1}\sqrt{2\nu-2\gamma_1-1}}. \quad \square
$$

Corollary 6.12. (to Theorem 6.6; $\lambda_\alpha = 0$, $p(t) = q(t) = 1$).
It holds

$$\int_{x_0}^x \left[|(D_{x_0}^{\gamma_2} f_2)(w)|^{\lambda_\beta} |(D_{x_0}^\nu f_1)(w)|^{\lambda_\nu} + |(D_{x_0}^{\gamma_2} f_1)(w)|^{\lambda_\beta} |(D_{x_0}^\nu f_2)(w)|^{\lambda_\nu} \right] dw$$

$$\leq C_2(x) \left(\int_{x_0}^x \left[|(D_{x_0}^\nu f_1)(w)|^p + |(D_{x_0}^\nu f_2)(w)|^p \right] dw \right)^{((\lambda_\nu + \lambda_\beta)/p)}, \quad (6.57)$$

all $x_0 \leq x \leq b$, where

$$C_2(x) := (A_0(x)|_{\lambda_\alpha=0}) \, 2^{(p-\lambda_\nu)/p} \left(\frac{\lambda_\nu}{\lambda_\beta + \lambda_\nu} \right)^{\lambda_\nu/p} \delta_3^{\lambda_\nu/p}, \quad (6.58)$$

$$\delta_3 := \left\{ \begin{array}{ll} 2^{\lambda_\beta/\lambda_\nu} - 1, & \text{if } \lambda_\beta \geq \lambda_\nu, \\ 1, & \text{if } \lambda_\beta \leq \lambda_\nu \end{array} \right\}.$$

We find that

$$(A_0(x)|_{\lambda_\alpha=0}) = \left\{ \left(\frac{(p-1)^{((\lambda_\beta p - \lambda_\beta)/p)}}{(\Gamma(\nu - \gamma_2))^{\lambda_\beta} (\nu p - \gamma_2 p - 1)^{((\lambda_\beta p - \lambda_\beta)/p)}} \right) \right.$$
$$\left. \cdot \left(\frac{(p-\lambda_\nu)^{(p-\lambda_\nu/p)}}{(\lambda_\beta \nu p - \lambda_\beta \gamma_2 p - \lambda_\beta + p - \lambda_\nu)^{((p-\lambda_\nu)/p)}} \right) \right\}$$
$$\cdot (x - x_0)^{(\lambda_\beta \nu p - \lambda_\beta \gamma_2 p - \lambda_\beta + p - \lambda_\nu/p)}. \quad (6.59)$$

Proof. Here we need to calculate

$$A_0(x)|_{\lambda_\alpha=0}.$$

From (6.17) we have

$$A_0(x)|_{\lambda_\alpha=0} = \left(\int_{x_0}^x (A(w)|_{\lambda_\alpha=0})^{p/p-\lambda_\nu} \, dw \right)^{(p-\lambda_\nu)/p}$$

where from (6.16) we find

$$A(w)|_{\lambda_\alpha=0} = \frac{(P_2(w))^{\lambda_\beta((p-1)/p)}}{(\Gamma(\nu - \gamma_2))^{\lambda_\beta}},$$

and here from (6.15) it is

$$P_2(w) = \int_{x_0}^w (w - t)^{(\nu-\gamma_2-1)p/(p-1)} \, dt.$$

Therefore we obtain

$$P_2(w) = \left(\frac{p-1}{\nu p - \gamma_2 p - 1} \right) (w - x_0)^{(\nu p - \gamma_2 p - 1)/(p-1)},$$

and

$$A(w)|_{\lambda_\alpha=0} = \frac{(p-1)^{(\lambda_\beta p - \lambda_\beta/p)}(w-x_0)^{(\lambda_\beta \nu p - \lambda_\beta \gamma_2 p - \lambda_\beta/p)}}{(\nu p - \gamma_2 p - 1)^{(\lambda_\beta p - \lambda_\beta/p)}(\Gamma(\nu-\gamma_2))^{\lambda_\beta}}.$$

That is,

$$A_0(x)|_{\lambda_\alpha=0} = \left(\frac{(p-1)^{(\lambda_\beta p - \lambda_\beta/p)}}{(\nu p - \gamma_2 p - 1)^{(\lambda_\beta p - \lambda_\beta/p)}(\Gamma(\nu-\gamma_2))^{\lambda_\beta}}\right)$$
$$\cdot \left(\int_{x_0}^x (w-x_0)^{(\lambda_\beta \nu p - \lambda_\beta \gamma_2 p - \lambda_\beta/p - \lambda_\nu)} dw\right)^{((p-\lambda_\nu)/p)}$$
$$= \left\{\left(\frac{(p-1)^{(\lambda_\beta p - \lambda_\beta/p)}}{(\nu p - \gamma_2 p - 1)^{(\lambda_\beta p - \lambda_\beta/p)}(\Gamma(\nu-\gamma_2))^{\lambda_\beta}}\right)\right.$$
$$\left.\cdot \left(\frac{(p-\lambda_\nu)^{(p-\lambda_\nu/p)}}{(\lambda_\beta \nu p - \lambda_\beta \gamma_2 p - \lambda_\beta + p - \lambda_\nu)^{((p-\lambda_\nu)/p)}}\right)\right\}$$
$$\cdot (x-x_0)^{(\lambda_\beta \nu p - \lambda_\beta \gamma_2 p - \lambda_\beta + p - \lambda_\nu/p)}.$$

□

Corollary 6.13. (to Theorem 6.6; $\lambda_\alpha=0$, $p(t)=q(t)=1$, $\lambda_\beta=\lambda_\nu=1$, $p=2$). *In detail*:
Let $\nu \geq 1$, $\gamma_1,\gamma_2 \geq 0$, such that $\nu-\gamma_1 \geq 1$, $\nu-\gamma_2 \geq 1$, and f_1, $f_2 \in C_{x_0}^\nu([a,b])$ with $f_1^{(i)}(x_0) = f_2^{(i)}(x_0) = 0$, $i=0,1,\ldots,n-1$, $n:=[\nu]$. Here $x,x_0 \in [a,b]$: $x \geq x_0$. Then

$$\int_{x_0}^x \left[|(D_{x_0}^{\gamma_2}f_2)(w)|\,|(D_{x_0}^\nu f_1)(w)| + |(D_{x_0}^{\gamma_2}f_1)(w)|\,|(D_{x_0}^\nu f_2)(w)|\right] dw$$
$$\leq \left(\frac{(x-x_0)^{(\nu-\gamma_2)}}{\sqrt{2}\Gamma(\nu-\gamma_2)\sqrt{\nu-\gamma_2}\,\sqrt{2\nu-2\gamma_2-1}}\right)$$
$$\cdot \left(\int_{x_0}^x \left(((D_{x_0}^\nu f_1)(w))^2 + ((D_{x_0}^\nu f_2(w))^2\right) dw\right), \tag{6.60}$$

all $x_0 \leq x \leq b$.

Proof. We apply Corollary 6.12. Here $\delta_3^{\lambda_\nu/p}=1$, $2^{p-\lambda_\nu/p}=\sqrt{2}$,

$$\left(\frac{\lambda_\nu}{\lambda_\beta+\lambda_\nu}\right)^{\lambda_\nu/p} = \frac{1}{\sqrt{2}}.$$

Furthermore we have

$$(A_0(x)|_{\lambda_\alpha=0}) = \left(\frac{(x-x_0)^{(\nu-\gamma_2)}}{\sqrt{2}\,\Gamma(\nu-\gamma_2)\sqrt{\nu-\gamma_2}\,\sqrt{2\nu-2\gamma_2-1}}\right).$$

Finally we get

$$C_2(x) = \left(\frac{(x - x_0)^{(\nu - \gamma_2)}}{\sqrt{2}\,\Gamma(\nu - \gamma_2)\sqrt{\nu - \gamma_2}\,\sqrt{2\nu - 2\gamma_2 - 1}} \right). \quad \square$$

Corollary 6.14. (to Theorem 6.7; $\lambda_\alpha = \lambda_\beta = \lambda_\nu = 1$, $p = 3$, $p(t) = q(t) = 1$).
  It holds

$$\int_{x_0}^{x} [[|(D_{x_0}^{\gamma_1} f_1)(w)||(D_{x_0}^{\gamma_2} f_2)(w)||(D_{x_0}^{\nu} f_1)(w)|$$

$$+ |(D_{x_0}^{\gamma_2} f_1)(w)||(D_{x_0}^{\gamma_1} f_2)(w)||(D_{x_0}^{\nu} f_2)(w)|]dw \le A_0(x)$$

$$\left(\sqrt[3]{2} + \frac{1}{\sqrt[3]{6}} \right) \cdot \left(\int_{x_0}^{x} (|(D_{x_0}^{\nu} f_1)(w)|^3 + |(D_{x_0}^{\nu} f_2)(w)|^3)dw \right), \qquad (6.61)$$

all $x_0 \le x \le b$. *Here*

$$A_0(x) =$$

$$\frac{4(x - x_0)^{(2\nu - \gamma_1 - \gamma_2)}}{\Gamma(\nu - \gamma_1)\Gamma(\nu - \gamma_2)[3(3\nu - 3\gamma_1 - 1)(3\nu - 3\gamma_2 - 1)(2\nu - \gamma_1 - \gamma_2)]^{2/3}}.$$

$$(6.62)$$

Proof. We apply inequality (6.34). Here $\tilde{\gamma}_1 = 3$, $\tilde{\gamma}_2 = 1$, and

$$\left(\frac{\lambda_\nu}{(\lambda_\alpha + \lambda_\beta)(\lambda_\alpha + \lambda_\beta + \lambda_\nu)} \right)^{\lambda_\nu/p} = \frac{1}{\sqrt[3]{6}}.$$

Furthermore

$$\left(\lambda_\alpha^{\lambda_\nu/p} \tilde{\gamma}_2 + 2^{(p - \lambda_\nu)/p} (\tilde{\gamma}_1 \lambda_\beta)^{\lambda_\nu/p} \right) = 1 + \sqrt[3]{12}.$$

The product of the last two quantities is $(\sqrt[3]{2} + 1/\sqrt[3]{6})$. It still remains to find $A_0(x)$ for this case.
  Here by (6.17),

$$A_0(x) = \left(\int_{x_0}^{x} A(w)^{3/2} dw \right)^{2/3},$$

and by (6.16),

$$A(w) = \frac{(P_1(w))^{2/3}(P_2(w))^{2/3}}{\Gamma(\nu - \gamma_1)\Gamma(\nu - \gamma_2)}.$$

Also by (6.15) we have for $k = 1, 2$,

$$P_k(w) = \int_{x_0}^{w} (w - t)^{((3\nu - 3\gamma_k - 3)/2)} dt.$$

That is,

$$P_k(w) = \frac{2(w - x_0)^{3\nu - 3\gamma_k - 1/2}}{(3\nu - 3\gamma_k - 1)}, \quad k = 1, 2,$$

and

$$A(w) = \frac{2^{4/3}(w - x_0)^{(6\nu - 3\gamma_1 - 3\gamma_2 - 2/3)}}{\Gamma(\nu - \gamma_1)\Gamma(\nu - \gamma_2)(3\nu - 3\gamma_1 - 1)^{2/3}(3\nu - 3\gamma_2 - 1)^{2/3}}.$$

Finally we find

$$
\begin{aligned}
A_0(x) &= \left(\frac{2^{4/3}}{\Gamma(\nu - \gamma_1)\Gamma(\nu - \gamma_2)(3\nu - 3\gamma_1 - 1)^{2/3}(3\nu - 3\gamma_2 - 1)^{2/3}} \right) \\
&\quad \cdot \left(\int_{x_0}^{x} (w - x_0)^{((6\nu - 3\gamma_1 - 3\gamma_2 - 2)/2)} \, dw \right)^{2/3} \\
&= \frac{4 \cdot (x - x_0)^{(2\nu - \gamma_1 - \gamma_2)}}{\Gamma(\nu - \gamma_1)\Gamma(\nu - \gamma_2)} \\
&\quad \frac{1}{[3(3\nu - 3\gamma_1 - 1)(3\nu - 3\gamma_2 - 1)(2\nu - \gamma_1 - \gamma_2)]^{2/3}}.
\end{aligned}
$$

$\square$

Corollary 6.15. (to Theorem 6.8; $\lambda_\alpha = 1$, $\lambda_{\alpha+1} = 1/2$, $p = 3/2$, $p(t) = q(t) = 1$). *In detail:*

Let $\nu \geq 2$ and $\gamma_1 \geq 0$ such that $\nu - \gamma_1 \geq 2$. Let $f_1, f_2 \in C_{x_0}^\nu([a, b])$ with $f_1^{(i)}(x_0) = f_2^{(i)}(x_0) = 0$, $i = 0, 1, \ldots, n - 1$, $n := [\nu]$. Here $x, x_0 \in [a, b]$: $x \geq x_0$. Then

$$\int_{x_0}^{x} \left[|(D_{x_0}^{\gamma_1} f_1)(w)| \sqrt{|(D_{x_0}^{\gamma_1 + 1} f_2)(w)|} + \right.$$

$$\left. |(D_{x_0}^{\gamma_1} f_2)(w)| \sqrt{|(D_{x_0}^{\gamma_1 + 1} f_1)(w)|} \right] dw$$

$$\leq \Phi(x) \cdot \left[\int_{x_0}^{x} (|(D_{x_0}^{\nu} f_1)(w)|^{3/2} + |(D_{x_0}^{\nu} f_2)(w)|^{3/2}) dw \right], \quad (6.63)$$

all $x_0 \leq x \leq b$. Here we find

$$\Phi(x) = \left(\frac{2}{\sqrt{3\nu - 3\gamma_1 - 2}} \right) \frac{(x - x_0)^{((3\nu - 3\gamma_1 - 1)/2)}}{(\Gamma(\nu - \gamma_1))^{3/2}}. \quad (6.64)$$

Proof. We apply inequality (6.45). Here $\omega_1 = \sqrt{2}$; that is, $\Phi(x) = \sqrt{2} \cdot T(x)$, and $\theta_3 = 3$. Furthermore we find $L(x) = \sqrt{2(x - x_0)}$, and

$$P_1(x) = \frac{(x - x_0)^{(3\nu - 3\gamma_1 - 2)}}{(3\nu - 3\gamma_1 - 2)}.$$

Thus

$$T(x) = \sqrt{2(x-x_0)} \cdot \left(\frac{(x-x_0)^{(3\nu-3\gamma_1-2/3)}}{\sqrt[3]{(3\nu-3\gamma_1-2)}\Gamma(\nu-\gamma_1)} \right)^{3/2}$$

$$= \frac{\sqrt{2} \cdot (x-x_0)^{(3\nu-3\gamma_1-1/2)}}{(\Gamma(\nu-\gamma_1))^{3/2} \cdot \sqrt{3\nu-3\gamma_1-2}}.$$

$\square$

Corollary 6.16. (to Theorem 6.9; $p = 2(\lambda_\alpha + \lambda_\nu) > 1$, $p(t) = q(t) = 1$).
It holds

$$\int_{x_0}^{x} [|(D_{x_0}^{\gamma_1} f_1)(w)|^{\lambda_\alpha} |(D_{x_0}^{\gamma_2} f_2)(w)|^{\lambda_\alpha+\lambda_\nu} |(D_{x_0}^{\nu} f_1)(w)|^{\lambda_\nu}$$

$$+ |(D_{x_0}^{\gamma_2} f_1)(w)|^{\lambda_\alpha+\lambda_\nu} |(D_{x_0}^{\gamma_1} f_2)(w)|^{\lambda_\alpha} |(D_{x_0}^{\nu} f_2)(w)|^{\lambda_\nu}] dw$$

$$\leq \tilde{T}(x) \left(\int_{x_0}^{x} \left[(|(D_{x_0}^{\nu} f_1)(w)|^{2(\lambda_\alpha+\lambda_\nu)} + (|(D_{x_0}^{\nu} f_2)(w)|)^{2(\lambda_\alpha+\lambda_\nu)} \right] dw \right),$$

(6.65)

all $x_0 \leq x \leq b$. Here

$$\tilde{T}(x) = A_0(x) \left(\frac{\lambda_\nu}{2(\lambda_\alpha+\lambda_\nu)} \right)^{\lambda_\nu/2(\lambda_\alpha+\lambda_\nu)}, \quad (6.66)$$

and

$$A_0(x) = \sigma\sigma^*(x-x_0)^\theta, \quad (6.67)$$

where

$$\sigma := \frac{1}{(\Gamma(\nu-\gamma_1))^{\lambda_\alpha} (\Gamma(\nu-\gamma_2))^{\lambda_\alpha+\lambda_\nu}}$$

$$\cdot \left(\frac{2\lambda_\alpha + 2\lambda_\nu - 1}{2\lambda_\alpha\nu + 2\lambda_\nu\nu - 2\lambda_\alpha\gamma_1 - 2\lambda_\nu\gamma_1 - 1} \right)^{(2\lambda_\alpha^2+2\lambda_\alpha\lambda_\nu-\lambda_\alpha)/(2\lambda_\alpha+2\lambda_\nu)}$$

$$\cdot \left(\frac{2\lambda_\alpha + 2\lambda_\nu - 1}{2\lambda_\alpha\nu + 2\lambda_\nu\nu - 2\lambda_\alpha\gamma_2 - 2\lambda_\nu\gamma_2 - 1} \right)^{((2\lambda_\alpha+2\lambda_\nu-1)/2)}$$

(6.68)

$$\sigma^* := \left[(2\lambda_\alpha + \lambda_\nu) \left(4\lambda_\alpha^2\nu + 6\lambda_\alpha\lambda_\nu\nu - 2\lambda_\alpha^2\gamma_1 - \right. \right.$$

$$\left. \left. -2\lambda_\alpha\lambda_\nu\gamma_1 - 2\lambda_\alpha^2\gamma_2 - 4\lambda_\alpha\lambda_\nu\gamma_2 + 2\lambda_\nu^2\nu - 2\lambda_\nu^2\gamma_2 \right)^{-1} \right]^{((2\lambda_\alpha+\lambda_\nu)/2(\lambda_\alpha+\lambda_\nu))},$$

(6.69)

and

$$\theta :=$$

$$\left(\frac{4\lambda_\alpha^2\nu + 6\lambda_\alpha\lambda_\nu\nu - 2\lambda_\alpha^2\gamma_1 - 2\lambda_\alpha\lambda_\nu\gamma_1 - 2\lambda_\alpha^2\gamma_2 - 4\lambda_\alpha\lambda_\nu\gamma_2 + 2\lambda_\nu^2\nu - 2\lambda_\nu^2\gamma_2}{2\lambda_\alpha + 2\lambda_\nu} \right).$$

(6.70)

Proof. We apply inequality (6.52). The constant $\tilde{T}(x)$ comes from (6.51) and the assumption on p here. Still we need to determine $A_0(x)$. Here from (6.17) we have

$$A_0(x) = \left(\int_{x_0}^{x} (A(w))^{2(\lambda_\alpha + \lambda_\nu)/2\lambda_\alpha + \lambda_\nu} \, dw \right)^{(2\lambda_\alpha + \lambda_\nu)/2(\lambda_\alpha + \lambda_\nu)},$$

where from (6.16) we find

$$A(w) = \frac{(P_1(w))^{\lambda_\alpha(2\lambda_\alpha + 2\lambda_\nu - 1/2\lambda_\alpha + 2\lambda_\nu)} (P_2(w))^{((2\lambda_\alpha + 2\lambda_\nu - 1)/2)}}{(\Gamma(\nu - \gamma_1))^{\lambda_\alpha} (\Gamma(\nu - \gamma_2))^{\lambda_\alpha + \lambda_\nu}},$$

and from (6.15) we find for $k = 1, 2$,

$$P_k(w) = \int_{x_0}^{w} (w - t)^{(\nu - \gamma_k - 1)((2\lambda_\alpha + 2\lambda_\nu)/2\lambda_\alpha + 2\lambda_\nu - 1)} \, dt.$$

Hence for $k = 1, 2$,

$$P_k(w) = \frac{(2\lambda_\alpha + 2\lambda_\nu - 1)(w - x_0)^{((2\lambda_\alpha\nu + 2\lambda_\nu\nu - 2\lambda_\alpha\gamma_k - 2\lambda_\nu\gamma_k - 1)/(2\lambda_\alpha + 2\lambda_\nu - 1))}}{(2\lambda_\alpha\nu + 2\lambda_\nu\nu - 2\lambda_\alpha\gamma_k - 2\lambda_\nu\gamma_k - 1)},$$

and

$$A(w) = \sigma(w - x_0)^{\left(\begin{array}{c} 4\lambda_\alpha^2\nu + 6\lambda_\alpha\lambda_\nu\nu - 2\lambda_\alpha^2\gamma_1 - 2\lambda_\alpha^2\gamma_2 - 2\lambda_\alpha\lambda_\nu\gamma_1 \\ - 4\lambda_\alpha\lambda_\nu\gamma_2 - 2\lambda_\nu^2\gamma_2 + 2\lambda_\nu^2\nu - 2\lambda_\alpha - \lambda_\nu \end{array} \right)/(2\lambda_\alpha + 2\lambda_\nu)}.$$

Finally we obtain

$$A_0(x) = \sigma \cdot$$

$$\left(\left[\int_{x_0}^{x} (w - x_0)^{\dfrac{\left(\begin{array}{c} 4\lambda_\alpha^2\nu + 6\lambda_\alpha\lambda_\nu\nu - 2\lambda_\alpha^2\gamma_1 - 2\lambda_\alpha^2\gamma_2 - 2\lambda_\alpha\lambda_\nu\gamma_1 \\ - 4\lambda_\alpha\lambda_\nu\gamma_2 - 2\lambda_\nu^2\gamma_2 + 2\lambda_\nu^2\nu - 2\lambda_\alpha - \lambda_\nu \end{array} \right)}{(2\lambda_\alpha + \lambda_\nu)}} \, dw \right]^{\dfrac{(2\lambda_\alpha + \lambda_\nu)}{2(\lambda_\alpha + \lambda_\nu)}} \right)$$

$$= \sigma\sigma^*(x - x_0)^\theta.$$

$\square$

Corollary 6.17. (to Theorem 6.9; $p = 4$, $\lambda_\alpha = \lambda_\nu = 1$, $p(t) = q(t) = 1$). *It holds*

$$\int_{x_0}^{x} \left[|(D_{x_0}^{\gamma_1} f_1)(w)|((D_{x_0}^{\gamma_2} f_2)(w))^2|(D_{x_0}^{\nu} f_1)(w)| + \right.$$

$$\left. ((D_{x_0}^{\gamma_2} f_1)(w))^2|(D_{x_0}^{\gamma_1} f_2)(w)| \, |(D_{x_0}^{\nu} f_2)(w)| \right] \, dw$$

$$\leq T^*(x) \left(\int_{x_0}^{x} (((D_{x_0}^{\nu} f_1)(w))^4 + ((D_{x_0}^{\nu} f_2)(w))^4) dw \right), \tag{6.71}$$

all $x_0 \leq x \leq b$. Here

$$T^*(x) = \frac{\tilde{A}_0(x)}{\sqrt{2}}, \tag{6.72}$$

and

$$\tilde{A}_0(x) = \tilde{\sigma} \, \tilde{\sigma}^* (x - x_0)^{\tilde{\theta}}, \tag{6.73}$$

where

$$\tilde{\sigma} := \frac{1}{\Gamma(\nu - \gamma_1)(\Gamma(\nu - \gamma_2))^2} \cdot \left(\frac{3}{4\nu - 4\gamma_1 - 1} \right)^{3/4} \cdot \left(\frac{3}{4\nu - 4\gamma_2 - 1} \right)^{3/2}, \tag{6.74}$$

$$\tilde{\sigma}^* := \left(\frac{3}{12\nu - 4\gamma_1 - 8\gamma_2} \right)^{3/4}, \tag{6.75}$$

and

$$\tilde{\theta} := 3\nu - \gamma_1 - 2\gamma_2. \tag{6.76}$$

Proof. By Corollary 6.16. $\square$

We continue with related results regarding the sup-norm $\| \cdot \|_{\infty}$.

Theorem 6.18. *Let* $\nu \geq 1$, $\gamma_1, \gamma_2 \geq 0$, *such that* $\nu - \gamma_1 \geq 1$, $\nu - \gamma_2 \geq 1$, *and* $f_1, f_2 \in C_{x_0}^{\nu}([a,b])$ *with* $f_1^{(i)}(x_0) = f_2^{(i)}(x_0) = 0$, $i = 0, 1, \ldots, n - 1$, $n := [\nu]$. *Here* $x, x_0 \in [a,b]: x \geq x_0$. *Consider* $p(x) \geq 0$ *a continuous function on* $[x_0, b]$. *Let* λ_α, λ_β, $\lambda_\nu \geq 0$. *Set*

$$\rho(x) := \frac{(x - x_0)^{(\nu\lambda_\alpha - \gamma_1\lambda_\alpha + \nu\lambda_\beta - \gamma_2\lambda_\beta + 1)} \|p(x)\|_{\infty}}{(\nu\lambda_\alpha - \gamma_1\lambda_\alpha + \nu\lambda_\beta - \gamma_2\lambda_\beta + 1)(\Gamma(\nu - \gamma_1 + 1))^{\lambda_\alpha} (\Gamma(\nu - \gamma_2 + 1))^{\lambda_\beta}}. \tag{6.77}$$

Then

$$\int_{x_0}^{x} p(w) \left[|(D_{x_0}^{\gamma_1} f_1)(w)|^{\lambda_\alpha} |(D_{x_0}^{\gamma_2} f_2)(w)|^{\lambda_\beta} |(D_{x_0}^{\nu} f_1)(w)|^{\lambda_\nu} \right.$$

$$\left. + |(D_{x_0}^{\gamma_2} f_1)(w)|^{\lambda_\beta} |(D_{x_0}^{\gamma_1} f_2)(w)|^{\lambda_\alpha} |(D_{x_0}^{\nu} f_2)(w)|^{\lambda_\nu} \right] dw$$

$$\leq \frac{\rho(x)}{2} \left[\|D_{x_0}^{\nu} f_1\|_{\infty}^{2(\lambda_\alpha + \lambda_\nu)} + \|D_{x_0}^{\nu} f_1\|_{\infty}^{2\lambda_\beta} + \|D_{x_0}^{\nu} f_2\|_{\infty}^{2\lambda_\beta} + \|D_{x_0}^{\nu} f_2\|_{\infty}^{2(\lambda_\alpha + \lambda_\nu)} \right], \tag{6.78}$$

all $x_0 \leq x \leq b$.

Proof. From (6.13) we get for $j = 1, 2$; $k = 1, 2$, that

$$|(D_{x_0}^{\gamma_k} f_j)(w)| \leq \frac{1}{\Gamma(\nu - \gamma_k)} \left(\int_{x_0}^{w} (w - t)^{\nu - \gamma_k - 1} dt \right) \|D_{x_0}^{\nu} f_j\|_{\infty}$$

$$= \frac{1}{\Gamma(\nu - \gamma_k)} \frac{(w - x_0)^{\nu - \gamma_k}}{(\nu - \gamma_k)} \|D_{x_0}^{\nu} f_j\|_{\infty}.$$

That is,

$$|(D_{x_0}^{\gamma_k} f_j)(w)| \le \frac{(w - x_0)^{(\nu - \gamma_k)}}{\Gamma(\nu - \gamma_k + 1)} \|D_{x_0}^{\nu} f_j\|_{\infty}, \tag{6.79}$$

all $x_0 \le w \le b$. Then we have

$$|(D_{x_0}^{\gamma_1} f_1)(w)|^{\lambda_\alpha} \le \frac{(w - x_0)^{(\nu - \gamma_1)\lambda_\alpha}}{(\Gamma(\nu - \gamma_1 + 1))^{\lambda_\alpha}} \|D_{x_0}^{\nu} f_1\|_{\infty}^{\lambda_\alpha}, \tag{6.80}$$

$$|(D_{x_0}^{\gamma_2} f_2)(w)|^{\lambda_\beta} \le \frac{(w - x_0)^{(\nu - \gamma_2)\lambda_\beta}}{(\Gamma(\nu - \gamma_2 + 1))^{\lambda_\beta}} \|D_{x_0}^{\nu} f_2\|_{\infty}^{\lambda_\beta}, \tag{6.81}$$

$$|(D_{x_0}^{\nu} f_1)(w)|^{\lambda_\nu} \le \|D_{x_0}^{\nu} f_1\|_{\infty}^{\lambda_\nu}. \tag{6.82}$$

Multiplying (6.80) through (6.82) we obtain

$$|(D_{x_0}^{\gamma_1} f_1)(w)|^{\lambda_\alpha} |(D_{x_0}^{\gamma_2} f_2)(w)|^{\lambda_\beta} |(D_{x_0}^{\nu} f_1)(w)|^{\lambda_\nu}$$

$$\le \frac{(w - x_0)^{(\nu\lambda_\alpha - \gamma_1\lambda_\alpha + \nu\lambda_\beta - \gamma_2\lambda_\beta)}}{(\Gamma(\nu - \gamma_1 + 1))^{\lambda_\alpha} (\Gamma(\nu - \gamma_2 + 1))^{\lambda_\beta}} \|D_{x_0}^{\nu} f_1\|_{\infty}^{\lambda_\alpha + \lambda_\nu} \|D_{x_0}^{\nu} f_2\|_{\infty}^{\lambda_\beta}. \tag{6.83}$$

Similarly we observe

$$|(D_{x_0}^{\gamma_2} f_1)(w)|^{\lambda_\beta} \le \frac{(w - x_0)^{(\nu - \gamma_2)\lambda_\beta}}{(\Gamma(\nu - \gamma_2 + 1))^{\lambda_\beta}} \|D_{x_0}^{\nu} f_1\|_{\infty}^{\lambda_\beta}, \tag{6.84}$$

$$|(D_{x_0}^{\gamma_1} f_2)(w)|^{\lambda_\alpha} \le \frac{(w - x_0)^{(\nu - \gamma_1)\lambda_\alpha}}{(\Gamma(\nu - \gamma_1 + 1))^{\lambda_\alpha}} \|D_{x_0}^{\nu} f_2\|_{\infty}^{\lambda_\alpha}, \tag{6.85}$$

$$|(D_{x_0}^{\nu} f_2)(w)|^{\lambda_\nu} \le \|D_{x_0}^{\nu} f_2\|_{\infty}^{\lambda_\nu}. \tag{6.86}$$

Multiplying (6.84) through (6.86) we find

$$|(D_{x_0}^{\gamma_2} f_1)(w)|^{\lambda_\beta} |(D_{x_0}^{\gamma_1} f_2)(w)|^{\lambda_\alpha} |(D_{x_0}^{\nu} f_2)(w)|^{\lambda_\nu}$$

$$\le \frac{(w - x_0)^{(\nu\lambda_\beta - \gamma_2\lambda_\beta + \nu\lambda_\alpha - \gamma_1\lambda_\alpha)}}{(\Gamma(\nu - \gamma_2 + 1))^{\lambda_\beta} (\Gamma(\nu - \gamma_1 + 1))^{\lambda_\alpha}} \|D_{x_0}^{\nu} f_1\|_{\infty}^{\lambda_\beta} \|D_{x_0}^{\nu} f_2\|_{\infty}^{\lambda_\alpha + \lambda_\nu}. \tag{6.87}$$

Adding (6.83) and (6.87) we have

$$|(D_{x_0}^{\gamma_1} f_1)(w)|^{\lambda_\alpha} |(D_{x_0}^{\gamma_2} f_2)(w)|^{\lambda_\beta} |(D_{x_0}^{\nu} f_1)(w)|^{\lambda_\nu}$$

$$+ |(D_{x_0}^{\gamma_2} f_1)(w)|^{\lambda_\beta} |(D_{x_0}^{\gamma_1} f_2)(w)|^{\lambda_\alpha} |(D_{x_0}^{\nu} f_2)(w)|^{\lambda_\nu}$$

$$\le \frac{(w - x_0)^{(\nu\lambda_\alpha - \gamma_1\lambda_\alpha + \nu\lambda_\beta - \gamma_2\lambda_\beta)}}{(\Gamma(\nu - \gamma_1 + 1))^{\lambda_\alpha} (\Gamma(\nu - \gamma_2 + 1))^{\lambda_\beta}}$$

$$\cdot \left[\|D_{x_0}^{\nu} f_1\|_{\infty}^{\lambda_\alpha + \lambda_\nu} \|D_{x_0}^{\nu} f_2\|_{\infty}^{\lambda_\beta} + \|D_{x_0}^{\nu} f_1\|_{\infty}^{\lambda_\beta} \|D_{x_0}^{\nu} f_2\|_{\infty}^{\lambda_\alpha + \lambda_\nu} \right]. \tag{6.88}$$

It follows that

$$\int_{x_0}^{x} p(w)[|(D_{x_0}^{\gamma_1} f_1)(w)|^{\lambda_\alpha}|(D_{x_0}^{\gamma_2} f_2)(w)|^{\lambda_\beta}|(D_{x_0}^{\nu} f_1)(w)|^{\lambda_\nu}$$

$$+|(D_{x_0}^{\gamma_2} f_1)(w)|^{\lambda_\beta}|(D_{x_0}^{\gamma_1} f_2)(w)|^{\lambda_\alpha}|(D_{x_0}^{\nu} f_2)(w)|^{\lambda_\nu}]dw$$

$$\leq \frac{\|p(x)\|_\infty \int_{x_0}^{x} (w-x_0)^{(\nu\lambda_\alpha-\gamma_1\lambda_\alpha+\nu\lambda_\beta-\gamma_2\lambda_\beta)} dw}{(\Gamma(\nu-\gamma_1+1))^{\lambda_\alpha}(\Gamma(\nu-\gamma_2+1))^{\lambda_\beta}}$$

$$\cdot \left[\|D_{x_0}^{\nu} f_1\|_\infty^{\lambda_\alpha+\lambda_\nu}\|D_{x_0}^{\nu} f_2\|_\infty^{\lambda_\beta} + \|(D_{x_0}^{\nu} f_1\|_\infty^{\lambda_\beta}\|D_{x_0}^{\nu} f_2\|_\infty^{\lambda_\alpha+\lambda_\nu} \right]$$

$$\leq \frac{\rho(x)}{2} \left[\|D_{x_0}^{\nu} f_1\|_\infty^{2(\lambda_\alpha+\lambda_\nu)} + \|D_{x_0}^{\nu} f_1\|_\infty^{2\lambda_\beta} + \|D_{x_0}^{\nu} f_2\|_\infty^{2\lambda_\beta} + \|D_{x_0}^{\nu} f_2\|_\infty^{2(\lambda_\alpha+\lambda_\nu)} \right].$$

We have established (6.78). □

Some special cases to the last theorem follow.

Theorem 6.19. (as in Theorem 6.18; $\lambda_\beta = 0$).
It holds

$$\int_{x_0}^{x} p(w) \left[|(D_{x_0}^{\gamma_1} f_1)(w)|^{\lambda_\alpha}|(D_{x_0}^{\nu} f_1)(w)|^{\lambda_\nu} + \right.$$

$$|(D_{x_0}^{\gamma_1} f_2)(w)|^{\lambda_\alpha}|(D_{x_0}^{\nu} f_2)(w)|^{\lambda_\nu} \Big] dw$$

$$\leq \left(\frac{(x-x_0)^{(\nu\lambda_\alpha-\gamma_1\lambda_\alpha+1)}\|p(x)\|_\infty}{(\nu\lambda_\alpha-\gamma_1\lambda_\alpha+1)(\Gamma(\nu-\gamma_1+1))^{\lambda_\alpha}} \right)$$

$$\cdot [\|D_{x_0}^{\nu} f_1\|_\infty^{\lambda_\alpha+\lambda_\nu} + \|D_{x_0}^{\nu} f_2\|_\infty^{\lambda_\alpha+\lambda_\nu}], \qquad (6.89)$$

all $x_0 \leq x \leq b$.

Proof. As in Theorem 6.18, similarly. □

Theorem 6.20. (as in Theorem 6.18; $\lambda_\beta = \lambda_\alpha + \lambda_\nu$).
It holds

$$\int_{x_0}^{x} p(w)[|(D_{x_0}^{\gamma_1} f_1)(w)|^{\lambda_\alpha}|(D_{x_0}^{\gamma_2} f_2)(w)|^{\lambda_\alpha+\lambda_\nu}|(D_{x_0}^{\nu} f_1)(w)|^{\lambda_\nu}$$

$$+|(D_{x_0}^{\gamma_2} f_1)(w)|^{\lambda_\alpha+\lambda_\nu}|(D_{x_0}^{\gamma_1} f_2)(w)|^{\lambda_\alpha}|(D_{x_0}^{\nu} f_2)(w)|^{\lambda_\nu}]dw$$

$$\leq \left(\frac{(x-x_0)^{(2\nu\lambda_\alpha-\gamma_1\lambda_\alpha+\nu\lambda_\nu-\gamma_2\lambda_\alpha-\gamma_2\lambda_\nu+1}}{(2\nu\lambda_\alpha-\gamma_1\lambda_\alpha+\nu\lambda_\nu-\gamma_2\lambda_\alpha-\gamma_2\lambda_\nu+1)} \right.$$

$$\cdot \left. \frac{\|p(x)\|_\infty}{(\Gamma(\nu-\gamma_1+1))^{\lambda_\alpha}(\Gamma(\nu-\gamma_2+1))^{\lambda_\alpha+\lambda_\nu}} \right)$$

$$\cdot \left(\|D_{x_0}^{\nu} f_1\|_\infty^{2(\lambda_\alpha+\lambda_\nu)} + \|(D_{x_0}^{\nu} f_2)\|_\infty^{2(\lambda_\alpha+\lambda_\nu)} \right), \text{ all } x_0 \leq x \leq b. \qquad (6.90)$$

Proof. As in Theorem 6.18, similarly. $\square$

Theorem 6.21. (as in Theorem 6.18; $\lambda_\nu = 0$, $\lambda_\alpha = \lambda_\beta$).
It holds

$$\int_{x_0}^{x} p(w) \left[|(D_{x_0}^{\gamma_1} f_1)(w)|^{\lambda_\alpha} |(D_{x_0}^{\gamma_2} f_2)(w)|^{\lambda_\alpha} + \right.$$

$$\left. |(D_{x_0}^{\gamma_2} f_1)(w)|^{\lambda_\alpha} |(D_{x_0}^{\gamma_1} f_2)(w)|^{\lambda_\alpha} \right] dw$$

$$\leq \rho^*(x) [\|D_{x_0}^{\nu} f_1\|_\infty^{2\lambda_\alpha} + \|D_{x_0}^{\nu} f_2\|_\infty^{2\lambda_\alpha}], \text{ all } x_0 \leq x \leq b. \qquad (6.91)$$

Here

$$\rho^*(x) := \left(\frac{(x - x_0)^{(2\nu\lambda_\alpha - \gamma_1\lambda_\alpha - \gamma_2\lambda_\alpha + 1)} \cdot \|p(x)\|_\infty}{(2\nu\lambda_\alpha - \gamma_1\lambda_\alpha - \gamma_2\lambda_\alpha + 1)(\Gamma(\nu - \gamma_1 + 1))^{\lambda_\alpha}(\Gamma(\nu - \gamma_2 + 1))^{\lambda_\alpha}} \right). \qquad (6.92)$$

Proof. As in Theorem 6.18, when $\lambda_\nu = 0$, we follow the proof and we obtain

$$\int_{x_0}^{x} p(w) \left[|(D_{x_0}^{\gamma_1} f_1)(w)|^{\lambda_\alpha} |(D_{x_0}^{\gamma_2} f_2)(w)|^{\lambda_\beta} + \right.$$

$$\left. |(D_{x_0}^{\gamma_2} f_1)(w)|^{\lambda_\beta} |(D_{x_0}^{\gamma_1} f_2)(w)|^{\lambda_\alpha} \right] dw$$

$$\leq \rho(x) [\|D_{x_0}^{\nu} f_1\|_\infty^{\lambda_\alpha} \|D_{x_0}^{\nu} f_2\|_\infty^{\lambda_\beta} + \|D_{x_0}^{\nu} f_1\|_\infty^{\lambda_\beta} \|D_{x_0}^{\nu} f_2\|_\infty^{\lambda_\alpha}], \qquad (6.93)$$

all $x_0 \leq x \leq b$, which inequality is by itself of interest. Then setting $\lambda_\alpha = \lambda_\beta$ into (6.93) and (6.77) we derive (6.91). $\square$

Theorem 6.22. (as in Theorem 6.18; $\lambda_\alpha = 0$, $\lambda_\beta = \lambda_\nu$).
It holds

$$\int_{x_0}^{x} p(w) \left[|(D_{x_0}^{\gamma_2} f_2)(w)|^{\lambda_\beta} |(D_{x_0}^{\nu} f_1)(w)|^{\lambda_\beta} + \right.$$

$$\left. (D_{x_0}^{\gamma_2} f_1)(w)|^{\lambda_\beta} |(D_{x_0}^{\nu} f_2)(w)|^{\lambda_\beta} \right] dw$$

$$\leq \left(\frac{(x - x_0)^{(\nu\lambda_\beta - \gamma_2\lambda_\beta + 1)} \cdot \|p(x)\|_\infty}{(\nu\lambda_\beta - \gamma_2\lambda_\beta + 1)(\Gamma(\nu - \gamma_2 + 1))^{\lambda_\beta}} \right) \cdot$$

$$[\|(D_{x_0}^{\nu} f_1)\|_\infty^{2\lambda_\beta} + \|(D_{x_0}^{\nu} f_2)\|_\infty^{2\lambda_\beta}], \qquad (6.94)$$

all $x_0 \leq x \leq b$.

Proof. Again we follow the proof of Theorem 6.18. $\square$

Corollary 6.23. (to Theorem 6.21; all are as in Theorem 6.18; $\lambda_\nu = 0$, $\lambda_\alpha = \lambda_\beta$, $\gamma_2 = \gamma_1 + 1$).
It holds

$$\int_{x_0}^{x} p(w) \left[|(D_{x_0}^{\gamma_1} f_1)(w)|^{\lambda_\alpha} |(D_{x_0}^{\gamma_1 + 1} f_2)(w)|^{\lambda_\alpha} + \right.$$

$$|(D_{x_0}^{\gamma_1+1}f_1)(w)|^{\lambda_\alpha}|(D_{x_0}^{\gamma_1}f_2)(w)|^{\lambda_\alpha}\big]\,dw$$

$$\leq \left(\frac{(x-x_0)^{(2\nu\lambda_\alpha-2\gamma_1\lambda_\alpha-\lambda_\alpha+1)}\|p(x)\|_\infty}{(2\nu\lambda_\alpha-2\gamma_1\lambda_\alpha-\lambda_\alpha+1)(\nu-\gamma_1)^{\lambda_\alpha}(\Gamma(\nu-\gamma_1))^{2\lambda_\alpha}}\right)$$

$$\cdot [\|D_{x_0}^\nu f_1\|_\infty^{2\lambda_\alpha}+\|D_{x_0}^\nu f_2\|_\infty^{2\lambda_\alpha}],\ \text{ all } x_0\leq x\leq b. \tag{6.95}$$

Proof. Obvious. $\square$

Corollary 6.24. (to Corollary 6.23). *In detail:*
Let $\nu\geq 2, \gamma_1\geq 0$ *such that* $\nu-\gamma_1\geq 2$ *and* $f_1, f_2\in C_{x_0}^\nu([a,b])$ *with* $f_1^{(i)}(x_0)=f_2^{(i)}(x_0)=0,\ i=0,1,\ldots,n-1,\ n:=[\nu].$ *Here* $x, x_0\in[a,b]: x\geq x_0$. *Then*

$$\int_{x_0}^x \big[|(D_{x_0}^{\gamma_1}f_1)(w)||(D_{x_0}^{\gamma_1+1}f_2)(w)|+$$

$$|(D_{x_0}^{\gamma_1+1}f_1)(w)|\,|(D_{x_0}^{\gamma_1}f_2)(w)|\big]\,dw$$

$$\leq \left(\frac{(x-x_0)^{2(\nu-\gamma_1)}}{2(\nu-\gamma_1)^2(\Gamma(\nu-\gamma_1))^2}\right)$$

$$[\|D_{x_0}^\nu f_1\|_\infty^2+\|D_{x_0}^\nu f_2\|_\infty^2],\ \text{ all } x_0\leq x\leq b. \tag{6.96}$$

Proof. Obvious. $\square$

Proposition 6.25. *Inequality (6.96) is sharp, in fact it is attained when* $f_1=f_2$.

Proof. Clearly (6.96) collapses to

$$\int_{x_0}^x |(D_{x_0}^{\gamma_1}f_1)(w)|\,|(D_{x_0}^{\gamma_1+1}f_1)(w)|dw \leq \left(\frac{(x-x_0)^{2(\nu-\gamma_1)}}{2(\nu-\gamma_1)^2(\Gamma(\nu-\gamma_1))^2}\right)\|D_{x_0}^\nu f_1\|_\infty^2, \tag{6.97}$$

all $x_0\leq x\leq b$.
In [17], see Propositions 9 and 10; there we proved that (6.97) is sharp. In fact it is attained. $\square$

6.4   Applications

We present a uniqueness of solution result for a system of fractional differential equations.

Theorem 6.26. *Let $\nu \geq 1, \gamma_i \geq 0$, $\nu - \gamma_i \geq 1$, $i = 1, \ldots, r \in \mathbb{N}$, $n := [\nu]$, $f_j \in C_a^\nu([a,b])$, $j = 1, 2$; $f_j^{(i)}(a) = a_{ij} \in \mathbb{R}$, $i = 0, 1, \ldots, n-1$. Furthermore we have for $j = 1, 2$ that*

$$(D_a^\nu f_j)(t) = F_j\left(t, \{(D_a^{\gamma_i} f_1)(t)\}_{i=1}^r, \{(D_a^{\gamma_i} f_2)(t)\}_{i=1}^r\right), \quad \text{all } t \in [a,b]. \quad (6.98)$$

Here F_j are continuous functions on $[a,b] \times \mathbb{R}^r \times \mathbb{R}^r$, and satisfy the Lipschitz condition

$$|F_j(t, z_1, \ldots, z_r, y_1, \ldots, y_r) - F_j(t, z_1', \ldots, z_r', y_1', \ldots, y_r')|$$

$$\leq \sum_{i=1}^r [q_{1,i,j}(t)|z_i - z_i'| + q_{2,i,j}(t)|y_i - y_i'|], \quad j = 1, 2, \quad (6.99)$$

where $q_{1,i,j}(t)$, $q_{2,i,j}(t) \geq 0$, $1 \leq i \leq r$ are continuous functions over $[a,b]$. Call

$$M_{1,i} := \max(\|q_{1,i,1}\|_\infty, \|q_{2,i,2}\|_\infty) \quad \text{and} \quad M_{2,i} := \max(\|q_{2,i,1}\|_\infty, \|q_{1,i,2}\|_\infty). \quad (6.100)$$

Assume here that

$$\phi^*(b) := \sum_{i=1}^r \left(\frac{M_{1,i}}{2} + \frac{M_{2,i}}{\sqrt{2}}\right) \cdot \left(\frac{(b-a)^{\nu - \gamma_i}}{\Gamma(\nu - \gamma_i)\sqrt{\nu - \gamma_i}\sqrt{2\nu - 2\gamma_i - 1}}\right) < 1. \quad (6.101)$$

Then if the system (6.98) has two pairs of solutions (f_1, f_2) and (f_1^*, f_2^*) we prove that $f_j = f_j^*$, $j = 1, 2$; that is, we have uniqueness of solution.

Proof. Assume there are two pairs of solutions (f_1, f_2), (f_1^*, f_2^*) to system (6.98). Set $g_j := f_j - f_j^*$, $j = 1, 2$. Then $g_j^{(i)} = f_j^{(i)} - f_j^{*(i)}$ and $g_j^{(i)}(a) = 0$, $i = 0, 1, \ldots, n-1$; $j = 1, 2$. It holds

$$
\begin{aligned}
(D_a^\nu g_j)(t) = {} & F_j(t, \{(D_a^{\gamma_i} f_1)(t)\}_{i=1}^r, \{(D_a^{\gamma_i} f_2)(t)\}_{i=1}^r) \\
& - F_j(t, \{(D_a^{\gamma_i} f_1^*)(t)\}_{i=1}^r, \{(D_a^{\gamma_i} f_2^*)(t)\}_{i=1}^r).
\end{aligned}
$$

Hence by (6.99) we have

$$|(D_a^\nu g_j)(t)| \leq \sum_{i=1}^r [q_{1,i,j}(t)|(D_a^{\gamma_i} g_1)(t)| + q_{2,i,j}(t)|(D_a^{\gamma_i} g_2)(t)|].$$

Thus

$$|(D_a^\nu g_j)(t)| \leq \sum_{i=1}^r [\|q_{1,i,j}\|_\infty |(D_a^{\gamma_i} g_1)(t)| + \|q_{2,i,j}\|_\infty |(D_a^{\gamma_i} g_2)(t)|].$$

The last implies that

$$
\begin{aligned}
|(D_a^\nu g_j)(t)|^2 \leq {} & \sum_{i=1}^r [\|q_{1,i,j}\|_\infty |(D_a^{\gamma_i} g_1)(t)||(D_a^\nu g_j)(t)| \\
& + \|q_{2,i,j}\|_\infty |(D_a^{\gamma_i} g_2)(t)| |(D_a^\nu g_j)(t)|.
\end{aligned}
$$

Consequently we observe

$$I := \int_a^b (((D_a^\nu g_1)(t))^2 + ((D_a^\nu g_2)(t))^2)dt$$

$$\leq \sum_{i=1}^r \left[\|q_{1,i,1}\|_\infty \int_a^b |(D_a^{\gamma_i} g_1)(t)|\, |(D_a^\nu g_1)(t)|dt + \right.$$

$$\left. \|q_{2,i,1}\|_\infty \int_a^b |(D_a^{\gamma_i} g_2)(t)|\, |(D_a^\nu g_1)(t)|dt \right]$$

$$+ \sum_{i=1}^r \left[\|q_{1,i,2}\|_\infty \int_a^b |(D_a^{\gamma_i} g_1)(t)|\, |(D_a^\nu g_2)(t)|dt + \right.$$

$$\left. \|q_{2,i,2}\|_\infty \int_a^b |(D_a^{\gamma_i} g_2)(t)|\, |(D_a^\nu g_2)(t)|dt \right]$$

$$\leq \sum_{i=1}^r M_{1,i} \int_a^b [|(D_a^{\gamma_i} g_1)(t)|\, |(D_a^\nu g_1)(t)| + |(D_a^{\gamma_i} g_2)(t)|\, |(D_a^\nu g_2)(t)|]dt$$

$$+ \sum_{i=1}^r M_{2,i} \int_a^b [|(D_a^{\gamma_i} g_2)(t)|\, |(D_a^\nu g_1)(t)| + |(D_a^{\gamma_i} g_1)(t)|\, |(D_a^\nu g_2)(t)|]dt =: (*).$$

However by Corollary 6.11 we obtain

$$\int_a^b [|(D_a^{\gamma_i} g_1)(t)|\, |(D_a^\nu g_1)(t)| + |(D_a^{\gamma_i} g_2)(t)|\, |(D_a^\nu g_2)(t)|]dt$$

$$\leq \left(\frac{(b-a)^{\nu-\gamma_i}}{2\Gamma(\nu-\gamma_i)\sqrt{\nu-\gamma_i}\,\sqrt{2\nu-2\gamma_i-1}} \right) \cdot I. \tag{6.102}$$

Also by Corollary 6.13 we find

$$\int_a^b [|(D_a^{\gamma_i} g_2)(t)|\, |(D_a^\nu g_1)(t)| + |(D_a^{\gamma_i} g_1)(t)|\, |(D_a^\nu g_2)(t)|]dt$$

$$\leq \left(\frac{(b-a)^{\nu-\gamma_i}}{\sqrt{2}\Gamma(\nu-\gamma_i)\sqrt{\nu-\gamma_i}\,\sqrt{2\nu-2\gamma_i-1}} \right) \cdot I. \tag{6.103}$$

Therefore by (6.102) and (6.103) we find

$$(*) \leq \sum_{i=1}^r \left[M_{1,i} \left(\frac{(b-a)^{\nu-\gamma_i}}{2\Gamma(\nu-\gamma_i)\sqrt{\nu-\gamma_i}\,\sqrt{2\nu-2\gamma_i-1}} \right) \cdot I \right.$$

$$\left. + M_{2,i} \left(\frac{(b-a)^{\nu-\gamma_i}}{\sqrt{2}\Gamma(\nu-\gamma_i)\sqrt{\nu-\gamma_i}\,\sqrt{2\nu-2\gamma_i-1}} \right) \cdot I \right] = \phi^*(b)I.$$

We have established that

$$I \le \phi^*(b)I. \tag{6.104}$$

If $I \ne 0$ then $\phi^*(b) \ge 1$, a contradiction by the assumption (6.101) that $\phi^*(b) < 1$. Therefore $I = 0$, implying that

$$((D_a^\nu g_1)(t))^2 + ((D_a^\nu g_2)(t))^2 = 0, \quad \text{a.e. } [a,b].$$

That is,

$$((D_a^\nu g_1)(t))^2 = 0, \quad ((D_a^\nu g_2)(t))^2 = 0, \quad \text{a.e. } [a,b].$$

That is,

$$(D_a^\nu g_1)(t) = 0, \quad (D_a^\nu g_2)(t) = 0, \quad \text{a.e. } [a,b].$$

But for $j = 1,2$ we get that

$$g_j^{(i)}(a) = 0, \quad 0 \le i \le n-1.$$

Hence from fractional Taylor's Theorem 6.1 (see (6.7)) we find that $g_j(t) = 0$ on $[a,b]$. That is $f_j \equiv f_j^*$, $j = 1,2$, proving the uniqueness claim of the theorem. $\square$

Next we give upper bounds on $D_a^\nu f_j$, solutions f_j, and so on, all involved in a system of fractional differential equations.

Theorem 6.27. *Let $\nu \ge 1, \gamma_i \ge 0$, $\nu - \gamma_i \ge 1$, $i = 1,\ldots,r \in \mathbb{N}$, $n := [\nu]$, $f_j \in C_a^\nu([a,b])$, $j = 1,2$; $f_j^{(i)}(a) = 0$, $i = 0,1,\ldots,n-1$, and*

$$(D_a^\nu f_j)(a) = A_j \in \mathbb{R}.$$

Furthermore for $a \le t \le b$ we have holding the system of fractional differential equations

$$(D_a^\nu f_j)'(t) = F_j(t, \{(D_a^{\gamma_i} f_1)(t)\}_{i=1}^r, (D_a^\nu f_1)(t), \{(D_a^{\gamma_i} f_2)(t)\}_{i=1}^r, (D_a^\nu f_2)(t)), \tag{6.105}$$

for $j = 1,2$. Here F_j are continuous functions on $[a,b] \times \mathbb{R}^{r+1} \times \mathbb{R}^{r+1}$ such that

$$|F_j(t, x_1, x_2, \ldots, x_r, x_{r+1}, y_1, \ldots, y_r, y_{r+1})|$$
$$\le \sum_{i=1}^r [q_{1,i,j}(t)|x_i| + q_{2,i,j}(t)|y_i|], \tag{6.106}$$

where $q_{1,i,j}(t)$, $q_{2,i,j}(t) \ge 0$, $1 \le i \le r$, $j = 1,2$, are continuous functions on $[a,b]$. Call

$$M_{1,i} := \max(\|q_{1,i,1}\|_\infty, \|q_{2,i,2}\|_\infty), \quad \text{and} \quad M_{2,i} := \max(\|q_{2,i,1}\|_\infty, \|q_{1,i,2}\|_\infty). \tag{6.107}$$

Also we set $(a \leq x \leq b)$

$$\theta(x) := \quad ((D_a^\nu f_1)(x))^2 + ((D_a^\nu f_2)(x))^2, \tag{6.108}$$
$$\rho := \quad A_1^2 + A_2^2, \tag{6.109}$$

$$Q(x) := \quad \sum_{i=1}^{r} (M_{1,i} + \sqrt{2}M_{2,i}) \cdot$$
$$\left(\frac{(x-a)^{\nu-\gamma_i}}{\Gamma(\nu-\gamma_i)\sqrt{\nu-\gamma_i}\sqrt{2\nu-2\gamma_i-1}} \right), \tag{6.110}$$

and

$$\chi(x) := \sqrt{\rho} \cdot \left\{ 1 + Q(x) \cdot e^{\left(\int_a^x Q(s)ds \right)} \left[\int_a^x \left(e^{-\left(\int_a^t Q(s)ds \right)} \right) dt \right] \right\}^{1/2}. \tag{6.111}$$

Then

$$\sqrt{\theta(x)} \leq \chi(x), \quad a \leq x \leq b. \tag{6.112}$$

Consequently it holds

$$|(D_a^\nu f_j)(x)| \quad \leq \quad \chi(x), \tag{6.113}$$
$$|f_j(x)| \quad \leq \quad \frac{1}{\Gamma(\nu)} \int_a^x (x-t)^{\nu-1}\chi(t)dt, \tag{6.114}$$

for $j = 1, 2$, $a \leq x \leq b$. *Also it holds that*

$$|(D_a^{\gamma_i} f_j)(x)| \leq \frac{1}{\Gamma(\nu-\gamma_i)} \int_a^x (x-t)^{\nu-\gamma_i-1}\chi(t)dt, \tag{6.115}$$

for $j = 1, 2$; $i = 1, \ldots, r$, $a \leq x \leq b$.

Proof. We see that

$$(D_a^\nu f_j)(t) \cdot (D_a^\nu f_j)'(t) \quad = \quad (D_a^\nu f_j)(t) \cdot F_j(t, \{(D_a^{\gamma_i} f_1)(t)\}_{i=1}^r, (D_a^\nu f_1)(t),$$
$$\{(D_a^{\gamma_i} f_2)(t)\}_{i=1}^r, (D_a^\nu f_2)(t)), \quad \text{all } a \leq x \leq b.$$

We then integrate the last equality

$$\int_a^x (D_a^\nu f_j)(t) \cdot ((D_a^\nu f_j)'(t))dt$$
$$= \int_a^x (D_a^\nu f_j)(t) \cdot F_j(t, \{(D_a^{\gamma_i} f_1)(t)\}_{i=1}^r,$$
$$(D_a^\nu f_1)(t), \{(D_a^{\gamma_i} f_2)(t)\}_{i=1}^r, (D_a^\nu f_2)(t))dt.$$

Hence we find

$$\frac{((D_a^\nu f_j)(t))^2}{2}\Big|_a^x \leq \int_a^x |(D_a^\nu f_j)(t)|\,|F_j \cdots|dt$$

$$\leq \int_a^x |(D_a^\nu f_j)(t)|$$

$$\left[\sum_{i=1}^r \{q_{1,i,j}(t)|(D_a^{\gamma_i} f_1)(t)| + q_{2,i,j}(t)|(D_a^{\gamma_i} f_2)(t)|\}\right]dt$$

$$\leq \sum_{i=1}^r \Bigg\{ \|q_{1,i,j}\|_\infty \int_a^x |(D_a^\nu f_j)(t)|\,|(D_a^{\gamma_i} f_1)(t)|dt$$

$$+ \|q_{2,i,j}\|_\infty \int_a^x |(D_a^\nu f_j)(t)|\,|(D_a^{\gamma_i} f_2)(t)|dt\Bigg\}.$$

Thus we obtain

$$((D_\alpha^\nu f_j)(x))^2 \leq A_j^2 + 2\sum_{i=1}^r \Bigg\{ \|q_{1,i,j}\|_\infty \cdot \int_a^x |(D_a^\nu f_j)(t)|\,|(D_a^{\gamma_i} f_1)(t)dt$$

$$+ \|q_{2,i,j}\|_\infty \cdot \int_a^x |(D_a^\nu f_j)(t)|\,|(D_a^{\gamma_i} f_2)(t)|dt\Bigg\}.$$

Consequently we write

$$\theta(x) \leq \rho + 2\sum_{i=1}^r$$

$$\Bigg\{ M_{1,i} \int_a^x [|(D_a^\nu f_1)(t)|\,|(D_a^{\gamma_i} f_1)(t)| + |(D_a^\nu f_2)(t)|\,|(D_a^{\gamma_i} f_2)(t)|]dt$$

$$+ M_{2,i} \int_a^x [|(D_a^\nu f_2)(t)|\,|(D_a^{\gamma_i} f_1)(t)| + |(D_a^\nu f_1)(t)|\,|(D_a^{\gamma_i} f_2)(t)|]dt\Bigg\}$$

$$\overset{(by\ (6.56),\ (6.60))}{\leq} \rho + 2\sum_{i=1}^r$$

$$\Bigg\{ M_{1,i}\left(\frac{(x-a)^{(\nu-\gamma_i)}}{2\Gamma(\nu-\gamma_i)\sqrt{\nu-\gamma_i}\sqrt{2\nu-2\gamma_i-1}}\right)\cdot\left(\int_a^x \theta(t)dt\right)$$

$$+ M_{2,i}\left(\frac{(x-a)^{(\nu-\gamma_i)}}{\sqrt{2}\Gamma(\nu-\gamma_i)\sqrt{\nu-\gamma_i}\sqrt{2\nu-2\gamma_i-1}}\right)\cdot\left(\int_a^x \theta(t)dt\right)\Bigg\}$$

$$= \rho + Q(x)\int_a^x \theta(t)dt.$$

That is, we get

$$\theta(x) \leq \rho + Q(x)\int_a^x \theta(t)dt, \qquad (6.116)$$

all $a \leq x \leq b$. Here $\rho \geq 0$, $Q(x) \geq 0$, $Q(a) = 0$, $\theta(x) \geq 0$, all $a \leq x \leq b$. Solving the integral inequality (6.116), exactly as in [17] (see pp. 224–225, Application 3.2 and inequalities (44), (47) there) we find that

$$\sqrt{\theta(x)} \leq \chi(x), \quad a \leq x \leq b.$$

Then (6.113) is obvious.

Next from (6.9) we have

$$|f_j(x)| \leq \frac{1}{\Gamma(\nu)} \int_a^x (x-t)^{\nu-1} |D_a^\nu f_j(t)| dt \leq \frac{1}{\Gamma(\nu)} \int_a^x (x-t)^{\nu-1} \chi(t) dt,$$

all $a \leq x \leq b$, $j = 1, 2$, proving (6.114).

Finally from (6.13) we find

$$\begin{aligned}
|(D_a^{\gamma_i} f_j)(x)| &\leq \frac{1}{\Gamma(\nu-\gamma_i)} \int_a^x (x-t)^{\nu-\gamma_i-1} |(D_a^\nu f_j)(t)| dt \\
&\leq \frac{1}{\Gamma(\nu-\gamma_i)} \int_a^x (x-t)^{\nu-\gamma_i-1} \chi(t) dt,
\end{aligned}$$

all $a \leq x \leq b$, $j = 1, 2$, $i = 1, \ldots, r$, proving (6.115). $\square$

In the final application we give similar upper bounds to Theorem 6.27, but under different conditions.

Theorem 6.28. *Let $a \neq b$, $\nu \geq 2$, $\gamma_i \geq 0$, $\nu - \gamma_i \geq 1$, $i = 1, \ldots, r \in \mathbb{N}$, $n := [\nu]$, $f_j \in C_a^\nu([a,b])$, $j = 1, 2$; $f_j^{(i)}(a) = 0$, $i = 0, 1, \ldots, n-1$, and*

$$(D_a^\nu f_j)(a) = A_j \in \mathbb{R}.$$

Furthermore for $a \leq t \leq b$ we have holding the system of fractional differential equations

$$\begin{aligned}
(D_a^\nu f_j)'(t) = \ & F_j(t, \{(D_a^{\gamma_i} f_1)(t)\}_{i=1}^r, (D_a^\nu f_1)(t), \\
& \{(D_a^{\gamma_i} f_2)(t)\}_{i=1}^r, (D_a^\nu f_2)(t)), \quad \text{for } j = 1, 2. \ (6.117)
\end{aligned}$$

For fixed $i_* \in \{1, \ldots, r\}$ we assume that $\gamma_{i_*+1} = \gamma_{i_*} + 1$, and $\nu - \gamma_{i_*} \geq 2$, where $\gamma_{i_*}, \gamma_{i_*+1} \in \{\gamma_1, \ldots, \gamma_r\}$. Call $k := \gamma_{i_*}$, $\gamma := \gamma_{i_*} + 1$ (i.e., $\gamma = k+1$). Here F_j are continuous functions on $[a,b] \times \mathbb{R}^{r+1} \times \mathbb{R}^{r+1}$ such that

$$\begin{aligned}
|F_j(t, x_1, \ldots, x_r, x_{r+1}, y_1, \ldots, y_r, y_{r+1})| \\
\leq q_{1,j}(t)|x_{i_*}|\sqrt{|y_{i_*+1}|} + q_{2,j}(t)|y_{i_*}|\sqrt{|x_{i_*+1}|}, \quad (6.118)
\end{aligned}$$

where

$$q_{1,j}, q_{2,j} \neq 0, q_{1,j}(t), q_{2,j}(t) \geq 0$$

are continuous functions over $[a,b]$. *Put*

$$M := \max(\|q_{1,1}\|_\infty, \|q_{2,1}\|_\infty, \|q_{1,2}\|_\infty, \|q_{2,2}\|_\infty), \tag{6.119}$$

$$\theta(x) := |(D_a^\nu f_1)(x)| + |(D_a^\nu f_2)(x)|, \quad a \le x \le b, \tag{6.120}$$

$$\rho := |A_1| + |A_2|. \tag{6.121}$$

Also we define

$$Q(x) := \left(\frac{4M}{\sqrt{3\nu - 3k - 2}}\right) \cdot \frac{(x-a)^{\left(\frac{3\nu - 3k - 1}{2}\right)}}{(\Gamma(\nu - k))^{3/2}}, \tag{6.122}$$

$$\sigma := \|Q(x)\|_\infty, \quad a \le x \le b. \tag{6.123}$$

We assume that

$$(b-a)\sigma\sqrt{\rho} < 2. \tag{6.124}$$

Call

$$\tilde{\varphi}(x) := \frac{(x-a) \cdot (\sigma - Q(x)) \cdot [\sigma\rho^2(x-a) - 4\rho^{3/2}] + 4\rho}{(2 - \sigma\sqrt{\rho}(x-a))^2}, \tag{6.125}$$

all $a \le x \le b$. *Then*

$$\theta(x) \le \tilde{\varphi}(x); \tag{6.126}$$

in particular, it holds

$$|(D_a^\nu f_j)(x)| \le \tilde{\varphi}(x), \quad j = 1, 2, \tag{6.127}$$

all $a \le x \le b$. *Furthermore we find*

$$|f_j(x)| \le \frac{1}{\Gamma(\nu)} \int_a^x (x-t)^{\nu-1} \tilde{\varphi}(t)dt, \tag{6.128}$$

$$|(D_a^{\gamma_i} f_j)(x)| \le \frac{1}{\Gamma(\nu - \gamma_i)} \int_a^x (x-t)^{\nu-\gamma_i-1} \tilde{\varphi}(t)dt, \tag{6.129}$$

$j = 1, 2$, $i = 1, \ldots, r$, *all* $a \le x \le b$.

Proof. For $a \le x \le b$ we find

$$\int_a^x (D_a^\nu f_j)'(t)dt = \int_a^x F_j(t, \{(D_a^{\gamma_i} f_1)(t)\}_{i=1}^r,$$
$$(D_a^\nu f_1)(t), \{(D_a^{\gamma_i} f_2)(t)\}_{i=1}^r, (D_a^\nu f_2)(t))dt.$$

That is,

$$(D_a^\nu f_j)(x) = A_j + \int_a^x F_j(t, \ldots)dt.$$

Then we observe

$$|(D_a^\nu f_j)(x)| \leq |A_j| + \int_a^x |F_j(t, \dots)| dt$$

$$\leq |A_j| + \int_a^x [q_{1,j}(t)|(D_a^k f_1)(t)|\sqrt{|(D_a^{k+1} f_2)(t)|}$$

$$+ q_{2,j}(t)|(D_a^k f_2)(t)|\sqrt{|(D_a^{k+1} f_1)(t)|}] dt$$

$$\leq |A_j| + M \left(\int_a^x \left[|(D_a^k f_1)(t)|\sqrt{|(D_a^{k+1} f_2)(t)|} + \right. \right.$$

$$\left. \left. |(D_a^k f_2)(t)|\sqrt{|(D_a^{k+1} f_1)(t)|} \right] dt \right).$$

Hence

$$\theta(x) \leq \rho + 2M \left(\int_a^x \left[|(D_a^k f_1)(t)|\sqrt{|(D_a^{k+1} f_2)(t)|} + \right. \right.$$

$$\left. \left. |(D_a^k f_2)(t)|\sqrt{|(D_a^{k+1} f_1)(t)|} \right] dt \right)$$

$$\overset{\text{(by (6.63), (6.64))}}{\leq} \rho + \left(\frac{4M}{\sqrt{3\nu - 3k - 2}} \right) \cdot \left(\frac{(x-a)^{((3\nu-3k-1)/2)}}{(\Gamma(\nu-k))^{3/2}} \right)$$

$$\cdot \left(\int_a^x (|(D_a^\nu f_1)(t)|^{3/2} + |(D_a^\nu f_2)(t)|^{3/2}) dt \right)$$

$$\overset{(6.24)}{\leq} \rho + \left(\frac{4M}{\sqrt{3\nu - 3k - 2}} \right) \cdot \left(\frac{(x-a)^{((3\nu-3k-1)/2)}}{(\Gamma(\nu-k))^{3/2}} \right)$$

$$\cdot \left(\int_a^x (|(D_a^\nu f_1)(t)| + |(D_a^\nu f_2)(t)|)^{3/2} dt \right).$$

We have proved that

$$\theta(x) \leq \rho + Q(x) \cdot \int_a^x (\theta(t))^{3/2} dt, \tag{6.130}$$

all $a \leq x \leq b$. Notice that $\theta(x) \geq 0$, $\rho \geq 0$, $Q(x) \geq 0$, and $Q(a) = 0$; also it is $\sigma > 0$.

Call

$$0 \leq w(x) := \int_a^x (\theta(t))^{3/2} dt, \quad w(a) = 0, \quad \text{all } a \leq x \leq b. \tag{6.131}$$

That is,

$$w'(x) = (\theta(x))^{3/2} \geq 0,$$

and

$$\theta(x) = (w'(x))^{2/3}, \quad \text{all } a \leq x \leq b. \tag{6.132}$$

Hence we rewrite (6.130) as follows,

$$(w'(x))^{2/3} \leq \rho + Q(x)w(x), \quad a \leq x \leq b. \qquad (6.133)$$

So that

$$(w'(x))^{2/3} \leq \rho + \sigma w(x) < \rho + \varepsilon + \sigma w(x),$$

for $\varepsilon > 0$ arbitrarily small. Hence

$$w'(x) < (\rho + \varepsilon + \sigma w(x))^{3/2}, \quad a \leq x \leq b.$$

Here $(\rho + \varepsilon + \sigma w(x))^{3/2} > 0$. In particular it holds

$$w'(t) < (\rho + \varepsilon + \sigma w(t))^{3/2}, \quad a \leq t \leq x,$$

and

$$\frac{w'(t)}{(\rho + \varepsilon + \sigma w(t))^{3/2}} < 1. \qquad (6.134)$$

The last is the same as

$$\left(-\frac{2}{\sigma}(\rho + \varepsilon + \sigma w(t))^{-1/2}\right)' < 1.$$

Therefore after integration we have

$$-\int_a^x d((\rho + \varepsilon + \sigma w(t))^{-1/2}) \leq \frac{\sigma}{2}(x - a),$$

and

$$(\rho + \varepsilon + \sigma w(t))^{-1/2}|_x^a \leq \frac{\sigma}{2}(x - a).$$

That is,

$$(\rho + \varepsilon)^{-1/2} - (\rho + \varepsilon + \sigma w(x))^{-1/2} \leq \frac{\sigma}{2}(x - a),$$

and

$$(\rho + \varepsilon)^{-1/2} - \frac{\sigma}{2}(x - a) \leq (\rho + \varepsilon + \sigma w(x))^{-1/2}.$$

That is,

$$\frac{2 - \sigma(\rho + \varepsilon)^{1/2}(x - a)}{2(\rho + \varepsilon)^{1/2}} \leq \frac{1}{(\rho + \varepsilon + \sigma w(x))^{1/2}}. \qquad (6.135)$$

By assumption (6.124) we get

$$(x - a)\sigma\sqrt{\rho} < 2, \quad \text{all } a \leq x \leq b.$$

Then for sufficiently small $\varepsilon > 0$ we still have

$$(x - a)\sigma(\rho + \varepsilon)^{1/2} < 2.$$

That is,
$$2 - (x - a)\sigma(\rho + \varepsilon)^{1/2} > 0, \quad \text{all } a \le x \le b. \tag{6.136}$$

From (6.135) and (6.136) we get

$$(\rho + \varepsilon + \sigma w(x))^{1/2} \le \frac{2(\rho + \varepsilon)^{1/2}}{2 - \sigma(\rho + \varepsilon)^{1/2}(x - a)}, \quad \text{all } a \le x \le b. \tag{6.137}$$

Solving (6.137) for $w(x)$ we find

$$w(x) \le \left(\frac{\rho + \varepsilon}{\sigma}\right) \left[\frac{4}{(2 - \sigma(\sqrt{\rho + \varepsilon})(x - a))^2} - 1\right], \tag{6.138}$$

for $\varepsilon > 0$ sufficiently small and for all $a \le x \le b$.

Letting $\varepsilon \to 0$ we obtain

$$w(x) \le \left(\frac{\rho}{\sigma}\right) \left[\frac{4}{(2 - \sigma\sqrt{\rho}(x - a))^2} - 1\right], \tag{6.139}$$

for all $a \le x \le b$.

That is,

$$w(x) \le \frac{4\rho^{3/2}(x - a) - \sigma\rho^2(x - a)^2}{(2 - \sigma\sqrt{\rho}(x - a))^2}, \tag{6.140}$$

for all $a \le x \le b$. Then by (6.130) and (6.140) we find

$$\theta(x) \le \rho + Q(x) \left[\frac{4\rho^{3/2}(x - a) - \sigma\rho^2(x - a)^2}{(2 - \sigma\sqrt{\rho}(x - a))^2}\right] = \tilde{\varphi}(x), \tag{6.141}$$

for all $a \le x \le b$. That is, (6.141) implies (6.126). Then by (6.120) and (6.126), inequality (6.127) is obvious.

Finally by (6.9), (6.13), and (6.127), the inequalities (6.128) and (6.129) are clear, respectively. □

7

Opial-Type Inequalities for Riemann–Liouville Fractional Derivatives of Two Functions with Applications

A wide variety of very general but basic $L_p(1 \leq p \leq \infty)$-form Opial-type inequalities [315] is established involving Riemann–Liouville fractional derivatives [17, 230, 295, 314] of two functions in different orders and powers.

From the developed results are derived several other concrete results of special interest. The sharpness of inequalities is established there. Finally applications of some of these special inequalities are given in establishing the uniqueness of solution and in giving upper bounds to solutions of initial value fractional problems involving a very general system of two fractional differential equations. Also upper bounds to various Riemann–Liouville fractional derivatives of the solutions that are involved in the above systems are presented. This treatment is based on [48].

7.1   Introduction

We start in Section 7.2 with background, we continue in Section 7.3 with the main results, and we finish in Section 7.4 with the applications.

To give the reader a flavor of the kind of inequalities we are dealing with, we briefly mention

$$\int_0^x q(w)[|(D^{\gamma_1} f_1)(w)|^{\lambda_\alpha}|(D^\nu f_1)(w)|^{\lambda_\nu} + |(D^{\gamma_1} f_2)(w)|^{\lambda_\alpha}|(D^\nu f_2)(w)|^{\lambda_\nu}]dw$$

G.A. Anastassiou, *Fractional Differentiation Inequalities*,   107
DOI 10.1007/978-0-387-98128-4_7, © Springer Science+Business Media, LLC 2009

$$\leq C(x, q(w), \gamma_1, \nu, \lambda_\alpha, \lambda_\nu, p(w), p)$$

$$\cdot \left[\int_0^x p(w)[|(D^\nu f_1)(w)^p + |(D^\nu f_2)(w)|^p] dw \right]^{\lambda_\alpha + \lambda_\nu / p}, x \geq 0,$$

for functions $f_1, f_2 \in L_1(0, x); p(w), q(w), 1/p(w) \in L_\infty(0, x)$, all exponents and orders are fractional, and so on. Here $D^\beta f$ stands for the Riemann–Liouville fractional derivative of order $\beta \geq 0$. Furthermore one system of fractional differential equations we are working on looks briefly like

$$(D^\nu f_j)(t) = F_j\left(t, \{(D^{\gamma_i} f_1)(t)\}_{i=1}^r, \{(D^{\gamma_i} f_2)(t)\}_{i=1}^r\right) \text{ all } t \in [0, x],$$

for $j = 1, 2$, and with $D^{\nu-i} f_j(0) = d_{ij} \in \mathbb{R}, i = 1, ..., n+1$, where $n := [\nu]$ is the integral part of $\nu > 0$, and so on.

7.2   Background

We need

Definition 7.1. (see [187, 295, 314]). Let $\alpha \in \mathbb{R}_+ - \{0\}$. For any $f \in L_1(0, x); x \in \mathbb{R}_+ - \{0\}$, the *Riemann–Liouville fractional integral* of f of order α is defined by

$$(J_\alpha f)(s) := \frac{1}{\Gamma(\alpha)} \int_0^s (s-t)^{\alpha-1} f(t) dt, \ \forall s \in [0, x], \qquad (7.1)$$

and the *Riemann–Liouville fractional derivative of f* of order α by

$$D^\alpha f(s) := \frac{1}{\Gamma(m-a)} \left(\frac{d}{ds}\right)^m \int_0^s (s-t)^{m-\alpha-1} f(t) dt, \qquad (7.2)$$

where $m := [\alpha] + 1, [\cdot]$ is the integral part. In addition, we set $D^0 f := f := J_0 f$, $J_{-\alpha} f = D^\alpha f$ if $\alpha > 0$, $D^{-\alpha} f := J_\alpha f$, if $0 < \alpha \leq 1$. If $\alpha \in \mathbb{N}$, then $D^\alpha f = f^{(\alpha)}$, the ordinary derivative.

Definition 7.2. [87]. We say that $f \in L_1(0, x)$ has an L_∞ *fractional derivative* $D^\alpha f$ in $[0, x], x \in \mathbb{R}_+ - \{0\}$, iff $D^{\alpha-k} f \in C([0, x]), k = 1, ..., m := [\alpha] + 1; \alpha \in \mathbb{R}_+ - \{0\}$, and $D^{\alpha-1} f \in AC([0, x])$ (absolutely continuous functions) and $D^\alpha f \in L_\infty(0, x)$.

We need

Lemma 7.3. [187]. *Let $\alpha \in \mathbb{R}_+$, $\beta > \alpha$, let $f \in L_1(0, x)$, $x \in \mathbb{R}_+ - \{0\}$, have an L_∞ fractional derivative $D^\beta f$ in $[0, x]$, and let $D^{\beta-k} f(0) = 0$ for $k = 1, \ldots, [\beta] + 1$. Then*

$$D^\alpha f(s) = \frac{1}{\Gamma(\beta - \alpha)} \int_0^s (s - t)^{\beta - \alpha - 1} D^\beta f(t) dt, \forall \ s \in [0, x]. \qquad (7.3)$$

Here $D^\alpha f \in AC([0, x])$ for $\beta - \alpha \geq 1$ and $D^\alpha f \in C([0, x])$ for $\beta - \alpha \in (0, 1)$, hence $D^\alpha f \in L_\infty(0, x)$ and $D^\alpha f \in L_1(0, x)$.

7.3 Main Results

We present our first main result.

Theorem 7.4. *Let $\alpha_1, \alpha_2 \in \mathbb{R}_+, \beta > \alpha_1, \alpha_2, \ \beta - \alpha_i > (1/p), p > 1, i = 1, 2$ and let $f_1, f_2 \in L_1(0, x), x \in \mathbb{R}_+ - \{0\}$ have, respectively, L_∞ fractional derivatives $D^\beta f_1, D^\beta f_2$ in $[0,x]$, and let $D^{\beta-k} f_i(0) = 0$, for $k = 1, \ldots, [\beta] + 1; i = 1, 2$. Consider also $p(t) > 0$ and $q(t) \geq 0$, with all $p(t), 1/p(t), q(t) \in L_\infty(0, x)$. Let $\lambda_\beta > 0$ and $\lambda_{\alpha_1}, \lambda_{\alpha_2} \geq 0$, such that $\lambda_\beta < p$, where $p > 1$. Put*

$$P_i(s) := \int_0^s (s - t)^{p(\beta - \alpha_i - 1)/p - 1} (p(t))^{-1/(p-1)} \, dt, \ i = 1, 2; 0 \leq s \leq x,$$

$$\qquad (7.4)$$

$$A(s) := \frac{q(s)(P_1(s))^{\lambda_{\alpha_1}(p-1/p)} (P_2(s))^{\lambda_{\alpha_2}((p-1)/p)} (p(s))^{-\lambda_\beta/p}}{(\Gamma(\beta - \alpha_1))^{\lambda_{\alpha_1}} (\Gamma(\beta - \alpha_2))^{\lambda_{\alpha_2}}} \qquad (7.5)$$

$$A_0(x) := \left(\int_0^x (A(s))^{p/(p-\lambda_\beta)} ds \right)^{(p-\lambda_\beta)/p}, \qquad (7.6)$$

and

$$\delta_1 := \begin{cases} 2^{1-((\lambda_{\alpha_1}+\lambda_\beta)/p)}, & if \quad \lambda_{\alpha_1} + \lambda_\beta \leq p, \\ 1, & if \quad \lambda_{\alpha_1} + \lambda_\beta \geq p. \end{cases} \qquad (7.7)$$

If $\lambda_{\alpha_2} = 0$, we obtain that

$$\int_0^x q(s) \left[|D^{\alpha_1} f_1(s)|^{\lambda_{\alpha_1}} \left| D^\beta f_1(s) \right|^{\lambda_\beta} + |D^{\alpha_1} f_2(s)|^{\lambda_{\alpha_1}} \left| D^\beta f_2(s) \right|^{\lambda_\beta} \right] ds \leq$$

$$(A_0(x)|_{\lambda_{\alpha_2}=0}) \left(\frac{\lambda_\beta}{\lambda_{\alpha_1} + \lambda_\beta} \right)^{\lambda_\beta/p}$$

$$\delta_1 \left[\int_0^x p(s) \left[|D^\beta f_1(s)|^p + |D^\beta f_2(s)|^p \right] ds \right]^{((\lambda_{\alpha_1}+\lambda_\beta)/p)}. \qquad (7.8)$$

Proof. From (7.3) we have

$$D^{\alpha_i} f_j(s) = \frac{1}{\Gamma(\beta - \alpha_i)} \int_0^s (s-t)^{\beta - \alpha_i - 1} D^\beta f_j(t) dt, \forall \, s \in [0, x], \; i = 1, 2; \; j = 1, 2.$$
(7.9)

Next, applying Hölder's inequality with indices $p, p/(p-1)$, we find

$$|D^{\alpha_i} f_j(s)| \le \frac{1}{\Gamma(\beta - \alpha_i)} \int_0^s (s-t)^{\beta - \alpha_i - 1} (p(t))^{-1/p} (p(t))^{1/p} |D^\beta f_j(t)| dt$$
(7.10)

$$\le \frac{1}{\Gamma(\beta - \alpha_i)} \left(\int_0^s \left((s-t)^{\beta - \alpha_i - 1} (p(t))^{-1/p} \right)^{p/(p-1)} dt \right)^{(p-1)/p}$$
(7.11)

$$\times \left(\int_0^s p(t) |D^\beta f_j(t)|^p dt \right)^{1/p} = \frac{1}{\Gamma(\beta - \alpha_i)} (P_i(s))^{(p-1)/p} \left(\int_0^s p(t) |D^\beta f_j(t)|^p dt \right)^{1/p}.$$
(7.12)

That is, it holds

$$|D^{\alpha_i} f_j(s)| \le \frac{1}{\Gamma(\beta - \alpha_i)} (P_i(s))^{p - 1/p} \left(\int_0^s p(t) |D^\beta f_j(t)|^p dt \right)^{1/p}.$$
(7.13)

Set

$$z_j(s) := \int_0^s p(t) |D^\beta f_j(t)|^p dt,$$
(7.14)

thus,

$$z_j'(s) = p(s) |D^\beta f_j(s)|^p, \text{a.e. in} (0, x), z_j(0) = 0; j = 1, 2.$$
(7.15)

Hence, we have

$$|D^{\alpha_i} f_j(s)| \le \frac{1}{\Gamma(\beta - \alpha_i)} (P_i(s))^{p - 1/p} (z_j(s))^{1/p},$$
(7.16)

and

$$|D^\beta f_j(s)|^{\lambda_\beta} = (p(s))^{-\lambda_\beta / p} (z_j'(s))^{\lambda_\beta / p}, \text{a.e. in} (0, x), i = 1, 2; j = 1, 2.$$
(7.17)

Therefore we obtain

$$q(s) |(D^{\alpha_1} f_1)(s)|^{\lambda_{\alpha_1}} |(D^{\alpha_2} f_2)(s)|^{\lambda_{\alpha_2}} |(D^\beta f_1)(s)|^{\lambda_\beta}$$

$$\le q(s) \frac{1}{(\Gamma(\beta - \alpha_1))^{\lambda_{\alpha_1}}} (P_1(s))^{(\lambda_{\alpha_1}(p-1)/p)} (z_1(s))^{\lambda_{\alpha_1}/p}$$
(7.18)

$$\frac{1}{(\Gamma(\beta - \alpha_2))^{\lambda_{\alpha_2}}} (P_2(s))^{\lambda_{\alpha_2}(p-1/p)} (z_2(s))^{\lambda_{\alpha_2}/p} (p(s))^{-\lambda_\beta/p} (z_1'(s))^{\lambda_\beta/p}$$
(7.19)

$$= A(s) (z_1(s))^{\lambda_{\alpha_1}/p} (z_2(s))^{\lambda_{\alpha_2}/p} (z_1'(s))^{\lambda_\beta/p}, a.e. in (0, x).$$
(7.20)

Consequently, by another Hölder's inequality application, we find (by $p/\lambda_\beta > 1$)

$$\int_0^x q(s) \left|(D^{\alpha_1} f_1)(s)\right|^{\lambda_{\alpha_1}} \left|(D^{\alpha_2} f_2)(s)\right|^{\lambda_{\alpha_2}} \left|(D^\beta f_1)(s)\right|^{\lambda_\beta} ds$$

$$\leq A_0(x) \left(\int_0^x (z_1(s))^{\lambda_{\alpha_1}/\lambda_\beta} (z_2(s))^{\lambda_{\alpha_2}/\lambda_\beta} z_1'(s) ds \right)^{\lambda_\beta/p}. \tag{7.21}$$

Similarly, one finds

$$\int_0^x q(s) \left|(D^{\alpha_1} f_2)(s)\right|^{\lambda_{\alpha_1}} \left|(D^{\alpha_2} f_1)(s)\right|^{\lambda_{\alpha_2}} \left|(D^\beta f_2)(s)\right|^{\lambda_\beta} ds$$

$$\leq A_0(x) \left(\int_0^x (z_1(s))^{\lambda_{\alpha_2}/\lambda_\beta} (z_2(s))^{\lambda_{\alpha_1}/\lambda_\beta} z_2'(s) ds \right)^{\lambda_\beta/p}. \tag{7.22}$$

Taking $\lambda_{\alpha_2} = 0$ and adding (7.21) and (7.22), we get

$$\int_0^x q(s) \left[\left|(D^{\alpha_1} f_1)(s)\right|^{\lambda_{\alpha_1}} \left|(D^\beta f_1)(s)\right|^{\lambda_\beta} + \left|(D^{\alpha_1} f_2)(s)\right|^{\lambda_{\alpha_1}} \left|(D^\beta f_2)(s)\right|^{\lambda_\beta} \right] ds$$

$$\leq (A_0(x)|_{\lambda_{\alpha_2}=0}) \left[\left(\int_0^x (z_1(s))^{\lambda_{\alpha_1}/\lambda_\beta} (z_1'(s)) ds \right)^{\lambda_\beta/p} \right.$$

$$\left. + \left(\int_0^x (z_2(s))^{\lambda_{\alpha_1}/\lambda_\beta} z_2'(s) ds \right)^{\lambda_\beta/p} \right] = \tag{7.23}$$

$$\left(A_0(x)\,|_{\lambda_{\alpha_2}=0} \right) \left[(z_1(x))^{((\lambda_{\alpha_1}+\lambda_\beta)/p)} + (z_2(x))^{((\lambda_{\alpha_1}+\lambda_\beta)/p)} \right] \left(\frac{\lambda_\beta}{\lambda_{\alpha_1} + \lambda_\beta} \right)^{\lambda_\beta/p} = \tag{7.24}$$

$$(A_0(x)|_{\lambda_{\alpha_2}=0}) \left(\frac{\lambda_\beta}{\lambda_{\alpha_1} + \lambda_\beta} \right)^{\lambda_\beta/p} \left[\left(\int_0^x p(t)|D^\beta f_1(t)|^p dt \right)^{((\lambda_{\alpha_1}+\lambda_\beta)/p)} \right.$$

$$\left. + \left(\int_0^x p(t)|D^\beta f_2(t)|^p dt \right)^{(\lambda_{\alpha_1}+\lambda_\beta/p)} \right] =: (*). \tag{7.25}$$

In this study, we frequently use the basic inequalities,

$$2^{r-1}(a^r + b^r) \leq (a+b)^r \leq a^r + b^r, \ a, b \geq 0, \ 0 \leq r \leq 1, \tag{7.26}$$

and

$$a^r + b^r \leq (a+b)^r \leq 2^{r-1}(a^r + b^r), \ a, b \geq 0, \ r \geq 1. \tag{7.27}$$

At last using (7.7), (7.26), and (7.27), we find

$$(*) \leq (A_0(x)|_{\lambda_{\alpha_2}=0}) \left(\frac{\lambda_\beta}{\lambda_{\alpha_1} + \lambda_\beta} \right)^{\lambda_\beta/p} \delta_1$$

$$\left[\int_0^x p(t) \left[|D^\beta f_1(t)|^p + |D^\beta f_2(t)|^p \right] dt \right]^{(\lambda_{\alpha_1} + \lambda_\beta/p)} . \tag{7.28}$$

Inequality (7.8) has been proved. □

The counterpart of the last theorem follows.

Theorem 7.5. *All here are as in Theorem 7.4. Denote*

$$\delta_3 := \begin{cases} 2^{\lambda_{\alpha_2}/\lambda_\beta} - 1, & if \quad \lambda_{\alpha_2} \geq \lambda_\beta, \\ 1, & if \quad \lambda_{\alpha_2} \leq \lambda_\beta. \end{cases} \tag{7.29}$$

If $\lambda_{\alpha_1} = 0$, *then*

$$\int_0^x q(s) \left[|(D^{\alpha_2} f_2)(s)|^{\lambda_{\alpha_2}} |(D^\beta f_1)(s)|^{\lambda_\beta} + |(D^{\alpha_2} f_1)(s)|^{\lambda_{\alpha_2}} |(D^\beta f_2)(s)|^{\lambda_\beta} \right] ds$$

$$\leq \left(A_0(x)|_{\lambda_{\alpha_1}=0} \right) 2^{(p-\lambda_\beta)/p} \left(\frac{\lambda_\beta}{\lambda_{\alpha_2} + \lambda_\beta} \right)^{\lambda_\beta/p} \delta_3^{\lambda_\beta/p}$$

$$\cdot \left(\int_0^x p(s) \left[|D^\beta f_1(s)|^p + |D^\beta f_2(s)|^p \right] ds \right)^{(\lambda_\beta + \lambda_{\alpha_2})/p} . \tag{7.30}$$

Proof. When $\lambda_{\alpha_1} = 0$ from (7.21) and (7.22) we obtain

$$\int_0^x q(s) \left[|D^{\alpha_2} f_2(s)|^{\lambda_{\alpha_2}} |D^\beta f_1(s)|^{\lambda_\beta} + |D^{\alpha_2} f_1(s)|^{\lambda_{\alpha_2}} |D^\beta f_2(s)|^{\lambda_\beta} \right] ds \leq$$

$$\left(A_0(x)|_{\lambda_{\alpha_1}=0} \right) \left[\left(\int_0^x (z_2(s))^{\lambda_{\alpha_2}/\lambda_\beta} (z_1'(s)) ds \right)^{\lambda_\beta/p} \right.$$

$$+ \left. \left(\int_0^x (z_1(s))^{\lambda_{\alpha_2}/\lambda_\beta} z_2'(s) ds \right)^{\lambda_\beta/p} \right] \leq \tag{7.31}$$

$$\left(A_0(x)|_{\lambda_{\alpha_1}=0} \right) 2^{1-(\lambda_\beta/p)} (M(x))^{\lambda_\beta/p} =: (*), \tag{7.32}$$

by $0 < \lambda_\beta/p < 1$ and by (7.26), where

$$M(x) := \int_0^x \left[(z_2(s))^{\lambda_{\alpha_2}/\lambda_\beta} z_1'(s) + (z_1(s))^{\lambda_{\alpha_2}/\lambda_\beta} z_2'(s) \right] ds. \tag{7.33}$$

Call

$$\delta_2 := \begin{cases} 1, & if \quad \lambda_{\alpha_2} \geq \lambda_\beta, \\ 2^{1-(\lambda_{\alpha_2}/\lambda_\beta)}, & if \quad \lambda_{\alpha_2} \leq \lambda_\beta. \end{cases} \tag{7.34}$$

We observe that

$$M(x) = \int_0^x ((z_1(s))^{\lambda_{\alpha_2}/\lambda_\beta} + (z_2(s))^{\lambda_{\alpha_2}/\lambda_\beta})(z_1'(s) + z_2'(s))ds$$

$$- \int_0^x \left[(z_1(s))^{\lambda_{\alpha_2}/\lambda_\beta} z_1'(s) + (z_2(s))^{\lambda_{\alpha_2}/\lambda_\beta} z_2'(s) \right] ds \qquad (7.35)$$

$$\overset{(by\ (7.26),(7.27))}{\leq} \delta_2 \int_0^x (z_1(s) + z_2(s))^{\lambda_{\alpha_2}/\lambda_\beta} (z_1(s) + z_2(s))' ds$$

$$- \left(\frac{\lambda_\beta}{\lambda_{\alpha_2} + \lambda_\beta} \right) \left[(z_1(x))^{((\lambda_{\alpha_2} + \lambda_\beta)/\lambda_\beta)} + (z_2(x))^{((\lambda_{\alpha_2} + \lambda_\beta)/\lambda_\beta)} \right] \qquad (7.36)$$

$$= \delta_2 (z_1(x) + z_2(x))^{(\lambda_{\alpha_2} + \lambda_\beta/\lambda_\beta)} \left(\frac{\lambda_\beta}{\lambda_{\alpha_2} + \lambda_\beta} \right)$$

$$- \left(\frac{\lambda_\beta}{\lambda_{\alpha_2} + \lambda_\beta} \right) \left[(z_1(x))^{((\lambda_{\alpha_2} + \lambda_\beta)/\lambda_\beta)} + (z_2(x))^{((\lambda_{\alpha_2} + \lambda_\beta)/\lambda_\beta)} \right] \qquad (7.37)$$

$$= \left(\frac{\lambda_\beta}{\lambda_{\alpha_2} + \lambda_\beta} \right) \left[\delta_2 (z_1(x) + z_2(x))^{(\lambda_{\alpha_2} + \lambda_\beta/\lambda_\beta)} - \right.$$

$$\left. \left[(z_1(x))^{(\lambda_{\alpha_2} + \lambda_\beta/\lambda_\beta)} + (z_2(x))^{(\lambda_{\alpha_2} + \lambda_\beta/\lambda_\beta)} \right] \right] \qquad (7.38)$$

$$\overset{(7.27)}{\leq} \left(\frac{\lambda_\beta}{\lambda_{\alpha_2} + \lambda_\beta} \right) \left[\delta_2\, 2^{\lambda_{\alpha_2}/\lambda_\beta} \left((z_1(x))^{(\lambda_{\alpha_2} + \lambda_\beta/\lambda_\beta)} + (z_2(x))^{(\lambda_{\alpha_2} + \lambda_\beta/\lambda_\beta)} \right) \right.$$

$$\left. - \left[(z_1(x))^{(\lambda_{\alpha_2} + \lambda_\beta/\lambda_\beta)} + (z_2(x))^{(\lambda_{\alpha_2} + \lambda_\beta)/\lambda_\beta} \right] \right] \qquad (7.39)$$

$$= \left(\frac{\lambda_\beta}{\lambda_{\alpha_2} + \lambda_\beta} \right) (\delta_2 2^{\lambda_{\alpha_2}/\lambda_\beta} - 1) \left[(z_1(x))^{(\lambda_{\alpha_2} + \lambda_\beta/\lambda_\beta)} + (z_2(x))^{(\lambda_{\alpha_2} + \lambda_\beta/\lambda_\beta)} \right] \qquad (7.40)$$

$$\text{(notice that } \delta_3 = \delta_2\, 2^{\lambda_{\alpha_2}/\lambda_\beta} - 1)$$

$$\overset{(7.27)}{\leq} \left(\frac{\lambda_\beta}{\lambda_{\alpha_2} + \lambda_\beta} \right) \delta_3 (z_1(x) + z_2(x))^{((\lambda_{\alpha_2} + \lambda_\beta)/\lambda_\beta)}. \qquad (7.41)$$

That is, we get that

$$M(x) \leq \left(\frac{\lambda_\beta}{\lambda_{\alpha_2} + \lambda_\beta} \right) \delta_3 (z_1(x) + z_2(x))^{((\lambda_{\alpha_2} + \lambda_\beta)/\lambda_\beta)}. \qquad (7.42)$$

Therefore, by (7.32) and (7.42), we find

$$(*) \leq (A_0(x)|_{\lambda_{\alpha_1}=0})\, 2^{(p-\lambda_\beta)/p} \left(\frac{\lambda_\beta}{\lambda_{\alpha_2} + \lambda_\beta} \right)^{\lambda_\beta/p} \delta_3^{\lambda_\beta/p} (z_1(x) + z_2(x))^{(\lambda_{\alpha_2} + \lambda_\beta)/p}$$

$$(7.43)$$

$$= \left(A_0(x)|_{\lambda_{\alpha_1}=0}\right) 2^{(p-\lambda_\beta/p)} \left(\frac{\lambda_\beta}{\lambda_{\alpha_2}+\lambda_\beta}\right)^{\lambda_\beta/p} \delta_3^{\lambda_\beta/p} \left[\int_0^x p(s)\left[|D^\beta f_1(s)|^p + \right.\right.$$

$$\left.\left.|D^\beta f_2(s)|^p\right]ds\right]^{((\lambda_{\alpha_2}+\lambda_\beta)/p)}. \tag{7.44}$$

We have established (7.30). □

The complete case λ_{α_1}, $\lambda_{\alpha_2} \neq 0$ follows.

Theorem 7.6. *All here are as in Theorem 7.4.*
Denote

$$\tilde{\gamma}_1 := \begin{cases} 2^{((\lambda_{\alpha_1}+\lambda_{\alpha_2})/\lambda_\beta)}-1, & if \quad \lambda_{\alpha_1}+\lambda_{\alpha_2} \geq \lambda_\beta, \\ 1, & if \quad \lambda_{\alpha_1}+\lambda_{\alpha_2} \leq \lambda_\beta, \end{cases} \tag{7.45}$$

and

$$\tilde{\gamma}_2 := \begin{cases} 1, & if \quad \lambda_{\alpha_1}+\lambda_{\alpha_2}+\lambda_\beta \geq p, \\ 2^{1-((\lambda_{\alpha_1}\lambda_{\alpha_2}+\lambda_\beta)/p)}, & if \quad \lambda_{\alpha_1}+\lambda_{\alpha_2}+\lambda_\beta \leq p. \end{cases} \tag{7.46}$$

Then

$$\int_0^x q(s)\left[|D^{\alpha_1}f_1(s)|^{\lambda_{\alpha_1}}|D^{\alpha_2}f_2(s)|^{\lambda_{\alpha_2}}|D^\beta f_1(s)|^{\lambda_\beta} + \right.$$

$$\left.|D^{\alpha_2}f_1(s)|^{\lambda_{\alpha_2}}|D^{\alpha_1}f_2(s)|^{\lambda_{\alpha_1}}|D^\beta f_2(s)|^{\lambda_\beta}\right]ds \leq$$

$$A_0(x)\left(\frac{\lambda_\beta}{(\lambda_{\alpha_1}+\lambda_{\alpha_2})(\lambda_{\alpha_1}+\lambda_{\alpha_2}+\lambda_\beta)}\right)^{\lambda_\beta/p}\cdot\left[\lambda_{\alpha_1}^{\lambda_\beta/p}\tilde{\gamma}_2 + 2^{(p-\lambda_\beta)/p}(\tilde{\gamma}_1\lambda_{\alpha_2})^{\lambda_\beta/p}\right]$$

$$\tag{(7.47)}$$

$$\cdot\left(\int_0^x p(s)\left(|D^\beta f_1(s)|^p + |D^\beta f_2(s)|^p\right)ds\right)^{(\lambda_{\alpha_1}+\lambda_{\alpha_2}+\lambda_\beta)/p}.$$

Proof. Here we use the known inequality

$$ab \leq \frac{a^p}{p} + \frac{b^q}{q}, \tag{7.48}$$

where $a,b \geq 0$ and $p,q > 1$ such that $1/p+1/q = 1$. From (7.21) we obtain

$$\int_0^x q(s)|D^{\alpha_1}f_1(s)|^{\lambda_{\alpha_1}}|D^{\alpha_2}f_2(s)|^{\lambda_{\alpha_2}}|D^\beta f_1(s)|^{\lambda_\beta}ds \leq$$

$$A_0(x)\left[\int_0^x\left[\left(\frac{\lambda_{\alpha_1}}{\lambda_{\alpha_1}+\lambda_{\alpha_2}}\right)(z_1(s))^{(\lambda_{\alpha_1}+\lambda_{\alpha_2}/\lambda_\beta)} + \right.\right.$$

$$\left.\left.\left(\frac{\lambda_{\alpha_2}}{\lambda_{\alpha_1}+\lambda_{\alpha_2}}\right)(z_2(s))^{(\lambda_{\alpha_1}+\lambda_{\alpha_2}/\lambda_\beta)}\right]z_1'(s)ds\right]^{\lambda_\beta/p} \tag{7.49}$$

$$\overset{(7.26)}{\leq} A_0(x) \left[\left(\left(\frac{\lambda_{\alpha_1}\lambda_\beta}{\lambda_{\alpha_1}+\lambda_{\alpha_2}} \right) \frac{(z_1(x))^{((\lambda_{\alpha_1}+\lambda_{\alpha_2}+\lambda_\beta)/\lambda_\beta)}}{(\lambda_{\alpha_1}+\lambda_{\alpha_2}+\lambda_\beta)} \right)^{\lambda_\beta/p} \right.$$

$$\left. + \left(\int_0^x \left(\frac{\lambda_{\alpha_2}}{\lambda_{\alpha_1}+\lambda_{\alpha_2}} \right) (z_2(s))^{((\lambda_{\alpha_1}+\lambda_{\alpha_2})/\lambda_\beta)} z_1'(s)ds \right)^{\lambda_\beta/p} \right]. \qquad (7.50)$$

Consequently,

$$\int_0^x q(s)|D^{\alpha_1}f_1(s)|^{\lambda_{\alpha_1}}|D^{\alpha_2}f_2(s)|^{\lambda_{\alpha_2}}|D^\beta f_1(s)|^{\lambda_\beta}ds \leq$$

$$A_0(x) \left\{ \left(\frac{\lambda_{\alpha_1}\lambda_\beta}{(\lambda_{\alpha_1}+\lambda_{\alpha_2})(\lambda_{\alpha_1}+\lambda_{\alpha_2}+\lambda_\beta)} \right)^{\lambda_\beta/p} (z_1(x))^{((\lambda_{\alpha_1}+\lambda_{\alpha_2}+\lambda_\beta)/p)} \right.$$

$$\qquad (7.51)$$

$$\left. + \left(\frac{\lambda_{\alpha_2}}{\lambda_{\alpha_1}+\lambda_{\alpha_2}} \right)^{\lambda_\beta/p} \left(\int_0^x (z_2(s))^{(\lambda_{\alpha_1}+\lambda_{\alpha_2}/\lambda_\beta)} z_1'(s)ds \right)^{\lambda_\beta/p} \right\}.$$

Working similarly, we get

$$\int_0^x q(s)|D^{\alpha_1}f_2(s)|^{\lambda_{\alpha_1}}|D^{\alpha_2}f_1(s)|^{\lambda_{\alpha_2}}|D^\beta f_2(s)|^{\lambda_\beta}ds$$

$$\leq A_0(x) \left\{ \left(\frac{\lambda_{\alpha_1}\lambda_\beta}{(\lambda_{\alpha_1}+\lambda_{\alpha_2})(\lambda_{\alpha_1}+\lambda_{\alpha_2}+\lambda_\beta)} \right)^{\lambda_\beta/p} (z_2(x))^{((\lambda_{\alpha_1}+\lambda_{\alpha_2}+\lambda_\beta)/p)} + \right.$$

$$\left. \left(\frac{\lambda_{\alpha_2}}{\lambda_{\alpha_1}+\lambda_{\alpha_2}} \right)^{\lambda_\beta/p} \left(\int_0^x (z_1(s))^{(\lambda_{\alpha_1}+\lambda_{\alpha_2}/\lambda_\beta)} z_2'(s)ds \right)^{\lambda_\beta/p} \right\}. \qquad (7.52)$$

Next, adding (7.51) and (7.52), we observe

$$\Omega := \int_0^x q(s) \left[|D^{\alpha_1}f_1(s)|^{\lambda_{\alpha_1}} |D^{\alpha_2}f_2(s)|^{\lambda_{\alpha_2}} |D^\beta f_1(s)|^{\lambda_\beta} + \right.$$

$$\left. |D^{\alpha_1}f_2(s)|^{\lambda_{\alpha_1}} |D^{\alpha_2}f_1(s)|^{\lambda_{\alpha_2}} |D^\beta f_2(s)|^{\lambda_\beta} \right] ds$$

$$\leq A_0(x) \left\{ \left(\frac{\lambda_{\alpha_1}\lambda_\beta}{(\lambda_{\alpha_1}+\lambda_{\alpha_2})(\lambda_{\alpha_1}+\lambda_{\alpha_2}+\lambda_\beta)} \right)^{\lambda_\beta/p} \right.$$

$$\left((z_1(x))^{(\lambda_{\alpha_1}+\lambda_{\alpha_2}+\lambda_\beta/p)} + (z_2(x))^{((\lambda_{\alpha_1}+\lambda_{\alpha_2}+\lambda_\beta)/p)} \right) +$$

$$\left(\frac{\lambda_{\alpha_2}}{\lambda_{\alpha_1}+\lambda_{\alpha_2}} \right)^{\lambda_\beta/p} \left[\left(\int_0^x (z_2(s))^{((\lambda_{\alpha_1}+\lambda_{\alpha_2})/\lambda_\beta)} z_1'(s)ds \right)^{\lambda_\beta/p} + \right.$$

$$\left. \left. \left(\int_0^x (z_1(s))^{((\lambda_{\alpha_1}+\lambda_{\alpha_2})/\lambda_\beta)} z_2'(s)ds \right)^{\lambda_\beta/p} \right] \right\} \qquad (7.53)$$

$$\overset{(7.26)}{\leq} A_0(x) \left\{ \left(\frac{\lambda_{\alpha_1}\lambda_\beta}{(\lambda_{\alpha_1}+\lambda_{\alpha_2})(\lambda_{\alpha_1}+\lambda_{\alpha_2}+\lambda_\beta)} \right)^{\lambda_\beta/p} \right.$$

$$\left((z_1(x))^{((\lambda_{\alpha_1}+\lambda_{\alpha_2}+\lambda_\beta)/p)} + (z_2(x))^{((\lambda_{\alpha_1}+\lambda_{\alpha_2}+\lambda_\beta)/p)} \right) +$$

$$\left. \left(\frac{\lambda_{\alpha_2}}{\lambda_{\alpha_1}+\lambda_{\alpha_2}} \right)^{\lambda_\beta/p} 2^{\,1-(\lambda_\beta/p)} \left(\Gamma^*(x) \right)^{\lambda_\beta/p} \right\}, \qquad (7.54)$$

where

$$\Gamma^*(x) := \int_0^x \left((z_2(s))^{((\lambda_{\alpha_1}+\lambda_{\alpha_2})/\lambda_\beta)} z_1'(s) + (z_1(s))^{((\lambda_{\alpha_1}+\lambda_{\alpha_2})/\lambda_\beta)} z_2'(s) \right) ds.$$
$$(7.55)$$

Again see (7.33) and (7.42). Similarly we obtain

$$\Gamma^*(x) \leq \left(\frac{\lambda_\beta \tilde{\gamma}_1}{\lambda_{\alpha_1}+\lambda_{\alpha_2}+\lambda_\beta} \right) (z_1(x)+z_2(x))^{((\lambda_{\alpha_1}+\lambda_{\alpha_2}+\lambda_\beta)/\lambda_\beta)}. \qquad (7.56)$$

So by (7.56) we find

$$\Omega \leq A_0(x) \left\{ \left(\frac{\lambda_{\alpha_1}\lambda_\beta}{(\lambda_{\alpha_1}+\lambda_{\alpha_2})(\lambda_{\alpha_1}+\lambda_{\alpha_2}+\lambda_\beta)} \right)^{\lambda_\beta/p} \right.$$

$$\left((z_1(x))^{(\lambda_{\alpha_1}+\lambda_{\alpha_2}+\lambda_\beta/p)} + (z_2(x))^{((\lambda_{\alpha_1}+\lambda_{\alpha_2}+\lambda_\beta)/p)} \right) +$$

$$\left. \left(\frac{\lambda_{\alpha_2}}{\lambda_{\alpha_1}+\lambda_{\alpha_2}} \right)^{\lambda_\beta/p} 2^{(p-\lambda_\beta)/p} \left(\frac{\lambda_\beta\tilde{\gamma}_1}{\lambda_{\alpha_1}+\lambda_{\alpha_2}+\lambda_\beta} \right)^{\lambda_\beta/p} (z_1(x)+z_2(x))^{(\lambda_{\alpha_1}+\lambda_{\alpha_2}+\lambda_\beta/p)} \right\}$$
$$(7.57)$$

$$\overset{by\ (7.26),(7.27)}{\leq} A_0(x) \left\{ \left(\frac{\lambda_{\alpha_1}\lambda_\beta}{(\lambda_{\alpha_1}+\lambda_{\alpha_2})(\lambda_{\alpha_1}+\lambda_{\alpha_2}+\lambda_\beta)} \right)^{\lambda_\beta/p} \right.$$

$$\tilde{\gamma}_2(z_1(x)+z_2(x))^{(\lambda_{\alpha_1}+\lambda_{\alpha_2}+\lambda_\beta/p)} +$$

$$\left. \left(\frac{\lambda_{\alpha_2}}{\lambda_{\alpha_1}+\lambda_{\alpha_2}} \right)^{\lambda_\beta/p} 2^{(p-\lambda_\beta/p)} \left(\frac{\lambda_\beta\tilde{\gamma}_1}{\lambda_{\alpha_1}+\lambda_{\alpha_2}+\lambda_\beta} \right)^{\lambda_\beta/p} (z_1(x)+z_2(x))^{(\lambda_{\alpha_1}+\lambda_{\alpha_2}+\lambda_\beta/p)} \right\}$$
$$(7.58)$$

$$= A_0(x) \left\{ \left(\frac{\lambda_{\alpha_1}\lambda_\beta}{(\lambda_{\alpha_1}+\lambda_{\alpha_2})(\lambda_{\alpha_1}+\lambda_{\alpha_2}+\lambda_\beta)} \right)^{\lambda_\beta/p} \tilde{\gamma}_2 + \right.$$

$$\left. \left(\frac{\lambda_{\alpha_2}}{\lambda_{\alpha_1}+\lambda_{\alpha_2}} \right)^{\lambda_\beta/p} 2^{(p-\lambda_\beta)/p} \left(\frac{\lambda_\beta\tilde{\gamma}_1}{\lambda_{\alpha_1}+\lambda_{\alpha_2}+\lambda_\beta} \right)^{\lambda_\beta/p} \right\} (z_1(x)+z_2(x))^{(\lambda_{\alpha_1}+\lambda_{\alpha_2}+\lambda_\beta/p)}$$
$$(7.59)$$

$$= A_0(x) \left(\frac{\lambda_\beta}{(\lambda_{\alpha_1} + \lambda_{\alpha_2})(\lambda_{\alpha_1} + \lambda_{\alpha_2} + \lambda_\beta)} \right)^{\lambda_\beta/p} \left[\lambda_{\alpha_1}^{\lambda_\beta/p} \, \tilde{\gamma}_2 + 2^{(p-\lambda_\beta)/p} (\tilde{\gamma}_1 \lambda_{\alpha_2})^{\lambda_\beta/p} \right]$$

$$\left[\int_0^x p(s)[|D^\beta f_1(s)|^p + |D^\beta f_2(s)|^p] ds \right]^{(\lambda_{\alpha_1} + \lambda_{\alpha_2} + \lambda_\beta)/p}. \tag{7.60}$$

We have proved that

$$\Omega \le A_0(x) \left(\frac{\lambda_\beta}{(\lambda_{\alpha_1} + \lambda_{\alpha_2})(\lambda_{\alpha_1} + \lambda_{\alpha_2} + \lambda_\beta)} \right)^{\lambda_\beta/p} \left[\lambda_{\alpha_1}^{\lambda_\beta/p} \, \tilde{\gamma}_2 + 2^{(p-\lambda_\beta)/p} (\tilde{\gamma}_1 \lambda_{\alpha_2})^{\lambda_\beta/p} \right]$$

$$\left[\int_0^x p(s)[|D^\beta f_1(s)|^p + |D^\beta f_2(s)|^p] ds \right]^{(\lambda_{\alpha_1} + \lambda_{\alpha_2} + \lambda_\beta)/p}. \tag{7.61}$$

We have proved (7.47). □

We need

Theorem 7.7. *Let $0 \le s \le x$ and $f \in L_\infty([0,x]), r > 0$. Define*

$$F(s) := \int_0^s (s-t)^r f(t) dt. \tag{7.62}$$

Then there exists

$$F'(s) = r. \int_0^s (s-t)^{r-1} f(t) dt, \text{ all } s \in [0,x]. \tag{7.63}$$

Proof. Fix $s_0 \in [0,x]$ and notice that

$$F(s_0) = \int_0^{s_0} (s_0 - t)^r f(t) dt = \int_0^x \chi_{[0,s_0]}(t)(s_0 - t)^r f(t) dt.$$

We call $g(s,t) := \chi_{[0,s]}(t)(s-t)^r f(t)$, which is a Lebesgue integrable function for every $s \in [0,x]$.

That is, $g(s_0,t) = \chi_{[0,s_0]}(t)(s_0 - t)^r f(t)$, all $t \in [0,x]$, and $F(s_0) = \int_0^x g(s_0,t) dt$.

We would like to study if there exists

$$\frac{\partial g(s_0,t)}{\partial s} = f(t) \left(\lim_{h \to 0} \frac{\chi_{[0,s_0+h]}(t)(s_0 + h - t)^r - \chi_{[0,s_0]}(t)(s_0 - t)^r}{h} \right). \tag{7.64}$$

We distinguish the following cases.

(1) Let $x \ge t > s_0$; then there exist small enough $h > 0$ such that $t \ge s_0 \pm h$.

That is,

$$\chi_{[0,s_0 \pm h]}(t) = \chi_{[0,s_0]}(t) = 0.$$

Hence, there exists

$$\frac{\partial g(s_0, t)}{\partial s} = 0, \quad \text{all} \quad t : s_0 < t \leq x. \tag{7.65}$$

(2) Let $0 \leq t < s_0$; then there exist small enough $h > 0$ such that $t < s_0 \pm h$.
That is,

$$\chi_{[0, s_0 \pm h]}(t) = \chi_{[0, s_0]}(t) = 1.$$

In that case

$$\frac{\partial g(s_0, t)}{\partial s} = f(t) \left(\lim_{h \to 0} \frac{(s_0 + h - t)^r - (s_0 - t)^r}{h} \right) = r(s_0 - t)^{r-1} f(t),$$
$$\tag{7.66}$$

exists for almost all $t : 0 \leq t < s_0$.
(3) Let $t = s_0$. Then we see that

$$\frac{\partial g_+(s_0, s_0)}{\partial s} = f(s_0) \left(\lim_{h \to 0+} \frac{h^r}{h} \right) = f(s_0) \left(\lim_{h \to 0+} h^{r-1} \right). \tag{7.67}$$

The last limit does not exist if $0 < r < 1$, equals $f(s_0)$ if $r = 1$, and may not exist, and equals 0 if $r > 1$.
Notice also that

$$\frac{\partial g_-(s_0, s_0)}{\partial s} = f(s_0) \left(\lim_{h \to 0-} \frac{\chi_{[0, s_0 + h]}(s_0) h^r}{h} \right) = f(s_0) \left(\lim_{h \to 0-} \chi_{[0, s_0 + h]} h^{r-1} \right) = 0,$$

by $\chi_{[0, s_0 + h]}(s_0) = 0$, $h < 0$.
That is,

$$\frac{\partial g_-(s_0, s_0)}{\partial s} = 0. \tag{7.68}$$

In general as a conclusion we get that $\partial g(s_0, t)/\partial s$ exists for almost all $t \in [0, x]$.
Next we define the difference quotient at s_0,

$$D_{s_0}(h, t) := f(t) \left(\frac{\chi_{[0, s_0 + h]}(t)(s_0 + h - t)^r - \chi_{[0, s_0]}(t)(s_0 - t)^r}{h} \right), \tag{7.69}$$

for $h \neq 0$, and $D_{s_0}(0, t) := 0$.
Again we distinguish the following cases.
(1) Let $x \geq t > s_0$; then there exist small enough $h > 0$ such that $t > s_0 \pm h$.
Clearly then $D_{s_0}(h, t) = 0$.
(2) Let $0 \leq t < s_0$; then there exist small enough $h > 0$ such that $t < s_0 \pm h$. Thus

$$D_{s_0}(\pm h, t) = f(t) \left(\frac{(s_0 \pm h - t)^r - (s_0 - t)^r}{\pm h} \right). \tag{7.70}$$

Call $\rho := s_0 - t > 0$; clearly $\rho \le x$. Define

$$\varphi(h) := \frac{(\rho + h)^r - \rho^r}{h} = \frac{(s_0 + h - t)^r - (s_0 - t)^r}{h} \qquad (7.71)$$

for h close to zero, $r > 0$.

That is, $D_{s_0}(h,t) = f(t)\varphi(h)$. If $r = 1$, then $\varphi(h) = 1$ and

$$D_{s_0}(h,t) = f(t). \qquad (7.72)$$

We now treat the following subcases.

$(2(i))$ If $r > 1$, then $\gamma(\rho) := \rho^r, 0 \le \rho \le x$, is convex and increasing. If $h > 0$, then by the mean value theorem we get that

$$\varphi(h) < rx^{r-1}.$$

That is,

$$|D_{s_0}(h,t)| \le rx^{r-1}|f(t)|. \qquad (7.73)$$

If $h < 0$, then, similarly, again we get

$$\varphi(h) = \frac{\rho^r - (\rho + h)^r}{-h} < rx^{r-1}.$$

That is, for small $|h|$ we have

$$|D_{s_0}(h,t)| \le rx^{r-1}|f(t)|, \quad r \ge 1. \qquad (7.74)$$

$(2(ii))$ If $0 < r < 1$, then $\gamma(\rho) = \rho^r$, $0 \le \rho \le x$ is concave and increasing. Let $h > 0$; then $\varphi(h) < \rho^{r-1} = (s_0 - t)^{r-1}$ and for $h < 0$, again $\varphi(h) \le \rho^{r-1} = (s_0 - t)^{r-1}$.

That is,

$$|D_{s_0}(h,t)| \le (s_0 - t)^{r-1}|f(t)|, \qquad (7.75)$$

for small $|h|$.

(3) Case of $t = s_0$; then

$$D_{s_0}(h, s_0) = f(s_0)\, h^{r-1}, \quad \text{for } h > 0, \qquad (7.76)$$

and

$$D_{s_0}(h, s_0) = 0, \quad \text{for } h < 0. \qquad (7.77)$$

So, if $r \ge 1$ we obtain

$$|D_{s_0}(h, s_0)| \le |f(s_0)|x^{r-1}, \qquad (7.78)$$

for small $|h|$.

If $0 < r < 1$, then for small $h > 0$ the function $D_{s_0}(h, s_0)$ may be unbounded.

In conclusion we get:

(*I*) For $r \geq 1$, that

$$|D_{s_0}(h,t)| \leq r x^{r-1} ||f(t)||_\infty < +\infty, \qquad (7.79)$$

for almost all $t \in [0,x]$.
   (*II*) For $0 < r < 1$, that

$$|D_{s_0}(h,t)| \leq \lambda(t) \text{ for almost all } t \in [0,x], \qquad (7.80)$$

where

$$\lambda(t) := \begin{cases} (s_0 - t)^{r-1}|f(t)|, & 0 \leq t < s_0, \\ 0, & for \quad s_0 \leq t \leq x. \end{cases} \qquad (7.81)$$

Clearly λ is integrable on $[0,x]$. Then by Theorem 24.5, pp. 193–194 of [10] we get that $\partial g(s_0, \cdot)/\partial s$ defines an integrable function, and there exists

$$F'(s_0) = \int_0^x \frac{\partial g(s_0,t)}{\partial s} dt = r \int_0^{s_0} (s_0 - t)^{r-1} f(t) dt +$$

$$\int_{s_0}^x 0 \, dt = r \int_0^{s_0} (s_0 - t)^{r-1} f(t) dt. \qquad (7.82)$$

That proves the claim. $\quad \square$

We proceed with a special important case.

Theorem 7.8. *Let $\beta > \alpha_1 + 1$, $\alpha_1 \in \mathbb{R}_+$, and let $f_1, f_2 \in L_1(0,x)$, $x \in \mathbb{R}_+ - \{0\}$ have, respectively, L_∞ fractional derivatives $D^\beta f_1, D^\beta f_2$ in $[0,x]$, and let $D^{\beta-k} f_i(0) = 0$, for $k = 1, \ldots, [\beta] + 1; i = 1, 2$. Consider also $p(t) > 0$ and $q(t) \geq 0$, with $p(t), 1/p(t), q(t) \in L_\infty(0,x)$. Let $\lambda_\alpha \geq 0$, $0 < \lambda_{\alpha+1} < 1$, and $p > 1$.*
   Denote

$$\theta_3 := \begin{cases} 2^{\lambda_\alpha/(\lambda_{\alpha+1})} - 1, & if \quad \lambda_\alpha \geq \lambda_{\alpha+1}, \\ 1, & if \quad \lambda_\alpha \leq \lambda_{\alpha+1}, \end{cases} \qquad (7.83)$$

$$L(x) := \left(2 \int_0^x (q(s))^{(1/(1-\lambda_{\alpha+1}))} ds \right)^{(1-\lambda_{\alpha+1})} \left(\frac{\theta_3 \lambda_{\alpha+1}}{\lambda_\alpha + \lambda_{\alpha+1}} \right)^{\lambda_{\alpha+1}}, \quad (7.84)$$

and

$$P_1(x) := \int_0^x (x - s)^{(\beta - \alpha_1 - 1)p/(p-1)} (p(s))^{-1/(p-1)} ds, \qquad (7.85)$$

$$T(x) := L(x) \left(\frac{P_1(x)^{((p-1)/p)}}{\Gamma(\beta - \alpha_1)} \right)^{(\lambda_\alpha + \lambda_{\alpha+1})}, \qquad (7.86)$$

and

$$\omega_1 := 2^{((p-1)/p)(\lambda_\alpha + \lambda_{\alpha+1})}, \qquad (7.87)$$

$$\Phi(x) := T(x)\omega_1. \qquad (7.88)$$

Then

$$\int_0^x q(s) \left[\left| D^{\alpha_1} f_1(s) \right|^{\lambda_\alpha} \left| D^{\alpha_1+1} f_2(s) \right|^{\lambda_{\alpha+1}} + \left| D^{\alpha_1} f_2(s) \right|^{\lambda_\alpha} \left| D^{\alpha_1+1} f_1(s) \right|^{\lambda_{\alpha+1}} \right] ds$$

$$\leq \Phi(x) \left[\int_0^x p(s) \left(\left| D^\beta f_1(s) \right|^p + \left| D^\beta f_2(s) \right|^p \right) ds \right]^{(\lambda_\alpha + \lambda_{\alpha+1})/p}. \qquad (7.89)$$

Proof. For convenience, we set

$$\alpha_2 := \alpha_1 + 1. \qquad (7.90)$$

By Lemma 7.3, we have

$$D^{\alpha_1} f_i(s) = \frac{1}{\Gamma(\beta - \alpha_1)} \int_0^s (s-t)^{\beta - \alpha_1 - 1} D^\beta f_i(t) dt, \qquad (7.91)$$

$\forall\, s \in [0, x]$, $i = 1, 2$, and

$$D^{\alpha_2} f_i(s) = D^{\alpha_1+1} f_i(s) = \frac{1}{\Gamma(\beta - \alpha_1 - 1)} \int_0^s (s-t)^{\beta - \alpha_1 - 2} D^\beta f_i(t) dt, \qquad (7.92)$$

$\forall\, s \in [0, x]$, $i = 1, 2$.

By Theorem 7.7 we obtain that

$$(D^{\alpha_1} f_i(s))' = D^{\alpha_1+1} f_i(s) = D^{\alpha_2} f_i(s), \qquad (7.93)$$

$\forall\, s \in [0, x]$, $i = 1, 2$.

So we have

$$|D^{\alpha_i} f_j(s)| \leq \frac{1}{\Gamma(\beta - \alpha_i)} \int_0^s (s-t)^{\beta - \alpha_i - 1} |D^\beta f_j(t)| dt =: g_{j,\alpha_i}(s), \qquad (7.94)$$

where $j = 1, 2$, $i = 1, 2$, $0 \leq s \leq x$.

We also have by Theorem 7.7 that

$$(g_{j,\alpha_1}(s))' = g_{j,\alpha_2}(s); \ g_{j,\alpha_i}(0) = 0. \qquad (7.95)$$

Notice that, if $\beta - \alpha_2 = 1$, then

$$g_{j,\alpha_2}(s) = \int_0^s |D^\beta f_j(t)| dt. \qquad (7.96)$$

Next, we apply Hölder's inequality with indices $1/\lambda_{\alpha+1}, 1/(1 - \lambda_{\alpha+1})$ to find

$$\int_0^x q(s) |D^{\alpha_1} f_1(s)|^{\lambda_\alpha} |D^{\alpha_1+1} f_2(s)|^{\lambda_{\alpha+1}} ds \leq$$

$$\int_0^x q(s)(g_{1,\alpha_1}(s))^{\lambda_\alpha}((g_{2,\alpha_1}(s))')^{\lambda_{\alpha+1}} ds \leq \tag{7.97}$$

$$\left(\int_0^x (q(s))^{(1/1-\lambda_{\alpha+1})} ds\right)^{(1-\lambda_{\alpha+1})} \left(\int_0^x (g_{1,\alpha_1}(s))^{\lambda_\alpha/\lambda_{\alpha+1}}(g_{2,\alpha_1}(s))' ds\right)^{\lambda_{\alpha+1}}. \tag{7.98}$$

Similarly, we get

$$\int_0^x q(s)\,|D^{\alpha_1} f_2(s)|^{\lambda_\alpha}\,|D^{\alpha_1+1} f_1(s)|^{\lambda_{\alpha+1}}\, ds \leq$$

$$\left(\int_0^x (q(s))^{(1/1-\lambda_{\alpha+1})} ds\right)^{(1-\lambda_{\alpha+1})} \left(\int_0^x (g_{2,\alpha_1}(s))^{\lambda_\alpha/\lambda_{\alpha+1}}(g_{1,\alpha_1}(s))' ds\right)^{\lambda_{\alpha+1}}. \tag{7.99}$$

Adding the last two inequalities (7.98), (7.99), we get

$$\int_0^x q(s)\left[|D^{\alpha_1} f_1(s)|^{\lambda_\alpha}|D^{\alpha_1+1} f_2(s)|^{\lambda_{\alpha+1}} + |D^{\alpha_1} f_2(s)|^{\lambda_\alpha}|D^{\alpha_1+1} f_1(s)|^{\lambda_{\alpha+1}}\right] ds \leq$$

$$\left(\int_0^x (q(s))^{(1/1-\lambda_{\alpha+1})} ds\right)^{(1-\lambda_{\alpha+1})} \left[\left(\int_0^x (g_{1,\alpha_1}(s))^{\lambda_\alpha/\lambda_{\alpha+1}}(g_{2,\alpha_1}(s))' ds\right)^{\lambda_{\alpha+1}} + \right.$$

$$\left.\left(\int_0^x (g_{2,\alpha_1}(s))^{\lambda_\alpha/\lambda_{\alpha+1}}(g_{1,\alpha_1}(s))' ds\right)^{\lambda_{\alpha+1}}\right] \tag{7.100}$$

$$\overset{(7.26)}{\leq} \left(2\int_0^x (q(s))^{(1/1-\lambda_{\alpha+1})} ds\right)^{(1-\lambda_{\alpha+1})} \times$$

$$\left[\int_0^x \left[(g_{1,\alpha_1}(s))^{\lambda_\alpha/\lambda_{\alpha+1}}(g_{2,\alpha_1}(s))' + (g_{2,\alpha_1}(s))^{\lambda_\alpha/\lambda_{\alpha+1}}(g_{1,\alpha_1}(s))'\right] ds\right]^{\lambda_{\alpha+1}} \tag{7.101}$$

(notice (7.33) and (7.42), so similarly we have)

$$\leq \left(2\int_0^x (q(s))^{(1/(1-\lambda_{\alpha+1}))} ds\right)^{(1-\lambda_{\alpha+1})} \left(\frac{\lambda_{\alpha+1}\theta_3}{\lambda_\alpha + \lambda_{\alpha+1}}\right)^{\lambda_{\alpha+1}}$$

$$(g_{1,\alpha_1}(x) + g_{2,\alpha_1}(x))^{(\lambda_\alpha+\lambda_{\alpha+1})} \tag{7.102}$$

$$= L(x)(g_{1,\alpha_1}(x) + g_{2,\alpha_1}(x))^{(\lambda_\alpha+\lambda_{\alpha+1})} \tag{7.103}$$

$$= \frac{L(x)}{(\Gamma(\beta - \alpha_1))^{(\lambda_\alpha+\lambda_{\alpha+1})}} \times$$

$$\left[\int_0^x (x-t)^{\beta-\alpha_1-1}(p(t))^{-1/p}(p(t))^{1/p}\left[|D^\beta f_1(t)| + |D^\beta f_2(t)|\right] dt\right]^{(\lambda_\alpha+\lambda_{\alpha+1})} \tag{7.104}$$

(applying Hölder's inequality with indices $p/(p-1)$ and p, we find)

$$\le \frac{L(x)}{(\Gamma(\beta-\alpha_1))^{(\lambda_\alpha+\lambda_\alpha+1)}} \left[\int_0^x (x-t)^{(\beta-\alpha_1-1)p/(p-1)}(p(t))^{-1/(p-1)}dt\right]^{(p-1/p)(\lambda_\alpha+\lambda_\alpha+1)}$$

$$\left[\int_0^x p(t)(|D^\beta f_1(t)|+|D^\beta f_2(t)|)^p dt\right]^{(\lambda_\alpha+\lambda_\alpha+1/p)} = \quad (7.105)$$

$$\frac{L(x)(P_1(x))^{(p-1/p)(\lambda_\alpha+\lambda_\alpha+1)}}{(\Gamma(\beta-\alpha_1))^{(\lambda_\alpha+\lambda_\alpha+1)}}\left[\int_0^x p(t)\left(\left|D^\beta f_1(t)\right|+\left|D^\beta f_2(t)\right|\right)^p dt\right]^{(\lambda_\alpha+\lambda_\alpha+1/p)} \quad (7.106)$$

$$= T(x)\left[\int_0^x p(t)\left(|D^\beta f_1(t)|+|D^\beta f_2(t)|\right)^p dt\right]^{(\lambda_\alpha+\lambda_\alpha+1/p)} \quad (7.107)$$

$$\le T(x)2^{(p-1/p)(\lambda_\alpha+\lambda_\alpha+1)}\left[\int_0^x p(t)\left(\left|D^\beta f_1(t)\right|^p+\left|D^\beta f_2(t)\right|^p\right)dt\right]^{(\lambda_\alpha+\lambda_\alpha+1/p)} \quad (7.108)$$

$$= \Phi(x)\left[\int_0^x p(t)\left(|D^\beta f_1(t)|^p+|D^\beta f_2(t)|^p\right)dt\right]^{(\lambda_\alpha+\lambda_\alpha+1)/p}. \quad (7.109)$$

We have proved (7.89). □

We continue with

Theorem 7.9. *All here are as in Theorem 7.4. Consider the special case of $\lambda_{\alpha_2} = \lambda_{\alpha_1}+\lambda_\beta$.*
Denote

$$\tilde{T}(x) := A_0(x)\left(\frac{\lambda_\beta}{\lambda_{\alpha_1}+\lambda_\beta}\right)^{\lambda_\beta/p}2^{(p-2\lambda_{\alpha_1}-3\lambda_\beta)/p}. \quad (7.110)$$

Then

$$\int_0^x q(s)\left[|D^{\alpha_1}f_1(s)|^{\lambda_{\alpha_1}}|D^{\alpha_2}f_2(s)|^{\lambda_{\alpha_1}+\lambda_\beta}|D^\beta f_1(s)|^{\lambda_\beta}+\right.$$

$$\left.|D^{\alpha_2}f_1(s)|^{\lambda_{\alpha_1}+\lambda_\beta}|D^{\alpha_1}f_2(s)|^{\lambda_{\alpha_1}}|D^\beta f_2(s)|^{\lambda_\beta}\right]ds \le$$

$$\tilde{T}(x)\left(\int_0^x p(s)\left(|D^\beta f_1(s)|^p+|D^\beta f_2(s)|^p\right)ds\right)^{2(\lambda_{\alpha_1}+\lambda_\beta)/p}. \quad (7.111)$$

Proof. We apply (7.21) and (7.22) for $\lambda_{\alpha_2}=\lambda_{\alpha_1}+\lambda_\beta$, and by adding we find

$$\int_0^x q(s)\left[|D^{\alpha_1}f_1(s)|^{\lambda_{\alpha_1}}|D^{\alpha_2}f_2(s)|^{\lambda_{\alpha_1}+\lambda_\beta}|D^\beta f_1(s)|^{\lambda_\beta}+\right.$$

$$|D^{\alpha_1} f_2(s)|^{\lambda_{\alpha_1}} |D^{\alpha_2} f_1(s)|^{\lambda_{\alpha_1}+\lambda_\beta} |D^\beta f_2(s)|^{\lambda_\beta} \Big] ds$$

$$\leq A_0(x) \left[\left(\int_0^x (z_1(s))^{\lambda_{\alpha_1}/\lambda_\beta} (z_2(s))^{\lambda_{\alpha_1}+\lambda_\beta/\lambda_\beta} z_1'(s) ds \right)^{\lambda_\beta/p} \right.$$

$$\left. + \left(\int_0^x (z_1(s))^{(\lambda_{\alpha_1}+\lambda_\beta)/\lambda_\beta} (z_2(s))^{\lambda_{\alpha_1}/\lambda_\beta} z_2'(s) ds \right)^{\lambda_\beta/p} \right] \qquad (7.112)$$

$$\overset{(7.26)}{\leq} A_0(x) 2^{1-(\lambda_\beta/p)} \times \left[\int_0^x (z_1(s))^{\lambda_{\alpha_1}/\lambda_\beta} (z_2(s))^{\lambda_{\alpha_1}/\lambda_\beta+1} z_1'(s) + \right.$$

$$\left. (z_1(s))^{(\lambda_{\alpha_1}/\lambda_\beta)+1} (z_2(s))^{\lambda_{\alpha_1}/\lambda_\beta} z_2'(s) \Big) ds \right]^{\lambda_\beta/p} = \qquad (7.113)$$

$$A_0(x) 2^{1-(\lambda_\beta/p)} \left[\int_0^x (z_1(s) z_2(s))^{\lambda_{\alpha_1}/\lambda_\beta} [z_2(s) z_1'(s) + z_1(s) z_2'(s)] ds \right]^{\lambda_\beta/p} = \qquad (7.114)$$

$$A_0(x) 2^{(p-\lambda_\beta)/p} \left[\int_0^x (z_1(s) z_2(s))^{(\lambda_{\alpha_1}/\lambda_\beta)} (z_1(s) z_2(s))' ds \right]^{\lambda_\beta/p} \qquad (7.115)$$

$$= A_0(x) 2^{(p-\lambda_\beta/p)} (z_1(x) z_2(x))^{(\lambda_{\alpha_1}+\lambda_\beta/p)} \left(\frac{\lambda_\beta}{\lambda_{\alpha_1}+\lambda_\beta} \right)^{\lambda_\beta/p} \qquad (7.116)$$

$$\leq A_0(x) 2^{(p-\lambda_\beta)/p} \left(\frac{\lambda_\beta}{\lambda_{\alpha_1}+\lambda_\beta} \right)^{\lambda_\beta/p} \left(\frac{z_1(x)+z_2(x)}{2} \right)^{2(\lambda_{\alpha_1}+\lambda_\beta)/p} \qquad (7.117)$$

$$= \tilde{T}(x) (z_1(x)+z_2(x))^{2(\lambda_{\alpha_1}+\lambda_\beta)/p} = \qquad (7.118)$$

$$\tilde{T}(x) \left[\int_0^x p(s) \left(|D^\beta f_1(s)|^p + |D^\beta f_2(s)|^p \right) ds \right]^{2(\lambda_{\alpha_1}+\lambda_\beta)/p} . \qquad (7.119)$$

We have proved (7.111).   □

Next follow special cases of the above theorems.

Corollary 7.10. (to Theorem 7.4) *Set* $\lambda_{\alpha_2} = 0$, $p(t) = q(t) = 1$. *Then*

$$\int_0^x \left[|D^{\alpha_1} f_1(s)|^{\lambda_{\alpha_1}} |D^\beta f_1(s)|^{\lambda_\beta} + |D^{\alpha_1} f_2(s)|^{\lambda_{\alpha_1}} |D^\beta f_2(s)|^{\lambda_\beta} \right] ds \leq$$

$$C_1(x) \left[\int_0^x \left[|D^\beta f_1(s)|^p + |D^\beta f_2(s)|^p \right] ds \right]^{(\lambda_{\alpha_1}+\lambda_\beta/p)}, \qquad (7.120)$$

where

$$C_1(x) := (A_0(x)|_{\lambda_{\alpha_2}=0}) \left(\frac{\lambda_\beta}{\lambda_{\alpha_1}+\lambda_\beta} \right)^{\lambda_\beta/p} \delta_1, \qquad (7.121)$$

$$\delta_1 := \begin{cases} 2^{1-((\lambda_{\alpha_1}+\lambda_\beta)/p)}, & if \quad \lambda_{\alpha_1} + \lambda_\beta \leq p, \\ 1, & if \quad \lambda_{\alpha_1} + \lambda_\beta \geq p. \end{cases} \qquad (7.122)$$

We find that

$$\left(A_0(x)|_{\lambda_{\alpha_2}=0} \right) = \left\{ \left(\frac{(p-1)^{((\lambda_{\alpha_1}p-\lambda_{\alpha_1})/p)}}{(\Gamma(\beta-\alpha_1))^{\lambda_{\alpha_1}} (\beta p - \alpha_1 p - 1)^{((\lambda_{\alpha_1}p-\lambda_{\alpha_1})/p)}} \right) \cdot \right.$$

$$\left. \left(\frac{(p-\lambda_\beta)^{((p-\lambda_\beta)/p)}}{(\lambda_{\alpha_1}\beta p - \lambda_{\alpha_1}\alpha_1 p - \lambda_{\alpha_1} + p - \lambda_\beta)^{((p-\lambda_\beta)/p)}} \right) \right\}$$

$$\times x^{((\lambda_{\alpha_1}\beta p - \lambda_{\alpha_1}\alpha_1 p - \lambda_{\alpha_1} + p - \lambda_\beta)/p)}. \qquad (7.123)$$

Proof. Here we need to calculate $\left(A_0(x)|_{\lambda_{\alpha_2}=0} \right)$. From (7.6) we have

$$A_0(x)|_{\lambda_{\alpha_2}=0} = \left(\int_0^x \left(A(s)|_{\lambda_{\alpha_2}=0} \right)^{p/(p-\lambda_\beta)} ds \right)^{(p-\lambda_\beta)/p}, \qquad (7.124)$$

where from (7.5) we get

$$A(s)|_{\lambda_{\alpha_2}=0} = \frac{(P_1(s))^{\lambda_{\alpha_1}(p-1)/p}}{(\Gamma(\beta-\alpha_1))^{\lambda_{\alpha_1}}}, \qquad (7.125)$$

and here, from (7.4), it is

$$P_1(s) = \int_0^s (s-t)^{p(\beta-\alpha_1-1)/(p-1)} dt. \qquad (7.126)$$

Therefore we obtain

$$P_1(s) = \left(\frac{p-1}{\beta p - \alpha_1 p - 1} \right) s^{((\beta p - \alpha_1 p - 1)/(p-1))}, \qquad (7.127)$$

and

$$A(s)|_{\lambda_{\alpha_2}=0} = \left(\frac{p-1}{\beta p - \alpha_1 p - 1} \right)^{(\lambda_{\alpha_1}p-\lambda_{\alpha_1}/p)} \frac{1}{(\Gamma(\beta-\alpha_1))^{\lambda_{\alpha_1}}} s^{(\lambda_{\alpha_1}\beta p - \lambda_{\alpha_1}\alpha_1 p - \lambda_{\alpha_1})/p}. \qquad (7.128)$$

That is,

$$A_0(x)|_{\lambda_{\alpha_2}=0} = \left(\frac{(p-1)^{((\lambda_{\alpha_1}p-\lambda_{\alpha_1})/p)}}{(\beta p - \alpha_1 p - 1)^{((\lambda_{\alpha_1}p-\lambda_{\alpha_1})/p)}(\Gamma(\beta-\alpha_1))^{\lambda_{\alpha_1}}} \right)$$

$$\left(\int_0^x s^{((\lambda_{\alpha_1}\beta p - \lambda_{\alpha_1}\alpha_1 p - \lambda_{\alpha_1})/(p-\lambda_\beta))} ds \right)^{((p-\lambda_\beta)/p)} =$$

$$\left\{ \left(\frac{(p-1)^{((\lambda_{\alpha_1}p - \lambda_{\alpha_1})/p)}}{(\Gamma(\beta - \alpha_1))^{\lambda_{\alpha_1}}(\beta\, p - \alpha_1 p - 1)^{((\lambda_{\alpha_1}p - \lambda_{\alpha_1})/p)}} \right) \times \right. \qquad (7.129)$$

$$\left. \left(\frac{(p - \lambda_\beta)^{((p - \lambda_\beta)/p)}}{(\lambda_{\alpha_1}\beta p - \lambda_{\alpha_1}\alpha_1 p - \lambda_{\alpha_1} + p - \lambda_\beta)^{((p - \lambda_\beta)/p)}} \right) \right\}$$

$$\times x^{((\lambda_{\alpha_1}\beta p - \lambda_{\alpha_1}\alpha_1 p - \lambda_{\alpha_1} + p - \lambda_\beta)/p)}.$$

□

We continue with

Corollary 7.11. (to Theorem 7.4: set $\lambda_{\alpha_2} = 0$, $p(t) = q(t) = 1, \lambda_{\alpha_1} = \lambda_\beta = 1, p = 2$.) *In detail: let $\alpha_1 \in \mathbb{R}_+, \beta > \alpha_1$, $\beta - \alpha_1 > (1/2)$, and let $f_1, f_2 \in L_1(0, x), x \in \mathbb{R}_+ - \{0\}$, have, respectively, L_∞ fractional derivatives $D^\beta f_1, D^\beta f_2$ in $[0, x]$, and let $D^{\beta - k} f_i(0) = 0$ for $k = 1, \ldots, [\beta] + 1; i = 1, 2$. Then*

$$\int_0^x \left[|D^{\alpha_1} f_1(s)| \, |D^\beta f_1(s)| + |D^{\alpha_1} f_2(s)| \, |D^\beta f_2(s)| \right] ds$$

$$\leq \left(\frac{x^{(\beta - \alpha_1)}}{2\Gamma(\beta - \alpha_1)\sqrt{\beta - \alpha_1}\sqrt{2\beta - 2\alpha_1 - 1}} \right) \left(\int_0^x \left[\left(D^\beta f_1(s) \right)^2 + \left(D^\beta f_2(s) \right)^2 \right] ds \right).$$

$$(7.130)$$

Proof. We apply Corollary 7.10. Here, $\delta_1 = 1$ and

$$\left(\frac{\lambda_\beta}{\lambda_{\alpha_1} + \lambda_\beta} \right)^{\lambda_\beta/p} = \frac{1}{\sqrt{2}}. \qquad (7.131)$$

Furthermore, we have

$$(A_0(x)|_{\lambda_{\alpha_2}=0}) = \left(\frac{1}{\Gamma(\beta - \alpha_1)\sqrt{2\beta - 2\alpha_1 - 1}} \right) \left(\frac{x^{(\beta - \alpha_1)}}{\sqrt{2}\sqrt{\beta - \alpha_1}} \right). \qquad (7.132)$$

Finally, we get

$$C(x) = \frac{x^{(\beta - \alpha_1)}}{2\,\Gamma(\beta - \alpha_1)\,\sqrt{\beta - \alpha_1}\sqrt{2\beta - 2\alpha_1 - 1}}. \qquad (7.133)$$

□

We continue with

Corollary 7.12. (to Theorem 7.5, $\lambda_{\alpha_1} = 0$, $p(t) = q(t) = 1$). *It holds*

$$\int_0^x \left[|D^{\alpha_2} f_2(s)|^{\lambda_{\alpha_2}} \, |D^\beta f_1(s)|^{\lambda_\beta} + |D^{\alpha_2} f_1(s)|^{\lambda_{\alpha_2}} \, |D^\beta f_2(s)|^{\lambda_\beta} \right] ds \leq$$

$$C_2(x) \left(\int_0^x \left[|D^\beta f_1(s)|^p + |D^\beta f_2(s)|^p \right] ds \right)^{(\lambda_\beta + \lambda_{\alpha_2})/p} . \qquad (7.134)$$

Here

$$C_2(x) := \left(A_0(x)|_{\lambda_{\alpha_1}=0} \right) 2^{(p-\lambda_\beta)/p} \left(\frac{\lambda_\beta}{\lambda_{\alpha_2} + \lambda_\beta} \right)^{\lambda_\beta/p} \delta_3^{\lambda_\beta/p}, \qquad (7.135)$$

$$\delta_3 := \begin{cases} 2^{\lambda_{\alpha_2}/\lambda_\beta} - 1, & if \quad \lambda_{\alpha_2} \geq \lambda_\beta, \\ 1, & if \quad \lambda_{\alpha_2} \leq \lambda_\beta. \end{cases} \qquad (7.136)$$

We find that

$$\left(A_0(x)|_{\lambda_{\alpha_1}=0} \right) = \left\{ \left(\frac{(p-1)^{((\lambda_{\alpha_2}p - \lambda_{\alpha_2})/p)}}{(\Gamma(\beta - \alpha_2))^{\lambda_{\alpha_2}} (\beta\, p - \alpha_2 p - 1)^{((\lambda_{\alpha_2}p - \lambda_{\alpha_2})/p)}} \right) \right.$$

$$\left. \left(\frac{(p - \lambda_\beta)^{((p-\lambda_\beta)/p)}}{(\lambda_{\alpha_2}\beta p - \lambda_{\alpha_2}\alpha_2 p - \lambda_{\alpha_2} + p - \lambda_\beta)^{((p-\lambda_\beta)/p)}} \right) \right\}$$

$$\cdot x^{((\lambda_{\alpha_2}\beta\, p\, -\, \lambda_{\alpha_2}\alpha_2\, p\, -\, \lambda_{\alpha_2} + p - \lambda_\beta)/p)}. \qquad (7.137)$$

Proof. Here we need to calculate $\left(A_0(x)|_{\lambda_{\alpha_1}=0} \right)$.
From (7.6) we have

$$\left(A_0(x)|_{\lambda_{\alpha_1}=0} \right) = \left(\int_0^x \left(A(s)|_{\lambda_{\alpha_1}=0} \right)^{p/(p-\lambda_\beta)} ds \right)^{(p-\lambda_\beta)/p}, \qquad (7.138)$$

and from (7.5) we get

$$\left(A(s)|_{\lambda_{\alpha_1}=0} \right) = \frac{(P_2(s))^{\lambda_{\alpha_2}(p-1)/p}}{(\Gamma(\beta - \alpha_2))^{\lambda_{\alpha_2}}}, \qquad (7.139)$$

also by (7.4) we have

$$P_2(s) = \int_0^s (s - t)^{(\beta - \alpha_2 - 1)p/(p-1)} dt, \ 0 \leq s \leq x. \qquad (7.140)$$

Therefore, we obtain

$$P_2(s) = \left(\frac{p-1}{\beta\, p - \alpha_2\, p - 1} \right) s^{((\beta\, p - \alpha_2\, p - 1)/(p-1))}, \qquad (7.141)$$

and

$$\left(A(s)|_{\lambda_{\alpha_1}=0} \right) = \frac{(p-1)^{((\lambda_{\alpha_2}p - \lambda_{\alpha_2})/p)} s^{((\lambda_{\alpha_2}\beta\, p - \lambda_{\alpha_2}\alpha_2\, p - \lambda_{\alpha_2})/p)}}{(\beta\, p - \alpha_2\, p - 1)^{((\lambda_{\alpha_2}p - \lambda_{\alpha_2})/p)} (\Gamma(\beta - \alpha_2))^{\lambda_{\alpha_2}}}. \qquad (7.142)$$

That is,

$$\left(A_0(x)|_{\lambda_{\alpha_1}=0}\right) = \left(\frac{(p-1)^{((\lambda_{\alpha_2}p-\lambda_{\alpha_2})/p)}}{(\beta\,p-\alpha_2 p-1)^{((\lambda_{\alpha_2}p-\lambda_{\alpha_2})/p)}(\Gamma(\beta-\alpha_2))^{\lambda_{\alpha_2}}}\right) \cdot$$

$$\left(\int_0^x s^{((\lambda_{\alpha_2}\beta\,p-\lambda_{\alpha_2}\alpha_2 p-\lambda_{\alpha_2})/(p-\lambda_\beta))}ds\right)^{((p-\lambda_\beta)/p)} \tag{7.143}$$

$$= \left\{\left(\frac{(p-1)^{((\lambda_{\alpha_2}p-\lambda_{\alpha_2})/p)}}{(\beta\,p-\alpha_2 p-1)^{((\lambda_{\alpha_2}p-\lambda_{\alpha_2})/p)}(\Gamma(\beta-\alpha_2))^{\lambda_{\alpha_2}}}\right) \cdot \right.$$

$$\left.\left(\frac{(p-\lambda_\beta)^{((p-\lambda_\beta)/p)}}{(\lambda_{\alpha_2}\beta p-\lambda_{\alpha_2}\alpha_2 p-\lambda_{\alpha_2}+p-\lambda_\beta)^{((p-\lambda_\beta)/p)}}\right)\right\}$$

$$\times x^{((\lambda_{\alpha_2}\beta\,p-\lambda_{\alpha_2}\alpha_2 p-\lambda_{\alpha_2}+p-\lambda_\beta)/p)}. \tag{7.144}$$

□

We give

Corollary 7.13. (to Theorem 7.5, $\lambda_{\alpha_1} = 0$, $p(t) = q(t) = 1$, $\lambda_{\alpha_2} = \lambda_\beta = 1$, $p = 2$). *In detail: let $\alpha_2 \in \mathbb{R}_+, \beta > \alpha_2$, $\beta - \alpha_2 > (1/2)$ and let $f_1, f_2 \in L_1(0,x), x \in \mathbb{R}_+ - \{0\}$, have, respectively, L_∞ fractional derivatives $D^\beta f_1, D^\beta f_2$ in $[0,x]$, and let $D^{\beta-k}f_i(0) = 0$, for $k = 1, \ldots, [\beta]+1; i = 1, 2$. Then*

$$\int_0^x \left[|D^{\alpha_2}f_2(s)|\,|D^\beta f_1(s)| + |D^{\alpha_2}f_1(s)|\,|D^\beta f_2(s)|\right] ds$$

$$\leq C_2^*(x)\left(\int_0^x \left[|D^\beta f_1(s)|^2 + |D^\beta f_2(s)|^2\right] ds\right), \tag{7.145}$$

where

$$C_2^*(x) := \frac{x^{(\beta-\alpha_2)}}{\sqrt{2}\;\Gamma(\beta-\alpha_2)\sqrt{\beta-\alpha_2}\;\sqrt{2\beta-2\alpha_2-1}}. \tag{7.146}$$

Proof. We apply Corollary 7.12. Here $\delta_3^{\lambda_\beta/p} = 1$, $2^{(p-\lambda_\beta)/p} = \sqrt{2}$,

$$\left(\frac{\lambda_\beta}{\lambda_{\alpha_2}+\lambda_\beta}\right)^{\lambda_\beta/p} = \frac{1}{\sqrt{2}}. \tag{7.147}$$

Furthermore, we have

$$\left(A_0(x)|_{\lambda_{\alpha_1}=0}\right) = \left(\frac{x^{(\beta-\alpha_2)}}{\sqrt{2}\;\Gamma(\beta-\alpha_2)\sqrt{\beta-\alpha_2}\;\sqrt{2\beta-2\alpha_2-1}}\right). \tag{7.148}$$

Finally, we find

$$C_2^*(x) = \left(A_0(x)|_{\lambda_{\alpha_1}=0}\right),\qquad(7.149)$$

proving the claim. □

We continue with

Corollary 7.14. (to Theorem 7.6, $\lambda_{\alpha_1} = \lambda_{\alpha_2} = \lambda_\beta = 1, p = 3, p(t) = q(t) = 1$). *It holds*

$$\int_0^x \left[|D^{\alpha_1} f_1(s)| \, |D^{\alpha_2} f_2(s)| \, \left|D^\beta f_1(s)\right| + |D^{\alpha_2} f_1(s)| \, |D^{\alpha_1} f_2(s)| \, \left|D^\beta f_2(s)\right| \right] ds$$

$$\leq A_0(x) \left(\sqrt[3]{2} + \frac{1}{\sqrt[3]{6}} \right) \left(\int_0^x \left(\left|D^\beta f_1(s)\right|^3 + \left|D^\beta f_2(s)\right|^3 \right) ds \right). \quad (7.150)$$

Here,

$$A_0(x) = \frac{4x^{(2\beta-\alpha_1-\alpha_2)}}{\Gamma(\beta-\alpha_1)\Gamma(\beta-\alpha_2)[3(3\beta-3\alpha_1-1)(3\beta-3\alpha_2-1)(2\beta-\alpha_1-\alpha_2)]^{2/3}}.$$
$$(7.151)$$

Proof. We apply inequality (7.47). Here $\tilde{\gamma}_1 = 3$, $\tilde{\gamma}_2 = 1$, and

$$\left(\frac{\lambda_\beta}{(\lambda_{\alpha_1} + \lambda_{\alpha_2})(\lambda_{\alpha_1} + \lambda_{\alpha_2} + \lambda_\beta)} \right)^{\lambda_\beta/p} = \frac{1}{\sqrt[3]{6}}. \quad (7.152)$$

Furthermore,

$$\left[\lambda_{\alpha_1}^{\lambda_\beta/p} \tilde{\gamma}_2 + 2^{(p-\lambda_\beta)/p} \left(\tilde{\gamma}_1 \lambda_{\alpha_2} \right)^{\lambda_\beta/p} \right] = 1 + \sqrt[3]{12}. \quad (7.153)$$

The product of the last two quantities is $\left(\sqrt[3]{2} + \frac{1}{\sqrt[3]{6}} \right)$. Next we find $A_0(x)$ in our case. Here, by (7.6),

$$A_0(x) = \left(\int_0^x (A(s))^{3/2} \, ds \right)^{2/3}, \quad (7.154)$$

and, by (7.5),

$$A(s) = \frac{(P_1(s))^{2/3}(P_2(s))^{2/3}}{\Gamma(\beta-\alpha_1)\Gamma(\beta-\alpha_2)}. \quad (7.155)$$

Also, by (7.4) we have

$$P_i(s) = \int_0^s (s-t)^{(\beta-\alpha_i-1)3/2} \, dt, \quad i = 1, 2. \quad (7.156)$$

That is,

$$P_i(s) = \frac{2 \, s^{(3\beta - 3\alpha_i - 1)/2}}{(3\beta - 3\alpha_i - 1)}, \quad i = 1, 2, \tag{7.157}$$

and

$$A(s) = \frac{2^{4/3} \quad s^{((6\beta - 3\alpha_1 - 3\alpha_2 - 2)/3)}}{\Gamma(\beta - \alpha_1)\Gamma(\beta - \alpha_2)(3\beta - 3\alpha_1 - 1)^{2/3}(3\beta - 3\alpha_2 - 1)^{2/3}}. \tag{7.158}$$

Finally, we get

$$A_0(x) = \left(\frac{2^{4/3}}{\Gamma(\beta - \alpha_1)\Gamma(\beta - \alpha_2)(3\beta - 3\alpha_1 - 1)^{2/3}(3\beta - 3\alpha_2 - 1)^{2/3}} \right) \cdot$$

$$\left(\int_0^x s^{((6\beta - 3\alpha_1 - 3\alpha_2 - 2)/2)} ds \right)^{2/3} =$$

$$\frac{4x^{(2\beta - \alpha_1 - \alpha_2)}}{\Gamma(\beta - \alpha_1)\Gamma(\beta - \alpha_2)[3(3\beta - 3\alpha_1 - 1)(3\beta - 3\alpha_2 - 1)(2\beta - \alpha_1 - \alpha_2)]^{2/3}}. \tag{7.159}$$

□

We give

Corollary 7.15. (to Theorem 7.8, here $\lambda_\alpha = 1, \lambda_{\alpha+1} = 1/2, p = 3/2$, $p(t) = q(t) = 1$). In detail: let $\beta > \alpha_1 + 1, \alpha_1 \in \mathbb{R}_+$ and let $f_1, f_2 \in L_1(0, x), \ x \in \mathbb{R}_+ - \{0\}$ have, respectively, L_∞ fractional derivatives $D^\beta f_1$, $D^\beta f_2$ in $[0, x]$, and let $D^{\beta-k} f_i(0) = 0$, for $k = 1, \ldots, [\beta] + 1; i = 1, 2$. Then

$$\int_0^x \left[|D^{\alpha_1} f_1(s)| \sqrt{|D^{\alpha_1 + 1} f_2(s)|} + |D^{\alpha_1} f_2(s)| \sqrt{|D^{\alpha_1 + 1} f_1(s)|} \right] ds$$

$$\leq \Phi^*(x) \left[\int_0^x \left(|D^\beta f_1(s)|^{3/2} + |D^\beta f_2(s)|^{3/2} \right) ds \right], \tag{7.160}$$

where

$$\Phi^*(x) = \frac{2 \, x^{(3\beta - 3\alpha_1 - 1)/2}}{(\Gamma(\beta - \alpha_1))^{3/2} \sqrt{3\beta - 3\alpha_1 - 2}}. \tag{7.161}$$

Proof. We apply here Theorem 7.8. We have $\theta_3 = 3, \ L(x) = \sqrt{2} \sqrt{x}$,

$$P_1(x) = \frac{x^{3\beta - 3\alpha_1 - 2}}{(3\beta - 3\alpha_1 - 2)},$$

and

$$T(x) = \sqrt{2} \left(\frac{x^{(3\beta - 3\alpha_1 - 1)/2}}{(\Gamma(\beta - \alpha_1))^{3/2} \sqrt{3\beta - 3\alpha_1 - 2}} \right), \tag{7.162}$$

also, $\omega_1 = \sqrt{2}$.

Finally, we find

$$\Phi^*(x) = \frac{2\, x^{(3\beta - 3\alpha_1 - 1)/2}}{(\Gamma(\beta - \alpha_1))^{3/2}\sqrt{3\beta - 3\alpha_1 - 2}}, \qquad (7.163)$$

which proves the claim. □

We continue with

Corollary 7.16. (to Theorem 7.9, here $p = 2(\lambda_{\alpha_1} + \lambda_\beta) > 1, p(t) = q(t) = 1$). *It holds*

$$\int_0^x \left[|D^{\alpha_1} f_1(s)|^{\lambda_{\alpha_1}} |D^{\alpha_2} f_2(s)|^{\lambda_{\alpha_1} + \lambda_\beta} |D^\beta f_1(s)|^{\lambda_\beta} \right.$$

$$\left. + |D^{\alpha_2} f_1(s)|^{\lambda_{\alpha_1} + \lambda_\beta} |D^{\alpha_1} f_2(s)|^{\lambda_{\alpha_1}} |D^\beta f_2(s)|^{\lambda_\beta} \right] ds \le$$

$$\tilde{\tilde{T}}(x) \left(\int_0^x \left(|D^\beta f_1(s)|^{2(\lambda_{\alpha_1} + \lambda_\beta)} + |D^\beta f_2(s)|^{2(\lambda_{\alpha_1} + \lambda_\beta)} \right) ds \right). \qquad (7.164)$$

Here,

$$\tilde{\tilde{T}}(x) := \tilde{A}_0(x) \left(\frac{\lambda_\beta}{2(\lambda_{\alpha_1} + \lambda_\beta)} \right)^{(\lambda_\beta / 2(\lambda_{\alpha_1} + \lambda_\beta))}, \qquad (7.165)$$

and

$$\tilde{A}_0(x) := \sigma\, \sigma^*\, x^\theta, \qquad (7.166)$$

where

$$\sigma := \frac{1}{(\Gamma(\beta - \alpha_1))^{\lambda_{\alpha_1}} (\Gamma(\beta - \alpha_2))^{\lambda_{\alpha_1} + \lambda_\beta}}$$

$$\cdot \left(\frac{2\lambda_{\alpha_1} + 2\lambda_\beta - 1}{2\lambda_{\alpha_1}\beta + 2\lambda_\beta\beta - 2\lambda_{\alpha_1}\alpha_1 - 2\lambda_\beta\alpha_1 - 1} \right)^{(2\lambda_{\alpha_1}^2 + 2\lambda_{\alpha_1}\lambda_\beta - \lambda_{\alpha_1})/(2\lambda_{\alpha_1} + 2\lambda_\beta)}$$

$$\cdot \left(\frac{2\lambda_{\alpha_1} + 2\lambda_\beta - 1}{2\lambda_{\alpha_1}\beta + 2\lambda_\beta\beta - 2\lambda_{\alpha_1}\alpha_2 - 2\lambda_\beta\alpha_2 - 1} \right)^{((2\lambda_{\alpha_1} + 2\lambda_\beta - 1)/2)}, \qquad (7.167)$$

$$\sigma_0 := [(2\lambda_{\alpha_1} + \lambda_\beta)$$

$$\left(4\lambda_{\alpha_1}^2\beta + 6\lambda_{\alpha_1}\lambda_\beta\beta - 2\lambda_{\alpha_1}^2\alpha_1 - 2\lambda_{\alpha_1}\lambda_\beta\alpha_1 - 2\lambda_{\alpha_1}^2\alpha_2 - 4\lambda_{\alpha_1}\lambda_\beta\alpha_2 + 2\lambda_\beta^2\beta - 2\lambda_\beta^2\alpha_2 \right)^{-1}],$$

$$\sigma^* := \sigma_0^{(2\lambda_{\alpha_1} + \lambda_\beta)/2(\lambda_{\alpha_1} + \lambda_\beta)}, \qquad (7.168)$$

and

$$\theta := \left(\frac{4\lambda_{\alpha_1}^2\beta + 6\lambda_{\alpha_1}\lambda_\beta\beta - 2\lambda_{\alpha_1}^2\alpha_1 - 2\lambda_{\alpha_1}\lambda_\beta\alpha_1}{2\lambda_{\alpha_1} + 2\lambda_\beta} + \right.$$

$$\frac{\left(-2\lambda_{\alpha_1}^2 \alpha_2 - 4\lambda_{\alpha_1}\lambda_\beta\alpha_2 + 2\lambda_\beta^2\beta - 2\lambda_\beta^2\alpha_2\right)}{2\lambda_{\alpha_1} + 2\lambda_\beta}\Bigg). \tag{7.169}$$

Proof. We apply Theorem 7.9. The constant $\tilde{\tilde{T}}(x)$ comes from (7.110) in our case.

Still, we need to determine $\tilde{A}_0(x)$.

From (7.6) we get here

$$\tilde{A}_0(x) = \left(\int_0^x (A(s))^{2(\lambda_{\alpha_1}+\lambda_\beta)/(2\lambda_{\alpha_1}+\lambda_\beta)} ds\right)^{((2\lambda_{\alpha_1}+\lambda_\beta)/2(\lambda_{\alpha_1}+\lambda_\beta))}, \tag{7.170}$$

where, from (7.5) we get also here

$$A(s) = \frac{\left\{P_1(s)^{\lambda_{\alpha_1}((2\lambda_{\alpha_1}+2\lambda_\beta-1)/2(\lambda_{\alpha_1}+\lambda_\beta))} P_2(s)^{((2\lambda_{\alpha_1}+2\lambda_\beta-1)/2)}\right\}}{(\Gamma(\beta-\alpha_1))^{\lambda_{\alpha_1}}(\Gamma(\beta-\alpha_2))^{\lambda_{\alpha_1}+\lambda_\beta}}, \tag{7.171}$$

and, from (7.4), we find for $i = 1, 2$ that

$$P_i(s) = \int_0^s (s-t)^{[(2\lambda_{\alpha_1}+2\lambda_\beta)(\beta-\alpha_i-1)/(2\lambda_{\alpha_1}+2\lambda_\beta-1)]} dt, \tag{7.172}$$

all $0 \le s \le x$.

Hence for $i = 1, 2$ we obtain

$$P_i(s) = \frac{(2\lambda_{\alpha_1} + 2\lambda_\beta - 1)s^{((2\lambda_{\alpha_1}\beta+2\lambda_\beta\beta-2\lambda_{\alpha_1}\alpha_i-2\lambda_\beta\alpha_i-1)/(2\lambda_{\alpha_1}+2\lambda_\beta-1))}}{(2\lambda_{\alpha_1}\beta + 2\lambda_\beta\beta - 2\lambda_{\alpha_1}\alpha_i - 2\lambda_\beta\alpha_i - 1)}, \tag{7.173}$$

and

$$A(s) = \sigma s \left\{\frac{(4\lambda_{\alpha_1}^2\beta+6\lambda_{\alpha_1}\lambda_\beta\beta-2\lambda_{\alpha_1}^2\alpha_1-2\lambda_{\alpha_1}^2\alpha_2-2\lambda_{\alpha_1}\lambda_\beta\alpha_1-4\lambda_{\alpha_1}\lambda_\beta\alpha_2-2\lambda_\beta^2\alpha_2+2\lambda_\beta^2\beta-2\lambda_{\alpha_1}-\lambda_\beta)}{(2\lambda_{\alpha_1}+2\lambda_\beta)}\right\}. \tag{7.174}$$

Finally, we find
$$\tilde{A}(x_0) = \sigma$$

$$\left(\int_0^x s^{\dfrac{(4\lambda_{\alpha_1}^2\beta + 6\lambda_{\alpha_1}\lambda_\beta\beta - 2\lambda_{\alpha_1}^2\alpha_1 - 2\lambda_{\alpha_1}^2\alpha_2 - 2\lambda_{\alpha_1}\lambda_\beta\alpha_1 - 4\lambda_{\alpha_1}\lambda_\beta\alpha_2 - 2\lambda_\beta^2\alpha_2 + 2\lambda_\beta^2\beta - 2\lambda_{\alpha_1} - \lambda_\beta)}{(2\lambda_{\alpha_1}+\lambda_\beta)}} ds\right)^{(2\lambda_{\alpha_1}+\lambda_\beta)/2(\lambda_{\alpha_1}+\lambda_\beta)}$$

$$= \sigma\sigma^* x^\theta. \tag{7.175}$$

□

We present the following interesting special case.

Corollary 7.17. (to Theorem 7.9, here $p = 4$, $\lambda_{\alpha_1} = \lambda_\beta = 1$, $p(t) = q(t) = 1$).

It holds

$$\int_0^x \left[|D^{\alpha_1} f_1(s)|(D^{\alpha_2} f_2(s))^2 |D^\beta f_1(s)| + (D^{\alpha_2} f_1(s))^2 |D^{\alpha_1} f_2(s)||D^\beta f_2(s)| \right] ds$$

$$\leq T^*(x) \left(\int_0^x \left(\left[D^\beta f_1(s) \right]^4 + \left(D^\beta f_2(s) \right)^4 \right) ds \right). \tag{7.176}$$

Here

$$T^*(x) = \frac{A_0^*(x)}{\sqrt{2}}, \tag{7.177}$$

and

$$A_0^*(x) = \tilde{\sigma}\tilde{\sigma}^* x^{\tilde{\theta}}, \tag{7.178}$$

where

$$\tilde{\sigma} \quad : \quad = \frac{1}{\Gamma(\beta - \alpha_1)(\Gamma(\beta - \alpha_2))^2} \left(\frac{3}{4\beta - 4\alpha_1 - 1} \right)^{3/4}$$

$$\left(\frac{3}{4\beta - 4\alpha_2 - 1} \right)^{3/2}, \tag{7.179}$$

$$\tilde{\sigma}^* := \left(\frac{3}{12\beta - 4\alpha_1 - 8\alpha_2} \right)^{3/4}, \tag{7.180}$$

and

$$\tilde{\theta} := 3\,\beta - \alpha_1 - 2\alpha_2. \tag{7.181}$$

Proof. By Corollary 7.16. □

We continue related results regarding the $L_\infty-$ norm $\| \cdot \|_\infty$.

Theorem 7.18. Let $\alpha_1, \alpha_2 \in \mathbb{R}_+$, $\beta > \alpha_1, \alpha_2$, and let $f_1, f_2 \in L_1(0, x)$, $x \in \mathbb{R}_+ - \{0\}$, have, respectively, L_∞ fractional derivatives $D^\beta f_1$, $D^\beta f_2$ in $[0, x]$, and let $D^{\beta-k} f_i(0) = 0$ for $k = 1, \ldots, [\beta] + 1$; $i = 1, 2$. Consider $p(s) \geq 0$, $p(s) \in L_\infty(0, x)$. Let $\lambda_{\alpha_1}, \lambda_{\alpha_2}, \lambda_\beta \geq 0$. Set

$$\rho(x) :=$$

$$\left\{ \frac{\|p(s)\|_\infty \quad x^{(\beta\lambda_{\alpha_1} - \alpha_1\lambda_{\alpha_1} + \beta\lambda_{\alpha_2} - \alpha_2\lambda_{\alpha_2} + 1)}}{(\Gamma(\beta - \alpha_1 + 1))^{\lambda_{\alpha_1}} (\Gamma(\beta - \alpha_2 + 1))^{\lambda_{\alpha_2}} [\beta\lambda_{\alpha_1} - \alpha_1\lambda_{\alpha_1} + \beta\lambda_{\alpha_2} - \alpha_2\lambda_{\alpha_2} + 1]} \right\}. \tag{7.182}$$

Then

$$\int_0^x p(s) \left[|D^{\alpha_1} f_1(s)|^{\lambda_{\alpha_1}} |D^{\alpha_2} f_2(s)|^{\lambda_{\alpha_2}} |D^\beta f_1(s)|^{\lambda_\beta} + \right.$$

$$\left. |D^{\alpha_2} f_1(s)|^{\lambda_{\alpha_2}} |D^{\alpha_1} f_2(s)|^{\lambda_{\alpha_1}} |D^\beta f_2(s)|^{\lambda_\beta} \right] ds \le$$

$$\frac{\rho(x)}{2} \left[\|D^\beta f_1\|_\infty^{2(\lambda_{\alpha_1} + \lambda_\beta)} + \|D^\beta f_1\|_\infty^{2\lambda_{\alpha_2}} + \|D^\beta f_2\|_\infty^{2\lambda_{\alpha_2}} + \|D^\beta f_2\|_\infty^{2(\lambda_{\alpha_1} + \lambda_\beta)} \right].$$

$$(7.183)$$

Proof. From (7.9) we get

$$|D^{\alpha_i} f_j(s)| \le \frac{1}{\Gamma(\beta - \alpha_i)} \int_0^s (s - t)^{\beta - \alpha_i - 1} |D^\beta f_j(t)| dt$$

$$\le \frac{1}{\Gamma(\beta - \alpha_i)} \left(\int_0^s (s - t)^{\beta - \alpha_i - 1} dt \right) \|D^\beta f_j\|_\infty \qquad (7.184)$$

$$= \frac{\|D^\beta f_j\|_\infty}{\Gamma(\beta - \alpha_i)} \frac{s^{\beta - \alpha_i}}{(\beta - \alpha_i)} = \frac{s^{\beta - \alpha_i}}{\Gamma(\beta - \alpha_i + 1)} \|D^\beta f_j\|_\infty. \qquad (7.185)$$

That is,

$$|D^{\alpha_i} f_j(s)| \le \frac{s^{\beta - \alpha_i}}{\Gamma(\beta - \alpha_i + 1)} \|D^\beta f_j\|_\infty, \qquad (7.186)$$

$\forall \quad s \in [0, x]$, $i = 1, 2$; $j = 1, 2$.
Then we have

$$|D^{\alpha_1} f_1(s)|^{\lambda_{\alpha_1}} \le \frac{s^{(\beta - \alpha_1)\lambda_{\alpha_1}}}{(\Gamma(\beta - \alpha_1 + 1))^{\lambda_{\alpha_1}}} \|D^\beta f_1\|_\infty^{\lambda_{\alpha_1}}, \qquad (7.187)$$

$$|D^{\alpha_2} f_2(s)|^{\lambda_{\alpha_2}} \le \frac{s^{(\beta - \alpha_2)\lambda_{\alpha_2}}}{(\Gamma(\beta - \alpha_2 + 1))^{\lambda_{\alpha_2}}} \|D^\beta f_2\|_\infty^{\lambda_{\alpha_2}}, \qquad (7.188)$$

$$|D^\beta f_1(s)|^{\lambda_\beta} \le \|D^\beta f_1\|_\infty^{\lambda_\beta}. \qquad (7.189)$$

Multiplying the last three inequalities (7.187) through (7.189), we obtain

$$|D^{\alpha_1} f_1(s)|^{\lambda_{\alpha_1}} |D^{\alpha_2} f_2(s)|^{\lambda_{\alpha_2}} |D^\beta f_1(s)|^{\lambda_\beta} \le$$

$$\frac{s^{(\beta\lambda_{\alpha_1} - \alpha_1\lambda_{\alpha_1} + \beta\lambda_{\alpha_2} - \alpha_2\lambda_{\alpha_2})}}{(\Gamma(\beta - \alpha_1 + 1))^{\lambda_{\alpha_1}} (\Gamma(\beta - \alpha_2 + 1))^{\lambda_{\alpha_2}}} \|D^\beta f_1\|_\infty^{\lambda_{\alpha_1} + \lambda_\beta} \|D^\beta f_2\|_\infty^{\lambda_{\alpha_2}}. \quad (7.190)$$

Similarly we see that

$$|D^{\alpha_2} f_1(s)|^{\lambda_{\alpha_2}} \le \frac{s^{(\beta - \alpha_2)\lambda_{\alpha_2}}}{(\Gamma(\beta - \alpha_2 + 1))^{\lambda_{\alpha_2}}} \|D^\beta f_1\|_\infty^{\lambda_{\alpha_2}}, \qquad (7.191)$$

$$|D^{\alpha_1} f_2(s)|^{\lambda_{\alpha_1}} \le \frac{s^{(\beta - \alpha_1)\lambda_{\alpha_1}}}{(\Gamma(\beta - \alpha_1 + 1))^{\lambda_{\alpha_1}}} \|D^\beta f_2\|_\infty^{\lambda_{\alpha_1}}, \qquad (7.192)$$

$$|D^\beta f_2(s)|^{\lambda_\beta} \le \|D^\beta f_2\|_\infty^{\lambda_\beta}. \tag{7.193}$$

Multiplying the last three inequalities (7.191) through (7.193), we get

$$|D^{\alpha_2} f_1(s)|^{\lambda_{\alpha_2}} |D^{\alpha_1} f_2(s)|^{\lambda_{\alpha_1}} |D^\beta f_2(s)|^{\lambda_\beta} \le$$

$$\frac{s^{(\beta\lambda_{\alpha_2} - \alpha_2\lambda_{\alpha_2} + \beta\lambda_{\alpha_1} - \alpha_1\lambda_{\alpha_1})}}{(\Gamma(\beta - \alpha_2 + 1))^{\lambda_{\alpha_2}} (\Gamma(\beta - \alpha_1 + 1))^{\lambda_{\alpha_1}}} \|D^\beta f_1\|_\infty^{\lambda_{\alpha_2}} \|D^\beta f_2\|_\infty^{\lambda_{\alpha_1} + \lambda_\beta}. \tag{7.194}$$

Adding (7.190), (7.194) produces

$$|D^{\alpha_1} f_1(s)|^{\lambda_{\alpha_1}} |D^{\alpha_2} f_2(s)|^{\lambda_{\alpha_2}} |D^\beta f_1(s)|^{\lambda_\beta} + |D^{\alpha_2} f_1(s)|^{\lambda_{\alpha_2}} |D^{\alpha_1} f_2(s)|^{\lambda_{\alpha_1}} |D^\beta f_2(s)|^{\lambda_\beta}$$

$$\le \frac{s^{(\beta\lambda_{\alpha_1} - \alpha_1\lambda_{\alpha_1} + \beta\lambda_{\alpha_2} - \alpha_2\lambda_{\alpha_2})}}{(\Gamma(\beta - \alpha_1 + 1))^{\lambda_{\alpha_1}} (\Gamma(\beta - \alpha_2 + 1))^{\lambda_{\alpha_2}}} \cdot$$

$$\left[\|D^\beta f_1\|_\infty^{\lambda_{\alpha_1} + \lambda_\beta} \|D^\beta f_2\|_\infty^{\lambda_{\alpha_2}} + \|D^\beta f_1\|_\infty^{\lambda_{\alpha_2}} \|D^\beta f_2\|_\infty^{\lambda_{\alpha_1} + \lambda_\beta} \right]. \tag{7.195}$$

It follows that

$$\int_0^x p(s) \left[|D^{\alpha_1} f_1(s)|^{\lambda_{\alpha_1}} |D^{\alpha_2} f_2(s)|^{\lambda_{\alpha_2}} |D^\beta f_1(s)|^{\lambda_\beta} + \right.$$

$$\left. |D^{\alpha_2} f_1(s)|^{\lambda_{\alpha_2}} |D^{\alpha_1} f_2(s)|^{\lambda_{\alpha_1}} |D^\beta f_2(s)|^{\lambda_\beta} \right] ds \le$$

$$\frac{\|p(s)\|_\infty \left(\int_0^x s^{(\beta\lambda_{\alpha_1} - \alpha_1\lambda_{\alpha_1} + \beta\lambda_{\alpha_2} - \alpha_2\lambda_{\alpha_2})} ds \right)}{(\Gamma(\beta - \alpha_1 + 1))^{\lambda_{\alpha_1}} (\Gamma(\beta - \alpha_2 + 1))^{\lambda_{\alpha_2}}} \cdot$$

$$\left[\|D^\beta f_1\|_\infty^{\lambda_{\alpha_1} + \lambda_\beta} \|D^\beta f_2\|_\infty^{\lambda_{\alpha_2}} + \|D^\beta f_1\|_\infty^{\lambda_{\alpha_2}} \|D^\beta f_2\|_\infty^{\lambda_{\alpha_1} + \lambda_\beta} \right] \tag{7.196}$$

$$= \frac{\|p(s)\|_\infty \; x^{(\beta\lambda_{\alpha_1} - \alpha_1\lambda_{\alpha_1} + \beta\lambda_{\alpha_2} - \alpha_2\lambda_{\alpha_2} + 1)}}{(\Gamma(\beta - \alpha_1 + 1))^{\lambda_{\alpha_1}} (\Gamma(\beta - \alpha_2 + 1))^{\lambda_{\alpha_2}} [\beta\lambda_{\alpha_1} - \alpha_1\lambda_{\alpha_1} + \beta\lambda_{\alpha_2} - \alpha_2\lambda_{\alpha_2} + 1]} \cdot$$

$$\left[\|D^\beta f_1\|_\infty^{\lambda_{\alpha_1} + \lambda_\beta} \|D^\beta f_2\|_\infty^{\lambda_{\alpha_2}} + \|D^\beta f_1\|_\infty^{\lambda_{\alpha_2}} \|D^\beta f_2\|_\infty^{\lambda_{\alpha_1} + \lambda_\beta} \right] \tag{7.197}$$

$$\le \frac{\rho(x)}{2} \left[\|D^\beta f_1\|_\infty^{2(\lambda_{\alpha_1} + \lambda_\beta)} + \|D^\beta f_2\|_\infty^{2\lambda_{\alpha_2}} + \|D^\beta f_1\|_\infty^{2\lambda_{\alpha_2}} + \|D^\beta f_2\|_\infty^{2(\lambda_{\alpha_1} + \lambda_\beta)} \right]. \tag{7.198}$$

We have proved (7.183). $\square$

We present some special cases of the last theorem.

Theorem 7.19. (all are as in Theorem 7.18; $\lambda_{\alpha_2} = 0$).
It holds

$$\int_0^x p(s) \left[|D^{\alpha_1} f_1(s)|^{\lambda_{\alpha_1}} |D^\beta f_1(s)|^{\lambda_\beta} + |D^{\alpha_1} f_2(s)|^{\lambda_{\alpha_1}} |D^\beta f_2(s)|^{\lambda_\beta} \right] ds \le$$

$$\frac{\|p(s)\|_\infty \; x^{(\beta\lambda_{\alpha_1} - \alpha_1\lambda_{\alpha_1} + 1)}}{(\Gamma(\beta - \alpha_1 + 1))^{\lambda_{\alpha_1}} [\beta\lambda_{\alpha_1} - \alpha_1\lambda_{\alpha_1} + 1]} \left[\|D^\beta f_1\|_\infty^{\lambda_{\alpha_1} + \lambda_\beta} + \|D^\beta f_2\|_\infty^{\lambda_{\alpha_1} + \lambda_\beta} \right]. \tag{7.199}$$

Proof. As in Theorem 7.18, similarly. □

We continue with

Theorem 7.20. (all are as in Theorem 7.18; $\lambda_{\alpha_2} = \lambda_{\alpha_1} + \lambda_\beta$).
It holds

$$\int_0^x p(s) \left[|D^{\alpha_1} f_1(s)|^{\lambda_{\alpha_1}} |D^{\alpha_2} f_2(s)|^{\lambda_{\alpha_1}+\lambda_\beta} |D^\beta f_1(s)|^{\lambda_\beta} + \right.$$

$$\left. |D^{\alpha_2} f_1(s)|^{\lambda_{\alpha_1}+\lambda_\beta} |D^{\alpha_1} f_2(s)|^{\lambda_{\alpha_1}} |D^\beta f_2(s)|^{\lambda_\beta} \right] ds \leq$$

$$\left\{ \frac{\|p(s)\|_\infty}{(\Gamma(\beta - \alpha_1 + 1))^{\lambda_{\alpha_1}} (\Gamma(\beta - \alpha_2 + 1))^{\lambda_{\alpha_1}+\lambda_\beta}} \right.$$

$$\left. \frac{x^{(2\beta\lambda_{\alpha_1} - \alpha_1\lambda_{\alpha_1} + \beta\lambda_\beta - \alpha_2\lambda_{\alpha_1} - \alpha_2\lambda_\beta + 1)}}{(2\beta\lambda_{\alpha_1} - \alpha_1\lambda_{\alpha_1} + \beta\lambda_\beta - \alpha_2\lambda_{\alpha_1} - \alpha_2\lambda_\beta + 1)} \right\}$$

$$\left[\|D^\beta f_1\|_\infty^{2(\lambda_{\alpha_1}+\lambda_\beta)} + \|D^\beta f_2\|_\infty^{2(\lambda_{\alpha_1}+\lambda_\beta)} \right]. \tag{7.200}$$

Proof. As in Theorem 7.18, similarly. □

We give

Theorem 7.21. (all are as in Theorem 7.18; $\lambda_\beta = 0$, $\lambda_{\alpha_1} = \lambda_{\alpha_2}$).
It holds

$$\int_0^x p(s) \left[|D^{\alpha_1} f_1(s)|^{\lambda_{\alpha_1}} |D^{\alpha_2} f_2(s)|^{\lambda_{\alpha_1}} + |D^{\alpha_2} f_1(s)|^{\lambda_{\alpha_1}} |D^{\alpha_1} f_2(s)|^{\lambda_{\alpha_1}} \right] ds$$

$$\leq \rho^*(x) \left[\|D^\beta f_1\|_\infty^{2\lambda_{\alpha_1}} + \|D^\beta f_2\|_\infty^{2\lambda_{\alpha_1}} \right], \tag{7.201}$$

where

$$\rho^*(x) := \left\{ \frac{\|p(s)\|_\infty \quad x^{(2\beta\lambda_{\alpha_1} - \alpha_1\lambda_{\alpha_1} - \alpha_2\lambda_{\alpha_1} + 1)}}{(\Gamma(\beta - \alpha_1 + 1)\Gamma(\beta - \alpha_2 + 1))^{\lambda_{\alpha_1}} (2\beta\lambda_{\alpha_1} - \alpha_1\lambda_{\alpha_1} - \alpha_2\lambda_{\alpha_1} + 1)} \right\}. \tag{7.202}$$

Proof. As in Theorem 7.18, when $\lambda_\beta = 0$, we follow the proof and we obtain

$$\int_0^x p(s) \left[|D^{\alpha_1} f_1(s)|^{\lambda_{\alpha_1}} |D^{\alpha_2} f_2(s)|^{\lambda_{\alpha_2}} + |D^{\alpha_2} f_1(s)|^{\lambda_{\alpha_2}} |D^{\alpha_1} f_2(s)|^{\lambda_{\alpha_1}} \right] ds$$

$$\leq \rho(x) \left[\|D^\beta f_1\|_\infty^{\lambda_{\alpha_1}} \|D^\beta f_2\|_\infty^{\lambda_{\alpha_2}} + \|D^\beta f_1\|_\infty^{\lambda_{\alpha_2}} \|D^\beta f_2\|_\infty^{\lambda_{\alpha_1}} \right], \tag{7.203}$$

which inequality is in itself of interest, and so on. Then setting $\lambda_{\alpha_1} = \lambda_{\alpha_2}$ into (7.203) and (7.182), we finally derive (7.201) and (7.202). $\square$

We give

Theorem 7.22. (all are as in Theorem 7.18; $\lambda_{\alpha_1} = 0$, $\lambda_{\alpha_2} = \lambda_\beta$). It holds

$$\int_0^x p(s) \left[|D^{\alpha_2} f_2(s)|^{\lambda_{\alpha_2}} |D^\beta f_1(s)|^{\lambda_{\alpha_2}} + |D^{\alpha_2} f_1(s)|^{\lambda_{\alpha_2}} |D^\beta f_2(s)|^{\lambda_{\alpha_2}} \right] ds$$

$$\leq \left(\frac{x^{(\beta\lambda_{\alpha_2} - \alpha_2\lambda_{\alpha_2} + 1)} \|p(s)\|_\infty}{(\beta\,\lambda_{\alpha_2} - \alpha_2\lambda_{\alpha_2} + 1)(\Gamma(\beta - \alpha_2 + 1))^{\lambda_{\alpha_2}}} \right) \left[\|D^\beta f_1\|_\infty^{2\lambda_{\alpha_2}} + \|D^\beta f_2\|_\infty^{2\lambda_{\alpha_2}} \right].$$

(7.204)

Proof. As in Theorem 7.18. $\square$

We continue with

Corollary 7.23. (to Theorem 7.21; all are as in Theorem 7.18; $\lambda_\beta = 0$, $\lambda_{\alpha_1} = \lambda_{\alpha_2}, \alpha_2 = \alpha_1 + 1$).
It holds

$$\int_0^x p(s) \left[|D^{\alpha_1} f_1(s)|^{\lambda_{\alpha_1}} |D^{\alpha_1+1} f_2(s)|^{\lambda_{\alpha_1}} + |D^{\alpha_1+1} f_1(s)|^{\lambda_{\alpha_1}} |D^{\alpha_1} f_2(s)|^{\lambda_{\alpha_1}} \right] ds$$

$$\leq \left(\frac{x^{(2\beta\lambda_{\alpha_1} - 2\alpha_1\lambda_{\alpha_1} - \lambda_{\alpha_1} + 1)} \|p(s)\|_\infty}{(2\beta\,\lambda_{\alpha_1} - 2\alpha_1\lambda_{\alpha_1} - \lambda_{\alpha_1} + 1)(\beta - \alpha_1)^{\lambda_{\alpha_1}} (\Gamma(\beta - \alpha_1))^{2\lambda_{\alpha_1}}} \right) \cdot$$

$$\left[\|D^\beta f_1\|_\infty^{2\lambda_{\alpha_1}} + \|D^\beta f_2\|_\infty^{2\lambda_{\alpha_1}} \right].$$

(7.205)

Proof. By Theorem 7.21. $\square$

We give

Corollary 7.24. (to Corollary 7.23) *In detail: let $\alpha_1 \in \mathbb{R}_+, \beta > \alpha_1 + 1$ and let $f_1, f_2 \in L_1(0,x)$, $x \in \mathbb{R}_+ - \{0\}$, have, respectively, L_∞ fractional derivative $D^\beta f_1, D^\beta f_2$ in $[0, x]$, and let $D^{\beta-k} f_i(0) = 0$, for $k = 1, \ldots, [\beta] + 1; i = 1, 2$.*
Then

$$\int_0^x \left[|D^{\alpha_1} f_1(s)| \, |D^{\alpha_1+1} f_2(s)| + |D^{\alpha_1+1} f_1(s)| \, |D^{\alpha_1} f_2(s)| \right] ds$$

$$\leq \left(\frac{x^{2(\beta-\alpha_1)}}{2(\beta-\alpha_1)^2(\Gamma(\beta-\alpha_1))^2} \right) \left[\|D^\beta f_1\|_\infty^2 + \|D^\beta f_2\|_\infty^2 \right]. \qquad (7.206)$$

Proof. In (7.205) set $\lambda_{\alpha_1} = 1$ and $p(s) = 1$.   □

Proposition 7.25. *Inequality (7.206) is sharp; in fact it is attained when* $f_1 = f_2$.

Proof. Clearly (7.206) when $f_1 = f_2$ collapses to

$$\int_0^x |D^{\alpha_1} f_1(s)| \, |D^{\alpha_1+1} f_1(s)| ds \leq \left(\frac{x^{2(\beta-\alpha_1)}}{2(\beta-\alpha_1)^2(\Gamma(\beta-\alpha_1))^2} \right) \|D^\beta f_1\|_\infty^2. \qquad (7.207)$$

In [64] (see Theorem 2.5, Remark 2.6) we proved that (7.207) is sharp; in fact it is attained by $f_1(s) = s^\beta$.   □

7.4 Applications

We present a uniqueness of solution result for a system of fractional differential equations involving Riemann–Liouville fractional derivatives.

Theorem 7.26. *Let* $\alpha_i \in \mathbb{R}_+$, $\beta > \alpha_i$, $\beta - \alpha_i > (1/2)$, $i = 1, \ldots, r \in \mathbb{N}$, *and let* $f_j \in L_1(0,x)$, $j = 1,2$; $x \in \mathbb{R}_+ - \{0\}$, *have, respectively,* L_∞ *fractional derivative* $D^\beta f_j$ *in* $[0,x]$, *and let* $D^{\beta-k} f_j(0) = a_{kj} \in \mathbb{R}$ *for* $k = 1, \ldots, [\beta]+1; j = 1,2$. *Furthermore for* $j = 1,2$ *we have that*

$$D^\beta f_j(s) = F_j \left(s, \{D^{\alpha_i} f_1(s)\}_{i=1}^r, \{D^{\alpha_i} f_2(s)\}_{i=1}^r \right), \qquad (7.208)$$

all $s \in [0,x]$.

Here $F_j(s, z_1, \ldots, z_r, y_1, \ldots, y_r)$ *are continuous for* $(z_1, \ldots, z_r, y_1, \ldots, y_r)$ $\in \mathbb{R}^{2r}$, *bounded for* $s \in [0,x]$, *and satisfy the Lipschitz condition*

$$|F_j(s, z_1, \ldots, z_r, y_1, \ldots, y_r) - F_j(s, z_1', \ldots, z_r', y_1', \ldots, y_r')| \leq$$

$$\sum_{i=1}^r [q_{1,i,j}(s)|z_i - z_i'| + q_{2,i,j}(s)|y_i - y_i'|], \qquad (7.209)$$

$j = 1,2$, *where* $q_{1,i,j}(s)$, $q_{2,i,j}(s) \geq 0$ *are bounded on* $[0,x]$, $1 \leq i \leq r$. *Call*

$$M_{1,i} := \max(\|q_{1,i,1}\|_\infty, \|q_{2,i,2}\|_\infty) \text{ and } M_{2,i} := \max(\|q_{2,i,1}\|_\infty, \|q_{1,i,2}\|_\infty). \qquad (7.210)$$

Assume here that

$$\phi^*(x) := \sum_{i=1}^{r} \left(\frac{M_{1,i}}{2} + \frac{M_{2,i}}{\sqrt{2}} \right) \left(\frac{x^{\beta-\alpha_i}}{\Gamma(\beta-\alpha_i)\sqrt{\beta-\alpha_i}\sqrt{2\beta-2\alpha_i-1}} \right) < 1.$$

$$(7.211)$$

Then, if the system (7.208) has two pairs of solutions (f_1, f_2) and (f_1^*, f_2^*), we prove that $f_j = f_j^*, j = 1, 2$; that is, we have the uniqueness of solution property for the fractional system (7.208).

Proof. Assume there are two pairs of solutions (f_1, f_2), (f_1^*, f_2^*) to system (7.208). Set $g_j := f_j - f_j^*$, $j = 1, 2$. Then

$$D^{\beta-k} g_j(0) = 0, k = 1, ..., [\beta] + 1; \; j = 1, 2. \quad (7.212)$$

It holds

$$D^{\beta} g_j(s) = F_j \left(s, \{D^{\alpha_i} f_1(s)\}_{i=1}^{r}, \{D^{\alpha_i} f_2(s)\}_{i=1}^{r} \right)$$
$$- F_j \left(s, \{D^{\alpha_i} f_1^*(s)\}_{i=1}^{r}, \{D^{\alpha_i} f_2^*(s)\}_{i=1}^{r} \right). \quad (7.213)$$

Hence by (7.209) we have

$$|D^{\beta} g_j(s)| \le \sum_{i=1}^{r} [g_{1,i,j}(s)|D^{\alpha_i} g_1(s)| + g_{2,i,j}(s)|D^{\alpha_i} g_2(s)|]. \quad (7.214)$$

Thus

$$|D^{\beta} g_j(s)| \le \sum_{i=1}^{r} [\|g_{1,i,j}\|_\infty |D^{\alpha_i} g_1(s)| + \|g_{2,i,j}\|_\infty |D^{\alpha_i} g_2(s)|]. \quad (7.215)$$

The last implies that

$$(D^{\beta} g_j(s))^2 \le \sum_{i=1}^{r} [\|g_{1,i,j}\|_\infty |D^{\alpha_i} g_1(s)| \; |D^{\beta} g_j(s)| +$$

$$\|g_{2,i,j}\|_\infty |D^{\alpha_i} g_2(s)| \; |D^{\beta} g_j(s)|], \quad j = 1, 2. \quad (7.216)$$

Consequently we see that

$$I := \int_0^x \left((D^{\beta} g_1(s))^2 + (D^{\beta} g_2(s))^2 \right) ds \le$$

$$\sum_{i=1}^{r} \left[\|g_{1,i,1}\|_\infty \int_0^x |D^{\alpha_i} g_1(s)| \; |D^{\beta} g_1(s)| ds + \|g_{2,i,1}\|_\infty \int_0^x |D^{\alpha_i} g_2(s)| \; |D^{\beta} g_1(s)| ds \right]$$

$$+ \sum_{i=1}^{r} \left[\|g_{1,i,2}\|_\infty \int_0^x |D^{\alpha_i} g_1(s)| \; |D^{\beta} g_2(s)| ds + \right.$$

$$\left. \|g_{2,i,2}\|_\infty \int_0^x |D^{\alpha_i} g_2(s)| \, |D^\beta g_2(s)| ds \right] \tag{7.217}$$

$$\leq \sum_{i=1}^r M_{1,i} \left(\int_0^x \left[|D^{\alpha_i} g_1(s)| \, |D^\beta g_1(s)| + |D^{\alpha_i} g_2(s)| \, |D^\beta g_2(s)| \right] ds \right)$$

$$+ \sum_{i=1}^r M_{2,i} \left(\int_0^x \left[|D^{\alpha_i} g_2(s)| \, |D^\beta g_1(s)| + |D^{\alpha_i} g_1(s)| \, |D^\beta g_2(s)| \right] ds \right) =: (*). \tag{7.218}$$

However by Corollary 7.11, (7.130) we have

$$\int_0^x \left[|D^{\alpha_i} g_1(s)| \, |D^\beta g_1(s)| + |D^{\alpha_i} g_2(s)| \, |D^\beta g_2(s)| \right] ds$$

$$\leq \left(\frac{x^{(\beta - \alpha_i)}}{2\Gamma(\beta - \alpha_i) \sqrt{\beta - \alpha_i} \sqrt{2\beta - 2\alpha_i - 1}} \right) I. \tag{7.219}$$

Also by Corollary 7.13, (7.145) and (7.146) we find

$$\int_0^x \left[|D^{\alpha_i} g_2(s)| \, |D^\beta g_1(s)| + |D^{\alpha_i} g_1(s)| \, |D^\beta g_2(s)| \right] ds$$

$$\leq \left(\frac{x^{(\beta - \alpha_i)}}{\sqrt{2}\Gamma(\beta - \alpha_i) \sqrt{\beta - \alpha_i} \sqrt{2\beta - 2\alpha_i - 1}} \right) I. \tag{7.220}$$

Therefore by (7.219) and (7.220) we get

$$(*) \leq \sum_{i=1}^r M_{1,i} \left(\frac{x^{(\beta - \alpha_i)}}{2\Gamma(\beta - \alpha_i) \sqrt{\beta - \alpha_i} \sqrt{2\beta - 2\alpha_i - 1}} \frac{I}{} \right) + \tag{7.221}$$

$$\sum_{i=1}^r M_{2,i} \left(\frac{x^{(\beta - \alpha_i)}}{\sqrt{2}\Gamma(\beta - \alpha_i) \sqrt{\beta - \alpha_i} \sqrt{2\beta - 2\alpha_i - 1}} \frac{I}{} \right) = \phi^*(x) I. \tag{7.222}$$

We have established that

$$I \leq \phi^*(x) \, I. \tag{7.223}$$

If $I \neq 0$ then $\phi^*(x) \geq 1$, a contradiction by assumption (7.211) that $\phi^*(x) < 1$. Therefore $I = 0$, implying that

$$(D^\beta g_1(s))^2 + (D^\beta g_2(s))^2 = 0, \text{ a.e. } in \, [0, x]. \tag{7.224}$$

That is,

$$(D^\beta g_1(s))^2 = 0, (D^\beta g_2(s))^2 = 0, \text{ a.e. } in \, [0, x]. \tag{7.225}$$

That is,

$$D^\beta g_1(s) = 0, \; D^\beta g_2(s) = 0, \text{ a.e. } in \, [0, x]. \tag{7.226}$$

By (7.212) and Lemma 7.3, apply (7.3) for $\alpha = 0$; we find $g_1(s) \equiv g_2(s) \equiv 0$ over $[0, x]$.

The last implies $f_j = f_j^*, j = 1, 2$, over $[0, x]$, thus proving the uniqueness of solution to the initial value problem of this theorem. $\square$

Next we give upper bounds on $D^\beta f_j$, solutions f_j, and so on, all involved in a system of fractional differential equations involving Riemann–Liouville fractional derivatives.

Theorem 7.27. *Let* $\alpha_i \in \mathbb{R}_+$, $\beta > \alpha_i$, $\beta - \alpha_i > (1/2)$, $i = 1, \ldots, r \in \mathbb{N}$, *and let* $f_j \in L_1(0, x), j = 1, 2$; $x \in \mathbb{R}_+ - \{0\}$, *have, respectively, fractional derivative* $D^\beta f_j$ *in* $[0, x]$ *that is absolutely continuous on* $[0, x]$, *and let* $D^{\beta-k} f_j(0) = 0$ *for* $k = 1, \ldots, [\beta] + 1; j = 1, 2$. *Furthermore* $(D^\beta f_j)(0) = A_j \in \mathbb{R}$.

Furthermore for $0 \leq s \leq x$ *we have holding the system of fractional differential equations*

$$\left(D^\beta f_j\right)'(s) = F_j\left(s, \{D^{\alpha_i} f_1(s)\}_{i=1}^r, \left(D^\beta f_1\right)(s), \{D^{\alpha_i} f_2(s)\}_{i=1}^r, \left(D^\beta f_2\right)(s)\right),$$

$$(7.227)$$

for $j = 1, 2$. *Here* F_j *is Lebesgue measurable on* $[0, x] \times \mathbb{R}^{r+1} \times \mathbb{R}^{r+1}$ *such that*

$$|F_j(s, x_1 x_2, \ldots, x_r, x_{r+1}, y_1, \ldots, y_r, y_{r+1})| \leq \sum_{i=1}^r \left[q_{1,i,j}(s) \cdot |x_i| + q_{2,i,j}(s)|y_i|\right],$$

$$(7.228)$$

where $q_{1,i,j}(s), q_{2,i,j}(s) \geq 0$, $1 \leq i \leq r$, $j = 1, 2$, *are bounded on* [0,x]. *Call*

$$M_{1,i} := \max(\|q_{1,i,1}\|_\infty, \|q_{2,i,2}\|_\infty), \tag{7.229}$$

and

$$M_{2,i} := \max(\|q_{2,i,1}\|_\infty, \|q_{1,i,2}\|_\infty). \tag{7.230}$$

Also we set $(0 \leq s < x)$

$$\theta(s) := ((D^\beta f_1)(s))^2 + ((D^\beta f_2)(s))^2, \tag{7.231}$$

$$\rho := A_1^2 + A_2^2, \tag{7.232}$$

$$Q(s) := \sum_{i=1}^r (M_{1,i} + \sqrt{2} \, M_{2,i}) \left(\frac{s^{\beta - \alpha_i}}{\Gamma(\beta - \alpha_i)\sqrt{\beta - \alpha_i}\sqrt{2\beta - 2\alpha_i - 1}}\right),$$

$$(7.233)$$

and

$$\chi(s) := \sqrt{\rho} \cdot \left\{1 + Q(s) \cdot e^{(\int_0^s Q(t)dt)} \cdot \left[\int_0^s \left(e^{-(\int_0^z Q(t)dt)}\right) dz\right]\right\}^{1/2}. \tag{7.234}$$

Then

$$\sqrt{\theta(s)} \leq \chi(s), \quad \text{all} \quad 0 \leq s \leq x. \tag{7.235}$$

Consequently it holds

$$|D^\beta f_j(s)| \le \chi(s), \tag{7.236}$$

$$|f_j(s)| \le \frac{1}{\Gamma(\beta)} \int_0^s (s-t)^{\beta-1}\chi(t)dt, \tag{7.237}$$

for $j = 1,2, \quad 0 \le s \le x.$
Also it holds that

$$|D^{\alpha_i} f_j(s)| \le \frac{1}{\Gamma(\beta-\alpha_i)} \int_0^s (s-t)^{\beta-\alpha_i-1}\chi(t)dt, \tag{7.238}$$

for $j = 1,2, \ i = 1,\ldots,r, \ \ all\ 0 \le s \le x.$

Proof. We observe that

$$(D^\beta f_j)(s)(D^\beta f_j)'(s) = \left(D^\beta f_j\right)(s) \cdot F_j\left(s, \{D^{\alpha_i} f_1(s)\}_{i=1}^r,\right.$$

$$\left(D^\beta f_1\right)(s), \{D^{\alpha_i} f_2(s)\}_{i=1}^r, \left(D^\beta f_2\right)(s)\right), \tag{7.239}$$

$j = 1,2, \ all\ 0 \le s \le x.$
We then integrate (7.239),

$$\int_0^y (D^\beta f_j)(s)(D^\beta f_j)'(s)ds = \int_0^y (D^\beta f_j)(s)\cdot F_j\left(s, \{D^{\alpha_i} f_1(s)\}_{i=1}^r, \left(D^\beta f_1\right)(s),\right.$$

$$\{D^{\alpha_i} f_2(s)\}_{i=1}^r, \left(D^\beta f_2\right)(s)\right) ds, \ 0 \le y \le x. \tag{7.240}$$

Hence we obtain

$$\left.\frac{((D^\beta f_j)(s))^2}{2}\right|_0^y \le \int_0^y |D^\beta f_j(s)|\ |F_j\cdots|ds \overset{(7.228)}{\le}$$

$$\int_0^y |D^\beta f_j(s)| \left[\sum_{i=1}^r [g_{1,i,j}(s)|D^{\alpha_i} f_1(s)| + g_{2,i,j}(s)|D^{\alpha_i} f_2(s)|]\right] ds \tag{7.241}$$

$$\le \sum_{i=1}^r \left\{\|g_{1,i,j}\|_\infty \left(\int_0^y |D^\beta f_j(s)|\ |D^{\alpha_i} f_1(s)|ds\right)\right.$$

$$\left. + \|g_{2,i,j}\|_\infty \left(\int_0^y |D^\beta f_j(s)|\ |D^{\alpha_i} f_2(s)|ds\right)\right\}. \tag{7.242}$$

Thus we find

$$((D^\beta f_j(y))^2 \le A_j^2 + 2\sum_{i=1}^r \left\{\|g_{1,i,j}\|_\infty \left(\int_0^y |D^\beta f_j(s)|\ |D^{\alpha_i} f_1(s)|ds\right)\right.$$

$$\left. + \|g_{2,i,j}\|_\infty \left(\int_0^y |D^\beta f_j(s)|\ |D^{\alpha_i} f_2(s)|ds\right)\right\}. \tag{7.243}$$

Consequently we write

$$\theta(y) \le \rho+2\sum_{i=1}^{r}\left\{M_{1,i}\int_{0}^{y}\left[|D^{\beta}f_1(s)|\,|D^{\alpha_i}f_1(s)|+|D^{\beta}f_2(s)|\,|D^{\alpha_i}f_2(s)|\right]ds\right.$$

$$\left.+M_{2,i}\int_{0}^{y}\left[|D^{\beta}f_2(s)|\,|D^{\alpha_i}f_1(s)|+|D^{\beta}f_1(s)|\,|D^{\alpha_i}f_2(s)|\right]ds\right\}\qquad(7.244)$$

$$\overset{(by\ (7.130),(7.145))}{\le}\rho+2\sum_{i=1}^{r}\left\{M_{1,i}\left(\frac{y^{(\beta-\alpha_i)}\int_{0}^{y}\theta(s)ds}{2\Gamma(\beta-\alpha_i)\sqrt{\beta-\alpha_i}\sqrt{2\beta-2\alpha_i-1}}\right)\right.$$

$$\left.+M_{2,i}\left(\frac{y^{(\beta-\alpha_i)}\int_{0}^{y}\theta(s)ds}{\sqrt{2}\Gamma(\beta-\alpha_i)\sqrt{\beta-\alpha_i}\sqrt{2\beta-2\alpha_i-1}}\right)\right\}\qquad(7.245)$$

$$=\rho+Q(y)\int_{0}^{y}\theta(s)ds.\qquad(7.246)$$

That is, we get

$$\theta(y)\le\rho+Q(y)\int_{0}^{y}\theta(s)ds,\qquad(7.247)$$

all $0\le y\le x$. Here $\rho\ge 0, Q(y)\ge 0, Q(0)=0, \theta(y)\ge 0$, all $0\le y\le x$.
Solving the integral inequality (7.247), exactly as in [17] (see pp. 224–225,
Application 3.2 and inequalities (44), (47) there) we find that

$$\sqrt{\theta(y)}\le\chi(y),\quad 0\le y\le x,\qquad(7.248)$$

proving (7.235).

Then (7.236) is obvious.

Next from (7.3), the case of $\alpha=0$, we obtain

$$|f_j(s)|\le\frac{1}{\Gamma(\beta)}\int_{0}^{s}(s-t)^{\beta-1}|D^{\beta}f_j(t)|dt\le\frac{1}{\Gamma(\beta)}\int_{0}^{s}(s-t)^{\beta-1}\chi(t)dt,$$
$$(7.249)$$

proving (7.237).

Finally, again from (7.3), we find

$$|D^{\alpha_i}f_j(s)|\le\frac{1}{\Gamma(\beta-\alpha_i)}\int_{0}^{s}(s-t)^{\beta-\alpha_i-1}|D^{\beta}f_j(t)|dt\qquad(7.250)$$

$$\le\frac{1}{\Gamma(\beta-\alpha_i)}\int_{0}^{s}(s-t)^{\beta-\alpha_i-1}\chi(t)dt,\qquad(7.251)$$

thus proving (7.238). $\square$

In the last application we give similar upper bounds as in Theorem 7.27,
but under different conditions.

Theorem 7.28. *Let $\alpha_i \in \mathbb{R}_+, \beta > \alpha_i \; i = 1, \ldots, r \in \mathbb{N}$, and $f_j \in L_1(0, x)$, $j = 1, 2$; $x \in \mathbb{R}_+ - \{0\}$, have, respectively, fractional derivative $D^\beta f_j$ in $[0, x]$ that is absolutely continuous on $[0, x]$, and let $D^{\beta - \mu} f_j(0) = 0$ for $\mu = 1, \ldots, [\beta] + 1; j = 1, 2$. And $(D^\beta f_j)(0) = A_j \in \mathbb{R}$. Furthermore for $0 \le s \le x$ we have holding the system of fractional differential equations*

$$\left(D^\beta f_j\right)'(s) = F_j\left(s, \{D^{\alpha_i} f_1(s)\}_{i=1}^r, (D^\beta f_1)(s),\right.$$

$$\left. \{D^{\alpha_i} f_2(s)\}_{i=1}^r, (D^\beta f_2)(s)\right), \; \text{for } j = 1, 2. \tag{7.252}$$

For fixed $i_* \in \{1, \ldots, r\}$ we assume that $\alpha_{i_*+1} = \alpha_{i_*} + 1$, where $\alpha_{i_*}, \alpha_{i_*+1} \in \{\alpha_1, \ldots, \alpha_r\}$. Call $k := \alpha_{i_*}, \gamma := \alpha_{i_*} + 1$; that is, $\gamma = k + 1$. Here F_j is Lebesgue measurable on $[0, x] \times \mathbb{R}^{r+1} \times \mathbb{R}^{r+1}$ such that

$$|F_j(s, x_1, \ldots, x_r, x_{r+1}, y_1, \ldots, y_r, y_{r+1})| \le$$

$$q_{1,j}(s)|x_{i_*}|\sqrt{|y_{i_*+1}|} + q_{2,j}(s)|y_{i_*}|\sqrt{|x_{i_*+1}|}, \tag{7.253}$$

where both $0 \not\equiv q_{1,j}, q_{2,j} \ge 0$, bounded functions over $[0, x]$. Put

$$M := \max(\|q_{1,1}\|_\infty, \|q_{2,1}\|_\infty, \|q_{1,2}\|_\infty, \|q_{2,2}\|_\infty), \tag{7.254}$$

$$\theta(s) := |(D^\beta f_1)(s)| + |(D^\beta f_2)(s)|, \quad 0 \le s \le x, \quad \rho := |A_1| + |A_2|. \tag{7.255}$$

Also we define

$$Q(s) := \left(\frac{4M}{\sqrt{3\beta - 3k - 2}}\right) \frac{s^{((3\beta - 3k - 1)/2)}}{(\Gamma(\beta - k))^{3/2}}, \tag{7.256}$$

$$\sigma := \left(\frac{4M}{\sqrt{3\beta - 3k - 2}}\right) \frac{x^{((3\beta - 3k - 1)/2)}}{(\Gamma(\beta - k))^{3/2}}, \tag{7.257}$$

all $0 \le s \le x$.
    We assume that

$$x\sigma\sqrt{\rho} < 2. \tag{7.258}$$

Call

$$\tilde{\varphi}(s) := \frac{s(\sigma - Q(s))\left[\sigma\rho^2 s - 4\rho^{3/2}\right] + 4\rho}{(2 - \sigma\sqrt{\rho}\, s)^2}, \tag{7.259}$$

all $0 \le s \le x$. Then

$$\theta(s) \le \tilde{\varphi}(s); \tag{7.260}$$

in particular, it holds

$$|D^\beta f_j(s)| \le \tilde{\varphi}(s), \quad j = 1, 2, \tag{7.261}$$

for all $0 \le s \le x$. Furthermore we find

$$|f_j(s)| \le \frac{1}{\Gamma(\beta)} \int_0^s (s - t)^{\beta - 1} \tilde{\varphi}(t) dt, \tag{7.262}$$

$$|D^{\alpha_i} f_j(s)| \le \frac{1}{\Gamma(\beta - \alpha_i)} \int_0^s (s-t)^{\beta - \alpha_i - 1} \tilde{\varphi}(t) dt, \tag{7.263}$$

$j = 1, 2;\ i = 1, \ldots, r,\ $ all $\ 0 \le s \le x.$

Proof. Clearly, here $D^\beta f_j$ are L_∞ fractional derivatives in $[0, x]$. For $0 \le s \le y \le x$, by (7.252) we get that

$$\int_0^y \left(D^\beta f_j \right)'(s) ds = \int_0^y F_j \left(s, \{ D^{\alpha_i} f_1(s) \}_{i=1}^r, \left(D^\beta f_1 \right)(s), \right.$$

$$\left. \{ D^{\alpha_i} f_2(s) \}_{i=1}^r, \left(D^\beta f_2 \right)(s) \right) ds. \tag{7.264}$$

That is,

$$(D^\beta f_j)(y) = A_j + \int_0^y F_j(s, \ldots) ds. \tag{7.265}$$

Then we observe

$$|(D^\beta f_j)(y)| \le |A_j| + \int_0^y |F_j(s, \ldots)| ds \le |A_j| +$$

$$\int_0^y \left[\|q_{1,j}\|_\infty |(D^k f_1)(s)| \sqrt{|(D^{k+1} f_2)(s)|} + \|q_{2,j}\|_\infty |(D^k f_2)(s)| \sqrt{|(D^{k+1} f_1)(s)|} \right] ds \tag{7.266}$$

$$\le |A_j| + M \left(\int_0^y \left[|(D^k f_1)(s)| \sqrt{|(D^{k+1} f_2)(s)|} + |D^k f_2(s)| \sqrt{|(D^{k+1} f_1)(s)|} \right] ds \right). \tag{7.267}$$

Therefore

$$\theta(y) \le \rho + 2M \left(\int_0^y \left[|D^k f_1(s)| \sqrt{|D^{k+1} f_2(s)|} + |D^k f_2(s)| \sqrt{|D^{k+1} f_1(s)|} \right] ds \right) \tag{7.268}$$

$$\overset{((7.160),(7.161))}{\le} \rho + \left(\frac{4M}{\sqrt{3\beta - 3k - 2}} \right) \cdot \frac{y^{(3\beta - 3k - 1/2)}}{(\Gamma(\beta - k))^{3/2}}$$

$$\cdot \left(\int_0^y (|D^\beta f_1)(s)|^{3/2} + |(D^\beta f_2(s)|^{3/2}) ds \right) \tag{7.269}$$

$$\overset{(7.27)}{\le} \rho + \left(\frac{4M}{\sqrt{3\beta - 3k - 2}} \right) \left(\frac{y^{(3\beta - 3k - 1/2)}}{(\Gamma(\beta - k))^{3/2}} \right) \left(\int_0^y (|D^\beta f_1(s)| + |D^\beta f_2(s)|)^{3/2} ds \right). \tag{7.270}$$

We have proved that

$$\theta(y) \le \rho + Q(y) \int_0^y (\theta(s))^{3/2} ds, \tag{7.271}$$

all $\ 0 \le y \le x.$
Notice that $\ \theta(y) \ge 0,\ \rho \ge 0,\ Q(y) \ge 0,$ and $Q(0) = 0,$ also it is $\ \sigma > 0.$

Call

$$0 \le w(y) := \int_0^y (\theta(s))^{3/2} ds, \quad w(0) = 0, \quad \text{all } 0 \le y \le x. \tag{7.272}$$

That is,

$$w'(y) = (\theta(y))^{3/2} \ge 0 \quad \text{over} \quad [0, x], \tag{7.273}$$

and

$$\theta(y) = (w'(y))^{2/3}, \quad \text{all} \quad 0 \le y \le x. \tag{7.274}$$

Hence we rewrite (7.271) as follows,

$$(w'(y))^{2/3} \le \rho + Q(y)w(y), \quad \text{all} \quad 0 \le y \le x. \tag{7.275}$$

So that

$$(w'(y))^{2/3} \le \rho + \sigma \, w(y) < \rho + \varepsilon + \sigma \, w(y), \tag{7.276}$$

for $\varepsilon > 0$ arbitrarily small. Thus

$$w'(y) < (\rho + \varepsilon + \sigma \, w(y))^{3/2}, \quad \text{all} \quad 0 \le y \le x. \tag{7.277}$$

Here $(\rho + \varepsilon + \sigma \, w(y))^{3/2} > 0$. In particular it holds

$$w'(s) < (\rho + \varepsilon + \sigma \, w(y))^{3/2}, \quad \text{all} \quad 0 \le s \le y, \tag{7.278}$$

and

$$\frac{w'(s)}{(\rho + \varepsilon + \sigma \, w(s))^{3/2}} < 1. \tag{7.279}$$

The last is the same as

$$\left(-\frac{2}{\sigma} (\rho + \varepsilon + \sigma \, w(s))^{-1/2} \right)' < 1. \tag{7.280}$$

Therefore after integration we get

$$-\int_0^y d((\rho + \varepsilon + \sigma \, w(s))^{-1/2}) \le \frac{\sigma}{2} y, \tag{7.281}$$

and

$$(\rho + \varepsilon + \sigma \, w(s))^{-1/2} \Big|_y^0 \le \frac{\sigma}{2} \, y. \tag{7.282}$$

That is,

$$(\rho + \varepsilon)^{-1/2} - (\rho + \varepsilon + \sigma \, w(y))^{-1/2} \le \frac{\sigma}{2} \, y, \tag{7.283}$$

and

$$(\rho + \varepsilon)^{-1/2} - \frac{\sigma}{2} \, y \le (\rho + \varepsilon + \sigma \, w(y))^{-1/2}. \tag{7.284}$$

That is,

$$\frac{(2 - \sigma(\rho + \varepsilon)^{1/2} y)}{2(\rho + \varepsilon)^{1/2}} \le \frac{1}{(\rho + \varepsilon + \sigma \, w(y))^{1/2}}. \tag{7.285}$$

By assumption (7.258) we have

$$y\,\sigma\sqrt{\rho} < 2, \quad \text{all} \quad 0 \le y \le x. \tag{7.286}$$

Then for sufficiently small $\varepsilon > 0$ we still have

$$y\,\sigma(\rho + \varepsilon)^{1/2} < 2. \tag{7.287}$$

That is,

$$2 - y\,\sigma(\rho + \varepsilon)^{1/2} > 0, \quad \text{all} \quad 0 \le y \le x. \tag{7.288}$$

From (7.285) and (7.288) we obtain

$$(\rho + \varepsilon + \sigma\,w(y))^{1/2} \le \frac{2(\rho + \varepsilon)^{1/2}}{(2 - \sigma(\rho + \varepsilon)^{1/2}y)}, \tag{7.289}$$

all $0 \le y \le x$.

Solving (7.289) for $w(y)$ we find

$$w(y) \le \left(\frac{\rho + \varepsilon}{\sigma}\right)\left[\frac{4}{(2 - \sigma(\sqrt{\rho + \varepsilon})y)^2} - 1\right], \tag{7.290}$$

for $\varepsilon > 0$ sufficiently small and for all $0 \le y \le x$.

Letting $\varepsilon \to 0$ we obtain

$$w(y) \le \left(\frac{\rho}{\sigma}\right)\left(\frac{4}{(2 - \sigma\,y\sqrt{\rho})^2} - 1\right), \tag{7.291}$$

for all $0 \le y < x$.

That is,

$$w(y) \le \frac{(4\,y\,\rho^{3/2} - \sigma\,y^2\rho^2)}{(2 - \sigma\,y\sqrt{\rho})^2}, \tag{7.292}$$

all $0 \le y \le x$.

Hence by (7.271) through (7.292) we derive

$$\theta(y) \le \rho + Q(y)\left[\frac{(4\,y\,\rho^{3/2} - \sigma\,y^2\rho^2)}{(2 - \sigma\,y\sqrt{\rho})^2}\right] = \tilde{\varphi}(y), \tag{7.293}$$

all $0 \le y \le x$.

We have proved (7.260) and (7.261).

Next by applying Lemma 7.3, formula (7.3) for $\alpha = 0$, and using (7.261), we obtain (7.262). Finally by again applying Lemma 7.3, formula (7.3) for α_i, and using (7.261), we get (7.263). The proof of the theorem now is complete. $\square$

8

Canavati Fractional Opial-Type Inequalities for Several Functions and Applications

A wide variety of very general L_p ($1 \leq p \leq \infty$)-form Opial-type inequalities [315] is presented involving generalized Canavati fractional derivatives [17, 101] of several functions in different orders and powers. The above are based on a generalization of Taylor's formula for generalized Canavati fractional derivatives [17]. From the established results are derived several other particular results of special interest. Applications of some of these special inequalities are given in proving the uniqueness of solution and in giving upper bounds to solutions of initial value problems involving a very general system of several fractional differential equations. Upper bounds to various fractional derivatives of the solutions that are involved in the above systems are given too. This treatment is based on [27].

8.1 Introduction

Here the author continues his study of Canavati fractional Opial inequalities now involving several different functions and produces a wide variety of corresponding results with important applications to systems of several fractional differential equations. This chapter is a generalization of Chapter 6.

We start in Section 8.2 with preliminaries, we continue in Section 8.3 with the main results, and we finish in Section 8.4 with applications.

G.A. Anastassiou, *Fractional Differentiation Inequalities*,
DOI 10.1007/978-0-387-98128-4_8, © Springer Science+Business Media, LLC 2009

To give the reader an idea of the kind of inequalities we are dealing with, we briefly mention a simple one

$$\int_a^x \left(\sum_{j=1}^M |(D_a^\gamma f_j)(w)| \, |(D_a^\nu f_j)(w)| \right) dw$$

$$\leq \left(\frac{(x-a)^{\nu-\gamma}}{2\Gamma(\nu-\gamma)\sqrt{\nu-\gamma}\sqrt{2\nu-2\gamma-1}} \right) \left\{ \int_a^x \left(\sum_{j=1}^M ((D_a^\nu f_j)(w))^2 \right) dw \right\},$$

all $a \leq x \leq b$, for a certain kind of continuous function f_j, $j = 1, \ldots, M \in \mathbb{N}$; $\nu \geq 1$, $\gamma \geq 0$, $\nu - \gamma \geq 1$, and so on. Furthermore one system of fractional differential equations we are dealing with is of the form

$$(D_a^\nu f_j)(t) = F_j \left(t, \{(D_a^{\gamma_i} f_1)(t)\}_{i=1}^r, \{(D_a^{\gamma_i} f_2)(t)\}_{i=1}^r, \right.$$

$$\ldots, \{(D_a^{\gamma_i} f_M)(t)\}_{i=1}^r \right), \quad \text{all } t \in [a, b],$$

for $j = 1, 2, \ldots, M \in \mathbb{N}$ and with $f_j^{(i)}(a) = a_{ij} \in \mathbb{R}$, $i = 0, 1, \ldots, n-1$, where $n := [\nu]$, $\nu \geq 2$, and so on.

In the literature there are many different definitions of fractional derivatives, some of them equivalent (see [197, 333]). In this chapter we use one of the most recent due to J. Canavati [101], generalized in [15] and [17] by the author.

One of the advantages of Canavati fractional derivatives is that in applications to fractional initial value problems we need only n initial conditions as with the ordinary derivative case, whereas with other definitions of fractional derivatives we need $n+1$ or more initial conditions (see [333]).

8.2   Preliminaries

In the sequel we follow [101]. Let $g \in C([0,1])$. Let ν be a positive number, $n := [\nu]$, and $\alpha := \nu - n$ $(0 < \alpha < 1)$. Define

$$(J_\nu g)(x) := \frac{1}{\Gamma(\nu)} \int_0^x (x-t)^{\nu-1} g(t) \, dt, \quad 0 \leq x \leq 1, \qquad (8.1)$$

the *Riemann–Liouville integral*, where Γ is the gamma function. We define the subspace $C^\nu([0,1])$ of $C^n([0,1])$ as follows,

$$C^\nu([0,1]) := \{g \in C^n([0,1]) : J_{1-\alpha} D^n g \in C^1([0,1])\},$$

where $D := d/dx$. So for $g \in C^\nu([0,1])$, we define the Canavati ν-*fractional derivative* of g as

$$D^\nu g := D J_{1-\alpha} D^n g. \qquad (8.2)$$

When $\nu \geq 1$ we have the Taylor's formula

$$g(t) = g(0) + g'(0)t + g''(0)\frac{t^2}{2!} + \cdots + g^{(n-1)}(0)\frac{t^{n-1}}{(n-1)!}$$

$$+ (J_\nu D^\nu g)(t), \quad \text{for all } t \in [0,1]. \tag{8.1}$$

When $0 < \nu < 1$ we find

$$g(t) = (J_\nu D^\nu g)(t), \quad \text{for all } t \in [0,1]. \tag{8.4}$$

Next we transfer the above notions over to arbitrary $[a,b] \subseteq \mathbb{R}$ (see [17]). Let $x, x_0 \in [a,b]$ such that $x \geq x_0$, where x_0 is fixed. Let $f \in C([a,b])$ and define

$$(J_\nu^{x_0} f)(x) := \frac{1}{\Gamma(\nu)} \int_{x_0}^{x} (x-t)^{\nu-1} f(t)\, dt, \quad x_0 \leq x \leq b, \tag{8.5}$$

the *generalized Riemann–Liouville integral*. We define the subspace $C_{x_0}^\nu([a,b])$ of $C^n([a,b])$:

$$C_{x_0}^\nu([a,b]) := \{f \in C^n([a,b]): J_{1-\alpha}^{x_0} D^n f \in C^1([x_0,b])\}.$$

For $f \in C_{x_0}^\nu([a,b])$, we define the *generalized ν-fractional derivative of f over $[x_0,b]$* as

$$D_{x_0}^\nu f := D J_{1-\alpha}^{x_0} f^{(n)} \quad (f^{(n)} := D^n f). \tag{8.6}$$

Observe that

$$(J_{1-\alpha}^{x_0} f^{(n)})(x) = \frac{1}{\Gamma(1-\alpha)} \int_{x_0}^{x} (x-t)^{-\alpha} f^{(n)}(t)\, dt$$

exists for $f \in C_{x_0}^\nu([a,b])$.

We mention the following fractional generalization of Taylor's formula (see [17, 101]); see also Theorem 2.1.

Theorem 8.1. *Let* $f \in C_{x_0}^\nu([a,b])$, $x_0 \in [a,b]$*, fixed.*

(i) *If* $\nu \geq 1$ *then*

$$f(x) = f(x_0) + f'(x_0)(x-x_0) + f''(x_0)\frac{(x-x_0)^2}{2}$$

$$+ \cdots + f^{(n-1)}(x_0)\frac{(x-x_0)^{n-1}}{(n-1)!}$$

$$+ (J_\nu^{x_0} D_{x_0}^\nu f)(x), \quad \text{for all } x \in [a,b]: x \geq x_0. \tag{8.2}$$

(ii) *If* $0 < \nu < 1$ *then*

$$f(x) = (J_\nu^{x_0} D_{x_0}^\nu f)(x), \quad \text{for all } x \in [a,b]: x \geq x_0. \tag{8.8}$$

We make

Remark 8.2. (1) $(D_{x_0}^n f) = f^{(n)}$, $n \in \mathbb{N}$.

(2) Let $f \in C_{x_0}^\nu([a, b])$, $\nu \geq 1$, and $f^{(i)}(x_0) = 0$, $i = 0, 1, \ldots, n - 1$; $n := [\nu]$. Then by (8.7)

$$f(x) = (J_\nu^{x_0} D_{x_0}^\nu f)(x).$$

That is,

$$f(x) = \frac{1}{\Gamma(\nu)} \int_{x_0}^x (x - t)^{\nu-1} (D_{x_0}^\nu f)(t)\, dt, \tag{8.9}$$

for all $x \in [a, b]$ with $x \geq x_0$. Notice that (8.9) is true, also when $0 < \nu < 1$.

We also make

Remark 8.3. Let $\nu \geq 1$, $\gamma \geq 0$, such that $\nu - \gamma \geq 1$, so that $\gamma < \nu$. Call $n := [\nu]$, $\alpha := \nu - n$; $m := [\gamma]$, $\rho := \gamma - m$. Note that $\nu - m \geq 1$ and $n - m \geq 1$. Let $f \in C_{x_0}^\nu([a, b])$ be such that $f^{(i)}(x_0) = 0$, $i = 0, 1, \ldots, n - 1$. Hence by (8.7)

$$f(x) = (J_\nu^{x_0} D_{x_0}^\nu f)(x), \quad \text{for all } x \in [a, b]: x \geq x_0.$$

Therefore by Leibnitz's formula and $\Gamma(p + 1) = p\Gamma(p)$, $p > 0$, we get that

$$f^{(m)}(x) = (J_{\nu-m}^{x_0} D_{x_0}^\nu f)(x), \quad \text{for all } x \geq x_0.$$

It follows that $f \in C_{x_0}^\gamma([a, b])$ and thus $(D_{x_0}^\gamma f)(x) := (DJ_{1-\rho}^{x_0} f^{(m)})(x)$ exists for all $x \geq x_0$.

We easily obtain

$$(D_{x_0}^\gamma f)(x) = D((J_{1-\rho}^{x_0} f^{(m)})(x)) = \frac{1}{\Gamma(\nu - \gamma)} \int_{x_0}^x (x - t)^{(\nu-\gamma)-1} (D_{x_0}^\nu f)(t)\, dt, \tag{8.10}$$

and thus

$$(D_{x_0}^\gamma f)(x) = (J_{\nu-\gamma}^{x_0} (D_{x_0}^\nu f))(x)$$

and is continuous in x on $[x_0, b]$.

8.3   Main Results

Here we often use the following basic inequalities.

Let $a_1, \ldots, a_n \geq 0$, $n \in \mathbb{N}$; then

$$a_1^r + \cdots + a_n^r \leq (a_1 + \cdots + a_n)^r, \quad r \geq 1, \tag{8.11}$$

and

$$a_1^r + \cdots + a_n^r \le n^{1-r}(a_1 + \cdots + a_n)^r, \quad 0 \le r \le 1. \tag{8.12}$$

Our first result follows next.

Theorem 8.4. *Let* $\nu \ge 1$, $\gamma_1, \gamma_2 \ge 0$, *such that* $\nu - \gamma_1 \ge 1$, $\nu - \gamma_2 \ge 1$, *and* $f_j \in C_{x_0}^\nu([a,b])$ *with* $f_j^{(i)}(x_0) = 0$, $i = 0, 1, \ldots, n-1$, $n := [\nu]$, $j = 1, \ldots, M \in \mathbb{N}$. *Here* $x, x_0 \in [a,b]$: $x \ge x_0$. *Also consider* $p(t) > 0$, *and* $q(t) \ge 0$ *continuous functions on* $[x_0, b]$. *Let* $\lambda_\nu > 0$ *and* $\lambda_\alpha, \lambda_\beta \ge 0$ *such that* $\lambda_\nu < p$, *where* $p > 1$. *Set*

$$P_k(w) := \int_{x_0}^w (w-t)^{(\nu-\gamma_k-1)p/p-1}(p(t))^{-1/p-1}\, dt, \; k = 1, 2, \tag{8.13}$$

$x_0 \le w \le b$;

$$A(w) := \frac{q(w) \cdot (P_1(w))^{\lambda_\alpha(p-1/p)} \cdot (P_2(w))^{\lambda_\beta((p-1)/p)}(p(w))^{-\lambda_\nu/p}}{(\Gamma(\nu-\gamma_1))^{\lambda_\alpha} \cdot (\Gamma(\nu-\gamma_2))^{\lambda_\beta}}; \tag{8.14}$$

$$A_0(x) := \left(\int_{x_0}^x A(w)^{p/p-\lambda_\nu}\, dw \right)^{(p-\lambda_\nu)/p}. \tag{8.15}$$

Call

$$\varphi_1(x) := \left(A_0(x)|_{\lambda_\beta=0}\right) \cdot \left(\frac{\lambda_\nu}{\lambda_\alpha + \lambda_\nu} \right)^{\lambda_\nu/p}, \tag{8.16}$$

$$\delta_1^* := \begin{cases} M^{1-(\lambda_\alpha+\lambda_\nu/p)}, & \text{if } \lambda_\alpha + \lambda_\nu \le p, \\ 2^{(\lambda_\alpha+\lambda_\nu/p)-1} & \text{if } \lambda_\alpha + \lambda_\nu \ge p. \end{cases} \tag{8.17}$$

If $\lambda_\beta = 0$, *we obtain that*

$$\int_{x_0}^x q(w) \left(\sum_{j=1}^M |(D_{x_0}^{\gamma_1} f_j)(w)|^{\lambda_\alpha} |(D_{x_0}^\nu f_j)(w)|^{\lambda_\nu} \right) dw$$

$$\le \delta_1^* \cdot \varphi_1(x) \cdot \left[\int_{x_0}^x p(w) \left(\sum_{j=1}^M |(D_{x_0}^\nu f_j)(w)|^p \right) dw \right]^{((\lambda_\alpha+\lambda_\nu)/p)}, \tag{8.18}$$

all $x_0 \le x \le b$.

Proof. From Theorem 2 of [26] (see here Theorem 6.5) we get

$$\int_{x_0}^x q(w) \big[|(D_{x_0}^{\gamma_1} f_j)(w)|^{\lambda_\alpha} |(D_{x_0}^\nu f_j)(w)|^{\lambda_\nu}$$

$$+ |(D_{x_0}^{\gamma_1} f_{j+1})(w)|^{\lambda_\alpha} |(D_{x_0}^\nu f_{j+1})(w)|^{\lambda_\nu} \big]\, dw \tag{8.19}$$

$$\le \delta_1 \varphi_1(x) \left[\int_{x_0}^x p(w) \big[|(D_{x_0}^\nu f_j)(w)|^p + |(D_{x_0}^\nu f_{j+1})(w)|^p \big]\, dw \right]^{((\lambda_\alpha+\lambda_\nu)/p)},$$

$j = 1, 2, \ldots, M - 1$, where

$$\delta_1 := \begin{cases} 2^{1-\left((\lambda_\alpha + \lambda_\nu)/p\right)}, & \text{if } \lambda_\alpha + \lambda_\nu \leq p, \\ 1, & \text{if } \lambda_\alpha + \lambda_\nu \geq p. \end{cases} \tag{8.20}$$

Hence by adding all the above we find

$$\int_{x_0}^{x} q(w) \left(\sum_{j=1}^{M-1} \left[|(D_{x_0}^{\gamma_1} f_j)(w)|^{\lambda_\alpha} |(D_{x_0}^{\nu} f_j)(w)|^{\lambda_\nu} \right. \right.$$

$$\left. \left. + |(D_{x_0}^{\gamma_1} f_{j+1})(w)|^{\lambda_\alpha} |(D_{x_0}^{\nu} f_{j+1})(w)|^{\lambda_\nu} \right] \right) dw$$

$$\leq \delta_1 \varphi_1(x) \cdot \left(\sum_{j=1}^{M-1} \left[\int_{x_0}^{x} p(w) [|(D_{x_0}^{\nu} f_j)(w)|^p \right. \right. \tag{8.21}$$

$$\left. \left. + |(D_{x_0}^{\nu} f_{j+1})(w)|^p \right] dw \right]^{\left((\lambda_\alpha + \lambda_\nu)/p\right)} \right).$$

Also it holds

$$\int_{x_0}^{x} q(w) \left[|(D_{x_0}^{\gamma_1} f_1)(w)|^{\lambda_\alpha} |(D_{x_0}^{\nu} f_1)(w)|^{\lambda_\nu} \right.$$

$$\left. + |(D_{x_0}^{\gamma_1} f_M)(w)|^{\lambda_\alpha} |(D_{x_0}^{\nu} f_M)(w)|^{\lambda_\nu} \right] dw \tag{8.22}$$

$$\leq \delta_1 \varphi_1(x) \left[\int_{x_0}^{x} p(w) [|(D_{x_0}^{\nu} f_1)(w)|^p + |(D_{x_0}^{\nu} f_M)(w)|^p] dw \right]^{\left((\lambda_\alpha + \lambda_\nu)/p\right)}.$$

Call

$$\varepsilon_1 = \begin{cases} 1, & \text{if } \lambda_\alpha + \lambda_\nu \geq p \\ M^{1-\left(\lambda_\alpha + \lambda_\nu/p\right)}, & \text{if } \lambda_\alpha + \lambda_\nu \leq p. \end{cases} \tag{8.23}$$

Adding (8.21) and (8.22), and using (8.11) and (8.12) we have

$$2 \int_{x_0}^{x} q(w) \left(\sum_{j=1}^{M} |(D_{x_0}^{\gamma_1} f_j)(w)|^{\lambda_\alpha} |(D_{x_0}^{\nu} f_j)(w)|^{\lambda_\nu} \right) dw$$

$$\leq \delta_1 \varphi_1(x) \left\{ \left\{ \sum_{j=1}^{M-1} \left[\int_{x_0}^{x} p(w) [|(D_{x_0}^{\nu} f_j)(w)|^p \right. \right. \right.$$

$$\left. \left. \left. + |(D_{x_0}^{\nu} f_{j+1})(w)|^p] dw \right]^{\left((\lambda_\alpha + \lambda_\nu)/p\right)} \right\} \right.$$

$$+\left\{\int_{x_0}^x p(w)[|(D_{x_0}^\nu f_1)(w)|^p + |(D_{x_0}^\nu f_M)(w)|^p]dw\right\}^{((\lambda_\alpha+\lambda_\nu)/p)}\right\}$$

$$\leq \delta_1\varepsilon_1\varphi_1(x)\left\{\int_{x_0}^x p(w)\left[2\sum_{j=1}^M |(D_{x_0}^\nu f_j)(w)|^p\right]dw\right\}^{((\lambda_\alpha+\lambda_\nu)/p)}.$$

We have proved

$$\int_{x_0}^x q(w)\left(\sum_{j=1}^M |(D_{x_0}^{\gamma_1} f_j)(w)|^{\lambda_\alpha}|(D_{x_0}^\nu f_j)(w)|^{\lambda_\nu}\right)dw$$

$$\leq \delta_1\left(2^{\left(\lambda_\alpha+\lambda_\nu/p\right)-1}\right)\varepsilon_1\varphi_1(x)$$

$$\cdot\left\{\int_{x_0}^x p(w)\left[\sum_{j=1}^M |(D_{x_0}^\nu f_j)(w)|^p\right]dw\right\}^{((\lambda_\alpha+\lambda_\nu)/p)}. \qquad (8.24)$$

Clearly here we have

$$\delta_1^* = \delta_1\left(2^{\left(\lambda_\alpha+\lambda_\nu/p\right)-1}\right)\varepsilon_1. \qquad (8.25)$$

From (8.24) and (8.25) we derive (8.18). □

Next we give

Theorem 8.5. *All here are as in Theorem 8.4. Denote*

$$\delta_3 := \begin{cases} 2^{\lambda_\beta/\lambda_\nu} - 1, & \text{if } \lambda_\beta \geq \lambda_\nu, \\ 1, & \text{if } \lambda_\beta \leq \lambda_\nu, \end{cases} \qquad (8.26)$$

$$\varepsilon_2 := \begin{cases} 1, & \text{if } \lambda_\nu + \lambda_\beta \geq p, \\ M^{1-\left(\lambda_\nu+\lambda_\beta/p\right)}, & \text{if } \lambda_\nu + \lambda_\beta \leq p, \end{cases} \qquad (8.27)$$

and

$$\varphi_2(x) := \left(A_0(x)\big|_{\lambda_\alpha=0}\right)2^{\left(p-\lambda_\nu/p\right)}\left(\frac{\lambda_\nu}{\lambda_\beta+\lambda_\nu}\right)^{\lambda_\nu/p}\delta_3^{\lambda_\nu/p}. \qquad (8.28)$$

If $\lambda_\alpha = 0$, then

$$\int_{x_0}^x q(w)\left\{\left\{\sum_{j=1}^{M-1}\left[|(D_{x_0}^{\gamma_2} f_{j+1})(w)|^{\lambda_\beta}|(D_{x_0}^\nu f_j)(w)|^{\lambda_\nu}\right.\right.$$

$$+ |(D_{x_0}^{\gamma_2} f_j)(w)|^{\lambda_\beta} |(D_{x_0}^\nu f_{j+1})(w)|^{\lambda_\nu}] \Big\}$$

$$+ \Big[|(D_{x_0}^{\gamma_2} f_M)(w)|^{\lambda_\beta} |(D_{x_0}^\nu f_1)(w)|^{\lambda_\nu}$$

$$+ |(D_{x_0}^{\gamma_2} f_1)(w)|^{\lambda_\beta} |(D_{x_0}^\nu f_M)(w)|^{\lambda_\nu}] \Big\} \, dw$$

$$\leq 2^{((\lambda_\nu + \lambda_\beta)/p)} \varepsilon_2 \varphi_2(x) \cdot \Big\{ \int_{x_0}^x p(w)$$

$$\cdot \Big[\sum_{j=1}^M |(D_{x_0}^\nu f_j)(w)|^p \Big] dw \Big\}^{((\lambda_\nu + \lambda_\beta)/p)} , \quad x \geq x_0. \tag{8.29}$$

Proof. From Theorem 3 of [26] (see here Theorem 6.6) we have

$$\int_{x_0}^x q(w) \big[|(D_{x_0}^{\gamma_2} f_{j+1})(w)|^{\lambda_\beta} |(D_{x_0}^\nu f_j)(w)|^{\lambda_\nu}$$

$$+ |(D_{x_0}^{\gamma_2} f_j)(w)|^{\lambda_\beta} |(D_{x_0}^\nu f_{j+1})(w)|^{\lambda_\nu} \big] \, dw$$

$$\leq \varphi_2(x) \Big(\int_{x_0}^x p(w) \big[|(D_{x_0}^\nu f_j)(w)|^p +$$

$$|(D_{x_0}^\nu f_{j+1})(w)|^p \big] dw \Big)^{((\lambda_\nu + \lambda_\beta)/p)}, \tag{8.30}$$

for $j = 1, \ldots, M - 1$. Hence by adding all the above we get

$$\int_{x_0}^x q(w) \Big(\sum_{j=1}^{M-1} \big[|(D_{x_0}^{\gamma_2} f_{j+1})(w)|^{\lambda_\beta} |(D_{x_0}^\nu f_j)(w)|^{\lambda_\nu}$$

$$+ |(D_{x_0}^{\gamma_2} f_j)(w)|^{\lambda_\beta} |(D_{x_0}^\nu f_{j+1})(w)|^{\lambda_\nu} \big] \Big) \, dw$$

$$\leq \varphi_2(x) \Big\{ \sum_{j=1}^{M-1} \Big(\int_{x_0}^x p(w) [|(D_{x_0}^\nu f_j)(w)|^p$$

$$+ |(D_{x_0}^\nu f_{j+1})(w)|^p \, dw \Big)^{((\lambda_\nu + \lambda_\beta)/p)} \Big\}. \tag{8.31}$$

Similarly it holds

$$
\int_{x_0}^{x} q(w) \left[|(D_{x_0}^{\gamma_2} f_M)(w)|^{\lambda_\beta} |(D_{x_0}^{\nu} f_1)(w)|^{\lambda_\nu} \right.
$$
$$
+ \left. |(D_{x_0}^{\gamma_2} f_1)(w)|^{\lambda_\beta} |(D_{x_0}^{\nu} f_M)(w)|^{\lambda_\nu} \right] dw
$$
$$
\leq \varphi_2(x) \left(\int_{x_0}^{x} p(w) \left[|(D_{x_0}^{\nu} f_1)(w)|^{p} + \right. \right.
$$
$$
\left. \left. |(D_{x_0}^{\nu} f_M)(w)|^{p} \right] dw \right)^{\left((\lambda_\nu + \lambda_\beta)/p \right)}. \tag{8.32}
$$

Adding (8.31) and (8.32) and using (8.11), (8.12) we produce (8.29). □

The general case follows.

Theorem 8.6. *All here are as in Theorem 8.4. Denote*

$$
\tilde{\gamma}_1 := \begin{cases} 2^{\left(\lambda_\alpha + \lambda_\beta / \lambda_\nu \right)} - 1, & \text{if } \lambda_\alpha + \lambda_\beta \geq \lambda_\nu, \\ 1, & \text{if } \lambda_\alpha + \lambda_\beta \leq \lambda_\nu, \end{cases} \tag{8.33}
$$

and

$$
\tilde{\gamma}_2 := \begin{cases} 1, & \text{if } \lambda_\alpha + \lambda_\beta + \lambda_\nu \geq p, \\ 2^{1 - \left(\lambda_\alpha + \lambda_\beta + \lambda_\nu / p \right)}, & \text{if } \lambda_\alpha + \lambda_\beta + \lambda_\nu \leq p. \end{cases} \tag{8.34}
$$

Set

$$
\varphi_3(x) := A_0(x) \cdot \left(\frac{\lambda_\nu}{(\lambda_\alpha + \lambda_\beta)(\lambda_\alpha + \lambda_\beta + \lambda_\nu)} \right)^{\lambda_\nu/p}
$$
$$
\cdot \left[\lambda_\alpha^{\lambda_\nu/p} \tilde{\gamma}_2 + 2^{\left(p - \lambda_\nu/p \right)} (\tilde{\gamma}_1 \lambda_\beta)^{\lambda_\nu/p} \right], \tag{8.35}
$$

and

$$
\varepsilon_3 := \begin{cases} 1, & \text{if } \lambda_\alpha + \lambda_\beta + \lambda_\nu \geq p, \\ M^{1 - \left((\lambda_\alpha + \lambda_\beta + \lambda_\nu)/p \right)}, & \text{if } \lambda_\alpha + \lambda_\beta + \lambda_\nu \leq p. \end{cases} \tag{8.36}
$$

Then

$$\int_{x_0}^{x} q(w) \left[\sum_{j=1}^{M-1} \left[|(D_{x_0}^{\gamma_1} f_j)(w)|^{\lambda_\alpha} |(D_{x_0}^{\gamma_2} f_{j+1})(w)|^{\lambda_\beta} |(D_{x_0}^{\nu} f_j)(w)|^{\lambda_\nu} \right. \right.$$

$$+ |(D_{x_0}^{\gamma_2} f_j)(w)|^{\lambda_\beta} |(D_{x_0}^{\gamma_1} f_{j+1})(w)|^{\lambda_\alpha} |(D_{x_0}^{\nu} f_{j+1})(w)|^{\lambda_\nu} \Big]$$

$$+ \left[|(D_{x_0}^{\gamma_1} f_1)(w)|^{\lambda_\alpha} |(D_{x_0}^{\gamma_2} f_M)(w)|^{\lambda_\beta} |(D_{x_0}^{\nu} f_1)(w)|^{\lambda_\nu} \right.$$

$$\left. \left. + |(D_{x_0}^{\gamma_2} f_1)(w)|^{\lambda_\beta} |(D_{x_0}^{\gamma_1} f_M)(w)|^{\lambda_\alpha} |(D_{x_0}^{\nu} f_M)(w)|^{\lambda_\nu} \right] \right] dw \quad (8.37)$$

$$\leq 2^{\left(\lambda_\alpha + \lambda_\beta + \lambda_\nu/p\right)} \varepsilon_3 \varphi_3(x) \cdot \left\{ \int_{x_0}^{x} p(w) \right.$$

$$\left. \left[\sum_{j=1}^{M} |(D_{x_0}^{\nu} f_j)(w)|^p \right] dw \right\}^{\left((\lambda_\alpha + \lambda_\beta + \lambda_\nu)/p \right)},$$

all $x_0 \leq x \leq b$.

Proof. From Theorem 4 of [26] (see here Theorem 6.7) and adding all together we have

$$\sum_{j=1}^{M-1} \int_{x_0}^{x} q(w) \left[|(D_{x_0}^{\gamma_1} f_j)(w)|^{\lambda_\alpha} |(D_{x_0}^{\gamma_2} f_{j+1})(w)|^{\lambda_\beta} |(D_{x_0}^{\nu} f_j)(w)|^{\lambda_\nu} \right.$$

$$\left. + |(D_{x_0}^{\gamma_2} f_j)(w)|^{\lambda_\beta} |(D_{x_0}^{\gamma_1} f_{j+1})(w)|^{\lambda_\alpha} |(D_{x_0}^{\nu} f_{j+1})(w)|^{\lambda_\nu} \right] dw \quad (8.38)$$

$$\leq \varphi_3(x) \sum_{j=1}^{M-1} \left(\int_{x_0}^{x} p(w) \left(|(D_{x_0}^{\nu} f_j)(w)|^p \right. \right.$$

$$\left. \left. + |(D_{x_0}^{\nu} f_{j+1})(w)|^p \right) dw \right)^{\left((\lambda_\alpha + \lambda_\beta + \lambda_\nu)/p \right)},$$

all $x_0 \leq x \leq b$.
Also it holds

$$\int_{x_0}^{x} q(w) \left[|(D_{x_0}^{\gamma_1} f_1)(w)|^{\lambda_\alpha} |(D_{x_0}^{\gamma_2} f_M)(w)|^{\lambda_\beta} |(D_{x_0}^{\nu} f_1)(w)|^{\lambda_\nu} \right.$$

$$\left. + |(D_{x_0}^{\gamma_2} f_1)(w)|^{\lambda_\beta} |(D_{x_0}^{\gamma_1} f_M)(w)|^{\lambda_\alpha} |(D_{x_0}^{\nu} f_M)(w)|^{\lambda_\nu} \right] dw \quad (8.39)$$

$$\leq \varphi_3(x) \left(\int_{x_0}^{x} p(w) \left(|(D_{x_0}^{\nu} f_1)(w)|^p + |(D_{x_0}^{\nu} f_M)(w)|^p \right) dw \right)^{\left((\lambda_\alpha + \lambda_\beta + \lambda_\nu/p) \right)},$$

all $x_0 \leq x \leq b$.
Adding (8.38) and (8.39), along with (8.11) and (8.12) we produce (8.37).
□

We continue with

Theorem 8.7. *Let $\nu \geq 2$ and $\gamma_1 \geq 0$ such that $\nu - \gamma_1 \geq 2$. Let $f_j \in C^\nu_{x_0}([a,b])$ with $f_j^{(i)}(x_0) = 0$, $i = 0, 1, \ldots, n-1$, $n := [\nu]$, $j = 1, \ldots, M \in \mathbb{N}$. Here $x, x_0 \in [a,b]: x \geq x_0$. Also consider $p(t) > 0$, and $q(t) \geq 0$ continuous functions on $[x_0, b]$. Let $\lambda_\alpha \geq 0$, $0 < \lambda_{\alpha+1} < 1$, and $p > 1$. Denote*

$$\theta_3 := \left\{ \begin{array}{ll} 2^{\left(\lambda_\alpha / \lambda_{\alpha+1}\right)} - 1, & \text{if } \lambda_\alpha \geq \lambda_{\alpha+1}, \\ 1, & \text{if } \lambda_\alpha \leq \lambda_{\alpha+1} \end{array} \right\} \tag{8.40}$$

$$L(x) := \left(2 \int_{x_0}^x (q(w))^{\left(1/1-\lambda_{\alpha+1}\right)} \, dw \right)^{(1-\lambda_{\alpha+1})}$$

$$\left(\frac{\theta_3 \lambda_{\alpha+1}}{\lambda_\alpha + \lambda_{\alpha+1}} \right)^{\lambda_{\alpha+1}}, \tag{8.41}$$

and

$$P_1(x) := \int_{x_0}^x (x - t)^{(\nu - \gamma_1 - 1)p/p-1} (p(t))^{-1/(p-1)} \, dt, \tag{8.42}$$

$$T(x) := L(x) \cdot \left(\frac{P_1(x)^{\left((p-1)/p\right)}}{\Gamma(\nu - \gamma_1)} \right)^{(\lambda_\alpha + \lambda_{\alpha+1})}, \tag{8.43}$$

and

$$\omega_1 := 2^{((p-1)/p)(\lambda_\alpha + \lambda_{\alpha+1})}, \tag{8.44}$$

$$\Phi(x) := T(x)\omega_1. \tag{8.45}$$

Also put

$$\varepsilon_4 := \left\{ \begin{array}{ll} 1, & \text{if } \lambda_\alpha + \lambda_{\alpha+1} \geq p, \\ M^{1-\left((\lambda_\alpha + \lambda_{\alpha+1})/p\right)}, & \text{if } \lambda_\alpha + \lambda_{\alpha+1} \leq p \end{array} \right\}. \tag{8.46}$$

Then

$$\int_{x_0}^{x} q(w) \left\{ \left\{ \sum_{j=1}^{M-1} \left[|(D_{x_0}^{\gamma_1} f_j)(w)|^{\lambda_\alpha} |(D_{x_0}^{\gamma_1+1} f_{j+1})(w)|^{\lambda_{\alpha+1}} \right. \right. \right.$$

$$\left. + |(D_{x_0}^{\gamma_1} f_{j+1})(w)|^{\lambda_\alpha} |(D_{x_0}^{\gamma_1+1} f_j)(w)|^{\lambda_{\alpha+1}} \right] \right\}$$

$$+ \left[|(D_{x_0}^{\gamma_1} f_1)(w)|^{\lambda_\alpha} |(D_{x_0}^{\gamma_1+1} f_M)(w)|^{\lambda_{\alpha+1}} \right.$$

$$\left. \left. + |(D_{x_0}^{\gamma_1} f_M)(w)|^{\lambda_\alpha} |(D_{x_0}^{\gamma_1+1} f_1)(w)|^{\lambda_{\alpha+1}} \right] \right\} dw$$

$$\leq 2^{\left(\lambda_\alpha + \lambda_{\alpha+1}/p \right)} \varepsilon_4 \Phi(x) \left[\int_{x_0}^{x} p(w) \right. \tag{8.47}$$

$$\left. \left(\sum_{j=1}^{M} |(D_{x_0}^{\nu} f_j)(w)|^p \right) dw \right]^{\left((\lambda_\alpha + \lambda_{\alpha+1})/p \right)},$$

all $x_0 \leq x \leq b$.

Proof. From Theorem 5 [26] (see here Theorem 6.8) we get

$$\int_{x_0}^{x} q(w) \sum_{j=1}^{M-1} \left[|(D_{x_0}^{\gamma_1} f_j)(w)|^{\lambda_\alpha} |(D_{x_0}^{\gamma_1+1} f_{j+1})(w)|^{\lambda_{\alpha+1}} \right.$$

$$\left. + |(D_{x_0}^{\gamma_1} f_{j+1})(w)|^{\lambda_\alpha} |(D_{x_0}^{\gamma_1+1} f_j)(w)|^{\lambda_{\alpha+1}} \right] dw$$

$$\leq \Phi(x) \sum_{j=1}^{M-1} \left[\int_{x_0}^{x} p(w) \left(|(D_{x_0}^{\nu} f_j)(w)|^p \right. \right. \tag{8.48}$$

$$\left. \left. + |(D_{x_0}^{\nu} f_{j+1})(w)|^p \right) dw \right]^{\left((\lambda_\alpha + \lambda_{\alpha+1})/p \right)}$$

all $x_0 \leq x \leq b$.
   Also it holds

$$\int_{x_0}^{x} q(w) \left[|(D_{x_0}^{\gamma_1} f_1)(w)|^{\lambda_\alpha} |(D_{x_0}^{\gamma_1+1} f_M)(w)|^{\lambda_{\alpha+1}} \right.$$

$$\left. + |(D_{x_0}^{\gamma_1} f_M)(w)|^{\lambda_\alpha} |(D_{x_0}^{\gamma_1+1} f_1)(w)|^{\lambda_{\alpha+1}} \right] dw$$

$$\leq \Phi(x) \left[\int_{x_0}^{x} p(w) \left(|(D_{x_0}^{\nu} f_1)(w)|^p \right. \right. \tag{8.49}$$

$$\left. \left. + |(D_{x_0}^{\nu} f_M)(w)|^p \right] dw \right]^{\left((\lambda_\alpha + \lambda_{\alpha+1})/p \right)},$$

all $x_0 \leq x \leq b$. Adding (8.48) and (8.49), along with (8.11) and (8.12) we
derive (8.47). $\square$

Next comes the following theorem.

Theorem 8.8. *All here are as in Theorem 8.4. Consider the special case* $\lambda_\beta = \lambda_\alpha + \lambda_\nu$. *Denote*

$$\tilde{T}(x) := A_0(x) \left(\frac{\lambda_\nu}{\lambda_\alpha + \lambda_\nu} \right)^{\lambda_\nu/p} 2^{\left((p-2\lambda_\alpha-3\lambda_\nu)/p\right)}, \qquad (8.50)$$

$$\varepsilon_5 := \left\{ \begin{array}{ll} 1, & \text{if } 2(\lambda_\alpha + \lambda_\nu) \geq p, \\ M^{1-\left(2(\lambda_\alpha+\lambda_\nu)/p\right)}, & \text{if } 2(\lambda_\alpha + \lambda_\nu) \leq p \end{array} \right\}. \qquad (8.51)$$

Then

$$\int_{x_0}^x q(w) \left\{ \left\{ \sum_{j=1}^{M-1} \left[|(D_{x_0}^{\gamma_1} f_j)(w)|^{\lambda_\alpha} |(D_{x_0}^{\gamma_2} f_{j+1})(w)|^{\lambda_\alpha+\lambda_\nu} |(D_{x_0}^\nu f_j)(w)|^{\lambda_\nu} \right. \right. \right.$$

$$\left. + |(D_{x_0}^{\gamma_2} f_j)(w)|^{\lambda_\alpha+\lambda_\nu} |(D_{x_0}^{\gamma_1} f_{j+1})(w)|^{\lambda_\alpha} |(D_{x_0}^\nu f_{j+1})(w)|^{\lambda_\nu} \right] \right\}$$

$$+ \left[|(D_{x_0}^{\gamma_1} f_1)(w)|^{\lambda_\alpha} |(D_{x_0}^{\gamma_2} f_M)(w)|^{\lambda_\alpha+\lambda_\nu} |(D_{x_0}^\nu f_1)(w)|^{\lambda_\nu} \right.$$

$$\left. \left. + |(D_{x_0}^{\gamma_2} f_1)(w)|^{\lambda_\alpha+\lambda_\nu} |(D_{x_0}^{\gamma_1} f_M)(w)|^{\lambda_\alpha} |(D_{x_0}^\nu f_M)(w)|^{\lambda_\nu} \right] \right\} dw$$

$$\leq 2^{2\left(\lambda_\alpha+\lambda_\nu/p\right)} \varepsilon_5 \tilde{T}(x) \left[\int_{x_0}^x p(w) \right. \qquad (8.52)$$

$$\left. \left(\sum_{j=1}^M |(D_{x_0}^\nu f_j)(w)|^p \right) dw \right]^{\left(2\left((\lambda_\alpha+\lambda_\nu)/p\right)\right)},$$

all $x_0 \leq x \leq b$.

Proof. Based on Theorem 6 [26] (see here Theorem 6.9). The rest is as in the proof of Theorem 8.7. □

Next we give special cases of the above theorems.

Corollary 8.9. (to Theorem 8.4; $\lambda_\beta = 0$, $p(t) = q(t) = 1$). *It holds*

$$\int_{x_0}^x \left(\sum_{j=1}^M |(D_{x_0}^{\gamma_1} f_j)(w)|^{\lambda_\alpha} |(D_{x_0}^\nu f_j)(w)|^{\lambda_\nu} \right) dw$$

$$\leq \delta_1^* \varphi_1(x) \left[\int_{x_0}^x \left[\sum_{j=1}^M |(D_{x_0}^\nu f_j)(w)|^p \right] dw \right]^{\left((\lambda_\alpha+\lambda_\nu)/p\right)} \qquad (8.53)$$

all $x_0 \leq x \leq b$.

In (8.53) $\left(A_0(x)\big|_{\lambda_\beta=0}\right)$ of $\varphi_1(x)$ is given in [26], Corollary 1, by Equation (55) there; see here (6.55).

Corollary 8.10. (to Theorem 8.4; $\lambda_\beta = 0$, $p(t) = q(t) = 1$, $\lambda_\alpha = \lambda_\nu = 1$, $p = 2$). *In detail: let $\nu \geq 1$, $\gamma_1 \geq 0$, such that $\nu - \gamma_1 \geq 1$, $f_j \in C_{x_0}^\nu([a,b])$ with $f_j^{(i)}(x_0) = 0$, $i = 1, \ldots, n-1$, $n := [\nu]$, $j = 1, \ldots, M \in \mathbb{N}$. Here $x, x_0 \in [a,b]: x \geq x_0$. Then*

$$\int_{x_0}^x \left(\sum_{j=1}^M |(D_{x_0}^{\gamma_1} f_j)(w)|\, |(D_{x_0}^\nu f_j)(w)| \right) dw \tag{8.54}$$

$$\leq \left(\frac{(x-x_0)^{\nu-\gamma_1}}{2\Gamma(\nu-\gamma_1)\sqrt{\nu-\gamma_1}\sqrt{2\nu-2\gamma_1-1}} \right)$$

$$\cdot \left\{ \int_{x_0}^x \left[\sum_{j=1}^M ((D_{x_0}^\nu f_j)(w))^2 \right] dw \right\},$$

all $x_0 \leq x \leq b$.

Proof. Based on our Corollary 8.9 and Corollary 1 of [26], especially Equation (55) there; see here (6.55). □

Corollary 8.11. (to Theorem 8.5, $\lambda_\alpha = 0$, $p(t) = q(t) = 1$). *It holds*

$$\int_{x_0}^x \left\{ \left\{ \sum_{j=1}^{M-1} \left[|(D_{x_0}^{\gamma_2} f_{j+1})(w)|^{\lambda_\beta} |(D_{x_0}^\nu f_j)(w)|^{\lambda_\nu} \right. \right. \right.$$

$$\left. + |(D_{x_0}^{\gamma_2} f_j)(w)|^{\lambda_\beta} |(D_{x_0}^\nu f_{j+1})(w)|^{\lambda_\nu} \right] \right\}$$

$$+ \left[|(D_{x_0}^{\gamma_2} f_M)(w)|^{\lambda_\beta} |(D_{x_0}^\nu f_1)(w)|^{\lambda_\nu} \right.$$

$$\left. \left. + |(D_{x_0}^{\gamma_2} f_1)(w)|^{\lambda_\beta} |(D_{x_0}^\nu f_M)(w)|^{\lambda_\nu} \right] \right\} dw$$

$$\leq 2^{\left(\lambda_\nu + \lambda_\beta/p\right)} \varepsilon_2 \varphi_2(x).$$

$$\left\{ \int_{x_0}^x \left[\sum_{j=1}^M |(D_{x_0}^\nu f_j)(w)|^p \right] dw \right\}^{\left((\lambda_\nu + \lambda_\beta)/p\right)}, \tag{8.55}$$

all $x_0 \leq x \leq b$.

In (8.55), $\left(A_0(x)\big|_{\lambda_\alpha=0}\right)$ of $\varphi_2(x)$ is given in [26], Corollary 3 by Equation (59); see here (6.59).

Corollary 8.12. (to Theorem 8.5, $\lambda_\alpha = 0$, $p(t) = q(t) = 1$, $\lambda_\beta = \lambda_\nu = 1$, $p = 2$). *In detail: let* $\nu \geq 1$, $\gamma_2 \geq 0$ *such that* $\nu - \gamma_2 \geq 1$ *and* $f_j \in C_{x_0}^\nu([a,b])$ *with* $f_j^{(i)}(x_0) = 0$, $i = 0, 1, \ldots, n-1$, $n := [\nu]$, $j = 1, \ldots, M \in \mathbb{N}$. *Here* $x, x_0 \in [a,b]$: $x \geq x_0$. *Then*

$$\int_{x_0}^x \left\{ \left\{ \sum_{j=1}^{M-1} \left[|(D_{x_0}^{\gamma_2} f_{j+1})(w)| \, |(D_{x_0}^\nu f_j)(w)| \right. \right. \right.$$

$$\left. + |(D_{x_0}^{\gamma_2} f_j)(w)| \, |(D_{x_0}^\nu f_{j+1})(w)| \right] \Big\} + \left[|(D_{x_0}^{\gamma_2} f_M)(w)| \, |(D_{x_0}^\nu f_1)(w)| \right.$$

$$\left. \left. + |(D_{x_0}^{\gamma_2} f_1)(w)| \, |(D_{x_0}^\nu f_M)(w)| \right] \right\} dw$$

$$\leq \left(\frac{\sqrt{2}(x - x_0)^{(\nu - \gamma_2)}}{\Gamma(\nu - \gamma_2)\sqrt{\nu - \gamma_2}\sqrt{2\nu - 2\gamma_2 - 1}} \right)$$

$$\left\{ \int_{x_0}^x \left[\sum_{j=1}^M ((D_{x_0}^\nu f_j)(w))^2 \right] dw \right\}, \tag{8.56}$$

all $x_0 \leq x \leq b$.

Proof. From our Corollary 8.11 and Corollary 3 of [26], especially Equation (59); see here (6.59). $\quad\square$

Corollary 8.13. (to Theorem 8.6, $\lambda_\alpha = \lambda_\beta = \lambda_\nu = 1$, $p = 3$, $p(t) = q(t) = 1$). *It holds*

$$\int_{x_0}^x \left[\sum_{j=1}^{M-1} \left[|(D_{x_0}^{\gamma_1} f_j)(w)| \, |(D_{x_0}^{\gamma_2} f_{j+1})(w)| |(D_{x_0}^\nu f_j)(w)| \right. \right.$$

$$\left. + |(D_{x_0}^{\gamma_2} f_j)(w)| \, |(D_{x_0}^{\gamma_1} f_{j+1})(w)| \, |(D_{x_0}^\nu f_{j+1})(w)| \right]$$

$$+ \left[|(D_{x_0}^{\gamma_1} f_1)(w)| \, |(D_{x_0}^{\gamma_2} f_M)(w)| \, |(D_{x_0}^\nu f_1)(w)| \right.$$

$$\left. \left. + |(D_{x_0}^{\gamma_2} f_1)(w)| \, |(D_{x_0}^{\gamma_1} f_M)(w)| \, |(D_{x_0}^\nu f_M)(w)| \right] \right] dw$$

$$\leq 2\varphi_3^*(x) \cdot \left[\int_{x_0}^x \left[\sum_{j=1}^M |(D_{x_0}^\nu f_j)(w)|^3 \, dw \right] \right], \tag{8.57}$$

all $x_0 \leq x \leq b$.
    Here

$$\varphi_3^*(x) := \left(\sqrt[3]{2} + \frac{1}{\sqrt[3]{6}} \right) A_0(x), \tag{8.58}$$

where in this special case

$$A_0(x) := \frac{4(x - x_0)^{(2\nu - \gamma_1 - \gamma_2)}}{\Gamma(\nu - \gamma_1)\Gamma(\nu - \gamma_2)[3(3\nu - 3\gamma_1 - 1)(3\nu - 3\gamma_2 - 1)(2\nu - \gamma_1 - \gamma_2)]^{2/3}} \cdot$$

(8.59)

Proof. From Theorem 8.6 and Equation (62) of [26], which is here Equation (8.59). □

Corollary 8.14. (to Theorem 8.7, $\lambda_\alpha = 1$, $\lambda_{\alpha+1} = 1/2$, $p = 3/2$, $p(t) = q(t) = 1$). *In detail :*

Let $\nu \geq 2$ and $\gamma_1 \geq 0$ such that $\nu - \gamma_1 \geq 2$. Let $f_j \in C_{x_0}^\nu([a,b])$ with $f_j^{(i)}(x_0) = 0$, $i = 0, 1, \ldots, n - 1$, $n := [\nu]$, $j = 1, \ldots, M \in \mathbb{N}$. Here $x, x_0 \in [a, b]$: $x \geq x_0$. *Call*

$$\Phi^*(x) := \left(\frac{2}{\sqrt{3\nu - 3\gamma_1 - 2}}\right)\frac{(x - x_0)^{(3\nu - 3\gamma_1 - 1/2)}}{(\Gamma(\nu - \gamma_1))^{3/2}}, \qquad (8.60)$$

all $x_0 \leq x \leq b$.
Then

$$\int_{x_0}^x \left\{ \left\{ \sum_{j=1}^{M-1} \left[|(D_{x_0}^{\gamma_1} f_j)(w)| \sqrt{|(D_{x_0}^{\gamma_1+1} f_{j+1})(w)|} \right. \right. \right.$$

$$\left. + |(D_{x_0}^{\gamma_1} f_{j+1})(w)| \sqrt{|(D_{x_0}^{\gamma_1+1} f_j)(w)|} \right] \right\}$$

$$+ \left[|(D_{x_0}^{\gamma_1} f_1)(w)| \sqrt{|(D_{x_0}^{\gamma_1+1} f_M)(w)|} \right.$$

$$\left. \left. + |(D_{x_0}^{\gamma_1} f_M)(w)| \sqrt{|(D_{x_0}^{\gamma_1+1} f_1)(w)|} \right] \right\} dw$$

$$\leq 2\Phi^*(x) \cdot \left[\int_{x_0}^x \left(\sum_{j=1}^M |(D_{x_0}^\nu f_j)(w)|^{3/2} \right) dw \right], \qquad (8.61)$$

all $x_0 \leq x \leq b$.

Proof. Based on Theorem 8.7 here, and Equation (64) of [26] to establish coefficient $\Phi^*(x)$ in (8.61). □

Corollary 8.15. (to Theorem 8.8, $p = 2(\lambda_\alpha + \lambda_\nu) > 1$, $p(t) = q(t) = 1$). *It holds*

$$\int_{x_0}^x \left\{ \left\{ \sum_{j=1}^{M-1} \left[|(D_{x_0}^{\gamma_1} f_j)(w)|^{\lambda_\alpha} |(D_{x_0}^{\gamma_2} f_{j+1})(w)|^{\lambda_\alpha + \lambda_\nu} |(D_{x_0}^\nu f_j)(w)|^{\lambda_\nu} \right. \right. \right.$$

$$+ |(D_{x_0}^{\gamma_2} f_j)(w)|^{\lambda_\alpha + \lambda_\nu} |(D_{x_0}^{\gamma_1} f_{j+1})(w)|^{\lambda_\alpha} |(D_{x_0}^{\nu} f_{j+1})(w)|^{\lambda_\nu}] \Big\}$$

$$+ \big[|(D_{x_0}^{\gamma_1} f_1)(w)|^{\lambda_\alpha} |(D_{x_0}^{\gamma_2} f_M)(w)|^{\lambda_\alpha + \lambda_\nu} |(D_{x_0}^{\nu} f_1)(w)|^{\lambda_\nu}$$

$$+ |(D_{x_0}^{\gamma_2} f_1)(w)|^{\lambda_\alpha + \lambda_\nu} |(D_{x_0}^{\gamma_1} f_M)(w)|^{\lambda_\alpha} |(D_{x_0}^{\nu} f_M)(w)|^{\lambda_\nu}] \Big\} \, dw$$

$$\leq 2\tilde{T}(x) \left[\int_{x_0}^{x} \left(\sum_{j=1}^{M} |(D_{x_0}^{\nu} f_j)(w)|^{2(\lambda_\alpha + \lambda_\nu)} \right) dw \right], \qquad (8.62)$$

all $x_0 \leq x \leq b$.

Here $\tilde{T}(x)$ in (8.62) is given precisely by Equations (66)–(70) of [26]; see here (6.66)–(6.70).

Corollary 8.16. (to Theorem 8.8, $p = 4$, $\lambda_\alpha = \lambda_\nu = 1$, $p(t) = q(t) = 1$). *It holds*

$$\int_{x_0}^{x} \Bigg\{ \Bigg\{ \sum_{j=1}^{M-1} \big[|(D_{x_0}^{\gamma_1} f_j)(w)| ((D_{x_0}^{\gamma_2} f_{j+1})(w))^2 |(D_{x_0}^{\nu} f_j)(w)|$$

$$+ ((D_{x_0}^{\gamma_2} f_j)(w))^2 |(D_{x_0}^{\gamma_1} f_{j+1})(w)| \, |(D_{x_0}^{\nu} f_{j+1})(w)|] \Bigg\}$$

$$+ \big[|(D_{x_0}^{\gamma_1} f_1)(w)| ((D_{x_0}^{\gamma_2} f_M)(w))^2 |(D_{x_0}^{\nu} f_1)(w)|$$

$$+ ((D_{x_0}^{\gamma_2} f_1)(w))^2 |(D_{x_0}^{\gamma_1} f_M)(w)| \, |(D_{x_0}^{\nu} f_M)(w)|] \Bigg\} \, dw$$

$$\leq 2\tilde{T}(x) \left[\int_{x_0}^{x} \left(\sum_{j=1}^{M} ((D_{x_0}^{\nu} f_j)(w))^4 \right) dw \right], \qquad (8.63)$$

all $x_0 \leq x \leq b$.

Here in (8.63) we have that $\tilde{T}(x) = T^*(x)$ of Corollary 8 in [26] for it (see there Equations (72)–(76), and here (6.72)–(6.76)) .

Next we present the supremum case.

Theorem 8.17. *Let* $\nu \geq 1$, $\gamma_1, \gamma_2 \geq 0$, *such that* $\nu - \gamma_1 \geq 1$, $\nu - \gamma_2 \geq 1$, *and* $f_j \in C_{x_0}^{\nu}([a,b])$ *with* $f_j^{(i)}(x_0) = 0$, $i = 0, 1, \ldots, n-1$, $n := [\nu]$, $j = 1, \ldots, M \in \mathbb{N}$. *Here* $x, x_0 \in [a,b]: x \geq x_0$. *Consider* $p(x) \geq 0$ *a continuous function on* $[x_0, b]$. *Let* $\lambda_\alpha, \lambda_\beta, \lambda_\nu \geq 0$. *Set*

$$\rho(x) := \frac{(x - x_0)^{(\nu \lambda_\alpha - \gamma_1 \lambda_\alpha + \nu \lambda_\beta - \gamma_2 \lambda_\beta + 1)} \|p(x)\|_\infty}{(\nu \lambda_\alpha - \gamma_1 \lambda_\alpha + \nu \lambda_\beta - \gamma_2 \lambda_\beta + 1)(\Gamma(\nu - \gamma_1 + 1))^{\lambda_\alpha} (\Gamma(\nu - \gamma_2 + 1))^{\lambda_\beta}}.$$

$$(8.64)$$

Then

$$\int_{x_0}^{x} p(w)\bigg\{\bigg\{\sum_{j=1}^{M-1}\big[|(D_{x_0}^{\gamma_1}f_j)(w)|^{\lambda_\alpha}|(D_{x_0}^{\gamma_2}f_{j+1})(w)|^{\lambda_\beta}|(D_{x_0}^{\nu}f_j)(w)|^{\lambda_\nu}$$

$$+|(D_{x_0}^{\gamma_2}f_j)(w)|^{\lambda_\beta}|(D_{x_0}^{\gamma_1}f_{j+1}(w)|^{\lambda_\alpha}|(D_{x_0}^{\nu}f_{j+1})(w)|^{\lambda_\nu}\big]\bigg\}$$

$$+\big[|(D_{x_0}^{\gamma_1}f_1)(w)|^{\lambda_\alpha}|(D_{x_0}^{\gamma_2}f_M)(w)|^{\lambda_\beta}|(D_{x_0}^{\nu}f_1)(w)|^{\lambda_\nu}$$

$$+|(D_{x_0}^{\gamma_2}f_1)(w)|^{\lambda_\beta}|(D_{x_0}^{\gamma_1}f_M)(w)|^{\lambda_\alpha}|(D_{x_0}^{\nu}f_M)(w)|^{\lambda_\nu}\big]\bigg\}\,dw$$

$$\leq \rho(x)\bigg\{\sum_{j=1}^{M}\{\|(D_{x_0}^{\nu}f_j)\|_{\infty}^{2(\lambda_\alpha+\lambda_\nu)}+\|(D_{x_0}^{\nu}f_j)\|_{\infty}^{2\lambda_\beta}\}\bigg\}, \qquad (8.65)$$

all $x_0 \leq x \leq b$.

Proof. Based on Theorem 7 of [26]; see here Theorem 6.18. $\quad\square$

Similarly we give

Theorem 8.18. (*as in Theorem 8.17,* $\lambda_\beta = 0$). *It holds*

$$\int_{x_0}^{x} p(w)\bigg(\sum_{j=1}^{M}|(D_{x_0}^{\gamma_1}f_j)(w)|^{\lambda_\alpha}|(D_{x_0}^{\nu}f_j)(w)|^{\lambda_\nu}\bigg)\,dw$$

$$\leq \bigg(\frac{(x-x_0)^{(\nu\lambda_\alpha-\gamma_1\lambda_\alpha+1)}\|p(x)\|_\infty}{(\nu\lambda_\alpha-\gamma_1\lambda_\alpha+1)(\Gamma(\nu-\gamma_1+1))^{\lambda_\alpha}}\bigg) \qquad (8.66)$$

$$\cdot\bigg(\sum_{j=1}^{M}\|D_{x_0}^{\nu}f_j\|_{\infty}^{\lambda_\alpha+\lambda_\nu}\bigg),$$

all $x_0 \leq x \leq b$.

Proof. Based on Theorem 8 of [26]; see here Theorem 6.19. $\quad\square$

It follows

Theorem 8.19. (*as in Theorem 8.17,* $\lambda_\beta = \lambda_\alpha + \lambda_\nu$). *It holds*

$$\int_{x_0}^{x} p(w)\bigg\{\bigg\{\sum_{j=1}^{M-1}\big[|(D_{x_0}^{\gamma_1}f_j)(w)|^{\lambda_\alpha}|(D_{x_0}^{\gamma_2}f_{j+1})(w)|^{\lambda_\alpha+\lambda_\nu}|(D_{x_0}^{\nu}f_j)(w)|^{\lambda_\nu}$$

$$+|(D_{x_0}^{\gamma_2} f_j)(w)|^{\lambda_\alpha+\lambda_\nu}|(D_{x_0}^{\gamma_1} f_{j+1})(w)|^{\lambda_\alpha}|(D_{x_0}^{\nu} f_{j+1})(w)|^{\lambda_\nu}]\Big\}$$

$$+\Big[|(D_{x_0}^{\gamma_1} f_1)(w)|^{\lambda_\alpha}|(D_{x_0}^{\gamma_2} f_M)(w)|^{\lambda_\alpha+\lambda_\nu}|(D_{x_0}^{\nu} f_1)(w)|^{\lambda_\nu}$$

$$+|(D_{x_0}^{\gamma_2} f_1)(w)|^{\lambda_\alpha+\lambda_\nu}|(D_{x_0}^{\gamma_1} f_M)(w)|^{\lambda_\alpha}|(D_{x_0}^{\nu} f_M)(w)|^{\lambda_\nu}]\Big\}\,dw$$

$$\leq \left(\frac{2(x-x_0)^{(2\nu\lambda_\alpha-\gamma_1\lambda_\alpha+\nu\lambda_\nu-\gamma_2\lambda_\alpha-\gamma_2\lambda_\nu+1)}}{(2\nu\lambda_\alpha-\gamma_1\lambda_\alpha+\nu\lambda_\nu-\gamma_2\lambda_\alpha-\gamma_2\lambda_\nu+1)(\Gamma(\nu-\gamma_1+1))^{\lambda_\alpha}}\right.$$

$$\left.\frac{\|p(x)\|_\infty}{(\Gamma(\nu-\gamma_2+1))^{(\lambda_\alpha+\lambda_\nu)}}\right)\cdot\left(\sum_{j=1}^{M}\|D_{x_0}^{\nu} f_j\|_\infty^{2(\lambda_\alpha+\lambda_\nu)}\right), \qquad (8.67)$$

all $x_0 \leq x \leq b$.

Proof. By Theorem 9 of [26]; see here Theorem 6.20. $\square$

We continue with

Theorem 8.20. *(as in Theorem 8.17, $\lambda_\nu = 0$, $\lambda_\alpha = \lambda_\beta$). It holds*

$$\int_{x_0}^{x} p(w)\Big\{\Big\{\sum_{j=1}^{M-1}[|(D_{x_0}^{\gamma_1} f_j)(w)|^{\lambda_\alpha}|(D_{x_0}^{\gamma_2} f_{j+1})(w)|^{\lambda_\alpha}$$

$$+|(D_{x_0}^{\gamma_2} f_j)(w)|^{\lambda_\alpha}|(D_{x_0}^{\gamma_1} f_{j+1})(w)|^{\lambda_\alpha}]\Big\}$$

$$+\Big[|(D_{x_0}^{\gamma_1} f_1)(w)|^{\lambda_\alpha}|(D_{x_0}^{\gamma_2} f_M)(w)|^{\lambda_\alpha}$$

$$+|(D_{x_0}^{\gamma_2} f_1)(w)|^{\lambda_\alpha}|(D_{x_0}^{\gamma_1} f_M)(w)|^{\lambda_\alpha}]\Big\}\,dw$$

$$\leq 2\rho^*(x)\left[\sum_{j=1}^{M}\|D_{x_0}^{\nu} f_j\|_\infty^{2\lambda_\alpha}\right], \qquad (8.68)$$

all $x_0 \leq x \leq b$.
Here we have

$$\rho^*(x) := \left(\frac{(x-x_0)^{(2\nu\lambda_\alpha-\gamma_1\lambda_\alpha-\gamma_2\lambda_\alpha+1)}\|p(x)\|_\infty}{(2\nu\lambda_\alpha-\gamma_1\lambda_\alpha-\gamma_2\lambda_\alpha+1)(\Gamma(\nu-\gamma_1+1))^{\lambda_\alpha}(\Gamma(\nu-\gamma_2+1))^{\lambda_\alpha}}\right).$$
$$(8.69)$$

Proof. Based on Theorem 10 of [26]; see here Theorem 6.21. $\square$

Next we give

Theorem 8.21. (as in Theorem 8.17, $\lambda_\alpha = 0$, $\lambda_\beta = \lambda_\nu$). It holds

$$
\int_{x_0}^{x} p(w) \Bigg\{ \Bigg\{ \sum_{j=1}^{M-1} \big[|(D_{x_0}^{\gamma_2} f_{j+1})(w)|^{\lambda_\beta} |(D_{x_0}^{\nu} f_j)(w)|^{\lambda_\beta}
$$

$$
+ |(D_{x_0}^{\gamma_2} f_j)(w)|^{\lambda_\beta} |(D_{x_0}^{\nu} f_{j+1})(w)|^{\lambda_\beta} \big] \Bigg\}
$$

$$
+ \big[|(D_{x_0}^{\gamma_2} f_M)(w)|^{\lambda_\beta} |(D_{x_0}^{\nu} f_1)(w)|^{\lambda_\beta}
$$

$$
+ |(D_{x_0}^{\gamma_2} f_1)(w)|^{\lambda_\beta} |(D_{x_0}^{\nu} f_M)(w)|^{\lambda_\beta} \big] \Bigg\} dw
$$

$$
\leq 2 \cdot \left(\frac{(x - x_0)^{(\nu\lambda_\beta - \gamma_2\lambda_\beta + 1)} \|p(x)\|_\infty}{(\nu\lambda_\beta - \gamma_2\lambda_\beta + 1)(\Gamma(\nu - \gamma_2 + 1))^{\lambda_\beta}} \right)
$$

$$
\left[\sum_{j=1}^{M} \|D_{x_0}^{\nu} f_j\|_\infty^{2\lambda_\beta} \right], \tag{8.70}
$$

all $x_0 \leq x \leq b$.

Proof. Based on Theorem 11 of [26]; see here Theorem 6.22. □

Some special cases follow.

Corollary 8.22. (to Theorem 8.20, all are as in Theorem 8.17, $\lambda_\nu = 0$, $\lambda_\alpha = \lambda_\beta$, $\gamma_2 = \gamma_1 + 1$). It holds

$$
\int_{x_0}^{x} p(w) \Bigg\{ \Bigg\{ \sum_{j=1}^{M-1} \big[|(D_{x_0}^{\gamma_1} f_j)(w)|^{\lambda_\alpha} |(D_{x_0}^{\gamma_1+1} f_{j+1})(w)|^{\lambda_\alpha}
$$

$$
+ |(D_{x_0}^{\gamma_1+1} f_j)(w)|^{\lambda_\alpha} |(D_{x_0}^{\gamma_1} f_{j+1})(w)|^{\lambda_\alpha} \big] \Bigg\}
$$

$$
+ \big[|(D_{x_0}^{\gamma_1} f_1)(w)|^{\lambda_\alpha} |(D_{x_0}^{\gamma_1+1} f_M)(w)|^{\lambda_\alpha}
$$

$$
+ |(D_{x_0}^{\gamma_1+1} f_1)(w)|^{\lambda_\alpha} |(D_{x_0}^{\gamma_1} f_M)(w)|^{\lambda_\alpha} \big] \Bigg\} dw
$$

$$
\leq 2 \cdot \left(\frac{(x - x_0)^{(2\nu\lambda_\alpha - 2\gamma_1\lambda_\alpha - \lambda_\alpha + 1)} \|p(x)\|_\infty}{(2\nu\lambda_\alpha - 2\gamma_1\lambda_\alpha - \lambda_\alpha + 1)(\nu - \gamma_1)^{\lambda_\alpha} (\Gamma(\nu - \gamma_1))^{2\lambda_\alpha}} \right)
$$

$$
\cdot \left[\sum_{j=1}^{M} \|D_{x_0}^{\nu} f_j\|_\infty^{2\lambda_\alpha} \right], \tag{8.71}
$$

all $x_0 \leq x \leq b$.

Proof. Based on Corollary 9 of [26]; see here Corollary 6.23. □

Corollary 8.23. (to Corollary 8.22). *In detail: let $\nu \geq 2$, $\gamma_1 \geq 0$, such that $\nu - \gamma_1 \geq 2$ and $f_j \in C_{x_0}^\nu([a,b])$ with $f_j^{(i)}(x_0) = 0$, $i = 0,1,\ldots,n-1$, $n := [\nu]$, $j = 1,\ldots,M \in \mathbb{N}$. Here $x, x_0 \in [a,b]: x \geq x_0$. Then*

$$\int_{x_0}^x \left\{ \left\{ \sum_{j=1}^{M-1} \left[|(D_{x_0}^{\gamma_1} f_j)(w)| \, |(D_{x_0}^{\gamma_1+1} f_{j+1})(w)| \right. \right. \right.$$

$$\left. + |(D_{x_0}^{\gamma_1+1} f_j)(w)| \, |(D_{x_0}^{\gamma_1} f_{j+1})(w)| \right] \right\}$$

$$+ \left[|(D_{x_0}^{\gamma_1} f_1)(w)| \, |(D_{x_0}^{\gamma_1+1} f_M)(w)| \right.$$

$$\left. \left. + |(D_{x_0}^{\gamma_1+1} f_1)(w)| \, |(D_{x_0}^{\gamma_1} f_M)(w)| \right] \right\} dw$$

$$\leq \left(\frac{(x-x_0)^{2(\nu-\gamma_1)}}{(\nu-\gamma_1)^2 (\Gamma(\nu-\gamma_1))^2} \right) \left(\sum_{j=1}^M \|D_{x_0}^\nu f_j\|_\infty^2 \right), \qquad (8.72)$$

all $x_0 \leq x \leq b$.

Proof. Based on Corollary 10 of [26]; see here Corollary 6.24. □

Corollary 8.24. (to Corollary 8.23). *It holds*

$$\int_{x_0}^x \left(\sum_{j=1}^M |(D_{x_0}^{\gamma_1} f_j)(w)| \, |(D_{x_0}^{\gamma_1+1} f_j)(w)| \right) dw$$

$$\leq \left(\frac{(x-x_0)^{2(\nu-\gamma_1)}}{2(\nu-\gamma_1)^2 (\Gamma(\nu-\gamma_1))^2} \right) \left(\sum_{j=1}^M \|D_{x_0}^\nu f_j\|_\infty^2 \right), \quad (8.73)$$

all $x_0 \leq x \leq b$.

Proof. Based on Equation (97) of [26]; see here (6.97). □

8.4 Applications

We present our first application.

Theorem 8.25. *Let $\nu \geq 1$, $\gamma_i \geq 0$, $\nu - \gamma_i \geq 1$, $i = 1,\ldots,r \in \mathbb{N}$, $n := [\nu]$, $f_j \in C_a^\nu([a,b])$, $j = 1,2,3,\ldots,M$, $f_j^{(i)}(a) = a_{ij} \in \mathbb{R}$, $i = 0,1,\ldots,n-1$. Furthermore we have for $j = 1,2,\ldots,M$ that*

$$(D_a^\nu f_j)(t) = F_j\big(t, \{(D_a^{\gamma_i} f_1)(t)\}_{i=1}^r, \{(D_a^{\gamma_i} f_2)(t)\}_{i=1}^r, \ldots, \{(D_a^{\gamma_i} f_M)(t)\}_{i=1}^r\big),$$
$$(8.74)$$

all $t \in [a, b]$.

Here F_j are continuous functions on $[a, b] \times (\mathbb{R}^r)^M$ and satisfy the Lipschitz condition

$$\left| F_j(t, x_{11}, x_{12}, \ldots, x_{1r}, x_{21}, \ldots, x_{2r}, x_{31}, \ldots, x_{3r}, \ldots, x_{M1}, \ldots, x_{Mr}) \right.$$
$$\left. - F_j(t, x'_{11}, x'_{12}, \ldots, x'_{1r}, x'_{21}, \ldots, x'_{2r}, x'_{31}, \ldots, x'_{3r}, x'_{M1}, \ldots, x'_{Mr}) \right|$$
$$\leq \sum_{i=1}^{r} \left(\sum_{\ell=1}^{M} q_{\ell,i,j}(t) |x_{\ell i} - x'_{\ell i}| \right), \tag{8.75}$$

$j = 1, 2, \ldots, M$, where all $q_{\ell,i,j} \geq 0$, $1 \leq i \leq r$, are continuous functions over $[a, b]$.

Call

$$W := \max\{\|q_{\ell,i,j}\|_\infty, \ell, j = 1, 2, \ldots, M, i = 1, \ldots, r\}. \tag{8.76}$$

Assume here that

$$\phi^*(b) := W \left(\frac{1}{2} + \frac{M-1}{\sqrt{2}} \right) \left(\sum_{i=1}^{r} \left(\frac{(b-a)^{\nu-\gamma_i}}{\Gamma(\nu-\gamma_i)\sqrt{\nu-\gamma_i}\sqrt{2\nu-2\gamma_i-1}} \right) \right) < 1. \tag{8.77}$$

Then if system (8.74) has two M-tuples of solutions $(f_1, f_2, \ldots, f_M)$ and $(f_1^*, f_2^*, \ldots, f_M^*)$ we prove that

$$f_j = f_j^*, \quad j = 1, 2, \ldots, M;$$

that is, we have uniqueness of solution.

Proof. Assume that there are two M-tuples of solutions $(f_1, f_2, \ldots, f_M)$ and $(f_1^*, \ldots, f_M^*)$ satisfying the system (8.74). Set $g_j := f_j - f_j^*$, $j = 1, 2, \ldots, M$. Then $g_j^{(i)} = f_j^{(i)} - f_j^{*(i)}$ and $g_j^{(i)}(a) = 0$, $i = 0, 1, \ldots, n-1$; $j = 1, 2, \ldots, M$. It holds

$$\begin{aligned}(D_a^\nu g_j)(t) &= F_j\left(t, \{(D_a^{\gamma_i} f_1)(t)\}_{i=1}^r, \ldots, \{(D_a^{\gamma_i} f_M)(t)\}_{i=1}^r\right) \\ &\quad - F_j\left(t, \{(D_a^{\gamma_i} f_1^*)(t)\}_{i=1}^r, \ldots, \{(D_a^{\gamma_i} f_M^*)(t)\}_{i=1}^r\right).\end{aligned}$$

Therefore by (8.75) we get

$$\begin{aligned}|(D_a^\nu g_j)(t)| \leq \sum_{i=1}^{r} &\big[q_{1,i,j}(t)|(D_a^{\gamma_i} g_1)(t)| + q_{2,i,j}(t)|(D_a^{\gamma_i} g_2)(t)| \\ &+ \cdots + q_{M,i,j}(t)|(D_a^{\gamma_i} g_M)(t)| \big].\end{aligned}$$

And thus

$$\begin{aligned}|(D_a^\nu g_j)(t)| \leq \sum_{i=1}^{r} &\big[\|q_{1,i,j}\|_\infty|(D_a^{\gamma_i} g_1)(t)| + \|q_{2,i,j}\|_\infty|(D_a^{\gamma_i} g_2)(t)| \\ &+ \cdots + \|q_{M,i,j}\|_\infty|(D_a^{\gamma_i} g_M)(t)| \big].\end{aligned}$$

Furthermore we have

$$|(D_a^\nu g_j)(t)| \leq W\left\{\sum_{i=1}^{r}\left[|(D_a^{\gamma_i}g_1)(t)| + |(D_a^{\gamma_i}g_2)(t)|\right.\right.$$

$$\left.\left. + \cdots + |(D_a^{\gamma_i}g_M)(t)|\right]\right\}. \tag{8.78}$$

Clearly (8.78) implies

$$\sum_{j=1}^{M}((D_a^\nu g_j)(t))^2 \leq W\left\{\sum_{i=1}^{r}\sum_{j=1}^{M}\left[|(D_a^{\gamma_i}g_1)(t)|\,|(D_a^\nu g_j)(t)|\right.\right.$$

$$+ |(D_a^{\gamma_i}g_2)(t)|\,|(D_a^\nu g_j)(t)|$$

$$\left.\left. + \cdots + |(D_a^{\gamma_i}g_M)(t)|\,|(D_a^\nu g_j)(t)|\right]\right\}. \tag{8.79}$$

Integrating (8.79) we observe

$$I := \int_a^b \left(\sum_{j=1}^{M}((D_a^\nu g_j)(t))^2\right)dt$$

$$\leq W\left\{\sum_{i=1}^{r}\sum_{j=1}^{M}\left[\int_a^b |(D_a^{\gamma_i}g_1)(t)|\,|(D_a^\nu g_j)(t)|\,dt\right.\right.$$

$$+ \int_a^b |(D_a^{\gamma_i}g_2)(t)|\,|(D_a^\nu g_j)(t)|\,dt$$

$$\left.\left. + \cdots + \int_a^b |(D_a^{\gamma_i}g_M)(t)|\,|(D_a^\nu g_j)(t)|\,dt\right]\right\}. \tag{8.80}$$

That is,

$$I \leq W\left\{\sum_{i=1}^{r}\left[\left(\int_a^b\left(\sum_{\lambda=1}^{M}|(D_a^{\gamma_i}g_\lambda)(t)|\,|(D_a^\nu g_\lambda)(t)|\right)dt\right)\right.\right.$$

$$+ \sum_{\substack{\tau,m\in\{1,\ldots,M\}\\ \tau\neq m}}\left(\int_a^b\left(|(D_a^{\gamma_i}g_m)(t)|\,|(D_a^\nu g_\tau)(t)|\right.\right.$$

$$\left.\left.\left.\left. + |(D_a^{\gamma_i}g_\tau)(t)|\,|(D_a^\nu g_m)(t)|\right)dt\right)\right]\right\}. \tag{8.81}$$

Using Corollary 2 from here and Corollary 4 of [26] (see here Corollary 6.11 and Corollary 6.13), we obtain

$$
I \leq W \left\{ \sum_{i=1}^{r} \left[\left(\frac{(b-a)^{\nu-\gamma_i}}{2\Gamma(\nu-\gamma_i)\sqrt{\nu-\gamma_i}\sqrt{2\nu-2\gamma_i-1}} \right) I \right. \right.
$$
$$
\left. \left. + \left(\frac{(b-a)^{\nu-\gamma_i}}{\sqrt{2}\Gamma(\nu-\gamma_i)\sqrt{\nu-\gamma_i}\sqrt{2\nu-2\gamma_i-1}} \right) (M-1)I \right] \right\}. \qquad (8.82)
$$

That is, we get that

$$
I \leq \phi^*(b) \cdot I. \qquad (8.83)
$$

If $I \neq 0$ then $\phi^*(b) \geq 1$, a contradiction by the assumption that $\phi^*(b) < 1$; see (8.77). Therefore $I = 0$, implying that

$$
\sum_{\lambda=1}^{M} \left((D_a^\nu g_\lambda)(t) \right)^2 = 0, \quad \text{a.e. in } [a,b].
$$

That is,

$$
(D_a^\nu g_\lambda)^2(t) = 0, \quad \text{a.e. in } [a,b].
$$

That is,

$$
(D_a^\nu g_\lambda)(t) = 0, \quad \lambda = 1,2,\ldots,M, \text{ a.e. in } [a,b].
$$

But for $\lambda = 1,2,\ldots,M$ we get that

$$
g_\lambda^{(i)}(a) = 0, \quad 0 \leq i \leq n-1.
$$

Hence from fractional Taylor's Theorem 8.1 we get that $g_\lambda(t) = 0$ on $[a,b]$. That is,

$$
f_\lambda = f_\lambda^*, \quad \lambda = 1,2,\ldots,M,
$$

proving the uniqueness argument of this theorem.   □

Another related application follows.

Theorem 8.26. *Let $\nu \geq 1$, $\gamma_i \geq 0$, $\nu - \gamma_i \geq 1$, $i = 1,\ldots,r \in \mathbb{N}$, $n := [\nu]$, $f_j \in C_a^\nu([a,b])$, $j = 1,2,\ldots,M$; $f_j^{(i)}(a) = 0$, $i = 0,1,\ldots,n-1$, and $(D_a^\nu f_j)(a) = A_j \in \mathbb{R}$. Furthermore for $a \leq t \leq b$ we have holding the system of fractional differential equations*

$$
(D_a^\nu f_j)'(x) = F_j\big(t, (\{(D_a^{\gamma_i} f_\lambda)(t)\}_{i=1}^r, (D_a^\nu f_\lambda)(t));
$$
$$
\lambda = 1,2,\ldots,M), \quad j = 1,2,\ldots,M. \qquad (8.84)
$$

Here F_j are continuous functions on $[a,b] \times (\mathbb{R}^{r+1})^M$ such that

$$
|F_j(t, x_{11}, x_{12}, \ldots, x_{1r}, x_{1,r+1}; x_{21}, x_{22}, \ldots, x_{2r}, x_{2,r+1}; x_{31}, x_{32}, \ldots, x_{3,r+1};
$$

$$x_{M1}, x_{M2}, \ldots, x_{M,r+1})| \leq \sum_{i=1}^{r} \left(\sum_{\ell=1}^{M} q_{\ell,i,j}(t) |x_{\ell i}| \right), \qquad (8.85)$$

where

$$q_{\ell,i,j}(t) \geq 0, \quad 1 \leq i \leq r; \quad \ell, j = 1, 2, \ldots, M,$$

are continuous functions on $[a, b]$.

Call

$$W := \max\{\|q_{\ell,i,j}\|_{\infty}; \ell, j = 1, 2, \ldots, M, i = 1, \ldots, r\}. \qquad (8.86)$$

Also we set $(a \leq x \leq b)$

$$\theta(x) := \sum_{\lambda=1}^{M} \left((D_a^\nu f_\lambda)(x) \right)^2, \qquad (8.87)$$

$$\rho := \sum_{\lambda=1}^{M} A_\lambda^2, \qquad (8.88)$$

$$Q(x) := W \left(1 + \sqrt{2}(M-1) \right)$$
$$\left(\sum_{i=1}^{r} \left(\frac{(x-a)^{\nu-\gamma_i}}{\Gamma(\nu-\gamma_i)\sqrt{\nu-\gamma_i}\sqrt{2\nu-2\gamma_i-1}} \right) \right) \qquad (8.89)$$

and

$$\chi(x) := \sqrt{\rho} \cdot \left\{ 1 + Q(x) \cdot e^{\left(\int_a^x Q(s)ds \right)} \cdot \left[\int_a^x \left(e^{-\left(\int_a^t Q(s)ds \right)} \right) dt \right] \right\}^{1/2}. \qquad (8.90)$$

Then

$$\sqrt{\theta(x)} \leq \chi(x), \quad a \leq x \leq b. \qquad (8.91)$$

Consequently we get

$$|(D_a^\nu f_j)(x)| \leq \chi(x), \qquad (8.92)$$

$$|f_j(x)| \leq \frac{1}{\Gamma(\nu)} \int_a^x (x-t)^{\nu-1} \chi(t)\, dt, \qquad (8.93)$$

all $a \leq x \leq b$, $j = 1, 2, \ldots, M$. *Also it holds*

$$|(D_a^{\gamma_i} f_j)(x)| \leq \frac{1}{\Gamma(\nu-\gamma_i)} \int_a^x (x-t)^{\nu-\gamma_i-1} \chi(t)\, dt, \qquad (8.94)$$

all $a \leq x \leq b$, $j = 1, 2, \ldots, M$, $i = 1, \ldots, r$.

Proof. We easily get that $(a \leq x \leq b)$,

$$\int_a^x (D_a^\nu f_j)(t)(D_a^\nu f_j)'(t)dt - \int_a^x (D_a^\nu f_j)(t) \cdot F_j \left(t, \left(\{(D_a^{\gamma_i} f_\lambda)(t)\}_{i=1}^r, \right. \right.$$
$$\left. \left. (D_a^\nu f_\lambda)(t) \right); \lambda = 1, 2, \ldots, M \right) dt. \qquad (8.95)$$

Hence we obtain

$$\left.\frac{((D_a^\nu f_j)(t))^2}{2}\right|_a^x \le \int_a^x |(D_a^\nu f_j)(t)| \, |F_j \cdots | dt$$

$$\le \int_a^x |(D_a^\nu f_j)(t)| \left[\sum_{i=1}^r \left(\sum_{\ell=1}^M q_{\ell,i,j}(t) |(D_a^{\gamma_i} f_\ell)(t)| \right)\right] dt$$

$$\le \sum_{i=1}^r \left(\sum_{\ell=1}^M \|q_{\ell,i,j}\|_\infty \int_a^x |(D_a^\nu f_j)(t)| \, |(D_a^{\gamma_i} f_\ell)(t)| \, dt \right)$$

$$\le W\left(\sum_{i=1}^r \sum_{\ell=1}^M \left(\int_a^x |(D_a^\nu f_j)(t)| \, |(D_a^{\gamma_i} f_\ell)(t)| \, dt \right)\right).$$

Thus we have for $j = 1, \ldots, M$ that

$$((D_a^\nu f_j)(x))^2 \le A_j^2 + 2W\Bigg\{\sum_{i=1}^r \sum_{\ell=1}^M$$
$$\cdot \left(\int_a^x |(D_a^\nu f_j)(t)| \, |(D_a^{\gamma_i} f_\ell)(t)| \, dt \right)\Bigg\}. \qquad (8.96)$$

Consequently it holds

$$\theta(x) \le \rho + 2W\Bigg\{\sum_{i=1}^r \left(\sum_{j=1}^M \sum_{\ell=1}^M \left(\int_a^x |(D_a^\nu f_j)(t)| \, |(D_a^{\gamma_i} f_\ell)(t)| \, dt \right)\right)\Bigg\}$$

$$= \rho + 2W\Bigg\{\sum_{i=1}^r \Bigg\{\int_a^x \left(\sum_{\lambda=1}^M |(D_a^{\gamma_i} f_\lambda)(t)| \, |(D_a^\nu f_\lambda)(t)| \right) dt$$

$$+ \sum_{\substack{\tau,m\in\{1,\ldots,M\} \\ \tau \ne m}} \left(\int_a^x (|(D_a^{\gamma_i} f_m)(t)| \, |(D_a^\nu f_\tau)(t)| \right.$$

$$\left. + |(D_a^{\gamma_i} f_\tau)(t)| \, |(D_a^\nu f_m)(t)|) \, dt \right)\Bigg\}\Bigg\}. \qquad (8.97)$$

Using Corollary 2 from here and Corollary 4 of [26] (see here Corollary 6.11 and Corollary 6.13) we obtain

$$\theta(x) \le \rho + 2W\Bigg\{\sum_{i=1}^r \Bigg\{\left(\frac{(x-a)^{\nu-\gamma_i}}{2\Gamma(\nu-\gamma_i)\sqrt{\nu-\gamma_i}\sqrt{2\nu-2\gamma_i-1}}\right)$$

$$\left(\int_a^x \theta(t) \, dt\right) + \left(\frac{(x-a)^{\nu-\gamma_i}}{\sqrt{2}\Gamma(\nu-\gamma_i)\sqrt{\nu-\gamma_i}\sqrt{2\nu-2\gamma_i-1}}\right) \qquad (8.98)$$

$$(M-1)\left(\int_a^x \theta(t) \, dt\right)\Bigg\}\Bigg\}.$$

Hence we have

$$\theta(x) \leq \rho + Q(x) \int_a^x \theta(t)\, dt, \quad \text{all } a \leq x \leq b. \tag{8.99}$$

Here $\rho \geq 0$, $Q(x) \geq 0$, $Q(a) = 0$, $\theta(x) \geq 0$, all $a \leq x \leq b$. As in the proof of Theorem 13 of [26] (see also [17]), we get (8.91) and (8.92). Using (8.9) we get (8.93), and using (8.10) we establish (8.94). □

Finally we give a specialized application.

Theorem 8.27. *Let $a \neq b$, $\nu \geq 2$, $\gamma_i \geq 0$, $\nu - \gamma_i \geq 1$, $i = 1, \ldots, r \in \mathbb{N}$, $n := [\nu]$, $f_j \in C_a^\nu([a,b])$, $j = 1, 2, \ldots, M$; $f_j^{(i)}(a) = 0$, $i = 0, 1, \ldots, n-1$, and*

$$(D_a^\nu f_j)(a) = A_j \in \mathbb{R}. \tag{8.100}$$

Furthermore for $a \leq t \leq b$ we have holding the system of fractional differential equations

$$(D_a^\nu f_j)'(t) \;=\; F_j\big(t, (\{(D_a^{\gamma_i} f_\ell)(t)\}_{i=1}^r, (D_a^\nu f_\ell)(t)); $$
$$\ell = 1, \ldots, M\big), \quad \text{for } j = 1, 2, \ldots, M. \tag{8.101}$$

For fixed $i_* \in \{1, \ldots, r\}$ we assume that $\gamma_{i_*+1} = \gamma_{i_*} + 1$, and $\nu - \gamma_{i_*} \geq 2$, where γ_{i_*}, $\gamma_{i_*+1} \in \{\gamma_1, \ldots, \gamma_r\}$. Call $k := \gamma_{i_*}$, $\gamma := \gamma_{i_*} + 1$; that is, $\gamma = k+1$.

Here F_j are continuous functions on $[a,b] \times (\mathbb{R}^{r+1})^M$ such that

$$|F_j(t, x_{11}, x_{12}, \ldots, x_{1r}, x_{1,r+1}; x_{21}, x_{22}, \ldots, x_{2r}, x_{2,r+1};$$

$$x_{31}, x_{32}, \ldots, x_{3r}, x_{3,r+1}; \ldots; x_{M1}, x_{M2}, \ldots, x_{Mr}, x_{M,r+1})|$$

$$\leq \left\{ \left\{ \sum_{\ell=1}^{M-1} \left(q_{\ell,1,j}(t)|x_{\ell i_*}|\sqrt{|x_{\ell+1,i_*+1}|} + q_{\ell,2,j}(t)|x_{\ell+1,i_*}|\sqrt{|x_{\ell,i_*+1}|} \right) \right\} \right. $$

$$\left. + \left(q_{M,1,j}(t)|x_{1i_*}|\sqrt{|x_{M,i_*+1}|} + q_{M,2,j}(t)|x_{Mi_*}|\sqrt{|x_{1,i_*+1}|} \right) \right\}, \tag{8.102}$$

where all $0 \leq q_{\ell,1,j}$, $q_{\ell,2,j} \not\equiv 0$ are continuous functions over $[a,b]$.
Put

$$W := \max\{\|q_{\ell,1,j}\|_\infty, \|q_{\ell,2,j}\|_\infty\}_{\ell,j=1}^M. \tag{8.103}$$

Also set

$$\theta(x) := \sum_{j=1}^M |(D_a^\nu f_j)(x)|, \quad a \leq x \leq b, \tag{8.104}$$

$$\rho := \sum_{j=1}^M |A_j|, \tag{8.105}$$

$$\Phi^*(x) := \left(\frac{2}{\sqrt{3\nu - 3k - 2}} \right) \frac{(x-a)^{((3\nu-3k-1)/2)}}{(\Gamma(\nu-k))^{3/2}}, \tag{8.106}$$

all $a \leq x \leq b$, and

$$Q(x) := 2MW\Phi^*(x), \quad a \leq x \leq b, \tag{8.107}$$

$$\sigma := \|Q(x)\|_\infty, \quad a \leq x \leq b. \tag{8.108}$$

We assume that

$$(b-a)\sigma\sqrt{\rho} < 2. \tag{8.109}$$

Call

$$\tilde{\varphi}(x) := \rho + Q(x) \cdot \left[\frac{4\rho^{3/2}(x-a) - \sigma\rho^2(x-a)^2}{(2 - \sigma\sqrt{\rho}(x-a))^2} \right], \quad \text{all } a \leq x \leq b. \tag{8.110}$$

Then

$$\theta(x) \leq \tilde{\varphi}(x), \quad \text{all } a \leq x \leq b; \tag{8.111}$$

in particular we have

$$|(D_a^\nu f_j)(x)| \leq \tilde{\varphi}(x), \quad j = 1, \ldots, M, \quad \text{all } a \leq x \leq b. \tag{8.112}$$

Furthermore we get

$$|f_j(x)| \leq \frac{1}{\Gamma(\nu)} \int_a^x (x-t)^{\nu-1} \tilde{\varphi}(t) \, dt, \tag{8.113}$$

and

$$|(D_a^{\gamma_i} f_j)(x)| \leq \frac{1}{\Gamma(\nu - \gamma_i)} \int_a^x (x-t)^{\nu-\gamma_i-1} \tilde{\varphi}(t) \, dt, \tag{8.114}$$

$j = 1, \ldots, M; i = 1, \ldots, r; \text{ all } a \leq x \leq b.$

Proof. Notice that $W > 0$ and $\sigma > 0$. For $a \leq x \leq b$ we get

$$\int_a^x (D_a^\nu f_j)'(t) dt = \int_a^x F_j\big(t, (\{D_a^{\gamma_i} f_\ell)(t)\}_{i=1}^r, (D_a^\nu f_\ell)(t));$$

$$\ell = 1, \ldots, M\big) \, dt, \quad j = 1, \ldots, M. \tag{8.115}$$

That is,

$$(D_a^\nu f_j)(x) = A_j + \int_a^x F_j(t, \ldots) \, dt. \tag{8.116}$$

Then we observe

$$|(D_a^\nu f_j)(x)| \leq |A_j| + \int_a^x |F_j(t, \ldots)| \, dt$$

$$\leq |A_j| + \int_a^x \left\{ \left\{ \sum_{\ell=1}^{M-1} (q_{\ell,1,j}(t)|(D_a^{\gamma_{i*}} f_\ell)(t)| \sqrt{|(D_a^{\gamma_{i*}+1} f_{\ell+1})(t)|} \right. \right.$$

$$+q_{\ell,2,j}(t)|(D_a^{\gamma_{i*}}f_{\ell+1})(t)|\sqrt{|(D_a^{\gamma_{i*}+1}f_\ell)(t)|}\Big\}$$

$$+\big(q_{M,1,j}(t)|(D_a^{\gamma_{i*}}f_1)(t)|\sqrt{|(D_a^{\gamma_{i*}+1}f_M)(t)|}$$

$$+\,q_{M,2,j}(t)|(D_a^{\gamma_{i*}}f_M)(t)|\sqrt{|(D_a^{\gamma_{i*}+1}f_1)(t)|}\big)\Big\}\,dt. \tag{8.117}$$

Thus

$$|(D_a^\nu f_j)(x)| \le |A_j| + W\left(\int_a^x\Big\{\Big\{\sum_{\ell=1}^{M-1}\big(|(D_a^k f_\ell)(t)|\sqrt{|(D_a^{k+1}f_{\ell+1})(t)|}\right.$$

$$+|(D_a^k f_{\ell+1})(t)|\sqrt{|(D_a^{k+1}f_\ell)(t)|}\big)\Big\}$$

$$+\big(|(D_a^k f_1)(t)|\sqrt{|(D_a^{k+1}f_M)(t)|}$$

$$+\,|(D_a^k f_M)(t)|\sqrt{|(D_a^{k+1}f_1)(t)|}\big)\Big\}\,dt. \tag{8.118}$$

By Corollary 8.14 we obtain

$$|(D_a^\nu f_j)(x)| \le |A_j| + 2W\Phi^*(x)\left(\int_a^x\left(\sum_{\ell=1}^M|(D_a^\nu f_\ell)(t)|^{3/2}\right)dt\right), \tag{8.119}$$

$j=1,2,\ldots,M$.

Therefore by adding all of the inequalities (8.119) we get

$$\theta(x) \le \rho + 2M\Phi^*(x)W\left(\int_a^x\left(\sum_{\ell=1}^M|(D_a^\nu f_\ell)(t)|^{3/2}\right)dt\right)$$

$$\overset{\text{(by (8.11))}}{\le} \rho + 2M\Phi^*(x)W\left(\int_a^x\left(\sum_{\ell=1}^M|(D_a^\nu f_\ell)(t)|\right)^{3/2}dt\right). \tag{8.120}$$

That is,

$$\theta(x) \le \rho + (2M\Phi^*(x)W)\left(\int_a^x(\theta(t))^{3/2}\,dt\right), \quad \text{all } a \le x \le b. \tag{8.121}$$

More precisely we get that

$$\theta(x) \le \rho + Q(x)\left(\int_a^x(\theta(t))^{3/2}\,dt\right), \quad a \le x \le b. \tag{8.122}$$

Notice that $\theta(x) \ge 0$, $\rho \ge 0$, $Q(x) \ge 0$, and $Q(a) = 0$ by $\Phi^*(a) = 0$. Acting here as in the proof of Theorem 14 of [26] (see here Theorem 6.28), we derive (8.111) and (8.112). Using (8.9) we get (8.113), and using (8.10) we establish (8.114). $\square$

9

Riemann–Liouville Fractional Opial-Type Inequalities for Several Functions and Applications

A wide variety of very general $L_p(1 \le p \le \infty)$-form Opial-type inequalities [315] is presented involving Riemann–Liouville fractional derivatives [17, 230, 295, 314] of several functions in different orders and powers.

Several other particular results of special interest from the established results are derived. Applications of some of these special inequalities are given in proving the uniqueness of solution and in giving upper bounds to solutions of initial value fractional problems involving a very general system of several fractional differential equations. Upper bounds to various Riemann–Liouville fractional derivatives of the solutions that are involved in the above systems are given too. This treatment is based on [46].

9.1 Introduction

Here the author continues his study of Riemann–Liouville fractional Opial-type inequalities now involving several different functions and produces a wide variety of corresponding results with important applications to systems of several fractional differential equations. This chapter continues Chapter 7.

We start in Section 9.2 with background, we continue in Section 9.3 with the main results, and we finish in Section 9.4 with applications.

G.A. Anastassiou, *Fractional Differentiation Inequalities*,   179
DOI 10.1007/978-0-387-98128-4_9, © Springer Science+Business Media, LLC 2009

To give the reader an idea of the kind of inequalities we are dealing with, we briefly mention a simple one,

$$\int_0^x \left(\sum_{j=1}^{M} |(D^\gamma f_j)(w)| \, |(D^\nu f_j)(w)| \right) dw \le$$

$$\left(\frac{x^{\nu-\gamma}}{2\Gamma(\nu-\gamma)\sqrt{\nu-\gamma}\sqrt{2\nu-2\gamma-1}} \right) \left\{ \int_0^x \left(\sum_{j=1}^{M} ((D^\nu f_j)(w))^2 \right) dw \right\},$$

$x \ge 0$, for functions $f_j \in L_1(0,x)$, $j = 1, \ldots, M \in \mathbb{N}$; $\nu > \gamma \ge 0$, and so on. Here $D^\beta f$ stands for the Riemann–Liouville fractional derivative of f of order $\beta \ge 0$. Furthermore one system of fractional differential equations we are dealing with briefly is of the form

$$(D^\nu f_j)(t) = F_j(t, \{(D^{\gamma_i} f_1)(t)\}_{i=1}^r,$$

$$\{(D^{\gamma_i} f_2)(t)\}_{i=1}^r, \ldots, \{(D^{\gamma_i} f_M)(t)\}_{i=1}^r), \text{ all } t \in [a, b],$$

$j = 1, \ldots, M; D^{\nu-k} f_j(0) = \alpha_{kj} \in \mathbb{R}, k = 1, \ldots, [\nu] + 1$. Here $[\nu]$ is the integral part of ν.

9.2   Background

We need

Definition 9.1. (see [187, 295, 314]). Let $\alpha \in \mathbb{R}_+ - \{0\}$. For any $f \in L_1(0,x); x \in \mathbb{R}_+ - \{0\}$, the *Riemann–Liouville fractional integral of* f of order α is defined by

$$(J_\alpha f)(s) := \frac{1}{\Gamma(\alpha)} \int_0^s (s-t)^{\alpha-1} f(t) dt, \; \forall s \in [0, x], \tag{9.1}$$

and the *Riemann–Liouville fractional derivative of* f of order α by

$$D^\alpha f(s) := \frac{1}{\Gamma(m-a)} \left(\frac{d}{ds} \right)^m \int_0^s (s-t)^{m-\alpha-1} f(t) dt, \tag{9.2}$$

where $m := [\alpha] + 1, [\cdot]$ is the integral part. In addition, we set $D^0 f := f := J_0 f$, $J_{-\alpha} f = D^\alpha f$ if $\alpha > 0$, $D^{-\alpha} f := J_\alpha f$, if $0 < \alpha \le 1$. If $\alpha \in \mathbb{N}$, then $D^\alpha f = f^{(\alpha)}$ the ordinary derivative.

Definition 9.2. [187]. We say that $f \in L_1(0,x)$ has an L_∞ *fractional derivative* $D^\alpha f$ in $[0, x], x \in \mathbb{R}_+ - \{0\}$, iff $D^{\alpha-k} f \in C([0, x])$,

$k = 1, \ldots, m := [\alpha] + 1; \alpha \in \mathbb{R}_+ - \{0\}$, and $D^{\alpha-1}f \in AC([0,x])$ (absolutely continuous functions) and $D^\alpha f \in L_\infty(0,x)$.

We need

Lemma 9.3. [187]. *Let $\alpha \in \mathbb{R}_+$, $\beta > \alpha$, let $f \in L_1(0,x)$, $x \in \mathbb{R}_+ - \{0\}$, have an L_∞ fractional derivative $D^\beta f$ in $[0,x]$, and let $D^{\beta-k}f(0) = 0$ for $k = 1, \ldots, [\beta] + 1$. Then*

$$D^\alpha f(s) = \frac{1}{\Gamma(\beta-\alpha)} \int_0^s (s-t)^{\beta-\alpha-1} D^\beta f(t)dt, \forall\, s \in [0,x]. \qquad (9.3)$$

Clearly here $D^\alpha f \in AC([0,x])$ for $\beta - \alpha \geq 1$ and in $C([0,x])$ for $\beta - \alpha \in (0,1)$, hence $D^\alpha f \in L_\infty(0,x)$ and $D^\alpha f \in L_1(0,x)$.

9.3   Main Results

Here we often use the following basic inequalities. Let $\alpha_1, \alpha_2, \ldots, \alpha_n \geq 0, n \in \mathbb{N}$; then

$$a_1^r + \ldots + a_n^r \leq (a_1 + \ldots + a_n)^r, \quad r \geq 1, \qquad (9.4)$$

and

$$a_1^r + \ldots + a_n^r \leq n^{1-r}(a_1 + \ldots + a_n)^r, \quad 0 \leq r \leq 1. \qquad (9.5)$$

Our first result follows.

Theorem 9.4. *Let $\alpha_1, \alpha_2 \in \mathbb{R}_+, \beta > \alpha_1, \alpha_2, \beta - \alpha_i > (1/p), p > 1, i = 1, 2$, and let $f_j \in L_1(0,x), j = 1, \ldots, M \in \mathbb{N}, x \in \mathbb{R}_+ - \{0\}$ have, respectively, L_∞ fractional derivatives $D^\beta f_j$ in $[0,x]$, and let $D^{\beta-k}f_j(0) = 0$, for $k = 1, \ldots, [\beta] + 1; j = 1, \ldots, M$. Consider also $p(t) > 0$ and $q(t) \geq 0$, with all $p(t), 1/p(t), q(t) \in L_\infty(0,x)$. Let $\lambda_\beta > 0$ and $\lambda_{\alpha_1}, \lambda_{\alpha_2} \geq 0$, such that $\lambda_\beta < p$. Set*

$$P_i(s) := \int_0^s (s-t)^{p(\beta-\alpha_i-1)/p-1}(p(t))^{-1/(p-1)}\, dt, \ i = 1,2; \ 0 \leq s \leq x, \qquad (9.6)$$

$$A(s) := \frac{q(s)(P_1(s))^{\lambda_{\alpha_1}(p-1/p)}(P_2(s))^{\lambda_{\alpha_2}((p-1)/p)}(p(s))^{-\lambda_\beta/p}}{(\Gamma(\beta-\alpha_1))^{\lambda_{\alpha_1}}(\Gamma(\beta-\alpha_2))^{\lambda_{\alpha_2}}}, \qquad (9.7)$$

$$A_0(x) := \left(\int_0^x (A(s))^{p/(p-\lambda_\beta)}ds\right)^{(p-\lambda_\beta)/p}, \qquad (9.8)$$

and

$$\delta_1^* := \begin{cases} M^{1-((\lambda_{\alpha_1}+\lambda_\beta)/p)}, & if \quad \lambda_{\alpha_1} + \lambda_\beta \leq p, \\ 2^{(\lambda_{\alpha_1}+\lambda_\beta/p)-1}, & if \quad \lambda_{\alpha_1} + \lambda_\beta \geq p. \end{cases} \tag{9.9}$$

Call

$$\varphi_1(x) := (A_0(x)|_{\lambda_{\alpha_2}=0}) \left(\frac{\lambda_\beta}{\lambda_{\alpha_1} + \lambda_\beta} \right)^{\lambda_\beta/p}. \tag{9.10}$$

If $\lambda_{\alpha_2} = 0$, we obtain that,

$$\int_0^x q(s) \left(\sum_{j=1}^{M} |D^{\alpha_1} f_j(s)|^{\lambda_{\alpha_1}} |D^\beta f_j(s)|^{\lambda_\beta} \right) ds \leq$$

$$\delta_1^* \varphi_1(x) \left[\int_0^x p(s) \left(\sum_{j=1}^{M} |D^\beta f_j(s)|^p \right) ds \right]^{(\lambda_{\alpha_1}+\lambda_\beta/p)}. \tag{9.11}$$

Proof. By Theorem 4 of [48] and here Theorem 7.4 we obtain

$$\int_0^x q(s) \left[|D^{\alpha_1} f_j(s)|^{\lambda_{\alpha_1}} |D^\beta f_j(s)|^{\lambda_\beta} + |D^{\alpha_1} f_{j+1}(s)|^{\lambda_{\alpha_1}} |D^\beta f_{j+1}(s)|^{\lambda_\beta} \right] ds \leq$$

$$(A_0(x)|_{\lambda_{\alpha_2}=0}) \left(\frac{\lambda_\beta}{\lambda_{\alpha_1} + \lambda_\beta} \right)^{(\lambda_\beta/p)} \delta_1 \left[\int_0^x p(s) \left[|D^\beta f_j(s)|^p + \right. \right.$$

$$\left. \left. |D^\beta f_{j+1}(s)|^p \right] ds \right]^{(\lambda_{\alpha_1}+\lambda_\beta/p)}, \quad j = 1, 2..., M-1, \tag{9.12}$$

where

$$\delta_1 := \begin{cases} 2^{1-((\lambda_{\alpha_1}+\lambda_\beta)/p)}, & if \quad \lambda_{\alpha_1} + \lambda_\beta \leq p, \\ 1, & if \quad \lambda_{\alpha_1} + \lambda_\beta \geq p. \end{cases} \tag{9.13}$$

Hence by adding all the above we get

$$\int_0^x q(s) \left\{ \sum_{j=1}^{M-1} \left[|D^{\alpha_1} f_j(s)|^{\lambda_{\alpha_1}} |D^\beta f_j(s)|^{\lambda_\beta} + |D^{\alpha_1} f_{j+1}(s)|^{\lambda_{\alpha_1}} |D^\beta f_{j+1}(s)|^{\lambda_\beta} \right] \right\} ds$$

$$\leq \delta_1 \varphi_1(x) \left\{ \sum_{j=1}^{M-1} \left[\int_0^x p(s) \left[|D^\beta f_j(s)|^p + |D^\beta f_{j+1}(s)|^p \right] ds \right]^{(\lambda_{\alpha_1}+\lambda_\beta)/p} \right\}. \tag{9.14}$$

Also it holds

$$\int_0^x q(s) \left[|D^{\alpha_1} f_1(s)|^{\lambda_{\alpha_1}} |D^\beta f_1(s)|^{\lambda_\beta} + |D^{\alpha_1} f_M(s)|^{\lambda_{\alpha_1}} |D^\beta f_M(s)|^{\lambda_\beta} \right] ds \leq$$

$$\delta_1 \varphi_1(x) \left[\int_0^x p(s) \left[|D^\beta f_1(s)|^p + |D^\beta f_M(s)|^p \right] ds \right]^{(\lambda_{\alpha_1}+\lambda_\beta)/p}. \tag{9.15}$$

Call

$$\varepsilon_1 = \begin{cases} 1, & if \quad \lambda_{\alpha_1} + \lambda_\beta \geq p, \\ M^{1-(\lambda_{\alpha_1}+\lambda_\beta/p)}, & if \quad \lambda_{\alpha_1} + \lambda_\beta \leq p. \end{cases} \tag{9.16}$$

Adding (9.14) and (9.15), and using (9.4) and (9.5) we have

$$2 \int_0^x q(s) \left(\sum_{j=1}^M |D^{\alpha_1} f_j(s)|^{\lambda_{\alpha_1}} |D^\beta f_j(s)|^{\lambda_\beta} \right) ds \leq \tag{9.17}$$

$$\delta_1 \varphi_1(x) \left\{ \sum_{j=1}^{M-1} \left[\int_0^x p(s) \left[|D^\beta f_j(s)|^p + |D^\beta f_{j+1}(s)|^p \right] ds \right]^{(\lambda_{\alpha_1}+\lambda_\beta/p)} + \right.$$

$$\left. \left[\int_0^x p(s) \left[|D^\beta f_1(s)|^p + |D^\beta f_M(s)|^p \right] ds \right]^{((\lambda_{\alpha_1}+\lambda_\beta)/p)} \right\} \leq$$

$$\delta_1 \varepsilon_1 \varphi_1(x) \left\{ \int_0^x p(s) \left(2 \sum_{j=1}^M |D^\beta f_j(s)|^p \right) ds \right\}^{((\lambda_{\alpha_1}+\lambda_\beta)/p)} . \tag{9.18}$$

We have derived

$$\int_0^x q(s) \left(\sum_{j=1}^M |D^{\alpha_1} f_j(s)|^{\lambda_{\alpha_1}} |D^\beta f_j(s)|^{\lambda_\beta} \right) ds \leq$$

$$\delta_1 \left(2^{(\lambda_{\alpha_1}+\lambda_\beta/p)-1} \right) \varepsilon_1 \varphi_1(x) \left\{ \int_0^x p(s) \left[\sum_{j=1}^M |D^\beta f_j(s)|^p \right] ds \right\}^{((\lambda_{\alpha_1}+\lambda_\beta)/p)} . \tag{9.19}$$

Clearly here we have

$$\delta_1^* = \delta_1 \left(2^{(\lambda_{\alpha_1}+\lambda_\beta/p)-1} \right) \varepsilon_1. \tag{9.20}$$

From (9.19) and (9.20) we derive (9.11). $\square$

Next we give

Theorem 9.5. *All here are as in Theorem 9.4. Denote*

$$\delta_3 := \begin{cases} 2^{\lambda_{\alpha_2}/\lambda_\beta} - 1, & if \quad \lambda_{\alpha_2} \geq \lambda_\beta, \\ 1, & if \quad \lambda_{\alpha_2} \leq \lambda_\beta, \end{cases} \tag{9.21}$$

$$\varepsilon_2 := \begin{cases} 1, & if \quad \lambda_\beta + \lambda_{\alpha_2} \geq p, \\ M^{1-(\lambda_\beta+\lambda_{\alpha_2}/p)}, & if \quad \lambda_\beta + \lambda_{\alpha_2} \leq p, \end{cases} \tag{9.22}$$

and

$$\varphi_2(x) := \left(A_0(x)\big|_{\lambda_{\alpha_1}=0}\right) 2^{(p-\lambda_\beta/p)} \left(\frac{\lambda_\beta}{\lambda_{\alpha_2}+\lambda_\beta}\right)^{\lambda_\beta/p} \delta_3^{(\lambda_\beta/p)}. \tag{9.23}$$

If $\lambda_{\alpha_1} = 0$, *then*

$$\int_0^x q(s) \left\{\left\{\sum_{j=1}^{M-1}\left[|D^{\alpha_2}f_{j+1}(s)|^{\lambda_{\alpha_2}}|D^\beta f_j(s)|^{\lambda_\beta}+|D^{\alpha_2}f_j(s)|^{\lambda_{\alpha_2}}|D^\beta f_{j+1}(s)|^{\lambda_\beta}\right]\right\}\right.$$

$$\left.+\left[|D^{\alpha_2}f_M(s)|^{\lambda_{\alpha_2}}|D^\beta f_1(s)|^{\lambda_\beta}+|D^{\alpha_2}f_1(s)|^{\lambda_{\alpha_2}}|D^\beta f_M(s)|^{\lambda_\beta}\right]\right\} ds \le$$

$$2^{(\lambda_\beta+\lambda_{\alpha_2}/p)}\varepsilon_2\varphi_2(x)\left\{\int_0^x p(s)\left[\sum_{j=1}^M |D^\beta f_j(s)|^p\right] ds\right\}^{((\lambda_\beta+\lambda_{\alpha_2})/p)}. \tag{9.24}$$

Proof. From Theorem 5 of [48] and here Theorem 7.5, we have

$$\int_0^x q(s)\left[|D^{\alpha_2}f_{j+1}(s)|^{\lambda_{\alpha_2}}|D^\beta f_j(s)|^{\lambda_\beta}+|D^{\alpha_2}f_j(s)|^{\lambda_{\alpha_2}}|D^\beta f_{j+1}(s)|^{\lambda_\beta}\right] ds \le$$

$$\varphi_2(x)\left(\int_0^x p(s)\left[|D^\beta f_j(s)|^p+|D^\beta f_{j+1}(s)|^p\right] ds\right)^{(\lambda_\beta+\lambda_{\alpha_2})/p}, \tag{9.25}$$

for $j = 1,\ldots,M-1$. Hence by adding all of the above we find

$$\int_0^x q(s)\left(\sum_{j=1}^{M-1}\left[|D^{\alpha_2}f_{j+1}(s)|^{\lambda_{\alpha_2}}|D^\beta f_j(s)|^{\lambda_\beta}+|D^{\alpha_2}f_j(s)|^{\lambda_{\alpha_2}}|D^\beta f_{j+1}(s)|^{\lambda_\beta}\right]\right) ds$$

$$\le \varphi_2(x)\left(\sum_{j=1}^{M-1}\left(\int_0^x p(s)\left[|D^\beta f_j(s)|^p+|D^\beta f_{j+1}(s)|^p\right] ds\right)^{(\lambda_\beta+\lambda_{\alpha_2})/p}\right). \tag{9.26}$$

Similarly it holds

$$\int_0^x q(s)\left[|D^{\alpha_2}f_M(s)|^{\lambda_{\alpha_2}}|D^\beta f_1(s)|^{\lambda_\beta}+|D^{\alpha_2}f_1(s)|^{\lambda_{\alpha_2}}|D^\beta f_M(s)|^{\lambda_\beta}\right] ds \le$$

$$\varphi_2(x)\left(\int_0^x p(s)\left[|D^\beta f_1(s)|^p+|D^\beta f_M(s)|^p\right] ds\right)^{(\lambda_\beta+\lambda_{\alpha_2})/p}. \tag{9.27}$$

Adding (9.26) and (9.27) and using (9.4), (9.5) we derive (9.24). $\square$

The general case follows.

Theorem 9.6. *All here are as in Theorem 9.4. Denote*

$$\tilde{\gamma}_1 := \begin{cases} 2^{((\lambda_{\alpha_1} + \lambda_{\alpha_2})/\lambda_\beta)} - 1, & if \quad \lambda_{\alpha_1} + \lambda_{\alpha_2} \geq \lambda_\beta, \\ 1, & if \quad \lambda_{\alpha_1} + \lambda_{\alpha_2} \leq \lambda_\beta, \end{cases} \tag{9.28}$$

and

$$\tilde{\gamma}_2 := \begin{cases} 1, & if \quad \lambda_{\alpha_1} + \lambda_{\alpha_2} + \lambda_\beta \geq p, \\ 2^{1-((\lambda_{\alpha_1} + \lambda_{\alpha_2} + \lambda_\beta)/p)}, & if \quad \lambda_{\alpha_1} + \lambda_{\alpha_2} + \lambda_\beta \leq p. \end{cases} \tag{9.29}$$

Set

$$\varphi_3(x) \quad : \; = A_0(x) \left(\frac{\lambda_\beta}{(\lambda_{\alpha_1} + \lambda_{\alpha_2})(\lambda_{\alpha_1} + \lambda_{\alpha_2} + \lambda_\beta)} \right)^{\lambda_\beta/p}$$
$$\left[\lambda_{\alpha_1}^{\lambda_\beta/p} \tilde{\gamma}_2 + 2^{(p-\lambda_\beta)/p} (\tilde{\gamma}_1 \lambda_{\alpha_2})^{\lambda_\beta/p} \right], \tag{9.30}$$

and

$$\varepsilon_3 := \begin{cases} 1, & if \quad \lambda_{\alpha_1} + \lambda_{\alpha_2} + \lambda_\beta \geq p, \\ M^{1-(\lambda_{\alpha_1} + \lambda_{\alpha_2} + \lambda_\beta/p)}, & if \quad \lambda_{\alpha_1} + \lambda_{\alpha_2} + \lambda_\beta \leq p. \end{cases} \tag{9.31}$$

Then

$$\int_0^x q(s) \left[\sum_{j=1}^{M-1} \left[|D^{\alpha_1} f_j(s)|^{\lambda_{\alpha_1}} |D^{\alpha_2} f_{j+1}(s)|^{\lambda_{\alpha_2}} |D^\beta f_j(s)|^{\lambda_\beta} + \right. \right.$$

$$|D^{\alpha_2} f_j(s)|^{\lambda_{\alpha_2}} |D^{\alpha_1} f_{j+1}(s)|^{\lambda_{\alpha_1}} |D^\beta f_{j+1}(s)|^{\lambda_\beta} \right] +$$
$$\left[|D^{\alpha_1} f_1(s)|^{\lambda_{\alpha_1}} |D^{\alpha_2} f_M(s)|^{\lambda_{\alpha_2}} |D^\beta f_1(s)|^{\lambda_\beta} \right.$$

$$\left. + |D^{\alpha_2} f_1(s)|^{\lambda_{\alpha_2}} |D^{\alpha_1} f_M(s)|^{\lambda_{\alpha_1}} |D^\beta f_M(s)|^{\lambda_\beta} \right] ds$$

$$\leq 2^{(\lambda_{\alpha_1} + \lambda_{\alpha_2} + \lambda_\beta/p)} \varepsilon_3 \varphi_3(x) \left\{ \int_0^x p(s) \left[\sum_{j=1}^M |D^\beta f_j(s)|^p \right] ds \right\}^{((\lambda_{\alpha_1} + \lambda_{\alpha_2} + \lambda_\beta)/p)}. \tag{9.32}$$

Proof. From Theorem 6 of [48] and here Theorem 7.6, and by adding we find

$$\sum_{j=1}^{M-1} \int_0^x q(s) \left[|D^{\alpha_1} f_j(s)|^{\lambda_{\alpha_1}} |D^{\alpha_2} f_{j+1}(s)|^{\lambda_{\alpha_2}} |D^\beta f_j(s)|^{\lambda_\beta} + \right.$$

$$|D^{\alpha_2} f_j(s)|^{\lambda_{\alpha_2}} |D^{\alpha_1} f_{j+1}(s)|^{\lambda_{\alpha_1}} |D^\beta f_{j+1}(s)|^{\lambda_\beta}] \, ds \le$$

$$\varphi_3(x) \sum_{j=1}^{M-1} \left(\int_0^x p(s)(|D^\beta f_j(s)|^p + |D^\beta f_{j+1}(s)|^p) ds \right)^{(\lambda_{\alpha_1} + \lambda_{\alpha_2} + \lambda_\beta)/p}.$$

$$(9.33)$$

Also it holds

$$\int_0^x q(s) \left[|D^{\alpha_1} f_1(s)|^{\lambda_{\alpha_1}} |D^{\alpha_2} f_M(s)|^{\lambda_{\alpha_2}} |D^\beta f_1(s)|^{\lambda_\beta} + \right.$$

$$\left. |D^{\alpha_2} f_1(s)|^{\lambda_{\alpha_2}} |D^{\alpha_1} f_M(s)|^{\lambda_{\alpha_1}} |D^\beta f_M(s)|^{\lambda_\beta} \right] ds \le$$

$$\varphi_3(x) \left(\int_0^x p(s) \left(|D^\beta f_1(s)|^p + |D^\beta f_M(s)|^p \right) ds \right)^{(\lambda_{\alpha_1} + \lambda_{\alpha_2} + \lambda_\beta)/p}. \quad (9.34)$$

Adding (9.33) and (9.34), along with (9.4), (9.5) we derive (9.32). □

We continue with

Theorem 9.7. *Let* $\beta > \alpha_1 + 1$, $\alpha_1 \in \mathbb{R}_+$ *and let* $f_j \in L_1(0, x)$, $j = 1, \ldots, M \in \mathbb{N}$, $x \in \mathbb{R}_+ - \{0\}$ *have, respectively,* L_∞ *fractional derivatives* $D^\beta f_j$, *in* $[0, x]$, *and let* $D^{\beta-k} f_j(0) = 0$, *for* $k = 1, \ldots, [\beta] + 1; j = 1, \ldots, M$. *Consider also* $p(t) > 0$ *and* $q(t) \ge 0$, *with* $p(t), 1/p(t), q(t) \in L_\infty(0, x)$. *Let* $\lambda_\alpha \ge 0$, $0 < \lambda_{\alpha+1} < 1$, *and* $p > 1$.
Denote

$$\theta_3 := \begin{cases} 2^{\lambda_\alpha/(\lambda_{\alpha+1})} - 1, & if \quad \lambda_\alpha \ge \lambda_{\alpha+1}, \\ 1, & if \quad \lambda_\alpha \le \lambda_{\alpha+1}, \end{cases} \quad (9.35)$$

$$L(x) := \left(2 \int_0^x (q(s))^{(1/(1-\lambda_{\alpha+1}))} ds \right)^{(1-\lambda_{\alpha+1})} \left(\frac{\theta_3 \lambda_{\alpha+1}}{\lambda_\alpha + \lambda_{\alpha+1}} \right)^{\lambda_{\alpha+1}}, \quad (9.36)$$

and

$$P_1(x) := \int_0^x (x-s)^{(\beta-\alpha_1-1)p/(p-1)} (p(s))^{-1/(p-1)} ds, \quad (9.37)$$

$$T(x) := L(x) \left(\frac{P_1(x)^{(p-1/p)}}{\Gamma(\beta - \alpha_1)} \right)^{(\lambda_\alpha + \lambda_{\alpha+1})}, \quad (9.38)$$

and

$$\omega_1 := 2^{(p-1/p)(\lambda_\alpha + \lambda_{\alpha+1})}, \quad (9.39)$$

$$\Phi(x) := T(x) \, \omega_1. \quad (9.40)$$

Also put

$$\varepsilon_4 := \begin{cases} 1, & if \quad \lambda_\alpha + \lambda_{\alpha+1} \ge p, \\ M^{1-(\lambda_\alpha + \lambda_{\alpha+1}/p)}, & if \quad \lambda_\alpha + \lambda_{\alpha+1} \le p. \end{cases} \quad (9.41)$$

Then

$$\int_0^x q(s) \left\{ \left\{ \sum_{j=1}^{M-1} \left[|D^{\alpha_1} f_j(s)|^{\lambda_\alpha} |D^{\alpha_1+1} f_{j+1}(s)|^{\lambda_{\alpha+1}} + \right. \right. \right.$$

$$\left. |D^{\alpha_1} f_{j+1}(s)|^{\lambda_\alpha} |D^{\alpha_1+1} f_j(s)|^{\lambda_{\alpha+1}} \right] \right\}$$

$$+ \left[|D^{\alpha_1} f_1(s)|^{\lambda_\alpha} |D^{\alpha_1+1} f_M(s)|^{\lambda_{\alpha+1}} + |D^{\alpha_1} f_M(s)|^{\lambda_\alpha} |D^{\alpha_1+1} f_1(s)|^{\lambda_{\alpha+1}} \right] \right\} ds \le$$

$$2^{(\lambda_\alpha+\lambda_{\alpha+1}/p)} \varepsilon_4 \Phi(x) \left[\int_0^x p(s) \left(\sum_{j=1}^M |D^\beta f_j(s)|^p \right) ds \right]^{((\lambda_\alpha+\lambda_{\alpha+1})/p)} . \tag{9.42}$$

Proof. From Theorem 8 of [48] and here Theorem 7.8, we find

$$\int_0^x q(s) \sum_{j=1}^{M-1} \left[|D^{\alpha_1} f_j(s)|^{\lambda_\alpha} |D^{\alpha_1+1} f_{j+1}(s)|^{\lambda_{\alpha+1}} + \right.$$

$$\left. |D^{\alpha_1} f_{j+1}(s)|^{\lambda_\alpha} |D^{\alpha_1+1} f_j(s)|^{\lambda_{\alpha+1}} \right] ds$$

$$\le \Phi(x) \sum_{j=1}^{M-1} \left[\int_0^x p(s)(|D^\beta f_j(s)|^p + |D^\beta f_{j+1}(s)|^p) ds \right]^{((\lambda_\alpha+\lambda_{\alpha+1})/p)} . \tag{9.43}$$

Similarly it holds

$$\int_0^x q(s) \left[|D^{\alpha_1} f_1(s)|^{\lambda_\alpha} |D^{\alpha_1+1} f_M(s)|^{\lambda_{\alpha+1}} + |D^{\alpha_1} f_M(s)|^{\lambda_\alpha} |D^{\alpha_1+1} f_1(s)|^{\lambda_{\alpha+1}} \right] ds$$

$$\le \Phi(x) \left[\int_0^x p(s)(|D^\beta f_1(s)|^p + |D^\beta f_M(s)|^p) ds \right]^{(\lambda_\alpha+\lambda_{\alpha+1})/p} . \tag{9.44}$$

Adding (9.43) and (9.44), along with (9.4), (9.5) we obtain (9.42). □

Next comes the following theorem.

Theorem 9.8. *All are as in Theorem 9.4. Consider the special case of* $\lambda_{\alpha_2} = \lambda_{\alpha_1} + \lambda_\beta$.
Denote

$$\tilde{T}(x) := A_0(x) \left(\frac{\lambda_\beta}{\lambda_{\alpha_1} + \lambda_\beta} \right)^{\lambda_\beta/p} 2^{(p-2\lambda_{\alpha_1}-3\lambda_\beta)/p}, \tag{9.45}$$

$$\varepsilon_5 := \begin{cases} 1, & if \quad 2(\lambda_{\alpha_1} + \lambda_\beta) \ge p, \\ M^{1-(2(\lambda_{\alpha_1}+\lambda_\beta)/p)}, & if \quad 2(\lambda_{\alpha_1} + \lambda_\beta) \le p. \end{cases} \tag{9.46}$$

Then

$$\int_0^x q(s) \left\{ \left\{ \sum_{j=1}^{M-1} \left[|D^{\alpha_1} f_j(s)|^{\lambda_{\alpha_1}} |D^{\alpha_2} f_{j+1}(s)|^{\lambda_{\alpha_1}+\lambda_\beta} |D^\beta f_j(s)|^{\lambda_\beta} \right. \right. \right.$$

$$+ |D^{\alpha_2} f_j(s)|^{\lambda_{\alpha_1}+\lambda_\beta} |D^{\alpha_1} f_{j+1}(s)|^{\lambda_{\alpha_1}} |D^\beta f_{j+1}(s)|^{\lambda_\beta} \Big] \Big\}$$

$$+ \Big[|D^{\alpha_1} f_1(s)|^{\lambda_{\alpha_1}} |D^{\alpha_2} f_M(s)|^{\lambda_{\alpha_1}+\lambda_\beta} |D^\beta f_1(s)|^{\lambda_\beta}$$

$$+ |D^{\alpha_2} f_1(s)|^{\lambda_{\alpha_1}+\lambda_\beta} |D^{\alpha_1} f_M(s)|^{\lambda_{\alpha_1}} |D^\beta f_M(s)|^{\lambda_\beta} \Big] \Big\} ds \le$$

$$2^{2(\lambda_{\alpha_1}+\lambda_\beta)/p} \, \varepsilon_5 \, \tilde{T}(x) \left[\int_0^x p(s) \left(\sum_{j=1}^M |D^\beta f_j(s)|^p \right) ds \right]^{2(\lambda_{\alpha_1}+\lambda_\beta)/p} . \qquad (9.47)$$

Proof. Based on Theorem 9 of [48] and here Theorem 7.9. The rest is as in the proof of Theorem 9.7. $\square$

Next we give a special case of the above theorems.

Corollary 9.9. (to Theorem 9.4, $\lambda_{\alpha_2} = 0$, $p(t) = q(t) = 1$). *It holds*

$$\int_0^x \left(\sum_{j=1}^M |D^{\alpha_1} f_j(s)|^{\lambda_{\alpha_1}} |D^\beta f_j(s)|^{\lambda_\beta} \right) ds \le$$

$$\delta_1^* \varphi_1(x) \left[\int_0^x \left[\sum_{j=1}^M |D^\beta f_j(s)|^p \right] ds \right]^{(\lambda_{\alpha_1}+\lambda_\beta/p)} . \qquad (9.48)$$

In (9.48), $(A_0(x) \mid_{\lambda_{\alpha_2}=0})$ *of* $\varphi_1(x)$ *is given in* [48], *Corollary 10, Equation* (123) *there; here see* (7.123).

Corollary 9.10. (to Theorem 9.4, $\lambda_{\alpha_2} = 0, p(t) = q(t) = 1, \lambda_{\alpha_1} = \lambda_\beta = 1, p = 2$). *In detail; let* $\alpha_1 \in \mathbb{R}_+, \beta > \alpha_1, \beta - \alpha_1 > (1/2)$, *and let* $f_j \in L_1(0, x), j = 1, \ldots, M \in \mathbb{N}, x \in \mathbb{R}_+ - \{0\}$, *have, respectively,* L_∞ *fractional derivatives* $D^\beta f_j$ *in* $[0, x]$, *and let* $D^{\beta-k} f_j(0) = 0$ *for* $k = 1, \ldots, [\beta] + 1; j = 1, \ldots, M$. *Then*

$$\int_0^x \left(\sum_{j=1}^M |D^{\alpha_1} f_j(s)| \, |D^\beta f_j(s)| \right) ds \le$$

$$\left(\frac{x^{(\beta-\alpha_1)}}{2\Gamma(\beta-\alpha_1)\sqrt{\beta-\alpha_1}\sqrt{2\beta-2\alpha_1-1}}\right)\left\{\int_0^x\left[\sum_{j=1}^M\left(D^\beta f_j(s)\right)^2\right]ds\right\}.$$
(9.49)

Proof. Based on our Corollary 9.9 and Corollary 11 of [48] (see inequality (130) there; also see here Corollary 7.11). □

Corollary 9.11. (to Theorem 9.5, $\lambda_{\alpha_1}=0$, $p(t)=q(t)=1$). It holds

$$\int_0^x\left\{\left\{\sum_{j=1}^{M-1}\left[\left|D^{\alpha_2}f_{j+1}(s)\right|^{\lambda_{\alpha_2}}\left|D^\beta f_j(s)\right|^{\lambda_\beta}+\left|D^{\alpha_2}f_j(s)\right|^{\lambda_{\alpha_2}}\left|D^\beta f_{j+1}(s)\right|^{\lambda_\beta}\right]\right\}\right.$$

$$\left.+\left[\left|D^{\alpha_2}f_M(s)\right|^{\lambda_{\alpha_2}}\left|D^\beta f_1(s)\right|^{\lambda_\beta}+\left|D^{\alpha_2}f_1(s)\right|^{\lambda_{\alpha_2}}\left|D^\beta f_M(s)\right|^{\lambda_\beta}\right]\right\}ds\le$$

$$2^{(\lambda_\beta+\lambda_{\alpha_2}/p)}\,\varepsilon_2\varphi_2(x)\left\{\int_0^x\left[\sum_{j=1}^M\left|D^\beta f_j(s)\right|^p\right]ds\right\}^{((\lambda_\beta+\lambda_{\alpha_2})/p)}.$$
(9.50)

In (9.50), $(A_0(x)\mid_{\lambda_{\alpha_1}=0})$ of $\varphi_2(x)$ is given in [48], Corollary 12, Equation (137) there; also see Corollary 7.12 here.

Corollary 9.12. (to Theorem 9.5, $\lambda_{\alpha_1}=0$, $p(t)=q(t)=1$, $\lambda_{\alpha_2}=\lambda_\beta=1$, $p=2$). In detail; let $\alpha_2\in\mathbb{R}_+$, $\beta>\alpha_2$, $\beta-\alpha_2>(1/2)$, and let $f_j\in L_1(0,x)$, $j=1,\ldots,M\in\mathbb{N}$, $x\in\mathbb{R}_+-\{0\}$, have, respectively, L_∞ fractional derivatives $D^\beta f_j$ in $[0,x]$, and let $D^{\beta-k}f_j(0)=0$, for $k=1,\ldots,[\beta]+1$; $j=1,\ldots,M$. Then

$$\int_0^x\left\{\left\{\sum_{j=1}^{M-1}\left[\left|D^{\alpha_2}f_{j+1}(s)\right|\left|D^\beta f_j(s)\right|+\left|D^{\alpha_2}f_j(s)\right|\left|D^\beta f_{j+1}(s)\right|\right]\right\}+\right.$$

$$\left.\left[\left|D^{\alpha_2}f_M(s)\right|\left|D^\beta f_1(s)\right|+\left|D^{\alpha_2}f_1(s)\right|\left|D^\beta f_M(s)\right|\right]\right\}ds$$

$$\le\left(\frac{\sqrt{2}\,x^{(\beta-\alpha_2)}}{\Gamma(\beta-\alpha_2)\,\sqrt{\beta-\alpha_2}\,\sqrt{2\beta-2\alpha_2-1}}\right)\left\{\int_0^x\left[\sum_{j=1}^M(D^\beta f_j(s))^2\right]ds\right\}.$$
(9.51)

Proof. From Corollary 9.11 and Corollary 13 of [48], especially Equation (146) there. Also see Corollary 7.13 here. □

Corollary 9.13. (to Theorem 9.6, $\lambda_{\alpha_1} = \lambda_{\alpha_2} = \lambda_\beta = 1, p = 3, p(t) = q(t) = 1$). *It holds*

$$\int_0^x \left[\sum_{j=1}^{M-1} \left[|D^{\alpha_1} f_j(s)| \ |D^{\alpha_2} f_{j+1}(s)| \ |D^\beta f_j(s)| + \right. \right.$$

$$|D^{\alpha_2} f_j(s)| \ |D^{\alpha_1} f_{j+1}(s)| \ |D^\beta f_{j+1}(s)|] +$$

$$\left. \left[|D^{\alpha_1} f_1(s)| \ |D^{\alpha_2} f_M(s)| \ |D^\beta f_1(s)| + |D^{\alpha_2} f_1(s)| \ |D^{\alpha_1} f_M(s)| \ |D^\beta f_M(s)| \right] \right] ds \le$$

$$2 \, \varphi_3^*(x) \left[\int_0^x \left[\sum_{j=1}^M \left| D^\beta f_j(s) \right|^3 ds \right] \right]. \tag{9.52}$$

Here

$$\varphi_3^* := \left(\sqrt[3]{2} + \frac{1}{\sqrt[3]{6}} \right) A_0(x), \tag{9.53}$$

where in this special case,

$$A_0(x) =$$

$$\frac{4x^{(2\beta - \alpha_1 - \alpha_2)}}{\Gamma(\beta - \alpha_1)\Gamma(\beta - \alpha_2)[3(3\beta - 3\alpha_1 - 1)(3\beta - 3\alpha_2 - 1)(2\beta - \alpha_1 - \alpha_2)]^{2/3}}. \tag{9.54}$$

Proof. From Theorem 9.6 and Corollary 14 of [48]; see there Equation (151) which is here (9.54); see also Corollary 7.14. $\square$

Corollary 9.14. (to Theorem 9.7, $\lambda_\alpha = 1, \lambda_{\alpha+1} = 1/2, p = 3/2, p(t) = q(t) = 1$). *In detail: let $\beta > \alpha_1 + 1, \alpha_1 \in \mathbb{R}_+$ and let $f_j \in L_1(0, x), j = 1, \ldots, M \in \mathbb{N}, x \in \mathbb{R}_+ - \{0\}$ have, respectively, L_∞ fractional derivatives $D^\beta f_j$ in $[0, x]$, and let $D^{\beta-k} f_j(0) = 0$, for $k = 1, \ldots, [\beta] + 1; j = 1, \ldots, M$. Set*

$$\Phi^*(x) := \left(\frac{2}{\sqrt{3\beta - 3\alpha_1 - 2}} \right) \frac{x^{(3\beta - 3\alpha_1 - 1/2)}}{(\Gamma(\beta - \alpha_1))^{3/2}}. \tag{9.55}$$

Then

$$\int_0^x \left\{ \left\{ \sum_{j=1}^{M-1} \left[|D^{\alpha_1} f_j(s)| \ \sqrt{|D^{\alpha_1+1} f_{j+1}(s)|} + |D^{\alpha_1} f_{j+1}(s)| \ \sqrt{|D^{\alpha_1+1} f_j(s)|} \right] \right\} + \right.$$

$$\left. \left[|D^{\alpha_1} f_1(s)| \ \sqrt{|D^{\alpha_1+1} f_M(s)|} + |D^{\alpha_1} f_M(s)| \ \sqrt{|D^{\alpha_1+1} f_1(s)|} \right] \right\} ds \le$$

$$2\Phi^*(x) \left[\int_0^x \left(\sum_{j=1}^M |D^\beta f_j(s)|^{3/2} \right) ds \right]. \tag{9.56}$$

Proof. Based on Theorem 9.7 here, and Corollary 15 of [48] (see there Equation (161) which is here (9.55)); see also Corollary 7.15 here. □

Corollary 9.15. (to Theorem 9.8, here $p = 2(\lambda_{\alpha_1} + \lambda_\beta) > 1, p(t) = q(t) = 1$). *It holds*

$$
\int_0^x \Bigg\{ \Bigg\{ \sum_{j=1}^{M-1} \Big[|D^{\alpha_1} f_j(s)|^{\lambda_{\alpha_1}} |D^{\alpha_2} f_{j+1}(s)|^{\lambda_{\alpha_1}+\lambda_\beta} |D^\beta f_j(s)|^{\lambda_\beta} +
$$

$$
|D^{\alpha_2} f_j(s)|^{\lambda_{\alpha_1}+\lambda_\beta} |D^{\alpha_1} f_{j+1}(s)|^{\lambda_{\alpha_1}} |D^\beta f_{j+1}(s)|^{\lambda_\beta} \Big] \Bigg\} +
$$

$$
\Big[|D^{\alpha_1} f_1(s)|^{\lambda_{\alpha_1}} |D^{\alpha_2} f_M(s)|^{\lambda_{\alpha_1}+\lambda_\beta} |D^\beta f_1(s)|^{\lambda_\beta} +
$$

$$
|D^{\alpha_2} f_1(s)|^{\lambda_{\alpha_1}+\lambda_\beta} |D^{\alpha_1} f_M(s)|^{\lambda_{\alpha_1}} |D^\beta f_M(s)|^{\lambda_\beta} \Big] \Bigg\} ds \le
$$

$$
2\tilde{T}(x) \left[\int_0^x \left(\sum_{j=1}^{M} |D^\beta f_j(s)|^{2(\lambda_{\alpha_1}+\lambda_\beta)} \right) ds \right]. \tag{9.57}
$$

Here $\tilde{T}(x)$ in (9.57) is given by (9.45) and in detail by $\tilde{\tilde{T}}(x)$ of [48]; see there Corollary 16 and Equations (165) – (169). Also see here Corollary 7.16 and (7.165)–(7.169) .

Corollary 9.16. (to Theorem 9.8, $p = 4$, $\lambda_{\alpha_1} = \lambda_\beta = 1, p(t) = q(t) = 1$). *It holds*

$$
\int_0^x \Bigg\{ \Bigg\{ \sum_{j=1}^{M-1} \Big[|D^{\alpha_1} f_j(s)| \, (D^{\alpha_2} f_{j+1}(s))^2 \, |D^\beta f_j(s)| +
$$

$$
(D^{\alpha_2} f_j(s))^2 \, |D^{\alpha_1} f_{j+1}(s)| \, |D^\beta f_{j+1}(s)| \Big] \Bigg\} +
$$

$$
\Big[|D^{\alpha_1} f_1(s)| \, (D^{\alpha_2} f_M(s))^2 \, |D^\beta f_1(s)| +
$$

$$
(D^{\alpha_2} f_1(s))^2 \, |D^{\alpha_1} f_M(s)| \, |D^\beta f_M(s)| \Big] \Bigg\} ds \le
$$

$$
2\tilde{T}(x) \left[\int_0^x \left(\sum_{j=1}^{M} (D^\beta f_j(s))^4 \right) ds \right]. \tag{9.58}
$$

Here in (9.58) we have that $\tilde{T}(x) = T^*(x)$ of Corollary 17 of [48]; for that see Equations (177)–(181) there. Also see here Corollary 7.17, Equations (7.177)–(7.181).

Next we present the L_∞ case.

Theorem 9.17. *Let* $\alpha_1, \alpha_2 \in \mathbb{R}_+, \beta > \alpha_1, \alpha_2$, *and let* $f_j \in L_1(0, x), j = 1, \ldots, M \in \mathbb{N}, x \in \mathbb{R}_+ - \{0\}$, *have, respectively,* L_∞ *fractional derivatives* $D^\beta f_j$ *in* $[0, x]$, *and let* $D^{\beta-k} f_j(0) = 0$ *for* $k = 1, \ldots, [\beta] + 1; j = 1, \ldots, M$. *Consider* $p(s) \geq 0, p(s) \in L_\infty(0, x)$. *Let* $\lambda_{\alpha_1}, \lambda_{\alpha_2}, \lambda_\beta \geq 0$. *Set*

$$\rho(x) := \left\{ \frac{\|p(s)\|_\infty}{(\Gamma(\beta - \alpha_1 + 1))^{\lambda_{\alpha_1}} (\Gamma(\beta - \alpha_2 + 1))^{\lambda_{\alpha_2}}} \right.$$

$$\left. \frac{x^{(\beta\lambda_{\alpha_1} - \alpha_1\lambda_{\alpha_1} + \beta\lambda_{\alpha_2} - \alpha_2\lambda_{\alpha_2} + 1)}}{[\beta\lambda_{\alpha_1} - \alpha_1\lambda_{\alpha_1} + \beta\lambda_{\alpha_2} - \alpha_2\lambda_{\alpha_2} + 1]} \right\}. \qquad (9.59)$$

Then

$$\int_0^x p(s) \left\{ \left\{ \sum_{j=1}^{M-1} \left[|D^{\alpha_1} f_j(s)|^{\lambda_{\alpha_1}} |D^{\alpha_2} f_{j+1}(s)|^{\lambda_{\alpha_2}} |D^\beta f_j(s)|^{\lambda_\beta} + \right. \right. \right.$$

$$\left. |D^{\alpha_2} f_j(s)|^{\lambda_{\alpha_2}} |D^{\alpha_1} f_{j+1}(s)|^{\lambda_{\alpha_1}} |D^\beta f_{j+1}(s)|^{\lambda_\beta} \right] \right\} +$$

$$\left[|D^{\alpha_1} f_1(s)|^{\lambda_{\alpha_1}} |D^{\alpha_2} f_M(s)|^{\lambda_{\alpha_2}} |D^\beta f_1(s)|^{\lambda_\beta} \right.$$

$$\left. \left. + |D^{\alpha_2} f_1(s)|^{\lambda_{\alpha_2}} |D^{\alpha_1} f_M(s)|^{\lambda_{\alpha_1}} |D^\beta f_M(s)|^{\lambda_\beta} \right] \right\} ds \leq$$

$$\rho(x) \left\{ \sum_{j=1}^M \left\{ \|D^\beta f_j\|_\infty^{2(\lambda_{\alpha_1} + \lambda_\beta)} + \|D^\beta f_j\|_\infty^{2\lambda_{\alpha_2}} \right\} \right\}. \qquad (9.60)$$

Proof. Based on Theorem 18 of [48]; see also Theorem 7.18 here. □

Similarly we give

Theorem 9.18. (*as in Theorem 9.17,* $\lambda_{\alpha_2} = 0$). *It holds*

$$\int_0^x p(s) \left(\sum_{j=1}^M |D^{\alpha_1} f_j(s)|^{\lambda_{\alpha_1}} |D^\beta f_j(s)|^{\lambda_\beta} \right) ds \leq$$

$$\left(\frac{x^{(\beta\lambda_{\alpha_1} - \alpha_1\lambda_{\alpha_1} + 1)}}{(\beta\lambda_{\alpha_1} - \alpha_1\lambda_{\alpha_1} + 1)(\Gamma(\beta - \alpha_1 + 1))^{\lambda_{\alpha_1}}} \frac{\|p(s)\|_\infty}{} \right) \left(\sum_{j=1}^M \|D^\beta f_j\|_\infty^{\lambda_{\alpha_1} + \lambda_\beta} \right). \qquad (9.61)$$

Proof. Based on Theorem 19 of [48]; see Theorem 7.19 here. □

It follows

Theorem 9.19. (as in Theorem 9.17, $\lambda_{\alpha_2} = \lambda_{\alpha_1} + \lambda_\beta$). *It holds*

$$\int_0^x p(s) \left\{ \left\{ \sum_{j=1}^{M-1} \left[|D^{\alpha_1} f_j(s)|^{\lambda_{\alpha_1}} |D^{\alpha_2} f_{j+1}(s)|^{\lambda_{\alpha_1}+\lambda_\beta} |D^\beta f_j(s)|^{\lambda_\beta} + \right. \right. \right.$$

$$\left. |D^{\alpha_2} f_j(s)|^{\lambda_{\alpha_1}+\lambda_\beta} |D^{\alpha_1} f_{j+1}(s)|^{\lambda_{\alpha_1}} |D^\beta f_{j+1}(s)|^{\lambda_\beta} \right] \right\} +$$

$$\left[|D^{\alpha_1} f_1(s)|^{\lambda_{\alpha_1}} |D^{\alpha_2} f_M(s)|^{\lambda_{\alpha_1}+\lambda_\beta} |D^\beta f_1(s)|^{\lambda_\beta} + \right.$$

$$\left. \left. |D^{\alpha_2} f_1(s)|^{\lambda_{\alpha_1}+\lambda_\beta} |D^{\alpha_1} f_M(s)|^{\lambda_{\alpha_1}} |D^\beta f_M(s)|^{\lambda_\beta} \right] \right\} ds \le$$

$$\left(\left(\frac{2 \, x^{(2\beta\lambda_{\alpha_1} - \alpha_1\lambda_{\alpha_1} + \beta\lambda_\beta - \alpha_2\lambda_{\alpha_1} - \alpha_2\lambda_\beta + 1)}}{(2\beta\lambda_{\alpha_1} - \alpha_1\lambda_{\alpha_1} + \beta\lambda_\beta - \alpha_2\lambda_{\alpha_1} - \alpha_2\lambda_\beta + 1)} \right) \right.$$

$$\left. \frac{\|p(s)\|_\infty}{(\Gamma(\beta - \alpha_1 + 1))^{\lambda_{\alpha_1}} (\Gamma(\beta - \alpha_2 + 1))^{(\lambda_{\alpha_1}+\lambda_\beta)}} \right)$$

$$\left(\sum_{j=1}^{M} \|D^\beta f_j\|_\infty^{2(\lambda_{\alpha_1}+\lambda_\beta)} \right). \tag{9.62}$$

Proof. By Theorem 20 of [48]; see Theorem 7.20 here. $\square$

We continue with

Theorem 9.20. (as in Theorem 9.17, $\lambda_\beta = 0, \lambda_{\alpha_1} = \lambda_{\alpha_2}$). *It holds*

$$\int_0^x p(s) \left\{ \left\{ \sum_{j=1}^{M-1} \left[|D^{\alpha_1} f_j(s)|^{\lambda_{\alpha_1}} |D^{\alpha_2} f_{j+1}(s)|^{\lambda_{\alpha_1}} + \right. \right. \right.$$

$$\left. \left. |D^{\alpha_2} f_j(s)|^{\lambda_{\alpha_1}} |D^{\alpha_1} f_{j+1}(s)|^{\lambda_{\alpha_1}} \right] \right\}$$

$$+ \left[|D^{\alpha_1} f_1(s)|^{\lambda_{\alpha_1}} |D^{\alpha_2} f_M(s)|^{\lambda_{\alpha_1}} + |D^{\alpha_2} f_1(s)|^{\lambda_{\alpha_1}} |D^{\alpha_1} f_M(s)|^{\lambda_{\alpha_1}} \right] \right\} ds$$

$$\le 2 \rho^*(x) \left[\sum_{j=1}^{M} \|D^\beta f_j\|_\infty^{2\lambda_{\alpha_1}} \right]. \tag{9.63}$$

Here we have

$$\rho^*(x) :=$$

$$\left(\frac{x^{(2\beta\lambda_{\alpha_1} - \alpha_1\lambda_{\alpha_1} - \alpha_2\lambda_{\alpha_1} + 1)} \|p(s)\|_\infty}{(2\beta\,\lambda_{\alpha_1} - \alpha_1\lambda_{\alpha_1} - \alpha_2\lambda_{\alpha_1} + 1)(\Gamma(\beta - \alpha_1 + 1))^{\lambda_{\alpha_1}} (\Gamma(\beta - \alpha_2 + 1))^{\lambda_{\alpha_1}}} \right).$$
(9.64)

Proof. Based on Theorem 21 of [48]; see here Theorem 7.21. □

Next we give

Theorem 9.21. (as in Theorem 9.17, $\lambda_{\alpha_1} = 0, \lambda_{\alpha_2} = \lambda_\beta$). *It holds*

$$\int_0^x p(s) \left\{ \left\{ \sum_{j=1}^{M-1} \left[|D^{\alpha_2} f_{j+1}(s)|^{\lambda_{\alpha_2}} |D^\beta f_j(s)|^{\lambda_{\alpha_2}} + \right. \right. \right.$$

$$\left. \left. |D^{\alpha_2} f_j(s)|^{\lambda_{\alpha_2}} |D^\beta f_{j+1}(s)|^{\lambda_{\alpha_2}} \right] \right\}$$

$$\left. \left[|D^{\alpha_2} f_M(s)|^{\lambda_{\alpha_2}} |D^\beta f_1(s)|^{\lambda_{\alpha_2}} + |D^{\alpha_2} f_1(s)|^{\lambda_{\alpha_2}} |D^\beta f_M(s)|^{\lambda_{\alpha_2}} \right] \right\} ds \leq$$

$$2 \left(\frac{x^{(\beta\lambda_{\alpha_2} - \alpha_2\lambda_{\alpha_2} + 1)} \|p(s)\|_\infty}{(\beta\lambda_{\alpha_2} - \alpha_2\lambda_{\alpha_2} + 1)(\Gamma(\beta - \alpha_2 + 1))^{\lambda_{\alpha_2}}} \right) \left(\sum_{j=1}^M \|D^\beta f_j\|_\infty^{2\lambda_{\alpha_2}} \right).$$
(9.65)

Proof. Based on Theorem 22 of [48]; see Theorem 7.22 here. □

Some special cases follow.

Corollary 9.22. (to Theorem 9.20, all are as in Theorem 9.17, $\lambda_\beta = 0, \lambda_{\alpha_1} = \lambda_{\alpha_2}, \alpha_2 = \alpha_1 + 1$). *It holds*

$$\int_0^x p(s) \left\{ \left\{ \sum_{j=1}^{M-1} \left[|D^{\alpha_1} f_j(s)|^{\lambda_{\alpha_1}} |D^{\alpha_1+1} f_{j+1}(s)|^{\lambda_{\alpha_1}} + \right. \right. \right.$$

$$\left. \left. |D^{\alpha_1+1} f_j(s)|^{\lambda_{\alpha_1}} |D^{\alpha_1} f_{j+1}(s)|^{\lambda_{\alpha_1}} \right] \right\}$$

$$+ \left[|D^{\alpha_1} f_1(s)|^{\lambda_{\alpha_1}} |D^{\alpha_1+1} f_M(s)|^{\lambda_{\alpha_1}} + |D^{\alpha_1+1} f_1(s)|^{\lambda_{\alpha_1}} |D^{\alpha_1} f_M(s)|^{\lambda_{\alpha_1}} \right] \right\} ds \leq$$

$$2 \left(\frac{x^{(2\beta\lambda_{\alpha_1} - 2\alpha_1\lambda_{\alpha_1} - \lambda_{\alpha_1} + 1)} \|p(s)\|_\infty}{(2\beta\lambda_{\alpha_1} - 2\alpha_1\lambda_{\alpha_1} - \lambda_{\alpha_1} + 1)(\beta - \alpha_1)^{\lambda_{\alpha_1}} (\Gamma(\beta - \alpha_1))^{2\lambda_{\alpha_1}}} \right)$$

$$\left[\sum_{j=1}^M \|D^\beta f_j\|_\infty^{2\lambda_{\alpha_1}} \right].$$
(9.66)

Proof. Based on Corollary 23 of [48]; see Corollary 7.23 here. □

Corollary 9.23. (to Corollary 9.22). *In detail: let* $\alpha_1 \in \mathbb{R}_+, \beta > \alpha_1 + 1$, *and let* $f_j \in L_1(0, x)$, $j = 1, \ldots, M \in \mathbb{N}$, $x \in \mathbb{R}_+ - \{0\}$, *have, respectively,* L_∞ *fractional derivatives* $D^\beta f_j$ *in* $[0, x]$, *and let* $D^{\beta-k} f_j(0) = 0$, *for* $k = 1, \ldots, [\beta] + 1; j = 1, \ldots, M$. *Then*

$$\int_0^x \left\{ \left\{ \sum_{j=1}^{M-1} \left[|D^{\alpha_1} f_j(s)| \, |D^{\alpha_1+1} f_{j+1}(s)| + |D^{\alpha_1+1} f_j(s)| \, |D^{\alpha_1} f_{j+1}(s)| \right] \right\} + \right.$$

$$\left. \left[|D^{\alpha_1} f_1(s)| \, |D^{\alpha_1+1} f_M(s)| + |D^{\alpha_1+1} f_1(s)| \, |D^{\alpha_1} f_M(s)| \right] \right\} ds$$

$$\leq \frac{x^{2(\beta-\alpha_1)}}{(\beta-\alpha_1)^2 (\Gamma(\beta-\alpha_1))^2} \left(\sum_{j=1}^{M} \|D^\beta f_j\|_\infty^2 \right). \tag{9.67}$$

Proof. Based on Corollary 24 of [48]; see Corollary 7.24 here. □

Corollary 9.24. (to Corollary 29.3). *It holds*

$$\int_0^x \left(\sum_{j=1}^{M} |D^{\alpha_1} f_j(s)| \, |D^{\alpha_1+1} f_j(s)| \right) ds \leq$$

$$\left(\frac{x^{2(\beta-\alpha_1)}}{2(\beta-\alpha_1)^2 (\Gamma(\beta-\alpha_1))^2} \right) \left(\sum_{j=1}^{M} \|D^\beta f_j\|_\infty^2 \right). \tag{9.68}$$

Proof. Based on inequality (207) of [48]; see also (7.207) here. □

9.4   Applications

We present our first application.

Theorem 9.25. *Let* $\alpha_i \in \mathbb{R}_+$, $\beta > \alpha_i$, $\beta - \alpha_i > (1/2)$, $i = 1, \ldots, r \in \mathbb{N}$, *and let* $f_j \in L_1(0, x)$, $j = 1, \ldots, M \in \mathbb{N}$; $x \in \mathbb{R}_+ - \{0\}$, *have, respectively,* L_∞ *fractional derivatives* $D^\beta f_j$ *in* $[0, x]$, *and let* $D^{\beta-k} f_j(0) = \alpha_{kj} \in \mathbb{R}$ *for* $k = 1, \ldots, [\beta] + 1; j = 1, \ldots, M$. *Furthermore for* $j = 1, \ldots, M$, *we have that*

$$D^\beta f_j(s) = F_j \left(s, \{D^{\alpha_i} f_j(s)\}_{i=1,j=1}^{r,M} \right), \tag{9.69}$$

all $s \in [0, x]$.

Here $F_j(s, \vec{z_1}, \vec{z_2}, \ldots, \vec{z_M})$ are continuous for $(\vec{z_1}, \vec{z_2}, \ldots, \vec{z_M}) \in (\mathbb{R}^r)^M$, bounded for $s \in [0, x]$, and satisfy the Lipschitz condition

$$|F_j(t; x_{11}, x_{12}, \ldots, x_{1r}; x_{21}, x_{22}, \ldots, x_{2r}; x_{31}, \ldots, x_{3r}; \ldots, x_{M1}, \ldots, x_{Mr})$$

$$-F_j(t, x'_{11}, x'_{12}, \ldots, x'_{1r}; x'_{21}, x'_{22}, \ldots, x'_{2r}; x'_{31}, \ldots, x'_{3r}; \ldots, x'_{M1}, \ldots, x'_{Mr})| \leq$$

$$\sum_{i=1}^{r} \left(\sum_{\ell=1}^{M} q_{\ell,i,j}(t) |x_{\ell i} - x'_{\ell i}| \right), \tag{9.70}$$

$j = 1, 2, \ldots, M$, *where all* $q_{\ell,i,j}(s) \geq 0$ *are bounded on* $[0, x]$, $1 \leq i \leq r, \ell = 1, \ldots, M$. *Call*

$$W := \max \left\{ \|q_{\ell,i,j}\|_\infty, \, | \, \ell, j = 1, 2, \ldots, M; i = 1, \ldots, r \right\}. \tag{9.71}$$

Assume here that

$$\phi^*(x) := W \left(\frac{1}{2} + \frac{M-1}{\sqrt{2}} \right) \left(\sum_{i=1}^{r} \left(\frac{x^{\beta - \alpha_i}}{\Gamma(\beta - \alpha_i)\sqrt{\beta - \alpha_i}\sqrt{2\beta - 2\alpha_i - 1}} \right) \right) < 1. \tag{9.72}$$

Then, if the system (9.69) has two pairs of solutions $(f_1, f_2, \ldots, f_M)$ and $(f_1^*, f_2^*, \ldots, f_M^*)$, we prove that $f_j = f_j^*, j = 1, 2, \ldots, M$; that is, we have uniqueness of solution for the fractional system (9.69).

Proof. Assume that there are two M -tuples of solutions $(f_1, \ldots, f_M)$ and $(f_1^*, \ldots, f_M^*)$ satisfying the system (9.69). Set $g_j := f_j - f_j^*$, $j = 1, 2, \ldots, M$. Then $D^{\beta-k} g_j(0) = 0, k = 1, \ldots, [\beta] + 1; \ j = 1, \ldots, M$. It holds

$$D^\beta g_j(s) = F_j \left(s, \{D^{\alpha_i} f_j(s)\}_{i=1,j=1}^{r,M} \right) - F_j \left(s, \{D^{\alpha_i} f_j^*(s)\}_{i=1,j=1}^{r,M} \right). \tag{9.73}$$

Hence by (9.70) we get

$$|D^\beta g_j(s)| \leq \sum_{i=1}^{r} \left(\sum_{\ell=1}^{M} q_{\ell,i,j}(t) |D^{\alpha_i} g_\ell(s)| \right). \tag{9.74}$$

Thus

$$|D^\beta g_j(s)| \leq W \sum_{i=1}^{r} \left(\sum_{\ell=1}^{M} |D^{\alpha_i} g_\ell(s)| \right). \tag{9.75}$$

The last implies

$$(D^\beta g_j(s))^2 \leq W \sum_{i=1}^{r} \left(\sum_{\ell=1}^{M} |D^\beta g_j(s)| \, |D^{\alpha_i} g_\ell(s)| \right), \tag{9.76}$$

and

$$\sum_{j=1}^{M}(D^{\beta}g_j(s))^2 \leq W \sum_{i=1}^{r}\sum_{j=1}^{M}\sum_{\ell=1}^{M}|D^{\beta}g_j(s)|\,|D^{\alpha_i}g_\ell(s)|. \qquad (9.77)$$

Integrating (9.77) we observe

$$I := \int_0^x \left(\sum_{j=1}^{M}(D^{\beta}g_j(s))^2\right)ds \leq$$

$$W\left\{\sum_{i=1}^{r}\sum_{j=1}^{M}\left(\sum_{\ell=1}^{M}\left(\int_0^x |D^{\beta}g_j(s)|\,|D^{\alpha_i}g_\ell(s)|ds\right)\right)\right\}. \qquad (9.78)$$

That is,

$$I \leq W\left\{\sum_{i=1}^{r}\left[\left(\int_0^x \left(\sum_{\lambda=1}^{M}|D^{\alpha_i}g_\lambda(s)|\,|D^{\beta}g_\lambda(s)|\right)ds\right)+\right.\right.$$

$$\left.\left.\sum_{\substack{\tau,m\in\{1,\dots,M\}\\ \tau\neq m}}\left(\int_0^x \left(|D^{\alpha_i}g_m(s)|\,|D^{\beta}g_\tau(s)|+|D^{\alpha_i}g_\tau(s)|\,|D^{\beta}g_m(s)|\right)ds\right)\right]\right\}.$$

$$(9.79)$$

Using Corollary 9.10 from here, and Corollary 13 of [48] (see also here Corollary 7.13) we find

$$I \leq W\left\{\sum_{i=1}^{r}\left[\left(\frac{x^{(\beta-\alpha_i)}\ \ I}{2\,\Gamma(\beta-\alpha_i)\sqrt{\beta-\alpha_i}\sqrt{2\beta-2\alpha_i-1}}\right)+\right.\right.$$

$$\left.\left.\left(\frac{x^{(\beta-\alpha_i)}\ (M-1)I}{\sqrt{2}\,\Gamma(\beta-\alpha_i)\sqrt{\beta-\alpha_i}\sqrt{2\beta-2\alpha_i-1}}\right)\right]\right\}. \qquad (9.80)$$

That is, we get that

$$I \leq \phi^*(x)\,I. \qquad (9.81)$$

If $I \neq 0$ then $\phi^*(x) \geq 1$, a contradiction by assumption that $\phi^*(x) < 1$; see (9.72). Therefore $I = 0$, implying that

$$\sum_{j=1}^{M}(D^{\beta}g_j(s))^2 = 0, \quad \text{a.e. in }[0,x]. \qquad (9.82)$$

That is,

$$(D^{\beta}g_j(s))^2 = 0, \quad \text{a.e. in }[0,x], \quad j=1,\dots,M, \qquad (9.83)$$

and

$$D^{\beta}g_j(s) = 0, \quad \text{a.e. in }[0,x], \quad j=1,\dots,M. \qquad (9.84)$$

By $D^{\beta-k}g_j(0) = 0$, $k = 1,\ldots,[\beta]+1; j = 1,\ldots,M$, and Lemma 9.3, apply (9.3) for $\alpha = 0$, we find $g_j(s) \equiv 0$, all $s \in [0,x]$, all $j = 1,\ldots,M$.

The last implies $f_j = f_j^*, j = 1,\ldots,M$, over $[0,x]$, thus proving the uniqueness of solution to the initial value problem of this theorem. □

Another related application follows.

Theorem 9.26. *Let* $\alpha_i \in \mathbb{R}_+, \beta > \alpha_i$, $\beta - \alpha_i > (1/2)$, $i = 1,\ldots,r \in \mathbb{N}$, *and let* $f_j \in L_1(0,x), j = 1,\ldots,M \in \mathbb{N}$; $x \in \mathbb{R}_+ - \{0\}$, *have, respectively, fractional derivatives* $D^\beta f_j$ *in* $[0,x]$, *that are absolutely continuous on* $[0,x]$, *and let* $D^{\beta-k}f_j(0) = 0$ *for* $k = 1,\ldots,[\beta]+1; j = 1,\ldots,M$. *And* $(D^\beta f_j)(0) = A_j \in \mathbb{R}$. *Furthermore for* $0 \le s \le x$ *we have holding the system of fractional differential equations*

$$\left(D^\beta f_j\right)'(s) = F_j\left(s, \{D^{\alpha_i}f_j(s)\}_{i=1,j=1}^{r,M}, \{D^\beta f_j(s)\}_{j=1}^{M}\right), \qquad (9.85)$$

for $j = 1,\ldots,M$. *Here* F_j *is Lebesgue measurable on* $[0,x] \times (\mathbb{R}^{r+1})^M$ *such that*

$$|F_j(t; x_{11}, x_{12}, \ldots, x_{1r}, x_{1,r+1}; x_{21}, x_{22}, \ldots, x_{2r}, x_{2,r+1}; \ldots$$

$$x_{M1}, x_{M2}, \ldots, x_{M,r+1})| \le \sum_{i=1}^{r}\left(\sum_{\ell=1}^{M} q_{\ell,i,j}(t)|x_{\ell i}|\right), \qquad (9.86)$$

where $q_{\ell,i,j} \ge 0$, $1 \le i \le r$; $\ell,j = 1,\ldots,M$, *are bounded on* $[0,x]$. *Set*

$$W := \max\{\|q_{\ell,i,j}\|_\infty; \ell,j = 1,\ldots,M; i = 1,\ldots,r\}. \qquad (9.87)$$

Also we put $(0 \le s \le x)$

$$\theta(s) := \sum_{\lambda=1}^{M}(D^\beta f_\lambda(s))^2, \qquad (9.88)$$

$$\rho := \sum_{\lambda=1}^{M} A_\lambda^2, \qquad (9.89)$$

$$Q(s) := W(1 + \sqrt{2}(M-1))\left(\sum_{i=1}^{r}\left(\frac{s^{\beta-\alpha_i}}{\Gamma(\beta-\alpha_i)\sqrt{\beta-\alpha_i}\sqrt{2\beta-2\alpha_i-1}}\right)\right), \qquad (9.90)$$

and

$$\chi(s) := \sqrt{\rho}\left\{1 + Q(s) \cdot e^{(\int_0^s Q(t)dt)} \cdot \left[\int_0^s \left(e^{-(\int_0^t Q(y)dy)}\right) dt\right]\right\}^{1/2}. \qquad (9.91)$$

Then

$$\sqrt{\theta(s)} \le \chi(s), \quad 0 \le s \le x. \qquad (9.92)$$

Consequently we get

$$|D^\beta f_j(s)| \le \chi(s), \tag{9.93}$$

$$|f_j(s)| \le \frac{1}{\Gamma(\beta)} \int_0^s (s-t)^{\beta-1} \chi(t)dt, \tag{9.94}$$

all $0 \le s \le x$, $j = 1, \ldots, M$. *Also it holds*

$$|D^{\alpha_i} f_j(s)| \le \frac{1}{\Gamma(\beta-\alpha_i)} \int_0^s (s-t)^{\beta-\alpha_i-1} \chi(t)dt, \tag{9.95}$$

all $0 \le s \le x$, $j = 1, \ldots, M$, $i = 1, \ldots, r$.

Proof. We see that

$$(D^\beta f_j)(s)(D^\beta f_j)'(s) = (D^\beta f_j)(s) F_j\left(s, \{D^{\alpha_i} f_j(s)\}_{i=1,j=1}^{r,M}, \{D^\beta f_j(s)\}_{j=1}^{M}\right), \tag{9.96}$$

$j = 1, \ldots, M$, all $0 \le s \le x$.
We then integrate (9.96),

$$\int_0^y (D^\beta f_j)(s)(D^\beta f_j)'(s)ds =$$

$$\int_0^y (D^\beta f_j)(s) F_j\left(s, \{D^{\alpha_i} f_j(s)\}_{i=1,j=1}^{r,M}, \{D^\beta f_j(s)\}_{j=1}^{M}\right) ds, \tag{9.97}$$

$0 \le y \le x$.
Hence we obtain

$$\frac{((D^\beta f_j)(s))^2}{2}\Big|_0^y \le \int_0^y |D^\beta f_j(s)|\, |F_j \cdots |ds \overset{(9.86)}{\underset{(9.87)}{\le}}$$

$$W \int_0^y |D^\beta f_j(s)| \left[\sum_{i=1}^r \left(\sum_{\ell=1}^M |D^{\alpha_i} f_\ell(s)|\right)\right] ds =$$

$$W\left(\sum_{i=1}^r \left(\sum_{\ell=1}^M \left(\int_0^y |D^{\alpha_i} f_\ell(s)||D^\beta f_j(s)|ds\right)\right)\right). \tag{9.98}$$

Thus we have for $j = 1, \ldots, M$ that

$$(D^\beta f_j(y))^2 \le A_j^2 + 2W\left(\sum_{i=1}^r \left(\sum_{\ell=1}^M \left(\int_0^y |D^{\alpha_i} f_\ell(s)||D^\beta f_j(s)|ds\right)\right)\right). \tag{9.99}$$

Consequently it holds

$$\theta(y) \le \rho + 2W\left(\sum_{i=1}^r \left(\sum_{j=1}^M \left(\sum_{\ell=1}^M \left(\int_0^y |D^{\alpha_i} f_\ell(s)||D^\beta f_j(s)|ds\right)\right)\right)\right) = \tag{9.100}$$

$$\rho + 2W \left\{ \sum_{i=1}^{r} \left\{ \left(\int_0^y \left(\sum_{\lambda=1}^{M} |D^{\alpha_i} f_\lambda(t)| \, |D^\beta f_\lambda(t)| \right) dt \right) + \right. \right.$$

$$\left. \left. \sum_{\substack{\tau, m \in \{1, \dots, M\} \\ \tau \neq m}} \left(\int_0^y \left(|D^{\alpha_i} f_m(t)| \, |D^\beta f_\tau(t)| + |D^{\alpha_i} f_\tau(t)| \, |D^\beta f_m(t)| \right) dt \right) \right\} \right\}.$$

$$(9.101)$$

Using Corollary 9.10 from here, and Corollary 13 of [48] (see here Corollary 7.13) we find

$$\theta(y) \le \rho + 2W \left\{ \sum_{i=1}^{r} \left\{ \left(\frac{y^{(\beta - \alpha_i)}}{2\,\Gamma(\beta - \alpha_i)\sqrt{\beta - \alpha_i}\sqrt{2\beta - 2\alpha_i - 1}} \right) \left(\int_0^y \theta(t) dt \right) + \right. \right.$$

$$\left. \left. \left(\frac{y^{(\beta - \alpha_i)}}{\sqrt{2}\,\Gamma(\beta - \alpha_i)\sqrt{\beta - \alpha_i}\sqrt{2\beta - 2\alpha_i - 1}} \right) (M - 1) \left(\int_0^y \theta(t) dt \right) \right\} \right\}.$$

$$(9.102)$$

Hence we have

$$\theta(y) \le \rho + Q(y) \int_0^y \theta(t) dt, \quad \text{all } 0 \le y \le x. \qquad (9.103)$$

Here $\rho \ge 0$, $Q(y) \ge 0$, $Q(0) = 0$, $\theta(y) \ge 0$, all $0 \le y \le x$. As in the proof of Theorem 27 of [48] (here also see the proof of Theorem 7.27; see also [17]) we get (9.92), (9.93). Using Lemma 9.3 (see (9.3)), for $\alpha = 0$, and (9.93) we obtain (9.94). Again using (9.3) and (9.93) we get (9.95). □

Finally we give a specialized application.

Theorem 9.27. *Let $\alpha_i \in \mathbb{R}_+, \beta > \alpha_i$, $i = 1, \dots, r \in \mathbb{N}$, and $f_j \in L_1(0, x)$, $j = 1, \dots, M \in \mathbb{N}$; $x \in \mathbb{R}_+ - \{0\}$, have, respectively, fractional derivatives $D^\beta f_j$ in $[0, x]$, that are absolutely continuous on $[0, x]$, and let $D^{\beta - \mu} f_j(0) = 0$ for $\mu = 1, \dots, [\beta] + 1; j = 1, \dots, M$. And $(D^\beta f_j)(0) = A_j \in \mathbb{R}$. Furthermore for $0 \le s \le x$ we have holding the system of fractional differential equations*

$$\left(D^\beta f_j \right)' (s) = F_j \left(s, \{ D^{\alpha_i} f_j(s) \}_{i=1, j=1}^{r, M}, \{ D^\beta f_j(s) \}_{j=1}^{M} \right), \qquad (9.104)$$

for $j = 1, \dots, M$.

For fixed $i_* \in \{1, \dots, r\}$ we assume that $\alpha_{i_*+1} = \alpha_{i_*} + 1$, where α_{i_*}, $\alpha_{i_*+1} \in \{\alpha_1, \dots, \alpha_r\}$. Call $k := \alpha_{i_*}$, $\gamma := \alpha_{i_*} + 1$; that is, $\gamma = k + 1$. Here F_j is Lebesgue measurable on $[0, x] \times (\mathbb{R}^{r+1})^M$ such that

$$|F_j(t, x_{11}, x_{12}, \dots, x_{1r}, x_{1,r+1}; x_{21}, x_{22}, \dots, x_{2r}, x_{2,r+1};$$

$$x_{31}, x_{32}, \dots, x_{3r}, x_{3,r+1}; \dots; x_{M1}, x_{M2}, \dots, x_{Mr}, x_{M,r+1})|$$

$$\leq \left\{ \left\{ \sum_{\ell=1}^{M-1} \left(q_{\ell,1,j}(t)|x_{\ell_{i*}}|\sqrt{|x_{\ell+1,i_*+1}|} + q_{\ell,2,j}(t)|x_{\ell+1,i_*}|\sqrt{|x_{\ell,i_*+1}|} \right) \right\} + \right.$$

$$\left. \left(q_{M,1,j}(t)|x_{1i_*}|\sqrt{|x_{M,i_*+1}|} + q_{M,2,j}(t)|x_{Mi_*}|\sqrt{|x_{1,i_*+1}|} \right) \right\}, \qquad (9.105)$$

where all $0 \leq q_{\ell,1,j}$, $q_{\ell,2,j} \not\equiv 0$, are bounded over $[0,x]$. Call

$$W := \max\{\|q_{\ell,1,j}\|_\infty, \|q_{\ell,2,j}\|_\infty\}_{\ell,j=1}^{M}. \qquad (9.106)$$

Also set

$$\theta(s) := \sum_{j=1}^{M} |(D^\beta f_j)(s)|, \quad 0 \leq s \leq x, \qquad (9.107)$$

$$\rho := \sum_{j=1}^{M} |A_j|,$$

$$\Phi^*(s) := \left(\frac{2}{\sqrt{3\beta - 3k - 2}} \right) \frac{s^{((3\beta-3k-1)/2)}}{(\Gamma(\beta-k))^{3/2}}, \qquad (9.108)$$

all $0 \leq s \leq x$, and

$$Q(s) := 2MW\Phi^*(s), \quad 0 \leq s \leq x, \qquad (9.109)$$

$$\sigma := \|Q(s)\|_\infty = \frac{4MW \; x^{((3\beta-3k-1)/2)}}{\sqrt{3\beta - 3k - 2} \; (\Gamma(\beta-k))^{3/2}}. \qquad (9.110)$$

We assume that

$$x\sigma\sqrt{\rho} < 2. \qquad (9.111)$$

Set

$$\breve{\varphi}(s) := \rho + Q(s) \left[\frac{4\rho^{3/2}s - \sigma\rho^2 s^2}{(2 - \sigma\sqrt{\rho}s)^2} \right], \quad \text{all } 0 \leq s \leq x. \qquad (9.112)$$

Then

$$\theta(s) \leq \breve{\varphi}(s), \quad \text{all} \quad 0 \leq s \leq x, \qquad (9.113)$$

in particular we have

$$|D^\beta f_j(s)| \leq \breve{\varphi}(s), \quad j = 1, ..., M, \text{ for all } 0 \leq s \leq x. \qquad (9.114)$$

Furthermore we get

$$|f_j(s)| \leq \frac{1}{\Gamma(\beta)} \int_0^s (s-t)^{\beta-1}\breve{\varphi}(t)dt, \qquad (9.115)$$

and

$$|D^{\alpha_i} f_j(s)| \leq \frac{1}{\Gamma(\beta-\alpha_i)} \int_0^s (s-t)^{\beta-\alpha_i-1}\breve{\varphi}(t)dt, \qquad (9.116)$$

$j = 1, \ldots, M; \ i = 1, \ldots, r, \ \text{all } 0 \le s \le x.$

Proof. Notice here that $W > 0$ and $\sigma > 0$. Clearly, here $D^\beta f_j$ are L_∞ fractional derivatives in $[0, x]$. For $0 \le s \le y \le x$, by (9.104) we get that

$$\int_0^y \left(D^\beta f_j\right)'(s) ds = \int_0^y F_j\left(s, \{D^{\alpha_i} f_j(s)\}_{i=1, j=1}^{r, M}, \{D^\beta f_j(s)\}_{j=1}^M\right) ds. \tag{9.117}$$

That is,

$$(D^\beta f_j)(y) = A_j + \int_0^y F_j(s, \ldots) ds. \tag{9.118}$$

Then we observe that

$$|D^\beta f_j(y)| \le |A_j| + \int_0^y |F_j(s, \ldots)| ds \le |A_j| +$$

$$W \int_0^y \left\{ \left\{ \sum_{\ell=1}^{M-1} (|D^{\alpha_{i*}} f_\ell(s)| \sqrt{|D^{\alpha_{i*}+1} f_{\ell+1}(s)|} + |D^{\alpha_{i*}} f_{\ell+1}(s)| \sqrt{|D^{\alpha_{i*}+1} f_\ell(s)|}) \right\} \right.$$

$$+ (|D^{\alpha_{i*}} f_1(s)| \sqrt{|D^{\alpha_{i*}+1} f_M(s)|} + |D^{\alpha_{i*}} f_M(s)| \sqrt{|D^{\alpha_{i*}+1} f_1(s)|}) \Big\} ds. \tag{9.119}$$

That is,

$$|D^\beta f_j(y)| \le |A_j| + W \left(\int_0^y \left\{ \left\{ \sum_{\ell=1}^{M-1} \left(|D^k f_\ell(s)| \sqrt{|D^{k+1} f_{\ell+1}(s)|} + \right. \right. \right. \right.$$

$$\left. |D^k f_{\ell+1}(s)| \sqrt{|D^{k+1} f_\ell(s)|} \right) \Big\} +$$

$$\left. \left(|D^k f_1(s)| \sqrt{|D^{k+1} f_M(s)|} + |D^k f_M(s)| \sqrt{|D^{k+1} f_1(s)|} \right) \right\} ds. \tag{9.120}$$

By Corollary 9.14 we obtain

$$|D^\beta f_j(s)| \le |A_j| + 2W \phi^*(s) \left(\int_0^s \left(\sum_{\ell=1}^M |D^\beta f_\ell(t)|^{3/2} \right) dt \right), \tag{9.121}$$

$j = 1, \ldots, M, \ \text{all } 0 \le s \le x.$

Therefore by adding all of inequalities (9.121) we find

$$\theta(s) \le \rho + 2MW \phi^*(s) \left(\int_0^s \left(\sum_{\ell=1}^M |D^\beta f_\ell(t)|^{3/2} \right) dt \right), \tag{9.122}$$

$$\overset{(9.4)}{\le} \rho + 2MW \phi^*(s) \left(\int_0^s \left(\sum_{\ell=1}^M |D^\beta f_\ell(t)| \right)^{3/2} dt \right). \tag{9.123}$$

That is,

$$\theta(s) \leq \rho + (2MW\phi^*(s)) \left(\int_0^s (\theta(t))^{3/2} dt \right), \tag{9.124}$$

all $0 \leq s \leq x$.

More precisely we get that

$$\theta(s) \leq \rho + Q(y) \left(\int_0^s (\theta(t))^{3/2} dt \right), \quad \text{all } 0 \leq s \leq x. \tag{9.125}$$

Notice that $\theta(s) \geq 0$, $\rho \geq 0$, $Q(s) \geq 0$, and $Q(0) = 0$ by $\Phi^*(0) = 0$. Acting here as in the proof of Theorem 28 of [48] (also see Theorem 7.28 here) we derive (9.113) and (9.114).

Using Lemma 9.3 (see (9.3)), along with (9.114), we obtain (9.115) and (9.116). □

10

Converse Canavati Fractional Opial-Type Inequalities for Several Functions

A collection of very general $L_p (0 < p < 1)$-form converse Opial-type inequalities [315] is presented involving generalized Canavati fractional derivatives [17, 101] of several functions in different orders and powers. Other particular results of special interest are derived from the established results. This treatment is based on [51].

10.1 Introduction

Opial inequalities appeared for the first time in [315] and since then many authors have dealt with them in different directions and for various cases. For a complete account of the recent activity in this field see [4], and it still remains a very active area of research. One of the main attractions to these inequalities is their applications, especially in proving the uniqueness and upper bounds of the solution of initial value problems in differential equations. The author was the first to present Opial-type inequalities involving fractional derivatives of functions in [15, 17] with applications to fractional differential equations.

Fractional derivatives come up naturally in a number of fields, especially in physics; see the recent books [197, 333]. Here the author continues his study of fractional Opial-type inequalities now involving several different functions and produces a wide variety of converse results. To give the reader an idea of the kind of inequalities we are dealing with, we briefly mention

G.A. Anastassiou, *Fractional Differentiation Inequalities*,   205
DOI 10.1007/978-0-387-98128-4_10, © Springer Science+Business Media, LLC 2009

a specific one (see Corollary 10.15).

$$\int_{x_0}^{x} \left(\sum_{j=1}^{M} \left| \left(D_{x_0}^{\gamma_1} f_j \right)(w) \right|^{\lambda_\alpha} \left| \left(D_{x_0}^{v} f_j \right)(w) \right|^{\lambda_v} \right) dw \geq$$

$$C(x) \left[\int_{x_0}^{x} \left[\sum_{j=1}^{M} \left| \left(D_{x_0}^{v} f_j \right)(w) \right|^{p} \right] dw \right]^{(\lambda_\alpha + \lambda_v / p)},$$

all $x_0 \leq x \leq b$.

In the last inequality $C(x)$ is a constant that depends on x_0, x, and the involved parameters, $\gamma_1 \geq 0$, $1 \leq v - \gamma_1 < 1/p$, $0 < p < 1$; $D_{x_0}^{v} f_j$ is of fixed sign on $[x_0, b]$, $j = 1, \ldots, M \in \mathbb{N}$. Also $\lambda_\alpha \geq 0$, $\lambda_v > p$. Here $f_j^{(i)}(x_0) = 0$, $i = 0, 1, \ldots, n - 1$, $n := [v]$ (integral part); $j = 1, \ldots, M$.

And $D_{x_0}^{\gamma_1} f_j$, $D_{x_0}^{v} f_j$ are the generalized (of Canavati) type [15, 101] fractional derivatives of f_j of orders γ_1, v, respectively.

10.2   Preliminaries

In the sequel we follow [101]. Let $g \in C([0, 1])$. Let v be a positive number, $n := [v]$, and $\alpha := v - n$ $(0 < \alpha < 1)$. Define

$$(J_v g)(x) := \frac{1}{\Gamma(v)} \int_0^x (x - t)^{v-1} g(t) \, dt, \quad 0 \leq x \leq 1, \qquad (10.1)$$

the *Riemann–Liouville integral,* where Γ is the gamma function. We define the subspace $C^v([0, 1])$ of $C^n([0, 1])$ as follows.

$$C^v([0, 1]) := \left\{ g \in C^n([0, 1]) : J_{1-\alpha} D^n g \in C^1([0, 1]) \right\},$$

where $D := d/dx$. So for $g \in C^v([0, 1])$, we define the Canavati v-*fractional derivative* of g as

$$D^v g := D J_{1-\alpha} D^n g. \qquad (10.2)$$

When $v \geq 1$ we have the fractional Taylor's formula

$$g(t) = g(0) + g'(0) t + g''(0) \frac{t^2}{2!} + \ldots + g^{(n-1)}(0) \frac{t^{n-1}}{(n-1)!}$$

$$+ (J_v D^v g)(t), \quad \text{for all } t \in [0, 1]. \qquad (10.3)$$

When $0 < v < 1$ we find

$$g(t) = (J_v D^v g)(t), \quad \text{for all } t \in [0, 1]. \qquad (10.4)$$

Next we carry the above notions over to arbitrary $[a, b] \subseteq \mathbb{R}$ (see [17]). Let x, $x_0 \in [a, b]$ such that $x > x_0$, where $x_0 < b$ is fixed. Let $f \in C([a, b])$ and define

$$(J_v^{x_0} f)(x) := \frac{1}{\Gamma(v)} \int_{x_0}^x (x - t)^{v-1} f(t) \, dt, \quad x_0 \leq x \leq b, \tag{10.5}$$

the *generalized Riemann–Liouville integral.* We define the subspace $C_{x_0}^v([a, b])$ of $C^n([a, b])$:

$$C_{x_0}^v([a, b]) := \left\{ f \in C^n([a, b]) : J_{1-\alpha}^{x_0} D^n f \in C^1([x_0, b]) \right\};$$

clearly $C_{x_0}^0([a, b]) = C([a, b])$; also $C_{x_0}^n([a, b]) = C^n([a, b])$, $n \in \mathbb{N}$.

For $f \in C_{x_0}^v([a, b])$, we define the *generalized v-fractional derivative of f over $[x_0, b]$* as

$$D_{x_0}^v f := D J_{1-\alpha}^{x_0} f^{(n)} \quad \left(f^{(n)} := D^n f \right). \tag{10.6}$$

Notice that

$$\left(J_{1-\alpha}^{x_0} f^{(n)} \right)(x) = \frac{1}{\Gamma(1-\alpha)} \int_{x_0}^x (x - t)^{-\alpha} f^{(n)}(t) \, dt$$

exists for $f \in C_{x_0}^v([a, b])$.

We recall the following fractional generalization of Taylor's formula (see [17, 101]).

Theorem 10.1. *Let $f \in C_{x_0}^v([a, b])$, $x_0 \in [a, b]$ fixed.*
(i) If $v \geq 1$ then

$$f(x) = f(x_0) + f'(x_0)(x - x_0) + f''(x_0) \frac{(x - x_0)^2}{2} + \ldots +$$

$$f^{(n-1)}(x_0) \frac{(x - x_0)^{n-1}}{(n-1)!} + \left(J_v^{x_0} D_{x_0}^v f \right)(x), \quad \text{for all } x \in [a, b] : x \geq x_0. \tag{10.7}$$

(ii) If $0 < v < 1$ then

$$f(x) = \left(J_v^{x_0} D_{x_0}^v f \right)(x), \quad \text{for all } x \in [a, b] : x \geq x_0. \tag{10.8}$$

We make

Remark 10.2. (1) $\left(D_{x_0}^n f \right) = f^{(n)}$, $n \in \mathbb{N}$.
(2) Let $f \in C_{x_0}^v([a, b])$, $v \geq 1$, and $f^{(i)}(x_0) = 0$, $i = 0, 1, \ldots, n - 1$, $n := [v]$. Then by (10.7),

$$f(x) = \left(J_v^{x_0} D_{x_0}^v f \right)(x).$$

That is,

$$f(x) = \frac{1}{\Gamma(v)} \int_{x_0}^{x} (x-t)^{v-1} \left(D_{x_0}^v f\right)(t)\, dt, \tag{10.9}$$

for all $x \in [a,b]$ with $x \geq x_0$. Notice that (10.9) is true, also when $0 < v < 1$.

We need the following from [17].

Lemma 10.3. *Let* $f \in C([a,b])$, $\mu, v > 0$. *Then*

$$J_\mu^{x_0} \left(J_v^{x_0} f\right) = J_{\mu+v}^{x_0}(f). \tag{10.10}$$

We also make

Remark 10.4. Let $v \geq \gamma + 1$, $\gamma \geq 0$, so that $\gamma < v$. Call $n := [v]$, $\alpha := v - n$; $m := [\gamma]$, $\rho := \gamma - m$. Note that $v - m \geq 1$ and $n - m \geq 1$. Let $f \in C_{x_0}^v([a,b])$ be such that $f^{(i)}(x_0) = 0$, $i = 0, 1, \ldots, n-1$. Hence by (10.7),

$$f(x) = \left(J_v^{x_0} D_{x_0}^v f\right)(x), \quad \text{for all } x \in [a,b] : x \geq x_0.$$

Therefore by Leibnitz's formula and $\Gamma(p+1) = p\Gamma(p)$, $p > 0$, we get that

$$f^{(m)}(x) = \left(J_{v-m}^{x_0} D_{x_0}^v f\right)(x), \quad \text{for all } x \geq x_0. \tag{10.11}$$

It follows that $f \in C_{x_0}^\gamma([a,b])$ and thus

$$\left(D_{x_0}^\gamma f\right)(x) := \left(D J_{1-\rho}^{x_0} f^{(m)}\right)(x) \quad \text{exists for all } x \geq x_0. \tag{10.12}$$

By the use of (10.11) we have on $[x_0, b]$,

$$J_{1-\rho}^{x_0}\left(f^{(m)}\right) = J_{1-\rho}^{x_0}\left(J_{v-m}^{x_0} D_{x_0}^v f\right) = \left(J_{1-\rho}^{x_0} \circ J_{v-m}^{x_0}\right)\left(D_{x_0}^v f\right)$$

$$= J_{v-m+1-\rho}^{x_0}\left(D_{x_0}^v f\right) = J_{v-\gamma+1}^{x_0}\left(D_{x_0}^v f\right),$$

by (10.10). That is,

$$\left(J_{1-\rho}^{x_0} f^{(m)}\right)(x) = \frac{1}{\Gamma(v-\gamma+1)} \int_{x_0}^{x} (x-t)^{v-\gamma} \left(D_{x_0}^v f\right)(t)\, dt.$$

Therefore

$$\left(D_{x_0}^\gamma f\right)(x) = D\left(\left(J_{1-\rho}^{x_0} f^{(m)}\right)(x)\right) = \frac{1}{\Gamma(v-\gamma)} \cdot$$
$$\int_{x_0}^{x} (x-t)^{(v-\gamma)-1} \left(D_{x_0}^v f\right)(t)\, dt; \tag{10.13}$$

hence

$$\left(D_{x_0}^\gamma f\right)(x) = \left(J_{v-\gamma}^{x_0}\left(D_{x_0}^v f\right)\right)(x)$$

and is continuous in x on $[x_0, b]$. In particular when $v \geq 2$ we have

$$\left(D_{x_0}^{v-1}f\right)(x) = \int_{x_0}^{x} \left(D_{x_0}^v f\right)(t)\,dt, \quad x \geq x_0. \tag{10.14}$$

That is,

$$\left(D_{x_0}^{v-1}f\right)' = D_{x_0}^v f, \quad \left(D_{x_0}^{v-1}f\right)(x_0) = 0.$$

10.3   Main Results

10.3.1   Results Involving Two Functions

We present our first main result.

Theorem 10.5. Let $\gamma_j \geq 0$, $1 \leq v - \gamma_j < 1/p$, $0 < p < 1$, $j = 1,2$, and $f_1, f_2 \in C_{x_0}^v([a,b])$ with $f_1^{(i)}(x_0) = f_2^{(i)}(x_0) = 0$, $i = 0,1,\ldots,n-1$, $n := [v]$. Here $x, x_0 \in [a,b] : x \geq x_0$. We assume here that $D_{x_0}^v f_i$ is of fixed sign on $[x_0, b]$, $i = 1, 2$. Consider also $p(t) > 0$ and $q(t) > 0$ continuous functions on $[x_0, b]$. Let $\lambda_v > 0$ and $\lambda_\alpha, \lambda_\beta \geq 0$ such that $\lambda_v > p$.
Set

$$P_k(w) := \int_{x_0}^{w} (w-t)^{(v-\gamma_k-1)p/p-1}(p(t))^{-1/p-1}\,dt, \quad k=1,2,\ x_0 \leq x \leq b;$$
$$\tag{10.15}$$

$$A(w) := \frac{q(w) \cdot (P_1(w))^{\lambda_\alpha(p-1/p)} \cdot (P_2(w))^{(\lambda_\beta(p-1)/p)}(p(w))^{-\lambda_v/p}}{(\Gamma(v-\gamma_1))^{\lambda_\alpha}(\Gamma(v-\gamma_2))^{\lambda_\beta}}, \tag{10.16}$$

$$A_0(x) := \left(\int_{x_0}^{x} (A(w))^{p/(p-\lambda_v)}\,dw\right)^{(p-\lambda_v)/p}, \tag{10.17}$$

and

$$\delta_1 := 2^{1-((\lambda_\alpha \,|\, \lambda_v)/p)}. \tag{10.18}$$

If $\lambda_\beta = 0$, we obtain that

$$\int_{x_0}^{x} q(w)\left[\left|\left(D_{x_0}^{\gamma_1}f_1\right)(w)\right|^{\lambda_\alpha} \cdot \left|\left(D_{x_0}^v f_1\right)(w)\right|^{\lambda_v} + \right.$$

$$\left. \left|\left(D_{x_0}^{\gamma_1}f_2\right)(w)\right|^{\lambda_\alpha} \cdot \left|\left(D_{x_0}^v f_2\right)(w)\right|^{\lambda_v}\right]dw$$

$$\geq \left.\left(A_0(x)\,\right|_{\lambda_\beta=0}\right) \cdot \left(\frac{\lambda_v}{\lambda_\alpha + \lambda_v}\right)^{\lambda_v/p} \cdot \delta_1 \cdot \left[\int_{x_0}^{x} p(w)\left[\left|\left(D_{x_0}^v f_1\right)(w)\right|^p + \right.\right.$$

$$\left.\left.\left|\left(D_{x_0}^v f_2\right)(w)\right|^p\right]dw\right]^{((\lambda_\alpha+\lambda_v)/p)}. \tag{10.19}$$

Proof. From (10.13) and assumption we get

$$\left|\left(D_{x_0}^{\gamma_k} f_j\right)(w)\right| = \frac{1}{\Gamma(v - \gamma_k)} \int_{x_0}^{w} (w - t)^{v - \gamma_k - 1} \left|\left(D_{x_0}^{v} f_j\right)(t)\right| dt,$$

for $k = 1, 2$, $j = 1, 2$ and for all $x_0 \leq w \leq b$.

Next applying Hölder's inequality with indices p, $p/p - 1$ we have

$$\left|\left(D_{x_0}^{\gamma_k} f_j\right)(w)\right| = \frac{1}{\Gamma(v - \gamma_k)} \int_{x_0}^{w} (w - t)^{v - \gamma_k - 1} (p(t))^{-1/p} (p(t))^{1/p}$$

$$\left|\left(D_{x_0}^{v} f_j\right)(t)\right| dt$$

$$\geq \frac{1}{\Gamma(v - \gamma_k)} \left(\int_{x_0}^{w} \left((w - t)^{v - \gamma_k - 1} (p(t))^{-1/p}\right)^{p/p - 1} dt\right)^{p - 1/p}$$

$$\left(\int_{x_0}^{w} p(t) \left|\left(D_{x_0}^{v} f_j\right)(t)\right|^p dt\right)^{1/p}$$

$$= \frac{1}{\Gamma(v - \gamma_k)} \left(P_k(w)\right)^{p - 1/p} \left(\int_{x_0}^{w} p(t) \left|\left(D_{x_0}^{v} f_j\right)(t)\right|^p dt\right)^{1/p}.$$

That is, it holds

$$\left|\left(D_{x_0}^{\gamma_k} f_j\right)(w)\right| \geq \frac{1}{\Gamma(v - \gamma_k)} \left(P_k(w)\right)^{p - 1/p} \left(\int_{x_0}^{w} p(t) \left|\left(D_{x_0}^{v} f_j\right)(t)\right|^p dt\right)^{1/p}.$$

$$(10.20)$$

Put

$$z_j(w) := \int_{x_0}^{w} p(t) \left|\left(D_{x_0}^{v} f_j\right)(t)\right|^p dt,$$

thus,

$$z_j'(w) = p(w) \left|\left(D_{x_0}^{v} f_j\right)(w)\right|^p, \quad z_j(x_0) = 0; \quad j = 1, 2.$$

Hence, we have

$$\left|\left(D_{x_0}^{\gamma_k} f_j\right)(w)\right| \geq \frac{1}{\Gamma(v - \gamma_k)} \left(P_k(w)\right)^{p - 1/p} \left(z_j(w)\right)^{1/p},$$

and

$$\left|\left(D_{x_0}^{v} f_j\right)(w)\right|^{\lambda_v} = p(w)^{-\lambda_v/p} \left(z_j'(w)\right)^{\lambda_v/p}, \quad j = 1, 2.$$

Therefore we obtain

$$q(w) \left|\left(D_{x_0}^{\gamma_1} f_1\right)(w)\right|^{\lambda_\alpha} \left|\left(D_{x_0}^{\gamma_2} f_2\right)(w)\right|^{\lambda_\beta} \left|\left(D_{x_0}^{v} f_1\right)(w)\right|^{\lambda_v}$$

$$\geq q(w) \frac{1}{\left(\Gamma(v - \gamma_1)\right)^{\lambda_\alpha}} \left(P_1(w)\right)^{\lambda_\alpha(p - 1/p)} \left(z_1(w)\right)^{\lambda_\alpha/p}$$

$$\frac{1}{\left(\Gamma\left(v-\gamma_2\right)\right)^{\lambda_\beta}}\left(P_2\left(w\right)\right)^{\lambda_\beta(p-1/p)}\left(z_2\left(w\right)\right)^{\lambda_\beta/p}\left(p\left(w\right)\right)^{-\lambda_v/p}\left(z_1^{'}\left(w\right)\right)^{\lambda_v/p}$$

$$= A\left(w\right)\left(z_1\left(w\right)\right)^{\lambda_\alpha/p}\left(z_2\left(w\right)\right)^{\lambda_\beta/p}\left(z_1^{'}\left(w\right)\right)^{\lambda_v/p}.$$

Consequently, by another Hölder's inequality application, we find (by $p/\lambda_v < 1$)

$$\int_{x_0}^{x} q\left(w\right)\left|\left(D_{x_0}^{\gamma_1}f_1\right)\left(w\right)\right|^{\lambda_\alpha}\left|\left(D_{x_0}^{\gamma_2}f_2\right)\left(w\right)\right|^{\lambda_\beta}\left|\left(D_{x_0}^{v}f_1\right)\left(w\right)\right|^{\lambda_v}dw$$

$$\geq A_0\left(x\right)\left[\int_{x_0}^{x}\left(z_1\left(w\right)\right)^{\lambda_\alpha/\lambda_v}\left(z_2\left(w\right)\right)^{\lambda_\beta/\lambda_v}z_1^{'}\left(w\right)dw\right]^{\lambda_v/p}.\qquad(10.21)$$

Similarly one derives

$$\int_{x_0}^{x} q\left(w\right)\left|\left(D_{x_0}^{\gamma_2}f_1\right)\left(w\right)\right|^{\lambda_\beta}\left|\left(D_{x_0}^{\gamma_1}f_2\right)\left(w\right)\right|^{\lambda_\alpha}\left|\left(D_{x_0}^{v}f_2\right)\left(w\right)\right|^{\lambda_v}dw$$

$$\geq A_0\left(x\right)\left[\int_{x_0}^{x}\left(z_1\left(w\right)\right)^{\lambda_\beta/\lambda_v}\left(z_2\left(w\right)\right)^{\lambda_\alpha/\lambda_v}z_2^{'}\left(w\right)dw\right]^{\lambda_v/p}.\qquad(10.22)$$

Taking $\lambda_\beta = 0$ and adding (10.21) and (10.22) we obtain

$$\int_{x_0}^{x} q\left(w\right)\left[\left|\left(D_{x_0}^{\gamma_1}f_1\right)\left(w\right)\right|^{\lambda_\alpha}\cdot\left|\left(D_{x_0}^{v}f_1\right)\left(w\right)\right|^{\lambda_v}+\right.$$

$$\left.\left|\left(D_{x_0}^{\gamma_1}f_2\right)\left(w\right)\right|^{\lambda_\alpha}\cdot\left|\left(D_{x_0}^{v}f_2\right)\left(w\right)\right|^{\lambda_v}\right]dw$$

$$\geq\left(A_0\left(x\right)|_{\lambda_\beta=0}\right)\left\{\left[\int_{x_0}^{x}\left(z_1\left(w\right)\right)^{\lambda_\alpha/\lambda_v}z_1^{'}\left(w\right)dw\right]^{\lambda_v/p}+\right.$$

$$\left.\left[\int_{x_0}^{x}\left(z_2\left(w\right)\right)^{\lambda_\alpha/\lambda_v}z_2^{'}\left(w\right)dw\right]^{\lambda_v/p}\right\}$$

$$=\left(A_0\left(x\right)|_{\lambda_\beta=0}\right)\left\{\left(z_1\left(x\right)\right)^{(\lambda_\alpha+\lambda_v)/p}+\right.$$

$$\left.\left(z_2\left(x\right)\right)^{(\lambda_\alpha+\lambda_v)/p}\right\}\left(\frac{\lambda_v}{\lambda_\alpha+\lambda_v}\right)^{\lambda_v/p}$$

$$=\left(A_0\left(x\right)|_{\lambda_\beta=0}\right)\left(\frac{\lambda_v}{\lambda_\alpha+\lambda_v}\right)^{\lambda_v/p}\left\{\left(\int_{x_0}^{x}p\left(t\right)\left|\left(D_{x_0}^{v}f_1\right)\left(t\right)\right|^{p}dt\right)^{(\lambda_\alpha+\lambda_v)/p}\right.$$

$$\left.+\left(\int_{x_0}^{x}p\left(t\right)\left|\left(D_{x_0}^{v}f_2\right)\left(t\right)\right|^{p}dt\right)^{(\lambda_\alpha+\lambda_v)/p}\right\}=:(*).$$

In this chapter we frequently use the basic inequalities

$$2^{r-1}\left(a^{r}+b^{r}\right)\leq\left(a+b\right)^{r}\leq a^{r}+b^{r},\quad a,b\geq 0,\ 0\leq r\leq 1, \tag{10.23}$$

$$a^{r}+b^{r}\leq\left(a+b\right)^{r}\leq 2^{r-1}\left(a^{r}+b^{r}\right),\quad a,b\geq 0,\ r\geq 1. \tag{10.24}$$

Finally using (10.23), (10.24), and (10.18) we find

$$(*)\geq\left(A_{0}\left(x\right)|_{\lambda_{\beta}=0}\right)\cdot\left(\frac{\lambda_{v}}{\lambda_{\alpha}+\lambda_{v}}\right)^{\lambda_{v}/p}\cdot\delta_{1}$$

$$\left\{\int_{x_{0}}^{x}p\left(t\right)\left[\left|\left(D_{x_{0}}^{v}f_{1}\right)\left(t\right)\right|^{p}+\left|\left(D_{x_{0}}^{v}f_{2}\right)\left(t\right)\right|^{p}\right]dt\right\}^{(\lambda_{\alpha}+\lambda_{v})/p}.$$

Inequality (10.19) has been established.

Here we see that $(p/p-1)\left(v-\gamma_{i}-1\right)+1>0$, $-1/(p-1)>0$ and $p\left(t\right)\in C\left(\left[x_{0},b\right]\right)$, thus (see (10.15)) $P_{i}\left(w\right)\in\mathbb{R}$ for every $w\in\left[x_{0},b\right]$, also $P_{i}\left(w\right)$ is continuous and bounded on $\left[x_{0},b\right]$ for $i=1,2$.

By $\lambda_{v}>p>0$, we have $0<p/\lambda_{v}<1$, $p/p-\lambda_{v}<0$.

We see that

$$\frac{1}{A\left(w\right)}=\frac{1}{q\left(w\right)}\left(\Gamma\left(v-\gamma_{1}\right)\right)^{\lambda_{\alpha}}\left(\Gamma\left(v-\gamma_{2}\right)\right)^{\lambda_{\beta}}\left(p\left(w\right)\right)^{\lambda_{v}/p}$$

$$\left(P_{1}\left(w\right)\right)^{\lambda_{\alpha}(1-p/p)}\left(P_{2}\left(w\right)\right)^{\lambda_{\beta}(1-p/p)}\in C\left(\left[x_{0},b\right]\right),$$

and $1/A\left(w\right)>0$ on $\left(x_{0},b\right]$, $1/A\left(x_{0}\right)=0$.

Therefore $0<A_{0}\left(x\right)<\infty$, and all we have done in this proof is valid. $\square$

The counterpart of the last theorem follows.

Theorem 10.6. *All here are as in Theorem* 10.5. *Further assume* $\lambda_{\beta}\geq\lambda_{v}$.

Denote

$$\delta_{2}:=2^{1-(\lambda_{\beta}/\lambda_{v})},\quad\delta_{3}:=\left(\delta_{2}-1\right)2^{-(\lambda_{\beta}/\lambda_{v})}. \tag{10.25}$$

If $\lambda_{\alpha}=0$, *then*

$$\int_{x_{0}}^{x}q\left(w\right)\left[\left|\left(D_{x_{0}}^{\gamma_{2}}f_{2}\right)\left(w\right)\right|^{\lambda_{\beta}}\cdot\left|\left(D_{x_{0}}^{v}f_{1}\right)\left(w\right)\right|^{\lambda_{v}}+\right.$$

$$\left.\left|\left(D_{x_{0}}^{\gamma_{2}}f_{1}\right)\left(w\right)\right|^{\lambda_{\beta}}\cdot\left|\left(D_{x_{0}}^{v}f_{2}\right)\left(w\right)\right|^{\lambda_{v}}\right]dw$$

$$\geq\left(A_{0}\left(x\right)|_{\lambda_{\alpha}=0}\right)2^{(p-\lambda_{v})/p}\left(\frac{\lambda_{v}}{\lambda_{\beta}+\lambda_{v}}\right)^{\lambda_{v}/p}\delta_{3}^{\lambda_{v}/p}$$

$$\cdot \left(\int_{x_0}^{x} p\left(w \right) \left[\left| \left(D_{x_0}^v f_1 \right) \left(w \right) \right|^p + \left| \left(D_{x_0}^v f_2 \right) \left(w \right) \right|^p \right] dw \right)^{((\lambda_v + \lambda_\beta)/p)},$$ (10.26)

all $x_0 \le x \le b$.

Proof. When $\lambda_\alpha = 0$ from (10.21) and (10.22) we have

$$\int_{x_0}^{x} q\left(w \right) \left| \left(D_{x_0}^{\gamma_2} f_2 \right) \left(w \right) \right|^{\lambda_\beta} \left| \left(D_{x_0}^v f_1 \right) \left(w \right) \right|^{\lambda_v} dw$$

$$\ge \left(A_0 \left(x \right) \big|_{\lambda_\alpha = 0} \right) \left[\int_{x_0}^{x} \left(z_2 \left(w \right) \right)^{\lambda_\beta / \lambda_v} z_1' \left(w \right) dw \right]^{\lambda_v / p},$$ (10.27)

and

$$\int_{x_0}^{x} q\left(w \right) \left| \left(D_{x_0}^{\gamma_2} f_1 \right) \left(w \right) \right|^{\lambda_\beta} \left| \left(D_{x_0}^v f_2 \right) \left(w \right) \right|^{\lambda_v} dw$$

$$\ge \left(A_0 \left(x \right) \big|_{\lambda_\alpha = 0} \right) \left[\int_{x_0}^{x} \left(z_1 \left(w \right) \right)^{\lambda_\beta / \lambda_v} z_2' \left(w \right) dw \right]^{\lambda_v / p},$$ (10.28)

all $x_0 \le x \le b$. Adding (10.27) and (10.28) we obtain

$$\int_{x_0}^{x} q\left(w \right) \left[\left| \left(D_{x_0}^{\gamma_2} f_2 \right) \left(w \right) \right|^{\lambda_\beta} \cdot \left| \left(D_{x_0}^v f_1 \right) \left(w \right) \right|^{\lambda_v} + \right.$$

$$\left. \left| \left(D_{x_0}^{\gamma_2} f_1 \right) \left(w \right) \right|^{\lambda_\beta} \cdot \left| \left(D_{x_0}^v f_2 \right) \left(w \right) \right|^{\lambda_v} \right] dw$$

$$\ge \left(A_0 \left(x \right) \big|_{\lambda_\alpha = 0} \right) \left\{ \left[\int_{x_0}^{x} \left(z_2 \left(w \right) \right)^{\lambda_\beta / \lambda_v} z_1' \left(w \right) dw \right]^{\lambda_v / p} + \right.$$

$$\left. \left[\int_{x_0}^{x} \left(z_1 \left(w \right) \right)^{\lambda_\beta / \lambda_v} z_2' \left(w \right) dw \right]^{\lambda_v / p} \right\}$$

$$\ge \left(A_0 \left(x \right) \big|_{\lambda_\alpha = 0} \right) \cdot 2^{p - \lambda_v / p} \cdot \left(M \left(x \right) \right)^{\lambda_v / p} =: (*) ,$$ (10.29)

by $\lambda_v / p > 1$ and (10.24), where

$$M \left(x \right) := \int_{x_0}^{x} \left(z_2 \left(w \right) \right)^{\lambda_\beta / \lambda_v} z_1' \left(w \right) + \left(z_1 \left(w \right) \right)^{\lambda_\beta / \lambda_v} z_2' \left(w \right) dw.$$ (10.30)

Next we work on $M \left(x \right)$. We have that

$$M \left(x \right) = \int_{x_0}^{x} \left(\left(z_1 \left(w \right) \right)^{\lambda_\beta / \lambda_v} + \left(z_2 \left(w \right) \right)^{\lambda_\beta / \lambda_v} \right) \left(z_1' \left(w \right) + z_2' \left(w \right) \right) dw$$

$$- \int_{x_0}^{x} \left[\left(z_1 \left(w \right) \right)^{\lambda_\beta / \lambda_v} z_1' \left(w \right) + \left(z_2 \left(w \right) \right)^{\lambda_\beta / \lambda_v} z_2' \left(w \right) \right] dw$$

$$\overset{(by\ (10.24))}{\geq}\ \delta_2 \int_{x_0}^{x} (z_1(w) + z_2(w))^{\lambda_\beta/\lambda_v} (z_1(w) + z_2(w))' \, dw$$

$$- \left(\frac{\lambda_v}{\lambda_\beta + \lambda_v}\right) \left[(z_1(x))^{((\lambda_v+\lambda_\beta)/\lambda_v)} + (z_2(x))^{((\lambda_v+\lambda_\beta)/\lambda_v)}\right]$$

$$= \delta_2 (z_1(x) + z_2(x))^{(\lambda_v+\lambda_\beta/\lambda_v)} \left(\frac{\lambda_v}{\lambda_v + \lambda_\beta}\right) - \left(\frac{\lambda_v}{\lambda_\beta + \lambda_v}\right)$$

$$\left[(z_1(x))^{(\lambda_v+\lambda_\beta/\lambda_v)} + (z_2(x))^{(\lambda_v+\lambda_\beta/\lambda_v)}\right]$$

$$= \left(\frac{\lambda_v}{\lambda_\beta + \lambda_v}\right) \left[\delta_2 (z_1(x) + z_2(x))^{(\lambda_v+\lambda_\beta/\lambda_v)} - \right.$$

$$\left. \left((z_1(x))^{(\lambda_v+\lambda_\beta/\lambda_v)} + (z_2(x))^{(\lambda_v+\lambda_\beta/\lambda_v)}\right)\right]$$

$$\overset{(10.24)}{\geq} \left(\frac{\lambda_v}{\lambda_\beta + \lambda_v}\right) \left[\delta_2 \left((z_1(x))^{(\lambda_v+\lambda_\beta/\lambda_v)} + (z_2(x))^{(\lambda_v+\lambda_\beta/\lambda_v)}\right)\right.$$

$$\left. - \left((z_1(x))^{(\lambda_v+\lambda_\beta/\lambda_v)} + (z_2(x))^{(\lambda_v+\lambda_\beta/\lambda_v)}\right)\right]$$

$$= \left(\frac{\lambda_v}{\lambda_\beta + \lambda_v}\right) (\delta_2 - 1) \left[(z_1(x))^{(\lambda_v+\lambda_\beta/\lambda_v)} + (z_2(x))^{(\lambda_v+\lambda_\beta/\lambda_v)}\right]$$

$$\overset{(10.24)}{\geq} \left(\frac{\lambda_v}{\lambda_\beta + \lambda_v}\right) \delta_3 (z_1(x) + z_2(x))^{(\lambda_v+\lambda_\beta/\lambda_v)}.$$

That is, we present that

$$M(x) \geq \left(\frac{\lambda_v}{\lambda_\beta + \lambda_v}\right) \delta_3 (z_1(x) + z_2(x))^{(\lambda_v+\lambda_\beta/\lambda_v)}. \tag{10.31}$$

Consequently, by (10.29) and (10.31) we obtain

$$(*) \geq (A_0(x)|_{\lambda_\alpha=0}) \, 2^{p-\lambda_v/p} \left(\frac{\lambda_v}{\lambda_\beta + \lambda_v}\right)^{\lambda_v/p}$$

$$\delta_3^{\lambda_v/p} (z_1(x) + z_2(x))^{(\lambda_v+\lambda_\beta/p)}$$

$$= (A_0(x)|_{\lambda_\alpha=0}) \, 2^{p-\lambda_v/p} \left(\frac{\lambda_v}{\lambda_\beta + \lambda_v}\right)^{\lambda_v/p} \delta_3^{\lambda_v/p}$$

$$\left(\int_{x_0}^{x} p(t) \left[|(D_{x_0}^v f_1)(t)|^p + |(D_{x_0}^v f_2)(t)|^p\right] dt\right)^{(\lambda_v+\lambda_\beta/p)}.$$

We have established (10.26). $\square$

A special important case follows.

Theorem 10.7. *Let $v \geq 2$ and $\gamma_1 \geq 0$ such that $2 \leq v - \gamma_1 < 1/p$, $0 < p < 1$. Let $f_1, f_2 \in C_{x_0}^v([a,b])$ with $f_1^{(i)}(x_0) = f_2^{(i)}(x_0) = 0$, $i = 0, 1, \ldots, n-1$, $n := [v]$. Here $x, x_0 \in [a,b] : x \geq x_0$. We assume here that $D_{x_0}^v f_j$ is of fixed sign on $[x_0, b]$, $j = 1, 2$. Consider also $p(t) > 0$ and $q(t) > 0$ continuous functions on $[x_0, b]$. Let $\lambda_\alpha \geq \lambda_{\alpha+1} > 1$. Denote*

$$\theta_3 := \left(2^{1-(\lambda_\alpha/\lambda_{\alpha+1})} - 1\right) 2^{-\lambda_\alpha/\lambda_{\alpha+1}}, \tag{10.32}$$

$$L(x) := \left(2 \int_{x_0}^x (q(w))^{(1/1-\lambda_{\alpha+1})} \, dw\right)^{(1-\lambda_{\alpha+1})} \left(\frac{\theta_3 \lambda_{\alpha+1}}{\lambda_\alpha + \lambda_{\alpha+1}}\right)^{\lambda_{\alpha+1}}, \tag{10.33}$$

and

$$P_1(x) := \int_{x_0}^x (x-t)^{(v-\gamma_1-1)p/p-1} (p(t))^{-1/p-1} \, dt, \tag{10.34}$$

$$T(x) := L(x) \cdot \left(\frac{P_1(x)^{(p-1/p)}}{\Gamma(v-\gamma_1)}\right)^{(\lambda_\alpha + \lambda_{\alpha+1})}, $$

$$\omega_1 := 2^{(p-1/p)(\lambda_\alpha + \lambda_{\alpha+1})}, \tag{10.35}$$

and

$$\Phi(x) := T(x)\,\omega_1. \tag{10.36}$$

Then

$$\int_{x_0}^x q(w) \left[\left|(D_{x_0}^{\gamma_1} f_1)(w)\right|^{\lambda_\alpha} \left|(D_{x_0}^{\gamma_1+1} f_2)(w)\right|^{\lambda_{\alpha+1}}\right.$$

$$\left. + \left|(D_{x_0}^{\gamma_1} f_2)(w)\right|^{\lambda_\alpha} \left|(D_{x_0}^{\gamma_1+1} f_1)(w)\right|^{\lambda_{\alpha+1}}\right] dw \tag{10.37}$$

$$\geq \Phi(x) \left[\int_{x_0}^x p(w) \left(\left|(D_{x_0}^v f_1)(w)\right|^p + \left|(D_{x_0}^v f_2)(w)\right|^p\right) dw\right]^{(\lambda_\alpha + \lambda_{\alpha+1})/p},$$

all $x_0 \leq x \leq b$.

Proof. For convenience we set $\gamma_2 := \gamma_1 + 1$. From (10.13) and assumption we obtain

$$\left|(D_{x_0}^{\gamma_k} f_j)(w)\right| = \frac{1}{\Gamma(v-\gamma_k)} \int_{x_0}^w (w-t)^{v-\gamma_k-1} \left|(D_{x_0}^v f_i)(t)\right| dt =: g_{j,\gamma_k}(w), \tag{10.38}$$

where $j = 1, 2$, $k = 1, 2$, all $x_0 \leq x \leq b$. We observe that

$$\left((D_{x_0}^{\gamma_1} f_j)(x)\right)' = (D_{x_0}^{\gamma_1+1} f_j)(x) = (D_{x_0}^{\gamma_2} f_j)(x), \tag{10.39}$$

all $x_0 \leq x \leq b$. And also

$$\left(g_{j,\gamma_1}(w)\right)' = g_{j,\gamma_2}(w); \quad g_{j,\gamma_k}(x_0) = 0. \tag{10.40}$$

Notice that if $v - \gamma_2 = 1$, then

$$g_{j,\gamma_2}(w) = \int_{x_0}^{w} \left| \left(D_{x_0}^{v} f_j \right)(t) \right| dt.$$

Next we apply Hölder's inequality with indices $1/\lambda_{\alpha+1} < 1$, $1/(1-\lambda_{\alpha+1}) < 0$; we obtain

$$\int_{x_0}^{x} q(w) \left| \left(D_{x_0}^{\gamma_1} f_1 \right)(w) \right|^{\lambda_\alpha} \left| \left(D_{x_0}^{\gamma_1+1} f_2 \right)(w) \right|^{\lambda_{\alpha+1}} dw$$

$$= \int_{x_0}^{x} q(w) \left(g_{1,\gamma_1}(w) \right)^{\lambda_\alpha} \left(\left(g_{2,\gamma_1}(w) \right)' \right)^{\lambda_{\alpha+1}} dw \qquad (10.41)$$

$$\geq \left(\int_{x_0}^{x} (q(w))^{(1/1-\lambda_{\alpha+1})} dw \right)^{(1-\lambda_{\alpha+1})}$$

$$\left(\int_{x_0}^{x} \left(g_{1,\gamma_1}(w) \right)^{\lambda_\alpha/\lambda_{\alpha+1}} \left(g_{2,\gamma_1}(w) \right)' dw \right)^{\lambda_{\alpha+1}}.$$

Similarly we have

$$\int_{x_0}^{x} q(w) \left| \left(D_{x_0}^{\gamma_1} f_2 \right)(w) \right|^{\lambda_\alpha} \left| \left(D_{x_0}^{\gamma_1+1} f_1 \right)(w) \right|^{\lambda_{\alpha+1}} dw$$

$$\geq \left(\int_{x_0}^{x} (q(w))^{(1/1-\lambda_{\alpha+1})} dw \right)^{(1-\lambda_{\alpha+1})}$$

$$\left(\int_{x_0}^{x} \left(g_{2,\gamma_1}(w) \right)^{\lambda_\alpha/\lambda_{\alpha+1}} \left(g_{1,\gamma_1}(w) \right)' dw \right)^{\lambda_{\alpha+1}}. \qquad (10.42)$$

Adding (10.41) and (10.42) we see that

$$\int_{x_0}^{x} q(w) \left[\left| \left(D_{x_0}^{\gamma_1} f_1 \right)(w) \right|^{\lambda_\alpha} \left| \left(D_{x_0}^{\gamma_1+1} f_2 \right)(w) \right|^{\lambda_{\alpha+1}} \right.$$

$$\left. + \left| \left(D_{x_0}^{\gamma_1} f_2 \right)(w) \right|^{\lambda_\alpha} \left| \left(D_{x_0}^{\gamma_1+1} f_1 \right)(w) \right|^{\lambda_{\alpha+1}} \right] dw$$

$$\geq \left(\int_{x_0}^{x} (q(w))^{(1/1-\lambda_{\alpha+1})} dw \right)^{(1-\lambda_{\alpha+1})}$$

$$\left[\left(\int_{x_0}^{x} \left(g_{1,\gamma_1}(w) \right)^{\lambda_\alpha/\lambda_{\alpha+1}} \left(g_{2,\gamma_1}(w) \right)' dw \right)^{\lambda_{\alpha+1}} \right.$$

$$\left. + \left(\int_{x_0}^{x} \left(g_{2,\gamma_1}(w) \right)^{\lambda_\alpha/\lambda_{\alpha+1}} \left(g_{1,\gamma_1}(w) \right)' dw \right)^{\lambda_{\alpha+1}} \right]$$

$$(10.24) \geq \left(2\int_{x_0}^{x} (q(w))^{(1/1-\lambda_{\alpha+1})} \, dw\right)^{(1-\lambda_{\alpha+1})}$$

$$\cdot \left[\int_{x_0}^{x} \left[(g_{1,\gamma_1}(w))^{\lambda_\alpha/\lambda_{\alpha+1}} (g_{2,\gamma_1}(w))' \right.\right.$$

$$\left.\left. + (g_{2,\gamma_1}(w))^{\lambda_\alpha/\lambda_{\alpha+1}} (g_{1,\gamma_1}(w))' \right] dw\right]^{\lambda_{\alpha+1}}$$

(note (10.30) and the proof of (10.31); accordingly here we have)

$$\geq \left(2\int_{x_0}^{x} (q(w))^{(1/1-\lambda_{\alpha+1})} \, dw\right)^{(1-\lambda_{\alpha+1})}$$

$$\left(\frac{\lambda_{\alpha+1}\theta_3}{\lambda_\alpha + \lambda_{\alpha+1}}\right)^{\lambda_{\alpha+1}} (g_{1,\gamma_1}(x) + g_{2,\gamma_1}(x))^{(\lambda_\alpha+\lambda_{\alpha+1})}$$

$$= L(x) (g_{1,\gamma_1}(x) + g_{2,\gamma_1}(x))^{(\lambda_\alpha+\lambda_{\alpha+1})}$$

$$= \frac{L(x)}{(\Gamma(v-\gamma_1))^{(\lambda_\alpha+\lambda_{\alpha+1})}} \left\{\int_{x_0}^{x} (x-t)^{v-\gamma_1-1} (p(t))^{-1/p} (p(t))^{1/p}\right.$$

$$\left. \left[|(D_{x_0}^v f_1)(t)| + |(D_{x_0}^v f_2)(t)|\right] dt\right\}^{(\lambda_\alpha+\lambda_{\alpha+1})}$$

(applying Hölder's inequality with indices $\dfrac{p}{p-1}$ and p we find)

$$\geq \frac{L(x)}{(\Gamma(v-\gamma_1))^{(\lambda_\alpha+\lambda_{\alpha+1})}} \cdot \left(\int_{x_0}^{x} (x-t)^{(v-\gamma_1-1)p/p-1}\right.$$

$$\left. (p(t))^{-1/p-1} \, dt\right)^{(p-1/p)(\lambda_\alpha+\lambda_{\alpha+1})}$$

$$\left(\int_{x_0}^{x} p(t) \left[|(D_{x_0}^v f_1)(t)| + |(D_{x_0}^v f_2)(t)|\right]^p dt\right)^{(\lambda_\alpha+\lambda_{\alpha+1}/p)}$$

$$= T(x) \cdot \left[\int_{x_0}^{x} p(t) \left(|(D_{x_0}^v f_1)(t)| + |(D_{x_0}^v f_2)(t)|\right)^p dt\right]^{(\lambda_\alpha+\lambda_{\alpha+1}/p)}$$

$$\geq \Phi(x) \cdot \left[\int_{x_0}^{x} p(t) \left(|(D_{x_0}^v f_1)(t)|^p + |(D_{x_0}^v f_2)(t)|^p\right) dt\right]^{(\lambda_\alpha+\lambda_{\alpha+1}/p)}.$$

We have proved (10.37). $\square$

Next we treat the case of exponents $\lambda_\beta = \lambda_\alpha + \lambda_v$.

Theorem 10.8. *All here are as in Theorem 10.5. Consider the special case of* $\lambda_\beta = \lambda_\alpha + \lambda_v$.

Assume here for $j = 1, 2$ that

$$z_j(x) := \int_{x_0}^x p(t) \left| (D_{x_0}^v f_j)(t) \right|^p dt \in [H, \Psi], 0 < H < \Psi,$$

$$h := \frac{\Psi}{H} > 1, \quad M_h(1) := \frac{(h-1) h^{1/h-1}}{e \ln h}. \tag{10.43}$$

Denote

$$\tilde{T}(x) := A_0(x) \left(\frac{\lambda_v}{\lambda_\alpha + \lambda_v} \right)^{\lambda_v/p} 2^{p-2\lambda_\alpha-3\lambda_v/p} (M_h(1))^{-2(\lambda_\alpha+\lambda_v)/p}. \tag{10.44}$$

Then

$$\int_{x_0}^x q(w) \left[\left| (D_{x_0}^{\gamma_1} f_1)(w) \right|^{\lambda_\alpha} \left| (D_{x_0}^{\gamma_2} f_2)(w) \right|^{\lambda_\alpha+\lambda_v} \left| (D_{x_0}^v f_1)(w) \right|^{\lambda_v} \right.$$

$$\left. + \left| (D_{x_0}^{\gamma_2} f_1)(w) \right|^{\lambda_\alpha+\lambda_v} \left| (D_{x_0}^{\gamma_1} f_2)(w) \right|^{\lambda_\alpha} \left| (D_{x_0}^v f_2)(w) \right|^{\lambda_v} \right] dw$$

$$\geq \tilde{T}(x) \left(\int_{x_0}^x p(w) \left(\left| (D_{x_0}^v f_1)(w) \right|^p + \left| (D_{x_0}^v f_2)(w) \right|^p \right) dw \right)^{2((\lambda_\alpha+\lambda_v)/p)}, \tag{10.45}$$

all $x_0 \leq x \leq b$.

Proof. We apply (10.21) and (10.22) for $\lambda_\beta = \lambda_\alpha + \lambda_v$ and add to find

$$\int_{x_0}^x q(w) \left[\left| (D_{x_0}^{\gamma_1} f_1)(w) \right|^{\lambda_\alpha} \left| (D_{x_0}^{\gamma_2} f_2)(w) \right|^{\lambda_\alpha+\lambda_v} \left| (D_{x_0}^v f_1)(w) \right|^{\lambda_v} \right.$$

$$\left. + \left| (D_{x_0}^{\gamma_2} f_1)(w) \right|^{\lambda_\alpha+\lambda_v} \left| (D_{x_0}^{\gamma_1} f_2)(w) \right|^{\lambda_\alpha} \left| (D_{x_0}^v f_2)(w) \right|^{\lambda_v} \right] dw$$

$$\geq A_0(x) \left\{ \left[\int_{x_0}^x (z_1(w))^{\lambda_\alpha/\lambda_v} (z_2(w))^{\lambda_\alpha/\lambda_v+1} z_1'(w) dw \right]^{\lambda_v/p} \right.$$

$$\left. + \left[\int_{x_0}^x (z_1(w))^{\lambda_\alpha/\lambda_v+1} (z_2(w))^{\lambda_\alpha/\lambda_v} z_2'(w) dw \right]^{\lambda_v/p} \right\}$$

$$\overset{(10.24)}{\geq} A_0(x) 2^{1-\lambda_v/p} \left\{ \int_{x_0}^x \left[(z_1(w))^{\lambda_\alpha/\lambda_v} (z_2(w))^{\lambda_\alpha/\lambda_v+1} z_1'(w) \right. \right.$$

$$\left. \left. + (z_1(w))^{\lambda_\alpha/\lambda_v+1} (z_2(w))^{\lambda_\alpha/\lambda_v} z_2'(w) \right] dw \right\}^{\lambda_v/p}$$

$$= A_0(x) 2^{1-\lambda_v/p} \left\{ \int_{x_0}^x (z_1(w) z_2(w))^{\lambda_\alpha/\lambda_v} \right.$$

$$\left. [z_2(w) z_1'(w) + z_1(w) z_2'(w)] dw \right\}^{\lambda_v/p}$$

$$= A_0\left(x\right) 2^{1-\lambda_v/p} \left\{ \int_{x_0}^{x} \left(z_1\left(w\right) z_2\left(w\right)\right)^{\lambda_\alpha/\lambda_v} \left(z_1\left(w\right) z_2\left(w\right)\right)' dw \right\}^{\lambda_v/p}$$

$$= A_0\left(x\right) 2^{1-\lambda_v/p} \left(\frac{\left(z_1\left(x\right) z_2\left(x\right)\right)^{\lambda_\alpha/\lambda_v+1}}{\frac{\lambda_\alpha}{\lambda_v} + 1} \right)^{\lambda_v/p}$$

$$= A_0\left(x\right) 2^{p-\lambda_v/p} \left(\frac{\lambda_v}{\lambda_\alpha + \lambda_v} \right)^{\lambda_v/p} \left(z_1\left(x\right) z_2\left(x\right)\right)^{(\lambda_\alpha+\lambda_v)/p}$$

(see [377])

$$\geq A_0\left(x\right) 2^{p-\lambda_v/p} \left(\frac{\lambda_v}{\lambda_\alpha + \lambda_v} \right)^{\lambda_v/p} \left(\frac{z_1\left(x\right) + z_2\left(x\right)}{2M_h\left(1\right)} \right)^{2(\lambda_\alpha+\lambda_v)/p}$$

$$= \tilde{T}\left(x\right) \left(z_1\left(x\right) + z_2\left(x\right)\right)^{2(\lambda_\alpha+\lambda_v)/p}$$

$$= \tilde{T}\left(x\right) \left(\int_{x_0}^{x} p\left(w\right) \left(\left|\left(D_{x_0}^v f_1\right)\left(w\right)\right|^p + \left|\left(D_{x_0}^v f_2\right)\left(w\right)\right|^p\right) dw \right)^{2(\lambda_\alpha+\lambda_v/p)}.$$

We have established (10.45). $\square$

Special cases of the above theorems follow next.

Corollary 10.9. (to Theorem 10.5; $\lambda_\beta = 0$, $p\left(t\right) = q\left(t\right) = 1$). *Then*

$$\int_{x_0}^{x} \left[\left|\left(D_{x_0}^{\gamma_1} f_1\right)\left(w\right)\right|^{\lambda_\alpha} \left|\left(D_{x_0}^v f_1\right)\left(w\right)\right|^{\lambda_v} + \left|\left(D_{x_0}^{\gamma_1} f_2\right)\left(w\right)\right|^{\lambda_\alpha} \left|\left(D_{x_0}^v f_2\right)\left(w\right)\right|^{\lambda_v} \right] dw$$

$$\geq C_1\left(x\right) \cdot \left(\int_{x_0}^{x} \left(\left|\left(D_{x_0}^v f_1\right)\left(w\right)\right|^p + \left|\left(D_{x_0}^v f_2\right)\left(w\right)\right|^p\right) dw \right)^{(\lambda_\alpha+\lambda_v/p)}, \qquad (10.46)$$

all $x_0 \leq x \leq b$, *where*

$$C_1\left(x\right) := \left(A_0\left(x\right)|_{\lambda_\beta=0}\right) \cdot \left(\frac{\lambda_v}{\lambda_\alpha + \lambda_v} \right)^{\lambda_v/p} \cdot \delta_1, \qquad (10.47)$$

$$\delta_1 := 2^{1-(\lambda_\alpha+\lambda_v/p)} \qquad (10.48)$$

We have that

$$\left(A_0\left(x\right)|_{\lambda_\beta=0}\right) = \left\{ \left(\frac{(p-1)^{((\lambda_\alpha p-\lambda_\alpha)/p)}}{\left(\Gamma\left(v-\gamma_1\right)\right)^{\lambda_\alpha} \left(vp-\gamma_1 p-1\right)^{((\lambda_\alpha p-\lambda_\alpha)/p)}} \right) \right. \qquad (10.49)$$

$$\left. \cdot \left(\frac{(p-\lambda_v)^{(p-\lambda_v/p)}}{\left(\lambda_\alpha vp - \lambda_\alpha \gamma_1 p - \lambda_\alpha + p - \lambda_v\right)^{((p-\lambda_v)/p)}} \right) \right\} \cdot$$

$$\left(x - x_0\right)^{((\lambda_\alpha vp-\lambda_\alpha \gamma_1 p-\lambda_\alpha+p-\lambda_v)/p)}.$$

Proof. By Theorem 10.5. The constant $\left(A_0\left(x\right)|_{\lambda_\beta=0}\right)$ was calculated in [26]; see here (6.55). $\square$

Corollary 10.10. (to Theorem 10.6; $\lambda_\alpha = 0$, $p\left(t\right) = q\left(t\right) = 1$, $\lambda_\beta \geq \lambda_v$). *Then*

$$\int_{x_0}^x \left[\left|\left(D_{x_0}^{\gamma_2} f_2\right)(w)\right|^{\lambda_\beta} \left|\left(D_{x_0}^v f_1\right)(w)\right|^{\lambda_v} + \left|\left(D_{x_0}^{\gamma_2} f_1\right)(w)\right|^{\lambda_\beta} \left|\left(D_{x_0}^v f_2\right)(w)\right|^{\lambda_v}\right] dw$$

$$\geq C_2\left(x\right) \left(\int_{x_0}^x \left[\left|\left(D_{x_0}^v f_1\right)(w)\right|^p + \left|\left(D_{x_0}^v f_2\right)(w)\right|^p\right] dw\right)^{(\lambda_v + \lambda_\beta/p)}, \qquad (10.50)$$

all $x_0 \leq x \leq b$, where

$$C_2\left(x\right) := \left(A_0\left(x\right)|_{\lambda_\alpha=0}\right) 2^{(p-\lambda_v)/p} \left(\frac{\lambda_v}{\lambda_\beta + \lambda_v}\right)^{\lambda_v/p} \delta_3^{\lambda_v/p}. \qquad (10.51)$$

We have that

$$\left(A_0\left(x\right)|_{\lambda_\alpha=0}\right) = \left\{\left(\frac{(p-1)^{((\lambda_\beta p - \lambda_\beta)/p)}}{(\Gamma\left(v - \gamma_2\right))^{\lambda_\beta} \left(vp - \gamma_2 p - 1\right)^{((\lambda_\beta p - \lambda_\beta)/p)}}\right) \qquad (10.52)$$

$$\cdot \left(\frac{(p-\lambda_v)^{(p-\lambda_v/p)}}{(\lambda_\beta vp - \lambda_\beta \gamma_2 p - \lambda_\beta + p - \lambda_v)^{((p-\lambda_v)/p)}}\right)\right\} \cdot$$

$$(x - x_0)^{((\lambda_\beta vp - \lambda_\beta \gamma_2 p - \lambda_\beta + p - \lambda_v)/p)}.$$

Proof. By Theorem 10.6. The constant $\left(A_0\left(x\right)|_{\lambda_\alpha=0}\right)$ was calculated in [26]; see here (6.59). $\square$

10.3.2   Results Involving Several Functions

Here we use the following basic inequality. Let $\alpha_1, \ldots, \alpha_n \geq 0$, $n \in \mathbb{N}$; then

$$a_1^r + \ldots + a_n^r \leq \left(a_1 + \ldots + a_n\right)^r \leq n^{r-1}\left(\sum_{i=1}^n a_i^r\right), r \geq 1. \qquad (10.53)$$

We present

Theorem 10.11. *Let $\gamma_1, \gamma_2 \geq 0$ such that $1 \leq v - \gamma_i < 1/p$, $0 < p < 1$, $i = 1, 2$, and $f_j \in C_{x_0}^v\left([a,b]\right)$ with $f_j^{(i)}\left(x_0\right) = 0$, $i = 0, 1, \ldots, n-1$, $n := [v]$, $j = 1, \ldots, M \in \mathbb{N}$. Here $x, x_0 \in [a,b] : x \geq x_0$. We assume that $D_{x_0}^v f_j$ is of fixed sign on $[x_0, b]$, $j = 1, \ldots, M$. Consider also $p\left(t\right) > 0$, and $q\left(t\right) > 0$ continuous functions on $[x_0, b]$. Let $\lambda_v > 0$ and $\lambda_\alpha, \lambda_\beta \geq 0$ such that $\lambda_v > p$.*

Set

$$P_k(w) := \int_{x_0}^{w} (w-t)^{(v-\gamma_k-1)p/p-1} (p(t))^{-1/p-1} dt, \ k = 1,2; \ x_0 \leq w \leq b;$$
(10.54)

$$A(w) := \frac{q(w) (P_1(w))^{(\lambda_\alpha(p-1)/p)} (P_2(w))^{(\lambda_\beta(p-1)/p)} (p(w))^{-\lambda_v/p}}{(\Gamma(v-\gamma_1))^{\lambda_\alpha} (\Gamma(v-\gamma_2))^{\lambda_\beta}};$$
(10.55)

$$A_0(x) := \left(\int_{x_0}^{x} (A(w))^{p/p-\lambda_v} dw \right)^{(p-\lambda_v)/p}.$$
(10.56)

Call

$$\varphi_1(x) := (A_0(x)|_{\lambda_\beta=0}) \cdot \left(\frac{\lambda_v}{\lambda_\alpha + \lambda_v} \right)^{\lambda_v/p},$$
(10.57)

$$\delta_1^* := M^{1-(\lambda_\alpha+\lambda_v/p)}.$$
(10.58)

If $\lambda_\beta = 0$, we obtain that

$$\int_{x_0}^{x} q(w) \left(\sum_{j=1}^{M} \left| (D_{x_0}^{\gamma_1} f_j)(w) \right|^{\lambda_\alpha} \left| (D_{x_0}^{v} f_j)(w) \right|^{\lambda_v} \right) dw$$

$$\geq \delta_1^* \cdot \varphi_1(x) \cdot \left[\int_{x_0}^{x} p(w) \left(\sum_{j=1}^{M} \left| (D_{x_0}^{v} f_j)(w) \right|^{p} \right) dw \right]^{((\lambda_\alpha+\lambda_v)/p)}, \quad (10.59)$$

all $x_0 \leq x \leq b$.

Proof. By Theorem 10.5 we have

$$\int_{x_0}^{x} q(w) \left[\left| (D_{x_0}^{\gamma_1} f_j)(w) \right|^{\lambda_\alpha} \left| (D_{x_0}^{v} f_j)(w) \right|^{\lambda_v} \right.$$

$$\left. + \left| (D_{x_0}^{\gamma_1} f_{j+1})(w) \right|^{\lambda_\alpha} \left| (D_{x_0}^{v} f_{j+1})(w) \right|^{\lambda_v} \right] dw \quad (10.60)$$

$$\geq \delta_1 \varphi_1(x) \left[\int_{x_0}^{x} p(w) \left[\left| (D_{x_0}^{v} f_j)(w) \right|^{p} + \left| (D_{x_0}^{v} f_{j+1})(w) \right|^{p} \right] dw \right]^{((\lambda_\alpha+\lambda_v)/p)},$$

$j = 1, 2, \ldots, M-1$.
Hence by adding all the above we derive

$$\int_{x_0}^{x} q(w) \left(\sum_{j=1}^{M-1} \left[\left| (D_{x_0}^{\gamma_1} f_j)(w) \right|^{\lambda_\alpha} \left| (D_{x_0}^{v} f_j)(w) \right|^{\lambda_v} \right. \right.$$

$$\left. \left. + \left| (D_{x_0}^{\gamma_1} f_{j+1})(w) \right|^{\lambda_\alpha} \left| (D_{x_0}^{v} f_{j+1})(w) \right|^{\lambda_v} \right] \right) dw \quad (10.61)$$

$$\geq \delta_1 \varphi_1 (x) \cdot \left(\sum_{j=1}^{M-1} \left[\int_{x_0}^{x} p(w) \left[\left| \left(D_{x_0}^v f_j \right) (w) \right|^p \right. \right. \right.$$

$$\left. \left. \left. + \left| \left(D_{x_0}^v f_{j+1} \right) (w) \right|^p \right] dw \right]^{(\lambda_\alpha + \lambda_v / p)} \right).$$

Also it holds

$$\int_{x_0}^{x} q(w) \left[\left| \left(D_{x_0}^{\gamma_1} f_1 \right) (w) \right|^{\lambda_\alpha} \left| \left(D_{x_0}^v f_1 \right) (w) \right|^{\lambda_v} \right.$$

$$\left. + \left| \left(D_{x_0}^{\gamma_1} f_M \right) (w) \right|^{\lambda_\alpha} \left| \left(D_{x_0}^v f_M \right) (w) \right|^{\lambda_v} \right] dw \qquad (10.62)$$

$$\geq \delta_1 \varphi_1 (x) \left[\int_{x_0}^{x} p(w) \left[\left| \left(D_{x_0}^v f_1 \right) (w) \right|^p + \left| \left(D_{x_0}^v f_M \right) (w) \right|^p \right] dw \right]^{((\lambda_\alpha + \lambda_v)/p)}.$$

Adding (10.61) and (10.62), and using (10.53) we have

$$2 \int_{x_0}^{x} q(w) \left(\sum_{j=1}^{M} \left| \left(D_{x_0}^{\gamma_1} f_j \right) (w) \right|^{\lambda_\alpha} \left| \left(D_{x_0}^v f_j \right) (w) \right|^{\lambda_v} \right) dw$$

$$\geq \delta_1 \varphi_1 (x) \left\{ \left\{ \sum_{j=1}^{M-1} \left[\int_{x_0}^{x} p(w) \left[\left| \left(D_{x_0}^v f_j \right) (w) \right|^p \right. \right. \right. \right. \qquad (10.63)$$

$$\left. \left. \left. + \left| \left(D_{x_0}^v f_{j+1} \right) (w) \right|^p \right] dw \right]^{(\lambda_\alpha + \lambda_v / p)} \right\}$$

$$+ \left\{ \int_{x_0}^{x} p(w) \left[\left| \left(D_{x_0}^v f_1 \right) (w) \right|^p + \left| \left(D_{x_0}^v f_M \right) (w) \right|^p \right] dw \right\}^{((\lambda_\alpha + \lambda_v)/p)} \right\} \geq$$

$$M^{1-(\lambda_\alpha + \lambda_v / p)} \delta_1 \varphi_1 (x) \left\{ \int_{x_0}^{x} p(w) \left[2 \sum_{j=1}^{M} \left| \left(D_{x_0}^v f_j \right) (w) \right|^p \right] dw \right\}^{((\lambda_\alpha + \lambda_v)/p)}.$$

$$(10.64)$$

We have proved

$$\int_{x_0}^{x} q(w) \left(\sum_{j=1}^{M} \left| \left(D_{x_0}^{\gamma_1} f_j \right) (w) \right|^{\lambda_\alpha} \left| \left(D_{x_0}^v f_j \right) (w) \right|^{\lambda_v} \right) dw \geq \qquad (10.65)$$

$$M^{1-(\lambda_\alpha + \lambda_v / p)} \delta_1 \left(2^{(\lambda_\alpha + \lambda_v / p) - 1} \right) \varphi_1 (x)$$

$$\cdot \left\{ \int_{x_0}^{x} p(w) \left[\sum_{j=1}^{M} \left| \left(D_{x_0}^v f_j \right) (w) \right|^p \right] dw \right\}^{((\lambda_\alpha + \lambda_v)/p)},$$

thus proving (10.59). □

Next we give

Theorem 10.12. *All here are as in Theorem 10.11. Assume* $\lambda_\beta \geq \lambda_v$. *Denote*

$$\varphi_2(x) := (A_0(x)|_{\lambda_\alpha=0}) \, 2^{(p-\lambda_v)/p} \left(\frac{\lambda_v}{\lambda_\beta + \lambda_v} \right)^{\lambda_v/p} \delta_3^{\lambda_v/p}. \tag{10.66}$$

If $\lambda_\alpha = 0$, *then*

$$\int_{x_0}^x q(w) \left\{ \left\{ \sum_{j=1}^{M-1} \left[\left| (D_{x_0}^{\gamma_2} f_{j+1})(w) \right|^{\lambda_\beta} \left| (D_{x_0}^v f_j)(w) \right|^{\lambda_v} \right. \right. \right.$$

$$+ \left| (D_{x_0}^{\gamma_2} f_j)(w) \right|^{\lambda_\beta} \left| (D_{x_0}^v f_{j+1})(w) \right|^{\lambda_v} \right] \right\}$$

$$+ \left[\left| (D_{x_0}^{\gamma_2} f_M)(w) \right|^{\lambda_\beta} \left| (D_{x_0}^v f_1)(w) \right|^{\lambda_v} \right.$$

$$\left. \left. + \left| (D_{x_0}^{\gamma_2} f_1)(w) \right|^{\lambda_\beta} \left| (D_{x_0}^v f_M)(w) \right|^{\lambda_v} \right] \right\} dw \geq$$

$$M^{1-((\lambda_v+\lambda_\beta)/p)} 2^{((\lambda_v+\lambda_\beta)/p)} \varphi_2(x) \cdot \left\{ \left[\int_{x_0}^x p(w) \right. \right.$$

$$\left. \left[\sum_{j=1}^M \left| (D_{x_0}^v f_j)(w) \right|^p \right] dw \right\}^{((\lambda_v+\lambda_\beta)/p)}, \quad x \geq x_0. \tag{10.67}$$

Proof. From Theorem 10.6 we obtain

$$\int_{x_0}^x q(w) \left[\left| (D_{x_0}^{\gamma_2} f_{j+1})(w) \right|^{\lambda_\beta} \left| (D_{x_0}^v f_j)(w) \right|^{\lambda_v} \right.$$

$$\left. + \left| (D_{x_0}^{\gamma_2} f_j)(w) \right|^{\lambda_\beta} \left| (D_{x_0}^v f_{j+1})(w) \right|^{\lambda_v} \right] dw$$

$$\geq \varphi_2(x) \left(\int_{x_0}^x p(w) \left[\left| (D_{x_0}^v f_j)(w) \right|^p + \left| (D_{x_0}^v f_{j+1})(w) \right|^p \right] dw \right)^{((\lambda_v+\lambda_\beta)/p)}, \tag{10.68}$$

for $j = 1, \ldots, M - 1$. Hence by adding all of the above we find

$$\int_{x_0}^x q(w) \left(\sum_{j=1}^{M-1} \left[\left| (D_{x_0}^{\gamma_2} f_{j+1})(w) \right|^{\lambda_\beta} \left| (D_{x_0}^v f_j)(w) \right|^{\lambda_v} \right. \right.$$

$$\left. \left. + \left| (D_{x_0}^{\gamma_2} f_j)(w) \right|^{\lambda_\beta} \left| (D_{x_0}^v f_{j+1})(w) \right|^{\lambda_v} \right] \right) dw$$

$$\geq \varphi_2\left(x\right)\left\{\sum_{j=1}^{M-1}\left(\int_{x_0}^{x}p\left(w\right)\left[\left|\left(D_{x_0}^{v}f_j\right)\left(w\right)\right|^{p}\right.\right.\right.$$

$$\left.\left.\left.+\left|\left(D_{x_0}^{v}f_{j+1}\right)\left(w\right)\right|^{p}\right]dw\right)^{((\lambda_v+\lambda_\beta)/p)}\right\}. \tag{10.69}$$

Similarly it holds

$$\int_{x_0}^{x}q\left(w\right)\left[\left|\left(D_{x_0}^{\gamma_2}f_M\right)\left(w\right)\right|^{\lambda_\beta}\left|\left(D_{x_0}^{v}f_1\right)\left(w\right)\right|^{\lambda_v}\right.$$

$$\left.+\left|\left(D_{x_0}^{\gamma_2}f_1\right)\left(w\right)\right|^{\lambda_\beta}\left|\left(D_{x_0}^{v}f_M\right)\left(w\right)\right|^{\lambda_v}\right]dw$$

$$\geq \varphi_2\left(x\right)\left(\int_{x_0}^{x}p\left(w\right)\left[\left|\left(D_{x_0}^{v}f_1\right)\left(w\right)\right|^{p}+\left|\left(D_{x_0}^{v}f_M\right)\left(w\right)\right|^{p}\right]dw\right)^{((\lambda_v+\lambda_\beta)/p)}. \tag{10.70}$$

Adding (10.69), (10.70), and using (10.53) we derive (10.67). $\square$

We continue with

Theorem 10.13. *Let* $v \geq 2$ *and* $\gamma_1 \geq 0$ *such that* $2 \leq v - \gamma_1 < 1/p$, $0 < p < 1$. *Let* $f_j \in C_{x_0}^{v}\left([a,b]\right)$ *with* $f_j^{(i)}\left(x_0\right) = 0$, $i = 0,1,\ldots,n-1$, $n := [v]$, $j = 1,\ldots,M \in \mathbb{N}$. *Here* $x, x_0 \in [a,b]: x \geq x_0$. *Assume that* $D_{x_0}^{v}f_j$ *is of fixed sign on* $[x_0,b]$, $j = 1,\ldots,M$. *Consider also* $p\left(t\right) > 0$, *and* $q\left(t\right) > 0$ *continuous functions on* $[x_0,b]$. *Let* $\lambda_\alpha \geq \lambda_{\alpha+1} > 1$; Φ *is as in Theorem 10.7.*
Then

$$\int_{x_0}^{x}q\left(w\right)\left\{\left\{\sum_{j=1}^{M-1}\left[\left|\left(D_{x_0}^{\gamma_1}f_j\right)\left(w\right)\right|^{\lambda_\alpha}\left|D_{x_0}^{\gamma_1+1}f_{j+1}\left(w\right)\right|^{\lambda_{\alpha+1}}\right.\right.\right.$$

$$\left.\left.+\left|\left(D_{x_0}^{\gamma_1}f_{j+1}\right)\left(w\right)\right|^{\lambda_\alpha}\left|D_{x_0}^{\gamma_1+1}f_j\left(w\right)\right|^{\lambda_{\alpha+1}}\right]\right\}$$

$$+\left[\left|\left(D_{x_0}^{\gamma_1}f_1\right)\left(w\right)\right|^{\lambda_\alpha}\left|D_{x_0}^{\gamma_1+1}f_M\left(w\right)\right|^{\lambda_{\alpha+1}}\right.$$

$$\left.\left.+\left|\left(D_{x_0}^{\gamma_1}f_M\right)\left(w\right)\right|^{\lambda_\alpha}\left|D_{x_0}^{\gamma_1+1}f_1\left(w\right)\right|^{\lambda_{\alpha+1}}\right]\right\}dw \geq$$

$$M^{1-((\lambda_\alpha+\lambda_{\alpha+1})/p)}2^{((\lambda_\alpha+\lambda_{\alpha+1})/p)}\Phi\left(x\right)$$

$$\left[\int_{x_0}^{x}p\left(w\right)\left(\sum_{j=1}^{M}\left|\left(D_{x_0}^{v}f_j\right)\left(w\right)\right|^{p}\right)dw\right]^{((\lambda_\alpha+\lambda_{\alpha+1})/p)}, \tag{10.71}$$

all $x_0 \leq x \leq b$.

Proof. From Theorem 10.7 we obtain

$$\int_{x_0}^{x} q\left(w\right) \sum_{j=1}^{M-1} \left[\left|\left(D_{x_0}^{\gamma_1} f_j\right)\left(w\right)\right|^{\lambda_\alpha} \left|D_{x_0}^{\gamma_1+1} f_{j+1}\left(w\right)\right|^{\lambda_{\alpha+1}}\right.$$

$$\left. + \left|\left(D_{x_0}^{\gamma_1} f_{j+1}\right)\left(w\right)\right|^{\lambda_\alpha} \left|D_{x_0}^{\gamma_1+1} f_j\left(w\right)\right|^{\lambda_{\alpha+1}}\right] dw$$

$$\geq \Phi\left(x\right) \sum_{j=1}^{M-1} \left[\int_{x_0}^{x} p\left(w\right) \left(\left|\left(D_{x_0}^{v} f_j\right)\left(w\right)\right|^{P} + \left|\left(D_{x_0}^{v} f_{j+1}\right)\left(w\right)\right|^{P}\right) dw\right]^{(\lambda_\alpha + \lambda_{\alpha+1}/p)},$$

(10.72)

all $x_0 \leq x \leq b$.

Also it holds

$$\int_{x_0}^{x} q\left(w\right) \left[\left|\left(D_{x_0}^{\gamma_1} f_1\right)\left(w\right)\right|^{\lambda_\alpha} \left|D_{x_0}^{\gamma_1+1} f_M\left(w\right)\right|^{\lambda_{\alpha+1}}\right.$$

$$\left. + \left|\left(D_{x_0}^{\gamma_1} f_M\right)\left(w\right)\right|^{\lambda_\alpha} \left|D_{x_0}^{\gamma_1+1} f_1\left(w\right)\right|^{\lambda_{\alpha+1}}\right] dw$$

$$\geq \Phi\left(x\right) \left[\int_{x_0}^{x} p\left(w\right) \left(\left|\left(D_{x_0}^{v} f_1\right)\left(w\right)\right|^{P} + \left|\left(D_{x_0}^{v} f_M\right)\left(w\right)\right|^{P}\right) dw\right]^{((\lambda_\alpha + \lambda_{\alpha+1})/p)},$$

(10.73)

all $x_0 \leq x \leq b$. Adding (10.72) and (10.73), along with (10.53) we derive
(10.71). $\square$

Next comes the following theorem.

Theorem 10.14. *All here are as in Theorem 10.11. Consider the special case of $\lambda_\beta = \lambda_\alpha + \lambda_v$. Here $\tilde{T}\left(x\right)$ is as in (10.44).*
Assume here for $j = 1, \ldots, M$ that

$$z_j\left(x\right) := \int_{x_0}^{x} p\left(t\right) \left|D_{x_0}^{v} f_j\left(t\right)\right|^{P} dt \in [H, \Psi], \ 0 < H < \Psi.$$

Then

$$\int_{x_0}^{x} q\left(w\right) \left\{\left\{\sum_{j=1}^{M-1} \left[\left|\left(D_{x_0}^{\gamma_1} f_j\right)\left(w\right)\right|^{\lambda_\alpha} \left|\left(D_{x_0}^{\gamma_2} f_{j+1}\right)\left(w\right)\right|^{\lambda_\alpha + \lambda_v} \left|\left(D_{x_0}^{v} f_j\right)\left(w\right)\right|^{\lambda_v}\right.\right.\right.$$

$$\left.\left.\left. + \left|\left(D_{x_0}^{\gamma_2} f_j\right)\left(w\right)\right|^{\lambda_\alpha + \lambda_v} \left|\left(D_{x_0}^{\gamma_1} f_{j+1}\right)\left(w\right)\right|^{\lambda_\alpha} \left|\left(D_{x_0}^{v} f_{j+1}\right)\left(w\right)\right|^{\lambda_v}\right]\right\}\right.$$

$$+ \left[\left|\left(D_{x_0}^{\gamma_1} f_1\right)(w)\right|^{\lambda_\alpha} \left|\left(D_{x_0}^{\gamma_2} f_M\right)(w)\right|^{\lambda_\alpha + \lambda_v} \left|\left(D_{x_0}^{v} f_1\right)(w)\right|^{\lambda_v} \right.$$

$$+ \left. \left|\left(D_{x_0}^{\gamma_2} f_1\right)(w)\right|^{\lambda_\alpha + \lambda_v} \left|\left(D_{x_0}^{\gamma_1} f_M\right)(w)\right|^{\lambda_\alpha} \left|\left(D_{x_0}^{v} f_M\right)(w)\right|^{\lambda_v} \right] \right\} dw \geq$$

$$\tag{10.74}$$

$$M^{(1-2(\lambda_\alpha+\lambda_v)/p)} 2^{2(\lambda_\alpha+\lambda_v/p)} \tilde{T}(x)$$

$$\times \left[\int_{x_0}^{x} p(w) \left(\sum_{j=1}^{M} \left|\left(D_{x_0}^{v} f_j\right)(w)\right|^{p} \right) dw \right]^{(2(\lambda_\alpha + \lambda_v)/p)},$$

all $x_0 \leq x \leq b$.

Proof. Based on Theorem 10.8. The rest is as in the proof of Theorem 10.13. $\square$

We continue with

Corollary 10.15. (to Theorem 10.11, $\lambda_\beta = 0$, $p(t) = q(t) = 1$). *Then*

$$\int_{x_0}^{x} \left(\sum_{j=1}^{M} \left|\left(D_{x_0}^{\gamma_1} f_j\right)(w)\right|^{\lambda_\alpha} \left|\left(D_{x_0}^{v} f_j\right)(w)\right|^{\lambda_v} \right) dw$$

$$\geq \delta_1^* \varphi_1(x) \left[\int_{x_0}^{x} \left[\sum_{j=1}^{M} \left|\left(D_{x_0}^{v} f_j\right)(w)\right|^{p} \right] dw \right]^{((\lambda_\alpha + \lambda_v)/p)}, \tag{10.75}$$

all $x_0 \leq x \leq b$.

In (10.75), $\left(A_0(x)|_{\lambda_\beta = 0}\right)$ *of* $\varphi_1(x)$ *is given by* (10.49).

Proof. Based on Theorem 10.11. $\square$

Corollary 10.16. (to Theorem 10.12, $\lambda_\alpha = 0$, $p(t) = q(t) = 1$). *It holds*

$$\int_{x_0}^{x} \left\{ \left\{ \sum_{j=1}^{M-1} \left[\left|\left(D_{x_0}^{\gamma_2} f_{j+1}\right)(w)\right|^{\lambda_\beta} \left|\left(D_{x_0}^{v} f_j\right)(w)\right|^{\lambda_v} \right. \right. \right.$$

$$+ \left. \left|\left(D_{x_0}^{\gamma_2} f_j\right)(w)\right|^{\lambda_\beta} \left|\left(D_{x_0}^{v} f_{j+1}\right)(w)\right|^{\lambda_v} \right] \right\}$$

$$+ \left[\left|\left(D_{x_0}^{\gamma_2} f_M\right)(w)\right|^{\lambda_\beta} \left|\left(D_{x_0}^{v} f_1\right)(w)\right|^{\lambda_v} \right.$$

$$+ \left. \left|\left(D_{x_0}^{\gamma_2} f_1\right)(w)\right|^{\lambda_\beta} \left|\left(D_{x_0}^{v} f_M\right)(w)\right|^{\lambda_v} \right] \right\} dw \geq$$

$$\left(M^{1-(\lambda_v+\lambda_\beta/p)}\right)2^{((\lambda_v+\lambda_\beta)/p)}\varphi_2\left(x\right)\left\{\int_{x_0}^x\left[\sum_{j=1}^M\left|\left(D_{x_0}^v f_j\right)(w)\right|^p\right]dw\right\}^{((\lambda_v+\lambda_\beta)/p)},$$

$$(10.76)$$

all $x_0 \leq x \leq b$.

In (10.76), $\left(A_0\left(x\right)\big|_{\lambda_\alpha=0}\right)$ *of* $\varphi_2\left(x\right)$ *is given by* (10.52).

Proof. Based on Theorem 10.12. $\square$

11

Converse Riemann–Liouville Fractional Opial-Type Inequalities for Several Functions

A collection of very general $L_p(0 < p < 1)$-form converse Opial-type inequalities [315] is presented involving Riemann – Liouville fractional derivatives [17, 230, 295, 314] of several functions in different orders and powers. Other particular results of special interest are derived from the established results. This treatment is based on [47].

11.1 Introduction

Opial inequalities appeared for the first time in [315] and since then many authors dealt with them in different directions and for various cases. For a complete account of the recent activity in this field see [4], and it still remains a very active area of research. One of the main attractions to these inequalities is their applications, especially in proving the uniqueness and upper bounds of the solution of initial value problems in differential equations. The author was the first to present Opial inequalities involving fractional derivatives of functions in [15, 17] with applications to fractional differential equations. See also [64, 65].

Fractional derivatives come up naturally in a number of fields, especially in physics, for example; see the recent books [197, 333]. Here the author continues his study of Riemann – Liouville fractional Opial-type inequalities now involving several different functions and produces a wide variety of reverse results. To give the reader an idea of the kind of converse inequalities we are dealing with, we briefly mention a specific one (see Corollary 11.15).

G.A. Anastassiou, *Fractional Differentiation Inequalities*, 229
DOI 10.1007/978-0-387-98128-4_11, © Springer Science+Business Media, LLC 2009

$$\int_0^x \left(\sum_{j=1}^M |D^{\alpha_1} f_j(s)|^{\lambda_{\alpha_1}} |D^\beta f_j(s)|^{\lambda_\beta} \right) ds \geq \left[\left(M^{1-((\lambda_{\alpha_1}+\lambda_\beta)/p)} \right) \right.$$

$$\left(\frac{\lambda_\beta}{\lambda_{\alpha_1}+\lambda_\beta} \right)^{\lambda_\beta/p} \frac{1}{(\Gamma(\beta-\alpha_1))^{\lambda_{\alpha_1}}} \left(\frac{p-1}{\beta p - \alpha_1 p - 1} \right)^{\lambda_{\alpha_1}(p-1/p)}$$

$$\left(\frac{p-\lambda_\beta}{\lambda_{\alpha_1}\beta p - \lambda_{\alpha_1}\alpha_1 p - \lambda_{\alpha_1} + p - \lambda_\beta} \right)^{(p-\lambda_\beta/p)} x^{((\lambda_{\alpha_1}\beta p - \lambda_{\alpha_1}\alpha_1 p - \lambda_{\alpha_1} + p - \lambda_\beta)/p)} \right]$$

$$\left[\int_0^x \sum_{j=1}^M |D^\beta f_j(s)|^p ds \right]^{((\lambda_{\alpha_1}+\lambda_\beta)/p)},$$

where $\alpha_1 \geq 0$, $0 < \beta - \alpha_1 < 1/p$, $0 < p < 1$, and $\lambda_{\alpha_1} \geq 0$, $\lambda_\beta > p$.

All fractional derivatives involved here are of Riemann – Liouville type. The L_∞ fractional derivatives $D^\beta f_j$ in $[0, x]$ are each of fixed sign a.e. on $[0, x]$, and $D^{\beta-k} f_j(0) = 0$, for $k = 1, \ldots, [\beta] + 1$; $j = 1, \ldots, M$.

Here $[\cdot]$ denotes the integral part of the number; Γ stands for the gamma function.

11.2   Background

We need

Definition 11.1. (see [187, 295, 314]). Let $\alpha \in \mathbb{R}_+ - \{0\}$. For any $f \in L_1(0, x)$; $x \in \mathbb{R}_+ - \{0\}$, the *Riemann – Liouville fractional integral* of f of order α is defined by

$$(J_\alpha f)(s) := \frac{1}{\Gamma(\alpha)} \int_0^s (s-t)^{\alpha-1} f(t)\, dt, \forall s \in [0, x], \qquad (11.1)$$

and the *Riemann – Liouville fractional derivative* of f of order α by

$$D^\alpha f(s) := \frac{1}{\Gamma(m-a)} \left(\frac{d}{ds} \right)^m \int_0^s (s-t)^{m-\alpha-1} f(t)\, dt, \qquad (11.2)$$

where $m := [\alpha] + 1$, $[\cdot]$ is the integral part. In addition, we set $D^0 f := f := J_0 f$, $J_{-\alpha} f = D^\alpha f$ if $\alpha > 0$, $D^{-\alpha} f := J_\alpha f$, if $0 < \alpha \leq 1$. If $\alpha \in \mathbb{N}$, then $D^\alpha f = f^{(\alpha)}$ the ordinary derivative.

Definition 11.2. [187]. We say that $f \in L_1(0, x)$ has an L_∞ *fractional derivative* $D^\alpha f$ in $[0, x]$, $x \in \mathbb{R}_+ - \{0\}$, iff $D^{\alpha-k} f \in C([0, x])$, $k = 1, \ldots, m := [\alpha] + 1$; $\alpha \in \mathbb{R}_+ - \{0\}$, and $D^{\alpha-1} f \in AC([0, x])$ (absolutely continuous functions) and $D^\alpha f \in L_\infty(o, x)$.

We need

Lemma 11.3. [187]. *Let* $\alpha \in \mathbb{R}_+$, $\beta > \alpha$, *let* $f \in L_1(o, x)$, $x \in \mathbb{R}_+ - \{0\}$, *have an* L_∞ *fractional derivative* $D^\beta f$ *in* $[0, x]$, *and let* $D^{\beta-k} f(0) = 0$ *for* $k = 1, \ldots, [\beta] + 1$. *Then*

$$D^\alpha f(s) = \frac{1}{\Gamma(\beta - \alpha)} \int_0^s (s - t)^{\beta - \alpha - 1} D^\beta f(t) \, dt, \forall s \in [0, x]. \qquad (11.3)$$

Here $D^\alpha f \in AC([0, x])$ *for* $\beta - \alpha \geq 1$ *and* $D^\alpha f \in C([0, x])$ *for* $\beta - \alpha \in (0, 1)$, *hence* $D^\alpha f \in L_\infty(o, x)$ *and* $D^\alpha f \in L_1(o, x)$.

11.3 Main Results

11.3.1 Results Involving Two Functions

We present our first main result.

Theorem 11.4. *Let* $\alpha_i \in \mathbb{R}_+$, $0 < \beta - \alpha_i < 1/p$, $0 < p < 1$, $i = 1, 2$, *and let* $f_1, f_2 \in L_1(0, x)$, $x \in \mathbb{R}_+ - \{0\}$ *have, respectively,* L_∞ *fractional derivatives* $D^\beta f_1, D^\beta f_2$ *in* $[0, x]$, *each of fixed sign a.e. on* $[0, x]$, *and let* $D^{\beta-k} f_i(0) = 0$, *for* $k = 1, \ldots, [\beta] + 1; i = 1, 2$.
Consider also $p(t) > 0$ *and* $q(t) \geq 0$, *with all* $p(t), 1/p(t), q(t), 1/q(t) \in L_\infty(0, x)$. *Let* $\lambda_\beta > 0$ *and* $\lambda_{\alpha_1}, \lambda_{\alpha_2} \geq 0$, *such that* $\lambda_\beta > p$.
Call

$$P_i(s) := \int_0^s (s - t)^{p(\beta - \alpha_i - 1)/p - 1} (p(t))^{-1/(p-1)} \, dt, \ i = 1, 2; 0 \leq s \leq x,$$

$$(11.4)$$

$$A(s) := \frac{q(s)(P_1(s))^{\lambda_{\alpha_1}((p-1)/p)} (P_2(s))^{\lambda_{\alpha_2}((p-1)/p)} (p(s))^{-\lambda_\beta/p}}{(\Gamma(\beta - \alpha_1))^{\lambda_{\alpha_1}} (\Gamma(\beta - \alpha_2))^{\lambda_{\alpha_2}}} \qquad (11.5)$$

$$A_0(x) := \left(\int_0^x (A(s))^{p/(p-\lambda_\beta)} \, ds \right)^{(p-\lambda_\beta)/p} \qquad (11.6)$$

and

$$\delta_1 := 2^{1 - ((\lambda_{\alpha_1} + \lambda_\beta)/p)}. \qquad (11.7)$$

If $\lambda_{\alpha_2} = 0$, *we obtain that*

$$\int_0^x q(s) \left[|D^{\alpha_1} f_1(s)|^{\lambda_{\alpha_1}} |D^\beta f_1(s)|^{\lambda_\beta} + |D^{\alpha_1} f_2(s)|^{\lambda_{\alpha_1}} |D^\beta f_2(s)|^{\lambda_\beta} \right] ds \geq$$

$$\left(A_0\left(x\right)|_{\lambda_{\alpha_2}=0}\right)\left(\frac{\lambda_\beta}{\lambda_{\alpha_1}+\lambda_\beta}\right)^{\lambda_\beta/p}\delta_1\left[\int_0^x p\left(s\right)\left[\left|D^\beta f_1\left(s\right)\right|^p+\right.\right.$$

$$\left.\left.\left|D^\beta f_2\left(s\right)\right|^p\right]ds\right]^{\left(\left(\lambda_{\alpha_1}+\lambda_\beta\right)/p\right)}. \tag{11.8}$$

Proof. From (11.3) and assumption we have

$$\left|D^{\alpha_i} f_j\left(s\right)\right|=\frac{1}{\Gamma\left(\beta-\alpha_i\right)}\int_0^s \left(s-t\right)^{\beta-\alpha_i-1}\left|D^\beta f_j\left(t\right)\right|dt, \tag{11.9}$$

$$\forall s\in\left[0,x\right],\ i=1,2;\ j=1,2.$$

Next, applying Hölder's inequality with indices p, $p/\left(p-1\right)$, we find

$$\left|D^{\alpha_i} f_j\left(s\right)\right|=\frac{1}{\Gamma\left(\beta-\alpha_i\right)}\int_0^s \left(s-t\right)^{\beta-\alpha_i-1}$$

$$\left(p\left(t\right)\right)^{-1/p}\left(p\left(t\right)\right)^{1/p}\left|D^\beta f_j\left(t\right)\right|dt \tag{11.10}$$

$$\geq\frac{1}{\Gamma\left(\beta-\alpha_i\right)}\left(\int_0^s\left(\left(s-t\right)^{\beta-\alpha_i-1}\left(p\left(t\right)\right)^{-1/p}\right)^{p/p-1}dt\right)^{\left(p-1\right)/p} \tag{11.11}$$

$$\times\left(\int_0^s p\left(t\right)\left|D^\beta f_j\left(t\right)\right|^p dt\right)^{1/p}=\frac{1}{\Gamma\left(\beta-\alpha_i\right)}\left(P_i\left(s\right)\right)^{p-1/p}$$

$$\left(\int_0^s p\left(t\right)\left|D^\beta f_j\left(t\right)\right|^p dt\right)^{1/p}. \tag{11.12}$$

That is, it holds

$$\left|D^{\alpha_i} f_j\left(s\right)\right|\geq\frac{1}{\Gamma\left(\beta-\alpha_i\right)}\left(P_i\left(s\right)\right)^{p-1/p}\left(\int_0^s p\left(t\right)\left|D^\beta f_j\left(t\right)\right|^p dt\right)^{1/p}. \tag{11.13}$$

Put

$$z_j\left(s\right):=\int_0^s p\left(t\right)\left|D^\beta f_j\left(t\right)\right|^p dt, \tag{11.14}$$

thus,

$$z_j'\left(s\right)=p\left(s\right)\left|D^\beta f_j\left(s\right)\right|^p,\ \text{a.e. in }\left(0,x\right),\ z_j\left(0\right)=0;\ j=1,2. \tag{11.15}$$

Hence, we have

$$\left|D^{\alpha_i} f_j\left(s\right)\right|\geq\frac{1}{\Gamma\left(\beta-\alpha_i\right)}\left(P_i\left(s\right)\right)^{p-1/p}\left(z_j\left(s\right)\right)^{1/p}, \tag{11.16}$$

and

$$\left| D^{\beta} f_j \left(s \right) \right|^{\lambda_{\beta}} = \left(p \left(s \right) \right)^{-\lambda_{\beta}/p} \left(z_j' \left(s \right) \right)^{\lambda_{\beta}/p}, \tag{11.17}$$

a.e. in $\left(o, x \right)$, $i = 1, 2$; $j = 1, 2$.

Therefore we obtain

$$q \left(s \right) \left| \left(D^{\alpha_1} f_1 \right) \left(s \right) \right|^{\lambda_{\alpha_1}} \left| \left(D^{\alpha_2} f_2 \right) \left(s \right) \right|^{\lambda_{\alpha_2}} \left| \left(D^{\beta} f_1 \right) \left(s \right) \right|^{\lambda_{\beta}}$$

$$\geq q \left(s \right) \frac{1}{\left(\Gamma \left(\beta - \alpha_1 \right) \right)^{\lambda_{\alpha_1}}} \left(P_1 \left(s \right) \right)^{\lambda_{\alpha_1} \left(p - 1/p \right)} \left(z_1 \left(s \right) \right)^{\lambda_{\alpha_1}/p} \tag{11.18}$$

$$\frac{1}{\left(\Gamma \left(\beta - \alpha_2 \right) \right)^{\lambda_{\alpha_2}}} \left(P_2 \left(s \right) \right)^{\lambda_{\alpha_2} \left(p - 1/p \right)} \left(z_2 \left(s \right) \right)^{\lambda_{\alpha_2}/p} \left(p \left(s \right) \right)^{-\lambda_{\beta}/p} \left(z_1' \left(s \right) \right)^{\lambda_{\beta}/p}$$
$$\tag{11.19}$$

$$= A \left(s \right) \left(z_1 \left(s \right) \right)^{\lambda_{\alpha_1}/p} \left(z_2 \left(s \right) \right)^{\lambda_{\alpha_2}/p} \left(z_1' \left(s \right) \right)^{\lambda_{\beta}/p}, \text{a.e. in} \left(0, x \right). \tag{11.20}$$

Consequently, by another Hölder's inequality application, we get (by $\left(p/\lambda_{\beta} \right) < 1$)

$$\int_0^x q \left(s \right) \left| \left(D^{\alpha_1} f_1 \right) \left(s \right) \right|^{\lambda_{\alpha_1}} \left| \left(D^{\alpha_2} f_2 \right) \left(s \right) \right|^{\lambda_{\alpha_2}} \left| \left(D^{\beta} f_1 \right) \left(s \right) \right|^{\lambda_{\beta}} ds$$

$$\geq A_0 \left(x \right) \left(\int_0^x \left(z_1 \left(s \right) \right)^{\lambda_{\alpha_1}/\lambda_{\beta}} \left(z_2 \left(s \right) \right)^{\lambda_{\alpha_2}/\lambda_{\beta}} z_1' \left(s \right) ds \right)^{\lambda_{\beta}/p}. \tag{11.21}$$

Similarly, one finds

$$\int_0^x q \left(s \right) \left| \left(D^{\alpha_1} f_2 \right) \left(s \right) \right|^{\lambda_{\alpha_1}} \left| \left(D^{\alpha_2} f_1 \right) \left(s \right) \right|^{\lambda_{\alpha_2}} \left| \left(D^{\beta} f_2 \right) \left(s \right) \right|^{\lambda_{\beta}} ds$$

$$\geq A_0 \left(x \right) \left(\int_0^x \left(z_1 \left(s \right) \right)^{\lambda_{\alpha_2}/\lambda_{\beta}} \left(z_2 \left(s \right) \right)^{\lambda_{\alpha_1}/\lambda_{\beta}} z_2' \left(s \right) ds \right)^{\lambda_{\beta}/p}. \tag{11.22}$$

Taking $\lambda_{\alpha_2} = 0$ and adding $\left(11.21 \right)$ and $\left(11.22 \right)$, we derive

$$\int_0^x q \left(s \right) \left[\left| \left(D^{\alpha_1} f_1 \right) \left(s \right) \right|^{\lambda_{\alpha_1}} \left| \left(D^{\beta} f_1 \right) \left(s \right) \right|^{\lambda_{\beta}} + \left| \left(D^{\alpha_1} f_2 \right) \left(s \right) \right|^{\lambda_{\alpha_1}} \left| \left(D^{\beta} f_2 \right) \left(s \right) \right|^{\lambda_{\beta}} \right] ds$$

$$\geq \left(A_0 \left(x \right) |_{\lambda_{\alpha_2} = 0} \right) \left[\left(\int_0^x \left(z_1 \left(s \right) \right)^{\lambda_{\alpha_1}/\lambda_{\beta}} \left(z_1' \left(s \right) \right) ds \right)^{\lambda_{\beta}/p} + \right.$$

$$\left(\int_0^x (z_2(s))^{\lambda_{\alpha_1}/\lambda_\beta} z_2'(s)\, ds \right)^{\lambda_\beta/p} \Bigg] = \qquad (11.23)$$

$$(A_0(x)\,|_{\lambda_{\alpha_2}=0}) \left[(z_1(x))^{(\lambda_{\alpha_1}+\lambda_\beta/p)} + \right.$$

$$\left. (z_2(x))^{(\lambda_{\alpha_1}+\lambda_\beta/p)} \right] \left(\frac{\lambda_\beta}{\lambda_{\alpha_1}+\lambda_\beta} \right)^{\lambda_\beta/p} = \qquad (11.24)$$

$$(A_0(x)\,|_{\lambda_{\alpha_2}=0}) \left(\frac{\lambda_\beta}{\lambda_{\alpha_1}+\lambda_\beta} \right)^{\lambda_\beta/p} \left[\left(\int_0^x p(t)\, |D^\beta f_1(t)|^p\, dt \right)^{(\lambda_{\alpha_1}+\lambda_\beta/p)} \right.$$

$$\left. + \left(\int_0^x p(t)\, |D^\beta f_2(t)|^p\, dt \right)^{(\lambda_{\alpha_1}+\lambda_\beta/p)} \right] =: (*). \qquad (11.25)$$

In this study, we frequently use the basic inequalities

$$2^{r-1}\,(a^r + b^r) \le (a+b)^r \le a^r + b^r, \ a,b \ge 0, \ 0 \le r \le 1, \qquad (11.26)$$

and

$$a^r + b^r \le (a+b)^r \le 2^{r-1}\,(a^r + b^r), \ a,b \ge 0, \ r \ge 1. \qquad (11.27)$$

At last using (11.7), (11.26), and (11.27), we get

$$(*) \ge (A_0(x)\,|_{\lambda_{\alpha_2}=0}) \left(\frac{\lambda_\beta}{\lambda_{\alpha_1}+\lambda_\beta} \right)^{\lambda_\beta/p} \delta_1$$

$$\left[\int_0^x p(t) \left[|D^\beta f_1(t)|^p + |D^\beta f_2(t)|^p \right] dt \right]^{(\lambda_{\alpha_1}+\lambda_\beta/p)}. \qquad (11.28)$$

Inequality (11.8) has been proved.

We remark the following.

Here we see that $(p/p-1)(\beta-\alpha_i-1)+1 > 0$, $-1/(p-1) > 0$ and $p(t) \in L_\infty(0,x)$; thus (see (11.4)) $P_i(s) \in \mathbb{R}$, for every $s \in [0,x]$; also $P_i(s)$ is continuous and bounded on $[0,x]$.

By $\lambda_\beta > p > 0$, we have $0 < p/\lambda_\beta < 1$, $p/p - \lambda_\beta < 0$.

We observe that

$$\frac{1}{A(s)} = \frac{1}{q(s)}\,(\Gamma(\beta-\alpha_1))^{\lambda_{\alpha_1}}\,(\Gamma(\beta-\alpha_2))^{\lambda_{\alpha_2}}\,(p(s))^{\lambda_\beta/p}$$

$$(P_1(s))^{\lambda_{\alpha_1}(1-p/p)}\,(P_2(s))^{\lambda_{\alpha_2}(1-p/p)} \in L_\infty(0,x),$$

and $1/A(s) > 0$, a.e. on $[0, x]$.

Therefore $0 < A_0(x) < \infty$, and all we have done in this proof is valid.

$\square$

The counterpart of the last theorem follows.

Theorem 11.5. *All here are as in Theorem 11.4. Further assume* $\lambda_{\alpha_2} \geq \lambda_\beta$.

Denote

$$\delta_2 := 2^{1-(\lambda_{\alpha_2}/\lambda_\beta)} \tag{11.29}$$

and

$$\delta_3 := (\delta_2 - 1) \, 2^{-\lambda_{\alpha_2}/\lambda_\beta} \tag{11.30}$$

If $\lambda_{\alpha_1} = 0$, *then*

$$\int_0^x q(s) \left[|(D^{\alpha_2} f_2)(s)|^{\lambda_{\alpha_2}} \left| \left(D^\beta f_1 \right)(s) \right|^{\lambda_\beta} + |(D^{\alpha_2} f_1)(s)|^{\lambda_{\alpha_2}} \left| \left(D^\beta f_2 \right)(s) \right|^{\lambda_\beta} \right] ds$$

$$\geq \left(A_0(x) |_{\lambda_{\alpha_1}=0} \right) 2^{(p-\lambda_\beta)/p} \left(\frac{\lambda_\beta}{\lambda_{\alpha_2} + \lambda_\beta} \right)^{\lambda_\beta/p} \delta_3^{\lambda_\beta/p}$$

$$\cdot \left(\int_0^x p(s) \left[|D^\beta f_1(s)|^p + |D^\beta f_2(s)|^p \right] ds \right)^{(\lambda_\beta + \lambda_{\alpha_2})/p}. \tag{11.31}$$

Proof. When $\lambda_{\alpha_1} = 0$ from (11.21) and (11.22) we find

$$\int_0^x q(s) \left[|D^{\alpha_2} f_2(s)|^{\lambda_{\alpha_2}} |D^\beta f_1(s)|^{\lambda_\beta} + |D^{\alpha_2} f_1(s)|^{\lambda_{\alpha_2}} |D^\beta f_2(s)|^{\lambda_\beta} \right] ds \geq \tag{11.32}$$

$$\left(A_0(x) |_{\lambda_{\alpha_1}=0} \right) \left[\left(\int_0^x (z_2(s))^{\lambda_{\alpha_2}/\lambda_\beta} \left(z_1'(s) \right) ds \right)^{\lambda_\beta/p} \right.$$

$$\left. \left(\int_0^x (z_1(s))^{\lambda_{\alpha_2}/\lambda_\beta} \left(z_2'(s) \right) ds \right)^{\lambda_\beta/p} \right] \geq$$

$$\left(A_0(x) |_{\lambda_{\alpha_1}=0} \right) 2^{1-\lambda_\beta/p} (M(x))^{\lambda_\beta/p} := (*), \tag{11.33}$$

by $\lambda_\beta/p > 1$ and by (11.27), where

$$M(x) := \int_0^x \left[(z_2(s))^{\lambda_{\alpha_2}/\lambda_\beta} z_1'(s) + (z_1(s))^{\lambda_{\alpha_2}/\lambda_\beta} z_2'(s) \right] ds. \tag{11.34}$$

We observe that

$$M(x) = \int_0^x \left((z_1(s))^{\lambda_{\alpha_2}/\lambda_\beta} + (z_2(s))^{\lambda_{\alpha_2}/\lambda_\beta} \right) \left(z_1'(s) + z_2'(s) \right) ds$$

$$- \int_0^x \left[(z_1(s))^{\lambda_{\alpha_2}/\lambda_\beta} z_1'(s) + (z_2(s))^{\lambda_{\alpha_2}/\lambda_\beta} z_2'(s) \right] ds \qquad (11.35)$$

$$\overset{(by\ (11.27))}{\geq} \delta_2 \int_0^x (z_1(s) + z_2(s))^{\lambda_{\alpha_2}/\lambda_\beta} (z_1(s) + z_2(s))' ds$$

$$- \left(\frac{\lambda_\beta}{\lambda_{\alpha_2} + \lambda_\beta} \right) \left[(z_1(x))^{((\lambda_{\alpha_2}+\lambda_\beta)/\lambda_\beta)} + (z_2(x))^{((\lambda_{\alpha_2}+\lambda_\beta)/\lambda_\beta)} \right] \qquad (11.36)$$

$$= \delta_2 (z_1(x) + z_2(x))^{(\lambda_{\alpha_2}+\lambda_\beta/\lambda_\beta)} \left(\frac{\lambda_\beta}{\lambda_{\alpha_2} + \lambda_\beta} \right)$$

$$- \left(\frac{\lambda_\beta}{\lambda_{\alpha_2} + \lambda_\beta} \right) \left[(z_1(x))^{((\lambda_{\alpha_2}+\lambda_\beta)/\lambda_\beta)} + (z_2(x))^{((\lambda_{\alpha_2}+\lambda_\beta)/\lambda_\beta)} \right] \qquad (11.37)$$

$$= \left(\frac{\lambda_\beta}{\lambda_{\alpha_2} + \lambda_\beta} \right)$$

$$\left[\delta_2 (z_1(x) + z_2(x))^{(\lambda_{\alpha_2}+\lambda_\beta/\lambda_\beta)} - \left[(z_1(x))^{(\lambda_{\alpha_2}+\lambda_\beta/\lambda_\beta)} + (z_2(x))^{(\lambda_{\alpha_2}+\lambda_\beta/\lambda_\beta)} \right] \right]$$

$$11.38$$

$$\overset{(11.27)}{\geq} \left(\frac{\lambda_\beta}{\lambda_{\alpha_2} + \lambda_\beta} \right) \left[\delta_2 \left((z_1(x))^{(\lambda_{\alpha_2}+\lambda_\beta/\lambda_\beta)} + (z_2(x))^{(\lambda_{\alpha_2}+\lambda_\beta/\lambda_\beta)} \right) \right.$$

$$\left. - \left[(z_1(x))^{(\lambda_{\alpha_2}+\lambda_\beta/\lambda_\beta)} + (z_2(x))^{(\lambda_{\alpha_2}+\lambda_\beta)/\lambda_\beta} \right] \right] \qquad (11.39)$$

$$= \left(\frac{\lambda_\beta}{\lambda_{\alpha_2} + \lambda_\beta} \right) (\delta_2 - 1) \left[(z_1(x))^{(\lambda_{\alpha_2}+\lambda_\beta/\lambda_\beta)} + (z_2(x))^{(\lambda_{\alpha_2}+\lambda_\beta/\lambda_\beta)} \right]$$

$$(11.40)$$

$$\overset{(11.27)}{\geq} \left(\frac{\lambda_\beta}{\lambda_{\alpha_2} + \lambda_\beta} \right) \delta_3 (z_1(x) + z_2(x))^{(\lambda_{\alpha_2}+\lambda_\beta/\lambda_\beta)} . \qquad (11.41)$$

That is, we get that

$$M(x) \geq \left(\frac{\lambda_\beta}{\lambda_{\alpha_2} + \lambda_\beta}\right) \delta_3 \left(z_1(x) + z_2(x)\right)^{(\lambda_{\alpha_2} + \lambda_\beta/\lambda_\beta)}. \tag{11.42}$$

Therefore, by $(11.32), (11.33),$ and $(11.42),$ we derive

$$(*) \geq \left(A_0(x)|_{\lambda_{\alpha_1}=0}\right) 2^{(p-\lambda_\beta)/p} \left(\frac{\lambda_\beta}{\lambda_{\alpha_2} + \lambda_\beta}\right)^{\lambda_\beta/p}$$

$$\delta_3^{\lambda_\beta/p} \left(z_1(x) + z_2(x)\right)^{(\lambda_{\alpha_2} + \lambda_\beta)/p} \tag{11.43}$$

$$= \left(A_0(x)|_{\lambda_{\alpha_1}=0}\right) 2^{(p-\lambda_\beta)/p} \left(\frac{\lambda_\beta}{\lambda_{\alpha_2} + \lambda_\beta}\right)^{\lambda_\beta/p} \delta_3^{\lambda_\beta/p}$$

$$\left[\int_0^x p(s) \left[\left|D^\beta f_1(s)\right|^p + \left|D^\beta f_2(s)\right|^p\right] ds\right]^{(\lambda_{\alpha_2} + \lambda_\beta/p)}. \tag{11.44}$$

We have established (11.31). $\quad\square$

We need

Theorem 11.6. (see [48]) *Let* $0 \leq s \leq x$ *and* $f \in L_\infty([0,x])$, $r > 0$. *Define*

$$F(s) := \int_0^s (s-t)^r f(t) \, dt. \tag{11.45}$$

Then there exists

$$F'(s) = r \int_0^s (s-t)^{r-1} f(t) \, dt, \text{ all } s \in [0,x] \tag{11.46}$$

We proceed with a special important case.

Theorem 11.7. *Let* $1 < \beta - \alpha_1 < 1/p$, $0 < p < 1$, $\alpha_1 \in \mathbb{R}_+$ *and let* $f_1, f_2 \in L_1(0,x)$, $x \in \mathbb{R}_+ - \{0\}$ *have, respectively,* L_∞ *fractional derivatives* $D^\beta f_1, D^\beta f_2$ *in* $[0,x]$, *each of fixed sign a.e. in* $[0,x]$, *and let* $D^{\beta-k} f_i(0) = 0$, *for* $k = 1, \ldots, [\beta] + 1$; $i = 1, 2$. *Consider also* $p(t) > 0$ *and* $q(t) \geq 0$, *with* $p(t), 1/p(t), q(t), 1/q(t) \in L_\infty(0,x)$. *Let* $\lambda_\alpha \geq \lambda_{\alpha+1} > 1$. *Denote*

$$\theta_3 := \left(2^{1-(\lambda_\alpha/\lambda_{\alpha+1})} - 1\right) 2^{-\lambda_\alpha/\lambda_{\alpha+1}} \tag{11.47}$$

$$L(x) := \left(2 \int_0^x (q(s))^{(1/(1-\lambda_{\alpha+1}))} ds\right)^{1-\lambda_{\alpha+1}} \left(\frac{\theta_3 \lambda_{\alpha+1}}{\lambda_\alpha + \lambda_{\alpha+1}}\right)^{\lambda_{\alpha+1}}. \tag{11.48}$$

and

$$P_1(x) := \int_0^x (x-s)^{(\beta-\alpha_1-1)p/(p-1)} (p(s))^{-1/(p-1)} ds, \qquad (11.49)$$

$$T(x) := L(x) \left(\frac{P_1(x)^{((p-1)/p)}}{\Gamma(\beta-\alpha_1)} \right)^{(\lambda_\alpha+\lambda_{\alpha+1})} \qquad (11.50)$$

and

$$\omega_1 := 2^{(p-1/p)(\lambda_\alpha+\lambda_{\alpha+1})}, \qquad (11.51)$$

$$\Phi(x) := T(x)\,\omega_1. \qquad (11.52)$$

Then

$$\int_0^x q(s) \left[|D^{\alpha_1} f_1(s)|^{\lambda_\alpha} |D^{\alpha_1+1} f_2(s)|^{\lambda_{\alpha+1}} + |D^{\alpha_1} f_2(s)|^{\lambda_\alpha} |D^{\alpha_1+1} f_1(s)|^{\lambda_{\alpha+1}} \right] ds$$

$$\geq \Phi(x) \left[\int_0^x p(s) \left(|D^{\beta} f_1(s)|^p + |D^{\beta} f_2(s)|^p \right) ds \right]^{(\lambda_\alpha+\lambda_{\alpha+1})/p}. \qquad (11.53)$$

Proof. For convenience, we set

$$\alpha_2 := \alpha_1 + 1. \qquad (11.54)$$

By Lemma 11.3 and assumption we have

$$|D^{\alpha_1} f_i(s)| = \frac{1}{\Gamma(\beta-\alpha_1)} \int_0^s (s-t)^{\beta-\alpha_1-1} |D^{\beta} f_i(t)| \, dt, \qquad (11.55)$$

$\forall s \in [0,x]$, $i = 1,2$, and

$$|D^{\alpha_2} f_i(s)| = |D^{\alpha_1+1} f_i(s)| =$$

$$\frac{1}{\Gamma(\beta-\alpha_1-1)} \int_0^s (s-t)^{\beta-\alpha_1-2} |D^{\beta} f_i(t)| \, dt, \qquad (11.56)$$

$\forall s \in [0,x]$, $i = 1,2$.

By Theorem 11.6 we obtain that

$$(D^{\alpha_1} f_i(s))' = D^{\alpha_1+1} f_i(s) = D^{\alpha_2} f_i(s), \qquad (11.57)$$

$\forall s \in [0,x]$, $i = 1,2$.

So we have

$$|D^{\alpha_i} f_j (s)| = \frac{1}{\Gamma(\beta - \alpha_i)} \int_0^s (s-t)^{\beta - \alpha_i - 1} |D^\beta f_j(t)| \, dt =: g_{j,\alpha_i}(s),$$

(11.58)

where $j = 1, 2$, $i = 1, 2$, $0 \leq s \leq x$.

We also have by Theorem 11.6 that

$$(g_{j,\alpha_1}(s))' = g_{j,\alpha_2}(s); \quad g_{j,\alpha_i}(0) = 0.$$

(11.59)

Notice that, if $\beta - \alpha_2 = 1$, then

$$g_{j,\alpha_2}(s) = \int_0^s |D^\beta f_j(t)| \, dt.$$

(11.60)

Next, we apply Hölder's inequality with indices $1/\lambda_{\alpha+1} < 1$, $1/(1 - \lambda_{\alpha+1}) < 0$ to obtain

$$\int_0^x q(s) |D^{\alpha_1} f_1(s)|^{\lambda_\alpha} |D^{\alpha_1 + 1} f_2(s)|^{\lambda_{\alpha+1}} \, ds =$$

(11.61)

$$\int_0^x q(s) (g_{1,\alpha_1}(s))^{\lambda_\alpha} \left((g_{2,\alpha_1}(s))'\right)^{\lambda_{\alpha+1}} \, ds \geq$$

$$\left(\int_0^x (q(s))^{(1/1 - \lambda_{\alpha+1})} \, ds\right)^{(1 - \lambda_{\alpha+1})}$$

$$\left(\int_0^x (g_{1,\alpha_1}(s))^{\lambda_\alpha / \lambda_{\alpha+1}} (g_{2,\alpha_1}(s))' \, ds\right)^{\lambda_{\alpha+1}}.$$

(11.62)

Similarly, we find

$$\int_0^x q(s) |D^{\alpha_1} f_2(s)|^{\lambda_\alpha} |D^{\alpha_1 + 1} f_1(s)|^{\lambda_{\alpha+1}} \, ds \geq$$

$$\left(\int_0^x (q(s))^{(1/1 - \lambda_{\alpha+1})} \, ds\right)^{(1 - \lambda_{\alpha+1})}$$

$$\left(\int_0^x (g_{2,\alpha_1}(s))^{\lambda_\alpha / \lambda_{\alpha+1}} (g_{1,\alpha_1}(s))' \, ds\right)^{\lambda_{\alpha+1}}.$$

(11.63)

Adding the last two inequalities (11.62), (11.63), we derive

$$\int_0^x q(s) \left[|D^{\alpha_1} f_1(s)|^{\lambda_\alpha} |D^{\alpha_1 + 1} f_2(s)|^{\lambda_{\alpha+1}} + |D^{\alpha_1} f_2(s)|^{\lambda_\alpha} |D^{\alpha_1 + 1} f_1(s)|^{\lambda_{\alpha+1}} \right] ds$$

$$\geq \left(\int_0^x \left(q\left(s\right) \right)^{\left(1/1-\lambda_{\alpha+1}\right)} ds\right)^{\left(1-\lambda_{\alpha+1}\right)} \left[\left(\int_0^x \left(g_{1,\alpha_1}\left(s\right) \right)^{\lambda_\alpha/\lambda_{\alpha+1}} \left(g_{2,\alpha_1}\left(s\right) \right)' ds\right)^{\lambda_{\alpha+1}} + \right.$$

$$\left. \left(\int_0^x \left(g_{2,\alpha_1}\left(s\right) \right)^{\lambda_\alpha/\lambda_{\alpha+1}} \left(g_{1,\alpha_1}\left(s\right) \right)' ds\right)^{\lambda_{\alpha+1}} \right] \qquad (11.64)$$

$$\overset{(11.27)}{\geq} \left(2 \int_0^x \left(q\left(s\right) \right)^{\left(1/1-\lambda_{\alpha+1}\right)} ds\right)^{\left(1-\lambda_{\alpha+1}\right)} \times$$

$$\left[\int_0^x \left[\left(g_{1,\alpha_1}\left(s\right) \right)^{\lambda_\alpha/\lambda_{\alpha+1}} \left(g_{2,\alpha_1}\left(s\right) \right)' + \left(g_{2,\alpha_1}\left(s\right) \right)^{\lambda_\alpha/\lambda_{\alpha+1}} \left(g_{1,\alpha_1}\left(s\right) \right)' \right] ds \right]^{\lambda_{\alpha+1}} \qquad (11.65)$$

(notice (11.34) and (11.42), so similarly we have)

$$\geq \left(2 \int_0^x \left(q\left(s\right) \right)^{\left(1/1-\lambda_{\alpha+1}\right)} ds\right)^{\left(1-\lambda_{\alpha+1}\right)}$$

$$\left(\frac{\lambda_{\alpha+1}\theta_3}{\lambda_\alpha + \lambda_{\alpha+1}}\right)^{\lambda_{\alpha+1}} \left(g_{1,\alpha_1}\left(x\right) + g_{2,\alpha_1}\left(x\right) \right)^{\left(\lambda_\alpha+\lambda_{\alpha+1}\right)} \qquad (11.66)$$

$$= L\left(x\right) \left(g_{1,\alpha_1}\left(x\right) + g_{2,\alpha_1}\left(x\right) \right)^{\left(\lambda_\alpha+\lambda_{\alpha+1}\right)} \qquad (11.67)$$

$$= \frac{L\left(x\right)}{\left(\Gamma\left(\beta - \alpha_1\right) \right)^{\left(\lambda_\alpha+\lambda_{\alpha+1}\right)}} \times$$

$$\left[\int_0^x \left(x - t\right)^{\beta-\alpha_1-1} \left(p\left(t\right) \right)^{-1/p} \left(p\left(t\right) \right)^{1/p} \left[\left| D^\beta f_1\left(t\right) \right| + \left| D^\beta f_2\left(t\right) \right| \right] dt \right]^{\left(\lambda_\alpha+\lambda_{\alpha+1}\right)} \qquad (11.68)$$

(applying Hölder's inequality with indices $p/(p-1)$ and p, we find)

$$\geq \frac{L\left(x\right)}{\left(\Gamma\left(\beta - \alpha_1\right) \right)^{\left(\lambda_\alpha+\lambda_{\alpha+1}\right)}} \left[\int_0^x \left(x - t\right)^{\left(\beta-\alpha_1-1\right) p/\left(p-1\right)} \left(p\left(t\right) \right)^{-1/\left(p-1\right)} dt \right]^{\left(\left(p-1\right)/p\right) \left(\lambda_\alpha+\lambda_{\alpha+1}\right)} \qquad (11.69)$$

$$\left[\int_0^x p\left(t\right) \left(\left| D^\beta f_1\left(t\right) \right| + \left| D^\beta f_2\left(t\right) \right| \right)^p dt \right]^{\left(\left(\lambda_\alpha+\lambda_{\alpha+1}\right)/p\right)} =$$

$$\frac{L\left(x\right) \left(P_1\left(x\right) \right)^{\left(p-1/p\right) \left(\lambda_\alpha+\lambda_{\alpha+1}\right)}}{\left(\Gamma\left(\beta - \alpha_1\right) \right)^{\left(\lambda_\alpha+\lambda_{\alpha+1}\right)}} \left[\int_0^x p\left(t\right) \left(\left| D^\beta f_1\left(t\right) \right| + \left| D^\beta f_2\left(t\right) \right| \right)^p dt \right]^{\left(\left(\lambda_\alpha+\lambda_{\alpha+1}\right)/p\right)} \qquad (11.70)$$

$$= T(x) \left[\int_0^x p(t) \left(\left| D^\beta f_1(t) \right| + \left| D^\beta f_2(t) \right| \right)^p dt \right]^{(\lambda_\alpha + \lambda_{\alpha+1}/p)}$$

$$\overset{(11.26)}{\geq} T(x) \, 2^{(p-1/p)(\lambda_\alpha + \lambda_{\alpha+1})} \left[\int_0^x p(t) \left(\left| D^\beta f_1(t) \right|^p + \left| D^\beta f_2(t) \right|^p \right) dt \right]^{(\lambda_\alpha + \lambda_{\alpha+1}/p)} \tag{11.71}$$

$$= \Phi(x) \left[\int_0^x p(t) \left(\left| D^\beta f_1(t) \right|^p + \left| D^\beta f_2(t) \right|^p \right) dt \right]^{(\lambda_\alpha + \lambda_{\alpha+1})/p}. \tag{11.72}$$

We have proved (11.53). $\square$

We continue with

Theorem 11.8. *All here are as in Theorem 11.4. Consider the special case of* $\lambda_{\alpha_2} = \lambda_{\alpha_1} + \lambda_\beta$.
Assume here for $j = 1, 2$ *that*

$$z_j(x) := \int_0^x p(t) \left| D^\beta f_j(t) \right|^p dt \in [H, \Psi], \, 0 < H < \Psi, \tag{11.73}$$

along with

$$h := \frac{\Psi}{H} > 1, \quad M_h(1) = \frac{(h-1) h^{1/h-1}}{e \ln h}. \tag{11.74}$$

Denote

$$\tilde{T}(x) := A_0(x) \left(\frac{\lambda_\beta}{\lambda_{\alpha_1} + \lambda_\beta} \right)^{\lambda_\beta/p} 2^{(p - 2\lambda_{\alpha_1} - 3\lambda_\beta)/p} \left(M_h(1) \right)^{-2(\lambda_{\alpha_1} + \lambda_\beta)/p}. \tag{11.75}$$

Then

$$\int_0^x q(s) \left[\left| D^{\alpha_1} f_1(s) \right|^{\lambda_{\alpha_1}} \left| D^{\alpha_2} f_2(s) \right|^{\lambda_{\alpha_1} + \lambda_\beta} \left| D^\beta f_1(s) \right|^{\lambda_\beta} + \right.$$

$$\left. \left| D^{\alpha_2} f_1(s) \right|^{\lambda_{\alpha_1} + \lambda_\beta} \left| D^{\alpha_1} f_2(s) \right|^{\lambda_{\alpha_1}} \left| D^\beta f_2(s) \right|^{\lambda_\beta} \right] ds \geq$$

$$\tilde{T}(x) \left(\int_0^x p(s) \left(\left| D^\beta f_1(s) \right|^p + \left| D^\beta f_2(s) \right|^p \right) ds \right)^{2(\lambda_{\alpha_1} + \lambda_\beta)/p}. \tag{11.76}$$

Proof. We apply (11.21) and (11.22) for $\lambda_{\alpha_2} = \lambda_{\alpha_1} + \lambda_\beta$, and by adding we obtain

$$\int_0^x q\,(s)\left[|D^{\alpha_1} f_1\,(s)|^{\lambda_{\alpha_1}}\,|D^{\alpha_2} f_2\,(s)|^{\lambda_{\alpha_1}+\lambda_\beta}\,|D^\beta f_1\,(s)|^{\lambda_\beta}\,+\right.$$

$$\left.|D^{\alpha_1} f_2\,(s)|^{\lambda_{\alpha_1}}\,|D^{\alpha_2} f_1\,(s)|^{\lambda_{\alpha_1}+\lambda_\beta}\,|D^\beta f_2\,(s)|^{\lambda_\beta}\right]ds$$

$$\geq A_0\,(x)\left[\left(\int_0^x (z_1\,(s))^{\lambda_{\alpha_1}/\lambda_\beta}\,(z_2\,(s))^{\lambda_{\alpha_1}+\lambda_\beta/\lambda_\beta}\,z_1'\,(s)\,ds\right)^{\lambda_\beta/p}+\right.$$

$$\left.\left(\int_0^x (z_1\,(s))^{(\lambda_{\alpha_1}+\lambda_\beta)/\lambda_\beta}\,(z_2\,(s))^{\lambda_{\alpha_1}/\lambda_\beta}\,z_2'\,(s)\,ds\right)^{\lambda_\beta/p}\right] \qquad (11.77)$$

$$\overset{(11.27)}{\geq} A_0\,(x)\,2^{1-(\lambda_\beta/p)}\times\left[\int_0^x (z_1\,(s))^{\lambda_{\alpha_1}/\lambda_\beta}\,(z_2\,(s))^{\lambda_{\alpha_1}/\lambda_\beta+1}\,z_1'\,(s)+\right.=$$

$$\left.(z_1\,(s))^{(\lambda_{\alpha_1}/\lambda_\beta)+1}\,(z_2\,(s))^{\lambda_{\alpha_1}/\lambda_\beta}\,z_2'\,(s)\,ds\right]^{\lambda_\beta/p}= \qquad (11.78)$$

$$A_0\,(x)\,2^{1-(\lambda_\beta/p)}\left[\int_0^x (z_1\,(s)\,z_2\,(s))^{\lambda_{\alpha_1}/\lambda_\beta}\right.$$

$$\left.\left[z_2\,(s)\,z_1'\,(s)+z_1\,(s)\,z_2'\,(s)\right]ds\right]^{\lambda_\beta/p}= \qquad (11.79)$$

$$A_0\,(x)\,2^{(p-\lambda_\beta)/p}\left[\int_0^x (z_1\,(s)\,z_2\,(s))^{(\lambda_{\alpha_1}/\lambda_\beta)}\,(z_1\,(s)\,z_2\,(s))'\,ds\right]^{\lambda_\beta/p}$$
$$\qquad (11.80)$$

$$=A_0\,(x)\,2^{(p-\lambda_\beta/p)}\,(z_1\,(x)\,z_2\,(x))^{(\lambda_{\alpha_1}+\lambda_\beta/p)}\left(\frac{\lambda_\beta}{\lambda_{\alpha_1}+\lambda_\beta}\right)^{\lambda_\beta/p} \qquad (11.81)$$

(see [377])

$$\geq A_0\,(x)\,2^{(p-\lambda_\beta/p)}\left(\frac{\lambda_\beta}{\lambda_{\alpha_1}+\lambda_\beta}\right)^{\lambda_\beta/p}\left(\frac{z_1\,(x)+z_2\,(x)}{2M_h\,(1)}\right)^{2(\lambda_{\alpha_1}+\lambda_\beta)/p}$$
$$\qquad (11.82)$$

$$=\tilde{T}\,(x)\,(z_1\,(x)+z_2\,(x))^{2(\lambda_{\alpha_1}+\lambda_\beta)/p}= \qquad (11.83)$$

$$\tilde{T}(x)\left[\int_0^x p(s)\left(\left|D^\beta f_1(s)\right|^p+\left|D^\beta f_2(s)\right|^p\right)ds\right]^{2\left(\lambda_{\alpha_1}+\lambda_\beta\right)/p}. \tag{11.84}$$

We have proved (11.76). $\square$

Next are special cases of the above theorems.

Corollary 11.9. (to Theorem 11.4) *Set* $\lambda_{\alpha_2}=0$, $p(t)=q(t)=1$. *Then*

$$\int_0^x\left[\left|D^{\alpha_1}f_1(s)\right|^{\lambda_{\alpha_1}}\left|D^\beta f_1(s)\right|^{\lambda_\beta}+\left|D^{\alpha_1}f_2(s)\right|^{\lambda_{\alpha_1}}\left|D^\beta f_2(s)\right|^{\lambda_\beta}\right]ds\geq$$

$$C_1(x)\left[\int_0^x\left[\left|D^\beta f_1(s)\right|^p+\left|D^\beta f_2(s)\right|^p\right]ds\right]^{\left(\lambda_{\alpha_1}+\lambda_\beta/p\right)},$$

where

$$C_1(x):=\left(A_0(x)|_{\lambda_{\alpha_2}=0}\right)\left(\frac{\lambda_\beta}{\lambda_{\alpha_1}+\lambda_\beta}\right)^{\lambda_\beta/p}\delta_1, \tag{11.85}$$

with

$$\delta_1=2^{1-\left(\left(\lambda_{\alpha_1}+\lambda_\beta\right)/p\right)}. \tag{11.86}$$

We have that

$$\left(A_0(x)|_{\lambda_{\alpha_2}=0}\right)=\left\{\left(\frac{1}{\left(\Gamma(\beta-\alpha_1)\right)^{\lambda_{\alpha_1}}}\right)\left(\frac{p-1}{\beta p-\alpha_1 p-1}\right)^{\left(\left(\lambda_{\alpha_1}p-\lambda_{\alpha_1}\right)/p\right)}\right.$$

$$\tag{11.87}$$

$$\left.\left(\frac{p-\lambda_\beta}{\lambda_{\alpha_1}\beta p-\lambda_{\alpha_1}\alpha_1 p-\lambda_{\alpha_1}+p-\lambda_\beta}\right)^{\left((p-\lambda_\beta)/p\right)}\right\}\times$$

$$x^{\left(\left(\lambda_{\alpha_1}\beta p-\lambda_{\alpha_1}\alpha_1 p-\lambda_{\alpha_1}+p-\lambda_\beta\right)/p\right)}.$$

Proof. By Theorem 11.4. The constant $\left(A_0(x)|_{\lambda_{\alpha_2}=0}\right)$ was calculated in [48]. $\square$

We continue with

Corollary 11.10. (to Theorem 11.5, $\lambda_{\alpha_1}=0$, $p(t)=q(t)=1$, $\lambda_{\alpha_2}\geq\lambda_\beta$). *Then*

$$\int_0^x \left[|D^{\alpha_2} f_2(s)|^{\lambda_{\alpha_2}} |D^{\beta} f_1(s)|^{\lambda_{\beta}} + |D^{\alpha_2} f_1(s)|^{\lambda_{\alpha_2}} |D^{\beta} f_2(s)|^{\lambda_{\beta}} \right] ds \geq \tag{11.88}$$

$$C_2(x) \left(\int_0^x \left[|D^{\beta} f_1(s)|^p + |D^{\beta} f_2(s)|^p \right] ds \right)^{(\lambda_{\beta} + \lambda_{\alpha_2})/p}.$$

Here

$$C_2(x) := \left(A_0(x)|_{\lambda_{\alpha_1}=0} \right) 2^{(p-\lambda_{\beta})/p} \left(\frac{\lambda_{\beta}}{\lambda_{\alpha_2} + \lambda_{\beta}} \right)^{\lambda_{\beta}/p} \delta_3^{\lambda_{\beta}/p}. \tag{11.89}$$

We have that

$$\left(A_0(x)|_{\lambda_{\alpha_1}=0} \right) = \left\{ \left(\frac{1}{(\Gamma(\beta - \alpha_2))^{\lambda_{\alpha_2}}} \right) \left(\frac{p-1}{\beta p - \alpha_2 p - 1} \right)^{((\lambda_{\alpha_2} p - \lambda_{\alpha_2})/p)} \right.$$

$$\left. \left(\frac{p - \lambda_{\beta}}{\lambda_{\alpha_2} \beta p - \lambda_{\alpha_2} \alpha_2 p - \lambda_{\alpha_2} + p - \lambda_{\beta}} \right)^{((p-\lambda_{\beta})/p)} \right\}. \tag{11.90}$$

$$x^{((\lambda_{\alpha_2} \beta p - \lambda_{\alpha_2} \alpha_2 p - \lambda_{\alpha_2} + p - \lambda_{\beta})/p)}.$$

Proof. By Theorem 11.5. The constant $\left(A_0(x)|_{\lambda_{\alpha_1}=0} \right)$ was calculated in [48]. $\square$

11.3.2   Results Involving Several Functions

Here we use the following basic inequalities. Let $\alpha_1, \alpha_2, \ldots, \alpha_n \geq 0$, $n \in \mathbb{N}$, then

$$a_1^r + \ldots + a_n^r \leq (a_1 + \ldots + a_n)^r \leq n^{r-1} \left(\sum_{i=1}^n a_i^r \right), \ r \geq 1, \tag{11.91}$$

and

$$n^{r-1} (a_1^r + \ldots + a_n^r) \leq (a_1 + \ldots + a_n)^r \leq \sum_{i=1}^n a_i^r, \ 0 \leq r \leq 1. \tag{11.92}$$

We present

Theorem 11.11. *Let $\alpha_i \in \mathbb{R}_+$, $0 < \beta - \alpha_i < 1/p$, $0 < p < 1$, $i = 1, 2$, and let $f_j \in L_1(0, x)$, $j = 1, \ldots, M \in \mathbb{N}$, $x \in \mathbb{R}_+ - \{0\}$ have, respectively, L_∞ fractional derivatives $D^\beta f_j$ in $[0, x]$, each of fixed sign a.e. on $[0, x]$, and let $D^{\beta-k} f_j(0) = 0$, for $k = 1, \ldots, [\beta] + 1$; $j = 1, \ldots, M$. Consider also $p(t) > 0$ and $q(t) \geq 0$, with all $p(t), 1/p(t), q(t), 1/q(t) \in L_\infty(0, x)$. Let $\lambda_\beta > 0$ and $\lambda_{\alpha_1}, \lambda_{\alpha_2} \geq 0$, such that $\lambda_\beta > p$.*
Put

$$P_i(s) := \int_0^s (s-t)^{p(\beta-\alpha_i-1)/p-1} (p(t))^{-1/(p-1)} \, dt, \ i = 1, 2; \ 0 \leq s \leq x,$$
(11.93)

$$A(s) := \frac{q(s)(P_1(s))^{\lambda_{\alpha_1}(p-1/p)}(P_2(s))^{\lambda_{\alpha_2}(p-1/p)}(p(s))^{-\lambda_\beta/p}}{(\Gamma(\beta-\alpha_1))^{\lambda_{\alpha_1}}(\Gamma(\beta-\alpha_2))^{\lambda_{\alpha_2}}},$$
(11.94)

$$A_0(x) := \left(\int_0^x (A(s))^{p/(p-\lambda_\beta)} \, ds \right)^{(p-\lambda_\beta)/p},$$
(11.95)

and

$$\delta_1^* := M^{1-((\lambda_{\alpha_1}+\lambda_\beta)/p)}.$$
(11.96)

Call

$$\varphi_1(x) := (A_0(x)|_{\lambda_{\alpha_2}=0}) \left(\frac{\lambda_\beta}{\lambda_{\alpha_1}+\lambda_\beta} \right)^{\lambda_\beta/p}.$$
(11.97)

If $\lambda_{\alpha_2} = 0$, we obtain that

$$\int_0^x q(s) \left(\sum_{j=1}^M |D^{\alpha_1} f_j(s)|^{\lambda_{\alpha_1}} |D^\beta f_j(s)|^{\lambda_\beta} \right) ds \geq$$

$$\delta_1^* \varphi_1(x) \left[\int_0^x p(s) \left(\sum_{j=1}^M |D^\beta f_j(s)|^p \right) ds \right]^{((\lambda_{\alpha_1}+\lambda_\beta)/p)}.$$
(11.98)

Proof. By Theorem 11.4 we get

$$\int_0^x q(s) \left[|D^{\alpha_1} f_j(s)|^{\lambda_{\alpha_1}} |D^\beta f_j(s)|^{\lambda_\beta} + |D^{\alpha_1} f_{j+1}(s)|^{\lambda_{\alpha_1}} |D^\beta f_{j+1}(s)|^{\lambda_\beta} \right] ds \geq$$

$$\left(A_0\left(x\right)|_{\lambda_{\alpha_2}=0}\right)\left(\frac{\lambda_\beta}{\lambda_{\alpha_1}+\lambda_\beta}\right)^{\lambda_\beta/p}\delta_1\left[\int_0^x p\left(s\right)\left[\left|D^\beta f_j\left(s\right)\right|^p+\right.\right.$$

$$\left.\left.\left|D^\beta f_{j+1}\left(s\right)\right|^p\right]ds\right]^{\left(\lambda_{\alpha_1}+\lambda_\beta/p\right)},\ j=1,2,\ldots,M-1. \tag{11.99}$$

Hence by adding all the above we obtain

$$\int_0^x q\left(s\right)\left\{\sum_{j=1}^{M-1}\left[\left|D^{\alpha_1}f_j\left(s\right)\right|^{\lambda_{\alpha_1}}\left|D^\beta f_j\left(s\right)\right|^{\lambda_\beta}+\right.\right.$$

$$\left.\left.\left|D^{\alpha_1}f_{j+1}\left(s\right)\right|^{\lambda_{\alpha_1}}\left|D^\beta f_{j+1}\left(s\right)\right|^{\lambda_\beta}\right]\right\}ds$$

$$\geq\delta_1\varphi_1\left(x\right)\left\{\sum_{j=1}^{M-1}\left[\int_0^x p\left(s\right)\left[\left|D^\beta f_j\left(s\right)\right|^p+\left|D^\beta f_{j+1}\left(s\right)\right|^p\right]ds\right]^{\left(\lambda_{\alpha_1}+\lambda_\beta/p\right)}\right\}. \tag{11.100}$$

Also it holds

$$\int_0^x q\left(s\right)\left[\left|D^{\alpha_1}f_1\left(s\right)\right|^{\lambda_{\alpha_1}}\left|D^\beta f_1\left(s\right)\right|^{\lambda_\beta}+\left|D^{\alpha_1}f_M\left(s\right)\right|^{\lambda_{\alpha_1}}\left|D^\beta f_M\left(s\right)\right|^{\lambda_\beta}\right]ds\geq$$

$$\delta_1\varphi_1\left(x\right)\left[\int_0^x p\left(s\right)\left[\left|D^\beta f_1\left(s\right)\right|^p+\left|D^\beta f_M\left(s\right)\right|^p\right]ds\right]^{\left(\left(\lambda_{\alpha_1}+\lambda_\beta\right)/p\right)} \tag{11.101}$$

Adding (11.100) and (11.101), and using (11.91) we have

$$2\int_0^x q\left(s\right)\left(\sum_{j=1}^M\left|D^{\alpha_1}f_j\left(s\right)\right|^{\lambda_{\alpha_1}}\left|D^\beta f_j\left(s\right)\right|^{\lambda_\beta}\right)ds\geq$$

$$\delta_1\varphi_1\left(x\right)\left\{\sum_{j=1}^{M-1}\left[\int_0^x p\left(s\right)\left[\left|D^\beta f_j\left(s\right)\right|^p+\left|D^\beta f_{j+1}\left(s\right)\right|^p\right]ds\right]^{\left(\left(\lambda_{\alpha_1}+\lambda_\beta\right)/p\right)}+\right.$$

$$\left.\left[\int_0^x p\left(s\right)\left[\left|D^\beta f_1\left(s\right)\right|^p+\left|D^\beta f_M\left(s\right)\right|^p\right]ds\right]^{\left(\lambda_{\alpha_1}+\lambda_\beta/p\right)}\right\}\geq \tag{11.102}$$

$$M^{1-\left(\lambda_{\alpha_1}+\lambda_\beta/p\right)}\delta_1\varphi_1\left(x\right)\left\{\int_0^x p\left(s\right)\left(2\sum_{j=1}^M\left|D^\beta f_j\left(s\right)\right|^p\right)ds\right\}^{\left(\left(\lambda_{\alpha_1}+\lambda_\beta\right)/p\right)}. \tag{11.103}$$

We have established

$$\int_0^x q\left(s\right) \left(\sum_{j=1}^M |D^{\alpha_1} f_j\left(s\right)|^{\lambda_{\alpha_1}} |D^\beta f_j\left(s\right)|^{\lambda_\beta}\right) ds \geq$$

$$M^{1-\left(\lambda_{\alpha_1}+\lambda_\beta/p\right)} \delta_1 \left(2^{\left(\lambda_{\alpha_1}+\lambda_\beta/p\right)-1}\right)$$

$$\varphi_1\left(x\right) \left\{\int_0^x p\left(s\right) \left[\sum_{j=1}^M |D^\beta f_j\left(s\right)|^p\right] ds\right\}^{\left(\lambda_{\alpha_1}+\lambda_\beta/p\right)}, \tag{11.104}$$

thus proving (11.98). □

Next we give

Theorem 11.12. *All here are as in Theorem 11.11. Further assume* $\lambda_{\alpha_2} \geq \lambda_\beta$.
Denote

$$\varphi_2\left(x\right) := \left(A_0\left(x\right)|_{\lambda_{\alpha_1}=0}\right) 2^{\left(p-\lambda_\beta\right)/p} \left(\frac{\lambda_\beta}{\lambda_{\alpha_2}+\lambda_\beta}\right)^{\lambda_\beta/p} \delta_3^{\lambda_\beta/p}. \tag{11.105}$$

If $\lambda_{\alpha_1} = 0$, *then*

$$\int_0^x q\left(s\right) \left\{\sum_{j=1}^{M-1} \left[|D^{\alpha_2} f_{j+1}\left(s\right)|^{\lambda_{\alpha_2}} |D^\beta f_j\left(s\right)|^{\lambda_\beta} + \right.\right.$$

$$\left.|D^{\alpha_2} f_j\left(s\right)|^{\lambda_{\alpha_2}} |D^\beta f_{j+1}\left(s\right)|^{\lambda_\beta}\right]\right\} +$$

$$\left[|D^{\alpha_2} f_M\left(s\right)|^{\lambda_{\alpha_2}} |D^\beta f_1\left(s\right)|^{\lambda_\beta} + |D^{\alpha_2} f_1\left(s\right)|^{\lambda_{\alpha_2}} |D^\beta f_M\left(s\right)|^{\lambda_\beta}\right]\right\} ds \geq \tag{11.106}$$

$$M^{1-\left(\lambda_\beta+\lambda_{\alpha_2}/p\right)} 2^{\left(\lambda_\beta+\lambda_{\alpha_2}/p\right)} \varphi_2\left(x\right) \left\{\int_0^x p\left(s\right) \left[\sum_{j=1}^M |D^\beta f_j\left(s\right)|^p\right] ds\right\}^{\left(\left(\lambda_\beta+\lambda_{\alpha_2}\right)/p\right)}.$$

Proof. From Theorem 11.5 we have

$$\int_0^x q\left(s\right) \left[|D^{\alpha_2} f_{j+1}\left(s\right)|^{\lambda_{\alpha_2}} |D^\beta f_j\left(s\right)|^{\lambda_\beta} + |D^{\alpha_2} f_j\left(s\right)|^{\lambda_{\alpha_2}} |D^\beta f_{j+1}\left(s\right)|^{\lambda_\beta}\right] ds \geq$$

$$\varphi_2\left(x\right)\left(\int_0^x p\left(s\right)\left[\left|D^\beta f_j\left(s\right)\right|^p+\left|D^\beta f_{j+1}\left(s\right)\right|^p\right]ds\right)^{\left(\lambda_\beta+\lambda_{\alpha_2}/p\right)}, \quad (11.107)$$

for $j=1,\ldots,M-1$. Hence by adding all of the above we derive

$$\int_0^x q\left(s\right)\left(\sum_{j=1}^{M-1}\left[\left|D^{\alpha_2}f_{j+1}\left(s\right)\right|^{\lambda_{\alpha_2}}\left|D^\beta f_j\left(s\right)\right|^{\lambda_\beta}+\right.\right.$$

$$\left.\left.\left|D^{\alpha_2}f_j\left(s\right)\right|^{\lambda_{\alpha_2}}\left|D^\beta f_{j+1}\left(s\right)\right|^{\lambda_\beta}\right]\right)ds\geq$$

$$\varphi_2\left(x\right)\left(\sum_{j=1}^{M-1}\left(\int_0^x p\left(s\right)\left[\left|D^\beta f_j\left(s\right)\right|^p+\left|D^\beta f_{j+1}\left(s\right)\right|^p\right]ds\right)^{\left(\lambda_\beta+\lambda_{\alpha_2}/p\right)}\right).$$

$$(11.108)$$

Similarly it holds

$$\int_0^x q\left(s\right)\left[\left|D^{\alpha_2}f_M\left(s\right)\right|^{\lambda_{\alpha_2}}\left|D^\beta f_1\left(s\right)\right|^{\lambda_\beta}+\left|D^{\alpha_2}f_1\left(s\right)\right|^{\lambda_{\alpha_2}}\left|D^\beta f_M\left(s\right)\right|^{\lambda_\beta}\right]ds\geq$$

$$\varphi_2\left(x\right)\left(\int_0^x p\left(s\right)\left[\left|D^\beta f_1\left(s\right)\right|^p+\left|D^\beta f_M\left(s\right)\right|^p\right]ds\right)^{\left(\lambda_\beta+\lambda_{\alpha_2}/p\right)}. \quad (11.109)$$

Adding (11.108) and (11.109) and using (11.91) we derive (11.106). $\quad\square$

We continue with

Theorem 11.13. *Let* $1<\beta-\alpha_1<1/p$, $0<p<1,\alpha_1\in\mathbb{R}_+$, *and let* $f_j\in L_1\left(0,x\right)$, $j=1,\ldots,M\in\mathbb{N}$, $x\in\mathbb{R}_+-\{0\}$ *have, respectively,* L_∞ *fractional derivatives* $D^\beta f_j$ *in* $[0,x]$, *each of fixed sign a.e. on* $[0,x]$, *and let* $D^{\beta-k}f_j\left(0\right)=0$, *for* $k=1,\ldots,[\beta]+1$; $j=1,\ldots,M$. *Consider also* $p\left(t\right)>0$ *and* $q\left(t\right)\geq0$, *with* $p\left(t\right),1/p\left(t\right),q\left(t\right),1/q\left(t\right)\in L_\infty\left(0,x\right)$. *Let* $\lambda_\alpha\geq\lambda_{\alpha+1}>1$; Φ *is as in Theorem 11.7.*
Then

$$\int_0^x q\left(s\right)\left\{\left\{\sum_{j=1}^{M-1}\left[\left|D^{\alpha_1}f_j\left(s\right)\right|^{\lambda_\alpha}\left|D^{\alpha_1+1}f_{j+1}\left(s\right)\right|^{\lambda_{\alpha+1}}+\right.\right.\right.$$

$$\left.\left.\left|D^{\alpha_1}f_{j+1}\left(s\right)\right|^{\lambda_\alpha}\left|D^{\alpha_1+1}f_j\left(s\right)\right|^{\lambda_{\alpha+1}}\right]\right\}$$

$$+\left[\left|D^{\alpha_1}f_1\left(s\right)\right|^{\lambda_\alpha}\left|D^{\alpha_1+1}f_M\left(s\right)\right|^{\lambda_{\alpha+1}}+\left|D^{\alpha_1}f_M\left(s\right)\right|^{\lambda_\alpha}\left|D^{\alpha_1+1}f_1\left(s\right)\right|^{\lambda_{\alpha+1}}\right]\right\}ds\geq$$

$$M^{1-(\lambda_\alpha+\lambda_{\alpha+1}/p)}2^{(\lambda_\alpha+\lambda_{\alpha+1}/p)}\Phi(x)\left[\int_0^x p(s)\left(\sum_{j=1}^M \left|D^\beta f_j(s)\right|^p\right)ds\right]^{((\lambda_\alpha+\lambda_{\alpha+1})/p)}.$$

(11.110)

Proof. From Theorem 11.7 we have

$$\int_0^x q(s)\sum_{j=1}^{M-1}\left[\left|D^{\alpha_1}f_j(s)\right|^{\lambda_\alpha}\left|D^{\alpha_1+1}f_{j+1}(s)\right|^{\lambda_{\alpha+1}}+\right.$$

$$\left.\left|D^{\alpha_1}f_{j+1}(s)\right|^{\lambda_\alpha}\left|D^{\alpha_1+1}f_j(s)\right|^{\lambda_{\alpha+1}}\right]ds$$

$$\geq \Phi(x)\sum_{j=1}^{M-1}\left[\int_0^x p(s)\left(\left|D^\beta f_j(s)\right|^p+\left|D^\beta f_{j+1}(s)\right|^p\right)ds\right]^{(\lambda_\alpha+\lambda_{\alpha+1}/p)}.$$

(11.111)

Similarly it holds

$$\int_0^x q(s)\left[\left|D^{\alpha_1}f_1(s)\right|^{\lambda_\alpha}\left|D^{\alpha_1+1}f_M(s)\right|^{\lambda_{\alpha+1}}+\right.$$

$$\left.\left|D^{\alpha_1}f_M(s)\right|^{\lambda_\alpha}\left|D^{\alpha_1+1}f_1(s)\right|^{\lambda_{\alpha+1}}\right]ds$$

$$\geq \Phi(x)\left[\int_0^x p(s)\left(\left|D^\beta f_1(s)\right|^p+\left|D^\beta f_M(s)\right|^p\right)ds\right]^{(\lambda_\alpha+\lambda_{\alpha+1}/p)}.$$

(11.112)

Adding (11.111) and (11.112), along with (11.91) we obtain (11.110). $\square$

We present

Theorem 11.14. *All are as in Theorem 11.11. Consider the special case of $\lambda_{\alpha_2}=\lambda_{\alpha_1}+\lambda_\beta$. Here $\tilde{T}(x)$ is as in (11.75).*
Assume here for $j=1,\ldots,M$ that

$$z_j(x):=\int_0^x p(t)\left|D^\beta f_j(t)\right|^p dt\in[H,\Psi],\ 0<H<\Psi.$$

Then

$$\int_0^x q(s)\left\{\left\{\sum_{j=1}^{M-1}\left[\left|D^{\alpha_1}f_j(s)\right|^{\lambda_{\alpha_1}}\left|D^{\alpha_2}f_{j+1}(s)\right|^{\lambda_{\alpha_1}+\lambda_\beta}\left|D^\beta f_j(s)\right|^{\lambda_\beta}\right.\right.\right.$$

$$+ |D^{\alpha_2} f_j (s)|^{\lambda_{\alpha_1} + \lambda_\beta} |D^{\alpha_1} f_{j+1} (s)|^{\lambda_{\alpha_1}} |D^\beta f_{j+1} (s)|^{\lambda_\beta} \Big] \Big\}$$

$$+ \Big[|D^{\alpha_1} f_1 (s)|^{\lambda_{\alpha_1}} |D^{\alpha_2} f_M (s)|^{\lambda_{\alpha_1} + \lambda_\beta} |D^\beta f_1 (s)|^{\lambda_\beta}$$

$$+ |D^{\alpha_2} f_1 (s)|^{\lambda_{\alpha_1} + \lambda_\beta} |D^{\alpha_1} f_M (s)|^{\lambda_{\alpha_1}} |D^\beta f_M (s)|^{\lambda_\beta} \Big] \Big\} ds \geq \qquad (11.113)$$

$$M^{(1 - (2(\lambda_{\alpha_1} + \lambda_\beta)/p))} 2^{2(\lambda_{\alpha_1} + \lambda_\beta)/p} \tilde{T}(x) \left[\int_0^x p(s) \left(\sum_{j=1}^M |D^\beta f_j (s)|^p \right) ds \right]^{2(\lambda_{\alpha_1} + \lambda_\beta)/p} .$$

Proof. Based on Theorem 11.8. The rest of the proof is as in the proof of Theorem 11.13. □

We continue with

Corollary 11.15. (to Theorem 11.11, $\lambda_{\alpha_2} = 0$, $p(t) = q(t) = 1$). *Then*

$$\int_0^x \left(\sum_{j=1}^M |D^{\alpha_1} f_j (s)|^{\lambda_{\alpha_1}} |D^\beta f_j (s)|^{\lambda_\beta} \right) ds \geq$$

$$\delta_1^* \varphi_1 (x) \left[\int_0^x \left[\sum_{j=1}^M |D^\beta f_j (s)|^p \right] ds \right]^{((\lambda_{\alpha_1} + \lambda_\beta)/p)} . \qquad (11.114)$$

Here $\left(A_0 (x) |_{\lambda_{\alpha_2} = 0} \right)$ *within* $\varphi_1 (x)$ *is given by* (11.87). □

Proof. Based on Theorem 11.11. □

We also give

Corollary 11.16. (to Theorem 11.12, $\lambda_{\alpha_1} = 0$, $p(t) = q(t) = 1$). *It holds*

$$\int_0^x \left\{ \left\{ \sum_{j=1}^{M-1} \Big[|D^{\alpha_2} f_{j+1} (s)|^{\lambda_{\alpha_2}} |D^\beta f_j (s)|^{\lambda_\beta} + \right. \right.$$

$$|D^{\alpha_2} f_j (s)|^{\lambda_{\alpha_2}} |D^\beta f_{j+1} (s)|^{\lambda_\beta} \Big] \Big\} +$$

$$\Big[|D^{\alpha_2} f_M (s)|^{\lambda_{\alpha_2}} |D^\beta f_1 (s)|^{\lambda_\beta} + |D^{\alpha_2} f_1 (s)|^{\lambda_{\alpha_2}} |D^\beta f_M (s)|^{\lambda_\beta} \Big] \Big\} ds \geq$$

$$\left(M^{1-(\lambda_\beta+\lambda_{\alpha_2}/p)}\right) 2^{(\lambda_\beta+\lambda_{\alpha_2}/p)} \varphi_2\left(x\right) \left\{\int_0^x \left[\sum_{j=1}^M \left|D^\beta f_j\left(s\right)\right|^p\right] ds\right\}^{((\lambda_\beta+\lambda_{\alpha_2})/p)}.$$

$$(11.115)$$

In (11.115), within $\varphi_2\left(x\right)$, $\left(A_0\left(x\right)|_{\lambda_{\alpha_1}=0}\right)$ is given by (11.90).

Proof. Based on Theorem 11.12. $\square$

11.3.3 Results with Respect to Generalized Riemann – Liouville Fractional Derivative

We give the notion of a generalized Riemann – Liouville fractional derivative at arbitrary anchor point $a \in \mathbb{R}$; see [64].

Definition 11.17. Let $v \geq 0$; define

$$\left(D_a^v f\right)\left(s\right) := \left(D^v f_a\right)\left(s - a\right), \quad s \geq a, \qquad (11.116)$$

where $f_a\left(t\right) := f\left(t+a\right)$ is the translate function, for $v = 0$ both sides equal to $f\left(s\right)$, and for $v = n \in \mathbb{N}$ we get $\left(D_a^n f\right)\left(s\right) = f^{(n)}\left(s\right)$, the ordinary derivative.

Clearly here

$$\left(D_a^v f\right)\left(z + a\right) = \left(D^v f_a\right)\left(z\right). \qquad (11.117)$$

For $p\left(s\right)$, $D_a^v f\left(s\right) \in L_\infty\left(a, x\right)$, $x > a$, $a, x \in \mathbb{R}$ we get

$$\int_a^w p\left(y\right)\left(D_a^v f\right)\left(y\right) dy = \int_a^{w-a} p\left(z+a\right)\left(D^v f_a\right)\left(z\right) dz, \qquad (11.118)$$

all $a \leq w \leq x$.

So here we transfer the results of Sections 11.3.1 and 11.3.2, using the above concepts, to arbitrary interval $[a, x]$. Thus earlier results are applied to f_a over $[0, x-a]$, for f over $[a, x]$ and use the generalized Riemann – Liouville fractional derivative. This method was applied extensively in [45]; see there for all details concerning this transfer.

We need

Definition 11.18. [45] We say that $f \in L_1\left(a, w\right)$, $a < w$; $a, w \in \mathbb{R}$ has an L_∞ fractional derivative $D_a^\beta f$ $[\beta > 0]$ in $[a, w]$, iff

(1) $D_a^{\beta-k} f \in C\left([a, w]\right)$, $k = 1, \ldots, m := [\beta] + 1$

$$(2) \ D_a^{\beta-1} f \in AC\left([a,w]\right),$$

and

$$(3) \ D_a^\beta f \in L_\infty\left(a,w\right).$$

We need

Lemma 11.19. (see [45]) *Let $\beta > \alpha \geq 0$ and let $f \in L_1\left(a,w\right)$ have an L_∞ fractional derivative $D_a^\beta f$ in $[a,w]$, and let $\left(D_a^{\beta-k}f\right)(a) = 0$, $k = 1,\ldots,[\beta]+1$.*
Then

$$D_a^\alpha f\left(s\right) = \frac{1}{\Gamma\left(\beta-\alpha\right)} \int_a^s \left(s-t\right)^{\beta-\alpha-1} D_a^\beta f\left(t\right) dt, \qquad (11.119)$$

all $a \leq s \leq w$.
Clearly here $D_a^\alpha f \in AC\left([a,w]\right)$ for $\beta - \alpha \geq 1$, and in $C\left([a,w]\right)$ for $\beta - \alpha \in (0,1)$, hence $D_a^\alpha f \in L_\infty\left(a,w\right)$, and $D_a^\alpha f \in L_1\left(a,w\right)$.
Notice that for $\beta > 0$ that

$$\left(D_a^{\beta-k}f\right)(a) = D^{\beta-k}f_a\left(0\right), \text{ all } k = 1,\ldots,[\beta]+1. \qquad (11.120)$$

Here we only show two such transfers; for the rest of the results so far we can get similar corresponding transfers, acting as in [45].

We give

Theorem 11.20. *Let $\alpha_i \geq 0$, $0 < \beta - \alpha_i < 1/p$, $0 < p < 1$, $i = 1,2$, and let*

$$f_1, f_2 \in L_1\left(a,x\right), \ a,x \in \mathbb{R}, a < x,$$

have, respectively, L_∞ fractional derivatives $D_a^\beta f_1, D_a^\beta f_2$ in $[a,x]$, each of fixed sign a.e. on $[a,x]$, and let $D_a^{\beta-k}f_i\left(a\right) = 0$, for $k = 1,\ldots,[\beta]+1$; $i = 1,2$.
Consider also

$$p\left(t\right) > 0 \text{ and } q\left(t\right) \geq 0,$$

with all

$$p\left(t\right), \frac{1}{p\left(t\right)}, \ q\left(t\right), \ \frac{1}{q\left(t\right)} \in L_\infty\left(a,x\right).$$

Let

$$\lambda_\beta > 0 \text{ and } \lambda_{\alpha_1}, \lambda_{\alpha_2} \geq 0,$$

such that

$$\lambda_\beta > p.$$

Put

$$P_i(s) := \int_0^s (s-t)^{p(\beta-\alpha_i-1)/p-1} (p(t+a))^{-1/(p-1)} dt, \; i = 1,2; \; 0 \le s \le x-a,$$

$$(11.121)$$

$$A(s) := \frac{q(s+a)(P_1(s))^{\lambda_{\alpha_1}(p-1/p)}(P_2(s))^{\lambda_{\alpha_2}(p-1/p)}(p(s+a))^{-\lambda_\beta/p}}{(\Gamma(\beta-\alpha_1))^{\lambda_{\alpha_1}}(\Gamma(\beta-\alpha_2))^{\lambda_{\alpha_2}}},$$

$$(11.122)$$

$$A_0(x-a) := \left(\int_0^{x-a} (A(s))^{p/(p-\lambda_\beta)} ds \right)^{(p-\lambda_\beta)/p}, \qquad (11.123)$$

and

$$\delta_1 := 2^{1-(\lambda_{\alpha_1}+\lambda_\beta/p)}. \qquad (11.124)$$

If $\lambda_{\alpha_2} = 0$, then

$$\int_a^x q(s) \left[|D_a^{\alpha_1} f_1(s)|^{\lambda_{\alpha_1}} |D_a^\beta f_1(s)|^{\lambda_\beta} + |D_a^{\alpha_1} f_2(s)|^{\lambda_{\alpha_1}} |D_a^\beta f_2(s)|^{\lambda_\beta} \right] ds \ge$$

$$\left(A_0(x-a)|_{\lambda_{\alpha_2}=0} \right) \left(\frac{\lambda_\beta}{\lambda_{\alpha_1}+\lambda_\beta} \right)^{\lambda_\beta/p} \delta_1 \left[\int_a^x p(s) \left[|D_a^\beta f_1(s)|^p + \right. \right.$$

$$\left. \left. |D_a^\beta f_2(s)|^p \right] ds \right]^{((\lambda_{\alpha_1}+\lambda_\beta)/p)}. \qquad (11.125)$$

Proof. We apply (11.8) for

$$f_{1a} := f_1(\cdot + a), \; f_{2a} := f_2(\cdot + a), \; p(\cdot + a), \; q(\cdot + a)$$

and the fractional derivatives

$$D^\beta f_{1a}, \; D^\beta f_{2a}, \; D^{\alpha_i} f_{1a}, \; D^{\alpha_i} f_{2a}, \; i = 1,2,$$

all on $[0, x-a]$.

Namely here we have that the functions $f_{1a}, f_{2a}, p(\cdot + a), q(\cdot + a)$ fulfill all the assumptions of Theorem 11.4 on the interval $[0, x-a]$, so for $\lambda_{\alpha_2} = 0$ we get by (11.8) that

$$\int_0^{x-a} q(s+a) \left[|D^{\alpha_1} f_{1a}(s)|^{\lambda_{\alpha_1}} |D^\beta f_{1a}(s)|^{\lambda_\beta} + |D^{\alpha_1} f_{2a}(s)|^{\lambda_{\alpha_1}} |D^\beta f_{2a}(s)|^{\lambda_\beta} \right] ds$$

$$\geq \left(A_0 \left(x - a \right) |_{\lambda_{\alpha_2} = 0} \right) \left(\frac{\lambda_\beta}{\lambda_{\alpha_1} + \lambda_\beta} \right)^{\lambda_\beta / p} \delta_1 \left[\int_0^{x-a} p\left(s+a\right) \left[\left| D^\beta f_{1a}\left(s\right) \right|^p + \right. \right.$$

$$\left. \left. \left| D^\beta f_{2a}\left(s\right) \right|^p \right] ds \right]^{\left(\lambda_{\alpha_1} + \lambda_\beta / p \right)}. \tag{11.126}$$

Using (11.117), and from (11.126) we find

$$\int_0^{x-a} q\left(s+a\right) \left[\left| D_a^{\alpha_1} f_1\left(s+a\right) \right|^{\lambda_{\alpha_1}} \left| D_a^\beta f_1\left(s+a\right) \right|^{\lambda_\beta} + \right.$$

$$\left. \left| D_a^{\alpha_1} f_2\left(s+a\right) \right|^{\lambda_{\alpha_1}} \left| D_a^\beta f_2\left(s+a\right) \right|^{\lambda_\beta} \right] ds \geq$$

$$\left(A_0 \left(x - a \right) |_{\lambda_{\alpha_2} = 0} \right) \left(\frac{\lambda_\beta}{\lambda_{\alpha_1} + \lambda_\beta} \right)^{\lambda_\beta / p} \delta_1 \left[\int_0^{x-a} p\left(s+a\right) \left[\left| D_a^\beta f_1\left(s+a\right) \right|^p + \right. \right.$$

$$\left. \left. \left| D_a^\beta f_2\left(s+a\right) \right|^p \right] ds \right]^{\left(\lambda_{\alpha_1} + \lambda_\beta / p \right)}. \tag{11.127}$$

Let $F_a\left(s\right) := F\left(s+a\right)$, $G_a\left(s\right) := G\left(s+a\right)$ denote the integrands of left and right integrals of (11.127). Notice easily that $F_a, G_a \in L_\infty \left(0, x-a\right)$, that is, $F, G \in L_\infty \left(a, x\right)$, hence $F, G \in L_1 \left(a, x\right)$.

Thus

$$\int_0^{x-a} F_a\left(s\right) ds = \int_a^x F\left(s\right) ds, \quad \int_0^{x-a} G_a\left(s\right) ds = \int_a^x G\left(s\right) ds,$$

proving (11.125). Notice for $a = 0$, inequality (11.125) collapses to inequality (11.8). $\square$

We finish this chapter with

Theorem 11.21. *Let $\alpha_i \geq 0$, $0 < \beta - \alpha_i < 1/p$, $0 < p < 1$, $i = 1, 2$, and let*

$$f_j \in L_1 \left(a, x\right), \quad j = 1, \ldots, M \in \mathbb{N}, a, x \in \mathbb{R}, \ a < x,$$

have, respectively, L_∞ fractional derivatives $D_a^\beta f_j$ in $[a, x]$, each of fixed sign a.e. on $[a, x]$, and let

$$D_a^{\beta - k} f_i\left(a\right) = 0, \ \text{for } k = 1, \ldots, [\beta] + 1; \ j = 1, \ldots, M.$$

Consider also

$$p\left(t\right) > 0 \ \text{and} \ q\left(t\right) \geq 0,$$

with all

$$p(t), \frac{1}{p(t)}, \ q(t), \ \frac{1}{q(t)} \in L_\infty(a, x).$$

Let

$$\lambda_\beta > 0 \text{ and } \lambda_{\alpha_1}, \lambda_{\alpha_2} \geq 0,$$

such that

$$\lambda_\beta > p.$$

Call

$$P_i(s) := \int_0^s (s - t)^{p(\beta - \alpha_i - 1)/p - 1} \, (p(t + a))^{-1/(p-1)} \, dt, \ i = 1, 2; \ 0 \leq s \leq x - a,$$

(11.128)

$$A(s) := \frac{q(s + a)(P_1(s))^{\lambda_{\alpha_1}(p-1/p)}(P_2(s))^{\lambda_{\alpha_2}((p-1)/p)}(p(s + a))^{-\lambda_\beta/p}}{(\Gamma(\beta - \alpha_1))^{\lambda_{\alpha_1}}(\Gamma(\beta - \alpha_2))^{\lambda_{\alpha_2}}},$$

(11.129)

$$A_0(x - a) := \left(\int_0^{x-a} (A(s))^{p/(p-\lambda_\beta)} \, ds \right)^{(p-\lambda_\beta)/p},$$

(11.130)

and

$$\delta_1^* := M^{\left(1 - \left(\lambda_{\alpha_1} + \lambda_\beta/p\right)\right)}.$$

(11.131)

Call

$$\varphi_1(x - a) := \left(A_0(x - a)|_{\lambda_{\alpha_2}=0} \right) \left(\frac{\lambda_\beta}{\lambda_{\alpha_1} + \lambda_\beta} \right)^{\lambda_\beta/p}.$$

(11.132)

If $\lambda_{\alpha_2} = 0$, then

$$\int_a^x q(s) \left(\sum_{j=1}^M |D_a^{\alpha_1} f_j(s)|^{\lambda_{\alpha_1}} |D_a^\beta f_j(s)|^{\lambda_\beta} \right) ds \geq$$

$$\delta_1^* \varphi_1(x - a) \left[\int_a^x p(s) \left(\sum_{j=1}^M |D_a^\beta f_j(s)|^p \right) ds \right]^{((\lambda_{\alpha_1} + \lambda_\beta)/p)}.$$

(11.133)

Proof. Based on Theorem 11.11 (see (11.98)) and similar to Theorem 11.20. □

12

Multivariate Canavati Fractional Taylor Formula

We present here is a multivariate fractional Taylor formula using the Canavati definition of fractional derivative. As related results we present that the order of fractional-ordinary partial differentiation is immaterial, we discuss fractional integration by parts, and we estimate the remainder of our multivariate fractional Taylor formula. This treatment is based on [40].

12.1 Introduction

The main motivation here comes from Canavati [101], Anastassiou [17], and Anastassiou [19], where there is presented a Taylor univariate fractional formula by using an appropriate definition of fractional derivative introduced first in Canavati [101].

So we extend this formula to the multivariate fractional case over a compact and convex subset of $\mathbb{R}^k$, $k \geq 2$, for all fractional orders $\nu > 0$.

We give an estimate to the remainder of our multivariate fractional Taylor formula. We give under mild and natural assumptions that the order of fractional-ordinary partial differentiation is immaterial. Also we present some fractional integration by parts results. The main overall ingredient here is the *Riemann–Liouville integral*.

G.A. Anastassiou, *Fractional Differentiation Inequalities*,   257
DOI 10.1007/978-0-387-98128-4_12, © Springer Science+Business Media, LLC 2009

12.2   Results

We make

Remark 12.1. We follow Anastassiou [19, p. 540] (see also Canavati [101] and Anastassiou [17]). Let $[a, b] \subseteq \mathbb{R}$. Let $x, x_0 \in [a, b]$ such that $x \geq x_0$, x_0 is fixed. Let $f \in C([a, b])$ and define

$$(J_\nu^{x_0} f)(x) = \frac{1}{\Gamma(\nu)} \int_{x_0}^{x} (x - t)^{\nu-1} f(t) dt, \ x_0 \leq x \leq b, \tag{12.1}$$

$\nu > 0$, the generalized *Riemann–Liouville integral*. We consider the subspace $C_{x_0}^\nu([a, b])$ of $C^n([a, b]), n := [\nu], \alpha := \nu - n (0 < \alpha < 1)$:

$$C_{x_0}^\nu([a, b]) := \{f \in C^n([a, b]) : J_{1-\alpha}^{x_0} f^{(n)} \in C^1([x_0, b])\}. \tag{12.2}$$

Hence, let $f \in C_{x_0}^\nu([a, b])$, we define the *generalized ν - fractional derivative of f over* $[x_0, b]$ (see also Canavati [101] and Anastassiou [17]) as

$$D_{x_0}^\nu f := (J_{1-\alpha}^{x_0} f^{(n)})'. \tag{12.3}$$

Notice that

$$\left(J_{1-\alpha}^{x_0} f^{(n)}\right)(x) = \frac{1}{\Gamma(1-\alpha)} \int_{x_0}^{x} (x - t)^{-\alpha} f^{(n)}(t) dt \tag{12.4}$$

exists for $f \in C_{x_0}^\nu([a, b])$.

Let $f_{x_0}(t) := f(x_0 + t), \ \ 0 \leq t \leq b - x_0, \ \ x \geq x_0$. By change of variable we obtain

$$(D_0^\nu f_{x_0})(x - x_0) = (D_{x_0}^\nu f)(x). \tag{12.5}$$

When $\nu \in \mathbb{N}$ then the fractional derivative collapses to the usual one.

We mention the fractional Taylor formula. See Anastassiou [19, p. 540], Canavati [101], and Anastassiou [17].

Theorem 12.2. *Let* $f \in C_{x_0}^\nu([a, b]), x_0 \in [a, b]$ *fixed.*
(i) If $\nu \geq 1$, *then it holds*

$$f(x) = f(x_0) + f'(x_0)(x - x_0) + f''(x_0)\frac{(x - x_0)^2}{2} + \cdots + f^{(n-1)}(x_0)\frac{(x - x_0)^{n-1}}{(n-1)!}$$

$$+ (J_\nu^{x_0} D_{x_0}^\nu f)(x), \quad all \quad x \in [a, b] : x \geq x_0. \tag{12.6}$$

(ii) If $0 < \nu < 1$ *we have*

$$f(x) = (J_\nu^{x_0} D_{x_0}^\nu f)(x), \quad all \quad x \in [a, b] : x \geq x_0. \tag{12.7}$$

We transfer Theorem 12.2 to the multivariate case.

We make

Remark 12.3. Let Q be a compact and convex subset of $\mathbb{R}^k$, $k \geq 2$; $z := (z_1, \ldots, z_k)$, $x_0 := (x_{01}, \ldots, x_{0k}) \in Q$. Let $f \in C^n(Q)$, $n \in \mathbb{N}$.
Call

$$g_z(t) := f(x_0 + t(z - x_0)), 0 \leq t \leq 1; g_z(0) = f(x_0), g_z(1) = f(z). \quad (12.8)$$

Then

$$g_z^{(j)}(t) = \left[\left(\sum_{i=1}^{k}(z_i - x_{0i})\frac{\partial}{\partial x_i}\right)^j f\right](x_0 + t(z - x_0)), \quad (12.9)$$

$j = 0, 1, 2, \ldots, n$,
and

$$g_z^{(n)}(0) = \left[\left(\sum_{i=1}^{k}(z_i - x_{0i})\frac{\partial}{\partial x_i}\right)^n f\right](x_0). \quad (12.10)$$

If all $f_\alpha(x_0) := \partial^\alpha f/\partial x^\alpha(x_0) = 0, \alpha := (\alpha_1, \ldots, \alpha_k)$, $\alpha_i \in \mathbb{Z}^+$, $i = 1, \ldots, k$; $|\alpha| := \sum_{i=1}^{k} \alpha_i =: l$, then $g_z^{(l)}(0) = 0$, where $l \in \{0, 1, \ldots, n\}$. We quote that

$$g_z'(t) = \sum_{i=1}^{k}(z_i - x_{0i})\frac{\partial f}{\partial x_i}(x_0 + t(z - x_0)). \quad (12.11)$$

First let $1 \leq \nu < 2$; then here $n := [\nu] - 1$ and $\alpha = \nu - 1$; $1 - \alpha = 2 - \nu$. Because $0 \leq \nu - 1 < 1$, then $n^* := [\nu - 1] = 0$, $\alpha^* = \nu - 1 - n^* = \nu - 1$, and $1 - \alpha^* = 2 - \nu$.
Set

$$(J_\nu g_z)(x) := \frac{1}{\Gamma(\nu)}\int_0^x (x - t)^{\nu-1} g_z(t)dt, \quad (12.12)$$

$0 \leq x \leq 1$.
Consider

$$C^\nu([0,1]) := \{g \in C^1([0,1]) : J_{2-\nu}g' \in C^1([0,1])\} \subseteq C^1([0,1]), 1 \leq \nu < 2. \quad (12.13)$$

Assume that as a function of t : $f_{x_i}(x_0 + t(z - x_0)) \in C^{\nu-1}([0,1])$, $i = 1, \ldots, k$, then there exists the fractional derivative $g_z^{(\nu)}$, $g_z^{(\nu)} = (J_{2-\nu}g_z')'$. The last comes by using (12.11) to have

$$(J_{2-\nu}g_z')(x) = \frac{1}{\Gamma(2-\nu)}\int_0^x (x - t)^{1-\nu} g_z'(t)dt \quad (12.14)$$

$$= \frac{\sum_{i=1}^{k}(z_i - x_{0i})}{\Gamma(2-\nu)} \int_0^x (x-t)^{1-\nu} \frac{\partial f}{\partial x_i}(x_0 + t(z-x_0))dt, \qquad (12.15)$$

$0 \leq x \leq 1$.

Hence it holds

$$(J_{2-\nu}g_z'(x))' =$$

$$\sum_{i=1}^{k}(z_i - x_{0i}) \left(\frac{1}{\Gamma(2-\nu)} \int_0^x (x-t)^{1-\nu} \frac{\partial f}{\partial x_i}(x_0 + t(z-x_0))dt \right)' \qquad (12.16)$$

$$= \sum_{i=1}^{k}(z_i - x_{0i})(J_{2-\nu}(f_{x_i}(x_0 + t(z-x_0))))'. \qquad (12.17)$$

That is,

$$g_z^{(\nu)}(t) = \sum_{i=1}^{k}(z_i - x_{0i}) \left(\frac{\partial f}{\partial x_i}(x_0 + t(z-x_0)) \right)^{(\nu-1)}, \qquad (12.18)$$

$0 \leq t \leq 1$, $1 \leq \nu < 2$.

Thus the remainder turns to

$$(J_\nu g_z^{(\nu)})(t) = \sum_{i=1}^{k}(z_i - x_{0i})[J_\nu(f_{x_i}(x_0 + t(z-x_0)))^{(\nu-1)}](t). \qquad (12.19)$$

By (12.6) applied on g_z we obtain

$$f(z) = g_z(1) = f(x_0) + (J_\nu g_z^{(\nu)})(1). \qquad (12.20)$$

That is, it holds

$$f(z) = f(x_0) + \sum_{i=1}^{k}(z_i - x_{0i})[J_\nu(f_{x_i}(x_0 + t(z-x_0)))^{(\nu-1)}](1). \qquad (12.21)$$

More precisely we find

$$f(z) = f(x_0) + \sum_{i=1}^{k}(z_i - x_{0i}) \frac{1}{\Gamma(\nu)}$$

$$\int_0^1 (1-t)^{\nu-1}(f_{x_i}(x_0 + t(z-x_0)))^{(\nu-1)}dt. \qquad (12.22)$$

From Remark 12.3 we have established the basic multivariate Canavati fractional Taylor formula.

Theorem 12.4. *Let $f \in C^1(Q)$, Q compact and convex $\subseteq \mathbb{R}^k$, $k \geq 2$. For fixed $x_0, z \in Q$, assume that as a function of t: $f_{x_i}(x_0 + t(z-x_0)) \in C^{\nu-1}([0,1])$, $1 \leq \nu < 2$, all $i = 1, \ldots, k$. Then*

(i)

$$f(z_1, \ldots, z_k) = f(x_{01}, \ldots, x_{0k}) +$$

$$\sum_{i=1}^{k} \frac{(z_i - x_{0i})}{\Gamma(\nu)} \int_0^1 (1-t)^{\nu-1} (f_{x_i}(x_0 + t(z - x_0)))^{(\nu-1)} dt. \qquad (12.23)$$

(ii) Given $f(x_0) = 0$, then

$$f(z) = \sum_{i=1}^{k} \frac{(z_i - x_{0i})}{\Gamma(\nu)} \int_0^1 (1-t)^{\nu-1} (f_{x_i}(x_0 + t(z - x_0)))^{(\nu-1)} dt. \qquad (12.24)$$

We make

Remark 12.5. Continuing from Remark 12.3. Here $f \in C^2(Q), Q \subseteq \mathbb{R}^2$, we have

$$g_z''(t) = (z_1 - x_{01})^2 \frac{\partial^2 f}{\partial x_1^2}(x_0 + t(z - x_0)) + 2(z_1 - x_{01})(z_2 - x_{02})$$

$$\frac{\partial^2 f}{\partial x_1 \partial x_2}(x_0 + t(z - x_0)) + (z_2 - x_{02})^2 \frac{\partial^2 f}{\partial x_2^2}(x_0 + t(z - x_0)). \qquad (12.25)$$

Let $2 \leq \nu < 3$; then $n := [\nu] = 2$, $\alpha := \nu - n = \nu - 2$, $1 - \alpha = 3 - \nu$. Set $\nu^* := \nu - 2$; then $n^* := [\nu - 2] = 0$, $\alpha^* = (\nu - 2) - n^* = \nu - 2$, $1 - \alpha^* = 3 - \nu$.
We have ($0 \leq x \leq 1$)

$$(J_{3-\nu} g_z'')(x) = \frac{1}{\Gamma(3-\nu)} \int_0^x (x-t)^{2-\nu} g_z''(t) dt =$$

$$(z_1 - x_{01})^2 \frac{1}{\Gamma(3-\nu)} \int_0^x (x-t)^{2-\nu} f_{x_1 x_1}(x_0 + t(z - x_0)) dt$$

$$+ 2(z_1 - x_{01})(z_2 - x_{02}) \frac{1}{\Gamma(3-\nu)} \int_0^x (x-t)^{2-\nu} f_{x_1 x_2}(x_0 + t(z - x_0)) dt$$

$$+ (z_2 - x_{02})^2 \frac{1}{\Gamma(3-\nu)} \int_0^x (x-t)^{2-\nu} f_{x_2 x_2}(x_0 + t(z - x_0)) dt. \qquad (12.26)$$

That is, it holds.

$$(J_{3-\nu} g_z'')(x) = (z_1 - x_{01})^2 (J_{3-\nu}(f_{x_1 x_1}(x_0 + t(z - x_0))))(x) +$$

$$2(z_1 - x_{01})(z_2 - x_{02})(J_{3-\nu}(f_{x_1 x_2}(x_0 + t(z - x_0))))(x) +$$

$$(z_2 - x_{02})^2 (J_{3-\nu}(f_{x_2 x_2}(x_0 + t(z - x_0))))(x). \qquad (12.27)$$

Assuming now that $f_{x_1x_1}(x_0 + t(z - x_0))$, $f_{x_1x_2}(x_0 + t(z - x_0))$, $f_{x_2x_2}(x_0 + t(z - x_0))$, as functions of t belong to $C^{(\nu-2)}([0,1])$ we obtain that there exists

$$g_z^{(\nu)}(t) = (z_1 - x_{01})^2 (f_{x_1x_1}(x_0 + t(z - x_0)))^{(\nu-2)} +$$

$$2(z_1 - x_{01})(z_2 - x_{02})(f_{x_1x_2}(x_0 + t(z - x_0)))^{(\nu-2)} +$$

$$(z_2 - x_{02})^2 (f_{x_2x_2}(x_0 + t(z - x_0)))^{(\nu-2)}. \tag{12.28}$$

Next we observe that

$$(J_\nu g_z^{(\nu)})(x) = (z_1 - x_{01})^2 (J_\nu(f_{x_1x_1}(x_0 + t(z - x_0)))^{(\nu-2)})(x)$$

$$+2(z_1 - x_{01})(z_2 - x_{02})(J_\nu(f_{x_1x_2}(x_0 + t(z - x_0)))^{(\nu-2)})(x)$$

$$+(z_2 - x_{02})^2 (J_\nu(f_{x_2x_2}(x_0 + t(z - x_0)))^{(\nu-2)})(x). \tag{12.29}$$

We have proved via (12.6) the next Taylor type result.

Theorem 12.6. Let $f \in C^2(Q)$, Q compact and convex $\subseteq \mathbb{R}^2$. For fixed $x_0, z \in Q$ assume that as functions of t: $f_{x_1x_1}(x_0 + t(z - x_0))$, $f_{x_1x_2}(x_0 + t(z - x_0))$, $f_{x_2x_2}(x_0 + t(z - x_0)) \in C^{(\nu-2)}([0,1])$, where $2 \le \nu < 3$. Then
(i)

$$f(z_1, z_2) = f(x_{01}, x_{02}) + (z_1 - x_{01})\frac{\partial f}{\partial x_1}(x_0) + (z_2 - x_{02})\frac{\partial f}{\partial x_2}(x_0) +$$

$$(z_1 - x_{01})^2 \frac{1}{\Gamma(\nu)} \int_0^1 (1 - t)^{\nu-1}(f_{x_1x_1}(x_0 + t(z - x_0)))^{(\nu-2)} dt$$

$$+2(z_1 - x_{01})(z_2 - x_{02})\frac{1}{\Gamma(\nu)} \int_0^1 (1 - t)^{\nu-1}(f_{x_1x_2}(x_0 + t(z - x_0)))^{(\nu-2)} dt +$$

$$(z_2 - x_{02})^2 \frac{1}{\Gamma(\nu)} \int_0^1 (1 - t)^{\nu-1}(f_{x_2x_2}(x_0 + t(z - x_0)))^{(\nu-2)} dt. \tag{12.30}$$

(ii) When $f(x_0) = \frac{\partial f}{\partial x_1}(x_0) = \frac{\partial f}{\partial x_2}(x_0) = 0$, then

$$f(z_1, z_2) = (z_1 - x_{01})^2 \frac{1}{\Gamma(\nu)} \int_0^1 (1 - t)^{\nu-1}(f_{x_1x_1}(x_0 + t(z - x_0)))^{(\nu-2)} dt$$

$$+2(z_1 - x_{01})(z_2 - x_{02})\frac{1}{\Gamma(\nu)} \int_0^1 (1 - t)^{\nu-1}(f_{x_1x_2}(x_0 + t(z - x_0)))^{(\nu-2)} dt +$$

$$(z_2 - x_{02})^2 \frac{1}{\Gamma(\nu)} \int_0^1 (1 - t)^{\nu-1}(f_{x_2x_2}(x_0 + t(z - x_0)))^{(\nu-2)} dt. \tag{12.31}$$

The following general multivariate Canavati fractional Taylor formula is valid.

Theorem 12.7. *Let* $f \in C^n(Q)$, Q *compact and convex* $\subseteq \mathbb{R}^k$, $k \geq 2$; *here* $\nu \geq 1$ *such that* $n = [\nu]$. *For fixed* $x_0, z \in Q$ *assume that as functions of* t: $f_\alpha(x_0 + t(z - x_0)) \in C^{(\nu-n)}([0,1])$, *for all* $\alpha := (\alpha_1, \ldots, \alpha_k)$, $\alpha_i \in \mathbb{Z}^+$, $i = 1, \ldots, k$; $|\alpha| := \sum_{i=1}^k \alpha_i = n$.
Then
(i)

$$f(z_1, \ldots, z_k) = f(x_{01}, \ldots, x_{0k}) + \sum_{i=1}^k (z_i - x_{0i})\frac{\partial f}{\partial x_i}(x_{01}, \ldots, x_{0k})$$

$$+ \frac{\left[\left(\sum_{i=1}^k (z_i - x_{0i})\frac{\partial}{\partial x_i}\right)^2 f\right](x_{01}, \ldots, x_{0k})}{2} + \ldots$$

$$\frac{\left[\left(\sum_{i=1}^k (z_i - x_{0i})\frac{\partial}{\partial x_i}\right)^{n-1} f\right](x_{01}, \ldots, x_{0k})}{(n-1)!} +$$

$$\frac{1}{\Gamma(\nu)}\int_0^1 (1-t)^{\nu-1}\left\{\left[\left(\sum_{i=1}^k (z_i - x_{0i})\frac{\partial}{\partial x_i}\right)^n f\right]^{(\nu-n)}(x_0 + t(z - x_0))\right\} dt.$$

$$(12.32)$$

(ii) If all $f_\alpha(x_0) = 0$, $\alpha := (\alpha_1, \ldots, \alpha_k)$, $\alpha_i \in \mathbb{Z}^+$, $i = 1, \ldots, k$, $|\alpha| := \sum_{i=1}^k \alpha_i = l$, $l = 0, \ldots, n-1$, *then*

$$f(z_1, \ldots, z_k) = \frac{1}{\Gamma(\nu)}\int_0^1 (1-t)^{\nu-1}$$

$$\left\{\left[\left(\sum_{i=1}^k (z_i - x_{0i})\frac{\partial}{\partial x_i}\right)^n f\right]^{(\nu-n)}(x_0 + t(z - x_0))\right\} dt. \qquad (12.33)$$

Proof. Use (12.6). $\square$

(Note that fractional differentiation is a linear operation.)
We make

Remark 12.8. Continuing from the previous remarks. Let here $0 < \nu < 1$. Assume that $f(x_0 + t(z - x_0)) \in C^\nu([0, 1])$ as a function of t. Then by $g_z(t) := f(x_0 + t(z - x_0))$ we have

$$g_z^{(\nu)}(t) = (f(x_0 + t(z - x_0)))^{(\nu)}, \quad \text{and} \quad (J_\nu \, g_z^{(\nu)})(t) = (J_\nu(f(x_0 + t(z - x_0)))^{(\nu)})(t),$$

$t \in [0, 1]$. Hence

$$(J_\nu \, g_z^{(\nu)})(1) = \frac{1}{\Gamma(\nu)} \int_0^1 (1 - t)^{\nu - 1} (f(x_0 + t(z - x_0)))^{(\nu)} dt. \qquad (12.34)$$

We have established the next multivariate Canavati fractional Taylor formula when $0 < \nu < 1$.

Theorem 12.9. *Let bounded $f : Q \to \mathbb{R}$, where Q convex $\subseteq \mathbb{R}^k, k \geq 2$, such that as a function of t : $f(x_0 + t(z - x_0)) \in C^\nu([0, 1]), 0 < \nu < 1, x_0, z \in Q$ being fixed.*
Then

$$f(z_1, \ldots, z_k) = \frac{1}{\Gamma(\nu)} \int_0^1 (1 - t)^{\nu - 1} (f(x_0 + t(z - x_0)))^{(\nu)} dt. \qquad (12.35)$$

Proof. Use (12.7). $\square$

We make

Remark 12.10. Next we study the ordinary partial derivatives of fractional derivatives. Let $0 < \alpha < 1, f \in C^1([0, 1]^2)$, $x \in [0, 1]$ fixed, and consider

$$\gamma(x, z) := \int_0^x (x - t)^{-\alpha} f(t, z) dt, \qquad (12.36)$$

$\forall \, z \in [0, 1]$.
We observe that

$$|\gamma(x, z)| \leq \int_0^x (x - t)^{-\alpha} |f(t, z)| dt \leq \|f\|_\infty \int_0^x (x - t)^{-\alpha} dt =$$

$$\|f\|_\infty \frac{x^{1-\alpha}}{1 - \alpha} \leq \frac{\|f\|_\infty}{1 - \alpha} < +\infty.$$

That is, the function

$$\rho(t) := (x - t)^{-\alpha} f(t, z) \qquad (12.37)$$

is Lebesgue integrable in $t \in [0, x]$, $\forall \, z \in [0, 1]$. Thus one can consider integration in (12.36) over $[0, x]$, $\forall \, z \in [0, 1]$.
Also the function

$$\lambda(z) := (x - t)^{-\alpha} f(t, z) \qquad (12.38)$$

is differentiable in $z \in [0,1]$, $\forall$ $t \in [0,x)$. That is, we have

$$\lambda'(z) = (x-t)^{-\alpha}\frac{\partial f(t,z)}{\partial z}, \forall t \in [0,x). \qquad (12.39)$$

Moreover,

$$|\lambda'(z)| \leq (x-t)^{-\alpha}\left\|\frac{\partial f}{\partial z}\right\|_{\infty}, \qquad (12.40)$$

$\forall\,(t,z) \in [0,x) \times [0,1]$.

The R.H.S (12.40) is integrable in $t \in [0,x]$ and nonnegative. Hence by H. Bauer [79, pp. 103–104] we obtain that $(x-t)^{-\alpha}\partial f(t,z)/\partial z$ is integrable in $t \in [0,x)$ and

$$\frac{\partial\gamma(x,z)}{\partial z} = \int_0^x (x-t)^{-\alpha}\frac{\partial f(t,z)}{\partial z}dt, \qquad (12.41)$$

$\forall\,z \in [0,1]$.

We have proved

Lemma 12.11. *Let* $0 < \alpha < 1$, $f \in C^1([0,1]^2)$, $0 \leq x \leq 1$. *Then*

$$\frac{\partial}{\partial z}\left(\int_0^x (x-t)^{-\alpha}f(t,z)dt\right) = \int_0^x (x-t)^{-\alpha}\frac{\partial f(t,z)}{\partial z}dt, \qquad (12.42)$$

$\forall\,z \in [0,1]$.

We make

Remark 12.12. Assume now $0 < \alpha < 1$, $f \in C^{n+1}([0,1]^2)$, $n \in \mathbb{N}$. Then by Lemma 12.11 we get

$$\frac{\partial}{\partial z}\left(\int_0^x (x-t)^{-\alpha}\frac{\partial^n f}{\partial t^n}(t,z)dt\right) = \int_0^x (x-t)^{-\alpha}\frac{\partial^{n+1} f}{\partial t^n \partial z}(t,z)dt. \qquad (12.43)$$

Let now $\nu > 0$, $n := [\nu]$, $\alpha := \nu - n$.
We suppose existence of

$$g^{(\nu)}(x,z) := \frac{\partial^\nu g(x,z)}{\partial x^\nu} = \frac{\partial}{\partial x}\left(J_{1-\alpha}\left(\frac{\partial^n g}{\partial t^n}(\cdot,z)\right)\right)(x,z). \qquad (12.44)$$

We also assume here that $g \in C^{n+1}([0,1]^2)$, and $(g^{(\nu)}(x,z))_z$, $g_z^{(\nu)}(x,z)$ exist and are jointly continuous in $(x,z) \in [0,1]^2$, $[\nu] = n \in \mathbb{N}$.
Then it holds

$$(g^{(\nu)}(x,z))_z = \frac{\partial}{\partial z}(g^{(\nu)}(x,z)) = \frac{\partial}{\partial z}\left(\frac{\partial}{\partial x}\left(J_{1-\alpha}\left(\frac{\partial^n g}{\partial t^n}(\cdot,z)\right)\right)\right)(x,z) =$$

$$\frac{\partial}{\partial x}\left(\frac{\partial}{\partial z}\left(J_{1-\alpha}\left(\frac{\partial^n g}{\partial t^n}(\cdot,z)\right)\right)\right)(x,z)(\text{by }(12.43))=$$

$$\frac{\partial}{\partial x}\left(J_{1-\alpha}\left(\frac{\partial^{n+1}}{\partial z\partial t^n}g(\cdot,z)\right)\right)(x,z)$$

$$=\frac{\partial}{\partial x}\left(J_{1-\alpha}\left(\frac{\partial^n}{\partial t^n}g_z(\cdot,z)\right)\right)=g_z^{(\nu)}(x,z).$$

That is,

$$(g^{(\nu)}(x,z))_z=g_z^{(\nu)}(x,z),\quad \forall (x,z)\in[0,1]^2,\ \nu>0. \tag{12.45}$$

In brief, it holds

$$(g^{(\nu)})_z=(g_z)^{(\nu)}. \tag{12.46}$$

Under more similar suitable assumptions one obtains

$$(g^{(\nu)})_{zz}=(g_{zz})^{(\nu)},\ (g^{(\nu)})_{z_1 z_2}=(g_{z_1 z_2})^{(\nu)},\ (g^{(\nu)})_{z_1 z_2 z_3}=(g_{z_1 z_2 z_3})^{(\nu)},\dots. \tag{12.47}$$

We have established that the order of fractional-ordinary partial differentiation is immaterial.

Theorem 12.13. *Let $g\in C^{n+1}([0,1]^2)$, and $\nu>0$ such that $[\nu]=n\in\mathbb{N}$. Assume the existence of $g^{(\nu)}(x,z)$, and $(g^{(\nu)}(x,z))_z$, $g_z^{(\nu)}(x,z)$ both exist and are jointly continuous in $(x,z)\in[0,1]^2$.*
Then

$$(g^{(\nu)}(x,z))_z=g_z^{(\nu)}(x,z), \tag{12.48}$$

$\forall(x,z)\in[0,1]^2.$

We make

Remark 12.14. Next comes fractional integration by parts.
Let $f,g\in C^\nu([0,1])$, $\nu>0$, $n:=[\nu]$, $\alpha:=\nu-n$. Here

$$g^{(\nu)}=\frac{d(J_{1-\alpha}g^{(n)})}{dx},f^{(\nu)}=\frac{d\left(J_{1-\alpha}f^{(n)}\right)}{dx}.$$

That is, $d\left(J_{1-\alpha}g^{(n)}\right)=g^{(\nu)}dx, d\left(J_{1-\alpha}f^{(n)}\right)=f^{(\nu)}dx$. We observe that

$$\int_0^1(J_{1-\alpha}f^{(n)})(x)g^{(\nu)}(x)dx=\int_0^1(J_{1-\alpha}f^{(n)})(x)d\left(J_{1-\alpha}g^{(n)}\right)(x)=$$

$$(J_{1-\alpha}f^{(n)})(1)(J_{1-\alpha}g^{(n)})(1)-\int_0^1(J_{1-\alpha}g^{(n)})(x)d\left(J_{1-\alpha}f^{(n)}\right)(x)=$$

$$(J_{1-\alpha}\, f^{(n)})(1)(J_{1-\alpha}g^{(n)})(1) - \int_0^1 (J_{1-\alpha}g^{(n)})(x)f^{(\nu)}(x)dx. \qquad (12.49)$$

Next let us take $g \in C^\nu([0,1])$, $1 \le \nu < 2$, $n := [\nu] = 1$, $\alpha := \nu - n = \nu - 1$, $1 - \alpha = 2 - \nu$, and $f \in C^1([0,1])$.

Then $g^{(\nu)}(x) = d\,(J_{2-\nu}\, g')(x)/dx$; that is, $d\,(J_{2-\nu}\, g')(x) = g^{(\nu)}(x)dx$. Hence

$$\int_0^1 f(x)g^{(\nu)}(x)dx = \int_0^1 f(x)d(J_{2-\nu}\, g')(x) =$$

$$f(1)(J_{2-\nu}\, g')(1) - \int_0^1 (J_{2-\nu}\, g')(x)f'(x)dx. \qquad (12.50)$$

We have established the following fractional integration by parts formulae.

Theorem 12.15. (i) Let $f, g \in C^\nu([0,1])$, $\nu > 0$, $n := [\nu]$, $\alpha := \nu - n$. Then

$$\int_0^1 (J_{1-\alpha}\, f^{(n)})(x)g^{(\nu)}(x)dx =$$

$$(J_{1-\alpha}\, f^{(n)})(1)(J_{1-\alpha}\, g^{(n)})(1) - \int_0^1 (J_{1-\alpha}\, g^{(n)})(x)f^{(\nu)}(x)dx. \qquad (12.51)$$

(ii) Let $g \in C^\nu([0,1])$, $1 \le \nu < 2$, $f \in C^1([0,1])$. Then

$$\int_0^1 f(x)g^{(\nu)}(x)dx = f(1)(J_{2-\nu}\, g')(1) - \int_0^1 (J_{2-\nu}\, g')(x)f'(x)dx. \quad (12.52)$$

We make the last

Remark 12.16. Here we estimate the remainder (12.32). By definition in this chapter (see (12.2), (12.3)), the fractional derivatives are continuous functions. So the function

$$G_\nu(t) := \left\{ \left[\left(\sum_{i=1}^k (z_i - x_{0i})\frac{\partial}{\partial x_i} \right)^n f \right]^{(\nu-n)} (x_0 + t(z - x_0)) \right\}, \qquad (12.53)$$

$t \in [0,1]$, that appears in the remainder of (12.32), is continuous in t. We write the remainder (12.32) as

$$R_\nu := \frac{1}{\Gamma(\nu)} \int_0^1 (1-t)^{\nu-1}$$

$$\left\{ \left[\left(\sum_{i=1}^{k} (z_i - x_{0i}) \frac{\partial}{\partial x_i} \right)^n f \right]^{(\nu - n)} (x_0 + t(z - x_0)) \right\} dt =$$

$$\frac{1}{\Gamma(\nu)} \int_0^1 (1 - t)^{\nu - 1} G_\nu(t) dt, \ \nu \geq 1. \tag{12.54}$$

We obtain

$$|R_\nu| \leq \frac{1}{\Gamma(\nu)} \int_0^1 (1 - t)^{\nu - 1} |G_\nu(t)| dt \leq \frac{1}{\Gamma(\nu)} \int_0^1 |G_\nu(t)| dt =$$

$$\frac{1}{\Gamma(\nu)} \|G_\nu\|_{L_1([0,1])}. \tag{12.55}$$

Also for $p, q > 1 : 1/p + 1/q = 1$, we get

$$|R_\nu| \leq \frac{1}{\Gamma(\nu)} \int_0^1 (1 - t)^{\nu - 1} |G_\nu(t)| dt \leq$$

$$\frac{1}{\Gamma(\nu)} \left(\int_0^1 ((1 - t)^{\nu - 1})^p dt \right)^{1/p} \left(\int_0^1 |G_\nu(t)|^q dt \right)^{1/q} =$$

$$\frac{1}{\Gamma(\nu)} \frac{1}{(p(\nu - 1) + 1)^{1/p}} \|G_\nu\|_{L_q([0,1])}. \tag{12.56}$$

In the case $p = q = 2$ we have

$$|R_\nu| \leq \frac{1}{\Gamma(\nu)} \frac{1}{\sqrt{2\nu - 1}} \|G_\nu\|_{L_2([0,1])}. \tag{12.57}$$

Finally we get that

$$|R_\nu| \leq \frac{1}{\Gamma(\nu)} \int_0^1 (1 - t)^{\nu - 1} |G_\nu(t)| dt \leq \frac{\|G_\nu\|_\infty}{\Gamma(\nu + 1)}. \tag{12.58}$$

We have established the following remainder estimate.

Theorem 12.17. *All here are as in Theorem 12.7. Let R_ν be the remainder in (12.32) (see (12.54)), and G_ν as in (12.53). Then*

$$|R_\nu| \leq min \left\{ \frac{\|G_\nu\|_{L_1([0,1])}}{\Gamma(\nu)}, \ \frac{\|G_\nu\|_{L_q([0,1])}}{\Gamma(\nu)(p(\nu - 1) + 1)^{1/p}}, \right.$$

$$\left. \frac{\|G_\nu\|_{L_2([0,1])}}{\Gamma(\nu)\sqrt{2\nu - 1}}, \ \frac{\|G_\nu\|_\infty}{\Gamma(\nu + 1)} \right\}, \tag{12.59}$$

where $p, q > 1 : 1/p + 1/q = 1$.

Comment. The chain rule as in ordinary differentiation is not possible in Canavati fractional differentiation. That limits us a lot from using the multivariate Canavati fractional Taylor formula, as we employ the usual one involving only ordinary partial derivatives of functions.

13
Multivariate Caputo Fractional Taylor Formula

This is a continuation of Chapter 12. We establish here a multivariate fractional Taylor formula via the Caputo fractional derivative. The fractional remainder is expressed as a composition of two Riemann–Liouville fractional integrals.

We estimate the remainder. This treatment is based on [53].

13.1 Background

We start with

Definition 13.1. [134] Let $\nu \geq 0$; the operator J_a^ν, defined on $L_1(a, b)$ by

$$J_a^\nu f(x) := \frac{1}{\Gamma(\nu)} \int_a^x (x - t)^{\nu-1} f(t) \, dt \tag{13.1}$$

for $a \leq x \leq b$, is called the Riemann–Liouville fractional integral operator of order ν. For $\nu = 0$, we set $J_a^0 := I$, the identity operator. Here Γ stands for the gamma function.

By Theorem 2.1 of [134, p. 13], $J_a^\nu f(x)$, $\nu > 0$, exists for almost all $x \in [a, b]$ and $J_a^\nu f \in L_1(a, b)$, where $f \in L_1(a, b)$.

Here $AC^n([a, b])$ is the space of functions with absolutely continuous $(n-1)$–st derivative.

G.A. Anastassiou, *Fractional Differentiation Inequalities*,   269
DOI 10.1007/978-0-387-98128-4_13, © Springer Science+Business Media, LLC 2009

We need to mention

Definition 13.2. [58, 134] Let $\nu \geq 0$; $n := \lceil \nu \rceil$, $\lceil \cdot \rceil$ is the ceiling of the number, $f \in AC^n\left([a,b]\right)$. We call the Caputo fractional derivative

$$D_{*a}^{\nu} f\left(x\right) := \frac{1}{\Gamma\left(n - \nu\right)} \int_a^x \left(x - t\right)^{n - \nu - 1} f^{(n)}\left(t\right) dt, \qquad (13.2)$$

$\forall x \in [a, b]$.

The above function $D_{*a}^{\nu} f\left(x\right)$ exists almost everywhere for $x \in [a, b]$.
If $\nu \in \mathbb{N}$, then $D_{*a}^{\nu} f = f^{(\nu)}$ the ordinary derivative; it is also $D_{*a}^0 f = f$.

We need

Theorem 13.3. (Taylor expansion for Caputo derivatives, [134, p. 40])
Assume $\nu \geq 0$, $n = \lceil \nu \rceil$, and $f \in AC^n\left([a,b]\right)$.
Then

$$f\left(x\right) = \sum_{k=0}^{n-1} \frac{f^{(k)}\left(a\right)}{k!} \left(x - a\right)^k + \frac{1}{\Gamma\left(\nu\right)} \int_a^x \left(x - t\right)^{\nu - 1} D_{*a}^{\nu} f\left(t\right) dt, \qquad (13.3)$$

$\forall x \in [a, b]$.

13.2   Results

We establish analogues of Theorem 13.3 to the multivariate case. We make

Remark 13.4. Let Q be a compact and convex subset of $\mathbb{R}^k$, $k \geq 2$; $z := \left(z_1, \ldots, z_k\right)$, $x_0 := \left(x_{01}, \ldots, x_{0k}\right) \in Q$. Let $f \in C^n\left(Q\right)$, $n \in \mathbb{N}$.
Set

$$g_z\left(t\right) := f\left(x_0 + t\left(z - x_0\right)\right),$$

$$0 \leq t \leq 1; \ g_z\left(0\right) = f\left(x_0\right), \ g_z\left(1\right) = f\left(z\right). \qquad (13.4)$$

Then

$$g_z^{(j)}\left(t\right) = \left[\left(\sum_{i=1}^k \left(z_i - x_{0i}\right) \frac{\partial}{\partial x_i}\right)^j f\right]\left(x_0 + t\left(z - x_0\right)\right), \qquad (13.5)$$

$j = 0, 1, 2, \ldots, n$,
and

$$g_z^{(n)}\left(0\right) = \left[\left(\sum_{i=1}^k \left(z_i - x_{0i}\right) \frac{\partial}{\partial x_i}\right)^n f\right]\left(x_0\right). \qquad (13.6)$$

If all $f_\alpha(x_0) := \partial^\alpha f/\partial x^\alpha(x_0) = 0$, $\alpha := (\alpha_1, \ldots, \alpha_k)$, $\alpha_i \in \mathbb{Z}^+$, $i = 1, \ldots, k$; $|\alpha| := \sum_{i=1}^k \alpha_i =: l$, then $g_z^{(l)}(0) = 0$, where $l \in \{0, 1, \ldots, n\}$. We quote that

$$g_z'(t) = \sum_{i=1}^k (z_i - x_{0i}) \frac{\partial f}{\partial x_i}(x_0 + t(z - x_0)). \tag{13.7}$$

When $f \in C^2(Q)$, $Q \subseteq \mathbb{R}^2$, we have

$$g_z''(t) = (z_1 - x_{01})^2 \frac{\partial^2 f}{\partial x_1^2}(x_0 + t(z - x_0)) +$$

$$2(z_1 - x_{01})(z_2 - x_{02}) \frac{\partial^2 f}{\partial x_1 \partial x_2}(x_0 + t(z - x_0)) +$$

$$(z_2 - x_{02})^2 \frac{\partial^2 f}{\partial x_2^2}(x_0 + t(z - x_0)), \tag{13.8}$$

and so on.

Clearly here $g_z \in C^n([0,1])$, hence $g_z \in AC^n([0,1])$.

Let now $\nu > 0$ with $\lceil \nu \rceil = n$.

By applying (13.3) for g_z we find

$$g_z(1) = \sum_{l=0}^{n-1} \frac{g_z^{(l)}(0)}{l!} + \frac{1}{\Gamma(\nu)} \int_0^1 (1-t)^{\nu-1} D_{*0}^\nu g_z(t)\, dt. \tag{13.9}$$

Here we observe by (13.2) that

$$D_{*0}^\nu g_z(t) = \frac{1}{\Gamma(n-\nu)} \int_0^t (t-s)^{n-\nu-1} g_z^{(n)}(s)\, ds. \tag{13.10}$$

Let us consider the case of $0 < \nu \le 1$; that is, $n = 1$. Then

$$D_{*0}^\nu g_z(t) - \frac{1}{\Gamma(1-\nu)} \int_0^t (t-s)^{-\nu} g_z'(s)\, ds$$

$$\overset{(13.7)}{=} \frac{1}{\Gamma(1-\nu)} \int_0^t (t-s)^{-\nu}$$

$$\left(\sum_{i=1}^k (z_i - x_{0i}) \frac{\partial f}{\partial x_i}(x_0 + s(z - x_0))\right) ds =$$

$$\frac{1}{\Gamma(1-\nu)} \left(\sum_{i=1}^k (z_i - x_{0i}) \int_0^t (t-s)^{-\nu} \frac{\partial f}{\partial x_i}(x_0 + s(z - x_0))\, ds\right)$$

$$= \sum_{i=1}^k (z_i - x_{0i}) J_0^{1-\nu}\left(\frac{\partial f}{\partial x_i}(x_0 + t(z - x_0))\right). \tag{13.11}$$

That is,

$$D_{*0}^{\nu} g_z (t) = \sum_{i=1}^{k} (z_i - x_{0i}) J_0^{1-\nu} \left(\frac{\partial f}{\partial x_i} (x_0 + t (z - x_0)) \right),$$ (13.12)

for all $t \in [0,1]$.

Consequently by (13.9) and (13.12) we get

$$f(z) = f(x_0) + \frac{1}{\Gamma(\nu)} \int_0^1 (1-t)^{\nu-1}$$

$$\left(\sum_{i=1}^{k} (z_i - x_{0i}) J_0^{1-\nu} \left(\frac{\partial f}{\partial x_i} (x_0 + t (z - x_0)) \right) \right) dt =$$

$$f(x_0) + \sum_{i=1}^{k} (z_i - x_{0i}) \left[\frac{1}{\Gamma(\nu)} \int_0^1 (1-t)^{\nu-1} \right.$$

$$\left. J_0^{1-\nu} \left(\frac{\partial f}{\partial x_i} (x_0 + t (z - x_0)) \right) dt \right].$$ (13.13)

Based on the last comments we present the following basic multivariate fractional fundamental theorem.

Theorem 13.5. *Let Q be a compact and convex subset of $\mathbb{R}^k$, $k \geq 2$; $z := (z_1, \dots, z_k)$, $x_0 := (x_{01}, \dots, x_{0k}) \in Q$, $f \in C^1(Q)$, $0 < \nu \leq 1$. Then*

$$f(z) = f(x_0) + \sum_{i=1}^{k} (z_i - x_{0i}) \left[\frac{1}{\Gamma(\nu)} \int_0^1 (1-t)^{\nu-1} \right.$$

$$\left. J_0^{1-\nu} \left(\frac{\partial f}{\partial x_i} (x_0 + t (z - x_0)) \right) dt \right].$$ (13.14)

We make

Remark 13.6. This is a continuation of Remark 13.4.
Let now $1 < \nu \leq 2$; that is, $n = 2$, $f \in C^2(Q)$, $Q \subseteq \mathbb{R}^2$.
Then

$$D_{*0}^{\nu} g_z (t) = \frac{1}{\Gamma(2-\nu)} \int_0^t (t-s)^{1-\nu} g_z''(s) ds \overset{(13.8)}{=}$$

$$\frac{1}{\Gamma(2-\nu)} \int_0^t (t-s)^{1-\nu} \left[(z_1 - x_{01})^2 \frac{\partial^2 f}{\partial x_1^2} (x_0 + s (z - x_0)) + \right.$$

$$2\left(z_1 - x_{01}\right)\left(z_2 - x_{02}\right)\frac{\partial^2 f}{\partial x_1 \partial x_2}\left(x_0 + s\left(z - x_0\right)\right) +$$

$$\left.\left(z_2 - x_{02}\right)^2 \frac{\partial^2 f}{\partial x_2^2}\left(x_0 + s\left(z - x_0\right)\right)\right] ds =$$

$$\left(z_1 - x_{01}\right)^2 \left[\frac{1}{\Gamma\left(2 - \nu\right)}\int_0^t \left(t - s\right)^{1-\nu}\frac{\partial^2 f}{\partial x_1^2}\left(x_0 + s\left(z - x_0\right)\right) ds\right]$$

$$+2\left(z_1 - x_{01}\right)\left(z_2 - x_{02}\right)$$

$$\left[\frac{1}{\Gamma\left(2 - \nu\right)}\int_0^t \left(t - s\right)^{1-\nu}\frac{\partial^2 f}{\partial x_1 \partial x_2}\left(x_0 + s\left(z - x_0\right)\right) ds\right]$$

$$+\left(z_2 - x_{02}\right)^2 \left(\frac{1}{\Gamma\left(2 - \nu\right)}\int_0^t \left(t - s\right)^{1-\nu}\frac{\partial^2 f}{\partial x_2^2}\left(x_0 + s\left(z - x_0\right)\right) ds\right)$$

$$= \left(z_1 - x_{01}\right)^2 \left(J_0^{2-\nu}\left(\frac{\partial^2 f}{\partial x_1^2}\left(x_0 + t\left(z - x_0\right)\right)\right)\right) \qquad (13.15)$$

$$+2\left(z_1 - x_{01}\right)\left(z_2 - x_{02}\right)\left(J_0^{2-\nu}\left(\frac{\partial^2 f}{\partial x_1 \partial x_2}\left(x_0 + t\left(z - x_0\right)\right)\right)\right)$$

$$+\left(z_2 - x_{02}\right)^2 \left(J_0^{2-\nu}\left(\frac{\partial^2 f}{\partial x_2^2}\left(x_0 + t\left(z - x_0\right)\right)\right)\right).$$

That is, we get

$$D_{*0}^\nu g_z\left(t\right) = \left(z_1 - x_{01}\right)^2 \left(J_0^{2-\nu}\left(\frac{\partial^2 f}{\partial x_1^2}\left(x_0 + t\left(z - x_0\right)\right)\right)\right)$$

$$+2\left(z_1 - x_{01}\right)\left(z_2 - x_{02}\right)\left(J_0^{2-\nu}\left(\frac{\partial^2 f}{\partial x_1 \partial x_2}\left(x_0 + t\left(z - x_0\right)\right)\right)\right) +$$

$$\left(z_2 - x_{02}\right)^2 \left(J_0^{2-\nu}\left(\frac{\partial^2 f}{\partial x_2^2}\left(x_0 + t\left(z - x_0\right)\right)\right)\right), \quad 0 \le t \le 1. \qquad (13.16)$$

Thus by (13.6), (13.7), (13.9), and (13.16) we obtain

$$f\left(z\right) = f\left(x_0\right) + \left(z_1 - x_{01}\right)\frac{\partial f}{\partial x_1}\left(x_0\right) +$$

$$\left(z_2 - x_{02}\right)\frac{\partial f}{\partial x_2}\left(x_0\right) + \frac{1}{\Gamma\left(\nu\right)}\int_0^1 \left(1 - t\right)^{\nu-1}$$

$$\left[(z_1 - x_{01})^2 \left(J_0^{2-\nu} \left(\frac{\partial^2 f}{\partial x_1^2} (x_0 + t(z - x_0)) \right) \right) + \right.$$

$$2(z_1 - x_{01})(z_2 - x_{02}) \left(J_0^{2-\nu} \left(\frac{\partial^2 f}{\partial x_1 \partial x_2} (x_0 + t(z - x_0)) \right) \right)$$

$$\left. + (z_2 - x_{02})^2 \left(J_0^{2-\nu} \left(\frac{\partial^2 f}{\partial x_2^2} (x_0 + t(z - x_0)) \right) \right) \right] dt \qquad (13.17)$$

$$= f(x_0) + (z_1 - x_{01}) \frac{\partial f}{\partial x_1} (x_0) + (z_2 - x_{02}) \frac{\partial f}{\partial x_2} (x_0) +$$

$$(z_1 - x_{01})^2 \left(\frac{1}{\Gamma(\nu)} \int_0^1 (1 - t)^{\nu - 1} \right.$$

$$\left. \left(J_0^{2-\nu} \left(\frac{\partial^2 f}{\partial x_1^2} (x_0 + t(z - x_0)) \right) \right) dt \right) +$$

$$+ 2(z_1 - x_{01})(z_2 - x_{02}) \left(\frac{1}{\Gamma(\nu)} \int_0^1 (1 - t)^{\nu - 1} \right.$$

$$\left. \left(J_0^{2-\nu} \left(\frac{\partial^2 f}{\partial x_1 \partial x_2} (x_0 + t(z - x_0)) \right) \right) dt \right) +$$

$$(z_2 - x_{02})^2 \left(\frac{1}{\Gamma(\nu)} \int_0^1 (1 - t)^{\nu - 1} \right.$$

$$\left. \left(J_0^{2-\nu} \left(\frac{\partial^2 f}{\partial x_2^2} (x_0 + t(z - x_0)) \right) \right) dt \right). \qquad (13.18)$$

We have established the following Caputo fractional bivariate Taylor formula.

Theorem 13.7. Let $f \in C^2(Q)$, $Q \subseteq \mathbb{R}^2$ compact and convex, $z := (z_1, z_2)$, $x_0 := (x_{01}, x_{02}) \in Q$, and $1 < \nu \leq 2$. Then

$$(1) \ f(z_1, z_2) = f(x_{01}, x_{02}) + (z_1 - x_{01}) \frac{\partial f}{\partial x_1} (x_{01}, x_{02}) +$$

$$(z_2 - x_{02}) \frac{\partial f}{\partial x_2} (x_{01}, x_{02}) + (z_1 - x_{01})^2$$

$$\left(\frac{1}{\Gamma(\nu)} \int_0^1 (1 - t)^{\nu - 1} \left(J_0^{2-\nu} \left(\frac{\partial^2 f}{\partial x_1^2} (x_0 + t(z - x_0)) \right) \right) dt \right)$$

$$+2\left(z_1 - x_{01}\right)\left(z_2 - x_{02}\right)\left(\frac{1}{\Gamma\left(\nu\right)}\int_0^1\left(1-t\right)^{\nu-1}\right)$$

$$\left(J_0^{2-\nu}\left(\frac{\partial^2 f}{\partial x_1\partial x_2}\left(x_0 + t\left(z - x_0\right)\right)\right)\right)dt\Bigg) +$$

$$\left(z_2 - x_{02}\right)^2\left(\frac{1}{\Gamma\left(\nu\right)}\int_0^1\left(1-t\right)^{\nu-1}\right)$$

$$\left(J_0^{2-\nu}\left(\frac{\partial^2 f}{\partial x_2^2}\left(x_0 + t\left(z - x_0\right)\right)\right)\right)dt\Bigg). \qquad (13.19)$$

Additionally assume that

$$f\left(x_0\right) = \frac{\partial f}{\partial x_1}\left(x_0\right) = \frac{\partial f}{\partial x_2}\left(x_0\right) = 0,$$

then

$$(2)\ f\left(z_1, z_2\right) = \left(z_1 - x_{01}\right)^2\left(\frac{1}{\Gamma\left(\nu\right)}\int_0^1\left(1-t\right)^{\nu-1}\right)$$

$$\left(J_0^{2-\nu}\left(\frac{\partial^2 f}{\partial x_1^2}\left(x_0 + t\left(z - x_0\right)\right)\right)\right)dt$$

$$+2\left(z_1 - x_{01}\right)\left(z_2 - x_{02}\right)\left(\frac{1}{\Gamma\left(\nu\right)}\int_0^1\left(1-t\right)^{\nu-1}\right)$$

$$\left(J_0^{2-\nu}\left(\frac{\partial^2 f}{\partial x_1\partial x_2}\left(x_0 + t\left(z - x_0\right)\right)\right)\right)dt\Bigg) +$$

$$\left(z_2 - x_{02}\right)^2\left(\frac{1}{\Gamma\left(\nu\right)}\int_0^1\left(1-t\right)^{\nu-1}\right)$$

$$\left(J_0^{2-\nu}\left(\frac{\partial^2 f}{\partial x_2^2}\left(x_0 + t\left(z - x_0\right)\right)\right)\right)dt\Bigg). \qquad (13.20)$$

We make

Remark 13.8. This is another continuation of Remark 13.4.
Let $\nu > 0$, $n = \lceil\nu\rceil$, $f \in C^n\left(Q\right)$. By (13.4), (13.6), and (13.9) we get

$$f\left(z\right) = f\left(x_0\right) + \sum_{i=1}^k\left(z_i - x_{0i}\right)\frac{\partial f}{\partial x_i}\left(x_0\right) +$$

$$\sum_{l=2}^{n-1} \frac{\left[\left(\sum_{i=1}^{k}(z_i - x_{0i})\frac{\partial}{\partial x_i}\right)^l f\right](x_0)}{l!}$$

$$+ \frac{1}{\Gamma(\nu)}\int_0^1 (1-t)^{\nu-1}D_{*0}^{\nu}g_z(t)\,dt. \tag{13.21}$$

But we have

$$D_{*0}^{\nu}g_z(t) = \frac{1}{\Gamma(n-\nu)}\int_0^t (t-s)^{n-\nu-1}g_z^{(n)}(s)\,ds =$$

$$\frac{1}{\Gamma(n-\nu)}\int_0^t (t-s)^{n-\nu-1}$$

$$\left[\left(\sum_{i=1}^{k}(z_i - x_{0i})\frac{\partial}{\partial x_i}\right)^n f\right](x_0 + s(z - x_0))\,ds$$

$$= J_0^{n-\nu}\left\{\left[\left(\sum_{i=1}^{k}(z_i - x_{0i})\frac{\partial}{\partial x_i}\right)^n f\right](x_0 + t(z - x_0))\right\}. \tag{13.22}$$

We have proved the following general multivariate Caputo fractional Taylor formula.

Theorem 13.9. *Let* $\nu > 0$, $n = \lceil \nu \rceil$, $f \in C^n(Q)$, *where* Q *is a compact and convex subset of* $\mathbb{R}^k$, $k \geq 2$; $z := (z_1, \ldots, z_k)$, $x_0 := (x_{01}, \ldots, x_{0k}) \in Q$. *Then*

$$1)\ f(z) = f(x_0) + \sum_{i=1}^{k}(z_i - x_{0i})\frac{\partial f(x_0)}{\partial x_i} +$$

$$\sum_{l=2}^{n-1} \frac{\left[\left(\sum_{i=1}^{k}(z_i - x_{0i})\frac{\partial}{\partial x_i}\right)^l f\right](x_0)}{l!}$$

$$+ \frac{1}{\Gamma(\nu)}\int_0^1 (1-t)^{\nu-1}\left[J_0^{n-\nu}\right.$$

$$\left.\left\{\left[\left(\sum_{i=1}^{k}(z_i - x_{0i})\frac{\partial}{\partial x_i}\right)^n f\right](x_0 + t(z - x_0))\right\}\right]dt. \tag{13.23}$$

Additionally assume that $f_\alpha(x_0) = 0$, $\alpha := (\alpha_1, \ldots, \alpha_k)$, $\alpha_i \in \mathbb{Z}^+$, $i = 1, \ldots, k$; $|\alpha| := \sum_{i=1}^k \alpha_i =: r$, $r = 0, \ldots, n-1$; *then*

$$2) \ f(z) = \frac{1}{\Gamma(\nu)} \int_0^1 (1-t)^{\nu-1} \left[J_0^{n-\nu} \right.$$

$$\left\{ \left[\left(\sum_{i=1}^k (z_i - x_{0i}) \frac{\partial}{\partial x_i} \right)^n f \right] (x_0 + t(z - x_0)) \right\} \right] dt =: R_\nu. \quad (13.24)$$

We continue with

Remark 13.10. Here we estimate the remainder of (13.23), which is the same as R_ν of (13.24).

The function

$$G_\nu(t) := J_0^{n-\nu} \left\{ \left[\left(\sum_{i=1}^k (z_i - x_{0i}) \frac{\partial}{\partial x_i} \right)^n f \right] \right.$$

$$(x_0 + t(z - x_0)) \right\}, \ t \in [0, 1], \quad (13.25)$$

which appears in R_ν is continuous; see Proposition 114 of [45] and $R_\nu \in \mathbb{R}$.

Similarly the remainder of (13.14) exists and the same holds for all the integrals of (13.19); they are all real numbers.

So we can write

$$R_\nu = \frac{1}{\Gamma(\nu)} \int_0^1 (1-t)^{\nu-1} G_\nu(t) dt, \ \nu > 0. \quad (13.26)$$

When $\nu \geq 1$ we obtain

$$|R_\nu| \leq \frac{1}{\Gamma(\nu)} \int_0^1 (1-t)^{\nu-1} |G_\nu(t)| dt \leq$$

$$\frac{1}{\Gamma(\nu)} \int_0^1 |G_\nu(t)| dt = \frac{1}{\Gamma(\nu)} \|G_\nu\|_{L_1(0,1)};$$

That is,

$$|R_\nu| \leq \frac{1}{\Gamma(\nu)} \|G_\nu\|_{L_1(0,1)}. \quad (13.27)$$

Also for $p, q > 1 : 1/p + 1/q = 1$ and with $p(\nu-1) + 1 > 0$ for $\nu > 0$, we derive

$$|R_\nu| \leq \frac{1}{\Gamma(\nu)} \int_0^1 (1-t)^{\nu-1} |G_\nu(t)| dt \leq$$

$$\frac{1}{\Gamma\left(\nu\right)}\left(\int_0^1\left(\left(1-t\right)^{\nu-1}\right)^p dt\right)^{1/p}\left(\int_0^1|G_\nu\left(t\right)|^q dt\right)^{1/q}=$$

$$\frac{1}{\Gamma\left(\nu\right)}\frac{1}{\left(p\left(\nu-1\right)+1\right)^{1/p}}\|G_\nu\|_{L_q\left([0,1]\right)}.$$

That is,

$$|R_\nu|\leq\frac{1}{\Gamma\left(\nu\right)}\frac{1}{\left(p\left(\nu-1\right)+1\right)^{1/p}}\|G_\nu\|_{L_q\left([0,1]\right)}.\qquad(13.28)$$

In the case of $p=q=2$ and $\nu>1/2$ we have

$$|R_\nu|\leq\frac{1}{\Gamma\left(\nu\right)}\frac{1}{\sqrt{2\nu-1}}\|G_\nu\|_{L_2\left([0,1]\right)}.\qquad(13.29)$$

Finally we get that

$$|R_\nu|\leq\frac{1}{\Gamma\left(\nu\right)}\int_0^1\left(1-t\right)^{\nu-1}|G_\nu\left(t\right)|\,dt\leq\frac{\|G_\nu\|_\infty}{\Gamma\left(\nu+1\right)};$$

that is,

$$|R_\nu|\leq\frac{\|G_\nu\|_\infty}{\Gamma\left(\nu+1\right)},\ \nu>0.\qquad(13.30)$$

We have established the following remainder estimate.

Theorem 13.11. *All here are as in Theorem* 13.9. *Let R_ν be the remainder in* (13.23), *and $G_\nu\left(t\right),t\in[0,1]$ as in* (13.25), $\nu\geq1$.
Then

$$|R_\nu|\leq\min\left\{\frac{\|G_\nu\|_{L_1\left([0,1]\right)}}{\Gamma\left(\nu\right)},\frac{\|G_\nu\|_{L_q\left([0,1]\right)}}{\Gamma\left(\nu\right)\left(p\left(\nu-1\right)+1\right)^{1/p}},\right.$$

$$\left.\frac{\|G_\nu\|_{L_2\left([0,1]\right)}}{\Gamma\left(\nu\right)\sqrt{2\nu-1}},\frac{\|G_\nu\|_\infty}{\Gamma\left(\nu+1\right)}\right\},\qquad(13.31)$$

where $p,q>1:1/p+1/q=1$.

14

Canavati Fractional Multivariate Opial-Type Inequalities on Spherical Shells

Here we introduce the concept of multivariate Canavati fractional differentiation especially of the fractional radial differentiation, by extending the univariate definition of [101]. Then we present Opial-type inequalities over compact and convex subsets of $\mathbb{R}^N$, $N \geq 2$, mainly over spherical shells, studying the problem in all possibilities. Our results involve one, two, or more functions. This treatment is based on [44].

14.1 Introduction

This chapter is motivated by the articles of Opial [315], Beesack [80], and Anastassiou [15, 17–19, 26, 27, 61].

We would like to mention

Theorem 14.1. (Opial [315, 1960]). *Let $c > 0$, and $y(x)$ be real, continuously differentiable on $[0, c]$, with $y(0) = y(c) = 0$. Then*

$$\int_0^c |y(x)y'(x)|dx \leq \frac{c}{4} \int_0^c (y'(x))^2 dx. \tag{14.1}$$

Equality holds for the function $y(x) = x$ on $[0, c/2]$, and $y(x) = c - x$ on $[c/2, c]$.

The next result implies Theorem 14.1 and is often used in applications.

Theorem 14.2. (Beesack [80,1962]). *Let* $b > 0$. *If* $y(x)$ *is real, continuously differentiable on* $[0, b]$, *and* $y(0) = 0$ *then*

$$\int_0^b |y(x)y'(x)| dx \le \frac{b}{2} \int_0^b (y'(x))^2 dx. \qquad (14.2)$$

Equality holds only for $y = mx$ *where* m *is a constant.*

We describe here our specific multivariate setting. Let the balls $B(0, R_1)$, $B(0, R_2)$; $0 < R_1 < R_2$. Here $B(0, R) := \{x \in \mathbb{R}^N : |x| < R\} \subseteq \mathbb{R}^N$, $N \ge 2$, $R > 0$, and the sphere $S^{N-1} := \{x \in \mathbb{R}^N : |x| = 1\}$, where $|\cdot|$ is the Euclidean norm. Let $d\omega$ be the element of surface measure on S^{N-1} and let $\omega_N = \int_{S^{N-1}} d\omega = 2\pi^{N/2}/\Gamma(N/2)$. For $x \in \mathbb{R}^N - \{0\}$ we can write uniquely $x = r\omega$, where $r = |x| > 0$, and $\omega = x/r \in S^{N-1}$, $|\omega| = 1$. Let the *spherical shell* $A := B(0, R_2) - \overline{B(0, R_1)}$. We have that $Vol(A) = \omega_N(R_2^N - R_1^N)/N$. Indeed $\bar{A} = [R_1, R_2] \times S^{N-1}$.

For $F \in C(\bar{A})$ it holds

$$\int_A F(x) dx = \int_{S^{N-1}} \left(\int_{R_1}^{R_2} F(r\omega) r^{N-1} dr \right) d\omega; \qquad (14.3)$$

we often use this formula here.

In this chapter we present a series of various fractional multivariate Opial-type inequalities over spherical shells and arbitrary domains. Opial-type inequalities find applications in establishing the uniqueness of solution of initial value problems for differential equations and their systems; see [406].

14.2   Results

We make

Remark 14.3. We introduce here the *partial Canavati type fractional derivatives.* Let $f : [0, 1]^2 \to \mathbb{R}$. Let $\nu > 0$, $n := [\nu]$, $\alpha := \nu - n$, $0 < \alpha < 1$; $\mu > 0$, $m := [\mu]$, $\beta := \mu - m$, $0 < \beta < 1$. Assume $\exists \, \partial^{m+n} f(t, s)/\partial x^n \partial y^m \in C([0, 1]^2)$; then $(x - t)^{-\alpha}(y - s)^{-\beta} \partial^{m+n} f(t, s)/\partial x^n \partial y^m$ is integrable over $[0, x] \times [0, y]; x, y \in [0, 1]$; that is,

$$F(x, y) := \int_0^x \int_0^y (x - t)^{-\alpha}(y - s)^{-\beta} \frac{\partial^{m+n} f(t, s)}{\partial x^n \partial y^m} dt \, ds \qquad (14.4)$$

is real-valued.

Thus, by Fubini's theorem, the order of integration in (14.4) does not matter.

Let now $g \in C([0,1])$; we define the *Riemann–Liouville integral*; Γ is the gamma function: $\Gamma(\nu) := \int_0^\infty e^{-t} t^{\nu-1} dt$, as

$$(\mathcal{J}_\nu g)(x) := \frac{1}{\Gamma(\nu)} \int_0^x (x-t)^{\nu-1} g(t) dt, \; 0 \le x \le 1. \tag{14.5}$$

We consider here the space

$$C^\nu([0,1]) := \{g \in C^n([0,1]) : \mathcal{J}_{1-\alpha} \, g^{(n)} \in C^1([0,1])\}; \tag{14.6}$$

then the ν – *fractional derivative of* g is defined by $g^{(\nu)} := (\mathcal{J}_{1-\alpha} \, g^{(n)})'$; see [101].

We assume here $f(\cdot, y) \in C^\nu([0,1])$, $\forall y \in [0,1]$; then we define the ν-*partial fractional derivate* of f with respect to $x: \partial f^\nu(\cdot, y)/\partial x^\nu$ as

$$\frac{\partial^\nu f(x,y)}{\partial x^\nu} := \frac{\partial}{\partial x} \left(\mathcal{J}_{1-\alpha} \frac{\partial^n f(x,y)}{\partial x^n} \right), \; \forall (x,y) \in [0,1]^2. \tag{14.7}$$

Also, we assume $f(x, \cdot) \in C^\mu([0,1])$, $\forall x \in [0,1]$, where

$$C^\mu([0,1]) := \{g \in C^m([0,1]) : \mathcal{J}_{1-\beta} \, g^{(m)} \in C^1([0,1])\}. \tag{14.8}$$

Then we define *the* μ – *partial fractional derivative of* f *with respect to* y: $\partial f^\mu / \partial y^\mu(x, \cdot)$ as

$$\frac{\partial f^\mu(x,y)}{\partial y^\mu} := \frac{\partial}{\partial y} \left(\mathcal{J}_{1-\beta} \frac{\partial^m f}{\partial y^m}(x,y) \right), \; \forall (x,y) \in [0,1]^2. \tag{14.9}$$

Define the space

$$C^{\nu+\mu}([0,1]^2) := \{f \in C^{n+m}([0,1]^2) :$$

$$\mathcal{J}_{1-\alpha} \left(\frac{\partial^n f(\cdot, y)}{\partial x^n} \right) \in C^1([0,1]), \; \forall y \in [0,1];$$

$$\mathcal{J}_{1-\beta} \left(\frac{\partial^m f(x, \cdot)}{\partial x^m} \right) \in C^1([0,1]), \; \forall x \in [0,1];$$

$$\exists \, F_x, F_y, F_{yx} \in C([0,1]^2)\}. \tag{14.10}$$

Define *the mixed Canavati fractional partial derivative*:

$$\frac{\partial^{\nu+\mu} f(x,y)}{\partial x^\nu \partial y^\mu} :=$$

$$\frac{1}{\Gamma(1-\alpha)\Gamma(1-\beta)} \frac{\partial^2}{\partial x \partial y} \int_0^x \int_0^y (x-t)^{-\alpha} (y-s)^{-\beta} \frac{\partial^{n+m} f(t,s)}{\partial x^n \partial y^m} dt\, ds. \tag{14.11}$$

One can have anchor points $x_0, y_0 \neq 0$; then all the above definitions go through for $x \geq x_0$, $y \geq y_0$.

Conclusion 1: Clearly then we have $F_{xy} = F_{yx}$, and

$$\frac{\partial^{\nu+\mu} f}{\partial x^\nu \partial y^\mu} = \frac{\partial^{\mu+\nu} f}{\partial y^\mu \partial x^\nu}, \tag{14.12}$$

so the order of fractional differentiation is immaterial.

Here, it is by definition

$$\frac{\partial^{\mu+\nu} f(x,y)}{\partial y^\mu \partial x^\nu} :=$$

$$\frac{1}{\Gamma(1-\alpha)\Gamma(1-\beta)} \frac{\partial^2}{\partial y \partial x} \int_0^x \int_0^y (x-t)^{-\alpha}(y-s)^{-\beta} \frac{\partial^{m+n} f(t,s)}{\partial y^m \partial x^n} \, dt \, ds. \tag{14.13}$$

Comment: (1) Let $\nu = 0$; then $n = \alpha = 0$, and (14.11) becomes

$$\frac{\partial^\mu f(x,y)}{\partial y^\mu} = \frac{1}{\Gamma(1-\beta)} \frac{\partial^2}{\partial x \partial y} \int_0^x \int_0^y (y-s)^{-\beta} \frac{\partial^m f(t,s)}{\partial y^m} \, dt \, ds =$$

$$\frac{1}{\Gamma(1-\beta)} \frac{\partial^2}{\partial y \partial x} \int_0^x \int_0^y (y-s)^{-\beta} \frac{\partial^m f(t,s)}{\partial y^m} \, dt \, ds =$$

$$\frac{1}{\Gamma(1-\beta)} \left(\frac{\partial}{\partial y} \left(\frac{\partial}{\partial x} \int_0^x \left(\int_0^y (y-s)^{-\beta} \frac{\partial^m f(t,s)}{\partial y^m} \, ds \right) dt \right) \right) =: (\star). \tag{14.14}$$

Notice for fixed y we have that $(y-s)^{-\beta} \partial^m f(t,s)/\partial y^m$ is integrable over $[0,y]$, so the function

$$\varphi(t) := \int_0^y (y-s)^{-\beta} \frac{\partial^m f(t,s)}{\partial y^m} \, ds \tag{14.15}$$

is real-valued for any $t \in [0,x]$.

By continuity of $\partial^m f/\partial y^m$ we have true that $\forall \varepsilon > 0 \; \exists \, \delta > 0$: whenever $|t_1 - t_2| < \delta$ we have $|\partial^m f(t_1,s)/\partial y^m - \partial^m f(t_2,s)/\partial y^m| < \varepsilon$. We further have

$$\varphi(t_1) - \varphi(t_2) = \int_0^y (y-s)^{-\beta} \left(\frac{\partial^m f(t_1,s)}{\partial y^m} - \frac{\partial^m f(t_2,s)}{\partial y^m} \right) ds.$$

Hence

$$|\varphi(t_1) - \varphi(t_2)| \leq \int_0^y (y-s)^{-\beta} \left| \frac{\partial^m f(t_1,s)}{\partial y^m} - \frac{\partial^m f(t_2,s)}{\partial y^m} \right| ds \leq$$

$$\varepsilon \int_0^y (y-s)^{-\beta} ds = \frac{\varepsilon\, y^{1-\beta}}{1-\beta}, \tag{14.16}$$

proving $\varphi(t)$ continuous.

Consequently

$$(\star) = \frac{1}{\Gamma(1-\beta)} \left(\frac{\partial}{\partial y} \left(\int_0^y (y-s)^{-\beta} \frac{\partial^m f(x,s)}{\partial y^m} ds \right) \right) =: \frac{\partial^\mu f(x,y)}{\partial y^\mu}. \tag{14.17}$$

Conclusion 2: When $\nu = 0$, the fractional mixed partial derivative collapses to the single fractional partial derivative.

(2) Let $\mu = 0$; then $m = \beta = 0$, and (14.11) becomes

$$\frac{\partial^\nu f(x,y)}{\partial x^\nu} = \frac{1}{\Gamma(1-\alpha)} \frac{\partial^2}{\partial x \partial y} \int_0^x \int_0^y (x-t)^{-\alpha} \frac{\partial^n f(t,s)}{\partial x^n} dt\, ds =$$

$$\frac{1}{\Gamma(1-\alpha)} \left(\frac{\partial}{\partial x} \left(\frac{\partial}{\partial y} \left(\int_0^y \left(\int_0^x (x-t)^{-\alpha} \frac{\partial^n f(t,s)}{\partial x^n} dt \right) ds \right) \right) \right) \tag{14.18}$$

$$\left(\text{notice} \int_0^x (x-t)^{-\alpha} \frac{\partial^n f(t,s)}{\partial x^n} dt \text{ is continuous in } s \in [0,y] \right) =$$

$$\frac{1}{\Gamma(1-\alpha)} \left(\frac{\partial}{\partial x} \left(\int_0^x (x-t)^{-\alpha} \frac{\partial^n f(t,y)}{\partial x^n} dt \right) \right) =: \frac{\partial^\nu f(x,y)}{\partial x^\nu}. \tag{14.19}$$

Conclusion 3: When $\mu = 0$, the mixed fractional derivative collapses again to the single one.

(3) Let now $n = \nu \in \mathbb{N}$ (i.e., $\alpha = 0$); then

$$\frac{\partial^\nu f(x,y)}{\partial x^\nu} = \frac{\partial}{\partial x} \left(\int_0^x \frac{\partial^n f(t,y)}{\partial x^n} dt \right) = \frac{\partial^n f(x,y)}{\partial x^n}, \tag{14.20}$$

the ordinary one.

(4) When $m = \mu \in \mathbb{N}$ (i.e., $\beta = 0$), then

$$\frac{\partial^\mu f(x,y)}{\partial y^\mu} = \frac{\partial}{\partial y} \int_0^y \frac{\partial^m f(x,s)}{\partial y^m} ds = \frac{\partial^m f(x,y)}{\partial y^m}, \tag{14.21}$$

the ordinary one.

(5) Furthermore, let finally both $\nu = n \in \mathbb{N}$ and $\mu = m \in \mathbb{N}$ (i.e., $\alpha = \beta = 0$). Then

$$\frac{\partial^{\nu+\mu} f(x,y)}{\partial x^\nu \partial y^\mu} = \frac{\partial^2}{\partial x \partial y} \int_0^x \int_0^y \frac{\partial^{n+m} f(t,s)}{\partial x^n \partial y^m} dt\, ds =$$

$$\frac{\partial}{\partial x} \left(\frac{\partial}{\partial y} \left(\int_0^y \left(\int_0^x \frac{\partial^{n+m} f(t,s)}{\partial x^n \partial y^m} dt \right) ds \right) \right)$$

$$= \frac{\partial}{\partial x}\left(\int_0^x \frac{\partial^{n+m} f(t,y)}{\partial x^n \partial y^m}\, dt\right) = \frac{\partial^{n+m} f(x,y)}{\partial x^n \partial y^m}, \qquad (14.22)$$

proving that the fractional mixed partial collapses to the ordinary one. Fractional differentiation is a linear operation.

Conclusion 4: The above definitions we gave for the *fractional partial derivatives* are natural extensions of the ordinary positive integer ones.

Having introduced the fractional partial derivatives we are ready to develop our Opial-type results.

We make

Remark 14.4. First we consider a general domain. Let Q be a compact and convex subset of $\mathbb{R}^N$, $N \geq 2$; $z := (z_1, ..., z_N)$, $x_0 := (x_{01}, ...x_{0N}) \in Q$ be fixed. Let $f \in C^n(Q)$, $n \in \mathbb{N}$. Set $g_z(t) = f(x_0 + t(z - x_0))$, $0 \leq t \leq 1$;

$$g_z(0) = f(x_0), \quad g_z(1) = f(z).$$

Then it holds

$$g_z^{(j)}(t) = \left[\left(\sum_{i=1}^N (z_i - x_{0i})\frac{\partial}{\partial x_i}\right)^j f\right](x_0 + t(z - x_0)), \qquad (14.23)$$

$j = 0, 1, 2, \ldots, n$, and in particular

$$g_z'(t) = \sum_{i=1}^N (z_i - x_{0i})\frac{\partial f}{\partial x_i}(x_0 + t(z - x_0)), \qquad (14.24)$$

$0 \leq t \leq 1$.

Clearly here $g_z \in C^n([0,1])$. Let first $1 \leq \nu < 2$; in that case we take $n := [\nu] = 1$. Following [40] and by assuming that as a function of t: $f_{x_i}(x_0 + t(z - x_0)) \in C^{\nu-1}([0,1])$, $i = 1, \ldots, N$, then there exists $g_z^{(\nu)} = (\mathcal{J}_{2-\nu}\, g_z')'$, and it holds

$$g_z^{(\nu)}(t) = \sum_{i=1}^N (z_i - x_{0i})\left(\frac{\partial f}{\partial x_i}(x_0 + t(z - x_0))\right)^{(\nu-1)}, \qquad (14.25)$$

$0 \leq t \leq 1$.
Also here we have

$$(\mathcal{J}_{2-\nu}\, g_z')(t) = \frac{\sum_{i=1}^N (z_i - x_{0i})}{\Gamma(2-\nu)}\int_0^t (t-s)^{1-\nu} f_{x_i}(x_0 + s(z - x_0))ds, \qquad (14.26)$$

$0 \leq t \leq 1$.
Clearly $(\mathcal{J}_{2-\nu}\, g_z')(t) \in C^1([0,1])$ and $(\mathcal{J}_{2-\nu}\, g_z')(0) = 0$. Therefore by (14.2) we get

$$\int_0^s |\mathcal{J}_{2-\nu}\, g_z'(t)|\, |D^\nu g_z(t)|\, dt \leq \frac{s}{2}\int_0^s (D^\nu g_z(t))^2\, dt, \; \forall\, s \in [0,1]. \quad (14.27)$$

We have established the following Opial-type result.

Theorem 14.5. *Let Q be a compact and convex subset of $\mathbb{R}^N$, $N \geq 2$; $z, x_0 \in Q$ be fixed; $1 \leq \nu < 2$. Let $f \in C^1(Q)$. Assume that as a function of t: $f_{x_i}(x_0 + t(z - x_0)) \in C^{\nu-1}([0, 1])$, $i = 1, \ldots, N$.*
Then

$$\frac{1}{\Gamma(2-\nu)} \int_0^s \left| \sum_{i=1}^N (z_i - x_{0i}) \left(\int_0^t (t-s)^{1-\nu} f_{x_i}(x_0 + s(z - x_0)) ds \right) \right|$$

$$\left| \sum_{i=1}^N (z_i - x_{0i})(f_{x_i}(x_0 + t(z - x_0)))^{(\nu-1)} \right| dt \leq$$

$$\frac{s}{2} \int_0^s \left(\sum_{i=1}^N (z_i - x_{0i})(f_{x_i}(x_0 + t(z - x_0)))^{(\nu-1)} \right)^2 dt, \tag{14.28}$$

$\forall s \in [0, 1]$.

Remark 14.6. (Continuation) Let here $\nu \geq 2$ and $n := [\nu]$, $\beta := \nu - n$. We assume that as functions of t: $f_\alpha(x_0 + t(z - x_0)) \in C^{(\nu-n)}([0, 1])$, for all $\alpha := (\alpha_1, \ldots, \alpha_k)$, $\alpha_i \in \mathbb{Z}^+$, $i = 1, \ldots, N$; $|\alpha| := \sum_{i=1}^N \alpha_i = n$. Clearly then there exists $g_z^{(\nu)} = (\mathcal{J}_{1-\beta} \, g_z^{(n)})'$, and it holds

$$g_z^{(\nu)}(t) = \left[\left(\sum_{i=1}^N (z_i - x_{0i}) \frac{\partial}{\partial x_i} \right)^n f \right]^{(\nu-n)} (x_0 + t(z - x_0)), \tag{14.29}$$

all $t \in [0, 1]$.
Of course, it holds

$$(\mathcal{J}_{1-\beta} \, g_z^{(n)})(t) \stackrel{(14.23)}{=} \frac{1}{\Gamma(1-\beta)} \int_0^t (t-s)^\beta$$

$$\left\{ \left[\left(\sum_{i=1}^N (z_i - x_{0i}) \frac{\partial}{\partial x_i} \right)^n f \right] (x_0 + s(z - x_0)) \right\} ds. \tag{14.30}$$

Notice $(\mathcal{J}_{1-\beta} \, g_z^{(n)})(0) = 0$. Hence again by (14.2) we get

$$\int_0^s |\mathcal{J}_{1-\beta} \, g_z^{(n)}(t)| \, |D^\nu g_z(t)| dt \leq \frac{s}{2} \int_0^s (D^\nu g_z(t))^2 dt, \forall s \in [0, 1]. \tag{14.31}$$

We have proved the following general Opial-type result.

Theorem 14.7. *Let Q be a compact and convex subset of $\mathbb{R}^N$, $N \geq 2$; $z, x_0 \in Q$ be fixed; $\nu \geq 2$, $n := [\nu]$, $\beta := \nu - n$. Let $f \in C^n(Q)$.*

Assume that as a function of $t : f_\alpha(x_0 + t(z - x_0)) \in C^{(\nu-n)}([0,1])$, for all $\alpha := (\alpha_1, ..., \alpha_k)$, $\alpha_i \in \mathbb{Z}^+$, $i = 1, \ldots, N$; $|\alpha| := \sum_{i=1}^{N} \alpha_i = n$. Then

$$\frac{1}{\Gamma(1-\beta)} \int_0^s \left| \int_0^t (t-s)^{-\beta} \left\{ \left[\left(\sum_{i=1}^N (z_i - x_{0i}) \frac{\partial}{\partial x_i} \right)^n f \right] (x_0 + s(z - x_0)) \right\} ds \right.$$

$$\left. \left| \left[\left(\sum_{i=1}^N (z_i - x_{0i}) \frac{\partial}{\partial x_i} \right)^n f \right]^{(\nu-n)} (x_0 + t(z - x_0)) \right| dt \leq$$

$$\frac{s}{2} \int_0^s \left\{ \left[\left(\sum_{i=1}^N (z_i - x_{0i}) \frac{\partial}{\partial x_i} \right)^n f \right]^{(\nu-n)} (x_0 + t(z - x_0)) \right\}^2 dt, \quad (14.32)$$

$\forall \, s \in [0,1]$.

Note. Following the last pattern one can transfer any univariate Opial-type inequality(see [4]) into this fractional multivariate general setting. Inasmuch as no chain rule is valid in the fractional differentiation, inequalities such as (14.31), (14.32) are not revealing themselves to totally decompose into all of their ingredients. Next, working over *spherical shells* we obtain a series of various Opial-type fractional multivariate inequalities that look nice and are very clear.

We give

Definition 14.8. (see [17; 19, p. 540]) In the following we carry earlier notions introduced in Remark 14.3 over to arbitrary $[a,b] \subseteq \mathbb{R}$. Let $x, x_0 \in [a,b]$ such that $x \geq x_0, x_0$ is fixed. Let $f \in C([a,b])$ and define

$$(\mathcal{J}_\nu^{x_0} f)(x) := \frac{1}{\Gamma(\nu)} \int_{x_0}^x (x-t)^{\nu-1} f(t) dt, \quad x_0 \leq x \leq b, \quad (14.33)$$

the generalized Riemann–Liouville integral. We consider the subspace $C_{x_0}^\nu([a,b])$ of $C^n([a,b])$:

$$C_{x_0}^\nu([a,b]) := \{ f \in C^n([a,b]) : \mathcal{J}_{1-\alpha}^{x_0} f^{(n)} \in C^1([x_0,b]) \} \quad (14.34)$$

Hence, let $f \in C_{x_0}^\nu([a,b])$; we define the *generalized ν - fractional derivative of f over $[x_0,b]$* as

$$D_{x_0}^\nu f := \left(\mathcal{J}_{1-\alpha}^{x_0} f^{(n)} \right)'. \quad (14.35)$$

Notice that

$$\left(\mathcal{J}_{1-\alpha}^{x_0} f^{(n)} \right)(x) = \frac{1}{\Gamma(1-\alpha)} \int_{x_0}^x (x-t)^{-\alpha} f^{(n)}(t) dt \quad (14.36)$$

exists for $f \in C_{x_0}^{\nu}([a, b])$.

Next we use

Theorem 14.9. [15; 19, p. 567] *Let* $\gamma_i \geq 0$, $\nu \geq 1$ *such that* $\nu - \gamma_i \geq$ 1; $i = 1, \ldots, l$ *and* $f \in C_{x_0}^{\nu}([a, b])$ *with* $f^{(j)}(x_0) = 0$, $j = 0, 1, \ldots, n - 1$, $n := [\nu]$. *Here* $x, x_0 \in [a, b] : x \geq x_0$. *Let* $q_1, q_2 > 0$ *be continuous functions on* $[a, b]$ *and* $r_i > 0 : \sum_{i=1}^{l} r_i = r$. *Let* $s_1, s_1' > 1 : 1/s_1 + 1/s_1' = 1$ *and* $s_2, s_2' > 1 : 1/s_2 + 1/s_2' = 1$ *and* $p > s_2$. *Furthermore suppose that*

$$Q_1 := \left(\int_{x_0}^{x} (q_1(\omega))^{s_1'} d\omega \right)^{1/s_1'} < +\infty \qquad (14.37)$$

and

$$Q_2 := \left(\int_{x_0}^{x} (q_2(\omega))^{-s_2'/p} d\omega \right)^{r/s_2'} < +\infty. \qquad (14.38)$$

Call $\sigma := p - s_2/ps_2$. *Then*

$$\int_{x_0}^{x} q_1(\omega) \prod_{i=1}^{l} |(D_{x_0}^{\gamma_i}(f))(\omega)|^{r_i} d\omega \leq Q_1 Q_2$$

$$\cdot \prod_{i=1}^{l} \left\{ \frac{\sigma^{r_i \sigma}}{(\Gamma(\nu - \gamma_i))^{r_i} (\nu - \gamma_i - 1 + \sigma)^{r_i \sigma}} \right\}$$

$$\cdot \frac{(x - x_0)^{(\sum_{i=1}^{l}(\nu - \gamma_i - 1)r_i + \sigma r + 1/s_1)}}{((\sum_{i=1}^{l}(\nu - \gamma_i - 1)r_i s_1) + rs_1 \sigma + 1)^{1/s_1}}$$

$$\cdot \left(\int_{x_0}^{x} q_2(\omega) |(D_{x_0}^{\nu} f)(\omega)|^p d\omega \right)^{r/p}. \qquad (14.39)$$

We next work in the setting of spherical shells introduced in Section 14.1, Introduction.

We need

Definition 14.10. Let $\nu > 0$, $n := [\nu]$, $\alpha := \nu - n$, $f \in C^n(\bar{A})$, and A is a spherical shell. Assume that there exists $\partial_{R_1}^{\nu} f(x)/\partial r^{\nu} \in C(\bar{A})$, given by

$$\frac{\partial_{R_1}^{\nu} f(x)}{\partial r^{\nu}} := \frac{1}{\Gamma(1 - \alpha)} \frac{\partial}{\partial r} \left(\int_{R_1}^{r} (r - t)^{-\alpha} \frac{\partial^n f(t\omega)}{\partial r^n} dt \right), \qquad (14.40)$$

where $x \in \bar{A}$; that is, $x = r\omega$, $r \in [R_1, R_2]$, and $\omega \in S^{N-1}$. We call $\partial_{R_1}^{\nu} f/\partial r^{\nu}$ the *radial fractional derivative of* f *of order* ν.

We need

Lemma 14.11. *Let* $\gamma \geq 0$, $\nu \geq 1$ *such that* $\nu - \gamma \geq 1$. *Let* $f \in C^n(\bar{A})$ *and there exists* $\partial^\nu_{R_1} f(x)/\partial r^\nu \in C(\bar{A})$, $x \in \bar{A}$, A *a spherical shell. Further assume that* $\partial^j f(R_1 \omega)/\partial r^j = 0$, $j = 0, 1, \ldots, n-1$, $n := [\nu]$, $\forall \, \omega \in S^{N-1}$. *Then there exists* $\partial^\gamma_{R_1} f(x)/\partial r^\gamma \in C(\bar{A})$.

Proof. The assumption implies that $\partial^\nu_{R_1} f(r\omega)/\partial r^\nu \in C([R_1, R_2])$, $\forall \, \omega \in S^{N-1}$; that is, $f(r\omega) \in C^\nu_{R_1}([R_1, R_2], \forall \, \omega \in S^{N-1}$. Following [17], and [19, pp. 544–545], we get that there exists $\partial^\gamma_{R_1} f(r\omega)/\partial r^\gamma$, given by

$$\frac{\partial^\gamma_{R_1} f(r\omega)}{\partial r^\gamma} = \frac{1}{\Gamma(\nu - \gamma)} \int_{R_1}^{r} (r-t)^{\nu-\gamma-1} \frac{\partial^\nu_{R_1} f(t\omega)}{\partial r^\nu} dt; \qquad (14.41)$$

indeed $f(r\omega) \in C^\gamma_{R_1}([R_1, R_2])$, $\forall \, \omega \in S^{N-1}$.

Hence

$$\frac{\partial^\gamma_{R_1} f(r\omega)}{\partial r^\gamma} = \frac{1}{\Gamma(\nu - \gamma)} \int_{R_1}^{R_2} \mathcal{X}_{[R_1, r]}(t)(r-t)^{\nu-\gamma-1} \frac{\partial^\nu_{R_1} f(t\omega)}{\partial r^\nu} dt. \quad (14.42)$$

Let $r_n \to r$, $\omega_n \to \omega$; then $\mathcal{X}_{[R_1, r_n]}(t) \to \mathcal{X}_{[R_1, r]}(t)$, a.e.; also $(r_n - t)^{\nu-\gamma-1} \to (r-t)^{\nu-\gamma-1}$, and

$$\frac{\partial^\nu_{R_1} f(t\omega_n)}{\partial r^\nu} \to \frac{\partial^\nu_{R_1} f(t\omega)}{\partial r^\nu}.$$

Furthermore it holds that

$$\mathcal{X}_{[R_1, r_n]}(t)(r_n - t)^{\nu-\gamma-1} \frac{\partial^\nu_{R_1} f(t\omega_n)}{\partial r^\nu} \longrightarrow$$

$$\mathcal{X}_{[R_1, r]}(t)(r - t)^{\nu-\gamma-1} \frac{\partial^\nu_{R_1} f(t\omega)}{\partial r^\nu}, \quad \text{a.e. on } [R_1, R_2]. \qquad (14.43)$$

However we have

$$\mathcal{X}_{[R_1, r_n]}(t)|r_n - t|^{\nu-\gamma-1} \left| \frac{\partial^\nu_{R_1} f(t\omega_n)}{\partial r^\nu} \right| \leq$$

$$(R_2 - R_1)^{\nu-\gamma-1} \left\| \frac{\partial^\nu_{R_1} f}{\partial r^\nu} \right\|_\infty < \infty. \qquad (14.44)$$

Thus, by the dominated convergence theorem we obtain

$$\int_{R_1}^{R_2} \mathcal{X}_{[R_1, r_n]}(t)(r_n - t)^{\nu-\gamma-1} \frac{\partial^\nu_{R_1} f(t\omega_n)}{\partial r^\nu} \, dt \to$$

$$\int_{R_1}^{R_2} \mathcal{X}_{[R_1, r]}(t)(r - t)^{\nu-\gamma-1} \frac{\partial^\nu_{R_1} f(t\omega)}{\partial r^\nu} \, dt, \qquad (14.45)$$

proving the claim.   □

We present the following very general result.

Theorem 14.12. *Let* $\gamma_i \geq 0,$ $\nu \geq 1,$ *such that* $\nu - \gamma_i \geq 1;$ $i = 1,\ldots,l,$ $n := [\nu].$ *Let* $f \in C^n(\bar{A})$ *and there exists* $\partial_{R_1}^{\nu} f(x)/\partial r^{\nu} \in C(\bar{A}),$ $x \in \bar{A};$ A *is a spherical shell:* $A := B(0,R_2) - \overline{B(0,R_1)} \subseteq \mathbb{R}^N,$ $N \geq 2.$ *Furthermore assume that* $\partial^j f/\partial r^j,$ $j = 0,1,\ldots,n-1,$ *vanish on* $\partial B(0,R_1).$ *Let* $r_i > 0 : \sum_{i=1}^{l} r_i = p.$ *Let* $s_1, s_1' > 1 : 1/s_1 + 1/s_1' = 1,$ *and* $s_2, s_2' > 1 : 1/s_2 + 1/s_2' = 1,$ *and* $p > s_2.$ *Denote*

$$Q_1 = \left(\frac{R_2^{(N-1)s_1'+1} - R_1^{(N-1)s_1'+1}}{(N-1)s_1' + 1} \right)^{1/s_1'}, \tag{14.46}$$

and

$$Q_2 = \left(\frac{R_2^{(1-N)s_2'/p+1} - R_1^{(1-N)s_2'/p+1}}{(1-N)\frac{s_2'}{p} + 1} \right)^{p/s_2'}. \tag{14.47}$$

Call $\sigma := p - s_2/p\, s_2.$
   Also call

$$C := Q_1 Q_2 \prod_{i=1}^{l} \left\{ \frac{\sigma^{r_i \sigma}}{(\Gamma(\nu - \gamma_i))^{r_i} (\nu - \gamma_i - 1 + \sigma)^{r_i \sigma}} \right\}$$

$$\frac{(R_2 - R_1)^{\left(\sum_{i=1}^{l} (\nu - \gamma_i - 1) r_i + \frac{p}{s_2} + \frac{1}{s_1} - 1 \right)}}{\left(\left(\sum_{i=1}^{l} (\nu - \gamma_i - 1) r_i s_1 \right) + s_1(\frac{p}{s_2} - 1) + 1 \right)^{1/s_1}}. \tag{14.48}$$

Then

$$\int_A \prod_{i=1}^{l} \left| \frac{\partial_{R_1}^{\gamma_i} f(x)}{\partial r^{\gamma_i}} \right|^{r_i} dx \leq C \int_A \left| \frac{\partial_{R_1}^{\nu} f(x)}{\partial r^{\nu}} \right|^p dx. \tag{14.49}$$

Proof. The assumption implies that $f(r\,\omega) \in C^n([R_1,R_2]), \partial_{R_1}^{\nu} f(r\,\omega)/\partial r^{\nu} \in C([R_1,R_2]),$ $\forall\, \omega \in S^{N-1}.$ By Theorem 14.9 we have

$$\int_{R_1}^{R_2} r^{N-1} \prod_{i=1}^{l} \left| \frac{\partial_{R_1}^{\gamma_i} f(r\,\omega)}{\partial r^{\gamma_i}} \right|^{r_i} dr \leq C \int_{R_1}^{R_2} r^{N-1} \left| \frac{\partial_{R_1}^{\nu} f(r\,\omega)}{\partial r^{\nu}} \right|^p dr, \ \forall\, \omega \in S^{N-1}. \tag{14.50}$$

Therefore it holds

$$\int_{S^{N-1}} \left(\int_{R_1}^{R_2} r^{N-1} \prod_{i=1}^{l} \left| \frac{\partial_{R_1}^{\gamma_i} f(r\,\omega)}{\partial r^{\gamma_i}} \right|^{r_i} dr \right) d\omega \leq$$

$$C \int_{S^{N-1}} \left(\int_{R_1}^{R_2} r^{N-1} \left| \frac{\partial_{R_1}^{\nu} f(r\,\omega)}{\partial r^{\nu}} \right|^p dr \right) d\omega. \qquad (14.51)$$

Using Lemma 14.11 and by (14.3) we derive (14.49). $\square$

We mention

Theorem 14.13. [15; 19, p. 573] *Let* $\gamma_i \geq 0$, $\nu \geq 1$ *such that* $\nu - \gamma_i \geq 1$; $i = 1, \ldots, l$ *and* $f \in C_{x_0}^{\nu}([a,b])$ *with* $f^{(j)}(x_0) = 0$, $j = 0, 1 \ldots, n-1$, $n := [\nu]$. *Here* $x, x_0 \in [a,b] : x \geq x_0$. *Let* $\tilde{q}(w) \geq 0$ *continuous on* $[a,b]$ *and* $r_i > 0 : \sum_{i=1}^{l} r_i = r$. *Then*

$$\int_{x_0}^{x} \tilde{q}(w) \cdot \prod_{i=1}^{l} (|D_{x_0}^{\nu} f|(w))^{r_i} dw$$

$$\leq \left\{ \frac{\|\tilde{q}\|_{\infty} (\|D_{x_0}^{\nu} f\|_{\infty})^r}{\prod_{i=1}^{l} (\Gamma(\nu - \gamma_i + 1))^{r_i}} \right\} \cdot \left\{ \frac{(x - x_0)^{r\nu - \sum_{i=1}^{l} r_i \gamma_i + 1}}{(r\nu - \sum_{i=1}^{l} r_i \gamma_i + 1)} \right\}. \qquad (14.52)$$

We give

Theorem 14.14. *Let* $\gamma_i \geq 0$, $\nu \geq 1$, *such that* $\nu - \gamma_i \geq 1$; $i = 1, \ldots, l$, $n := [\nu]$. *Let* $f \in C^n(\bar{A})$ *and there exists* $\partial_{R_1}^{\nu} f(x)/\partial r^{\nu} \in C(\bar{A})$, $x \in \bar{A}$; A *is a spherical shell:* $A := B(0, R_2) - \overline{B(0, R_1)} \subseteq \mathbb{R}^N$, $N \geq 2$. *Furthermore assume that* $\partial^j f / \partial r^j$, $j = 0, 1, \ldots, n-1$, *vanish on* $\partial B(0, R_1)$. *Let* $r_i > 0 : \sum_{i=1}^{l} r_i = r$. *Call*

$$M := \frac{R_2^{N-1} (R_2 - R_1)^{r\nu - \sum_{i=1}^{l} r_i \gamma_i + 1}}{\prod_{i=1}^{l} (\Gamma(\nu - \gamma_i + 1))^{r_i} (r\nu - \sum_{i=1}^{l} r_i \gamma_i + 1)} > 0. \qquad (14.53)$$

Then

$$\int_A \left(\prod_{i=1}^{l} \left| \frac{\partial_{R_1}^{\gamma_i} f(x)}{\partial r^{\gamma_i}} \right|^{r_i} \right) dx \leq M \frac{2\pi^{N/2}}{\Gamma(N/2)} \left\| \frac{\partial_{R_1}^{\nu} f}{\partial r^{\nu}} \right\|_{\infty, \bar{A}}^r. \qquad (14.54)$$

Proof. By Theorem 14.13 we get that

$$\int_{R_1}^{R_2} r^{N-1} \left(\prod_{i=1}^{l} \left| \frac{\partial_{R_1}^{\gamma_i} f(r\omega)}{\partial r^{\gamma_i}} \right|^{r_i} \right) dr \leq M \left\| \frac{\partial_{R_1}^{\nu} f}{\partial r^{\nu}} \right\|_{\infty, \bar{A}}^r. \qquad (14.55)$$

Hence it holds

$$\int_{S^{N-1}} \left(\int_{R_1}^{R_2} r^{N-1} \left(\prod_{i=1}^{l} \left| \frac{\partial_{R_1}^{\gamma_i} f(r\,\omega)}{\partial r^{\gamma_i}} \right|^{r_i} \right) dr \right) d\omega \leq M\, \omega_N \left| \frac{\partial_{R_1}^{\nu} f}{\partial r^{\nu}} \right|_{\infty, \bar{A}}^r. \qquad (14.56)$$

Using (14.3) and Lemma 14.11, we establish the claim.   □

We need

Theorem 14.15. [61]. *Let $\gamma \geq 0$, $\nu \geq 1$, $\nu - \gamma \geq 1$, $\alpha, \beta > 0$, $r > \alpha$, $r > 1$; let $p > 0$, $q \geq 0$ be continuous functions on $[a,b]$. Let $f \in C^\nu_{x_0}([a,b])$ with $f^{(i)}(x_0) = 0$, $i = 0, 1, \ldots, n-1$, $n := [\nu]$. Let $x, x_0 \in [a,b]$ with $x \geq x_0$. Then*

$$\int_{x_0}^x q(w)|D^\gamma_{x_0} f(w)|^\beta |D^\nu_{x_0} f(w)|^\alpha dw \leq K(p,q,\gamma,\nu,\alpha,\beta,r,x,x_0)$$

$$\cdot \left(\int_{x_0}^x p(w)\left|D^\nu_{x_0} f(w)\right|^r dw\right)^{(\alpha+\beta/r)}. \tag{14.57}$$

Here

$$K(p,q,\gamma,\nu,\alpha,\beta,r,x,x_0) := \left(\frac{\alpha}{\alpha+\beta}\right)^{\alpha/r} \cdot \frac{1}{(\Gamma(\nu-\gamma))^\beta}$$

$$\cdot \left(\int_{x_0}^x (q(w))^r \cdot (p(w))^{-\alpha})^{1/(r-\alpha)} \cdot (P_1(w))^{(\beta(r-1)/r-\alpha)} \cdot dw\right)^{r-\alpha/r}, \tag{14.58}$$

with

$$P_1(w) := \int_{x_0}^w (p(t))^{-1/r-1} \cdot (w-t)^{(\nu-\gamma-1)(r/r-1)} \cdot dt. \tag{14.59}$$

We present

Theorem 14.16. *Let $\gamma \geq 0$, $\nu \geq 1$, $n := [\nu]$, $\nu - \gamma \geq 1$, $\alpha, \beta > 0$, $\alpha + \beta > 1$. Let $f \in C^n(\bar{A})$ and there exists $\partial^\nu_{R_1} f(x)/\partial r^\nu \in C(\bar{A})$, $x \in \bar{A}$; A is a spherical shell: $A := B(0, R_2) - \overline{B(0, R_1)} \subseteq \mathbb{R}^N$, $N \geq 2$. Furthermore assume that $\partial^j f/\partial r^j = 0$, for $j = 0, 1, \ldots, n-1$, on $\partial B(0, R_1)$. Then*

$$\int_A \left|\frac{\partial^\gamma_{R_1} f(x)}{\partial r^\gamma}\right|^\beta \left|\frac{\partial^\nu_{R_1} f(x)}{\partial r^\nu}\right|^\alpha dx \leq K \int_A \left|\frac{\partial^\nu_{R_1} f(x)}{\partial r^\nu}\right|^{\alpha+\beta} dx. \tag{14.60}$$

Here

$$K = \left(\frac{\alpha}{\alpha+\beta}\right)^{\alpha/(\alpha+\beta)} \frac{1}{(\Gamma(\nu-\gamma))^\beta} \int_{R_1}^{R_2} r^{N-1} \left((P_1(r))^{(\alpha+\beta-1)}dr\right)^{\beta/(\alpha+\beta)}, \tag{14.61}$$

with

$$P_1(r) := \int_{R_1}^r t^{(1-N)/(\alpha+\beta-1)}(r-t)^{(\nu-\gamma-1)(\alpha+\beta)/(\alpha+\beta-1)}dr. \tag{14.62}$$

Proof. The assumption implies that $f(r\omega) \in C^n([R_1, R_2])$ and $\partial_{R_1}^\nu f(r\omega)/\partial r^\nu \in C([R_1, R_2])$, $\forall\, \omega \in S^{N-1}$. Hence by Theorem 14.15, $\forall\, \omega \in S^{N-1}$, we get that

$$\int_{R_1}^{R_2} r^{N-1} \left|\frac{\partial_{R_1}^\gamma f(r\omega)}{\partial r^\gamma}\right|^\beta \left|\frac{\partial_{R_1}^\nu f(r\omega)}{\partial r^\nu}\right|^\alpha dr \le K \int_{R_1}^{R_2} r^{N-1} \left|\frac{\partial_{R_1}^\nu f(r\omega)}{\partial r^\nu}\right|^{\alpha+\beta} dr. \tag{14.63}$$

Therefore it holds

$$\int_{S^{N-1}} \left(\int_{R_1}^{R_2} r^{N-1} \left|\frac{\partial_{R_1}^\gamma f(r\omega)}{\partial r^\gamma}\right|^\beta \left|\frac{\partial_{R_1}^\nu f(r\omega)}{\partial r^\nu}\right|^\alpha dr\right) d\omega$$

$$= K \left(\int_{S^{N-1}} \left(\int_{R_1}^{R_2} r^{N-1} \left|\frac{\partial_{R_1}^\nu f(r\omega)}{\partial r^\nu}\right|^{\alpha+\beta} dr\right) d\omega\right). \tag{14.64}$$

Using Lemma 14.11 and by (14.3) we derive (14.60). □

We need

Theorem 14.17. [61]. Let $\nu \ge 1$, $\alpha, \beta > 0$, $r > \alpha$, $r > 1$; let $p > 0$, $q \ge 0$ be continuous functions on $[a, b]$. Let $f \in C_{x_0}^\nu([a, b])$ with $f^{(i)}(x_0) = 0$, $i = 0, 1, \ldots, n-1$, $n := [\nu]$. Let $x, x_0 \in [a, b]$ with $x \ge x_0$. Then

$$\int_{x_0}^x g(w) |f(w)|^\beta |D_{x_0}^\nu f(x)|^\alpha dw \le K^*(p, q, \nu, \alpha, \beta, r, x, x_0)$$

$$\cdot \left(\int_{x_0}^x p(w) |D_{x_0}^\nu f(w)|^r dw\right)^{(\alpha+\beta/r)}. \tag{14.65}$$

Here

$$K^*(p, q, \nu, \alpha, \beta, r, x, x_0) := \left(\frac{\alpha}{\alpha+\beta}\right)^{\alpha/r} \cdot 1/(\Gamma(\nu))^\beta$$

$$\cdot \left(\int_{x_0}^x ((q(w))^r \cdot (p(w))^{-\alpha})^{1/(r-\alpha)} \cdot (P_1^*(w))^{(\beta(r-1)/r-\alpha)} \cdot dw\right)^{(r-\alpha)/r}, \tag{14.66}$$

with

$$P_1^*(w) := \int_{x_0}^w (p(t))^{-1/r-1} \cdot (w-t)^{(\nu-1)(r/(r-1))} dt. \tag{14.67}$$

Based on Theorem 14.17 we give similarly:

Theorem 14.18. Let $\nu \ge 1$, $n := [\nu]$, $\alpha, \beta > 0$, $\alpha + \beta > 1$. Let $f \in C^n(\bar{A})$ and there exists $\partial_{R_1}^\nu f(x)/\partial r^\nu \in C(\bar{A})$, $x \in \bar{A}$; A is a spherical *shell:* $A := B(0, R_2) - \overline{B(0, R_1)} \subseteq \mathbb{R}^N$, $N \geq 2$. *Furthermore assume that* $\partial^j f / \partial r^j = 0$, *for* $j = 0, 1, \ldots, n-1$, *on* $\partial B(0, R_1)$. *Then*

$$\int_A |f(x)|^\beta \left| \frac{\partial^\nu_{R_1} f(x)}{\partial r^\nu} \right|^\alpha dx \leq K^* \int_A \left| \frac{\partial^\nu_{R_1} f(x)}{\partial r^\nu} \right|^{\alpha+\beta} dx. \tag{14.68}$$

Here

$$K^* := \left(\frac{\alpha}{\alpha+\beta} \right)^{(\alpha/\alpha+\beta)} \frac{1}{(\Gamma(\nu))^\beta} \left(\int_{R_1}^{R_2} r^{N-1} (P_1^*(r))^{(\alpha+\beta-1)} dr \right)^{(\beta/(\alpha+\beta))}, \tag{14.69}$$

with

$$P_1^*(r) := \int_{R_1}^r t^{(1-N/\alpha+\beta-1)} (r-t)^{(\alpha+\beta)(\nu-1)/(\alpha+\beta-1)} dt. \tag{14.70}$$

Next we present a set of multivariate fractional Opial-type inequalities involving two functions over the shell.

We need

Theorem 14.19. [26]. *Let* $\nu \geq 1$, $\gamma_1, \gamma_2 \geq 0$, *such that* $\nu - \gamma_1 \geq 1$, $\nu - \gamma_2 \geq 1$, *and* $f_1, f_2 \in C^\nu_{x_0}([a,b])$ *with*

$$f_1^{(i)}(x_0) = f_2^{(i)}(x_0) = 0, \quad i = 0, 1, \ldots, n-1, \quad n := [\nu]. \tag{14.71}$$

Here, $x, x_0, \in [a,b] : x \geq x_0$. *Consider also* $p(t) > 0$, *and* $q(t) \geq 0$ *continuous functions on* $[x_0, b]$.
Let $\lambda_\nu > 0$ *and* $\lambda_\alpha, \lambda_\beta \geq 0$, *such that* $\lambda_\nu < p$, *where* $p > 1$. *Set*

$$P_k(w) := \int_{x_0}^w (w-t)^{(\nu-\gamma_k-1)p/(p-1)} (p(t))^{-1/(p-1)} dt, \quad k=1,2, \quad x_0 \leq w \leq b, \tag{14.72}$$

$$A(w) := \frac{q(w) \cdot (P_1(w))^{\lambda_\alpha((p-1)/p)} \cdot (P_2(w))^{\lambda_\beta((p-1)/p)} (p(w))^{-\lambda_\nu/p}}{(\Gamma(\nu-\gamma_1))^{\lambda_\alpha} \cdot (\Gamma(\nu-\gamma_2))^{\lambda_\beta}}, \tag{14.73}$$

$$A_0(x) := \left(\int_{x_0}^x A(w)^{p/(p-\lambda_\nu)} dw \right)^{(p-\lambda_\nu)/p}, \tag{14.74}$$

and

$$\delta_1 := \begin{cases} 2^{1-((\lambda_\alpha+\lambda_\nu)/p)}, & \text{if} \quad \lambda_\alpha + \lambda_\nu \leq p, \\ 1, & \text{if} \quad \lambda_\alpha + \lambda_\nu \geq p. \end{cases} \tag{14.75}$$

If $\lambda_\beta = 0$, *we obtain that*

$$\int_{x_0}^x q(w) \left[\left| (D_{x_0}^{\gamma_1} f_1)(w) \right|^{\lambda_\alpha} \cdot \left| (D_{x_0}^\nu f_1)(w) \right|^{\lambda_\nu} + \right.$$

$$\left| \left(D_{x_0}^{\lambda_1} f_2 \right)(w) \right|^{\lambda_\alpha} \cdot \left| \left(D_{x_0}^{\nu} f_2 \right)(w) \right|^{\lambda_\nu} \right] dw$$

$$\leq \left(A_0(x) \big|_{\lambda_\beta=0} \right) \cdot \left(\frac{\lambda_\nu}{\lambda_\alpha + \lambda_\nu} \right)^{\lambda_\nu/p} \cdot \delta_1$$

$$\cdot \left[\int_{x_0}^{x} p(w) \left[\left| \left(D_{x_0}^{\nu} f_1 \right)(w) \right|^p + \left| \left(D_{x_0}^{\nu} f_2 \right)(w) \right|^p \right] dw \right]^{((\lambda_\alpha+\lambda_\nu)/p)}. \quad (14.76)$$

Similarly, by (14.76), we derive

Theorem 14.20. *Let $\nu \geq 1$, $\gamma_1, \gamma_2 \geq 0$, such that $\nu - \gamma_1 \geq 1$, $\nu - \gamma_2 \geq 1$, $n := [\nu]$, and $f_1, f_2 \in C^n(\bar{A})$, and there exist $\partial_{R_1}^{\nu} f_1(x)/\partial r^{\nu}$, $\partial_{R_1}^{\nu} f_2(x)/\partial r^{\nu} \in C(\bar{A})$, $A := B(0, R_2) - \overline{B(0, R_1)} \subseteq \mathbb{R}^N$, $N \geq 2$. Furthermore assume $\partial^j f_1/\partial r^j = \partial^j f_2/\partial r^j = 0$, for $j = 0, 1, \ldots, n-1$, on $\partial B(0, R_1)$. Let $\lambda_\nu > 0$ and $\lambda_\alpha > 0$; $\lambda_\beta \geq 0$, $p := \lambda_\alpha + \lambda_\nu > 1$. Call*

$$P_k(w) := \int_{R_1}^{w} (w - t)^{(\nu-\gamma_k-1)p/(p-1)} t^{(1-N)/(p-1)} dt, \quad (14.77)$$

$k = 1, 2$, $R_1 \leq w \leq R_2$,

$$A(w) := \frac{w^{(N-1)(1-(\lambda_\nu/p))} (P_1(w))^{\lambda_\alpha((p-1)/p)} (P_2(w))^{\lambda_\beta((p-1)/p)}}{(\Gamma(\nu - \gamma_1))^{\lambda_\alpha} (\Gamma(\nu - \gamma_2))^{\lambda_\beta}}, \quad (14.78)$$

$$A_0(R_2) := \left(\int_{R_1}^{R_2} (A(w))^{p/\lambda_\alpha} dw \right)^{\lambda_\alpha/p}. \quad (14.79)$$

Take the case of $\lambda_\beta = 0$. Then

$$\int_A \left[\left| \frac{\partial_{R_1}^{\gamma_1}}{\partial r^{\gamma_1}} f_1(x) \right|^{\lambda_\alpha} \left| \left(\frac{\partial_{R_1}^{\nu} f_1}{\partial r^{\nu}} \right)(x) \right|^{\lambda_\nu} + \right.$$

$$\left. \left| \left(\frac{\partial_{R_1}^{\gamma_1} f_2}{\partial r^{\gamma_1}} \right)(x) \right|^{\lambda_\alpha} \left| \left(\frac{\partial_{R_1}^{\nu} f_2}{\partial r^{\nu}} \right)(x) \right|^{\lambda_\nu} \right] dx \leq$$

$$\left(A_0(R_2)\big|_{\lambda_\beta=0} \right) \left(\frac{\lambda_\nu}{p} \right)^{(\lambda_\nu/p)} \int_A \left[\left| \left(\frac{\partial_{R_1}^{\nu} f_1}{\partial r^{\nu}} \right)(x) \right|^p + \left| \left(\frac{\partial_{R_1}^{\nu} f_2}{\partial r^{\nu}} \right)(x) \right|^p \right] dx. \quad (14.80)$$

We need

Theorem 14.21. [26]. *All here are as in Theorem* 14.19. *Denote*

$$\delta_3 := \begin{cases} 2^{\lambda_\beta/\lambda_\nu} - 1, & if \quad \lambda_\beta \ge \lambda_\nu, \\ 1, & if \quad \lambda_\beta \le \lambda_\nu. \end{cases}$$

If $\lambda_\alpha = 0$, *then*

$$\int_{x_0}^x q(w) \left[\left| \left(D_{x_0}^{\gamma_2} f_2 \right)(w) \right|^{\lambda_\beta} \cdot \left| \left(D_{x_0}^\nu f_1 \right)(w) \right|^{\lambda_\nu} + \right.$$

$$\left. \left| \left(D_{x_0}^{\gamma_2} f_1 \right)(w) \right|^{\lambda_\beta} \cdot \left| \left(D_{x_0}^\nu f_2 \right)(w) \right|^{\lambda_\nu} \right] dw$$

$$\le \left(A_0(x)|_{\lambda_\alpha=0} \right) 2^{p-\lambda_\nu/p} \left(\frac{\lambda_\nu}{\lambda_\beta + \lambda_\nu} \right)^{\lambda_\nu/p} \delta_3^{\lambda_\nu/p}$$

$$\cdot \left(\int_{x_0}^x p(w) \left[\left| \left(D_{x_0}^\nu f_1 \right)(w) \right|^p + \left| \left(D_{x_0}^\nu f_2 \right)(w) \right|^p \right] dw \right)^{((\lambda_\nu+\lambda_\beta)/p)}, \qquad (14.81)$$

all $x_0 \le x \le b$.

Similarly, by (14.81), we derive

Theorem 14.22. *All basic assumptions are as in Theorem* 14.20. *Let* $\lambda_\nu > 0$, $\lambda_\alpha = 0$, $\lambda_\beta > 0$, $p := \lambda_\nu + \lambda_\beta > 1$, P_2 *defined by* (14.77). *Now it is*

$$A(w) := \frac{w^{(N-1)(1-\lambda_\nu/p)}(P_2(w))^{\lambda_\beta((p-1)/p)}}{(\Gamma(\nu-\gamma_2))^{\lambda_\beta}}, \qquad (14.82)$$

$$A_0(R_2) := \left(\int_{R_1}^{R_2} (A(w))^{p/\lambda_\beta} dw \right)^{\lambda_\beta/p}. \qquad (14.83)$$

Denote

$$\delta_3 := \begin{cases} 2^{\lambda_\beta/\lambda_\nu} - 1, & if \quad \lambda_\beta \ge \lambda_\nu, \\ 1, & if \quad \lambda_\beta \le \lambda_\nu. \end{cases} \qquad (14.84)$$

Then

$$\int_A \left[\left| \frac{\partial_{R_1}^{\gamma_2} f_2(x)}{\partial r^{\gamma_2}} \right|^{\lambda_\beta} \left| \frac{\partial_{R_1}^\nu f_1(x)}{\partial r^\nu} \right|^{\lambda_\nu} + \left| \frac{\partial_{R_1}^{\gamma_2} f_1(x)}{\partial r^{\gamma_2}} \right|^{\lambda_\beta} \left| \frac{\partial_{R_1}^\nu f_2(x)}{\partial r^\nu} \right|^{\lambda_\nu} \right] dx \le$$

$$A_0(R_2) 2^{\lambda_\beta/p} \left(\frac{\lambda_\nu}{p} \right)^{(\lambda_\nu/p)} \delta_3^{\lambda_\nu/p} \int_A \left(\left| \frac{\partial_{R_1}^\nu f_1(x)}{\partial r^\nu} \right|^p + \left| \frac{\partial_{R_1}^\nu f_2(x)}{\partial r^\nu} \right|^p \right) dx. \qquad (14.85)$$

We need

Theorem 14.23. [26]. *All here are as in Theorem 14.19 , $(\lambda_\alpha, \lambda_\beta \neq 0)$.*
Denote

$$
\tilde{\gamma}_1 := \begin{cases} 2^{((\lambda_\alpha+\lambda_\beta)/\lambda_\nu)-1}, & if \quad \lambda_\alpha + \lambda_\beta \geq \lambda_\nu, \\ 1, & if \quad \lambda_\alpha + \lambda_\beta \leq \lambda_\nu, \end{cases} \tag{14.86}
$$

and

$$
\tilde{\gamma}_2 := \begin{cases} 1, & if \quad \lambda_\alpha + \lambda_\beta + \lambda_\nu \geq p, \\ 2^{1-((\lambda_\alpha+\lambda_\beta+\lambda_\nu)/p)}, & if \quad \lambda_\alpha + \lambda_\beta + \lambda_\nu \leq p. \end{cases} \tag{14.87}
$$

Then

$$
\int_{x_0}^{x} q(w) \left[\left| \left(D_{x_0}^{\gamma_1} f_1 \right)(w) \right|^{\lambda_\alpha} \cdot \left| \left(D_{x_0}^{\gamma_2} f_2 \right)(w) \right|^{\lambda_\beta} \cdot \left| \left(D_{x_0}^{\nu} f_1 \right)(w) \right|^{\lambda_\nu} \right.
$$

$$
\left. + \left| \left(D_{x_0}^{\gamma_2} f_1 \right)(w) \right|^{\lambda_\beta} \cdot \left| \left(D_{x_0}^{\gamma_1} f_2 \right)(w) \right|^{\lambda_\alpha} \cdot \left| \left(D_{x_0}^{\nu} f_2 \right)(w) \right|^{\lambda_\nu} \right] dw
$$

$$
\leq A_0(x) \left(\frac{\lambda_\nu}{(\lambda_\alpha + \lambda_\beta)(\lambda_\alpha + \lambda_\beta + \lambda_\nu)} \right)^{\lambda_\nu/p} \left[\lambda_\alpha^{\lambda_\nu/p} \tilde{\gamma}_2 + 2^{(p-\lambda_\nu)/p} (\tilde{\gamma}_1 \lambda_\beta)^{\lambda_\nu/p} \right]
$$

$$
\cdot \left(\int_{x_0}^{x} p(w) \left(\left| \left(D_{x_0}^{\nu} f_1 \right)(w) \right|^{p} + \left| \left(D_{x_0}^{\nu} f_2 \right)(w) \right|^{p} \right) dw \right)^{((\lambda_\alpha+\lambda_\beta+\lambda_\nu)/p)},
$$

$$
\tag{14.88}
$$

all $x_0 \leq x \leq b$.

Similarly, by (14.88), we obtain

Theorem 14.24. *Let all basics be as in Theorem 14.20. Here, $\lambda_\nu, \lambda_\alpha, \lambda_\beta$*
> 0, $p := \lambda_\alpha + \lambda_\beta + \lambda_\nu > 1$. Also P_k, $k = 1, 2$ as in (14.77), and $A(w)$
is as in (14.78).
Here it is

$$
A_0(R_2) := \left(\int_{R_1}^{R_2} (A(w))^{p/\lambda_\alpha+\lambda_\beta} dw \right)^{\lambda_\alpha+\lambda_\beta/p}, \tag{14.89}
$$

$$
\tilde{\gamma}_1 := \begin{cases} 2^{((\lambda_\alpha+\lambda_\beta)/\lambda_\nu)} - 1, & if \quad \lambda_\alpha + \lambda_\beta \geq \lambda_\nu, \\ 1, & if \quad \lambda_\alpha + \lambda_\beta \leq \lambda_\nu. \end{cases} \tag{14.90}
$$

Then

$$
\int_A \left[\left| \frac{\partial_{R_1}^{\gamma_1} f_1(x)}{\partial r^{\gamma_1}} \right|^{\lambda_\alpha} \left| \frac{\partial_{R_1}^{\gamma_2} f_2(x)}{\partial r^{\gamma_2}} \right|^{\lambda_\beta} \left| \frac{\partial_{R_1}^{\nu} f_1(x)}{\partial r^{\nu}} \right|^{\lambda_\nu} + \right.
$$

$$\left|\frac{\partial_{R_1}^{\gamma_2} f_1(x)}{\partial r^{\gamma_2}}\right|^{\lambda_\beta} \left|\frac{\partial_{R_1}^{\gamma_1} f_2(x)}{\partial r^{\gamma_1}}\right|^{\lambda_\alpha} \left|\frac{\partial_{R_1}^{\nu} f_2(x)}{\partial r^{\nu}}\right|^{\lambda_\nu}\right] dx \leq$$

$$A_0(R_2) \left(\frac{\lambda_\nu}{(\lambda_\alpha + \lambda_\beta)p}\right)^{(\lambda_\nu/p)} \left[\lambda_\alpha^{\lambda_\nu/p} + 2^{(\lambda_\alpha+\lambda_\beta)/p}(\tilde{\gamma_1}\lambda_\beta)^{\lambda_\nu/p}\right]$$

$$\int_A \left(\left|\frac{\partial_{R_1}^{\nu} f_1(x)}{\partial r^{\nu}}\right|^p + \left|\frac{\partial_{R_1}^{\nu} f_2(x)}{\partial r^{\nu}}\right|^p\right) dx. \tag{14.91}$$

We need

Theorem 14.25. [26]. Let $\nu \geq 2$ and $\gamma_1 \geq 0$, such that $\nu - \gamma_1 \geq 2$. Let $f_1, f_2 \in C_{x_0}^{\nu}([a, b])$ with

$$f_1^{(i)}(x_0) = f_2^{(i)}(x_0) = 0, \quad i = 0, 1, \dots, n - 1,$$

$n := [\nu]$. Here $x, x_0 \in [a, b] : x \geq x_0$. Consider also, $p(t) > 0$, and $q(t) \geq 0$ continuous functions on $[x_0, b]$. Let

$$\lambda_\alpha \geq 0, \quad 0 < \lambda_{\alpha+1} < 1,$$

and $p > 1$. Denote

$$\theta_3 := \begin{cases} 2^{\lambda_\alpha/(\lambda_{\alpha+1})} - 1, & if \quad \lambda_\alpha \geq \lambda_{\alpha+1}, \\ 1, & if \quad \lambda_\alpha \leq \lambda_{\alpha+1}, \end{cases}$$

$$L(x) := \left(2 \int_{x_0}^{x} (q(w))^{(1/1-(\lambda_{\alpha+1}))} dw\right)^{(1-\lambda_{\alpha+1})} \left(\frac{\theta_3 \lambda_{\alpha+1}}{\lambda_\alpha + \lambda_{\alpha+1}}\right)^{\lambda_{\alpha+1}}, \tag{14.92}$$

and

$$P_1(x) := \int_{x_0}^{x} (x - t)^{(\nu-\gamma_1-1)p/(p-1)}(p(t))^{-1/(p-1)} dt, \tag{14.93}$$

$$T(x) := L(x) \cdot \left(\frac{P_1(x)^{((p-1)/p)}}{\Gamma(\nu - \gamma_1)}\right)^{(\lambda_\alpha+\lambda_{\alpha+1})}, \tag{14.94}$$

and

$$\omega_1 := 2^{(p-1/p)(\lambda_\alpha+\lambda_{\alpha+1})}, \tag{14.95}$$

$$\Phi(x) := T(x) \, \omega_1. \tag{14.96}$$

Then

$$\int_{x_0}^{x} q(w) \left[\left|(D_{x_0}^{\gamma_1} f_1)(w)\right|^{\lambda_\alpha} \cdot \left|(D_{x_0}^{\gamma_1+1} f_2)(w)\right|^{\lambda_{\alpha+1}} +\right.$$

$$\left. \left| \left(D_{x_0}^{\gamma_1} f_2 \right)(w) \right|^{\lambda_\alpha} \cdot \left| \left(D_{x_0}^{\gamma_1 + 1} f_1 \right)(w) \right|^{\lambda_{\alpha+1}} \right] dw \leq$$

$$\Phi(x) \left[\int_{x_0}^{x} p(w) \cdot \left| \left(D_{x_0}^{\nu} f_1 \right)(w) \right|^{p} + \left| \left(D_{x_0}^{\nu} f_2 \right)(w) \right|^{p} dw \right]^{((\lambda_\alpha + \lambda_{\alpha+1})/p)},$$

$$\tag{14.97}$$

all $x_0 \leq x \leq b$.

Similarly, by (14.97), we obtain

Theorem 14.26. *Let* $\nu \geq 2$, $\gamma_1 \geq 0$, *such that* $\nu - \gamma_1 \geq 2$, $n := [\nu]$. *Let* $f_1, f_2 \in C^n(\bar{A})$ *and there exist* $\partial_{R_1}^{\nu} f_1(x)/\partial r^{\nu}, \partial_{R_1}^{\nu} f_2(x)/\partial r^{\nu} \in C(\bar{A})$, $A := B(0, R_2) - \overline{B(0, R_1)} \subseteq \mathbb{R}^N$, $N \geq 2$. *Furthermore assume* $\partial^j f_1/\partial r^j = \partial^j f_2/\partial r^j = 0$, $j = 0, 1, \ldots, n-1$, *on* $\partial B(0, R_1)$. *Let* $\lambda_\alpha > 0$, $0 < \lambda_{\alpha+1} < 1$, *such that* $p := \lambda_\alpha + \lambda_{\alpha+1} > 1$.
Denote

$$\theta_3 := \begin{cases} 2^{(\lambda_\alpha/\lambda_{\alpha+1})} - 1 & \text{if} \quad \lambda_\alpha \geq \lambda_{\alpha+1} \\ 1, & \text{if} \quad \lambda_\alpha \leq \lambda_{\alpha+1}, \end{cases} \tag{14.98}$$

$$L(R_2) :=$$

$$\left[2 \frac{(1 - \lambda_{\alpha+1})}{(N - \lambda_{\alpha+1})} \left(R_2^{(N-\lambda_{\alpha+1})/(1-\lambda_{\alpha+1})} - R_1^{(N-\lambda_{\alpha+1})/(1-\lambda_{\alpha+1})} \right) \right]^{(1-\lambda_{\alpha+1})} \left(\frac{\theta_3 \lambda_{\alpha+1}}{p} \right)^{\lambda_{\alpha+1}},$$

$$\tag{14.99}$$

and

$$P_1(R_2) := \int_{R_1}^{R_2} (R_2 - t)^{(\nu-\gamma_1-1)p/(p-1)} t^{(1-N)/(p-1)} \, dt, \tag{14.100}$$

$$\Phi(R_2) := L(R_2) \left(\frac{P_1(R_2)^{(p-1)}}{(\Gamma(\nu - \gamma_1))^p} \right) 2^{p-1}. \tag{14.101}$$

Then

$$\int_A \left[\left| \frac{\partial_{R_1}^{\gamma_1} f_1(x)}{\partial r^{\gamma_1}} \right|^{\lambda_\alpha} \left| \frac{\partial_{R_1}^{\gamma_1+1} f_2(x)}{\partial r^{\gamma_1+1}} \right|^{\lambda_{\alpha+1}} + \left| \frac{\partial_{R_1}^{\gamma_1} f_2(x)}{\partial r^{\gamma_1}} \right|^{\lambda_\alpha} \left| \frac{\partial_{R_1}^{\gamma_1+1} f_1(x)}{\partial r^{\gamma_1+1}} \right|^{\lambda_{\alpha+1}} \right] dx$$

$$\leq \Phi(R_2) \int_A \left(\left| \frac{\partial_{R_1}^{\nu} f_1(x)}{\partial r^{\nu}} \right|^p + \left| \frac{\partial_{R_1}^{\nu} f_2(x)}{\partial r^{\nu}} \right|^p \right) dx. \tag{14.102}$$

We need

Theorem 14.27. [26]. *All here are as in Theorem 14.19. Consider the special case* $\lambda_\beta = \lambda_\alpha + \lambda_\nu$. *Denote*

$$\tilde{T}(x) := A_0(x) \left(\frac{\lambda_\nu}{\lambda_\alpha + \lambda_\nu} \right)^{\lambda_\nu/p} 2^{(p-2\lambda_\alpha-3\lambda_\nu)/p}. \tag{14.103}$$

Then

$$\int_{x_0}^{x} q(w) \left[\left| \left(D_{x_0}^{\gamma_1} f_1 \right) (w) \right|^{\lambda_\alpha} \left| \left(D_{x_0}^{\gamma_2} f_2 \right) (w) \right|^{\lambda_\alpha + \lambda_\nu} \left| \left(D_{x_0}^{\nu} f_1 \right) (w) \right|^{\lambda_\nu} \right.$$

$$\left. + \left| \left(D_{x_0}^{\lambda_2} f_1 \right) (w) \right|^{\lambda_\alpha + \lambda_\nu} \left| \left(D_{x_0}^{\gamma_1} f_2 \right) (w) \right|^{\lambda_\alpha} \left| \left(D_{x_0}^{\nu} f_2 \right) (w) \right|^{\lambda_\nu} \right] dw \leq$$

$$\tilde{T}(x) \left(\int_{x_0}^{x} p(w) \left(\left| \left(D_{x_0}^{\nu} f_1 \right) (w) \right|^{p} + \left| \left(D_{x_0}^{\nu} f_2 \right) (w) \right|^{p} \right) dw \right)^{2((\lambda_\alpha + \lambda_\nu)/p)} ,$$

(14.104)

all $x_0 \leq x \leq b$.

Similarly, by (14.104), we obtain

Theorem 14.28. *Here all are as in Theorem 14.20. Consider the case* $\lambda_\beta = \lambda_\alpha + \lambda_\nu$; $\lambda_\alpha \geq 0$, $\lambda_\nu > 0$, $\lambda_\beta > 1/2$, $p := 2\lambda_\beta$. *Here* P_k, $k = 1, 2$, *as in (14.77) and* $A(w)$ *is as in (14.78). Call*

$$A_0(R_2) := \left(\int_{R_1}^{R_2} (A(w))^{p/(2\lambda_\alpha + \lambda_\nu)} \right)^{((2\lambda_\alpha + \lambda_\nu)/p)} .$$

(14.105)

Also set

$$\tilde{T}(R_2) := A_0(R_2) \left(\frac{\lambda_\nu}{\lambda_\beta} \right)^{\lambda_\nu/p} 2^{(-\lambda_\nu/p)} .$$

(14.106)

Then

$$\int_{A} \left[\left| \frac{\partial_{R_1}^{\gamma_1} f_1(x)}{\partial r^{\gamma_1}} \right|^{\lambda_\alpha} \left| \frac{\partial_{R_1}^{\gamma_2} f_2(x)}{\partial r^{\gamma_2}} \right|^{\lambda_\beta} \left| \frac{\partial_{R_1}^{\nu} f_1(x)}{\partial r^{\nu}} \right|^{\lambda_\nu} + \right.$$

$$\left. \left| \frac{\partial_{R_1}^{\gamma_2} f_1(x)}{\partial r^{\gamma_2}} \right|^{\lambda_\beta} \left| \frac{\partial_{R_1}^{\gamma_1} f_2(x)}{\partial r^{\gamma_1}} \right|^{\lambda_\alpha} \left| \frac{\partial_{R_1}^{\nu} f_2(x)}{\partial r^{\nu}} \right|^{\lambda_\nu} \right] dx \leq$$

$$\tilde{T}(R_2) \int_{A} \left(\left| \frac{\partial_{R_1}^{\nu} f_1(x)}{\partial r^{\nu}} \right|^{p} + \left| \frac{\partial_{R_1}^{\nu} f_2(x)}{\partial r^{\nu}} \right|^{p} \right) dx.$$

(14.107)

We need

Theorem 14.29. [26]. *Let* $\nu \geq 1$, $\gamma_1, \gamma_2 \geq 0$, *such that* $\nu - \gamma_1 \geq 1$, $\nu - \gamma_2 \geq 1$ *and* $f_1, f_2 \in C_{x_0}^{\nu}([a, b])$ *with* $f_1^{(i)}(x_0) = f_2^{(i)}(x_0) = 0$, $i = 0, 1, \ldots, n - 1$, $n := [\nu]$. *Here,* $x, x_0 \in [a, b] : x \geq x_0$. *Consider* $p(x) \geq 0$ *continuous functions on* $[x_0, b]$. *Let* $\lambda_\alpha, \lambda_\beta, \lambda_\nu \geq 0$. *Set*

$$\rho(x) := \frac{(x - x_0)^{(\nu\lambda_\alpha - \gamma_1\lambda_\alpha + \nu\lambda_\beta - \gamma_2\lambda_\beta + 1)} \|p(x)\|_\infty}{(\nu\lambda_\alpha - \gamma_1\lambda_\alpha + \nu\lambda_\beta - \gamma_2\lambda_\beta + 1)(\Gamma(\nu - \gamma_1 + 1))^{\lambda_\alpha} (\Gamma(\nu - \gamma_2 + 1))^{\lambda_\beta}} .$$

(14.108)

Then

$$\int_{x_0}^{x} q(w) \left[\left| \left(D_{x_0}^{\gamma_1} f_1 \right)(w) \right|^{\lambda_\alpha} \left| \left(D_{x_0}^{\gamma_2} f_2 \right)(w) \right|^{\lambda_\beta} \left| \left(D_{x_0}^{\nu} f_1 \right)(w) \right|^{\lambda_\nu} + \right.$$

$$\left. \left| \left(D_{x_0}^{\lambda_2} f_1 \right)(w) \right|^{\lambda_\beta} \left| \left(D_{x_0}^{\gamma_1} f_2 \right)(w) \right|^{\lambda_\alpha} \left| \left(D_{x_0}^{\nu} f_2 \right)(w) \right|^{\lambda_\nu} \right] dw$$

$$\leq \frac{\rho(x)}{2} \left[\| D_{x_0}^{\nu} f_1 \|_\infty^{2(\lambda_\alpha + \lambda_\nu)} + \| D_{x_0}^{\nu} f_1 \|_\infty^{2\lambda_\beta} + \| D_{x_0}^{\nu} f_2 \|_\infty^{2\lambda_\beta} + \| D_{x_0}^{\nu} f_2 \|_\infty^{2(\lambda_\alpha + \lambda_\nu)} \right],$$

(14.109)

all $x_0 \leq x \leq b$.

Similarly, by (14.109), we get

Theorem 14.30. *Same basic assumptions as in Theorem 14.20. Let* $\lambda_\alpha, \lambda_\beta, \lambda_\nu \geq 0$. *Call*

$$\rho(R_2) := \frac{R_2^{N-1}(R_2 - R_1)^{(\nu\lambda_\alpha - \gamma_1\lambda_\alpha + \nu\lambda_\beta - \gamma_2\lambda_\beta + 1)}}{(\nu\lambda_\alpha - \gamma_1\lambda_\alpha + \nu\lambda_\beta - \gamma_2\lambda_\beta + 1)(\Gamma(\nu - \gamma_1 + 1))^{\lambda_\alpha}(\Gamma(\nu - \gamma_2 + 1))^{\lambda_\beta}}.$$

(14.110)

Then

$$\int_A \left[\left| \frac{\partial_{R_1}^{\gamma_1} f_1(x)}{\partial r^{\gamma_1}} \right|^{\lambda_\alpha} \left| \frac{\partial_{R_1}^{\gamma_2} f_2(x)}{\partial r^{\gamma_2}} \right|^{\lambda_\beta} \left| \frac{\partial_{R_1}^{\nu} f_1(x)}{\partial r^{\nu}} \right|^{\lambda_\nu} + \right.$$

$$\left. \left| \frac{\partial_{R_1}^{\gamma_2} f_1(x)}{\partial r^{\gamma_2}} \right|^{\lambda_\beta} \left| \frac{\partial_{R_1}^{\gamma_1} f_2(x)}{\partial r^{\gamma_1}} \right|^{\lambda_\alpha} \left| \frac{\partial_{R_1}^{\nu} f_2(x)}{\partial r^{\nu}} \right|^{\lambda_\nu} \right] dx \leq$$

$$\rho(R_2) \frac{\pi^{N/2}}{\Gamma(N/2)} \left[\left\| \frac{\partial_{R_1}^{\nu} f_1}{\partial r^{\nu}} \right\|_\infty^{2(\lambda_\alpha + \lambda_\nu)} + \left\| \frac{\partial_{R_1}^{\nu} f_1}{\partial r^{\nu}} \right\|_\infty^{2\lambda_\beta} + \right.$$

$$\left. \left\| \frac{\partial_{R_1}^{\nu} f_2}{\partial r^{\nu}} \right\|_\infty^{2\lambda_\beta} + \left\| \frac{\partial_{R_1}^{\nu} f_2}{\partial r^{\nu}} \right\|_\infty^{2(\lambda_\alpha + \lambda_\nu)} \right].$$

(14.111)

We need

Theorem 14.31. [26]. *Assume, as in Theorem 14.29,* $\lambda_\beta = 0$. *It holds*

$$\int_{x_0}^{x} p(w) \left[\left| \left(D_{x_0}^{\gamma_1} f_1 \right)(w) \right|^{\lambda_\alpha} \cdot \left| \left(D_{x_0}^{\nu} f_1 \right)(w) \right|^{\lambda_\nu} + \right.$$

$$\left. \left| \left(D_{x_0}^{\gamma_1} f_2 \right)(w) \right|^{\lambda_\alpha} \cdot \left| \left(D_{x_0}^{\nu} f_2 \right)(w) \right|^{\lambda_\nu} \right] dw \leq$$

$$\left(\frac{(x-x_0)^{(\nu\lambda_\alpha - \gamma_1\lambda_\alpha + 1)} \|p(x)\|_\infty}{(\nu\lambda_\alpha - \gamma_1\lambda_\alpha + 1)(\Gamma(\nu - \gamma_1 + 1))^{\lambda_\alpha}} \right) \cdot \left[\left\| D_{x_0}^\nu f_1 \right\|_\infty^{\lambda_\alpha + \lambda_\nu} + \left\| D_{x_0}^\nu f_2 \right\|_\infty^{\lambda_\alpha + \lambda_\nu} \right],$$

$$(14.112)$$

all $x_0 \le x \le b$.

Similarly, by (14.112), we derive

Theorem 14.32. *All are as in Theorem 14.30. Assume* $\lambda_\beta = 0$. *Then*

$$\int_A \left[\left| \frac{\partial_{R_1}^{\gamma_1} f_1(x)}{\partial r^{\gamma_1}} \right|^{\lambda_\alpha} \left| \frac{\partial_{R_1}^\nu f_1(x)}{\partial r^\nu} \right|^{\lambda_\nu} + \left| \frac{\partial_{R_1}^{\gamma_1} f_2(x)}{\partial r^{\gamma_1}} \right|^{\lambda_\alpha} \left| \frac{\partial_{R_1}^\nu f_2(x)}{\partial r^\nu} \right|^{\lambda_\nu} \right] dx \le$$

$$\frac{2\pi^{N/2}}{\Gamma(N/2)} \frac{R_2^{N-1}(R_2 - R_1)^{(\nu\lambda_\alpha - \gamma_1\lambda_\alpha + 1)}}{(\nu\lambda_\alpha - \gamma_1\lambda_\alpha + 1)(\Gamma(\nu - \gamma_1 + 1))^{\lambda_\alpha}} \left(\left\| \frac{\partial_{R_1}^\nu f_1}{\partial r^\nu} \right\|_\infty^{\lambda_\alpha + \lambda_\nu} + \left\| \frac{\partial_{R_1}^\nu f_2}{\partial r^\nu} \right\|_\infty^{\lambda_\alpha + \lambda_\nu} \right).$$

$$(14.113)$$

We need

Theorem 14.33. [26]. *(In relationship to Theorem 14.29,* $\lambda_\beta = \lambda_\alpha + \lambda_\nu$.) *It holds*

$$\int_{x_0}^x p(w) \left[\left| \left(D_{x_0}^{\gamma_1} f_1 \right)(w) \right|^{\lambda_\alpha} \left| \left(D_{x_0}^{\gamma_2} f_2 \right)(w) \right|^{\lambda_\alpha + \lambda_\nu} \left| \left(D_{x_0}^\nu f_1 \right)(w) \right|^{\lambda_\nu} + \right.$$

$$\left. \left| \left(D_{x_0}^{\gamma_2} f_1 \right)(w) \right|^{\lambda_\alpha + \lambda_\nu} \left| \left(D_{x_0}^{\gamma_1} f_2 \right)(w) \right|^{\lambda_\alpha} \left| \left(D_{x_0}^\nu f_2 \right)(w) \right|^{\lambda_\nu} \right] dw \le$$

$$\left(\frac{(x-x_0)^{(2\nu\lambda_\alpha - \gamma_1\lambda_\alpha + \nu\lambda_\nu - \gamma_2\lambda_\alpha - \gamma_2\lambda_\nu + 1)}}{(2\nu\lambda_\alpha - \gamma_1\lambda_\alpha + \nu\lambda_\nu - \gamma_2\lambda_\alpha - \gamma_2\lambda_\nu + 1)} \right.$$

$$\left. \cdot \frac{\|p(x)\|_\infty}{(\Gamma(\nu - \gamma_1 + 1))^{\lambda_\alpha} (\Gamma(\nu - \gamma_2 + 1))^{\lambda_\alpha + \lambda_\nu}} \right)$$

$$\cdot \left(\left\| D_{x_0}^\nu f_1 \right\|_\infty^{2(\lambda_\alpha + \lambda_\nu)} + \left\| \left(D_{x_0}^\nu f_2 \right) \right\|_\infty^{2(\lambda_\alpha + \lambda_\nu)} \right),$$

$$(14.114)$$

all $x_0 \le x \le b$.

Similarly, by (14.114), we derive

Theorem 14.34. *All are as in Theorem 14.30. Assume* $\lambda_\beta = \lambda_\alpha + \lambda_\nu$. *Then*

$$\int_A \left[\left| \frac{\partial_{R_1}^{\gamma_1} f_1(x)}{\partial r^{\gamma_1}} \right|^{\lambda_\alpha} \left| \frac{\partial_{R_1}^{\gamma_2} f_2(x)}{\partial r^{\gamma_2}} \right|^{\lambda_\alpha + \lambda_\nu} \left| \frac{\partial_{R_1}^\nu f_1(x)}{\partial r^\nu} \right|^{\lambda_\nu} + \right.$$

$$\left| \frac{\partial_{R_1}^{\gamma_2} f_1(x)}{\partial r^{\gamma_2}} \right|^{\lambda_\alpha + \lambda_\nu} \left| \frac{\partial_{R_1}^{\gamma_1} f_2(x)}{\partial r^{\gamma_1}} \right|^{\lambda_\alpha} \left| \frac{\partial_{R_1}^{\nu} f_2(x)}{\partial r^{\nu}} \right|^{\lambda_\nu} \right] dx \le \left(\frac{2\pi^{N/2}}{\Gamma(N/2)} \right)$$

$$\left(\frac{R_2^{N-1}(R_2 - R_1)^{(2\nu\lambda_\alpha - \gamma_1\lambda_\alpha + \nu\lambda_\nu - \gamma_2\lambda_\alpha - \gamma_2\lambda_\nu + 1)}}{(2\nu\lambda_\alpha - \gamma_1\lambda_\alpha + \nu\lambda_\nu - \gamma_2\lambda_\alpha - \gamma_2\lambda_\nu + 1)(\Gamma(\nu - \gamma_1 + 1))^{\lambda_\alpha}(\Gamma(\nu - \gamma_2 + 1))^{\lambda_\alpha + \lambda_\nu}} \right)$$

$$\left(\left\| \frac{\partial_{R_1}^{\nu} f_1}{\partial r^{\nu}} \right\|_\infty^{2(\lambda_\alpha + \lambda_\nu)} + \left\| \frac{\partial_{R_1}^{\nu} f_2}{\partial r^{\nu}} \right\|_\infty^{2(\lambda_\alpha + \lambda_\nu)} \right). \qquad (14.115)$$

We need

Theorem 14.35. [26]. (In relationship to Theorem 14.29, $\lambda_\nu = 0$, $\lambda_\alpha = \lambda_\beta$.) *It holds*

$$\int_{x_0}^{x} p(w) \left[\left| (D_{x_0}^{\gamma_1} f_1)(w) \right|^{\lambda_\alpha} \cdot \left| (D_{x_0}^{\gamma_2} f_2)(w) \right|^{\lambda_\alpha} + \right.$$

$$\left. \left| (D_{x_0}^{\gamma_2} f_1)(w) \right|^{\lambda_\alpha} \cdot \left| (D_{x_0}^{\gamma_1} f_2)(w) \right|^{\lambda_\alpha} \right] dw \le$$

$$\rho^*(x) \left[\left\| D_{x_0}^{\nu} f_1 \right\|_\infty^{2\lambda_\alpha} + \left\| D_{x_0}^{\nu} f_2 \right\|_\infty^{2\lambda_\alpha} \right], \qquad (14.116)$$

all $x_0 \le x \le b$.
Here

$$\rho^*(x) := \left(\frac{(x - x_0)^{(2\nu\lambda_\alpha - \gamma_1\lambda_\alpha - \gamma_2\lambda_\alpha + 1)} \cdot \|p(x)\|_\infty}{(2\nu\lambda_\alpha - \gamma_1\lambda_\alpha - \gamma_2\lambda_\alpha + 1)(\Gamma(\nu - \gamma_1 + 1))^{\lambda_\alpha}(\Gamma(\nu - \gamma_2 + 1))^{\lambda_\alpha}} \right).$$
$$(14.117)$$

We get, by (14.116), the result

Theorem 14.36. *All are as in Theorem* 14.30. *Assume* $\lambda_\nu = 0$, $\lambda_\alpha = \lambda_\beta$. *Then*

$$\int_A \left[\left| \frac{\partial_{R_1}^{\gamma_1} f_1(x)}{\partial r^{\gamma_1}} \right|^{\lambda_\alpha} \left| \frac{\partial_{R_1}^{\gamma_2} f_2(x)}{\partial r^{\gamma_2}} \right|^{\lambda_\alpha} + \left| \frac{\partial_{R_1}^{\gamma_2} f_1(x)}{\partial r^{\gamma_2}} \right|^{\lambda_\alpha} \left| \frac{\partial_{R_1}^{\gamma_1} f_2(x)}{\partial r^{\gamma_1}} \right|^{\lambda_\alpha} \right] dx \le$$

$$\left(\frac{2\pi^{N/2}}{\Gamma(N/2)} \right) \rho^*(R_2) \left[\left\| \frac{\partial_{R_1}^{\nu} f_1}{\partial r^{\nu}} \right\|_\infty^{2\lambda_\alpha} + \left\| \frac{\partial_{R_1}^{\nu} f_2}{\partial r^{\nu}} \right\|_\infty^{2\lambda_\alpha} \right], \qquad (14.118)$$

where

$$\rho^*(R_2) := \left(\frac{R_2^{N-1}(R_2 - R_1)^{(2\nu\lambda_\alpha - \gamma_1\lambda_\alpha - \gamma_2\lambda_\alpha + 1)}}{(2\nu\lambda_\alpha - \gamma_1\lambda_\alpha - \gamma_2\lambda_\alpha + 1)(\Gamma(\nu - \gamma_1 + 1))^{\lambda_\alpha}(\Gamma(\nu - \gamma_2 + 1))^{\lambda_\alpha}} \right).$$
$$(14.119)$$

We need

Theorem 14.37. [26]. (In relationship to Theorem 14.29, $\lambda_\alpha = 0$, $\lambda_\beta = \lambda_\nu$.) It holds

$$\int_{x_0}^x p(w) \left[\left| (D_{x_0}^{\gamma_2} f_2)(w) \right|^{\lambda_\beta} \cdot \left| (D_{x_0}^\nu f_1)(w) \right|^{\lambda_\beta} + \right.$$

$$\left. \left| (D_{x_0}^{\gamma_2} f_1)(w) \right|^{\lambda_\beta} \cdot \left| (D_{x_0}^\nu f_2)(w) \right|^{\lambda_\beta} \right] dw \le$$

$$\left(\frac{(x - x_0)^{(\nu\lambda_\beta - \gamma_2\lambda_\beta + 1)} \cdot \|p(x)\|_\infty}{(\nu\lambda_\beta - \gamma_2\lambda_\beta + 1)(\Gamma(\nu - \gamma_2 + 1))^{\lambda_\beta}} \right) \cdot \left[\left\| D_{x_0}^\nu f_1 \right\|_\infty^{2\lambda_\beta} + \left\| (D_{x_0}^\nu f_2) \right\|_\infty^{2\lambda_\beta} \right],$$

$$(14.120)$$

all $x_0 \le x \le b$.

We get, by (14.120), the next result.

Theorem 14.38. All are as in Theorem 14.30. Assume $\lambda_\alpha = 0$ and $\lambda_\beta = \lambda_\nu$. Then

$$\int_A \left[\left| \frac{\partial_{R_1}^{\gamma_2} f_2(x)}{\partial r^{\gamma_2}} \right|^{\lambda_\beta} \left| \frac{\partial_{R_1}^\nu f_1(x)}{\partial r^\nu} \right|^{\lambda_\beta} + \left| \frac{\partial_{R_1}^{\gamma_2} f_1(x)}{\partial r^{\gamma_2}} \right|^{\lambda_\beta} \left| \frac{\partial_{R_1}^\nu f_2(x)}{\partial r^\nu} \right|^{\lambda_\beta} \right] dx \le$$

$$\left(\frac{2\pi^{N/2}}{\Gamma(N/2)} \right) \left(\frac{R_2^{N-1}(R_2 - R_1)^{(\nu\lambda_\beta - \gamma_2\lambda_\beta + 1)}}{(\nu\lambda_\beta - \gamma_2\lambda_\beta + 1)(\Gamma(\nu - \gamma_2 + 1))^{\lambda_\beta}} \right) \left[\left\| \frac{\partial_{R_1}^\nu f_1}{\partial r^\nu} \right\|_\infty^{2\lambda_\beta} \right.$$

$$\left. + \left\| \frac{\partial_{R_1}^\nu f_2}{\partial r^\nu} \right\|_\infty^{2\lambda_\beta} \right]. \qquad (14.121)$$

We make

Assumption 14.39. Let $\nu \ge 1$, $n := [\nu]$, $f_j \in C^n(\bar{A})$, $j = 1, \ldots, M \in \mathbb{N}$, and there exist $\partial_{R_1}^\nu f_j / \partial r^\nu \in C(\bar{A})$, $A := B(0, R_2) - \overline{B(0, R_1)} \subseteq \mathbb{R}^N$, $N \ge 2$. Furthermore assume that $\partial^i f_j / \partial r^i = 0$, $i = 0, 1, \ldots, n - 1$, on $\partial B(0, R_1)$, for all $j = 1, \ldots, M$.

Next we present a set of multivariate fractional Opial-type inequalities involving several functions over the shell.
We need

Theorem 14.40. [27] Let $\nu \ge 1$, $\gamma_1, \gamma_2 \ge 0$, such that $\nu - \gamma_1 \ge 1$, $\nu - \gamma_2 \ge 1$ and $f_j \in C_{x_0}^\nu([a, b])$ with $f_j^{(i)}(x_0) = 0$, $i = 0, 1, \ldots, n - 1$, $n := [\nu]$, $j =$

$1, \ldots, M \in \mathbb{N}$. Here, $x, x_0 \in [a, b] : x \geq x_0$. Consider also $p(t) > 0$, and $q(t) \geq 0$ continuous functions on $[x_0, b]$. Let $\lambda_\nu > 0$ and $\lambda_\alpha, \lambda_\beta \geq 0$ such that $\lambda_\nu < p$, where $p > 1$. Call

$$P_k(w) := \int_{x_0}^{w} (w - t)^{(\nu - \gamma_k - 1)p/p - 1} (p(t))^{-1/p - 1} \, dt, \quad k = 1, 2, \ x_0 \leq w \leq b; \tag{14.122}$$

$$A(w) := \frac{q(w) \cdot (P_1(w))^{\lambda_\alpha((p-1)/p)} \cdot (P_2(w))^{\lambda_\beta((p-1)/p)} (p(w))^{-\lambda_\nu/p}}{(\Gamma(\nu - \gamma_1))^{\lambda_\alpha} \cdot (\Gamma(\nu - \gamma_2))^{\lambda_\beta}}; \tag{14.123}$$

$$A_0(x) := \left(\int_{x_0}^{x} A(w)^{p/(p-\lambda_\nu)} dw \right)^{(p-\lambda_\nu)/p}. \tag{14.124}$$

Set

$$\varphi_1(x) := \left(A_0(x) \big|_{\lambda_\beta = 0} \right) \cdot \left(\frac{\lambda_\nu}{\lambda_\alpha + \lambda_\nu} \right)^{\lambda_\nu/p}, \tag{14.125}$$

$$\delta_1^* := \begin{cases} M^{1 - (\lambda_\alpha + \lambda_\nu/p)}, & \text{if } \lambda_\alpha + \lambda_\nu \leq p, \\ 2^{(\lambda_\alpha + \lambda_\nu/p)} - 1, & \text{if } \lambda_\alpha + \lambda_\nu \geq p. \end{cases} \tag{14.126}$$

If $\lambda_\beta = 0$, we obtain that

$$\int_{x_0}^{x} q(w) \left(\sum_{j=1}^{M} \left| \left(D_{x_0}^{\gamma_1} f_j \right) (w) \right|^{\lambda_\alpha} \cdot \left| \left(D_{x_0}^{\nu} f_j \right) (w) \right|^{\lambda_\nu} \right) dw \leq$$

$$\delta_1^* \cdot \varphi_1(x) \cdot \left[\int_{x_0}^{x} p(w) \left(\sum_{j=1}^{M} \left| \left(D_{x_0}^{\nu} f_j \right) (w) \right|^p \right) dw \right]^{((\lambda_\alpha + \lambda_\nu)/p)}, \tag{14.127}$$

all $x_0 \leq x \leq b$.

Similarly, by (14.127), we derive

Theorem 14.41. Let $f_j, j = 1, \ldots, M,$ as in Assumption 14.39. Let $\gamma_1, \gamma_2 \geq 0$, such that $\nu - \gamma_1 \geq 1$, $\nu - \gamma_2 \geq 1$. Let $\lambda_\nu > 0$, and $\lambda_\alpha > 0$; $\lambda_\beta \geq 0$, $p := \lambda_\alpha + \lambda_\nu > 1$. Set

$$P_k(w) := \int_{R_1}^{w} (w - t)^{(\nu - \gamma_k - 1)p/(p-1)} t^{(1-N)/(p-1)} \, dt, \tag{14.128}$$

$k = 1, 2, \quad R_1 \leq w \leq R_2,$

$$A(w) := \frac{w^{(N-1)(1 - (\lambda_\nu/p))} (P_1(w))^{\lambda_\alpha((p-1)/p)} (P_2(w))^{\lambda_\beta((p-1)/p)}}{(\Gamma(\nu - \gamma_1))^{\lambda_\alpha} (\Gamma(\nu - \gamma_2))^{\lambda_\beta}}, \tag{14.129}$$

$$A_0(R_2) := \left(\int_{R_1}^{R_2} (A(w))^{p/\lambda_\alpha} dw \right)^{\lambda_\alpha/p}. \tag{14.130}$$

Take the case of $\lambda_\beta = 0$. *Then*

$$\sum_{j=1}^{M} \int_A \left| \frac{\partial_{R_1}^{\gamma_1} f_j(x)}{\partial r^{\gamma_1}} \right|^{\lambda_\alpha} \left| \frac{\partial_{R_1}^{\nu} f_j}{\partial r^{\nu}} \right|^{\lambda_\nu} dx$$

$$\leq \left(A_0(R_2) |_{\lambda_\beta = 0} \right) \left(\frac{\lambda_\nu}{p} \right)^{(\lambda_\nu/p)} \left[\sum_{j=1}^{M} \left(\int_A \left| \frac{\partial_{R_1}^{\nu} f_j(x)}{\partial r^{\nu}} \right|^p dx \right) \right]. \qquad (14.131)$$

We need

Theorem 14.42. [27]. *All here are as in Theorem 14.40. Denote*

$$\delta_3 := \begin{cases} 2^{\lambda_\beta/\lambda_\nu} - 1, & \text{if } \lambda_\beta \geq \lambda_\nu, \\ 1, & \text{if } \lambda_\beta \leq \lambda_\nu, \end{cases} \qquad (14.132)$$

$$\varepsilon_2 := \begin{cases} 1, & \text{if } \lambda_\nu + \lambda_\beta \geq p, \\ M^{1-((\lambda_\nu+\lambda_\beta)/p)}, & \text{if } \lambda_\nu + \lambda_\beta \leq p, \end{cases} \qquad (14.133)$$

and

$$\varphi_2(x) := \left(A_0(x) |_{\lambda_\alpha=0} \right) 2^{(p-\lambda_\nu/p)} \left(\frac{\lambda_\nu}{\lambda_\beta + \lambda_\nu} \right)^{\lambda_\nu/p} \delta_3^{\lambda_\nu/p}. \qquad (14.134)$$

If $\lambda_\alpha = 0$, *then*

$$\int_{x_0}^{x} q(w) \left\{ \left\{ \sum_{j=1}^{M-1} \left[\left| \left(D_{x_0}^{\gamma_2} f_{j+1} \right)(w) \right|^{\lambda_\beta} \left| \left(D_{x_0}^{\nu} f_j \right)(w) \right|^{\lambda_\nu} \right. \right. \right.$$

$$+ \left| \left(D_{x_0}^{\gamma_2} f_j \right)(w) \right|^{\lambda_\beta} \left| \left(D_{x_0}^{\nu} f_{j+1} \right)(w) \right|^{\lambda_\nu} \right] \right\}$$

$$+ \left[\left| \left(D_{x_0}^{\gamma_2} f_M \right)(w) \right|^{\lambda_\beta} \left| \left(D_{x_0}^{\nu} f_1 \right)(w) \right|^{\lambda_\nu} \right.$$

$$\left. \left. + \left| \left(D_{x_0}^{\gamma_2} f_1 \right)(w) \right|^{\lambda_\beta} \left| \left(D_{x_0}^{\nu} f_M \right)(w) \right|^{\lambda_\nu} \right] \right\} dw \leq$$

$$2^{((\lambda_\nu+\lambda_\beta)/p)} \varepsilon_2 \varphi_2(x) \cdot \left\{ \left[\int_{x_0}^{x} p(w) \cdot \left[\sum_{j=1}^{M} \left| \left(D_{x_0}^{\nu} f_j \right)(w) \right|^p \right] dw \right] \right\}^{((\lambda_\nu+\lambda_\beta)/p)},$$

$$\qquad (14.135)$$

$x \geq x_0.$

Similarly, by (14.135), we obtain

Theorem 14.43. *All basic assumptions are as in Theorem* 14.41. *Let* $\lambda_\nu > 0$, $\lambda_\alpha = 0$; $\lambda_\beta > 0$, $p := \lambda_\nu + \lambda_\beta > 1$, P_2 *defined by* (14.128).
Now it is

$$A(w) := \frac{w^{(N-1)(1-\lambda_\nu/p)}(P_2(w))^{\lambda_\beta(p-1/p)}}{(\Gamma(\nu-\gamma_2))^{\lambda_\beta}}, \tag{14.136}$$

$$A_0(R_2) := \left(\int_{R_1}^{R_2} (A(w))^{p/\lambda_\beta}\, dw\right)^{\lambda_\beta/p}. \tag{14.137}$$

Denote

$$\delta_3 := \begin{cases} 2^{\lambda_\beta/\lambda_\nu} - 1, & \text{if } \lambda_\beta \geq \lambda_\nu, \\ 1, & \text{if } \lambda_\beta \leq \lambda_\nu. \end{cases} \tag{14.138}$$

Call

$$\varphi_2(R_2) := A_0(R_2) 2^{\lambda_\beta/p} \left(\frac{\lambda_\nu}{p}\right)^{\lambda_\nu/p} \delta_3^{\lambda_\nu/p}. \tag{14.139}$$

Then

$$\int_A \Bigg\{ \Bigg\{ \sum_{j=1}^{M-1} \left[\left|\frac{\partial_{R_1}^{\gamma_2} f_{j+1}(x)}{\partial r^{\gamma_2}}\right|^{\lambda_\beta} \left|\frac{\partial_{R_1}^{\nu} f_j(x)}{\partial r^\nu}\right|^{\lambda_\nu} + \right. $$
$$\left. \left|\frac{\partial_{R_1}^{\gamma_2} f_j(x)}{\partial r^{\gamma_2}}\right|^{\lambda_\beta} \left|\frac{\partial_{R_1}^{\nu} f_{j+1}(x)}{\partial r^\nu}\right|^{\lambda_\nu} \right] \Bigg\} + $$
$$\left[\left|\frac{\partial_{R_1}^{\gamma_2} f_M(x)}{\partial r^{\gamma_2}}\right|^{\lambda_\beta} \left|\frac{\partial_{R_1}^{\nu} f_1(x)}{\partial r^\nu}\right|^{\lambda_\nu} + \right. $$
$$\left. \left|\frac{\partial_{R_1}^{\gamma_2} f_1(x)}{\partial r^{\gamma_2}}\right|^{\lambda_\beta} \left|\frac{\partial_{R_1}^{\nu} f_M(x)}{\partial r^\nu}\right|^{\lambda_\nu} \right] \Bigg\} \, dx \leq $$
$$2\varphi_2(R_2) \left[\sum_{j=1}^M \left(\int_A \left|\frac{\partial_{R_1}^{\nu} f_j(x)}{\partial r^\nu}\right|^p dx \right) \right]. \tag{14.140}$$

We need

Theorem 14.44. [27]. *All here are as in Theorem* 14.40, $(\lambda_\alpha, \lambda_\beta \neq 0)$.

$$\tilde{\gamma}_1 := \begin{cases} 2^{((\lambda_\alpha+\lambda_\beta)/\lambda_\nu)} - 1, & \text{if } \lambda_\alpha + \lambda_\beta \geq \lambda_\nu, \\ 1, & \text{if } \lambda_\alpha + \lambda_\beta \leq \lambda_\nu, \end{cases} \tag{14.141}$$

and

$$\tilde{\gamma}_2 := \begin{cases} 1, & \text{if } \lambda_\alpha + \lambda_\beta + \lambda_\nu \geq p, \\ 2^{1-((\lambda_\alpha+\lambda_\beta+\lambda_\nu)/p)}, & \text{if } \lambda_\alpha + \lambda_\beta + \lambda_\nu \leq p. \end{cases} \tag{14.142}$$

Set

$$\varphi_3(x) := A_0(x) \cdot \left(\frac{\lambda_\nu}{(\lambda_\alpha + \lambda_\beta)(\lambda_\alpha + \lambda_\beta + \lambda_\nu)} \right)^{\lambda_\nu/p} \tag{14.143}$$

and

$$\varepsilon_3 := \begin{cases} 1, & \text{if } \lambda_\alpha + \lambda_\beta + \lambda_\nu \geq p, \\ M^{1-(\lambda_\alpha + \lambda_\beta + \lambda_\nu/p)}, & \text{if } \lambda_\alpha + \lambda_\beta + \lambda_\nu \leq p, \end{cases} \tag{14.144}$$

Then

$$\int_{x_0}^{x} q(w) \left[\sum_{j=1}^{M-1} \left[\left| \left(D_{x_0}^{\gamma_1} f_j \right)(w) \right|^{\lambda_\alpha} \left| \left(D_{x_0}^{\gamma_2} f_{j+1} \right)(w) \right|^{\lambda_\beta} \left| \left(D_{x_0}^{\nu} f_j \right)(w) \right|^{\lambda_\nu} \right. \right.$$

$$+ \left| \left(D_{x_0}^{\gamma_2} f_j \right)(w) \right|^{\lambda_\beta} \left| \left(D_{x_0}^{\gamma_1} f_{j+1} \right)(w) \right|^{\lambda_\alpha} \left| \left(D_{x_0}^{\nu} f_{j+1} \right)(w) \right|^{\lambda_\nu} \right]$$

$$+ \left[\left| \left(D_{x_0}^{\gamma_1} f_1 \right)(w) \right|^{\lambda_\alpha} \left| \left(D_{x_0}^{\gamma_2} f_M \right)(w) \right|^{\lambda_\beta} \left| \left(D_{x_0}^{\nu} f_1 \right)(w) \right|^{\lambda_\nu} \right.$$

$$+ \left| \left(D_{x_0}^{\gamma_2} f_1 \right)(w) \right|^{\lambda_\beta} \left| \left(D_{x_0}^{\gamma_1} f_M \right)(w) \right|^{\lambda_\alpha} \left| \left(D_{x_0}^{\nu} f_M \right)(w) \right|^{\lambda_\nu} \right] \bigg] \, dw$$

$$\leq 2^{(\lambda_\alpha + \lambda_\beta + \lambda_\nu/p)} \varepsilon_3 \varphi_3(x) \cdot \left\{ \int_{x_0}^{x} p(w) \left[\sum_{j=1}^{M} \left| \left(D_{x_0}^{\nu} f_j \right)(w) \right|^{p} \right] dw \right\}^{((\lambda_\alpha + \lambda_\beta + \lambda_\nu)/p)}, \tag{14.145}$$

all $x_0 \leq x \leq b$.

Similarly, by (14.145), we obtain

Theorem 14.45. *All basic assumptions are as in Theorem 14.41. Here* $\lambda_\nu, \ \lambda_\alpha, \lambda_\beta > 0$, $p := \lambda_\alpha + \lambda_\beta + \lambda_\nu > 1$, P_k *as in* (14.128). *A is as in* (14.129). *Here*

$$A_0(R_2) := \left(\int_{R_1}^{R_2} (A(w))^{p/(\lambda_\alpha + \lambda_\beta)}, \right)^{(\lambda_\alpha + \lambda_\beta)/p}, \tag{14.146}$$

$$\tilde{\gamma}_1 := \begin{cases} 2^{(\lambda_\alpha + \lambda_\beta/\lambda_\nu)} - 1, & \text{if } \lambda_\alpha + \lambda_\beta \geq \lambda_\nu, \\ 1, & \text{if } \lambda_\alpha + \lambda_\beta \leq \lambda_\nu. \end{cases} \tag{14.147}$$

Set

$$\varphi_3(R_2) := A_0(R_2) \left(\frac{\lambda_\nu}{(\lambda_\alpha + \lambda_\beta)p} \right)^{(\lambda_\nu/p)} \left[\lambda_\alpha^{(\lambda_\nu/p)} + 2^{((\lambda_\alpha + \lambda_\beta)/p)} (\tilde{\gamma}_1 \lambda_\beta)^{(\lambda_\nu/p)} \right]. \tag{14.148}$$

Then

$$\int_A \left[\sum_{j=1}^{M-1} \left[\left| \frac{\partial_{R_1}^{\gamma_1} f_j(x)}{\partial r^{\gamma_1}} \right|^{\lambda_\alpha} \left| \frac{\partial_{R_1}^{\gamma_2} f_{j+1}(x)}{\partial r^{\gamma_1}} \right|^{\lambda_\beta} \left| \frac{\partial_{R_1}^\nu f_j(x)}{\partial r^\nu} \right|^{\lambda_\nu} + \right. \right.$$

$$\left. \left| \frac{\partial_{R_1}^{\gamma_2} f_j(x)}{\partial r^{\gamma_2}} \right|^{\lambda_\beta} \left| \frac{\partial_{R_1}^{\gamma_1} f_{j+1}(x)}{\partial r^{\gamma_1}} \right|^{\lambda_\alpha} \left| \frac{\partial_{R_1}^\nu f_{j+1}(x)}{\partial r^\nu} \right|^{\lambda_\nu} \right] + $$

$$\left[\left| \frac{\partial_{R_1}^{\gamma_1} f_1(x)}{\partial r^{\gamma_1}} \right|^{\lambda_\alpha} \left| \frac{\partial_{R_1}^{\gamma_2} f_M(x)}{\partial r^{\gamma_2}} \right|^{\lambda_\beta} \left| \frac{\partial_{R_1}^\nu f_1(x)}{\partial r^\nu} \right|^{\lambda_\nu} + \right.$$

$$\left. \left. \left| \frac{\partial_{R_1}^{\gamma_2} f_1(x)}{\partial r^{\gamma_2}} \right|^{\lambda_\beta} \left| \frac{\partial_{R_1}^{\gamma_1} f_M(x)}{\partial r^{\gamma_1}} \right|^{\lambda_\alpha} \left| \frac{\partial_{R_1}^\nu f_M(x)}{\partial r^\nu} \right|^{\lambda_\nu} \right] \right] dx \leq$$

$$2\varphi_3(R_2) \left[\sum_{j=1}^M \left(\int_A \left| \frac{\partial_{R_1}^\nu f_j(x)}{\partial r^\nu} \right|^p dx \right) \right]. \tag{14.149}$$

We need

Theorem 14.46. [27]. *Let* $\nu \geq 2$, *and* $\gamma_1 \geq 0$, *such that* $\nu - \gamma_1 \geq 2$. *Let* $f_j \in C_{x_0}^\nu([a,b])$ *with* $f_j^{(i)}(x_0) = 0$, $i = 0, 1, \ldots, n-1$, $n := [\nu]$, $j = 1, \ldots, M \in \mathbb{N}$. *Here,* $x, x_0 \in [a,b] : x \geq x_0$. *Consider also* $p(t) > 0$, , *and* $q(t) \geq 0$ *continuous functions on* $[x_0, b]$. *Let* $\lambda_\alpha \geq 0$, $0 < \lambda_{\alpha+1} < 1$, *and* $p > 1$. *Denote*

$$\theta_3 := \begin{cases} 2^{(\lambda_\alpha/\lambda_{\alpha+1})} - 1, & \text{if} \quad \lambda_\alpha \geq \lambda_{\alpha+1}, \\ 1, & \text{if} \quad \lambda_\alpha \leq \lambda_{\alpha+1}, \end{cases} \tag{14.150}$$

$$L(x) := \left(2 \int_{x_0}^x (q(w))^{(1/1-\lambda_{\alpha+1})} dw \right)^{(1-\lambda_{\alpha+1})} \left(\frac{\theta_3 \lambda_{\alpha+1}}{\lambda_\alpha + \lambda_{\alpha+1}} \right)^{\lambda_{\alpha+1}}, \tag{14.151}$$

and

$$P_1(x) := \int_{x_0}^x (x-t)^{((\nu-\gamma_1-1)p/(p-1))} (p(t))^{-1/(p-1)} dt, \tag{14.152}$$

$$T(x) := L(x) \cdot \left(\frac{P_1(x)^{(p-1/p)}}{\Gamma(\nu - \gamma_1)} \right)^{(\lambda_\alpha + \lambda_{\alpha+1})}, \tag{14.153}$$

and

$$\omega_1 := 2^{((p-1)/p)(\lambda_\alpha + \lambda_{\alpha+1})}, \tag{14.154}$$

$$\Phi(x) := T(x)\, \omega_1.$$

Also put

$$\varepsilon_4 := \left\{ \begin{array}{ll} 1, & if \quad \lambda_\alpha + \lambda_{\alpha+1} \geq p, \\ M^{1-(\lambda_\alpha+\lambda_{\alpha+1}/p)}, & if \quad \lambda_\alpha + \lambda_{\alpha+1} \leq p \end{array} \right\}. \tag{14.155}$$

Then

$$\int_{x_0}^x q(w) \left\{ \left\{ \sum_{j=1}^{M-1} \left[\left| \left(D_{x_0}^{\gamma_1} f_j \right)(w) \right|^{\lambda_\alpha} \left| \left(D_{x_0}^{\gamma_1+1} f_{j+1} \right)(w) \right|^{\lambda_{\alpha+1}} \right. \right. \right.$$

$$+ \left| \left(D_{x_0}^{\gamma_1} f_{j+1} \right)(w) \right|^{\lambda_\alpha} \left| \left(D_{x_0}^{\gamma_1+1} f_j \right)(w) \right|^{\lambda_{\alpha+1}} \right] \Bigg\}$$

$$+ \left[\left| \left(D_{x_0}^{\gamma_1} f_1 \right)(w) \right|^{\lambda_\alpha} \left| \left(D_{x_0}^{\gamma_1+1} f_M \right)(w) \right|^{\lambda_{\alpha+1}} \right.$$

$$+ \left| \left(D_{x_0}^{\gamma_1} f_M \right)(w) \right|^{\lambda_\alpha} \left| \left(D_{x_0}^{\gamma_1+1} f_1 \right)(w) \right|^{\lambda_{\alpha+1}} \right] \Bigg\} dw$$

$$\leq 2^{(\lambda_\alpha+\lambda_{\alpha+1}/p)} \varepsilon_4 \Phi(x) \left[\int_{x_0}^x p(w) \left(\sum_{j=1}^M \left| \left(D_{x_0}^\nu f_j \right)(w) \right|^p \right) dw \right]^{((\lambda_\alpha+\lambda_{\alpha+1})/p)}, \tag{14.156}$$

all $x_0 \leq x \leq b$.

Similarly, by (14.156), we get

Theorem 14.47. *Let all be as in Assumption* 14.39. *Here* $\nu \geq 2$, $\gamma_1 \geq 0$ *such that* $\nu - \gamma_1 \geq 2$. *Let* $\lambda_\alpha > 0$, $0 < \lambda_{\alpha+1} < 1$, *such that* $p := \lambda_\alpha + \lambda_{\alpha+1} > 1$. *Denote*

$$\theta_3 := \left\{ \begin{array}{ll} 2^{(\lambda_\alpha/\lambda_{\alpha+1})} - 1, & if \quad \lambda_\alpha > \lambda_{\alpha+1}, \\ 1, & if \quad \lambda_\alpha \leq \lambda_{\alpha+1}, \end{array} \right. \tag{14.157}$$

$$L(R_2) := \left[2 \left(\frac{1-\lambda_{\alpha+1}}{N-\lambda_{\alpha+1}} \right) \left(R_2^{(N-\lambda_{\alpha+1})/(1-\lambda_{\alpha+1})} \right. \right.$$

$$\left. \left. R_1^{(N-\lambda_{\alpha+1})/(1-\lambda_{\alpha+1})} \right) \right]^{(1-\lambda_{\alpha+1})} \left(\frac{\theta_3 \lambda_{\alpha+1}}{\lambda_\alpha + \lambda_{\alpha+1}} \right)^{\lambda_{\alpha+1}}, \tag{14.158}$$

and

$$P(R_2) := \int_{R_1}^{R_2} (R_2 - t)^{(\nu-\gamma_1-1)(p/(p-1))} t^{(1-N)/(p-1)} dt, \tag{14.159}$$

$$\Phi(R_2) := L(R_2) \left(\frac{P_1(R_2)^{(p-1)}}{(\Gamma(\nu-\gamma_1))^p} \right) 2^{(p-1)}. \tag{14.160}$$

Then

$$\int_A \left\{ \left\{ \sum_{j=1}^{M-1} \left[\left| \frac{\partial_{R_1}^{\gamma_1} f_j(x)}{\partial r^{\gamma_1}} \right|^{\lambda_\alpha} \left| \frac{\partial_{R_1}^{\gamma_1+1} f_{j+1}(x)}{\partial r^{\gamma_1+1}} \right|^{\lambda_{\alpha+1}} + \right. \right. \right.$$

$$\left. \left. \left| \frac{\partial_{R_1}^{\gamma_1} f_{j+1}(x)}{\partial r^{\gamma_1}} \right|^{\lambda_\alpha} \left| \frac{\partial_{R_1}^{\gamma_1+1} f_j(x)}{\partial r^{\gamma_1+1}} \right|^{\lambda_{\alpha+1}} \right] \right\} +$$

$$\left[\left| \frac{\partial_{R_1}^{\gamma_1} f_1(x)}{\partial r^{\gamma_1}} \right|^{\lambda_\alpha} \left| \frac{\partial_{R_1}^{\gamma_1+1} f_M(x)}{\partial r^{\gamma_1+1}} \right|^{\lambda_{\alpha+1}} + \right.$$

$$\left. \left. \left| \frac{\partial_{R_1}^{\gamma_1} f_M(x)}{\partial r^{\gamma_1}} \right|^{\lambda_\alpha} \left| \frac{\partial_{R_1}^{\gamma_1+1} f_1(x)}{\partial r^{\gamma_1+1}} \right|^{\lambda_{\alpha+1}} \right] \right\} dx \le$$

$$2\Phi(R_2) \left[\sum_{j=1}^{M} \left(\int_A \left| \frac{\partial_{R_1}^{\nu} f_j(x)}{\partial r^{\nu}} \right|^p dx \right) \right]. \tag{14.161}$$

We need

Theorem 14.48. [27]. *All here are as in Theorem 14.40. Consider the special case* $\lambda_\beta = \lambda_\alpha + \lambda_\nu$. *Denote*

$$\tilde{T}(x) := A_0(x) \left(\frac{\lambda_\nu}{\lambda_\alpha + \lambda_\nu} \right)^{\lambda_\nu/p} 2^{(p-2\lambda_\alpha-3\lambda_\nu/p)}, \tag{14.162}$$

$$\varepsilon_5 := \left\{ \begin{array}{ll} 1, & \text{if } 2(\lambda_\alpha + \lambda_\nu) \ge p, \\ M^{1-(2(\lambda_\alpha+\lambda_\nu)/p)}, & \text{if } 2(\lambda_\alpha + \lambda_\nu) \le p \end{array} \right\}. \tag{14.163}$$

Then

$$\int_{x_0}^x q(w) \left\{ \left\{ \sum_{j=1}^{M-1} \left[\left| (D_{x_0}^{\gamma_1} f_j)(w) \right|^{\lambda_\alpha} \left| (D_{x_0}^{\gamma_2} f_{j+1})(w) \right|^{\lambda_\alpha+\lambda_\nu} \left| (D_{x_0}^{\nu} f_j)(w) \right|^{\lambda_\nu} \right. \right. \right.$$

$$\left. \left. + \left| (D_{x_0}^{\gamma_2} f_j)(w) \right|^{\lambda_\alpha+\lambda_\nu} \left| (D_{x_0}^{\gamma_1} f_{j+1})(w) \right|^{\lambda_\alpha} \left| (D_{x_0}^{\nu} f_{j+1})(w) \right|^{\lambda_\nu} \right] \right\}$$

$$+ \left[\left| (D_{x_0}^{\gamma_1} f_1)(w) \right|^{\lambda_\alpha} \left| (D_{x_0}^{\gamma_2} f_M)(w) \right|^{\lambda_\alpha+\lambda_\nu} \left| (D_{x_0}^{\nu} f_1)(w) \right|^{\lambda_\nu} \right.$$

$$\left. \left. + \left| (D_{x_0}^{\gamma_2} f_1)(w) \right|^{\lambda_\alpha+\lambda_\nu} \left| (D_{x_0}^{\gamma_1} f_M)(w) \right|^{\lambda_\alpha} \left| (D_{x_0}^{\nu} f_M)(w) \right|^{\lambda_\nu} \right] \right\} dw$$

$$\le 2^{2((\lambda_\alpha+\lambda_\nu)/p)} \varepsilon_5 \, \tilde{T}(x) \left[\int_{x_0}^x p(w) \left(\sum_{j=1}^{M} \left| (D_{x_0}^{\nu} f_j)(w) \right|^p \right) dw \right]^{(2(\lambda_\alpha+\lambda_\nu)/p)},$$

$$\tag{14.164}$$

all $x_0 \leq x \leq b.$

Similarly, by (14.164), we have

Theorem 14.49. *Here all are as in Theorem 14.41. Consider the case* $\lambda_\beta = \lambda_\alpha + \lambda_\nu$; $\lambda_\alpha \geq 0$, $\lambda_\nu > 0$, $\lambda_\beta > 1/2$, $p := 2\lambda_\beta$. *Here* P_k, $k = 1, 2$, *as in (14.128) and A is as in (14.129). Set*

$$A_0(R_2) := \left(\int_{R_1}^{R_2} (A(w))^{p/(2\lambda_\alpha + \lambda_\nu)} dw \right)^{(2\lambda_\alpha + \lambda_\nu/p)}. \tag{14.165}$$

Also put

$$\tilde{T}(R_2) := A_0(R_2) \left(\frac{\lambda_\nu}{\lambda_\beta} \right)^{(\lambda_\nu/p)} 2^{-(\lambda_\nu/p)}. \tag{14.166}$$

Then

$$\int_A \left\{ \left\{ \sum_{j=1}^{M-1} \left[\left| \frac{\partial_{R_1}^{\gamma_1} f_j(x)}{\partial r^{\gamma_1}} \right|^{\lambda_\alpha} \left| \frac{\partial_{R_1}^{\gamma_2} f_{j+1}(x)}{\partial r^{\gamma_2}} \right|^{\lambda_\alpha + \lambda_\nu} \left| \frac{\partial_{R_1}^{\nu} f_j(x)}{\partial r^\nu} \right|^{\lambda_\nu} + \right. \right. \right.$$

$$\left. \left| \frac{\partial_{R_1}^{\gamma_2} f_j(x)}{\partial r^{\gamma_2}} \right|^{\lambda_\alpha + \lambda_\nu} \left| \frac{\partial_{R_1}^{\gamma_1} f_{j+1}(x)}{\partial r^{\gamma_1}} \right|^{\lambda_\alpha} \left| \frac{\partial_{R_1}^{\nu} f_{j+1}(x)}{\partial r^\nu} \right|^{\lambda_\nu} \right] \right\} +$$

$$\left[\left| \frac{\partial_{R_1}^{\gamma_1} f_1(x)}{\partial r^{\gamma_1}} \right|^{\lambda_\alpha} \left| \frac{\partial_{R_1}^{\gamma_2} f_M(x)}{\partial r^{\gamma_2}} \right|^{\lambda_\alpha + \lambda_\nu} \left| \frac{\partial_{R_1}^{\nu} f_1(x)}{\partial r^\nu} \right|^{\lambda_\nu} + \right.$$

$$\left. \left. \left| \frac{\partial_{R_1}^{\gamma_2} f_1(x)}{\partial r^{\gamma_2}} \right|^{\lambda_\alpha + \lambda_\nu} \left| \frac{\partial_{R_1}^{\gamma_1} f_M(x)}{\partial r^{\gamma_1}} \right|^{\lambda_\alpha} \left| \frac{\partial_{R_1}^{\nu} f_M(x)}{\partial r^\nu} \right|^{\lambda_\nu} \right] \right\} dx$$

$$\leq 2 \tilde{T}(R_2) \left[\sum_{j=1}^{M} \left(\int_A \left| \frac{\partial_{R_1}^{\nu} f_j(x)}{\partial r^\nu} \right|^p dx \right) \right]. \tag{14.167}$$

We need

Theorem 14.50. [27]. *Let* $\nu \geq 1$, $\gamma_1, \gamma_2 \geq 0$, *such that* $\nu - \gamma_1 \geq 1$, $\nu - \gamma_2 \geq 1$ *and* $f_j \in C_{x_0}^\nu([a,b])$ *with* $f_j^{(i)}(x_0) = 0$, $i = 0, 1, \ldots, n-1$, $n := [\nu]$, $j = 1, \ldots, M \in \mathbb{N}$. *Here,* $x, x_0 \in [a, b] : x \geq x_0$. *Consider* $p(x) \geq 0$ *continuous functions on* $[x_0, b]$. *Let* $\lambda_\alpha, \lambda_\beta, \lambda_\nu \geq 0$. *Set*

$$\rho(x) := \frac{(x - x_0)^{(\nu\lambda_\alpha - \gamma_1\lambda_\alpha + \nu\lambda_\beta - \gamma_2\lambda_\beta + 1)} \|p(x)\|_\infty}{(\nu\lambda_\alpha - \gamma_1\lambda_\alpha + \nu\lambda_\beta - \gamma_2\lambda_\beta + 1)(\Gamma(\nu - \gamma_1 + 1))^{\lambda_\alpha}(\Gamma(\nu - \gamma_2 + 1))^{\lambda_\beta}}. \tag{14.168}$$

Then

$$\int_{x_0}^{x} p(w) \left\{ \left\{ \sum_{j=1}^{M-1} \left[\left| \left(D_{x_0}^{\gamma_1} f_j\right)(w)\right|^{\lambda_\alpha} \left| \left(D_{x_0}^{\gamma_2} f_{j+1}\right)(w)\right|^{\lambda_\beta} \left| \left(D_{x_0}^{\nu} f_j\right)(w)\right|^{\lambda_\nu} + \right. \right. \right.$$

$$\left| \left(D_{x_0}^{\gamma_2} f_j\right)(w)\right|^{\lambda_\beta} \left| \left(D_{x_0}^{\gamma_1} f_{j+1}\right)(w)\right|^{\lambda_\alpha} \left| \left(D_{x_0}^{\nu} f_{j+1}\right)(w)\right|^{\lambda_\nu} \right] \right\} +$$

$$\left[\left| \left(D_{x_0}^{\gamma_1} f_1\right)(w)\right|^{\lambda_\alpha} \left| \left(D_{x_0}^{\gamma_2} f_M\right)(w)\right|^{\lambda_\beta} \left| \left(D_{x_0}^{\nu} f_1\right)(w)\right|^{\lambda_\nu} + \right.$$

$$\left. \left| \left(D_{x_0}^{\gamma_2} f_1\right)(w)\right|^{\lambda_\beta} \left| \left(D_{x_0}^{\gamma_1} f_M\right)(w)\right|^{\lambda_\alpha} \left| \left(D_{x_0}^{\nu} f_M\right)(w)\right|^{\lambda_\nu} \right] \right\} dw$$

$$\le \rho(x) \left\{ \sum_{j=1}^{M} \left\{ \left\| \left(D_{x_0}^{\nu} f_j\right) \right\|_{\infty}^{2(\lambda_\alpha + \lambda_\nu)} + \left\| \left(D_{x_0}^{\nu} f_j\right) \right\|_{\infty}^{2\lambda_\beta} \right\} \right\}, \qquad (14.169)$$

all $\quad x_0 \le x \le b.$

Similarly, by (14.169), we have

Theorem 14.51. *All are as in Assumption 14.39. Let* $\gamma_1, \gamma_2 \ge 0$, *such that*
$\nu - \gamma_1 \ge 1, \ \nu - \gamma_2 \ge 1; \ \lambda_\alpha, \lambda_\beta, \lambda_\nu \ge 0.$ *Set*

$$\rho(R_2) := \frac{R_2^{N-1}(R_2 - R_1)^{(\nu\lambda_\alpha - \gamma_1\lambda_\alpha + \nu\lambda_\beta - \gamma_2\lambda_\beta + 1)}}{(\nu\lambda_\alpha - \gamma_1\lambda_\alpha + \nu\lambda_\beta - \gamma_2\lambda_\beta + 1)(\Gamma(\nu - \gamma_1 + 1))^{\lambda_\alpha}(\Gamma(\nu - \gamma_2 + 1))^{\lambda_\beta}}.$$
$$(14.170)$$

Then

$$\int_{A} \left\{ \left\{ \sum_{j=1}^{M-1} \left[\left| \frac{\partial_{R_1}^{\gamma_1} f_j(x)}{\partial r^{\gamma_1}} \right|^{\lambda_\alpha} \left| \frac{\partial_{R_1}^{\gamma_2} f_{j+1}(x)}{\partial r^{\gamma_2}} \right|^{\lambda_\beta} \left| \frac{\partial_{R_1}^{\nu} f_j(x)}{\partial r^{\nu}} \right|^{\lambda_\nu} + \right. \right. \right.$$

$$\left| \frac{\partial_{R_1}^{\gamma_2} f_j(x)}{\partial r^{\gamma_2}} \right|^{\lambda_\beta} \left| \frac{\partial_{R_1}^{\gamma_1} f_{j+1}(x)}{\partial r^{\gamma_1}} \right|^{\lambda_\alpha} \left| \frac{\partial_{R_1}^{\nu} f_{j+1}(x)}{\partial r^{\nu}} \right|^{\lambda_\nu} \right] \right\} +$$

$$\left[\left| \frac{\partial_{R_1}^{\gamma_1} f_1(x)}{\partial r^{\gamma_1}} \right|^{\lambda_\alpha} \left| \frac{\partial_{R_1}^{\gamma_2} f_M(x)}{\partial r^{\gamma_2}} \right|^{\lambda_\beta} \left| \frac{\partial_{R_1}^{\nu} f_1(x)}{\partial r^{\nu}} \right|^{\lambda_\nu} + \right.$$

$$\left. \left| \frac{\partial_{R_1}^{\gamma_2} f_1(x)}{\partial r^{\gamma_2}} \right|^{\lambda_\beta} \left| \frac{\partial_{R_1}^{\gamma_1} f_M(x)}{\partial r^{\gamma_1}} \right|^{\lambda_\alpha} \left| \frac{\partial_{R_1}^{\nu} f_M(x)}{\partial r^{\nu}} \right|^{\lambda_\nu} \right] \right\} dx \le$$

$$\frac{2\pi^{N/2}}{\Gamma(N/2)} \rho(R_2) \left\{ \sum_{j=1}^{M} \left\{ \left\| \frac{\partial_{R_1}^{\nu} f_j}{\partial r^{\nu}} \right\|_{\infty}^{2(\lambda_\alpha + \lambda_\nu)} + \left\| \frac{\partial_{R_1}^{\nu} f_j}{\partial r^{\nu}} \right\|_{\infty}^{2\lambda_\beta} \right\} \right\}. \qquad (14.171)$$

We need

Theorem 14.52. [27]. (*As in Theorem 14.50,* $\lambda_\beta = 0$.) *It holds*

$$\int_{x_0}^{x} p(w) \left(\sum_{j=1}^{M} \left| \left(D_{x_0}^{\gamma_1} f_j \right)(w) \right|^{\lambda_\alpha} \left| \left(D_{x_0}^{\nu} f_j \right)(w) \right|^{\lambda_\nu} \right) dw \le$$

$$\left(\frac{(x - x_0)^{(\nu\lambda_\alpha - \gamma_1\lambda_\alpha + 1)} \|p(x)\|_\infty}{(\nu\lambda_\alpha - \gamma_1\lambda_\alpha + 1)(\Gamma(\nu - \gamma_1 + 1))^{\lambda_\alpha}} \right) \cdot \left(\sum_{j=1}^{M} \left\| D_{x_0}^{\nu} f_j \right\|_\infty^{\lambda_\alpha + \lambda_\nu} \right), \quad (14.172)$$

all $x_0 \le x \le b$.

Similarly, by (14.172), we obtain

Theorem 14.53. *Here all are as in Theorem 14.51. Case of* $\lambda_\beta = 0$.
Then

$$\sum_{j=1}^{M} \left(\int_{A} \left| \frac{\partial_{R_1}^{\gamma_1} f_j(x)}{\partial r^{\gamma_1}} \right|^{\lambda_\alpha} \left| \frac{\partial_{R_1}^{\nu} f_j(x)}{\partial r^{\nu}} \right|^{\lambda_\nu} dx \right) \le$$

$$\left(\frac{2\pi^{N/2}}{\Gamma(N/2)} \right) \left(\frac{R_2^{N-1}(R_2 - R_1)^{(\nu\lambda_\alpha - \gamma_1\lambda_\alpha + 1)}}{(\nu\lambda_\alpha - \gamma_1\lambda_\alpha + 1)(\Gamma(\nu - \gamma_1 + 1))^{\lambda_\alpha}} \right) \left(\sum_{j=1}^{M} \left\| \frac{\partial_{R_1}^{\nu} f_j}{\partial r^{\nu}} \right\|_\infty^{\lambda_\alpha + \lambda_\nu} \right).$$

$$(14.173)$$

We need

Theorem 14.54. [27]. (*As in Theorem 14.50,* $\lambda_\beta = \lambda_\alpha + \lambda_\nu$.) *It holds*

$$\int_{x_0}^{x} p(w) \left\{ \left\{ \sum_{j=1}^{M-1} \left[\left| \left(D_{x_0}^{\gamma_1} f_j \right)(w) \right|^{\lambda_\alpha} \left| \left(D_{x_0}^{\gamma_2} f_{j+1} \right)(w) \right|^{\lambda_\alpha + \lambda_\nu} \left| \left(D_{x_0}^{\nu} f_j \right)(w) \right|^{\lambda_\nu} + \right. \right. \right.$$

$$\left| \left(D_{x_0}^{\gamma_2} f_j \right)(w) \right|^{\lambda_\alpha + \lambda_\nu} \left| \left(D_{x_0}^{\gamma_1} f_{j+1} \right)(w) \right|^{\lambda_\alpha} \left| \left(D_{x_0}^{\nu} f_{j+1} \right)(w) \right|^{\lambda_\nu} \right] \right\} +$$

$$\left[\left| \left(D_{x_0}^{\gamma_1} f_1 \right)(w) \right|^{\lambda_\alpha} \left| \left(D_{x_0}^{\gamma_2} f_M \right)(w) \right|^{\lambda_\alpha + \lambda_\nu} \left| \left(D_{x_0}^{\nu} f_1 \right)(w) \right|^{\lambda_\nu} + \right.$$

$$\left| \left(D_{x_0}^{\gamma_2} f_1 \right)(w) \right|^{\lambda_\alpha + \lambda_\nu} \left| \left(D_{x_0}^{\gamma_1} f_M \right)(w) \right|^{\lambda_\alpha} \left| \left(D_{x_0}^{\nu} f_M \right)(w) \right|^{\lambda_\nu} \right] \right\} dw \le$$

$$\left(\frac{2(x - x_0)^{(2\nu\lambda_\alpha - \gamma_1\lambda_\alpha + \nu\lambda_\nu - \gamma_2\lambda_\alpha - \gamma_2\lambda_\nu + 1)}}{(2\nu\lambda_\alpha - \gamma_1\lambda_\alpha + \nu\lambda_\nu - \gamma_2\lambda_\alpha - \gamma_2\lambda_\nu + 1)} \right.$$

$$\left. \frac{\|p(x)\|_\infty}{(\Gamma(\nu - \gamma_1 + 1))^{\lambda_\alpha}(\Gamma(\nu - \gamma_2 + 1))^{(\lambda_\alpha + \lambda_\nu)}} \right)$$

$$\cdot \left(\sum_{j=1}^{M} \left\| D_{x_0}^{\nu} f_j \right\|_{\infty}^{2(\lambda_\alpha + \lambda_\nu)} \right), \tag{14.174}$$

all $x_0 \le x \le b$.

Similarly, by (14.174), we derive

Theorem 14.55. *Here all are as in Theorem 14.51. Case of* $\lambda_\beta = \lambda_\alpha + \lambda_\nu$. *Then*

$$\int_A \left\{ \left\{ \sum_{j=1}^{M-1} \left[\left| \frac{\partial_{R_1}^{\gamma_1} f_j(x)}{\partial r^{\gamma_1}} \right|^{\lambda_\alpha} \left| \frac{\partial_{R_1}^{\gamma_2} f_{j+1}(x)}{\partial r^{\gamma_2}} \right|^{\lambda_\alpha + \lambda_\nu} \left| \frac{\partial_{R_1}^{\nu} f_j(x)}{\partial r^{\nu}} \right|^{\lambda_\nu} + \right. \right. \right.$$

$$\left. \left. \left| \frac{\partial_{R_1}^{\gamma_2} f_j(x)}{\partial r^{\gamma_2}} \right|^{\lambda_\alpha + \lambda_\nu} \left| \frac{\partial_{R_1}^{\gamma_1} f_{j+1}(x)}{\partial r^{\gamma_1}} \right|^{\lambda_\alpha} \left| \frac{\partial_{R_1}^{\nu} f_{j+1}(x)}{\partial r^{\nu}} \right|^{\lambda_\nu} \right] \right\} +$$

$$\left[\left| \frac{\partial_{R_1}^{\gamma_1} f_1(x)}{\partial r^{\gamma_1}} \right|^{\lambda_\alpha} \left| \frac{\partial_{R_1}^{\gamma_2} f_M(x)}{\partial r^{\gamma_2}} \right|^{\lambda_\alpha + \lambda_\nu} \left| \frac{\partial_{R_1}^{\nu} f_1(x)}{\partial r^{\nu}} \right|^{\lambda_\nu} + \right.$$

$$\left. \left. \left| \frac{\partial_{R_1}^{\gamma_2} f_1(x)}{\partial r^{\gamma_2}} \right|^{\lambda_\alpha + \lambda_\nu} \left| \frac{\partial_{R_1}^{\gamma_1} f_M(x)}{\partial r^{\gamma_1}} \right|^{\lambda_\alpha} \left| \frac{\partial_{R_1}^{\nu} f_M(x)}{\partial r^{\nu}} \right|^{\lambda_\nu} \right] \right\} dx \le$$

$$\frac{4\pi^{N/2}}{\Gamma(N/2)} \cdot \left(\frac{R_2^{N-1}}{(2\nu\lambda_\alpha - \gamma_1\lambda_\alpha + \nu\lambda_\nu - \gamma_2\lambda_\alpha - \gamma_2\lambda_\nu + 1)} \right.$$

$$\left. \frac{(R_2 - R_1)^{(2\nu\lambda_\alpha - \gamma_1\lambda_\alpha + \nu\lambda_\nu - \gamma_2\lambda_\alpha - \gamma_2\lambda_\nu + 1)}}{(\Gamma(\nu - \gamma_1 + 1))^{\lambda_\alpha} (\Gamma(\nu - \gamma_2 + 1))^{(\lambda_\alpha + \lambda_\nu)}} \right)$$

$$\cdot \left(\sum_{j=1}^{M} \left\| \frac{\partial_{R_1}^{\nu} f_j}{\partial r^{\nu}} \right\|_{\infty}^{2(\lambda_\alpha + \lambda_\nu)} \right). \tag{14.175}$$

We need

Theorem 14.56. [27]. *(As in Theorem 14.50,* $\lambda_\nu = 0$, $\lambda_\alpha = \lambda_\beta$.) *It holds*

$$\int_{x_0}^{x} p(w) \left\{ \left\{ \sum_{j=1}^{M-1} \left[\left| (D_{x_0}^{\gamma_1} f_j)(w) \right|^{\lambda_\alpha} \left| (D_{x_0}^{\gamma_2} f_{j+1})(w) \right|^{\lambda_\alpha} \right. \right. \right.$$

$$+ \left|\left(D_{x_0}^{\gamma_2} f_j\right)(w)\right|^{\lambda_\alpha} \left|\left(D_{x_0}^{\gamma_1} f_{j+1}\right)(w)\right|^{\lambda_\alpha}\right]\Big\}$$

$$+ \left[\left|\left(D_{x_0}^{\gamma_1} f_1\right)(w)\right|^{\lambda_\alpha} \left|\left(D_{x_0}^{\gamma_2} f_M\right)(w)\right|^{\lambda_\alpha}\right.$$

$$+ \left|\left(D_{x_0}^{\gamma_2} f_1\right)(w)\right|^{\lambda_\alpha} \left|\left(D_{x_0}^{\gamma_1} f_M\right)(w)\right|^{\lambda_\alpha}\right]\Big\} dw$$

$$\le 2\,\rho^*(x) \left[\sum_{j=1}^{M} \left\|D_{x_0}^\nu f_j\right\|_\infty^{2\lambda_\alpha}\right], \tag{14.176}$$

all $x_0 \le x \le b$.
Here we have

$$\rho^*(x) := \left(\frac{(x-x_0)^{(2\nu\lambda_\alpha - \gamma_1\lambda_\alpha - \gamma_2\lambda_\alpha + 1)}\|p(x)\|_\infty}{(2\nu\lambda_\alpha - \gamma_1\lambda_\alpha - \gamma_2\lambda_\alpha + 1)(\Gamma(\nu-\gamma_1+1))^{\lambda_\alpha}(\Gamma(\nu-\gamma_2+1))^{\lambda_\alpha}}\right). \tag{14.177}$$

Similarly, by (14.177), we derive

Theorem 14.57. *Here all are as in Theorem 14.51. Case of $\lambda_\nu = 0$, $\lambda_\alpha = \lambda_\beta$.*
Then

$$\int_A \left\{\left\{\sum_{j=1}^{M-1}\left[\left|\frac{\partial_{R_1}^{\gamma_1} f_j(x)}{\partial r^{\gamma_1}}\right|^{\lambda_\alpha}\left|\frac{\partial_{R_1}^{\gamma_2} f_{j+1}(x)}{\partial r^{\gamma_2}}\right|^{\lambda_\alpha}\right.\right.\right.$$

$$+ \left.\left|\frac{\partial_{R_1}^{\gamma_2} f_j(x)}{\partial r^{\gamma_2}}\right|^{\lambda_\alpha}\left|\frac{\partial_{R_1}^{\gamma_1} f_{j+1}(x)}{\partial r^{\gamma_1}}\right|^{\lambda_\alpha}\right]\right\}$$

$$+ \left[\left|\frac{\partial_{R_1}^{\gamma_1} f_1(x)}{\partial r^{\gamma_1}}\right|^{\lambda_\alpha}\left|\frac{\partial_{R_1}^{\gamma_2} f_M(x)}{\partial r^{\gamma_2}}\right|^{\lambda_\alpha}\right.$$

$$+ \left.\left.\left|\frac{\partial_{R_1}^{\gamma_2} f_1(x)}{\partial r^{\gamma_2}}\right|^{\lambda_\alpha}\left|\frac{\partial_{R_1}^{\gamma_1} f_M(x)}{\partial r^{\gamma_1}}\right|^{\lambda_\alpha}\right]\right\} dx$$

$$\le \left(\frac{4\pi^{N/2}}{\Gamma(N/2)}\right)\rho^*(R_2)\left[\sum_{j=1}^{M}\left\|\frac{\partial_{R_1}^\nu f_j}{\partial r^\nu}\right\|_\infty^{2\lambda_\alpha}\right]. \tag{14.178}$$

Here we have

$$\rho^*(R_2) := \frac{R_2^{N-1}(R_2-R_1)^{(2\nu\lambda_\alpha - \gamma_1\lambda_\alpha - \gamma_2\lambda_\alpha + 1)}}{(2\nu\lambda_\alpha - \gamma_1\lambda_\alpha - \gamma_2\lambda_\alpha + 1)(\Gamma(\nu-\gamma_1+1))^{\lambda_\alpha}(\Gamma(\nu-\gamma_2+1))^{\lambda_\alpha}}. \tag{14.179}$$

We need

Theorem 14.58. [27]. (*As in Theorem 14.50,* $\lambda_\alpha = 0$, $\lambda_\beta = \lambda_\nu$.) *It holds*

$$\int_{x_0}^{x} p(w) \left\{ \left\{ \sum_{j=1}^{M-1} \left[\left| \left(D_{x_0}^{\gamma_2} f_{j+1} \right)(w) \right|^{\lambda_\beta} \left| \left(D_{x_0}^{\nu} f_j \right)(w) \right|^{\lambda_\beta} \right. \right. \right.$$

$$+ \left. \left| \left(D_{x_0}^{\gamma_2} f_j \right)(w) \right|^{\lambda_\beta} \left| \left(D_{x_0}^{\nu} f_{j+1} \right)(w) \right|^{\lambda_\beta} \right] \right\}$$

$$+ \left[\left| \left(D_{x_0}^{\gamma_2} f_M \right)(w) \right|^{\lambda_\beta} \left| \left(D_{x_0}^{\nu} f_1 \right)(w) \right|^{\lambda_\beta} \right.$$

$$+ \left. \left| \left(D_{x_0}^{\gamma_2} f_1 \right)(w) \right|^{\lambda_\beta} \left| \left(D_{x_0}^{\nu} f_M \right)(w) \right|^{\lambda_\beta} \right] \right\} dw$$

$$\leq 2 \cdot \left(\frac{(x - x_0)^{(\nu\lambda_\beta - \gamma_2\lambda_\beta + 1)} \| p(x) \|_\infty}{(\nu\lambda_\beta - \gamma_2\lambda_\beta + 1)(\Gamma(\nu - \gamma_2 + 1))^{\lambda_\beta}} \right) \left[\sum_{j=1}^{M} \left\| D_{x_0}^{\nu} f_j \right\|_\infty^{2\lambda_\beta} \right], \quad (14.180)$$

all $x_0 \leq x \leq b.$

Similarly, by (14.180), we give the following.

Theorem 14.59. *Here all are as in Theorem 14.51. Case of* $\lambda_\alpha = 0$, $\lambda_\beta = \lambda_\nu$. *Then*

$$\int_A \left\{ \left\{ \sum_{j=1}^{M-1} \left[\left| \frac{\partial_{R_1}^{\gamma_2} f_{j+1}(x)}{\partial r^{\gamma_2}} \right|^{\lambda_\beta} \left| \frac{\partial_{R_1}^{\nu} f_j(x)}{\partial r^{\nu}} \right|^{\lambda_\beta} \right. \right. \right.$$

$$+ \left. \left| \frac{\partial_{R_1}^{\gamma_2} f_j(x)}{\partial r^{\gamma_2}} \right|^{\lambda_\beta} \left| \frac{\partial_{R_1}^{\nu} f_{j+1}(x)}{\partial r^{\nu}} \right|^{\lambda_\beta} \right] \right\}$$

$$+ \left[\left| \frac{\partial_{R_1}^{\gamma_2} f_M(x)}{\partial r^{\gamma_2}} \right|^{\lambda_\beta} \left| \frac{\partial_{R_1}^{\nu} f_1(x)}{\partial r^{\nu}} \right|^{\lambda_\beta} \right.$$

$$+ \left. \left| \frac{\partial_{R_1}^{\gamma_2} f_1(x)}{\partial r^{\gamma_2}} \right|^{\lambda_\beta} \left| \frac{\partial_{R_1}^{\nu} f_M(x)}{\partial r^{\nu}} \right|^{\lambda_\beta} \right] \right\} dx \leq$$

$$\left(\frac{4\pi^{N/2}}{\Gamma(N/2)} \right) \left(\frac{R_2^{N-1} (R_2 - R_1)^{(\nu\lambda_\beta - \gamma_2\lambda_\beta + 1)}}{(\nu\lambda_\beta - \gamma_2\lambda_\beta + 1)(\Gamma(\nu - \gamma_2 + 1))^{\lambda_\beta}} \right) \left[\sum_{j=1}^{M} \left\| \frac{\partial_{R_1}^{\nu} f_j}{\partial r^{\nu}} \right\|_\infty^{2\lambda_\beta} \right].$$

$$(14.181)$$

To extend the above research we give a motivational result regarding fractional integration by parts.

Proposition 14.60. *Let* $f \in C^1([0,1])$ *and* $g \in C^\nu([0,1])$, $\nu > 0$, $n := [\nu]$, $\alpha := \nu - n$. *Then*

$$\int_0^1 f(x) g^{(\nu)}(x) dx = f(1)(\mathcal{J}_{1-\alpha} \, g^{(n)})(1) - \int_0^1 (\mathcal{J}_{1-\alpha} \, g^{(n)})(x) f'(x) dx.$$
$$(14.182)$$

Proof. Here $g^{(\nu)} = d \, (\mathcal{J}_{1-\alpha} \, g^{(n)})(x)/dx$; that is, $d \, (\mathcal{J}_{1-\alpha} \, g^{(n)})(x) = g^{(\nu)}(x) \, dx$.

Hence, by ordinary integration by parts we have:

$$\int_0^1 f(x) g^{(\nu)}(x) \, dx = \int_0^1 f(x) \, d \, (\mathcal{J}_{1-\alpha} \, g^{(n)})(x) =$$

$$f(1)(\mathcal{J}_{1-\alpha} \, g^{(n)})(1) - \int_0^1 (\mathcal{J}_{1-\alpha} \, g^{(n)})(x) f'(x) \, dx,$$

by $(\mathcal{J}_{1-\alpha} \, g^{(n)})(0) = 0$. $\square$

Now we are ready to give

Definition 14.61. Let $\nu > 0$, $n := [\nu]$, $\alpha := \nu - n$, $g : [0,1] \to \mathbb{R}$ such that there exists $g^{(n)}$ which is measurable. Assume that $(\mathcal{J}_{1-\alpha} \, g^{(n)}) \in L^1([0,1])$. We say that $g^{(\nu)} \in L^1([0,1])$ is a *weak fractional derivative of order* ν *for* g, iff

$$\int_0^1 u(x) \, g^{(\nu)}(x) \, dx = - \int_0^1 (\mathcal{J}_{1-\alpha} \, g^{(n)})(x) u'(x) \, dx, \qquad (14.183)$$

$\forall \, u \in C^\infty([0,1]) : \; u(1) = 0$.

Based on the above Definition 14.61, we can extend the concept of weak fractional differentiation to anchor points $x_0 \neq 0$, and to the multivariate case, especially to the radial case. Then we try to generalize the results of this chapter.

15
Riemann–Liouville Fractional Multivariate Opial-Type Inequalities over a Spherical Shell

Here we introduce the concept of the Riemann–Liouville fractional radial derivative for a function defined on a spherical shell. Using polar coordinates we are able to derive multivariate Opial-type inequalities over a spherical shell of $\mathbb{R}^N$, $N \geq 2$, by studying the topic in all possibilities. Our results involve one, two, or more functions. We also produce several generalized univariate fractional Opial-type inequalities, many of which are used to achieve the main goals. This treatment is based on [45].

15.1 Introduction

This chapter is motivated by articles of Opial [315], Bessack [80], Anastassiou, Koliha, and Pecaric [64, 65], and Anastassiou [46, 48].

Opial-type inequalities usually find applications in establishing the uniqueness of solution of initial value problems for differential equations and their systems; see Willett [406]. In this chapter we present a series of various Riemann–Liouville fractional multivariate Opial-type inequalities over spherical shells. To achieve our goal we use polar coordinates, and we introduce and use the Riemann–Liouville fractional radial derivative. We work on the spherical shell, and not on the ball, because a radial derivative cannot be defined at zero. So, we reduce the problem to a univariate one.

Consequently we use a large array of univariate Opial-type inequalities involving generalized Riemann–Liouville fractional derivatives; these are Riemann–Liouville fractional derivatives defined at the arbitrary anchor point $a \in \mathbb{R}$. So we also present a very large set of generalized univariate Riemann–Liouville fractional Opial-type inequalities transferred from earlier ones, proved at anchor point zero for the standard Riemann–Liouville fractional derivative. In our results we involve one, two, or several functions. But first we need to develop an extensive background in three parts, and then follow the main results in three subsections.

15.2   Background—I

Here we follow [356, pp. 149–150] and [383, pp. 87–88]. Also here $\mathbb{R}^N$, $N > 1$ denotes the N-tuple of reals $\mathbb{R}$, and $\mathbb{N}$ denotes the natural numbers. Let us denote by $dx \equiv \lambda_{\mathbb{R}^N}(dx)$ the Lebesgue measure on $\mathbb{R}^N$, $N > 1$, and $S^{N-1} := \{x \in \mathbb{R}^N : |x| = 1\}$ the unit sphere on $\mathbb{R}^N$, where $|\cdot|$ stands for the Euclidean norm in $\mathbb{R}^N$. Also denote the ball

$$B(0, R) := \{x \in \mathbb{R}^N : |x| < R\} \subseteq \mathbb{R}^N, \; R > 0,$$

and the spherical shell

$$A := B(0, R_2) - \overline{B(0, R_1)}, \;\; 0 < R_1 < R_2.$$

For $x \in \mathbb{R}^N - \{0\}$ we can write uniquely $x = rw$, where $r = |x| > 0$, and $w = x/r \in S^{N-1}$, $|w| = 1$. Clearly here

$$\mathbb{R}^N - \{0\} = (0, \infty) \times S^{N-1},$$

also the map

$$\Phi : \mathbb{R}^N - \{0\} \to S^{N-1} : \; \Phi(x) = \frac{x}{|x|}$$

is continuous. In addition, $\overline{A} = [R_1, R_2] \times S^{N-1}$. Let us denote by $dw \equiv \lambda_{S^{N-1}}(w)$ the surface measure on S^{N-1} to be defined as the image under Φ of $N \cdot \lambda_{\mathbb{R}^N}$ restricted to the Borel class of $B(0, 1) - \{0\}$. More precisely, the last definition is as follows. Let $A \subset S^{N-1}$ be a Borel set, and let

$$\tilde{A} := \{ru : \; 0 < r < 1, \; u \in A\} \subset \mathbb{R}^N;$$

we define

$$\lambda_{S^{N-1}}(A) = N \cdot \lambda_{\mathbb{R}^N}(\tilde{A}).$$

Noting $\Phi(rx) = \Phi(x)$, all $r > 0$ and $x \in \mathbb{R}^N - \{0\}$, one can conclude that

$$\int_{B(0,r)-\{0\}} f \circ \Phi(x) dx = r^N \int_{B(0,1)-\{0\}} f \circ \Phi(x) dx$$

and thus

$$\int_{B(0,r)-\{0\}} f \circ \Phi(x) dx = \frac{r^N}{N} \int_{S^{N-1}} f(w) \lambda_{S^{N-1}}(dw),$$

for all f nonnegative and measurable functions on $(S^{N-1}, \mathcal{B}_{S^{N-1}})$; $\mathcal{B}$ stands for the Borel class.

We denote by

$$w_N \equiv \lambda_{S^{N-1}}(S^{N-1}) = \int_{S^{N-1}} dw = \frac{2\pi^{N/2}}{\Gamma(N/2)}$$

the surface area of S^{N-1} and we get the volume

$$|B(0,r)| = \frac{w_N \, r^N}{N} = \frac{2\pi^{N/2} \, r^N}{N \, \Gamma(N/2)},$$

so that $|B(0,1)| = 2\pi^{N/2}/(N \, \Gamma(N/2))$. Clearly here

$$\mathrm{Vol}(A) = |A| = \frac{w_N(R_2^N - R_1^N)}{N} = \frac{2\pi^{N/2}(R_2^N - R_1^N)}{N \, \Gamma(N/2)}.$$

Next, define

$$\psi : (0,\infty) \times S^{N-1} \to \mathbb{R}^N - \{0\}$$

by $\psi(r,w) := rw$; ψ is a one-to-one and onto function. Thus

$$(r,w) \equiv \psi^{-1}(x) = (|x|, \Phi(x))$$

are called the polar coordinates of $x \in \mathbb{R}^N - \{0\}$.

Finally, define the measure R_N on $\big((0,\infty), \mathcal{B}_{(0,\infty)}\big)$ by

$$R_N(\Gamma) = \int_{\Gamma} r^{N-1} dr, \ \text{any} \ \Gamma \in \mathcal{B}_{(0,\infty)}.$$

We mention the following very important theorem.

Theorem 15.1. (See exercise 6, pp. 149–150 in [356] and Theorem 5.2.2 pp. 87–88 of [383].) *We have that* $\lambda_{\mathbb{R}^N} = (R_N \times \lambda_{S^{N-1}}) \circ \psi^{-1}$ *on* $\mathcal{B}_{\mathbb{R}^N - \{0\}}$.

In particular, if f is a nonnegative Borel measurable function on $(\mathbb{R}^N, \mathcal{B}_{\mathbb{R}^N})$, then the Lebesgue integral

$$\int_{\mathbb{R}^N} f(x)dx = \int_{(0,\infty)} r^{N-1} \left(\int_{S^{N-1}} f(rw)\lambda_{S^{N-1}}(dw) \right) dr$$

$$= \int_{S^{N-1}} \left(\int_{(0,\infty)} f(rw)r^{N-1} dr \right) \lambda_{S^{N-1}}(dw). \tag{15.1}$$

Clearly 15.1 is true for f a Borel integrable function taking values in $\mathbb{R}$.

Using the facts that:

(i) The Lebesgue measure of a Lebesgue measurable set K equals the Lebesgue measure of a Borel set (i.e., there exist M, an F_σ, and T a G_δ sets: $M \subset K \subset T$ with $\lambda_{\mathbb{R}^N}(K) = \lambda_{\mathbb{R}^N}(M) = \lambda_{\mathbb{R}^N}(T)$; see [354], p. 62).

(ii) For each g Lebesgue measurable function, there exists an f Borel measurable function such that $g = f$ a.e. (see [356, p. 145]), we get valid that (15.1) is true for Lebesgue integrable functions $f : \mathbb{R}^N \to \mathbb{R}$.

We give the important

Proposition 15.2. *Let $f : B(0, R) \to \mathbb{R}$, $R > 0$, be a Lebesgue integrable function. Then*

$$\int_{B(0,R)} f(x)dx = \int_{S^{N-1}} \left(\int_0^R f(rw)r^{N-1}dr \right) dw. \qquad (15.2)$$

Proof. Call

$$F(x) := \begin{cases} f(x), & x \in B(0, R), \\ \\ 0, & x \in \mathbb{R}^N - B(0, R). \end{cases}$$

Then apply (15.1) for F to get (15.2) easily.

At last here we give the main tool for this chapter.

Proposition 15.3. *Let $f : A \to \mathbb{R}$ be a Lebesgue integrable function, where*

$$A := B(0, R_2) - \overline{B(0, R_1)}, \quad 0 < R_1 < R_2.$$

Then

$$\int_A f(x)dx = \int_{S^{N-1}} \left(\int_{R_1}^{R_2} f(rw)r^{N-1}dr \right) dw. \qquad (15.3)$$

Proof. Apply (15.1) for

$$F(x) := \begin{cases} f(x), & x \in A, \\ \\ 0, & x \in \mathbb{R}^N - A; \end{cases}$$

then (15.3) is valid. $\square$

We also need the following well-known result.

Proposition 15.4. *Let $f : [a, b] \to \mathbb{R}$, be a Lebesgue integrable function. Then*

$$\int_a^b f(z)dz = \int_0^{b-a} f(t+a)dt. \qquad (15.4)$$

So if $f_a(t) := f(a+t)$, the translation of f, then

$$\int_a^b f(z)dz = \int_0^{b-a} f_a(t)dt. \qquad (15.5)$$

15.3   Background—II

Here we define the Riemann–Liouville fractional derivative that we use.

Definition 15.5. (See [64, 65, 187].) Let $\alpha > 0$. For any $f \in L_1(0, x)$; $x > 0$, the Riemann–Liouville fractional integral of f of order α is defined by

$$(J_\alpha f)(s) := \frac{1}{\Gamma(\alpha)} \int_0^s (s-t)^{\alpha-1} f(t)dt, \qquad (15.6)$$

all $s \in [0, x]$, and the Riemann–Liouville fractional derivative of f of order α by

$$D^\alpha f(s) := \frac{1}{\Gamma(m-\alpha)} \left(\frac{d}{ds}\right)^m \int_0^s (s-t)^{m-\alpha-1} f(t)dt, \qquad (15.7)$$

where $m = [\alpha] + 1$, $[\cdot]$ is the integer part.
    In addition, we set

$$D^0 f := f := J_0 f,$$

$$J_{-\alpha} f = D^\alpha f, \quad \text{if } \alpha > 0,$$

$$D^{-\alpha} f := J_\alpha f, \quad \text{if } 0 < \alpha \le 1.$$

If $\alpha \in \mathbb{N}$, then $D^\alpha f = f^{(\alpha)}$ the ordinary derivative.

Definition 15.6. (See [187].) We say that $f \in L_1(0, x)$ has an L_∞ fractional derivative $D^\alpha f$ in $[0, x]$ if $x > 0$, if and only if

$$D^{\alpha-k} f \in C([0, x]), \quad k = 1, \dots, m := [\alpha] + 1; \; \alpha > 0,$$

and

$$D^{\alpha-1} f \in AC([0, x]) \text{ (absolutely continuous functions)},$$

and $D^\alpha f \in L_\infty(0, x)$.

We mention

Lemma 15.7. (See [187].) *Let $\beta > \alpha \geq 0$. Let $f \in L_1(0, x)$, $x > 0$, have an L_∞ fractional derivative $D^\beta f$ in $[0, x]$ and let*

$$D^{\beta-k} f(0) = 0 \text{ for } k = 1, \ldots, [\beta] + 1.$$

Then

$$D^\alpha f(s) := \frac{1}{\Gamma(\beta - \alpha)} \int_0^s (s - t)^{\beta-\alpha-1} D^\beta f(t) dt, \qquad (15.8)$$

all $s \in [0, x]$. Clearly here

$$D^\alpha f \in AC([0, x]) \text{ for } \beta - \alpha \geq 1$$

and

$$D^\alpha f \in C([0, x]), \text{ for } \beta - \alpha \in (0, 1),$$

hence $D^\alpha f \in L_\infty(0, x)$, and $D^\alpha f \in L_1(0, x)$.

Next, we define the generalized Riemann–Liouville fractional derivative with arbitrary anchor point $a \in \mathbb{R}$; see [64].

Definition 15.8. Let $v \geq 0$, define

$$(D_a^v f)(s) := (D^v f_a)(s - a), \quad s \geq a, \qquad (15.9)$$

for $v = 0$ both sides equal to $f(s)$, and for $v = n \in \mathbb{N}$ we easily get that $(D_a^n f)(s) = f^{(n)}(s)$, the ordinary derivative. Clearly here

$$(D_a^v f)(z + a) = (D^v f_a)(z). \qquad (15.10)$$

We use $p(s)$ and $D_a^v f(s)$ in $L_\infty(a, x)$, $x > a$, $a, x \in \mathbb{R}$. In that case by using (15.5) we obtain

$$\int_a^w p(y)(D_a^v f)(y) dy = \int_0^{w-a} p(z + a)(D^v f_a)(z) dz, \qquad (15.11)$$

for all $a \leq w \leq x$, $a, x \in \mathbb{R}$, which identity we use a lot in this chapter.

Our initial intention is to transfer Riemann–Liouville fractional Opial inequalities, [46, 48, 64, 65] applied to f_a over $[0, w - a]$, for f over $[a, w]$ and use the generalized Riemann–Liouville fractional derivative. For that we observe the following.

Lemma 15.9. $f \in L_1(a, w)$ *if and only if $f_a \in L_1(0, w - a)$, where $w \geq a$, $a, w \in \mathbb{R}$; $f_a(t) := f(a + t)$.*

Proof. We see that

$$\int_a^w |f(z)|dz = \int_0^{w-a} |f_a(t)|dt. \quad \square$$

We need

Lemma 15.10. *Let* $F(s) := f(s-a)$, $a \in \mathbb{R}$ *be fixed. Here* $f : [0, w-a] \to \mathbb{R}$, *where* $w > a$ *and* $F : [a, w] \to \mathbb{R}$. *Then*

(i) $F \in C([a, w])$ *if and only if* $f \in C([0, w-a])$.

(ii) $F \in L_\infty(a, w)$ *if and only if* $f \in L_\infty(0, w-a)$.

(iii) $F \in AC([a, w])$ *if and only if* $f \in AC([0, w-a])$.

Proof. It is based on the fact that the map $g : [a, w] \to [0, w-a]$, such that $g(s) := s - a$ is one to one and onto.

(i) ($\Rightarrow$) Let F be continuous, and let z_n, $z \in [0, w-a]$: $z_n \to z$; that is, $z_n + a \to z + a$, here $z_n + a$, $z + a \in [a, w]$. Hence $F(z_n + a) \to F(z + a)$ (i.e., $f(z_n) \to f(z)$), proving continuity of f.

($\Leftarrow$) Let f be continuous, and let

$$s_n \to s; \; s_n, \; s \in [a, w] \iff s_n - a, \; s - a \in [0, w-a],$$

and $s_n - a \to s - a$. Hence $f(s_n - a) \to f(s - a)$; that is, $F(s_n) \to F(s)$. That is, F is continuous.

(ii) We see that

$$|F(s)| = |f(s - a)| \leq \|f\|_{\infty, [0,w-a]}$$

a.e. in $s \in [a, w]$. Hence

$$\|F\|_{\infty, [a,w]} \leq \|f\|_{\infty, [0,w-a]}.$$

Also

$$|f(s - a)| = |F(s)| \leq \|F\|_{\infty, [a,w]}$$

a.e. in $s \in [a, w]$. So that

$$\|f\|_{\infty, [0,w-a]} \leq \|F\|_{\infty, [a,w]};$$

that is,

$$\|F\|_{\infty, [a,w]} = \|f\|_{\infty, [0,w-a]},$$

proving the claim.

(iii) ($\Rightarrow$) Let F be absolutely continuous; that is, for every $\epsilon > 0$ there is a $\delta > 0$ such that whenever $(a_1, b_1), \ldots, (a_n, b_n)$ are disjoint open subintervals of $[a, w]$, then

$$\sum_{i=1}^{n}(b_i - a_i) < \delta \;\Rightarrow\; \sum_{i=1}^{n}|F(b_i) - F(a_i)| < \epsilon.$$

Here $(a_i - a, b_i - a) \subset [0, w - a]$, $i = 1, \ldots, n$ and also disjoint. Rewriting the last statement we have

$$\sum_{i=1}^{n}\left((b_i - a) - (a_i - a)\right) < \delta \;\Rightarrow\; \sum_{i=1}^{n}|f(b_i - a) - f(a_i - a)| < \epsilon;$$

that is, f is absolutely continuous. Notice that any open subinterval $(a_i', b_i') \subset [0, w - a]$ has the form $(a_i - a, b_i - a)$, where $(a_i, b_i) \subset [a, w]$, for each $i = 1, \ldots, n$; by $a_i' = a_i - a$, $b_i' = b_i - a$.

($\Leftarrow$) Assume now f is absolutely continuous; that is, for every $\epsilon > 0$ there is a $\delta > 0$: for any $(a_1, b_1), \ldots, (a_n, b_n)$ that are disjoint subintervals of $[0, w - a]$, then

$$\sum_{i=1}^{n}(b_i - a_i) < \delta \;\Rightarrow\; \sum_{i=1}^{n}|f(b_i) - f(a_i)| < \epsilon.$$

The last statement is rewritten as

$$(\forall\, \epsilon > 0)\; (\exists\, \delta > 0) : \; (a_1 + a, b_1 + a), \ldots, (a_n + a, b_n + a) \subset [a, w]$$

(they are disjoint open subintervals); then

$$\sum_{i=1}^{n}\left((b_i + a) - (a_i + a)\right) < \delta \;\Rightarrow\; \sum_{i=1}^{n}|F(b_i + a) - F(a_i + a)| < \epsilon,$$

by

$$f(b_i) = F(b_i + a), \quad f(a_i) = F(a_i + a).$$

Therefore F is absolutely continuous. Notice again here that any open subinterval $(a_i', b_i') \subset [a, w]$ has the form $(a_i + a, b_i + a)$, where $(a_i, b_i) \subset [0, w - a]$ all $i = 1, \ldots, n$; by $a_i' = a_i + a$, $b_i' = b_i + a$. $\square$

We need

Lemma 15.11. *Here $a < w$, $a, w \in \mathbb{R}$. Then*

$$p(s) \in L_\infty(a, w) \text{ if and only if } \delta(z) := p(a + z) \in L_\infty(0, w - a).$$

In fact $\|\delta\|_{\infty,\, [0, w-a]} = \|p(s)\|_{\infty,\, [a, w]}.$

Proof. Let $\mathcal{L}^1$ stand for the class of Lebesgue measurable sets. Assume $p(s)$ is Lebesgue measurable on $[a, w]$. Then for any $c \in \mathbb{R}$ we have

$$\mathcal{L}^1 ([a, w]) \ni \{x \in [a, w] : p(x) \leq c\} = \{a + (x - a) \in [a, w] : p(a + (x - a)) \leq c\} =$$

$$a + \{(x - a) \in [0, w - a] : p(a + (x - a)) \leq c\} = a + \{u \in [0, w - a] : p(a + u) \leq c\}.$$

Therefore

$$\{u \in [0, w - a] : \delta(u) \leq c\} = \{u \in [0, w - a] : p(a + u) \leq c\} =$$

$$-a + \{x \in [a, w] : p(x) \leq c\} \in \mathcal{L}^1 ([0, w - a]),$$

for all $c \in \mathbb{R}$. Hence δ is Lebesgue measurable on $[0, w - a]$.

Assume now that δ is Lebesgue measurable on $[0, w - a]$. Then for any $c \in \mathbb{R}$ we have

$$\mathcal{L}^1 ([0, w - a]) \ni \{z \in [0, w - a] : \delta(z) \leq c\} =$$

$$\{z \in [0, w - a] : p(a + z) \leq c\} = \{(a + z) - a \in [0, w - a] : p(a + z) \leq c\} =$$

$$-a + \{(a + z) \in [a, w] : p(a + z) \leq c\} = -a + \{x \in [a, w] : p(x) \leq c\};$$

that is,

$$\{x \in [a, w] : p(x) \leq c\} = a + \{z \in [0, w - a] : \delta(z) \leq c\} \in \mathcal{L}^1 ([a, w]),$$

for all $c \in \mathbb{R}$. Hence $p(s)$ is Lebesgue measurable on $[a, w]$. We do have that

$$|\delta(z)| = |p(a + z)| \leq \|p(s)\|_{\infty, [a,w]},$$

a.e. $z \in [0, w - a]$. Hence $\|\delta\|_{\infty, [0,w-a]} \leq \|p(s)\|_{\infty, [a,w]}$. Also we have

$$|p(a + z)| = |\delta(z)| \leq \|\delta\|_{\infty, [0,w-a]},$$

a.e. $z \in [0, w - a]$. Hence $\|p(s)\|_{\infty, [a,w]} \leq \|\delta\|_{\infty, [0,w-a]}$, proving the claim.
$\square$

We continue with

Definition 15.12. We say that $f \in L_1(a, w)$, $a < w$; $a, w \in \mathbb{R}$ has an L_∞ fractional derivative $D_a^\beta f$ ($\beta > 0$) in $[a, w]$, if and only if (1) $D_a^{\beta-k} f \in C([a, w])$, $k = 1, \ldots, m := [\beta] + 1$; (2)$D_a^{\beta-1} f \in AC([a, w])$; and (3)$D_a^\beta f \in L_\infty(a, w)$.

Based on Lemmas 15.9 and 15.10, the last definition 15.12 is equivalent step by step to

Definition 15.13. We say that $f_a(s) := f(a+s) \in L_1(0, w-a)$ has an L_∞ fractional derivative $D^\beta f_a$ in $[0, w-a]$, $\beta > 0$, $w > a$; $a, w \in \mathbb{R}$, if and only if (1) $D^{\beta-k} f_a \in C([0, w-a])$, $k = 1, \ldots, m := [\beta] + 1$; (2) $D^{\beta-1} f_a \in AC([0, w-a])$; and (3) $D^\beta f_a \in L_\infty(0, w-a)$.

Definition 15.14. Here we define for $s \geq a$,

$$D_a^{\beta-m} f(s) = D_a^{\beta-([\beta]+1)} f(s) := D^{\beta-([\beta]+1)} f_a(s-a) = J_{([\beta]+1-\beta)} f_a(s-a) =$$

$$\frac{1}{\Gamma([\beta]+1-\beta)} \int_0^{s-a} (s-(a+t))^{[\beta]-\beta} f(a+t) dt, \qquad (15.12)$$

where $f \in L_1(a, w)$, $a < w$; $a, w \in \mathbb{R}$. Notice that $0 < [\beta] + 1 - \beta \leq 1$. If $f \in L_\infty(a, w)$, then

$$D_a^{\beta-m} f(s) = \frac{1}{\Gamma(m-\beta)} \int_a^s (s-t)^{[\beta]-\beta} f(t) dt.$$

Remark 15.15. Notice that

$$\left(D_a^{\beta-k} f\right)(a) = D^{\beta-k} f_a(0), \quad for \ k = 1, \ldots, [\beta] + 1. \qquad (15.13)$$

Based on Lemma 15.7 we get

Lemma 15.16. *Let $\beta > \alpha \geq 0$ and let $f \in L_1(a, w)$ (if and only if $f_a \in L_1(0, w-a)$, $a < w$; $a, w \in \mathbb{R}$) have an L_∞ fractional derivative $D_a^\beta f$ in $[a, w]$ (if and only if f_a have an L_∞ fractional derivative $D^\beta f_a$ in $[0, w-a]$), and let*

$$\left(D_a^{\beta-k} f\right)(a) = 0, \quad k = 1, \ldots, [\beta] + 1$$

(which is the same as $D^{\beta-k} f_a(0) = 0$, for $k = 1, \ldots, [\beta] + 1$). Then

(i) $D^\alpha f_a(s) = \dfrac{1}{\Gamma(\beta-\alpha)} \displaystyle\int_0^s (s-t)^{\beta-\alpha-1} D^\beta f_a(t) dt, \qquad (15.14)$

all $s \in [0, w-a]$.

(ii) $D_a^\alpha f(s) = \dfrac{1}{\Gamma(\beta-\alpha)} \displaystyle\int_a^s (s-t)^{\beta-\alpha-1} D_a^\beta f(t) dt, \qquad (15.15)$

all $a \leq s \leq w$.
   Clearly here

$D_a^\alpha f \in AC([a, w])$ for $\beta - \alpha \geq 1$, and $D_a^\alpha f \in C([a, w])$ for $\beta - \alpha \in (0, 1)$;

hence $D_a^\alpha f \in L_\infty(a, w)$ and $D_a^\alpha f \in L_1(a, w)$, and likewise for $D^\alpha f_a$ on $[0, w - a]$.

Proof. By Lemma 15.7, and by Definition 15.8 and (15.14), we have

$$(D_a^\alpha f)(s) = (D^\alpha f_a)(s-a) \overset{(15.14)}{=} \frac{1}{\Gamma(\beta - \alpha)} \int_0^{s-a} ((s-a)-t)^{\beta-\alpha-1} \left(D_a^\beta f\right)(t+a)dt$$

$$= \frac{1}{\Gamma(\beta - \alpha)} \int_0^{s-a} (s - (t + a))^{\beta-\alpha-1} \left(D_a^\beta f\right)(t + a)dt$$

$$\overset{(15.4)}{=} \frac{1}{\Gamma(\beta - \alpha)} \int_a^s (s - t)^{\beta-\alpha-1} \left(D_a^\beta f\right)(t)dt, \qquad (15.16)$$

proving (15.15). $\square$

15.4 Background—III

We make

Remark 15.17. Let $f \in L_1(a, w)$, where $a < w$; $a, w \in \mathbb{R}$. Let $\beta > 0$, $a \le s \le w$; by Definition 15.8 we have

$$(D_a^\beta f)(s) = \frac{1}{\Gamma(m - \beta)} \left(\frac{d}{ds}\right)^m \int_0^{s-a} (s - a - t)^{m-\beta-1} f(a+t)dt, \quad (15.17)$$

where $m := [\beta] + 1$.

If $\beta = 0$, then $(D_a^\beta f)(s) = f(s)$. Let now

$$F \in L_1(A) = L_1\left([R_1, R_2] \times S^{N-1}\right).$$

For a fixed $w \in S^{N-1}$, define $g_w(r) := F(rw) = F(x)$, where

$$x \in A := B(0, R_2) - \overline{B(0, R_1)},$$

$$0 < R_1 \le r \le R_2, \quad r = |x|, \quad w = \frac{x}{r} \in S^{N-1}.$$

By Fubini's theorem $g_w \in L_1\left([R_1, R_2], \mathcal{B}_{[R_1, R_2]}, R_N\right)$, for $\lambda_{S^{N-1}}$-almost every $w \in S^{N-1}$.

Call

$$K(F) := \{w \in S^{N-1} : g_w \notin L_1\left([R_1, R_2], \mathcal{B}_{[R_1, R_2]}, R_N\right)\}$$

$$= \{w \in S^{N-1} : F(\cdot \ w) \notin L_1\left([R_1, R_2], \mathcal{B}_{[R_1, R_2]}, R_N\right)\}. \qquad (15.18)$$

That is, $\lambda_{S^{N-1}}(K(F)) = 0$. Of course, $\Theta(F) := [R_1, R_2] \times K(F) \subset A$ and $\lambda_{\mathbb{R}^N}(\Theta(F)) = 0$.

By (15.17) then we have

$$\left(D_{R_1}^{\beta} g_w\right)(r) = \frac{1}{\Gamma(m-\beta)} \left(\frac{d}{dr}\right)^m \int_0^{r-R_1} (r-R_1-t)^{m-\beta-1} g_w(R_1+t)dt,$$

$$(15.19)$$

where $\beta > 0$, $m := [\beta] + 1$, $r \in [R_1, R_2]$. If $\beta = 0$, then $\left(D_{R_1}^{\beta} g_w\right)(r) = g_w(r)$. Formula (15.19) is written for all $w \in S^{N-1} - K(F)$. We set

$$\left(D_{R_1}^{\beta} g_w\right)(r) \equiv 0, \text{ for all } w \in K(F), \ r \in [R_1, R_2], \text{ any } \beta > 0.$$

The above leads to the following definition.

Definition 15.18. Let $\beta > 0$, $m := [\beta] + 1$, $F \in L_1(A)$; A is the spherical shell. We define

$$\frac{\partial_{R_1}^{\beta} F(x)}{\partial r^{\beta}} := \begin{cases} \frac{1}{\Gamma(m-\beta)} \left(\frac{d}{dr}\right)^m \int_0^{r-R_1} (r-R_1-t)^{m-\beta-1} F((R_1+t)w)dt, \\ \qquad\qquad\qquad\qquad\qquad\qquad \text{for } w \in S^{N-1} - K(F), \\ 0, \text{ for } w \in K(F), \end{cases}$$

$$(15.20)$$

where $x = rw \in A$, $r \in [R_1, R_2]$, $w \in S^{N-1}$.
  If $\beta = 0$, define

$$\frac{\partial_{R_1}^{\beta} F(x)}{\partial r^{\beta}} := F(x).$$

We call

$$\frac{\partial_{R_1}^{\beta} F(x)}{\partial r^{\beta}}$$

the Riemann–Liouville radial fractional derivative of F of order β.

We make

Remark 15.19. If $f \in L_\infty(a, w)$, then (15.17) becomes

$$\left(D_a^{\beta} f\right)(s) = \frac{1}{\Gamma(m-\beta)} \left(\frac{d}{ds}\right)^m \int_a^s (s-t)^{m-\beta-1} f(t)dt, \qquad (15.21)$$

$\beta > 0$, $m := [\beta] + 1$, $s \in [a, w]$. If F is a Lebesgue measurable function from A into $\mathbb{R}$ and bounded, that is, there exists

$$M^* > 0: \ |F(x)| \leq M^*, \text{ all } x \in A,$$

then of course $F \in L_1(A)$. Clearly then $|g_w(r)| \leq M^*$, all $r \in [R_1, R_2]$, and all $w \in S^{N-1}$.

Therefore (15.19) becomes

$$\left(D^{\beta}_{R_1} g_w\right)(r) = \frac{1}{\Gamma(m-\beta)}\left(\frac{d}{dr}\right)^m \int_{R_1}^r (r-t)^{m-\beta-1} g_w(t) dt, \quad (15.22)$$

where $m := [\beta] + 1$, $\beta > 0$, $r \in [R_1, R_2]$, for all $w \in S^{N-1} - K(F)$. In this last case, (15.20) becomes

$$\frac{\partial^{\beta}_{R_1} F(x)}{\partial r^{\beta}} := \begin{cases} \frac{1}{\Gamma(m-\beta)}\left(\frac{d}{dr}\right)^m \int_{R_1}^r (r-t)^{m-\beta-1} F(tw) dt, \\ \qquad\qquad\qquad\qquad\qquad \text{for } w \in S^{N-1} - K(F), \\ 0, \quad \text{for } w \in K(F), \end{cases}$$

$$(15.23)$$

where $x = rw \in A$, $r \in [R_1, R_2]$, $w \in S^{N-1}$.

We need

Theorem 15.20. *Let $\beta > \alpha > 0$ and $F \in L_1(A)$. Assume that*

$$\frac{\partial^{\beta}_{R_1} F(x)}{\partial r^{\beta}} \in L_{\infty}(A).$$

Further assume that $D^{\beta}_{R_1} F(rw)$ takes real values for almost all $r \in [R_1, R_2]$, for each $w \in S^{N-1}$, and for these $|D^{\beta}_{R_1} F(rw)| \leq M_1$ for some $M_1 > 0$. For each $w \in S^{N-1} - K(F)$, we assume that $F(\cdot\, w)$ have an L_{∞} fractional derivative $D^{\beta}_{R_1} F(\cdot\, w)$ in $[R_1, R_2]$, and that

$$D^{\beta-k}_{R_1} F(R_1 w) = 0, \quad k = 1, \ldots, [\beta] + 1.$$

Then

$$\frac{\partial^{\alpha}_{R_1} F(x)}{\partial r^{\alpha}} = \left(D^{\alpha}_{R_1} F\right)(rw) = \frac{1}{\Gamma(\beta-\alpha)} \int_{R_1}^r (r-t)^{\beta-\alpha-1} \left(D^{\beta}_{R_1} F\right)(tw) dt,$$

$$(15.24)$$

is true for all $x \in A$, that is, is true for all $r \in [R_1, R_2]$ and for all $w \in S^{N-1}$. Here

$$\left(D^{\alpha}_{R_1} F\right)(\cdot\, w) \in AC([R_1, R_2]) \text{ for } \beta - \alpha \geq 1$$

and

$$\left(D^{\alpha}_{R_1} F\right)(\cdot\, w) \in C([R_1, R_2]) \text{ for } \beta - \alpha \in (0, 1),$$

for all $w \in S^{N-1}$. Furthermore

$$\frac{\partial^{\alpha}_{R_1} F(x)}{\partial r^{\alpha}} \in L_{\infty}(A).$$

In particular, it holds

$$F(x) = F(rw) = \frac{1}{\Gamma(\beta)} \int_{R_1}^{r} (r - t)^{\beta - 1} \left(D_{R_1}^{\beta} F \right) (tw) dt, \qquad (15.25)$$

for all $r \in [R_1, R_2]$ and $w \in S^{N-1} - K(F)$; $x = rw$, and

$$F(\cdot \, w) \in AC([R_1, R_2]) \text{ for } \beta \geq 1$$

and

$$F(\cdot \, w) \in C([R_1, R_2]) \text{ for } \beta \in (0, 1),$$

for all $w \in S^{N-1} - K(F)$.

Proof. Here we observe that for each $w \in S^{N-1} - K(F)$, we have

$$F(\cdot \, w) \in L_1 \left([R_1, R_2], \mathcal{B}_{[R_1, R_2]}, R_N \right).$$

By our assumptions and Lemma 15.16, we have

$$\left(D_{R_1}^{\alpha} F \right) (rw) = \frac{1}{\Gamma(\beta - \alpha)} \int_{R_1}^{r} (r - t)^{\beta - \alpha - 1} D_{R_1}^{\beta} F(tw) dt, \qquad (15.26)$$

for all $r \in [R_1, R_2]$ and $w \in S^{N-1} - K(F)$, for $\beta > \alpha \geq 0$, thus initially proving (15.25) by setting $\alpha = 0$ in (15.26). Here

$$D_{R_1}^{\alpha} F(\cdot \, w) \in AC([R_1, R_2]), \text{ for all } w \in S^{N-1} - K(F), \; \beta - \alpha \geq 1,$$

and

$$D_{R_1}^{\alpha} F(\cdot \, w) \in C([R_1, R_2]), \text{ for } \beta - \alpha \in (0, 1).$$

Formula (15.26) for $\alpha > 0$, is true for all $r \in [R_1, R_2]$ and $w \in S^{N-1}$, and

$$\left(D_{R_1}^{\alpha} F \right) (\cdot \, w) \in AC([R_1, R_2]), \text{ for all } w \in S^{N-1}, \; \beta - \alpha \geq 1$$

and

$$D_{R_1}^{\alpha} F(\cdot \, w) \in C([R_1, R_2]), \text{ for } \beta - \alpha \in (0, 1).$$

Thus proving (15.24). Fixing $r \in [R_1, R_2]$, the function

$$\delta_r(t, w) := (r - t)^{\beta - \alpha - 1} D_{R_1}^{\beta} F(tw)$$

is measurable on

$$\left([R_1, r] \times S^{N-1}, \overline{\mathcal{B}_{[R_1, r]} \times \mathcal{B}_{S^{N-1}}} \right).$$

Here $\overline{\mathcal{B}_{[R_1, r]} \times \mathcal{B}_{S^{N-1}}}$ stands for the complete σ- algebra generated by $\mathcal{B}_{[R_1, r]} \times \mathcal{B}_{S^{N-1}}$, where $\overline{\mathcal{B}_X}$ stands for the completion of $\mathcal{B}_X$. Then we get that

$$\int_{S^{N-1}} \left(\int_{R_1}^{r} |\delta_r(t, w)| dt \right) dw = \int_{S^{N-1}} \left(\int_{R_1}^{r} (r - t)^{\beta - \alpha - 1} |D_{R_1}^{\beta} F(tw)| dt \right) dw$$

$$\leq \left\| \frac{\partial_{R_1}^\beta F(x)}{\partial r^\beta} \right\|_{\infty,\ ([R_1,r]\times S^{N-1})} \left(\int_{S^{N-1}} \left(\int_{R_1}^r (r-t)^{\beta-\alpha-1} dt \right) dw \right)$$

(15.27)

$$= \left\| \frac{\partial_{R_1}^\beta F(x)}{\partial r^\beta} \right\|_{\infty,\ ([R_1,r]\times S^{N-1})} \left(\frac{2\pi^{N/2}}{\Gamma(N/2)} \right) \frac{(r-R_1)^{\beta-\alpha}}{(\beta-\alpha)} \leq$$

$$\left\| \frac{\partial_{R_1}^\beta F(x)}{\partial r^\beta} \right\|_{\infty,\ A} \left(\frac{2\pi^{N/2}}{\Gamma(N/2)} \right) \frac{(R_2-R_1)^{\beta-\alpha}}{(\beta-\alpha)} < \infty.$$

(15.28)

Hence $\delta_r(t,w)$ is integrable on

$$\left([R_1,r] \times S^{N-1}, \overline{\mathcal{B}}_{[R_1,r]} \times \overline{\mathcal{B}}_{S^{N-1}} \right).$$

Consequently, by Fubini's theorem and (15.24), we obtain that

$$\left(D_{R_1}^\alpha F \right)(rw),\ \beta > \alpha > 0,$$

is integrable in w over $\left(S^{N-1},\ \overline{\mathcal{B}}_{S^{N-1}} \right)$. So we have that

$$\left(D_{R_1}^\alpha F \right)(rw),\ \beta > \alpha > 0,$$

is continuous in $r \in [R_1, R_2]$ for each $w \in S^{N-1}$, and measurable in $w \in S^{N-1}$ for each $r \in [R_1, R_2]$. So, it is a Carathéodory function. Here $[R_1, R_2]$ is a separable metric space and S^{N-1} is a measurable space, and the function takes values in $\mathbb{R}^* = \mathbb{R} \cup \{\pm\infty\}$, which is a metric space. Therefore by Theorem 20.15, p. 156 of [10], $\left(D_{R_1}^\alpha F \right)(rw)$, $\beta > \alpha > 0$ is jointly $\left(\mathcal{B}_{[R_1,R_2]} \times \overline{\mathcal{B}}_{S^{N-1}} \right)$-measurable on $[R_1, R_2] \times S^{N-1} = A$, that is, Lebesgue measurable on A. Indeed then we have that

$$\left| \left(D_{R_1}^\alpha F \right)(rw) \right| \leq \frac{1}{\Gamma(\beta-\alpha)} \int_{R_1}^r (r-t)^{\beta-\alpha-1} \left| \left(D_{R_1}^\beta F \right)(tw) \right| dt$$

$$\leq \frac{\left\| D_{R_1}^\beta F(\cdot\ w) \right\|_{\infty,\ [R_1,R_2]}}{\Gamma(\beta-\alpha)} \left(\int_{R_1}^r (r-t)^{\beta-\alpha-1} dt \right) \leq \frac{M_1}{\Gamma(\beta-\alpha)} \frac{(r-R_1)^{\beta-\alpha}}{(\beta-\alpha)}$$

(15.29)

$$\leq \frac{M_1}{\Gamma(\beta-\alpha+1)} (R_2-R_1)^{\beta-\alpha} := \tau < \infty,$$

for all $w \in S^{N-1}$ and all $r \in [R_1, R_2]$; that is, we proved that

$$\left| \left(D_{R_1}^\alpha F \right)(rw) \right| \leq \tau < \infty,$$

(15.30)

for all $w \in S^{N-1}$ and all $r \in [R_1, R_2]$, hence proving that

$$\frac{\partial_{R_1}^\alpha F(x)}{\partial r^\alpha} \in L_\infty(A).$$

We have finished our proof. □

We have built the machinery to do Riemann–Liouville fractional Opial-type inequalities on the spherical shell.

Now we are ready to present our main results next.

15.5   Main Results

15.5.1   Riemann–Liouville Fractional Opial-Type Inequalities Involving One Function

We mention

Theorem 15.21. (See [64].) *Let* $1/p + 1/q = 1$ *with* $p, q > 1$, *let* $\gamma \geq 0$, $v > \gamma + 1 - 1/p$, *and* $f \in L_1(0, x)$ *have an* L_∞ *fractional derivative* $D^v f$ *in* $[0, x]$, $x > 0$, *such that*

$$D^{v-j} f(0) = 0, \ for \ j = 1, \ldots, [v] + 1.$$

Then

$$\int_0^x |D^\gamma f(s)| \, |D^v f(s)| ds \leq \Omega(x) \left(\int_0^x |D^v f(s)|^q ds \right)^{2/q} \tag{15.31}$$

where

$$\Omega(x) := \frac{x^{(rp+2)/p}}{2^{1/q} \, \Gamma(r+1) \, ((rp+1)(rp+2))^{1/p}} \tag{15.32}$$

and

$$r := v - \gamma - 1. \tag{15.33}$$

We transfer Theorem 15.21 to arbitrary anchor point $a \in \mathbb{R}$. We present

Theorem 15.22. *Let* $1/p + 1/q = 1$ *with* $p, q > 1$, *let* $\gamma \geq 0$, $v > \gamma + 1 - 1/p$, *and* $f \in L_1(a, x)$ *have an* L_∞ *fractional derivative* $D_a^v f$ *in* $[a, x]$, a, $x \in \mathbb{R}$, $a < x$, *such that*

$$D_a^{v-j} f(a) = 0, \ for \ j = 1, \ldots, [v] + 1.$$

Then

$$\int_a^x |D_a^\gamma f(s)| \, |D_a^v f(s)| ds \leq \Omega(x - a) \left(\int_a^x |D_a^v f(s)|^q ds \right)^{2/q}, \tag{15.34}$$

where Ω *is as in* (15.32).

Proof. By Lemma 15.9, $f_a \in L_1(0, x-a)$ with L_∞ fractional derivative $D^v f_a$ in $[0, x-a]$; see Definitions 15.12 and 15.13. Furthermore it holds (see (15.13)),

$$D^{v-j} f_a(0) = 0, \text{ for } j = 1, \ldots, [v] + 1.$$

Therefore by (15.31) we have

$$\int_0^{x-a} |D^\gamma f_a(s)| \, |D^v f_a(s)| ds \le \Omega(x-a) \left(\int_0^{x-a} |D^v f_a(s)|^q ds \right)^{2/q}. \quad (15.35)$$

Using (15.10) we have

$$\int_0^{x-a} |(D_a^\gamma f)(s+a)| \, |(D_a^v f)(s+a)| ds \le \Omega(x-a) \left(\int_0^{x-a} |(D_a^v f)(s+a)|^q ds \right)^{2/q}.$$

By Lemma 15.16, we have that

$$D_a^\gamma f \in AC([a, x]) \text{ for } v - \gamma \ge 1 \text{ and } D_a^\gamma f \in C([a, x]) \text{ for } v - \gamma \in (0, 1).$$

Clearly then by Proposition 15.4 we get

$$\int_0^{x-a} |(D_a^\gamma f)(s+a)| \, |(D_a^v f)(s+a)| ds = \int_a^x |D_a^\gamma f(s)| \, |(D_a^v f)(s)| ds \quad (15.36)$$

and

$$\int_0^{x-a} |(D_a^v f)(s+a)|^q ds = \int_a^x |(D_a^v f)(s)|^q ds,$$

notice here functions under right-hand side integrations are integrable. That proves (15.34). □

We mention

Theorem 15.23. (See [64].) *Let $v > \gamma \ge 0$, and let $f \in L_1(0, x)$ have an L_∞ fractional derivative $D^v f$ in $[0, x]$, $x > 0$, such that*

$$D^{v-j} f(0) = 0, \text{ for } j = 1, \ldots, [v] + 1.$$

Then

$$\int_0^x |D^\gamma f(s)| \, |D^v f(s)| ds \le \Omega_1(x) \text{ ess } \sup_{s \in [0,x]} |D^v f(s)|^2, \quad (15.37)$$

where

$$\Omega_1(x) = \frac{x^{r+2}}{\Gamma(r+3)}, \quad r = v - \gamma - 1. \quad (15.38)$$

We give the general transfer.

Theorem 15.24. *Let $v > \gamma \geq 0$, and let $f \in L_1(a, x)$ have an L_∞ fractional derivative $D_a^v f$ in $[a, x]$, a, $x \in \mathbb{R}$, $a < x$, such that*

$$D_a^{v-j} f(a) = 0, \ \text{for } j = 1, \ldots, [v] + 1.$$

Then

$$\int_a^x |D_a^\gamma f(s)| \, |D_a^v f(s)| ds \leq \Omega_1 (x - a) \text{ ess sup}_{s \in [a,x]} |D_a^v f(s)|^2, \quad (15.39)$$

where Ω_1 is as in (15.38).

Proof. By Lemma 15.9, $f_a \in L_1(0, x - a)$ with L_∞ fractional derivative $D^v f_a$ in $[0, x - a]$; see Definitions 15.12 and 15.13. Furthermore it holds (see (15.13)),

$$D^{v-j} f_a(0) = 0, \ \text{for } j = 1, \ldots, [v] + 1.$$

Therefore by (15.37) we have

$$\int_0^{x-a} |D^\gamma f_a(s)| \, |D^v f_a(s)| ds \leq \Omega_1 (x - a) \text{ ess sup}_{s \in [0, x-a]} |D^v f_a(s)|^2.$$
$$(15.40)$$

Using (15.10) we get

$$\int_0^{x-a} |D_a^\gamma f(s+a)| \, |(D_a^v f)(s+a)| ds \leq \Omega_1 (x-a) \text{ess sup}_{s \in [0, x-a]} |D_a^v f(s+a)|^2.$$

We have again by Proposition 15.4 that

$$\int_0^{x-a} |D_a^\gamma f(s + a)| \, |D_a^v f(s + a)| ds = \int_a^x |D_a^\gamma f(s)| \, |D_a^v f(s)| ds. \quad (15.41)$$

Also by Lemma 15.11 we obtain

$$\text{ess sup}_{s \in [0, x-a]} |D_a^v f(s + a)|^2 = \text{ess sup}_{s \in [a,x]} |D_a^v f(s)|^2,$$

thus proving the claim. $\square$

We give the transfer

Theorem 15.25. *Let $1/p + 1/q = 1$ with $0 < p < 1$, let $v > \gamma \geq 0$, and let $f \in L_1(a, x)$ have an L_∞ fractional derivative $D_a^v f$ in $[a, x]$, a, $x \in \mathbb{R}$, $a < x$, such that*

$$D_a^{v-j} f(a) = 0, \ \text{for } j = 1, \ldots, [v] + 1.$$

Additionally assume $(1/D_a^v f) \in L_\infty(a, x)$ *and that* $D_a^v f$ *has the same sign a.e. in* (a, x). *Then*

$$\int_a^x |D_a^\gamma f(s)| \, |D_a^v f(s)| ds \geq \Omega(x - a) \left(\int_a^x |D_a^v f(s)|^q ds \right)^{2/q} \qquad (15.42)$$

where Ω *is defined by* (15.32).

Proof. Based on Theorem 2.3, of [64], special case of $a = 0$. It is a similar method of proving as in Theorem 15.22. □

We give the transfer.

Theorem 15.26. *Let* $1/p + 1/q = 1$ *with* $p, q > 1$ *let* $\gamma \geq 0$, $v \geq \gamma + 2 - 1/p$, *and* $f \in L_1(a, x)$ *have an* L_∞ *fractional derivative* $D_a^v f$ *in* $[a, x]$, $a, x \in \mathbb{R}$, $a < x$, *such that*

$$D_a^{v-j} f(a) = 0, \; for \; j = 1, \ldots, [v] + 1.$$

Then

$$\int_a^x |D_a^\gamma f(s)| \, |D_a^{\gamma+1} f(s)| ds \leq \Omega_2 (x - a) \left(\int_a^x |D_a^v f(s)|^q ds \right)^{2/q}, \qquad (15.43)$$

where

$$\Omega_2(t) := \frac{t^{2 \, (rp+1)/p}}{2 \, (\Gamma(r + 1))^2 (rp + 1)^{2/p}}, \quad r = v - \gamma - 1, \; t \geq 0. \qquad (15.44)$$

Proof. Based on Theorem 2.4 of [64], special case of $a = 0$. It is a similar method of proving as in Theorem 15.22. □

We present the transfer.

Theorem 15.27. *Let* $\gamma \geq 0$, $v > \gamma + 1$, *and let* $f \in L_1(a, x)$ *have an* L_∞ *fractional derivative* $D_a^v f$ *in* $[a, x]$, $a, x \in \mathbb{R}$, $a < x$, *such that*

$$D_a^{v-j} f(a) = 0, \; for \; j = 1, \ldots, [v] + 1.$$

Then

$$\int_a^x |D_a^\gamma f(s)| \, |D_a^{\gamma+1} f(s)| ds \leq \Omega_3 (x - a) \; ess \; \sup_{t \in (a,x)} |D_a^v f(t)|^2, \qquad (15.45)$$

where

$$\Omega_3(t) := \frac{t^{2(v-\gamma)}}{2 \, (\Gamma(v - \gamma + 1))^2}, \quad t \geq 0. \qquad (15.46)$$

Proof. Based on Theorem 2.5, of [64], special case of $a = 0$. It is a similar method of proving as in Theorem 15.24. □

We further give

Proposition 15.28. *Inequality (15.45) is sharp; namely it is attained by*

$$f_*(s) := (s - a)^v, \quad a \leq s \leq x, \quad v > \gamma + 1, \quad \gamma \geq 0.$$

Proof. Here we are acting as in Remark 2.6 of [64]. We use the known formula

$$\int_a^s (s - t)^{u-1}(t - a)^{v-1} dt = \frac{\Gamma(u)\,\Gamma(v)}{\Gamma(u + v)}(s - a)^{u+v-1}, \quad u, v > 0. \quad (15.47)$$

Let

$$0 \leq j \leq [v] + 1, \quad m := [v] - j + 1, \quad \alpha := v - [v].$$

We have $1 - \alpha > 0$, $v + 1 > 0$, and by (15.21) we obtain

$$D_a^{v-j} f_*(s) = D_a^{v-j}(s-a)^v = \frac{1}{\Gamma(1 - \alpha)}\left(\frac{d}{ds}\right)^m \int_a^s (s-t)^{(1-\alpha)-1}(t-a)^{(v+1)-1} dt \quad (15.48)$$

$$= \frac{1}{\Gamma(1 - \alpha)}\,\frac{\Gamma(1 - \alpha)\,\Gamma(v + 1)}{\Gamma(m + j + 1)}\left(\frac{d}{ds}\right)^m (s - a)^{m+j} = \frac{\Gamma(v + 1)}{j!}(s - a)^j; \quad (15.49)$$

that is,

$$D_a^{v-j}(s - a)^v = \frac{\Gamma(v + 1)}{j!}(s - a)^j. \quad (15.50)$$

Hence

$$D_a^{v-j} f_*(a) = 0, \quad \text{for } j = 1, \ldots, [v] + 1 \quad (15.51)$$

and

$$D_a^v f_*(s) = D_a^v(s - a)^v = \Gamma(v + 1). \quad (15.52)$$

Using Lemma 15.16, in particular applying (15.15), we obtain

$$D_a^\gamma(s - a)^v = \frac{\Gamma(v + 1)(s - a)^{v-\gamma}}{\Gamma(v - \gamma + 1)}, \quad D_a^{\gamma+1}(s - a)^v = \frac{\Gamma(v + 1)}{\Gamma(v - \gamma)}(s - a)^{v-\gamma-1}. \quad (15.53)$$

Therefore

$$L.H.S. \ (15.45) = \int_a^x |D_a^\gamma f_*(s)| \, |D_a^{\gamma+1} f_*(s)| ds = \frac{(\Gamma(v + 1))^2}{\Gamma(v - \gamma)\,\Gamma(v - \gamma + 1)}$$

$$\int_a^x (s-a)^{2(v-\gamma)-1}ds = \frac{1}{2}\left(\frac{\Gamma(v+1)}{\Gamma(v-\gamma+1)}\right)^2 (x-a)^{2(v-\gamma)} = R.H.S. \ (15.45),$$
(15.54)

thus proving the claim. □

We present

Theorem 15.29. *Let* $1/p+1/q = 1$ *with* $0 < p < 1$, *let* $\gamma \geq 0$, $v > \gamma+1$, *and let* $f \in L_1(a,x)$ *have an* L_∞ *fractional derivative* $D_a^v f$ *in* $[a,x]$, $a, \ x \in \mathbb{R}$, $a < x$, *such that*

$$D_a^{v-j} f(a) = 0, \ for \ j = 1, \ldots, [v] + 1.$$

Also assume $(1/D_a^v f) \in L_\infty(a,x)$ *and that* $D_a^v f$ *has the same sign a.e. in* (a,x). *Then*

$$\int_a^x |D_a^\gamma f(s)| \ |D_a^{\gamma+1} f(s)|ds \geq \Omega_2(x-a)\left(\int_a^x |D_a^v f(s)|^q ds\right)^{2/q}, \quad (15.55)$$

where Ω_2 *is given by* (15.44).

Proof. We transfer here for arbitrary anchor point $a \in \mathbb{R}$, Theorem 2.7 of [64]. We apply the earlier established method; see Theorem 15.22. □

We present

Theorem 15.30. *Let* $1/p + 1/q = 1$ *with* $p,q > 1$, *let* $\gamma \geq 0$, $v > \gamma + 1 - 1/p$, *and let* $f \in L_1(a,x)$ *have an* L_∞ *fractional derivative* $D_a^v f$ *in* $[a,x]$, $a, \ x \in \mathbb{R}$, $a < x$, *such that*

$$D_a^{v-j} f(a) = 0, \ for \ j = 1, \ldots, [v] + 1;$$

let $m > 0$. *Then*

$$\int_a^x |D_a^\gamma f(s)|^m ds \leq \Omega_4(x-a)\left(\int_a^x |D_a^v f(s)|^q ds\right)^{m/q}, \quad (15.56)$$

where

$$\Omega_4(t) := \frac{t^{(rm+1+(m/p))}}{(\Gamma(r+1))^m \left(rm+1+\left(\frac{m}{p}\right)\right)(rp+1)^{m/p}},$$

$$r := v - \gamma - 1, \quad t \geq 0. \quad (15.57)$$

Proof. Based on Theorem 2.8 of [64] and its transfer to arbitrary anchor point $a \in \mathbb{R}$. □

We give

Theorem 15.31. *Let $v > \gamma \geq 0$, and let $f \in L_1(a, x)$ have an L_∞ fractional derivative $D_a^v f$ in $[a, x]$, a, $x \in \mathbb{R}$, $a < x$, such that*

$$D_a^{v-j} f(a) = 0, \text{ for } j = 1, \ldots, [v] + 1;$$

let $m > 0$. Then

$$\int_a^x |D_a^\gamma f(s)|^m ds \leq \Omega_5(x - a) \mathrm{esssup}_{t \in [a,x]} |D_a^v f(t)|^m, \qquad (15.58)$$

where

$$\Omega_5(t) := \frac{t^{(v-\gamma)m+1}}{(\Gamma(v - \gamma + 1))^m \left((v - \gamma - 1)m + 1 + \left(\frac{m}{p}\right)\right) ((v - \gamma)m + 1)}, \quad t \geq 0.$$

$$(15.59)$$

Proof. Based on Theorem 2.9 of [64], and so on. $\square$

We next give the notation valid for the rest of Section 15.5.1; we follow [65].

Notation 15.32. Here we call

- l: a positive integer.

- v, r_i: positive real numbers, $i = 1, \ldots, l$, $r = \sum_{i=1}^l r_i$.

- μ_i: real numbers satisfying $0 \leq \mu_i < v$, $i = 1, \ldots, l$.

- $\alpha_i = v - \mu_i - 1$, $i = 1, \ldots, l$.

- $\alpha = \max\{(\alpha_i)_- : i = 1, \ldots, l\}$, where $(\alpha_i)_- := (-\alpha_i)_+$.

- $\beta = \max\{(\alpha_i)_+ : i = 1, \ldots, l\}$, where $(\alpha_i)_+ := \max(\alpha_i, 0)$.

- w_1, w_2: continuous positive weight functions on $[a, x]$, a, $x \in \mathbb{R}$, $a < x$.

- w: continuous nonnegative weight function on $[a, x]$.

- $s_k, s_k' : s_k > 0$ and $1/s_k + 1/s_k' = 1$ $k = 1, 2$.

We write $\bar{\mu} = (\mu_1, \ldots, \mu_l)$ for a selection of the orders μ_i of fractional derivatives, and $\bar{r} = (r_1, \ldots, r_l)$ for a selection of the constants r_i.

We mention

Theorem 15.33. [65] *Let* $f \in L_1(0, x)$ *have an* L_∞ *fractional derivative* $D^v f$ *in* $[0, x]$, $x > 0$, *such that*

$$D^{v-j} f(0) = 0, \ for \ j = 1, \ldots, [v] + 1.$$

Here $a = 0$. *For* $k = 1, 2$, *let* $s_k > 1$ *and* $p > 0$ *satisfy*

$$\alpha s_2 < 1, \quad p > \frac{s_2}{1 - \alpha s_2} \tag{15.60}$$

and let $\sigma = 1/s_2 - 1/p$. *Finally, let*

$$Q_1 := \left(\int_0^x w_1(\tau)^{s_1'} d\tau \right)^{1/s_1'}; \quad Q_2 := \left(\int_0^x w_2(\tau)^{-s_2'/p} d\tau \right)^{r/s_2'}. \tag{15.61}$$

Then

$$\int_0^x w_1(\tau) \prod_{i=1}^l |D^{\mu_i} f(\tau)|^{r_i} d\tau \le$$

$$Q_1 Q_2 C_1 \ x^{\rho + (1/s_1)} \left(\int_0^x w_2(\tau) \ |D^v f(\tau)|^p d\tau \right)^{r/p}, \tag{15.62}$$

where $\rho := \sum_{i=1}^l \alpha_i r_i + \sigma r$, *and*

$$C_1 := C_1(v, \overline{\mu}, \overline{r}, p, s_1, s_2) := \frac{\sigma^{r\sigma}}{\prod_{i=1}^l \Gamma(v - \mu_i)^{r_i} (\alpha_i + \sigma)^{r_i \sigma} (\rho s_1 + 1)^{1/s_1}}. \tag{15.63}$$

We transfer the last theorem to arbitrary anchor point $a \in \mathbb{R}$.

Theorem 15.34. *Here all constant and parameter notation is as in Theorem 15.33. Let* $f \in L_1(a, x)$, $a < x$, $a, x \in \mathbb{R}$, *have an* L_∞ *fractional derivative* $D_a^v f$ *in* $[a, x]$, *such that*

$$D_a^{v-j} f(a) = 0, \ for \ j = 1, \ldots, [v] + 1.$$

Let

$$Q_1(a) := \left(\int_a^x w_1(\tau)^{s_1'} d\tau \right)^{1/s_1'}, \quad Q_2(a) := \left(\int_a^x w_2(\tau)^{-s_2'/p} d\tau \right)^{r/s_2'}. \tag{15.64}$$

Then

$$\int_a^x w_1(\tau) \prod_{i=1}^l |D_a^{\mu_i} f(\tau)|^{r_i} d\tau \le$$

$$Q_1(a)Q_2(a)C_1(x-a)^{\rho+(1/s_1)}\left(\int_a^x w_2(\tau)\,|D_a^v f(\tau)|^p d\tau\right)^{r/p}. \qquad (15.65)$$

Proof. By Lemma 15.9, $f_a \in L_1(0, x-a)$ with an L_∞ fractional derivative $D^v f_a$ in $[0, x-a]$; see Definitions 15.12, and 15.13. Furthermore it holds (see (15.13))

$$D^{v-j}f_a(0) = 0, \text{ for } j = 1, \ldots, [v] + 1.$$

Notice that

$$Q_1(a) := \left(\int_0^{x-a} w_1(\tau+a)^{s_1'} d\tau\right)^{1/s_1'}, \quad Q_2(a) := \left(\int_a^{x-a} w_2(\tau+a)^{-s_2'/p} d\tau\right)^{r/s_2'}. \qquad (15.66)$$

Next we apply (15.62) on $[0, x-a]$ to f_a with respect to

$$w_1(a+\tau), \ w_2(a+\tau), \ \tau \in [0, x-a].$$

We have

$$\int_0^{x-a} w_1(a+\tau)\prod_{i=1}^l |D^{\mu_i}f_a(\tau)|^{r_i} d\tau \le$$

$$Q_1(a)Q_2(a)C_1\,(x-a)^{\rho+(1/s_1)}\left(\int_0^{x-a} w_2(a+\tau)\,|D^v f_a(\tau)|^p d\tau\right)^{r/p}. \qquad (15.67)$$

Equivalently, via (15.10), we write

$$\int_0^{x-a} w_1(a+\tau)\prod_{i=1}^l |D_a^{\mu_i}f(a+\tau)|^{r_i} d\tau \le$$

$$Q_1(a)Q_2(a)C_1\,(x-a)^{\rho+(1/s_1)}\left(\int_0^{x-a} w_2(a+\tau)\,|D_a^v f(a+\tau)|^p d\tau\right)^{r/p}. \qquad (15.68)$$

By Lemma 15.16, we have that $D_a^{\mu_i}f \in AC([a, x])$. Hence, by Proposition 15.4, we get (15.65). $\square$

Next, we apply Theorem 15.34 to the spherical shell A. We give

Theorem 15.35. *Here all constant and parameter notation is as in Theorem 15.33. Let $f \in L_1(A)$ with*

$$\frac{\partial_{R_1}^v f(x)}{\partial r^v} \in L_\infty(A), \ x \in A;$$

$$A := B(0, R_2) - \overline{B(0, R_1)} \subseteq \mathbb{R}^N, \ N \ge 2, \ 0 < R_1 < R_2.$$

Further assume that $D_{R_1}^v f(rw) \in \mathbb{R}$ for almost all $r \in [R_1, R_2]$, for each $w \in S^{N-1}$, and for these $|D_{R_1}^v f(rw)| \leq M_1$ for some $M_1 > 0$. For each $w \in S^{N-1} - K(F)$, we assume that $f(\cdot\, w)$ has an L_∞ fractional derivative $D_{R_1}^v f(\cdot\, w)$ in $[R_1, R_2]$, and that

$$D_{R_1}^{v-j} f(R_1 w) = 0, \quad j = 1, \ldots, [v] + 1.$$

We take $p = \sum_{i=1}^l r_i$, and $0 \leq \mu_1 < \mu_2 \leq \mu_3 \leq \ldots \leq \mu_l < v$. If $\mu_1 = 0$ we set $r_1 = 1$. Denote

$$Q_1(R_1) := \left(\frac{R_2^{(N-1)s_1'+1} - R_1^{(N-1)s_1'+1}}{(N-1)s_1' + 1} \right)^{1/s_1'},$$

$$Q_2(R_1) := \left(\frac{R_2^{(1-N)(s_2'/p)+1} - R_1^{(1-N)(s_2'/p)+1}}{(1-N)(\frac{s_2'}{p}) + 1} \right)^{p/s_2'}. \tag{15.69}$$

Then

$$\int_A \prod_{i=1}^l \left| \frac{\partial_{R_1}^{\mu_i} f(x)}{\partial r^{\mu_i}} \right|^{r_i} dx \leq C^* \left(\int_A \left| \frac{\partial_{R_1}^v f(x)}{\partial r^v} \right|^p dx \right), \tag{15.70}$$

where

$$C^* := Q_1(R_1) Q_2(R_1) C_1 (R_2 - R_1)^{p+(1/s_1)}. \tag{15.71}$$

Proof. By Theorem 15.20 for $\mu_i > 0$ we get that

$$\frac{\partial_{R_1}^{\mu_i} f(x)}{\partial r^{\mu_i}} \in L_\infty(A).$$

In general here we get that

$$\prod_{i=1}^l \left| \frac{\partial_{R_1}^{\mu_i} f(x)}{\partial r^{\mu_i}} \right|^{r_i} \in L_1(A).$$

Thus, by Proposition 15.3 we have

$$I_1 := \int_A \prod_{i=1}^l \left| \frac{\partial_{R_1}^{\mu_i} f(x)}{\partial r^{\mu_i}} \right|^{r_i} dx = \int_{S^{N-1}} \left(\int_{R_1}^{R_2} \prod_{i=1}^l |D_{R_1}^{\mu_i} f(rw)|^{r_i} r^{N-1} dr \right) dw$$

$$= \int_{(S^{N-1}-K(f))} \left(\int_{R_1}^{R_2} \prod_{i=1}^l |D_{R_1}^{\mu_i} f(rw)|^{r_i} r^{N-1} dr \right) dw. \tag{15.72}$$

Because $\left| \frac{\partial_{R_1}^v f(x)}{\partial r^v} \right|^p \in L_1(A)$, we also obtain

$$I_2 := \int_A \left| \frac{\partial_{R_1}^v f(x)}{\partial r^v} \right|^p dx = \int_{S^{N-1}} \left(\int_{R_1}^{R_2} |D_{R_1}^r f(rw)|^p \, r^{N-1} dr \right) dw$$

$$= \int_{(S^{N-1}-K(f))} \left(\int_{R_1}^{R_2} |D_{R_1}^r f(rw)|^p \, r^{N-1} dr \right) dw. \qquad (15.73)$$

Notice here

$$f(\cdot \, w) \in L_1([R_1, R_2]), \text{ for all } w \in S^{N-1} - K(f),$$

and $\lambda_{S^{N-1}}(K(f)) = 0$.

Setting $w_1(r) = w_2(r) := r^{N-1}$, $r \in [R_1, R_2]$, we use Theorem 15.34, for every $w \in S^{N-1} - K(f)$. We get

$$\int_{R_1}^{R_2} \prod_{i=1}^{l} |D_{R_1}^{\mu_i} f(rw)|^{r_i} \, r^{N-1} dr$$

$$\leq Q_1(R_1) Q_2(R_1) C_1 (R_2 - R_1)^{\rho + (1/s_1)} \left(\int_{R_1}^{R_2} |D_{R_1}^v f(rw)|^p \, r^{N-1} dr \right);$$
$$(15.74)$$

that is, we find that

$$\int_{R_1}^{R_2} \prod_{i=1}^{l} |D_{R_1}^{\mu_i} f(rw)|^{r_i} \, r^{N-1} dr \leq$$

$$C^* \int_{R_1}^{R_2} |D_{R_1}^v f(rw)|^p \, r^{N-1} dr, \text{ for all } w \in S^{N-1} - K(f). \qquad (15.75)$$

Therefore

$$\int_{(S^{N-1}-K(f))} \left(\int_{R_1}^{R_2} \prod_{i=1}^{l} |D_{R_1}^{\mu_i} f(rw)|^{r_i} \, r^{N-1} dr \right) dw$$

$$\leq C^* \left(\int_{(S^{N-1}-K(f))} \left(\int_{R_1}^{R_2} |D_{R_1}^v f(rw)|^p \, r^{N-1} dr \right) dw \right). \qquad (15.76)$$

That is,

$$I_1 \leq C^* I_2, \qquad (15.77)$$

thus proving (15.70). $\square$

We continue with

Theorem 15.36. *Let* $f \in L_1(a, x)$, $a < x$, $a, x \in \mathbb{R}$ *have an* L_∞ *fractional derivative* $D_a^v f$ *in* $[a, x]$ *such that*

$$D_a^{v-j} f(a) = 0, \ \text{for } j = 1, \ldots, [v] + 1.$$

Then

$$\int_a^x w(\tau) \prod_{i=1}^l |D_a^{\mu_i} f(\tau)|^{r_i} d\tau \le \frac{\|w\|_\infty (x - a)^\rho}{\rho \prod_{i=1}^l \Gamma(v - \mu_i + 1)^{r_i}} \|D_a^v f\|_\infty^r, \quad (15.78)$$

where $\rho := \sum_{i=1}^l (v - \mu_i) r_i + 1.$

Proof. This is a transfer of Theorem 2.2 of [65] and its proof. By (15.15) we have

$$|D_a^{\mu_i} f(\tau)| \le \frac{1}{\Gamma(v - \mu_i)} \int_a^\tau (\tau - t)^{\alpha_i} |D_a^v f(\tau)| dt, \quad (15.79)$$

implying that

$$|D_a^{\mu_i} f(\tau)| \le \frac{\|D_a^v f\|_\infty (\tau - a)^{v - \mu_i}}{\Gamma(v - \mu_i + 1)}. \quad (15.80)$$

Hence

$$|D_a^{\mu_i} f(\tau)|^{r_i} \le \frac{\|D_a^v f\|_\infty^{r_i} (\tau - a)^{(v - \mu_i) r_i}}{(\Gamma(v - \mu_i + 1))^{r_i}}, \quad (15.81)$$

and

$$w(\tau) \prod_{i=1}^l |D_a^{\mu_i} f(\tau)|^{r_i} \le \frac{\|w\|_\infty \|D_a^v f\|_\infty^r (\tau - a)^{\sum_{i=1}^l (v - \mu_i) r_i}}{\prod_{i=1}^l (\Gamma(v - \mu_i + 1))^{r_i}}. \quad (15.82)$$

Integrating (15.82) over $[a, x]$ we get

$$\int_a^x w(\tau) \prod_{i=1}^l |D_a^{\mu_i} f(\tau)|^{r_i} d\tau \le \frac{\|w\|_\infty \|D_a^v f\|_\infty^r}{\prod_{i=1}^l (\Gamma(v - \mu_i + 1))^{r_i}} \int_a^x (\tau - a)^{\sum_{i=1}^l (v - \mu_i) r_i} d\tau =$$

$$\frac{\|w\|_\infty \|D_a^v f\|_\infty^r}{\prod_{i=1}^l (\Gamma(v - \mu_i + 1))^{r_i}} \frac{(x - a)^\rho}{\rho}, \quad (15.83)$$

proving (15.78). □

We apply Theorem 15.36 to the spherical shell A case.

Theorem 15.37. *Let* $f \in L_1(A)$ *with*

$$\frac{\partial_{R_1}^v f}{\partial r^v} \in L_\infty(A).$$

Assume $D_{R_1}^v f(rw) \in \mathbb{R}$ for almost all $r \in [R_1, R_2]$, for each $w \in S^{N-1}$, and for these $|D_{R_1}^v f(rw)| \leq M_1$ for some $M_1 > 0$. For each $w \in S^{N-1} - K(f)$, we assume that $f(\cdot\, w)$ has an L_∞ fractional derivative $D_{R_1}^v f(\cdot\, w)$ in $[R_1, R_2]$, and that

$$D_{R_1}^{v-j} f(R_1 w) = 0, \; for \; j = 1, \ldots, [v] + 1.$$

We take

$$0 \leq \mu_1 < \mu_2 \leq \mu_3 \leq \ldots \leq \mu_l < v.$$

If $\mu_1 = 0$ we set $r_1 = 1$. Then

$$\int_A \prod_{i=1}^l \left| \frac{\partial_{R_1}^{\mu_i} f(x)}{\partial r^{\mu_i}} \right|^{r_i} dx \leq \frac{R_2^{N-1}(R_2 - R_1)^\rho M_1^r}{\rho \, \prod_{i=1}^l (\Gamma(v - \mu_i + 1))^{r_i}} \, \frac{2\pi^{N/2}}{\Gamma(N/2)}, \qquad (15.84)$$

where $\rho := \sum_{i=1}^l (v - \mu_i) r_i + 1$.

Proof. Here

$$\prod_{i=1}^l \left| \frac{\partial_{R_1}^{\mu_i} f(x)}{\partial r^{\mu_i}} \right|^{r_i} \in L_1(A).$$

Hence as before

$$I_1 := \int_A \prod_{i=1}^l \left| \frac{\partial_{R_1}^{\mu_i} f(x)}{\partial r^{\mu_i}} \right|^{r_i} dx = \int_{(S^{N-1} - K(f))} \left(\int_{R_1}^{R_2} \prod_{i=1}^l |D_{R_1}^{\mu_i} f(rw)|^{r_i} \, r^{N-1} dr \right) dw.$$
$$(15.85)$$

Here we set $w(r) := r^{N-1}$, $r \in [R_1, R_2]$. For each $w \in S^{N-1} - K(f)$ we apply Theorem 15.36. From (15.78) we obtain

$$\int_{R_1}^{R_2} \prod_{i=1}^l |D_{R_1}^{\mu_i} f(rw)|^{r_i} \, r^{N-1} dr \leq \frac{R_2^{N-1}(R_2 - R_1)^\rho}{\rho \, \prod_{i=1}^l (\Gamma(v - \mu_i + 1))^{r_i}} \, \|D_{R_1}^v f(\cdot\, w)\|_{\infty, \, [R_1, R_2]}^r$$
$$(15.86)$$

$$\leq \frac{R_2^{N-1}(R_2 - R_1)^\rho \, M_1^r}{\rho \, \prod_{i=1}^l (\Gamma(v - \mu_i + 1))^{r_i}} := \theta, \; for \; all \; w \in S^{N-1} - K(f). \qquad (15.87)$$

Therefore

$$I_1 = \int_{(S^{N-1} - K(f))} \left(\int_{R_1}^{R_2} \prod_{i=1}^l |D_{R_1}^{\mu_i} f(rw)|^{r_i} \, r^{N-1} dr \right) dw$$

$$\leq \theta \int_{(S^{N-1} - K(f))} dw = \theta \int_{S^{N-1}} dw \qquad (15.88)$$

$$= \theta \, \frac{2\pi^{N/2}}{\Gamma(N/2)}, \qquad (15.89)$$

proving (15.84).   □

We continue with

Theorem 15.38. *Let $f \in L_1(a,x)$, $a < x$, $a,x \in \mathbb{R}$ have an L_∞ fractional derivative $D_a^v f$ in $[a,x]$ such that*

$$D_a^{v-j} f(a) = 0, \text{ for } j = 1,\dots,[v]+1.$$

Assume also that $D_a^v f$ has the same sign a.e. in (a,x). For $k = 1,2$, let $0 < s_k < 1$, let $p < 0$, and let $\sigma := 1/s_2 - 1/p$, $Q_1(a)$ and $Q_2(a)$ as in (15.64), C_1 as in (15.63). Then

$$\int_a^x w(\tau) \prod_{i=1}^l |D_a^{\mu_i} f(\tau)|^{r_i} d\tau \geq$$

$$Q_1(a)Q_2(a)C_1 \, (x-a)^{\rho+(1/s_1)} \left(\int_a^x w_2(\tau)|D_a^v f(\tau)|^p d\tau \right)^{r/p}, \quad (15.90)$$

where

$$\rho := \sum_{i=1}^l \alpha_i r_i + \sigma r. \quad (15.91)$$

Proof. Similar to Theorem 15.34 with transfer of Theorem 2.3 of [65] to anchor point $a \in \mathbb{R}$.   □

We give

Theorem 15.39. *Let $s_1, s_2, p \in (0,1)$, $rs_1 \leq 1$ and*

$$\frac{s_2}{1-\alpha s_2 + s_2} < p < \frac{s_2}{1+\beta s_2}, \quad \sigma = \frac{1}{s_2} - \frac{1}{p},$$

ρ as in (15.91), $Q_1(a)$ and $Q_2(a)$ as in (15.64), and C_1 as in (15.63). Let $f \in L_1(a,x)$, $a < x$, $a,x \in \mathbb{R}$ have an L_∞ fractional derivative $D_a^v f$ in $[a,x]$, such that

$$D_a^{v-j} f(a) = 0, \text{ for } j = 1,\dots,[v]+1.$$

Assume also that $D_a^v f$ has the same sign a.e. in (a,x). Then

$$\int_a^x w_1(\tau) \prod_{i=1}^l |D_a^{\mu_i} f(\tau)|^{r_i} d\tau \geq$$

$$Q_1(a)Q_2(a)C_1 \, (x-a)^{\rho+(1/s_1)} \left(\int_a^x w_2(\tau)|D_a^v f(\tau)|^p d\tau \right)^{r/p}. \quad (15.92)$$

Proof. Similar transfer of Theorem 2.4 from [65]. □

We apply Theorem 15.39 on the spherical shell A.

Theorem 15.40. *All parameters and constants are as in Theorem 15.39. We take $p = \sum_{i=1}^{l} r_i$, and $0 \le \mu_1 < \mu_2 \le \mu_3 \le \cdots \le \mu_l < v$. If $\mu_1 = 0$ we set $r_1 = 1$. Let $f \in L_1(A)$ with*

$$\frac{\partial_{R_1}^v f}{\partial r^v} \in L_\infty(A).$$

Assume that $D_{R_1}^v f(rw) \in \mathbb{R}$ for almost all $r \in [R_1, R_2]$, for each $w \in S^{N-1}$, and for these $|D_{R_1}^v f(rw)| \le M_1$ for some $M_1 > 0$. For each $w \in S^{N-1} - K(F)$, we assume that $f(\cdot\ w)$ has an L_∞ fractional derivative $D_{R_1}^v f(\cdot\ w)$ in $[R_1, R_2]$, and that

$$D_{R_1}^{v-j} f(R_1 w) = 0, \ j = 1, \ldots, [v] + 1,$$

also $D_{R_1}^v (\cdot\ w)$ has the same sign a.e. in $[R_1, R_2]$. Then

$$\int_A \prod_{i=1}^{l} \left| \frac{\partial_{R_1}^{\mu_i} f(x)}{\partial r^{\mu_i}} \right|^{r_i} dx \ge C^* \left(\int_A \left| \frac{\partial_{R_1}^v f(x)}{\partial r^v} \right|^p dx \right), \tag{15.93}$$

where

$$C^* := Q_1(R_1) Q_2(R_1) C_1 (R_2 - R_1)^{\rho + (1/s_1)}. \tag{15.94}$$

Proof. Similar to Theorem 15.35 by using (15.92). □

We continue with

Theorem 15.41. *Let $f \in L_1(a, x)$, $a < x$, $a, x \in \mathbb{R}$, have an L_∞ fractional derivative $D_a^v f$ in $[a, x]$, such that*

$$D_a^{v-j} f(a) = 0, \ for \ j = 1, \ldots, [v] + 1.$$

Let $v > \mu_2 \ge \mu_1 + 1 \ge 1$. If $p, q > 1 : \ 1/p + 1/q = 1$; then

$$\int_a^x |D_a^{\mu_1} f(\tau)| \, |D_a^{\mu_2} f(\tau)| d\tau \le C_2 (x - a)^{2v - \mu_1 - \mu_2 - 1 + (2/q)}$$

$$\left(\int_a^x |D_a^v f(\tau)|^p d\tau \right)^{2/p}, \tag{15.95}$$

where $C_2 = C_2(v, \mu_1, \mu_2, p)$ is given by

$$C_2 := \frac{\left(\frac{1}{2}\right)^{(1/p)}}{\Gamma(v - \mu_1)\Gamma(v - \mu_2 + 1)\left((v - \mu_1)q + 1\right)^{1/q}\left((2v - \mu_1 - \mu_2 - 1)q + 2\right)^{1/q}}.$$

$$(15.96)$$

Proof. Transfer to $a \in \mathbb{R}$ of Theorem 2.5 [65]. $\square$

We give

Theorem 15.42. *Let* $f \in L_1(a, x)$, $a < x$, $a, x \in \mathbb{R}$, *have an* L_∞
fractional derivative $D_a^v f$ *in* $[a, x]$, *such that*

$$D_a^{v-j} f(a) = 0, \text{ for } j = 1, \ldots, [v] + 1;$$

also $|D_a^v f|$ *is decreasing on* $[a, x]$. *Let* $l \geq 2$. *If* $p, q > 1 : 1/p + 1/q = 1$ *and*
$\left(\sum_{i=1}^l \alpha_i\right) p > -1$, *then*

$$\int_a^x \prod_{i=1}^l |D_a^{\mu_i} f(\tau)| d\tau \leq C_3 (x - a)^{(\gamma p + lp + 1)/p} \left(\int_a^x |D_a^v f(t)|^{lq} dt\right)^{1/q},$$

$$(15.97)$$

where $\gamma := \sum_{i=1}^l \alpha_i$ *and*

$$C_3 = C_3(v, \overline{\mu}, p) := \frac{p}{(\gamma p + 1)^{1/p}(\gamma p + p + 1) \prod_{i=1}^l \Gamma(v - \mu_i)}. \quad (15.98)$$

Proof. Transfer of Theorem 2.6 of [65]. $\square$

We finish this subsection with

Theorem 15.43. *All are us in Theorem 15.42 with* $p = 1$ *and* $q = \infty$.
Then

$$\int_a^x \prod_{i=1}^l |D_a^{\mu_i} f(\tau)| d\tau \leq C_4 (x - a)^{\gamma + l + 1} \|D_a^v f\|_\infty, \quad (15.99)$$

where $\gamma := \sum_{i=1}^l \alpha_i$ *and*

$$C_4 = C_4(v, \overline{\mu}) := \frac{1}{(\gamma + 1)(\gamma + l + 1) \prod_{i=1}^l \Gamma(v - \mu_i)}. \quad (15.100)$$

Proof. Transfer of Theorem 2.7 [65]. $\square$

15.5.2 Riemann–Liouville Fractional Opial-Type Inequalities Involving Two Functions

We present

Theorem 15.44. *Let*

$$\alpha_1, \alpha_2 \in \mathbb{R}_+, \ \beta > \alpha_1, \alpha_2, \ \beta - \alpha_i > (1/p), p > 1, i = 1, 2,$$

and let

$$f_1, f_2 \in L_1(a, x), \ a, \ x \in \mathbb{R}, \ a < x$$

have, respectively, L_∞ fractional derivatives $D_a^\beta f_1, D_a^\beta f_2$ in $[a, x]$, and let

$$D_a^{\beta-k} f_i(a) = 0, \ for \ k = 1, \ldots, [\beta] + 1; \ i = 1, 2.$$

Consider also $p(t) > 0$ and $q(t) \geq 0$, with all $p(t), 1/p(t), q(t) \in L_\infty(a, x)$. Let $\lambda_\beta > 0$ and $\lambda_{\alpha_1}, \lambda_{\alpha_2} \geq 0$, such that $\lambda_\beta < p$. Set

$$P_i(s) := \int_0^s (s-t)^{p(\beta-\alpha_i-1)/(p-1)} (p(t+a))^{-1/(p-1)} \, dt, \ i = 1, 2; \ 0 \leq s \leq x-a,$$
$$(15.101)$$

$$A(s) := \frac{q(s+a) (P_1(s))^{\lambda_{\alpha_1}((p-1)/p)} (P_2(s))^{\lambda_{\alpha_2}((p-1)/p)} (p(s+a))^{-\lambda_\beta/p}}{(\Gamma(\beta-\alpha_1))^{\lambda_{\alpha_1}} (\Gamma(\beta-\alpha_2))^{\lambda_{\alpha_2}}},$$
$$(15.102)$$

$$A_0(x-a) := \left(\int_0^{x-a} (A(s))^{p/(p-\lambda_\beta)} \, ds \right)^{(p-\lambda_\beta)/p}, \qquad (15.103)$$

and

$$\delta_1 := \begin{cases} 2^{1-((\lambda_{\alpha_1}+\lambda_\beta)/p)}, & if \ \lambda_{\alpha_1} + \lambda_\beta \leq p, \\ 1, & if \ \lambda_{\alpha_1} + \lambda_\beta \geq p. \end{cases} \qquad (15.104)$$

If $\lambda_{\alpha_2} = 0$ we obtain that

$$\int_a^x q(s) \left[|D_a^{\alpha_1} f_1(s)|^{\lambda_{\alpha_1}} |D_a^\beta f_1(s)|^{\lambda_\beta} + | D_a^{\alpha_1} f_2(s)|^{\lambda_{\alpha_1}} |D_a^\beta f_2(s)|^{\lambda_\beta} \right] ds \leq$$

$$\left(A_0(x-a) \Big|_{\lambda_{\alpha_2}=0} \right) \left(\frac{\lambda_\beta}{\lambda_{\alpha_1} + \lambda_\beta} \right)^{\lambda_\beta/p} \delta_1$$

$$\left[\int_a^x p(s) \left[|D_a^\beta f_1(s)|^p + |D_a^\beta f_2(s)|^p \right] ds \right]^{((\lambda_{\alpha_1}+\lambda_\beta)/p)}. \qquad (15.105)$$

Proof. Similar proof to those of Theorems 15.22 and 15.34. Here we transfer Theorem 4 of [48] to an arbitrary anchor point $a \in \mathbb{R}$. In fact for

$a = 0$ inequality (15.105) is identical to inequality (8) of Theorem 4 of [48]. We apply it here for the translates

$$f_{1a} := f_1(\cdot + a), \ f_{2a} := f_2(\cdot + a), \ p(\cdot + a), \ q(\cdot + a)$$

and the fractional derivatives

$$D^\beta f_{1a}, \ D^\beta f_{2a}, \ D^{\alpha_i} f_{1a}, \ D^{\alpha_i} f_{2a}, \ i = 1, 2,$$

all over $[0, x - a]$.

We use Lemma 15.9, the equivalent definitions 15.12, 15.13, and (15.13). We also use (15.9) through (15.11). We get the result by Proposition 15.4 applied at the end. $\square$

We continue with

Theorem 15.45. *All here are as in Theorem* 15.44. *Denote*

$$\delta_3 := \begin{cases} 2^{\lambda_{\alpha_2}/\lambda_\beta} - 1, & \text{if } \lambda_{\alpha_2} \geq \lambda_\beta, \\ 1, & \text{if } \lambda_{\alpha_2} \leq \lambda_\beta. \end{cases} \tag{15.106}$$

If $\lambda_{\alpha_1} = 0$, *then*

$$\int_a^x q(s) \left[|D_a^{\alpha_2} f_2(s)|^{\lambda_{\alpha_2}} \, |D_a^\beta f_1(s)|^{\lambda_\beta} + |D_a^{\alpha_2} f_1(s)|^{\lambda_{\alpha_2}} \, |D_a^\beta f_2(s)|^{\lambda_\beta} \right] ds \leq$$

$$\left(A_0(x-a) \Big|_{\lambda_{\alpha_1}=0} \right) 2^{(p-\lambda_\beta)/p} \left(\frac{\lambda_\beta}{\lambda_{\alpha_2} + \lambda_\beta} \right)^{\lambda_\beta/p} \delta_3^{\lambda_\beta/p}$$

$$\left(\int_a^x p(s) \left[|D_a^\beta f_1(s)|^p + |D_a^\beta f_2(s)|^p \right] ds \right)^{(\lambda_{\alpha_2}+\lambda_\beta)/p}. \tag{15.107}$$

Proof. Transfer of Theorem 5 of [48] to $a \in \mathbb{R}$. The proof is similar to that of Theorem 15.44. $\square$

The complete case $\lambda_{\alpha_1}, \lambda_{\alpha_2} \neq 0$ follows.

Theorem 15.46. *All here are as in Theorem* 15.44. *Denote*

$$\tilde{\gamma}_1 := \begin{cases} 2^{((\lambda_{\alpha_1}+\lambda_{\alpha_2})/\lambda_\beta)} - 1, & \text{if } \lambda_{\alpha_1} + \lambda_{\alpha_2} \geq \lambda_\beta, \\ 1, & \text{if } \lambda_{\alpha_1} + \lambda_{\alpha_2} \leq \lambda_\beta, \end{cases} \tag{15.108}$$

and

$$\tilde{\gamma}_2 := \begin{cases} 1, & \text{if } \lambda_{\alpha_1} + \lambda_{\alpha_2} + \lambda_\beta \geq p, \\ 2^{1-((\lambda_{\alpha_1}+\lambda_{\alpha_2}+\lambda_\beta)/p)}, & \text{if } \lambda_{\alpha_1} + \lambda_{\alpha_2} + \lambda_\beta \leq p. \end{cases} \tag{15.109}$$

Then, it holds

$$\int_a^x q(s) \Big[|D_a^{\alpha_1} f_1(s)|^{\lambda_{\alpha_1}} \, |D_a^{\alpha_2} f_2(s)|^{\lambda_{\alpha_2}} \, |D_a^{\beta} f_1(s)|^{\lambda_{\beta}}$$

$$+| \, D_a^{\alpha_2} f_1(s)|^{\lambda_{\alpha_2}} \, |D_a^{\alpha_1} f_2(s)|^{\lambda_{\alpha_1}} \, |D_a^{\beta} f_2(s)|^{\lambda_{\beta}} \Big] ds \le A_0 (x - a)$$

$$\left(\frac{\lambda_{\beta}}{(\lambda_{\alpha_1} + \lambda_{\alpha_2})(\lambda_{\alpha_1} + \lambda_{\alpha_2} + \lambda_{\beta})} \right)^{\lambda_{\beta}/p} \cdot [\lambda_{\alpha_1}^{\lambda_{\beta}/p} \tilde{\gamma}_2 + 2^{(p-\lambda_{\beta})/p} \, (\tilde{\gamma}_1 \lambda_{\alpha_2})^{\lambda_{\beta}/p}].$$

$$\left[\int_a^x p(s) \left(|D_a^{\beta} f_1(s)|^p + |D_a^{\beta} f_2(s)|^p \right) ds \right]^{(\lambda_{\alpha_1} + \lambda_{\alpha_2} + \lambda_{\beta})/p}. \qquad (15.110)$$

Proof. Similar transfer to $a \in \mathbb{R}$ of Theorem 6 of [48]. $\square$

We proceed with a special important case.

Theorem 15.47. *Let $\beta > \alpha_1 + 1$, $\alpha_1 \in \mathbb{R}_+$ and let*

$$f_1, f_2 \in L_1(a, x), \ a, \ x \in \mathbb{R}, \ a < x$$

have, respectively, L_∞ fractional derivatives $D_a^{\beta} f_1, D_a^{\beta} f_2$ in $[0, x]$, and let

$$D_a^{\beta - k} f_i(a) = 0, \ for \ k = 1, \ldots, [\beta] + 1; \ i = 1, 2.$$

Consider also $p(t) > 0$ and $q(t) \ge 0$, with $p(t)$, $1/p(t)$, $q(t) \in L_\infty(a, x)$. Let $\lambda_\alpha \ge 0$, $0 < \lambda_{\alpha+1} < 1$, and $p > 1$. Denote

$$\theta_3 := \begin{cases} 2^{\lambda_\alpha / \lambda_{\alpha+1}} - 1, & if \ \lambda_\alpha \ge \lambda_{\alpha+1}, \\ 1, & if \ \lambda_\alpha \le \lambda_{\alpha+1}, \end{cases} \qquad (15.111)$$

$$L(x - a) := \left(2 \int_a^x (q(s))^{(1/(1-\lambda_{\alpha+1}))} ds \right)^{(1-\lambda_{\alpha+1})} \left(\frac{\theta_3 \lambda_{\alpha+1}}{\lambda_\alpha + \lambda_{\alpha+1}} \right)^{\lambda_{\alpha+1}}, \qquad (15.112)$$

and

$$P_1(x - a) := \int_a^x (x - s)^{(\beta - \alpha_1 - 1)p/(p-1)} \, (p(s))^{-1/(p-1)} ds, \qquad (15.113)$$

$$T(x - a) := L(x - a) \left(\frac{P_1(x - a)^{(p-1/p)}}{\Gamma(\beta - \alpha_1)} \right)^{(\lambda_\alpha + \lambda_{\alpha+1})}, \qquad (15.114)$$

and

$$w_1 := 2^{((p-1)/p)(\lambda_\alpha + \lambda_{\alpha+1})}, \qquad (15.115)$$

with

$$\Phi(x - a) := T(x - a)w_1. \tag{15.116}$$

Then

$$\int_a^x q(s) \left[|D_a^{\alpha_1} f_1(s)|^{\lambda_\alpha} \, |D_a^{\alpha_1+1} f_2(s)|^{\lambda_{\alpha+1}} + |D_a^{\alpha_1} f_2(s)|^{\lambda_\alpha} \, |D_a^{\alpha_1+1} f_1(s)|^{\lambda_{\alpha+1}} \right] ds$$

$$\leq \Phi(x-a) \left[\int_a^x p(s) \left(|D_a^\beta f_1(s)|^p + |D_a^\beta f_2(s)|^p \right) ds \right]^{((\lambda_\alpha + \lambda_{\alpha+1})/p)}. \tag{15.117}$$

Proof. Similar transfer to $a \in \mathbb{R}$ of Theorem 8 of [48]. $\square$

We give

Theorem 15.48. *All here are as in Theorem* 15.44. *Consider the special case of*

$$\lambda_{\alpha_2} = \lambda_{\alpha_1} + \lambda_\beta.$$

Denote

$$\tilde{T}(x - a) := A_0(x - a) \left(\frac{\lambda_\beta}{\lambda_{\alpha_1} + \lambda_\beta} \right)^{\lambda_\beta/p} 2^{(p - 2\lambda_{\alpha_1} - 3\lambda_\beta)/p}. \tag{15.118}$$

Then

$$\int_a^x q(s) \left[|D_a^{\alpha_1} f_1(s)|^{\lambda_{\alpha_1}} \, |D_a^{\alpha_2} f_2(s)|^{\lambda_{\alpha_1} + \lambda_\beta} \, |D_a^\beta f_1(s)|^{\lambda_\beta} \right.$$

$$\left. + |D_a^{\alpha_2} f_1(s)|^{\lambda_{\alpha_1} + \lambda_\beta} \, |D_a^{\alpha_1} f_2(s)|^{\lambda_{\alpha_1}} \, |D_a^\beta f_2(s)|^{\lambda_\beta} \right] ds \leq$$

$$\tilde{T}(x - a) \left(\int_a^x p(s) \left(|D_a^\beta f_1(s)|^p + |D_a^\beta f_2(s)|^p \right) ds \right)^{2(\lambda_{\alpha_1} + \lambda_\beta)/p}. \tag{15.119}$$

Proof. Transfer of Theorem 9 of [48]. $\square$

Next follow special cases of the last theorems.

Corollary 15.49. (to Theorem 15.44) *Set* $\lambda_{\alpha_2} = 0$, $p(t) = q(t) = 1$. *Then*

$$\int_a^x \left[|D_a^{\alpha_1} f_1(s)|^{\lambda_{\alpha_1}} \, |D_a^\beta f_1(s)|^{\lambda_\beta} + |D_a^{\alpha_1} f_2(s)|^{\lambda_{\alpha_1}} \, |D_a^\beta f_2(s)|^{\lambda_\beta} \right] ds \leq$$

$$C_1(x - a) \left[\int_a^x \left[|D_a^\beta f_1(s)|^p + |D_a^\beta f_2(s)|^p \right] ds \right]^{(\lambda_{\alpha_1} + \lambda_\beta)/p}, \tag{15.120}$$

where

$$C_1(x-a) := \left(A_0(x-a) \Big|_{\lambda_{\alpha_2}=0} \right) \left(\frac{\lambda_\beta}{\lambda_{\alpha_1}+\lambda_\beta} \right)^{\lambda_\beta/p} \delta_1, \qquad (15.121)$$

$$\delta_1 := \begin{cases} 2^{1-((\lambda_{\alpha_1}+\lambda_\beta)/p)}, & \text{if } \lambda_{\alpha_1}+\lambda_\beta \le p, \\ 1, & \text{if } \lambda_{\alpha_1}+\lambda_\beta \ge p. \end{cases} \qquad (15.122)$$

We find that

$$\left(A_0(x-a) \Big|_{\lambda_{\alpha_2}=0} \right) = \left\{ \left(\frac{(p-1)^{((\lambda_{\alpha_1}p-\lambda_{\alpha_1})/p)}}{(\Gamma(\beta-\alpha_1))^{\lambda_{\alpha_1}} (\beta p - \alpha_1 p - 1)^{((\lambda_{\alpha_1}p-\lambda_{\alpha_1})/p)}} \right) \times \right.$$

$$\left. \left(\frac{(p-\lambda_\beta)^{((p-\lambda_\beta)/p)}}{(\lambda_{\alpha_1}\beta p - \lambda_{\alpha_1}\alpha_1 p - \lambda_{\alpha_1} + p - \lambda_\beta)^{((p-\lambda_\beta)/p)}} \right) \right\} \times$$

$$(x-a)^{((\lambda_{\alpha_1}\beta p - \lambda_{\alpha_1}\alpha_1 p - \lambda_{\alpha_1} + p - \lambda_\beta)/p)}. \qquad (15.123)$$

Proof. Transfer of Corollary 10 of [48]. □

We continue with

Corollary 15.50. (To Theorem 15.44: set $\lambda_{\alpha_2} = 0$, $p(t) = q(t) = 1$, $\lambda_{\alpha_1} = \lambda_\beta = 1$, $p = 2$.) *In detail; let $\alpha_1 \in \mathbb{R}_+$, $\beta > \alpha_1$, $\beta - \alpha_1 > (1/2)$, and let*

$$f_1, f_2 \in L_1(a,x), \quad a, \ x \in \mathbb{R}, \ a < x$$

have, respectively, L_∞ fractional derivatives $D_a^\beta f_1, D_a^\beta f_2$ in $[a,x]$, and let

$$D_a^{\beta-k} f_i(a) = 0, \ \text{for } k = 1, \ldots, [\beta]+1; \ i = 1, 2.$$

Then

$$\int_a^x \left[|D_a^{\alpha_1} f_1(s)| \ |D_a^\beta f_1(s)| + |D_a^{\alpha_1} f_2(s)| \ |D_a^\beta f_2(s)| \right] ds \le$$

$$\left(\frac{(x-a)^{(\beta-\alpha_1)}}{2\Gamma(\beta-\alpha_1)\sqrt{\beta-\alpha_1} \ \sqrt{2\beta-2\alpha_1-1}} \right) \left(\int_a^x \left[(D_a^\beta f_1(s))^2 + (D_a^\beta f_2(s))^2 \right] ds \right). \qquad (15.124)$$

Proof. Transfer of Corollary 11 of [48]. □

We continue with

Corollary 15.51. (To Theorem 15.45; $\lambda_{\alpha_1} = 0$, $p(t) = q(t) = 1$.) *It holds*

$$\int_a^x \left[|D_a^{\alpha_2} f_2(s)|^{\lambda_{\alpha_2}} \, |D_a^\beta f_1(s)|^{\lambda_\beta} + |D_a^{\alpha_2} f_1(s)|^{\lambda_{\alpha_2}} \, |D_a^\beta f_2(s)|^{\lambda_\beta} \right] ds \leq$$

$$C_2(x-a) \left[\int_a^x \left[|D_a^\beta f_1(s)|^p + |D_a^\beta f_2(s)|^p \right] ds \right]^{(\lambda_\beta + \lambda_{\alpha_2})/p}. \tag{15.125}$$

Here

$$C_2(x-a) := \left(A_0(x-a) \Big|_{\lambda_{\alpha_1}=0} \right) 2^{(p-\lambda_\beta/p)} \left(\frac{\lambda_\beta}{\lambda_{\alpha_2} + \lambda_\beta} \right)^{\lambda_\beta/p} \delta_3^{\lambda_\beta/p}, \tag{15.126}$$

$$\delta_3 := \begin{cases} 2^{\lambda_{\alpha_2}/\lambda_\beta} - 1, & \text{if } \lambda_{\alpha_2} \geq \lambda_\beta, \\ 1, & \text{if } \lambda_{\alpha_2} \leq \lambda_\beta. \end{cases} \tag{15.127}$$

We find that

$$\left(A_0(x-a) \Big|_{\lambda_{\alpha_1}=0} \right) = \left\{ \left(\frac{(p-1)^{(\lambda_{\alpha_2}p - \lambda_{\alpha_2})/p}}{(\Gamma(\beta - \alpha_2))^{\lambda_{\alpha_2}} (\beta p - \alpha_2 p - 1)^{(\lambda_{\alpha_2}p - \lambda_{\alpha_2})/p)}} \right) \times \right.$$

$$\left. \left(\frac{(p-\lambda_\beta)^{((p-\lambda_\beta)/p)}}{(\lambda_{\alpha_2}\beta p - \lambda_{\alpha_2}\alpha_2 p - \lambda_{\alpha_2} + p - \lambda_\beta)^{((p-\lambda_\beta)/p)}} \right) \right\} \times$$

$$(x-a)^{((\lambda_{\alpha_2}\beta p - \lambda_{\alpha_2}\alpha_2 p - \lambda_{\alpha_2} + p - \lambda_\beta)/p)}. \tag{15.128}$$

Proof. Transfer of Corollary 12 of [48]. $\square$

We give

Corollary 15.52. (To Theorem 15.45; $\lambda_{\alpha_1} = 0$, $p(t) = q(t) = 1$, $\lambda_{\alpha_2} = \lambda_\beta = 1$, $p = 2$.) *In detail; let* $\alpha_2 \in \mathbb{R}_+$, $\beta > \alpha_2$, $\beta \ \ \alpha_2 > (1/2)$, *and let*

$$f_1, f_2 \in L_1(a, x), \ a, \ x \in \mathbb{R}, \ a < x$$

have, respectively, L_∞ fractional derivatives $D_a^\beta f_1, D_a^\beta f_2$ in $[a, x]$, and let

$$D_a^{\beta-k} f_i(a) = 0, \ \text{for } k = 1, \ldots, [\beta] + 1; \ i = 1, 2.$$

Then

$$\int_a^x \left[|D_a^{\alpha_2} f_2(s)| \, |D_a^\beta f_1(s)| + |D_a^{\alpha_2} f_1(s)| \, |D_a^\beta f_2(s)| \right] ds \leq$$

$$C_2^*(x-a) \left(\int_a^x \left[(D_a^\beta f_1(s))^2 + (D_a^\beta f_2(s))^2 \right] ds \right), \tag{15.129}$$

where

$$C_2^*(x-a) := \frac{(x-a)^{(\beta-\alpha_2)}}{\sqrt{2}\,\Gamma(\beta-\alpha_2)\sqrt{\beta-\alpha_2}\,\sqrt{2\beta-2\alpha_2-1}}. \qquad (15.130)$$

Proof. Transfer of Corollary 13 of [48].   □

We continue with

Corollary 15.53. (To Theorem 15.46; $\lambda_{\alpha_1} = \lambda_{\alpha_2} = \lambda_\beta = 1$, $p = 3$, $p(t) = q(t) = 1$.) *It holds*

$$\int_a^x \Big[|D_a^{\alpha_1}f_1(s)|\,|D_a^{\alpha_2}f_2(s)|\,|D_a^\beta f_1(s)| + |D_a^{\alpha_2}f_1(s)|\,|D_a^{\alpha_1}f_2(s)|\,|D_a^\beta f_2(s)|\Big]ds \le$$

$$A_0(x-a)\left(\sqrt[3]{2}+\frac{1}{\sqrt[3]{6}}\right)\left(\int_a^x \big(|D_a^\beta f_1(s)|^3 + |D_a^\beta f_2(s)|^3\big)\,ds\right). \qquad (15.131)$$

Here,

$$A_0(x-a) = 4(x-a)^{(2\beta-\alpha_1-\alpha_2)} \times$$

$$\frac{1}{\Gamma(\beta-\alpha_1)\,\Gamma(\beta-\alpha_2)[3(3\beta-3\alpha_1-1)(3\beta-3\alpha_2-1)(2\beta-\alpha_1-\alpha_2)]^{2/3}}. \qquad (15.132)$$

Proof. Transfer of Corollary 14 of [48].   □

We give

Corollary 15.54. (To Theorem 15.47; here $\lambda_\alpha = 1$, $\lambda_{\alpha+1} = 1/2$, $p = 3/2$, $p(t) = q(t) = 1$.) *In detail: let $\beta > \alpha_1 + 1$, $\alpha_1 \in \mathbb{R}_+$, and let*

$$f_1, f_2 \in L_1(a,x), \quad a,\ x \in \mathbb{R}, \ a < x$$

have, respectively, L_∞ fractional derivatives $D_a^\beta f_1, D_a^\beta f_2$ in $[a,x]$, and let

$$D_a^{\beta-k}f_i(a) = 0, \ for\ k = 1,\dots,[\beta]+1;\ i = 1,2.$$

Then

$$\int_a^x \Big[|D_a^{\alpha_1}f_1(s)|\,\sqrt{|D_a^{\alpha_1+1}f_2(s)|} + |D_a^{\alpha_1}f_2(s)|\,\sqrt{|D_a^{\alpha_1+1}f_1(s)|}\Big]ds \le$$

$$\Phi^*(x-a)\left[\int_a^x \big(|D_a^\beta f_1(s)|^{3/2} + |D_a^\beta f_2(s)|^{3/2}\big)\,ds\right], \qquad (15.133)$$

where

$$\Phi^*(x-a) := \frac{2(x-a)^{(3\beta-3\alpha_1-1)/2}}{(\Gamma(\beta-\alpha_1))^{3/2}\sqrt{3\beta-3\alpha_1-2}}. \qquad (15.134)$$

Proof. Transfer of Corollary 15 of [48].   □

We continue

Corollary 15.55. (To Theorem 15.48; here $p = 2(\lambda_{\alpha_1} + \lambda_\beta) > 1$, $p(t) = q(t) = 1$.) *It holds*

$$\int_a^x \Big[|D_a^{\alpha_1} f_1(s)|^{\lambda_{\alpha_1}} \, |D_a^{\alpha_2} f_2(s)|^{\lambda_{\alpha_1}+\lambda_\beta} \, |D_a^\beta f_1(s)|^{\lambda_\beta}$$

$$+|D_a^{\alpha_2} f_1(s)|^{\lambda_{\alpha_1}+\lambda_\beta} \, |D_a^{\alpha_1} f_2(s)|^{\lambda_{\alpha_1}} \, |D_a^\beta f_2(s)|^{\lambda_\beta}\Big] ds \le$$

$$\tilde{\tilde{T}}(x - a)\Big(\int_a^x \Big(|D_a^\beta f_1(s)|^{2(\lambda_{\alpha_1}+\lambda_\beta)} + |D_a^\beta f_2(s)|^{2(\lambda_{\alpha_1}+\lambda_\beta)}\Big) ds\Big). \quad (15.135)$$

Here,

$$\tilde{\tilde{T}}(x - a) := \tilde{A}_0(x - a) \left(\frac{\lambda_\beta}{2(\lambda_{\alpha_1} + \lambda_\beta)}\right)^{(\lambda_\beta/2(\lambda_{\alpha_1}+\lambda_\beta))}, \quad (15.136)$$

and

$$\tilde{A}_0(x - a) := \sigma\sigma^*(x - a)^\theta, \quad (15.137)$$

where

$$\sigma := \frac{1}{(\Gamma(\beta - \alpha_1))^{\lambda_{\alpha_1}} (\Gamma(\beta - \alpha_2))^{\lambda_{\alpha_1}+\lambda_\beta}}$$

$$\times \left(\frac{2\lambda_{\alpha_1} + 2\lambda_\beta - 1}{2\lambda_{\alpha_1}\beta + 2\lambda_\beta\beta - 2\lambda_{\alpha_1}\alpha_1 - 2\lambda_\beta\alpha_1 - 1}\right)^{(2\lambda_{\alpha_1}^2+2\lambda_{\alpha_1}\lambda_\beta-\lambda_{\alpha_1})/(2\lambda_{\alpha_1}+2\lambda_\beta)}$$

$$\times \left(\frac{2\lambda_{\alpha_1} + 2\lambda_\beta - 1}{2\lambda_{\alpha_1}\beta + 2\lambda_\beta\beta - 2\lambda_{\alpha_1}\alpha_2 - 2\lambda_\beta\alpha_2 - 1}\right)^{((2\lambda_{\alpha_1}+2\lambda_\beta-1)/2)}, \quad (15.138)$$

$$\sigma^* := \left(\frac{2\lambda_{\alpha_1} + \lambda_\beta}{S}\right)^{(2\lambda_{\alpha_1}+\lambda_\beta)/2(\lambda_{\alpha_1}+\lambda_\beta)},$$

where

$$S := 4\lambda_{\alpha_1}^2\beta + 6\lambda_{\alpha_1}\lambda_\beta\beta - 2\lambda_{\alpha_1}^2\alpha_1 - 2\lambda_{\alpha_1}\lambda_\beta\alpha_1$$

$$-2\lambda_{\alpha_1}^2\alpha_2 - 4\lambda_{\alpha_1}\lambda_\beta\alpha_2 + 2\lambda_\beta^2\beta - 2\lambda_\beta^2\alpha_2$$

and

$$\theta := \left(\frac{S}{2\lambda_{\alpha_1} + 2\lambda_\beta}\right). \quad (15.139)$$

Proof. Transfer of Corollary 16 of [48]. □

We give the following interesting special case.

Corollary 15.56. (To Theorem 15.48; here $p = 4, \lambda_{\alpha_1} = \lambda_\beta = 1, p(t) = q(t) = 1$.) It holds

$$\int_a^x \left[|D_a^{\alpha_1} f_1(s)| \, (D_a^{\alpha_2} f_2(s))^2 \, |D_a^\beta f_1(s)| + (D_a^{\alpha_2} f_1(s))^2 \, |D_a^{\alpha_1} f_2(s)| \, |D_a^\beta f_2(s)| \right] ds \leq$$

$$T^*(x - a) \left(\int_a^x (D_a^\beta f_1(s))^4 + (D_a^\beta f_2(s)^4) \, ds \right). \tag{15.140}$$

Here,

$$T^*(x - a) := \frac{A_0^*(x - a)}{\sqrt{2}}, \tag{15.141}$$

and

$$A^*(x - a) := \tilde{\sigma}\tilde{\sigma}^*(x - a)^{\tilde{\theta}}, \tag{15.142}$$

where

$$\tilde{\sigma} := \frac{1}{\Gamma(\beta - \alpha_1)(\Gamma(\beta - \alpha_2))^2} \left(\frac{3}{4\beta - 4\alpha_1 - 1} \right)^{3/4} \left(\frac{3}{4\beta - 4\alpha_2 - 1} \right)^{3/2}, \tag{15.143}$$

$$\tilde{\sigma}^* := \left(\frac{3}{12\beta - 4\alpha_1 - 8\alpha_2} \right)^{3/4}, \tag{15.144}$$

and

$$\tilde{\theta} := 3\beta - \alpha_1 - 2\alpha_2. \tag{15.145}$$

Proof. Transfer of Corollary 17 of [48]. □

We continue with related results regarding the L_∞-norm $\| \cdot \|_\infty$.

Theorem 15.57. Let $\alpha_1, \alpha_2 \in \mathbb{R}_+, \ \beta > \alpha_1, \alpha_2,$ and let

$$f_1, f_2 \in L_1(a, x), \ a, \ x \in \mathbb{R}, \ a < x$$

have, respectively, L_∞ fractional derivatives $D_a^\beta f_1, D_a^\beta f_2$ in $[a, x]$, and let

$$D_a^{\beta-k} f_i(a) = 0, \ for \ k = 1, \ldots, \lceil \beta \rceil + 1; \ i = 1, 2.$$

Consider also $p(s) \geq 0, \ p(s) \in L_\infty(a, x)$. Let $\lambda_{\alpha_1}, \lambda_{\alpha_2}, \lambda_\beta \geq 0$. Set

$$\rho(x - a) :=$$

$$\frac{\|p(s)\|_\infty \, (x-a)^{(\beta\lambda_{\alpha_1} - \alpha_1\lambda_{\alpha_1} + \beta\lambda_{\alpha_2} - \alpha_2\lambda_{\alpha_2} + 1)}}{(\Gamma(\beta - \alpha_1 + 1))^{\lambda_{\alpha_1}} (\Gamma(\beta - \alpha_2 + 1))^{\lambda_{\alpha_2}} [\beta\lambda_{\alpha_1} - \alpha_1\lambda_{\alpha_1} + \beta\lambda_{\alpha_2} - \alpha_2\lambda_{\alpha_2} + 1]}.$$

$$(15.146)$$

Then

$$\int_a^x p(s) \Big[|D_a^{\alpha_1} f_1(s)|^{\lambda_{\alpha_1}} \, |D_a^{\alpha_2} f_2(s)|^{\lambda_{\alpha_2}} \, |D_a^{\beta} f_1(s)|^{\lambda_\beta}$$

$$+ |D_a^{\alpha_2} f_1(s)|^{\lambda_{\alpha_2}} \, |D_a^{\alpha_1} f_2(s)|^{\lambda_{\alpha_1}} \, |D_a^{\beta} f_2(s)|^{\lambda_\beta} \Big] ds \le$$

$$\frac{\rho(x-a)}{2} \Big[\|D_a^\beta f_1\|_\infty^{2(\lambda_{\alpha_1} + \lambda_\beta)} + \|D_a^\beta f_1\|_\infty^{2\lambda_{\alpha_2}}$$

$$+ \|D_a^\beta f_2\|_\infty^{2\lambda_{\alpha_2}} + \|D_a^\beta f_2\|_\infty^{2(\lambda_{\alpha_1} + \lambda_\beta)} \Big].$$

$$(15.147)$$

Proof. Transfer of Theorem 18 of [48]. It is similar to the proof of Theorem 15.24, and so on. $\quad\square$

We give special cases of the last theorem.

Theorem 15.58. (*All are as in Theorem 15.57;* $\lambda_{\alpha_2} = 0$.) *It holds*

$$\int_a^x p(s) \Big[|D_a^{\alpha_1} f_1(s)|^{\lambda_{\alpha_1}} \, |D_a^{\beta} f_1(s)|^{\lambda_\beta} + |D_a^{\alpha_1} f_2(s)|^{\lambda_{\alpha_1}} \, |D_a^{\beta} f_2(s)|^{\lambda_\beta} \Big] ds \le$$

$$\frac{\|p(s)\|_\infty \, (x-a)^{(\beta\lambda_{\alpha_1} - \alpha_1\lambda_{\alpha_1} + 1)}}{(\Gamma(\beta - \alpha_1 + 1))^{\lambda_{\alpha_1}} [\beta\lambda_{\alpha_1} - \alpha_1\lambda_{\alpha_1} + 1]} \cdot \Big[\|D_a^\beta f_1\|_\infty^{\lambda_{\alpha_1} + \lambda_\beta} + \|D_a^\beta f_2\|_\infty^{\lambda_{\alpha_1} + \lambda_\beta} \Big].$$

$$(15.148)$$

Proof. Transfer of Theorem 19 of [48]. It is similar to the proof of Theorem 15.57. $\quad\square$

We continue with

Theorem 15.59. (*All are as in Theorem 15.57;* $\lambda_{\alpha_2} = \lambda_{\alpha_1} + \lambda_\beta$.) *It holds*

$$\int_a^x p(s) \Big[|D_a^{\alpha_1} f_1(s)|^{\lambda_{\alpha_1}} \, |D_a^{\alpha_2} f_2(s)|^{\lambda_{\alpha_1} + \lambda_\beta} \, |D_a^{\beta} f_1(s)|^{\lambda_\beta}$$

$$+ |D_a^{\alpha_2} f_1(s)|^{\lambda_{\alpha_1} + \lambda_\beta} \, |D_a^{\alpha_1} f_2(s)|^{\lambda_{\alpha_1}} \, |D_a^{\beta} f_2(s)|^{\lambda_\beta} \Big] ds$$

$$\le \Bigg\{ \frac{\|p(s)\|_\infty}{(\Gamma(\beta - \alpha_1 + 1))^{\lambda_{\alpha_1}} (\Gamma(\beta - \alpha_2 + 1))^{\lambda_{\alpha_1} + \lambda_\beta}}$$

$$\cdot \frac{(x-a)^{(2\beta\lambda_{\alpha_1} - \alpha_1\lambda_{\alpha_1} + \beta\lambda_\beta - \alpha_2\lambda_{\alpha_1} - \alpha_2\lambda_\beta + 1)}}{(2\beta\lambda_{\alpha_1} - \alpha_1\lambda_{\alpha_1} + \beta\lambda_\beta - \alpha_2\lambda_{\alpha_1} - \alpha_2\lambda_\beta + 1)} \Bigg\}$$

$$\cdot \Big[\|D_a^\beta f_1\|_\infty^{2(\lambda_{\alpha_1} + \lambda_\beta)} + \|D_a^\beta f_2\|_\infty^{2(\lambda_{\alpha_1} + \lambda_\beta)} \Big].$$

$$(15.149)$$

Proof. Transfer of Theorem 20 of [48]. □

We give

Theorem 15.60. (All are as in Theorem 15.57; $\lambda_\beta = 0$, $\lambda_{\alpha_1} = \lambda_{\alpha_2}$.) *It holds*

$$\int_a^x p(s) \left[|D_a^{\alpha_1} f_1(s)|^{\lambda_{\alpha_1}} \, |D_a^{\alpha_2} f_2(s)|^{\lambda_{\alpha_1}} + |D_a^{\alpha_2} f_1(s)|^{\lambda_{\alpha_1}} \, |D_a^{\alpha_1} f_2(s)|^{\lambda_{\alpha_1}} \right] ds \leq$$

$$\rho^*(x-a) \left[\|D_a^\beta f_1\|_\infty^{2\lambda_{\alpha_1}} + \|D_a^\beta f_2\|_\infty^{2\lambda_{\alpha_1}} \right], \qquad (15.150)$$

where

$$\rho^*(x-a) = \left\{ \frac{\|p(s)\|_\infty \, (x-a)^{(2\beta\lambda_{\alpha_1} - \alpha_1\lambda_{\alpha_1} - \alpha_2\lambda_{\alpha_1} + 1)}}{(\Gamma(\beta-\alpha_1+1) \, \Gamma(\beta-\alpha_2+1))^{\lambda_{\alpha_1}} \, (2\beta\lambda_{\alpha_1} - \alpha_1\lambda_{\alpha_1} - \alpha_2\lambda_{\alpha_1} + 1)} \right\}.$$
$$(15.151)$$

Proof. Transfer of Theorem 21 of [48]. □

We give

Theorem 15.61. (All are as in Theorem 15.57; $\lambda_{\alpha_1} = 0$, $\lambda_{\alpha_2} = \lambda_\beta$.) *It holds*

$$\int_a^x p(s) \left[|D_a^{\alpha_2} f_2(s)|^{\lambda_{\alpha_2}} \, |D_a^\beta f_1(s)|^{\lambda_{\alpha_2}} + |D_a^{\alpha_2} f_1(s)|^{\lambda_{\alpha_2}} \, |D_a^\beta f_2(s)|^{\lambda_{\alpha_2}} \right] ds \leq$$

$$\left(\frac{(x-a)^{(\beta\lambda_{\alpha_2} - \alpha_2\lambda_{\alpha_2} + 1)} \, \|p(s)\|_\infty}{(\beta\lambda_{\alpha_2} - \alpha_2\lambda_{\alpha_2} + 1)(\Gamma(\beta-\alpha_2+1))^{\lambda_{\alpha_2}}} \right) \cdot \left[\|D_a^\beta f_1\|_\infty^{2\lambda_{\alpha_2}} + \|D_a^\beta f_2\|_\infty^{2\lambda_{\alpha_2}} \right].$$
$$(15.152)$$

Proof. Transfer of Theorem 22 of [48]. □

We continue with

Corollary 15.62. (To Theorem 15.60; all are as in Theorem 15.57; $\lambda_\beta = 0$, $\lambda_{\alpha_1} = \lambda_{\alpha_2}$, $\alpha_2 = \alpha_1 + 1$.) *It holds*

$$\int_a^x p(s) \left[|D_a^{\alpha_1} f_1(s)|^{\lambda_{\alpha_1}} \, |D_a^{\alpha_1+1} f_2(s)|^{\lambda_{\alpha_1}} + |D_a^{\alpha_1+1} f_1(s)|^{\lambda_{\alpha_1}} \, |D_a^{\alpha_1} f_2(s)|^{\lambda_{\alpha_1}} \right] ds \leq$$

$$\left(\frac{(x-a)^{(2\beta\lambda_{\alpha_1} - 2\alpha_1\lambda_{\alpha_1} - \lambda_{\alpha_1} + 1)} \, \|p(s)\|_\infty}{(2\beta\lambda_{\alpha_1} - 2\alpha_1\lambda_{\alpha_1} - \lambda_{\alpha_1} + 1)(\beta-\alpha_1)^{\lambda_{\alpha_1}} \, (\Gamma(\beta-\alpha_1))^{2\lambda_{\alpha_1}}} \right)$$

$$\cdot \left[\|D_a^\beta f_1\|_\infty^{2\lambda_{\alpha_1}} + \|D_a^\beta f_2\|_\infty^{2\lambda_{\alpha_1}} \right]. \qquad (15.153)$$

Proof. Transfer of Corollary 23 of [48]. □

We give

Corollary 15.63. (to Corollary 15.62) *In detail: let $\alpha_1 \in \mathbb{R}_+$, $\beta > \alpha_1 + 1$, and let*

$$f_1, f_2 \in L_1(a, x), \quad a, \ x \in \mathbb{R}, \ a < x$$

have, respectively, L_∞ fractional derivatives $D_a^\beta f_1, D_a^\beta f_2$ in $[a, x]$, and let

$$D_a^{\beta-k} f_i(a) = 0, \ \text{for } k = 1, \ldots, [\beta] + 1; \ i = 1, 2.$$

Then

$$\int_a^x \left[|D_a^{\alpha_1} f_1(s)| \, |D_a^{\alpha_1+1} f_2(s)| + |D_a^{\alpha_1+1} f_1(s)| \, |D_a^{\alpha_1} f_2(s)| \right] ds \leq$$

$$\left(\frac{(x-a)^{2(\beta-\alpha_1)}}{2(\beta-\alpha_1)^2 (\Gamma(\beta-\alpha_1))^2} \right) \cdot \left[\|D_a^\beta f_1\|_\infty^2 + \|D_a^\beta f_2\|_\infty^2 \right]. \tag{15.154}$$

Proof. Transfer of Corollary 24 of [48]. □

We finally give

Proposition 15.64. *Inequality* (15.154) *is sharp; in fact it is attained when $f_1 = f_2$, by*

$$f_1(s) = (s-a)^\beta, \ a \leq s \leq x, \ \beta > \alpha_1 + 1, \ \alpha_1 \geq 0.$$

Proof. Clearly (15.154), when $f_1 = f_2$, collapses to

$$\int_a^x |D_a^{\alpha_1} f_1(s)| \, |D_a^{\alpha_1+1} f_1(s)| ds \leq \left(\frac{(x-u)^{2(\beta-\alpha_1)}}{2(\Gamma(\beta-\alpha_1+1))^2} \right) \cdot \|D_a^\beta f_1\|_\infty^2;$$

$$\tag{15.155}$$

see Theorem 15.27 and Proposition 15.28. □

Next we apply the above results on the spherical shell A.
We make

Assumption 15.65. Let

$$\alpha_1, \alpha_2 \in \mathbb{R}_+, \ \beta > \alpha_1, \alpha_2, \ \beta - \alpha_i > (1/p), p > 1, i = 1, 2,$$

and let $f_1, f_2 \in L_1(A)$ with

$$\frac{\partial_{R_1}^\beta f_1(x)}{\partial r^\beta}, \ \frac{\partial_{R_1}^\beta f_2(x)}{\partial r^\beta} \in L_\infty(A), \ x \in A;$$

$$A := B(0, R_2) - \overline{B(0, R_1)} \subseteq \mathbb{R}^N, \ N \geq 2, \ 0 < R_1 < R_2.$$

Further assume that each $D^\beta_{R_1} f_i(rw) \in \mathbb{R}$ for almost all $r \in [R_1, R_2]$, for each $w \in S^{N-1}$, and for these $|D^\beta_{R_1} f_i(rw)| \leq M_i$ for some $M_i > 0$; $i = 1, 2$. For each $w \in S^{N-1} - (K(f_1) \cup K(f_2))$, we assume that $f_i(\cdot \ w)$ has an L_∞ fractional derivative $D^\beta_{R_1} f_i(\cdot \ w)$ in $[R_1, R_2]$, and that

$$D^{\beta-k}_{R_1} f_i(R_1 w) = 0, \ k = 1, \ldots, [\beta] + 1;$$

$i = 1, 2$. Let $\lambda_\beta > 0$ and $\lambda_{\alpha_1}, \lambda_{\alpha_2} \geq 0$, such that $\lambda_\beta < p$. If $\alpha_1 = 0$ we set $\lambda_{\alpha_1} = 1$, and if $\alpha_2 = 0$ we set $\lambda_{\alpha_2} = 1$.

We need

Notation 15.66. (on Assumption 15.65) Set

$$P_i(s) := \int_0^s (s-r)^{p(\beta-\alpha_i-1)/p-1} (r+R_1)^{1-N/p-1} \, dr, \ i = 1, 2; \ 0 \leq s \leq R_2 - R_1, \tag{15.156}$$

$$A(s) := \frac{(s+R_1)^{(N-1)(1-(\lambda_\beta/p))} (P_1(s))^{\lambda_{\alpha_1}((p-1)/p)} (P_2(s))^{\lambda_{\alpha_2}((p-1)/p)}}{(\Gamma(\beta-\alpha_1))^{\lambda_{\alpha_1}} (\Gamma(\beta-\alpha_2))^{\lambda_{\alpha_2}}}, \tag{15.157}$$

and

$$A_0(R_2 - R_1) := \left(\int_0^{R_2-R_1} (A(s))^{p/(p-\lambda_\beta)} \, ds \right)^{(p-\lambda_\beta)/p}. \tag{15.158}$$

We present

Theorem 15.67. (All are as in Assumption 15.65 and Notation 15.66.) Here $\lambda_{\alpha_1} > 0$, $\lambda_{\alpha_2} = 0$, and $p = \lambda_{\alpha_1} + \lambda_\beta > 1$. Then

$$\int_A \left[\left| \frac{\partial^{\alpha_1}_{R_1} f_1(x)}{\partial r^{\alpha_1}} \right|^{\lambda_{\alpha_1}} \left| \frac{\partial^\beta_{R_1} f_1(x)}{\partial r^\beta} \right|^{\lambda_\beta} + \left| \frac{\partial^{\alpha_1}_{R_1} f_2(x)}{\partial r^{\alpha_1}} \right|^{\lambda_{\alpha_1}} \left| \frac{\partial^\beta_{R_1} f_2(x)}{\partial r^\beta} \right|^{\lambda_\beta} \right] dx \leq$$

$$\left(A_0(R_2 - R_1) \Big|_{\lambda_{\alpha_2}=0} \right) \left(\frac{\lambda_\beta}{\lambda_{\alpha_1} + \lambda_\beta} \right)^{(\lambda_\beta/(\lambda_{\alpha_1}+\lambda_\beta))}$$

$$\left[\int_A \left[\left| \frac{\partial^\beta_{R_1} f_1(x)}{\partial r^\beta} \right|^p + \left| \frac{\partial^\beta_{R_1} f_2(x)}{\partial r^\beta} \right|^p \right] dx \right]. \tag{15.159}$$

Proof. By Theorem 15.20 for $\alpha_1 > 0$ we get that

$$\frac{\partial^{\alpha_1}_{R_1} f_i(x)}{\partial r^{\alpha_1}} \in L_\infty(A), \ i = 1, 2.$$

In general here the integrands of both integrals of (15.159) are in $L_1(A)$. Thus, by Proposition 15.3 we have

$$I_1 := L.H.S(15.159) = \int_{S^{N-1}} \left(\int_{R_1}^{R_2} [\ |D_{R_1}^{\alpha_1} f_1(rw)|^{\lambda_{\alpha_1}} \ |D_{R_1}^{\beta} f_1(rw)|^{\lambda_\beta} + \right.$$

$$\left. |D_{R_1}^{\alpha_1} f_2(rw)|^{\lambda_{\alpha_1}} \ |D_{R_1}^{\beta} f_2(rw)|^{\lambda_\beta}]r^{N-1}dr \right) dw =$$

$$\int_{(S^{N-1}-(K(f_1)\cup K(f_2)))} \left(\int_{R_1}^{R_2} [\ |D_{R_1}^{\alpha_1} f_1(rw)|^{\lambda_{\alpha_1}} \ |D_{R_1}^{\beta} f_1(rw)|^{\lambda_\beta} + \right.$$

$$\left. |D_{R_1}^{\alpha_1} f_2(rw)|^{\lambda_{\alpha_1}} \ |D_{R_1}^{\beta} f_2(rw)|^{\lambda_\beta}]r^{N-1}dr \right) dw. \tag{15.160}$$

Similarly we have

$$I_2 := \int_A \left[\left| \frac{\partial_{R_1}^{\beta} f_1(x)}{\partial r^{\beta}} \right|^p + \left| \frac{\partial_{R_1}^{\beta} f_2(x)}{\partial r^{\beta}} \right|^p \right] dx$$

$$= \int_{S^{N-1}} \left(\int_{R_1}^{R_2} [\ |D_{R_1}^{\beta} f_1(rw)|^p + |D_{R_1}^{\beta} f_2(rw)|^p\ ]r^{N-1}dr \right) dw =$$

$$\int_{(S^{N-1}-(K(f_1)\cup K(f_2)))} \left(\int_{R_1}^{R_2} [\ |D_{R_1}^{\beta} f_1(rw)|^p + |D_{R_1}^{\beta} f_2(rw)|^p\ ]r^{N-1}dr \right) dw. \tag{15.161}$$

Notice here $\lambda_{S^{N-1}}(K(f_1)\cup K(f_2)) = 0$. Here for every $w \in S^{N-1}-(K(f_1)\cup K(f_2))$ and for $p(r) = q(r) = r^{N-1}$, $r \in [R_1, R_2]$, $N \geq 2$ we apply Theorem 15.44. We obtain

$$\int_{R_1}^{R_2} [\ |D_{R_1}^{\alpha_1} f_1(rw)|^{\lambda_{\alpha_1}} \ |D_{R_1}^{\beta} f_1(rw)|^{\lambda_\beta} + |D_{R_1}^{\alpha_1} f_2(rw)|^{\lambda_{\alpha_1}} \ |D_{R_1}^{\beta} f_2(rw)|^{\lambda_\beta}]r^{N-1}dr \leq$$

$$\left(A_0(R_2 - R_1)\Big|_{\lambda_{\alpha_2}=0} \right) \left(\frac{\lambda_\beta}{\lambda_{\alpha_1} + \lambda_\beta} \right)^{(\lambda_\beta/\lambda_{\alpha_1}+\lambda_\beta)} \times$$

$$\left(\int_{R_1}^{R_2} [\ |D_{R_1}^{\beta} f_1(rw)|^p + |D_{R_1}^{\beta} f_2(rw)|^p\ ]r^{N-1} \right) dr. \tag{15.162}$$

Integrating now (15.162) over $S^{N-1} - (K(f_1) \cup K(f_2))$ and taking into account (15.160) and (15.161) we derive (15.159). $\square$

We continue with

Theorem 15.68. (All are as in Assumption 15.65 and Notation 15.66.) Here $\lambda_{\alpha_1} = 0$, $\lambda_{\alpha_2} > 0$, and $p = \lambda_\beta + \lambda_{\alpha_2} > 1$. Denote

$$\delta_3 := \begin{cases} 2^{\lambda_{\alpha_2}/\lambda_\beta} - 1, & \text{if } \lambda_{\alpha_2} \geq \lambda_\beta, \\ 1, & \text{if } \lambda_{\alpha_2} \leq \lambda_\beta. \end{cases} \tag{15.163}$$

Then

$$\int_A \left[\left| \frac{\partial_{R_1}^{\alpha_2} f_2(x)}{\partial r^{\alpha_2}} \right|^{\lambda_{\alpha_2}} \left| \frac{\partial_{R_1}^{\beta} f_1(x)}{\partial r^{\beta}} \right|^{\lambda_\beta} + \left| \frac{\partial_{R_1}^{\alpha_2} f_1(x)}{\partial r^{\alpha_2}} \right|^{\lambda_{\alpha_2}} \left| \frac{\partial_{R_1}^{\beta} f_2(x)}{\partial r^{\beta}} \right|^{\lambda_\beta} \right] dx \le$$

$$\left(A_0(R_2 - R_1)\Big|_{\lambda_{\alpha_1}=0} \right) 2^{\lambda_{\alpha_2}/p} \left(\frac{\lambda_\beta}{p} \right)^{\lambda_\beta/p} \delta_3^{\lambda_\beta/p}$$

$$\left(\int_A \left[\left| \frac{\partial_{R_1}^{\beta} f_1(x)}{\partial r^{\beta}} \right|^p + \left| \frac{\partial_{R_1}^{\beta} f_2(x)}{\partial r^{\beta}} \right|^p \right] dx \right). \tag{15.164}$$

Proof. Based on Theorem 15.45 and similar to the proof of Theorem 15.67. □

The complete case λ_{α_1}, $\lambda_{\alpha_2} > 0$ follows.

Theorem 15.69. (All are as in Assumption 15.65 and Notation 15.66.) *Here* λ_{α_1}, $\lambda_{\alpha_2} > 0$, $p = \lambda_{\alpha_1} + \lambda_{\alpha_2} + \lambda_\beta > 1$. *Denote*

$$\tilde{\gamma}_1 := \begin{cases} 2^{((\lambda_{\alpha_1}+\lambda_{\alpha_2})/\lambda_\beta)} - 1, & if \ \lambda_{\alpha_1} + \lambda_{\alpha_2} \ge \lambda_\beta, \\ 1, & if \ \lambda_{\alpha_1} + \lambda_{\alpha_2} \le \lambda_\beta. \end{cases} \tag{15.165}$$

Then

$$\int_A \left[\left| \frac{\partial_{R_1}^{\alpha_1} f_1(x)}{\partial r^{\alpha_1}} \right|^{\lambda_{\alpha_1}} \left| \frac{\partial_{R_1}^{\alpha_2} f_2(x)}{\partial r^{\alpha_2}} \right|^{\lambda_{\alpha_2}} \left| \frac{\partial_{R_1}^{\beta} f_1(x)}{\partial r^{\beta}} \right|^{\lambda_\beta} + \right.$$

$$\left. \left| \frac{\partial_{R_1}^{\alpha_2} f_1(x)}{\partial r^{\alpha_2}} \right|^{\lambda_{\alpha_2}} \left| \frac{\partial_{R_1}^{\alpha_1} f_2(x)}{\partial r^{\alpha_1}} \right|^{\lambda_{\alpha_1}} \left| \frac{\partial_{R_1}^{\beta} f_2(x)}{\partial r^{\beta}} \right|^{\lambda_\beta} \right] dx \le$$

$$A_0(R_2 - R_1) \left(\frac{\lambda_\beta}{(\lambda_{\alpha_1} + \lambda_{\alpha_2})p} \right)^{(\lambda_\beta/p)} [\lambda_{\alpha_1}^{(\lambda_\beta/p)} + 2^{(p-\lambda_\beta)/p} (\tilde{\gamma}_1\lambda_{\alpha_2})^{(\lambda_\beta/p)}].$$

$$\left(\int_A \left[\left| \frac{\partial_{R_1}^{\beta} f_1(x)}{\partial r^{\beta}} \right|^p + \left| \frac{\partial_{R_1}^{\beta} f_2(x)}{\partial r^{\beta}} \right|^p \right] dx \right). \tag{15.166}$$

Proof. Based on Theorem 15.46 and similar to the proof of Theorem 15.67. □

A special important case is next.

Theorem 15.70. (All are as in Assumption 15.65.) *Here* $\alpha_2 = \alpha_1 + 1$, $\lambda_\alpha := \lambda_{\alpha_1} \ge 0$, $\lambda_{\alpha+1} := \lambda_{\alpha_2} \in (0,1)$, *and* $p = \lambda_\alpha + \lambda_{\alpha+1} > 1$. *Denote*

$$\theta_3 := \begin{cases} 2^{(\lambda_\alpha/\lambda_{\alpha+1})} - 1, & if \ \lambda_\alpha \ge \lambda_{\alpha+1}, \\ 1, & if \ \lambda_\alpha \le \lambda_{\alpha+1}, \end{cases} \tag{15.167}$$

$$L(R_2 - R_1) :=$$

$$\left[2\, \frac{(1-\lambda_{\alpha+1})}{(N-\lambda_{\alpha+1})} \left(R_2^{\,N-\lambda_{\alpha+1}/1-\lambda_{\alpha+1}} - R_1^{\,N-\lambda_{\alpha+1}/1-\lambda_{\alpha+1}} \right) \right]^{(1-\lambda_{\alpha+1})}$$

$$\left(\frac{\theta_3\, \lambda_{\alpha+1}}{p} \right)^{\lambda_{\alpha+1}}, \tag{15.168}$$

and

$$P_1(R_2 - R_1) := \int_{R_1}^{R_2} (R_2 - r)^{(\beta-\alpha_1-1)p/(p-1)}\, r^{(1-N)/(p-1)} dr, \tag{15.169}$$

$$\Phi(R_2 - R_1) := L(R_2 - R_1) \left(\frac{(P_1(R_2 - R_1))^{(p-1)}}{(\Gamma(\beta-\alpha_1))^p} \right) 2^{(p-1)}. \tag{15.170}$$

Then

$$\int_A \left[\left| \frac{\partial_{R_1}^{\alpha_1} f_1(x)}{\partial r^{\alpha_1}} \right|^{\lambda_\alpha} \left| \frac{\partial_{R_1}^{\alpha_1+1} f_2(x)}{\partial r^{\alpha_1+1}} \right|^{\lambda_{\alpha+1}} + \left| \frac{\partial_{R_1}^{\alpha_1} f_2(x)}{\partial r^{\alpha_1}} \right|^{\lambda_\alpha} \left| \frac{\partial_{R_1}^{\alpha_1+1} f_1(x)}{\partial r^{\alpha_1+1}} \right|^{\lambda_{\alpha+1}} \right] dx$$

$$\leq \Phi(R_2 - R_1) \left(\int_A \left[\left| \frac{\partial_{R_1}^{\beta} f_1(x)}{\partial r^{\beta}} \right|^{p} + \left| \frac{\partial_{R_1}^{\beta} f_2(x)}{\partial r^{\beta}} \right|^{p} \right] dx \right). \tag{15.171}$$

Proof. Based on Theorem 15.47 and similar to the proof of Theorem 15.67. □

We also give

Theorem 15.71. (All are as in Assumption 15.65 and Notation 15.66.) Here $\lambda_{\alpha_2} = \lambda_{\alpha_1} + \lambda_\beta$ and $p = 2(\lambda_{\alpha_1} + \lambda_\beta) > 1$. Denote

$$\tilde{T}(R_2 - R_1) := A_0(R_2 - R_1) \left(\frac{\lambda_\beta}{\lambda_{\alpha_1} + \lambda_\beta} \right)^{\lambda_\beta/p} 2^{-\lambda_\beta/p}. \tag{15.172}$$

Then

$$\int_A \left[\left| \frac{\partial_{R_1}^{\alpha_1} f_1(x)}{\partial r^{\alpha_1}} \right|^{\lambda_{\alpha_1}} \left| \frac{\partial_{R_1}^{\alpha_2} f_2(x)}{\partial r^{\alpha_2}} \right|^{\lambda_{\alpha_1}+\lambda_\beta} \left| \frac{\partial_{R_1}^{\beta} f_1(x)}{\partial r^{\beta}} \right|^{\lambda_\beta} + \right.$$

$$\left. \left| \frac{\partial_{R_1}^{\alpha_2} f_1(x)}{\partial r^{\alpha_2}} \right|^{\lambda_{\alpha_1}+\lambda_\beta} \left| \frac{\partial_{R_1}^{\alpha_1} f_2(x)}{\partial r^{\alpha_1}} \right|^{\lambda_{\alpha_1}} \left| \frac{\partial_{R_1}^{\beta} f_2(x)}{\partial r^{\beta}} \right|^{\lambda_\beta} \right] dx \leq$$

$$\tilde{T}(R_2 - R_1) \left(\int_A \left[\left| \frac{\partial_{R_1}^{\beta} f_1(x)}{\partial r^{\beta}} \right|^{p} + \left| \frac{\partial_{R_1}^{\beta} f_2(x)}{\partial r^{\beta}} \right|^{p} \right] dx \right). \tag{15.173}$$

Proof. Based on Theorem 15.48. □

We continue with related L_∞ results on the shell A. We make

Assumption 15.72. Let $\alpha_1, \alpha_2 \in \mathbb{R}_+$, $\beta > \alpha_1, \alpha_2$ and let $f_1, f_2 \in L_1(A)$ with

$$\frac{\partial_{R_1}^\beta f_1(x)}{\partial r^\beta}, \ \frac{\partial_{R_1}^\beta f_2(x)}{\partial r^\beta} \in L_\infty(A), \ x \in A;$$

$$A := B(0, R_2) - \overline{B(0, R_1)} \subseteq \mathbb{R}^N, \ N \geq 2, \ 0 < R_1 < R_2.$$

Further assume that each $D_{R_1}^\beta f_i(rw) \in \mathbb{R}$ for almost all $r \in [R_1, R_2]$, for each $w \in S^{N-1}$, and for these $|D_{R_1}^\beta f_i(rw)| \leq M_i$ for some $M_i > 0$; $i = 1, 2$. For each $w \in S^{N-1} - (K(f_1) \cup K(f_2))$, we assume that $f_i(\cdot \ w)$ has an L_∞ fractional derivative $D_{R_1}^\beta f_i(\cdot \ w)$ in $[R_1, R_2]$, and that

$$D_{R_1}^{\beta-k} f_i(R_1 w) = 0, \ k = 1, \ldots, [\beta] + 1;$$

$i = 1, 2$. Let $\lambda_{\alpha_1}, \lambda_{\alpha_2}, \lambda_\beta \geq 0$. If $\alpha_1 = 0$ we set $\lambda_{\alpha_1} = 1$, and if $\alpha_2 = 0$ we set $\lambda_{\alpha_2} = 1$.

We present

Theorem 15.73. (All here are as in Assumption 15.72.) *Set*

$$\rho(R_2 - R_1) =$$

$$\frac{R_2^{N-1}(R_2 - R_1)^{(\beta\lambda_{\alpha_1} - \alpha_1\lambda_{\alpha_1} + \beta\lambda_{\alpha_2} - \alpha_2\lambda_{\alpha_2} + 1)}}{(\Gamma(\beta - \alpha_1 + 1))^{\lambda_{\alpha_1}} (\Gamma(\beta - \alpha_2 + 1))^{\lambda_{\alpha_2}} (\beta\lambda_{\alpha_1} - \alpha_1\lambda_{\alpha_1} + \beta\lambda_{\alpha_2} - \alpha_2\lambda_{\alpha_2} + 1)}.$$

$$(15.174)$$

Then

$$\int_A \Big[\Big| \frac{\partial_{R_1}^{\alpha_1} f_1(x)}{\partial r^{\alpha_1}} \Big|^{\lambda_{\alpha_1}} \Big| \frac{\partial_{R_1}^{\alpha_2} f_2(x)}{\partial r^{\alpha_2}} \Big|^{\lambda_{\alpha_2}} \Big| \frac{\partial_{R_1}^\beta f_1(x)}{\partial r^\beta} \Big|^{\lambda_\beta} +$$

$$\Big| \frac{\partial_{R_1}^{\alpha_2} f_1(x)}{\partial r^{\alpha_2}} \Big|^{\lambda_{\alpha_2}} \Big| \frac{\partial_{R_1}^{\alpha_1} f_2(x)}{\partial r^{\alpha_1}} \Big|^{\lambda_{\alpha_1}} \Big| \frac{\partial_{R_1}^\beta f_2(x)}{\partial r^\beta} \Big|^{\lambda_\beta} \Big] dx \leq$$

$$\rho(R_2 - R_1)[M_1^{2(\lambda_{\alpha_1} + \lambda_\beta)} + M_1^{2\lambda_{\alpha_2}} + M_2^{2\lambda_{\alpha_2}} + M_2^{2(\lambda_{\alpha_1} + \lambda_\beta)}] \frac{\pi^{N/2}}{\Gamma(N/2)}.$$

$$(15.175)$$

Proof. It is based on Theorem 15.57. By Theorem 15.20 for $\alpha_j > 0$ we get that

$$\frac{\partial_{R_1}^{\alpha_j} f_i}{\partial r^{\alpha_j}} \in L_\infty(A), \ i = 1, 2; \ j = 1, 2.$$

In general here the integrand of the integral of (15.175) belongs to $L_1(A)$. Thus by Proposition 15.3 we have

$$L.H.S(15.175) = \int_{S^{N-1}} \Big(\int_{R_1}^{R_2} [|D_{R_1}^{\alpha_1} f_1(rw)|^{\lambda_{\alpha_1}} |D_{R_1}^{\alpha_2} f_2(rw)|^{\lambda_{\alpha_2}} |D_{R_1}^\beta f_1(rw)|^{\lambda_\beta} +$$

$$|D_{R_1}^{\alpha_2} f_1(rw)|^{\lambda_{\alpha_2}} |D_{R_1}^{\alpha_1} f_2(rw)|^{\lambda_{\alpha_1}} |D_{R_1}^\beta f_2(rw)|^{\lambda_\beta}] r^{N-1} dr \Big) dw \quad (15.176)$$

$$= \int_{(S^{N-1}-(K(f_1)\cup K(f_2)))} \left(\int_{R_1}^{R_2} [\,|D_{R_1}^{\alpha_1} f_1(rw)|^{\lambda_{\alpha_1}}\,|D_{R_1}^{\alpha_2} f_2(rw)|^{\lambda_{\alpha_2}}\,|D_{R_1}^{\beta} f_1(rw)|^{\lambda_{\beta}}+ \right.$$

$$\left. |D_{R_1}^{\alpha_2} f_1(rw)|^{\lambda_{\alpha_2}}\,|D_{R_1}^{\alpha_1} f_2(rw)|^{\lambda_{\alpha_1}}\,|D_{R_1}^{\beta} f_2(rw)|^{\lambda_{\beta}}]r^{N-1}dr \right) dw \qquad (15.177)$$

(by (15.147) for $p(r) = r^{N-1}$, $r \in [R_1, R_2]$)

$$\leq \frac{\rho(R_2 - R_1)}{2} \int_{(S^{N-1}-(K(f_1)\cup K(f_2)))} \left[\|D_{R_1}^{\beta} f_1(\cdot\ w)\|_{\infty,\ [R_1,R_2]}^{2(\lambda_{\alpha_1}+\lambda_{\beta})}+ \right.$$

$$\left. \|D_{R_1}^{\beta} f_1(\cdot\ w)\|_{\infty,\ [R_1,R_2]}^{2\lambda_{\alpha_2}} + \|D_{R_1}^{\beta} f_2(\cdot\ w)\|_{\infty,\ [R_1,R_2]}^{2\lambda_{\alpha_2}} + \|D_{R_1}^{\beta} f_2(\cdot\ w)\|_{\infty,\ [R_1,R_2]}^{2(\lambda_{\alpha_1}+\lambda_{\beta})} \right] dw$$
$$(15.178)$$

$$\leq \frac{\rho(R_2 - R_1)}{2} \left[M_1^{2(\lambda_{\alpha_1}+\lambda_{\beta})} + M_1^{2\lambda_{\alpha_2}} + M_2^{2\lambda_{\alpha_2}} + M_2^{2(\lambda_{\alpha_1}+\lambda_{\beta})} \right] \cdot$$
$$(15.179)$$

$$\int_{(S^{N-1}-(K(f_1)\cup K(f_2)))} dw =$$

$$\frac{\rho(R_2 - R_1)}{2} \left[M_1^{2(\lambda_{\alpha_1}+\lambda_{\beta})} + M_1^{2\lambda_{\alpha_2}} + M_2^{2\lambda_{\alpha_2}} + M_2^{2(\lambda_{\alpha_1}+\lambda_{\beta})} \right] \frac{2\pi^{N/2}}{\Gamma(N/2)},$$
$$(15.180)$$

by

$$\int_{(S^{N-1}-(K(f_1)\cup K(f_2)))} dw = \int_{S^{N-1}} dw = \frac{2\pi^{N/2}}{\Gamma(N/2)}, \qquad (15.181)$$

because $\lambda_{S^{N-1}}(K(f_1) \cup K(f_2)) = 0$, thus proving inequality (15.175). $\quad\square$

We give special cases of the last theorem.

Theorem 15.74. (All here are as in Assumption 15.72; here $\lambda_{\alpha_2} = 0$.)
It holds

$$\int_A \left[\left| \frac{\partial_{R_1}^{\alpha_1} f_1(x)}{\partial r^{\alpha_1}} \right|^{\lambda_{\alpha_1}} \left| \frac{\partial_{R_1}^{\beta} f_1(x)}{\partial r^{\beta}} \right|^{\lambda_{\beta}} + \left| \frac{\partial_{R_1}^{\alpha_1} f_2(x)}{\partial r^{\alpha_1}} \right|^{\lambda_{\alpha_1}} \left| \frac{\partial_{R_1}^{\beta} f_2(x)}{\partial r^{\beta}} \right|^{\lambda_{\beta}} \right] dx \leq$$

$$\frac{R_2^{N-1}(R_2 - R_1)^{(\beta\lambda_{\alpha_1}-\alpha_1\lambda_{\alpha_1}+1)}}{(\Gamma(\beta - \alpha_1 + 1))^{\lambda_{\alpha_1}}\ [\beta\lambda_{\alpha_1} - \alpha_1\lambda_{\alpha_1} + 1]} \times$$

$$\left[M_1^{(\lambda_{\alpha_1}+\lambda_{\beta})} + M_2^{(\lambda_{\alpha_1}+\lambda_{\beta})} \right] \frac{2\pi^{N/2}}{\Gamma(N/2)}. \qquad (15.182)$$

Proof. Based on Theorem 15.58 and similar to the proof of Theorem 15.73. $\quad\square$

We continue with

Theorem 15.75. (All here are as in Assumption 15.72; here $\lambda_{\alpha_2} = \lambda_{\alpha_1} + \lambda_\beta$.) *It holds*

$$
\int_A \left[\left| \frac{\partial_{R_1}^{\alpha_1} f_1(x)}{\partial r^{\alpha_1}} \right|^{\lambda_{\alpha_1}} \left| \frac{\partial_{R_1}^{\alpha_2} f_2(x)}{\partial r^{\alpha_2}} \right|^{\lambda_{\alpha_1}+\lambda_\beta} \left| \frac{\partial_{R_1}^{\beta} f_1(x)}{\partial r^\beta} \right|^{\lambda_\beta} + \right.
$$

$$
\left. \left| \frac{\partial_{R_1}^{\alpha_2} f_1(x)}{\partial r^{\alpha_2}} \right|^{\lambda_{\alpha_1}+\lambda_\beta} \left| \frac{\partial_{R_1}^{\alpha_1} f_2(x)}{\partial r^{\alpha_1}} \right|^{\lambda_{\alpha_1}} \left| \frac{\partial_{R_1}^{\beta} f_2(x)}{\partial r^\beta} \right|^{\lambda_\beta} \right] dx \le
$$

$$
\frac{R_2^{N-1}}{(\Gamma(\beta - \alpha_1 + 1))^{\lambda_{\alpha_1}} (\Gamma(\beta - \alpha_2 + 1))^{\lambda_{\alpha_1}+\lambda_\beta}}
$$

$$
\frac{(R_2 - R_1)^{(2\beta\lambda_{\alpha_1} - \alpha_1\lambda_{\alpha_1} + \beta\lambda_\beta - \alpha_2\lambda_{\alpha_1} - \alpha_2\lambda_\beta + 1)}}{[2\beta\lambda_{\alpha_1} - \alpha_1\lambda_{\alpha_1} + \beta\lambda_\beta - \alpha_2\lambda_{\alpha_1} - \alpha_2\lambda_\beta + 1]}
$$

$$
\times \left[M_1^{2(\lambda_{\alpha_1}+\lambda_\beta)} + M_2^{2(\lambda_{\alpha_1}+\lambda_\beta)} \right] \frac{2\pi^{N/2}}{\Gamma(N/2)}. \tag{15.183}
$$

Proof. Based on Theorem 15.59 and similar to the proof of Theorem 15.73. □

We give

Theorem 15.76. (All here are as in Assumption 15.72; here $\lambda_\beta = 0$, $\lambda_{\alpha_1} = \lambda_{\alpha_2}$.) *It holds*

$$
\int_A \left[\left| \frac{\partial_{R_1}^{\alpha_1} f_1(x)}{\partial r^{\alpha_1}} \right|^{\lambda_{\alpha_1}} \left| \frac{\partial_{R_1}^{\alpha_2} f_2(x)}{\partial r^{\alpha_2}} \right|^{\lambda_{\alpha_1}} + \left| \frac{\partial_{R_1}^{\alpha_2} f_1(x)}{\partial r^{\alpha_2}} \right|^{\lambda_{\alpha_1}} \left| \frac{\partial_{R_1}^{\alpha_1} f_2(x)}{\partial r^{\alpha_1}} \right|^{\lambda_{\alpha_1}} \right] dx \le
$$

$$
\frac{R_2^{N-1}(R_2 - R_1)^{(2\beta\lambda_{\alpha_1} - \alpha_1\lambda_{\alpha_1} - \alpha_2\lambda_{\alpha_1} + 1)}}{(\Gamma(\beta - \alpha_1 + 1)\,\Gamma(\beta - \alpha_2 + 1))^{\lambda_{\alpha_1}} (2\beta\lambda_{\alpha_1} - \alpha_1\lambda_{\alpha_1} - \alpha_2\lambda_{\alpha_1} + 1)} \times
$$

$$
\left[M_1^{2\lambda_{\alpha_1}} + M_2^{2\lambda_{\alpha_1}} \right] \frac{2\pi^{N/2}}{\Gamma(N/2)}. \tag{15.184}
$$

Proof. Based on Theorem 15.60 and similar to the proof of Theorem 15.73. □

We give

Theorem 15.77. (All here are as in Assumption 15.72; here $\lambda_{\alpha_1} = 0$, $\lambda_{\alpha_2} = \lambda_\beta$.) *It holds*

$$
\int_A \left[\left| \frac{\partial_{R_1}^{\alpha_2} f_2(x)}{\partial r^{\alpha_2}} \right|^{\lambda_{\alpha_2}} \left| \frac{\partial_{R_1}^{\beta} f_1(x)}{\partial r^\beta} \right|^{\lambda_{\alpha_2}} + \left| \frac{\partial_{R_1}^{\alpha_2} f_1(x)}{\partial r^{\alpha_2}} \right|^{\lambda_{\alpha_2}} \left| \frac{\partial_{R_1}^{\beta} f_2(x)}{\partial r^\beta} \right|^{\lambda_{\alpha_2}} \right] dx \le
$$

$$\frac{R_2^{N-1}(R_2 - R_1)^{(\beta\lambda_{\alpha_2} - \alpha_2\lambda_{\alpha_2} + 1)}}{(\beta\lambda_{\alpha_2} - \alpha_2\lambda_{\alpha_2} + 1)(\Gamma(\beta - \alpha_2 + 1))^{\lambda_{\alpha_2}}} \left(M_1^{2\lambda_{\alpha_2}} + M_2^{2\lambda_{\alpha_2}} \right) \frac{2\pi^{N/2}}{\Gamma(N/2)}.$$
$$(15.185)$$

Proof. Based on Theorem 15.61, and so on. $\square$

We finish this section with

Corollary 15.78. (To Theorem 15.76. All here are as in Assumption 15.72, $\lambda_\beta = 0$, $\lambda_{\alpha_1} = \lambda_{\alpha_2}$, $\alpha_2 = \alpha_1 + 1$.) *It holds*

$$\int_A \left[\left| \frac{\partial_{R_1}^{\alpha_1} f_1(x)}{\partial r^{\alpha_1}} \right|^{\lambda_{\alpha_1}} \left| \frac{\partial_{R_1}^{\alpha_1+1} f_2(x)}{\partial r^{\alpha_1+1}} \right|^{\lambda_{\alpha_1}} + \left| \frac{\partial_{R_1}^{\alpha_1+1} f_1(x)}{\partial r^{\alpha_1+1}} \right|^{\lambda_{\alpha_1}} \left| \frac{\partial_{R_1}^{\alpha_1} f_2(x)}{\partial r^{\alpha_1}} \right|^{\lambda_{\alpha_1}} \right] dx \le$$

$$\frac{R_2^{N-1}(R_2 - R_1)^{(2\beta\lambda_{\alpha_1} - 2\alpha_1\lambda_{\alpha_1} - \lambda_{\alpha_1} + 1)}}{(2\beta\lambda_{\alpha_1} - 2\alpha_1\lambda_{\alpha_1} - \lambda_{\alpha_1} + 1)(\beta - \alpha_1)^{\lambda_{\alpha_1}} (\Gamma(\beta - \alpha_1))^{2\lambda_{\alpha_1}}} \times$$

$$\left[M_1^{2\lambda_{\alpha_1}} + M_2^{2\lambda_{\alpha_1}} \right] \frac{2\pi^{N/2}}{\Gamma(N/2)}.$$
$$(15.186)$$

Proof. Based on Corollary 15.62, and so on. $\square$

15.5.3 Riemann–Liouville Fractional Opial-Type Inequalities Involving Several Functions

We present the following theorem.

Theorem 15.79. *Let*

$$\alpha_1, \alpha_2 \in \mathbb{R}_+, \ \beta > \alpha_1, \alpha_2, \ \beta - \alpha_i > (1/p), p > 1, i = 1, 2,$$

and let

$$f_j \in L_1(a, x), \ j = 1, \ldots, M \in \mathbb{N}, \ a, \ x \in \mathbb{R}, \ a < x,$$

have, respectively, L_∞ fractional derivatives $D_a^\beta f_j$ in $[a, x]$, and let

$$D_a^{\beta-k} f_j(a) = 0, \ for \ k = 1, \ldots, [\beta] + 1; \ j = 1, \ldots, M.$$

Consider also $p(t) > 0$ and $q(t) \ge 0$, with all $p(t), 1/p(t), q(t) \in L_\infty(a, x)$. Let $\lambda_\beta > 0$ and $\lambda_{\alpha_1}, \lambda_{\alpha_2} \ge 0$, such that $\lambda_\beta < p$. Set

$$P_i(s) := \int_0^s (s-t)^{p(\beta-\alpha_i-1)/(p-1)} (p(t+a))^{-1/(p-1)} \, dt, \ i = 1, 2; \ 0 \le s \le x-a,$$
$$(15.187)$$

$$A(s) := \frac{q(s+a)\,(P_1(s))^{\lambda_{\alpha_1}((p-1)/p)}\,(P_2(s))^{\lambda_{\alpha_2}((p-1)/p)}\,(p(s+a))^{-\lambda_\beta/p}}{(\Gamma(\beta-\alpha_1))^{\lambda_{\alpha_1}}\,(\Gamma(\beta-\alpha_2))^{\lambda_{\alpha_2}}},$$

(15.188)

$$A_0(x-a) := \left(\int_0^{x-a} (A(s))^{p/(p-\lambda_\beta)}\,ds \right)^{(p-\lambda_\beta)/p},$$

(15.189)

and

$$\delta_1^* := \begin{cases} M^{1-((\lambda_{\alpha_1}+\lambda_\beta)/p)}, & \text{if } \lambda_{\alpha_1}+\lambda_\beta \le p, \\ 2^{(\lambda_{\alpha_1}+\lambda_\beta/p)-1}, & \text{if } \lambda_{\alpha_1}+\lambda_\beta \ge p. \end{cases}$$

(15.190)

Call

$$\varphi_1(x-a) := \left(A_0(x-a)\Big|_{\lambda_{\alpha_2}=0} \right) \left(\frac{\lambda_\beta}{\lambda_{\alpha_1}+\lambda_\beta} \right)^{\lambda_\beta/p}.$$

(15.191)

If $\lambda_{\alpha_2}=0$, we obtain that

$$\int_a^x q(s) \left(\sum_{j=1}^M |D_a^{\alpha_1} f_j(s)|^{\lambda_{\alpha_1}}\,|D_a^\beta f_j(s)|^{\lambda_\beta} \right) ds \le$$

$$\delta_1^* \varphi_1(x-a) \left[\int_a^x p(s) \left(\sum_{j=1}^M |D_a^\beta f_j(s)|^p \right) ds \right]^{((\lambda_{\alpha_1}+\lambda_\beta)/p)}.$$

(15.192)

Proof. Similar to Theorem 15.44 and transfer of Theorem 4 of [46]. □

Next we give

Theorem 15.80. (All here are as in Theorem 15.79). *Denote*

$$\delta_3 := \begin{cases} 2^{\lambda_{\alpha_2}/\lambda_\beta} - 1, & \text{if } \lambda_{\alpha_2} \ge \lambda_\beta, \\ 1, & \text{if } \lambda_{\alpha_2} \le \lambda_\beta. \end{cases}$$

(15.193)

$$\epsilon_2 := \begin{cases} 1, & \text{if } \lambda_\beta + \lambda_{\alpha_2} \ge p, \\ M^{1-((\lambda_\beta+\lambda_{\alpha_2})/p)}, & \text{if } \lambda_\beta + \lambda_{\alpha_2} \le p, \end{cases}$$

(15.194)

and

$$\varphi_2(x-a) := \left(A_0(x-a)\Big|_{\lambda_{\alpha_1}=0} \right) 2^{((p-\lambda_\beta)/p)} \left(\frac{\lambda_\beta}{\lambda_{\alpha_2}+\lambda_\beta} \right)^{\lambda_\beta/p} \delta_3^{(\lambda_\beta/p)}.$$

(15.195)

If $\lambda_{\alpha_1}=0$, then

$$\int_a^x q(s) \Big\{ \{ \sum_{j=1}^M [\,|D_a^{\alpha_2} f_{j+1}(s)|^{\lambda_{\alpha_2}}\,|D_a^\beta f_j(s)|^{\lambda_\beta} + |D_a^{\alpha_2} f_j(s)|^{\lambda_{\alpha_2}}\,|D_a^\beta f_{j+1}(s)|^{\lambda_\beta}\,] \} +$$

$$[\,|D_a^{\alpha_2} f_M(s)|^{\lambda_{\alpha_2}}\,|D_a^{\beta} f_1(s)|^{\lambda_\beta} + |D_a^{\alpha_2} f_1(s)|^{\lambda_{\alpha_2}}\,|D_a^{\beta} f_M(s)|^{\lambda_\beta}\,]\}ds \leq$$

$$2^{((\lambda_\beta+\lambda_{\alpha_2})/p)}\epsilon_2\varphi_2(x-a)\Big\{\int_a^x p(s)\Big(\sum_{j=1}^M |D_a^{\beta} f_j(s)|^p\Big)ds\Big\}^{((\lambda_\beta+\lambda_{\alpha_2})/p)}.$$

$$(15.196)$$

Proof. Usual transfer of Theorem 5 of [46]. □

The general case follows.

Theorem 15.81. (All here are as in Theorem 15.79). *Denote*

$$\tilde{\gamma}_1 := \begin{cases} 2^{((\lambda_{\alpha_1}+\lambda_{\alpha_2})/\lambda_\beta)} - 1, & if \ \lambda_{\alpha_1} + \lambda_{\alpha_2} \geq \lambda_\beta, \\ 1, & if \ \lambda_{\alpha_1} + \lambda_{\alpha_2} \leq \lambda_\beta, \end{cases} \qquad (15.197)$$

and

$$\tilde{\gamma}_2 := \begin{cases} 1, & if \ \lambda_{\alpha_1} + \lambda_{\alpha_2} + \lambda_\beta \geq p, \\ 2^{1-((\lambda_{\alpha_1}+\lambda_{\alpha_2}+\lambda_\beta)/p)}, & if \ \lambda_{\alpha_1} + \lambda_{\alpha_2} + \lambda_\beta \leq p. \end{cases} \qquad (15.198)$$

Set

$$\varphi_3(x-a) := A_0(x-a)\left(\frac{\lambda_\beta}{(\lambda_{\alpha_1}+\lambda_{\alpha_2})(\lambda_{\alpha_1}+\lambda_{\alpha_2}+\lambda_\beta)}\right)^{\lambda_\beta/p}$$

$$\cdot [\lambda_{\alpha_1}^{\lambda_\beta/p}\tilde{\gamma}_2 + 2^{(p-\lambda_\beta)/p}\,(\tilde{\gamma}_1\lambda_{\alpha_2})^{\lambda_\beta/p}], \qquad (15.199)$$

and

$$\epsilon_3 := \begin{cases} 1, & if \ \lambda_{\alpha_1} + \lambda_{\alpha_2} + \lambda_\beta \geq p, \\ M^{1-((\lambda_{\alpha_1}+\lambda_{\alpha_2}+\lambda_\beta)/p)}, & if \ \lambda_{\alpha_1} + \lambda_{\alpha_2} + \lambda_\beta \leq p. \end{cases} \qquad (15.200)$$

Then

$$\int_a^x q(s)\Big[\sum_{j=1}^{M-1}[\,|D_a^{\alpha_1} f_j(s)|^{\lambda_{\alpha_1}}\,|D_a^{\alpha_2} f_{j+1}(s)|^{\lambda_{\alpha_2}}\,|D_a^{\beta} f_j(s)|^{\lambda_\beta} +$$

$$|D_a^{\alpha_2} f_j(s)|^{\lambda_{\alpha_2}}\,|D_a^{\alpha_1} f_{j+1}(s)|^{\lambda_{\alpha_1}}\,|D_a^{\beta} f_{j+1}(s)|^{\lambda_\beta}\,] +$$

$$[\,|D_a^{\alpha_1} f_1(s)|^{\lambda_{\alpha_1}}\,|D_a^{\alpha_2} f_M(s)|^{\lambda_{\alpha_2}}\,|D_a^{\beta} f_1(s)|^{\lambda_\beta} +$$

$$|D_a^{\alpha_2} f_1(s)|^{\lambda_{\alpha_2}}\,|D_a^{\alpha_1} f_M(s)|^{\lambda_{\alpha_1}}\,|D_a^{\beta} f_M(s)|^{\lambda_\beta}\,]\Big]ds \leq \qquad (15.201)$$

$$2^{((\lambda_{\alpha_1}+\lambda_{\alpha_2}+\lambda_\beta)/p)}\epsilon_3\varphi_3(x-a)\Big\{\int_a^x p(s)\Big(\sum_{j=1}^M |D_a^{\beta} f_j(s)|^p\Big)ds\Big\}^{((\lambda_{\alpha_1}+\lambda_{\alpha_2}+\lambda_\beta)/p)}.$$

Proof. Usual transfer of Theorem 6 of [46]. □

We continue.

Theorem 15.82. *Let* $\beta > \alpha_1 + 1$, $\alpha_1 \in \mathbb{R}_+$ *and let*

$$f_j \in L_1(a, x), \ j = 1, \ldots, M \in \mathbb{N}, \ a, \ x \in \mathbb{R}, \ a < x,$$

have, respectively, L_∞ *fractional derivatives* $D_a^\beta f_j$ *in* $[a, x]$, *and let*

$$D_a^{\beta-k} f_j(a) = 0, \ for \ k = 1, \ldots, [\beta] + 1; \ j = 1, \ldots, M.$$

Consider also $p(t) > 0$ *and* $q(t) \geq 0$, *with all* $p(t), 1/p(t), q(t) \in L_\infty(a, x)$. *Let* $\lambda_\alpha \geq 0$, $0 < \lambda_{\alpha+1} < 1$, *and* $p > 1$. *Denote*

$$\theta_3 := \begin{cases} 2^{\lambda_\alpha/(\lambda_{\alpha+1})} - 1, & if \ \lambda_\alpha \geq \lambda_{\alpha+1}, \\ 1, & if \ \lambda_\alpha \leq \lambda_{\alpha+1}, \end{cases}$$

$$L(x - a) := \left(2 \int_a^x (q(s))^{(1/(1-\lambda_{\alpha+1}))} ds \right)^{(1-\lambda_{\alpha+1})} \left(\frac{\theta_3 \, \lambda_{\alpha+1}}{\lambda_\alpha + \lambda_{\alpha+1}} \right)^{\lambda_{\alpha+1}},$$
(15.202)

and

$$P_1(x - a) := \int_a^x (x - s)^{(\beta - \alpha_1 - 1)p/(p-1)} (p(s))^{-1/(p-1)} ds,$$
(15.203)

$$T(x - a) := L(x - a) \left(\frac{(P_1(x - a))^{(p-1/p)}}{\Gamma(\beta - \alpha_1)} \right)^{(\lambda_\alpha + \lambda_{\alpha+1})},$$
(15.204)

and

$$w_1 := 2^{(p-1/p)(\lambda_\alpha + \lambda_{\alpha+1})},$$
(15.205)

$$\Phi(x - a) := T(x - a) w_1.$$
(15.206)

Also put

$$\epsilon_4 := \begin{cases} 1, & if \ \lambda_\alpha + \lambda_{\alpha+1} \geq p, \\ M^{1-(\lambda_\alpha + \lambda_{\alpha+1}/p)}, & if \ \lambda_\alpha + \lambda_{\alpha+1} \leq p. \end{cases}$$
(15.207)

Then

$$\int_a^x q(s) \Bigg\{ \Bigg\{ \sum_{j=1}^{M-1} \big[\, |D_a^{\alpha_1} f_j(s)|^{\lambda_\alpha} \, |D_a^{\alpha_1+1} f_{j+1}(s)|^{\lambda_{\alpha+1}} +$$

$$|D_a^{\alpha_1} f_{j+1}(s)|^{\lambda_\alpha} \, |D_a^{\alpha_1+1} f_j(s)|^{\lambda_{\alpha+1}} \big] \Bigg\} +$$

$$\big[\, |D_a^{\alpha_1} f_1(s)|^{\lambda_\alpha} \, |D_a^{\alpha_1+1} f_M(s)|^{\lambda_{\alpha+1}} + |D_a^{\alpha_1} f_M(s)|^{\lambda_\alpha} \, |D_a^{\alpha_1+1} f_1(s)|^{\lambda_{\alpha+1}} \, \big] \Bigg\} ds \leq$$

$$2^{((\lambda_\alpha + \lambda_{\alpha+1})/p)} \epsilon_4 \phi(x - a) \left[\int_a^x p(s) \left(\sum_{j=1}^M |D_a^\beta f_j(s)|^p \right) ds \right]^{((\lambda_\alpha + \lambda_{\alpha+1})/p)}.$$
(15.208)

Proof. Transfer of Theorem 7 of [46]. □

Next comes

Theorem 15.83. (All are as in Theorem 15.79). *Consider the special case of* $\lambda_{\alpha_2} = \lambda_{\alpha_1} + \lambda_\beta$. *Denote*

$$\tilde{T}(x-a) := A_0(x-a) \left(\frac{\lambda_\beta}{\lambda_{\alpha_1}+\lambda_\beta} \right)^{\lambda_\beta/p} 2^{(p-2\lambda_{\alpha_1}-3\lambda_\beta)/p}. \qquad (15.209)$$

$$\epsilon_5 := \begin{cases} 1, & if\ 2(\lambda_{\alpha_1}+\lambda_\beta) \geq p, \\ M^{1-(2(\lambda_{\alpha_1}+\lambda_\beta)/p)}, & if\ 2(\lambda_{\alpha_1}+\lambda_\beta) \leq p. \end{cases} \qquad (15.210)$$

Then

$$\int_a^x q(s) \Big\{ \Big\{ \sum_{j=1}^{M-1} [\ |D_a^{\alpha_1} f_j(s)|^{\lambda_{\alpha_1}}\ |D_a^{\alpha_2} f_{j+1}(s)|^{\lambda_{\alpha_1}+\lambda_\beta}\ |D_a^\beta f_j(s)|^{\lambda_\beta} +$$

$$|D_a^{\alpha_2} f_j(s)|^{\lambda_{\alpha_1}+\lambda_\beta}\ |D_a^{\alpha_1} f_{j+1}(s)|^{\lambda_{\alpha_1}}\ |D_a^\beta f_{j+1}(s)|^{\lambda_\beta}\ ]\Big\} +$$

$$[\ |D_a^{\alpha_1} f_1(s)|^{\lambda_{\alpha_1}}\ |D_a^{\alpha_2} f_M(s)|^{\lambda_{\alpha_1}+\lambda_\beta}\ |D_a^\beta f_1(s)|^{\lambda_\beta} +$$

$$|D_a^{\alpha_2} f_1(s)|^{\lambda_{\alpha_1}+\lambda_\beta}\ |D_a^{\alpha_1} f_M(s)|^{\lambda_{\alpha_1}}\ |D_a^\beta f_M(s)|^{\lambda_\beta}\ ]\Big\} ds \leq$$

$$2^{\left(2(\lambda_{\alpha_1}+\lambda_\beta)/p\right)} \epsilon_5 \tilde{T}(x-a) \Big[\int_a^x p(s) \Big(\sum_{j=1}^M |D_a^\beta f_j(s)|^p \Big) ds \Big]^{\left(2(\lambda_{\alpha_1}+\lambda_\beta)/p\right)}.$$

$$(15.211)$$

Proof. Transfer of Theorem 8 of [46]. □

Special cases follow.

Corollary 15.84. (To Theorem 15.79; $\lambda_{\alpha_2} = 0$, $p(t) = q(t) = 1$.) *It holds*

$$\int_a^x \Big(\sum_{j=1}^M |D_a^{\alpha_1} f_j(s)|^{\lambda_{\alpha_1}}\ |D_a^\beta f_j(s)|^{\lambda_\beta} \Big) ds \leq$$

$$\delta_1^* \varphi_1(x-a) \Big[\int_a^x \sum_{j=1}^M [\ |D_a^\beta f_j(s)|^p\ ] ds \Big]^{\left((\lambda_{\alpha_1}+\lambda_\beta)/p\right)}. \qquad (15.212)$$

In (15.212)

$$\Big(A_0(x-a) \Big|_{\lambda_{\alpha_2}=0} \Big)$$

of $\varphi_1(x)$ *is given in Corollary* 15.49, *Equation* (15.123).

Proof. Transfer of Corollary 9 of [46]. □

Corollary 15.85. (To Theorem 15.79; $\lambda_{\alpha_2} = 0$, $p(t) = q(t) = 1, \lambda_{\alpha_1} = \lambda_\beta = 1$, $p = 2$.) *In detail: let $\beta > \alpha_1$, $\alpha_1 \in \mathbb{R}_+$, $\beta - \alpha_1 > (1/2)$, and let*

$$f_j \in L_1(a, x), \ j = 1, \ldots, M \in \mathbb{N}, \ a, \ x \in \mathbb{R}, \ a < x,$$

have, respectively, L_∞ fractional derivatives $D_a^\beta f_j$ in $[a, x]$, and let

$$D_a^{\beta-k} f_j(a) = 0, \ \text{for } k = 1, \ldots, [\beta] + 1; \ j = 1, \ldots, M.$$

Then

$$\int_a^x \left(\sum_{j=1}^M |D_a^{\alpha_1} f_j(s)| \ |D_a^\beta f_j(s)| \right) ds \leq$$

$$\frac{(x-a)^{(\beta-\alpha_1)}}{2\Gamma(\beta - \alpha_1) \sqrt{\beta - \alpha_1} \sqrt{2\beta - 2\alpha_1 - 1}} \left\{ \int_a^x \left[\sum_{j=1}^M (D_a^\beta f_j(s))^2 \right] ds \right\}. \quad (15.213)$$

Proof. Transfer of Corollary 10 of [46]. □

Corollary 15.86. (To Theorem 15.80; $\lambda_{\alpha_1} = 0$, $p(t) = q(t) = 1$.) *It holds*

$$\int_a^x \left\{ \left\{ \sum_{j=1}^{M-1} [\ |D_a^{\alpha_2} f_{j+1}(s)|^{\lambda_{\alpha_2}} \ |D_a^\beta f_j(s)|^{\lambda_\beta} + |D_a^{\alpha_2} f_j(s)|^{\lambda_{\alpha_2}} \ |D_a^\beta f_{j+1}(s)|^{\lambda_\beta}] \right\} + \right.$$

$$\left. [\ |D_a^{\alpha_2} f_M(s)|^{\lambda_{\alpha_2}} \ |D_a^\beta f_1(s)|^{\lambda_\beta} + |D_a^{\alpha_2} f_1(s)|^{\lambda_{\alpha_2}} \ |D_a^\beta f_M(s)|^{\lambda_\beta} \ ] \right\} ds \leq$$

$$2^{(\lambda_\beta+\lambda_{\alpha_2}/p)} \epsilon_2 \varphi_2(x-a) \left\{ \int_a^x \left(\sum_{j=1}^M |D_a^\beta f_j(s)|^p \right) ds \right\}^{((\lambda_\beta+\lambda_{\alpha_2})/p)}. \quad (15.214)$$

In (15.214),

$$\left(A_0(x-a) \Big|_{\lambda_{\alpha_1}=0} \right)$$

of $\varphi_2(x-a)$ is given in Corollary 15.51; see Equation (15.128).

Proof. Transfer of Corollary 11 of [46]. □

Corollary 15.87. (To Theorem 15.80, $\lambda_{\alpha_1} = 0$, $p(t) = q(t) = 1, \lambda_{\alpha_2} = \lambda_\beta = 1$, $p = 2$.) *In detail: let $\alpha_2 \in \mathbb{R}_+$, $\beta > \alpha_2$, $\beta - \alpha_2 > (1/2)$, and let*

$$f_j \in L_1(a, x), \ j = 1, \ldots, M \in \mathbb{N}, \ a, \ x \in \mathbb{R}, \ a < x,$$

have, respectively, L_∞ fractional derivatives $D_a^\beta f_j$ in $[a, x]$, and let

$$D_a^{\beta-k} f_j(a) = 0, \ \text{for } k = 1, \ldots, [\beta] + 1; \ j = 1, \ldots, M.$$

Then

$$\int_a^x \Big\{ \Big\{ \sum_{j=1}^{M-1} [\,|D_a^{\alpha_2} f_{j+1}(s)|\,|D_a^\beta f_j(s)| + |D_a^{\alpha_2} f_j(s)|\,|D_a^\beta f_{j+1}(s)||\,] \Big\} +$$

$$[\,|D_a^{\alpha_2} f_M(s)|\,|D_a^\beta f_1(s)| + |D_a^{\alpha_2} f_1(s)|\,|D_a^\beta f_M(s)|\,]\,\Big\} ds$$

$$\le \frac{\sqrt{2}(x-a)^{(\beta-\alpha_2)}}{\Gamma(\beta-\alpha_2)\,\sqrt{\beta-\alpha_2}\,\sqrt{2\beta-2\alpha_2-1}} \Big\{ \int_a^x \Big[\sum_{j=1}^M (D_a^\beta f_j(s))^2 \Big] ds \Big\}.$$

$$(15.215)$$

Proof. Transfer of Corollary 12 of [46]. □

Corollary 15.88. (To Theorem 15.81; $\lambda_{\alpha_1} = \lambda_{\alpha_2} = \lambda_\beta = 1$, $p(t) = q(t) = 1$, $p = 3$.) *It holds*

$$\int_a^x \Big[\sum_{j=1}^{M-1} [\,|D_a^{\alpha_1} f_j(s)|\,|D_a^{\alpha_2} f_{j+1}(s)|\,\,|D_a^\beta f_j(s)| +$$

$$|D_a^{\alpha_2} f_j(s)|\,|D_a^{\alpha_1} f_{j+1}(s)|\,|D_a^\beta f_{j+1}(s)|\,] + [\,|D_a^{\alpha_1} f_1(s)|\,|D_a^{\alpha_2} f_M(s)|\,|D_a^\beta f_1(s)| +$$

$$|D_a^{\alpha_2} f_1(s)|\,|D_a^{\alpha_1} f_M(s)|\,|D_a^\beta f_M(s)|\,] \Big] ds \le 2\varphi_3^*(x-a) \Big[\int_a^x \Big(\sum_{j=1}^M |D_a^\beta f_j(s)|^3 \Big) ds \Big].$$

$$(15.216)$$

Here

$$\varphi_3^*(x-a) := \left(\sqrt[3]{2} + \frac{1}{\sqrt[3]{6}} \right) A_0(x-a), \qquad (15.217)$$

where in this special case

$$A_0(x-a) =$$

$$\frac{4(x-a)^{(2\beta-\alpha_1-\alpha_2)}}{\Gamma(\beta-\alpha_1)\,\Gamma(\beta-\alpha_2)\,[3(3\beta-3\alpha_1-1)(3\beta-3\alpha_2-1)(2\beta-\alpha_1-\alpha_2)\,]^{2/3}}.$$

$$(15.218)$$

Proof. Transfer of Corollary 13 of [46]. □

Corollary 15.89. (To Theorem 15.82; $\lambda_\alpha = 1$, $\lambda_{\alpha+1} = 1/2$, $p(t) = q(t) = 1$, $p = 3/2$.) *In detail: let $\alpha_1 \in \mathbb{R}_+$, $\beta > \alpha_1 + 1$, and let*

$$f_j \in L_1(a,x), \ j=1,\dots,M \in \mathbb{N}, \ a, \ x \in \mathbb{R}, \ a < x,$$

have, respectively, L_∞ fractional derivatives $D_a^\beta f_j$ in $[a,x]$, and let

$$D_a^{\beta-k} f_j(a) = 0, \ \text{for } k = 1,\dots,[\beta]+1; \ j=1,\dots,M.$$

Set

$$\Phi^*(x - a) := \left(\frac{2}{\sqrt{3\beta - 3\alpha_1 - 2}}\right) \cdot \frac{(x - a)^{(3\beta - 3\alpha_1 - 1/2)}}{(\Gamma(\beta - \alpha_1))^{3/2}}. \qquad (15.219)$$

Then

$$\int_a^x \left\{\left\{\sum_{j=1}^{M-1}[\,|D_a^{\alpha_1} f_j(s)|\,\sqrt{|D_a^{\alpha_1+1} f_{j+1}(s)|} + |D_a^{\alpha_1} f_{j+1}(s)|\,\sqrt{|D_a^{\alpha_1+1} f_j(s)|}\,]\right\} + \right.$$

$$\left. [\,|D_a^{\alpha_1} f_1(s)|\,\sqrt{|D_a^{\alpha_1+1} f_M(s)|} + |D_a^{\alpha_1} f_M(s)|\,\sqrt{|D_a^{\alpha_1+1} f_1(s)|}\,]\right\} ds \leq$$

$$2\Phi^*(x - a)\left[\int_a^x \left(\sum_{j=1}^M |D_a^\beta f_j(s)|^{3/2}\right) ds\right]. \qquad (15.220)$$

Proof. Transfer of Corollary 14 of [46]. □

Corollary 15.90. (To Theorem 15.83; $p = 2(\lambda_{\alpha_1} + \lambda_\beta) > 1$, $p(t) = q(t) = 1$.) *It holds*

$$\int_a^x \left\{\left\{\sum_{j=1}^{M-1}[\,|D_a^{\alpha_1} f_j(s)|^{\lambda_{\alpha_1}}\,|D_a^{\alpha_2} f_{j+1}(s)|^{\lambda_{\alpha_1}+\lambda_\beta}\,|D_a^\beta f_j(s)|^{\lambda_\beta} + \right.\right.$$

$$|D_a^{\alpha_2} f_j(s)|^{\lambda_{\alpha_1}+\lambda_\beta}\,|D_a^{\alpha_1} f_{j+1}(s)|^{\lambda_{\alpha_1}}\,|D_a^\beta f_{j+1}(s)|^{\lambda_\beta}\,]\right\} +$$

$$[\,|D_a^{\alpha_1} f_1(s)|^{\lambda_{\alpha_1}}\,|D_a^{\alpha_2} f_M(s)|^{\lambda_{\alpha_1}+\lambda_\beta}\,|D_a^\beta f_1(s)|^{\lambda_\beta} +$$

$$\left. |D_a^{\alpha_2} f_1(s)|^{\lambda_{\alpha_1}+\lambda_\beta}\,|D_a^{\alpha_1} f_M(s)|^{\lambda_{\alpha_1}}\,|D_a^\beta f_M(s)|^{\lambda_\beta}\,]\right\} ds \leq$$

$$2\tilde{T}(x - a)\left[\int_a^x \left(\sum_{j=1}^M |D_a^\beta f_j(s)|^{2(\lambda_{\alpha_1}+\lambda_\beta)}\right) ds\right]. \qquad (15.221)$$

Here $\tilde{T}(x - a)$ in (15.221) is given by (15.209) and in detail by $\tilde{T}(x - a)$ of Corollary 15.55 and Equations (15.136)–(15.139).

Proof. Transfer of Corollary 15 of [46]. □

Corollary 15.91. (To Theorem 15.83; $p = 4$, $\lambda_{\alpha_1} = \lambda_\beta = 1$, $p(t) = q(t) = 1$.) *It holds*

$$\int_a^x \left\{\left\{\sum_{j=1}^{M-1}[\,|D_a^{\alpha_1} f_j(s)|\,(D_a^{\alpha_2} f_{j+1}(s))^2\,|D_a^\beta f_j(s)| + \right.\right.$$

$$(D_a^{\alpha_2} f_j(s))^2|D_a^{\alpha_1} f_{j+1}(s)|\,|D_a^\beta f_{j+1}(s)|\,]\right\} + [\,|D_a^{\alpha_1} f_1(s)|\,(D_a^{\alpha_2} f_M(s))^2\,|D_a^\beta f_1(s)| +$$

$$(D_a^{\alpha_2} f_1(s))^2 \, |D_a^{\alpha_1} f_M(s)| \, |D_a^{\beta} f_M(s)| \, | \big\} ds \le 2\tilde{T}(x-a) \Big[\int_a^x \Big(\sum_{j=1}^M (D_a^{\beta} f_j(s))^4 \Big) ds \Big].$$

$$(15.222)$$

Here in (15.222) we have that $\tilde{T}(x-a) = T^*(x-a)$ of Corollary 15.56; see the Equations (15.141)–(15.145).

Proof. Transfer of Corollary 16 of [46]. □

Next we present the L_∞ case.

Theorem 15.92. *Let $\alpha_1, \alpha_2 \in \mathbb{R}_+$, $\beta > \alpha_1, \alpha_2$, and let*

$$f_j \in L_1(a,x), \quad j = 1, \ldots, M \in \mathbb{N}, \ a, \ x \in \mathbb{R}, \ a < x,$$

have, respectively, L_∞ fractional derivatives $D_a^{\beta} f_j$ in $[a,x]$, and let

$$D_a^{\beta-k} f_j(a) = 0, \ \text{for } k = 1, \ldots, [\beta]+1; \ j = 1, \ldots, \ M.$$

Consider also $p(s) \ge 0$, $p(s) \in L_\infty(a,x)$. Let $\lambda_{\alpha_1}, \lambda_{\alpha_2}, \lambda_{\beta} \ge 0$. Set

$$\rho(x-a) =$$

$$\frac{\|p(s)\|_\infty \, (x-a)^{(\beta\lambda_{\alpha_1} - \alpha_1\lambda_{\alpha_1} + \beta\lambda_{\alpha_2} - \alpha_2\lambda_{\alpha_2} + 1)}}{(\Gamma(\beta-\alpha_1+1))^{\lambda_{\alpha_1}} (\Gamma(\beta-\alpha_2+1))^{\lambda_{\alpha_2}} [\beta\lambda_{\alpha_1} - \alpha_1\lambda_{\alpha_1} + \beta\lambda_{\alpha_2} - \alpha_2\lambda_{\alpha_2} + 1]}.$$

$$(15.223)$$

Then

$$\int_a^x p(s) \Big\{ \Big\{ \sum_{j=1}^{M-1} [\, |D_a^{\alpha_1} f_j(s)|^{\lambda_{\alpha_1}} \, |D_a^{\alpha_2} f_{j+1}(s)|^{\lambda_{\alpha_2}} \, |D_a^{\beta} f_j(s)|^{\lambda_{\beta}} +$$

$$|D_a^{\alpha_2} f_j(s)|^{\lambda_{\alpha_2}} \, |D_a^{\alpha_1} f_{j+1}(s)|^{\lambda_{\alpha_1}} \, |D_a^{\beta} f_{j+1}(s)|^{\lambda_{\beta}} \,]\} +$$

$$[\, |D_a^{\alpha_1} f_1(s)|^{\lambda_{\alpha_1}} \, |D_a^{\alpha_2} f_M(s)|^{\lambda_{\alpha_2}} \, |D_a^{\beta} f_1(s)|^{\lambda_{\beta}} +$$

$$|D_a^{\alpha_2} f_1(s)|^{\lambda_{\alpha_2}} \, |D_a^{\alpha_1} f_M(s)|^{\lambda_{\alpha_1}} \, |D_a^{\beta} f_M(s)|^{\lambda_{\beta}} \,]\} ds$$

$$\le \rho(x-a) \Big\{ \sum_{j=1}^M \{ \|D_a^{\beta} f_j\|_\infty^{2(\lambda_{\alpha_1}+\lambda_{\beta})} + \|D_a^{\beta} f_j\|_\infty^{2\lambda_{\alpha_2}} \} \Big\}.$$

$$(15.224)$$

Proof. Transfer of Corollary 17 of [46]. □

Similarly we give

Theorem 15.93. (As in Theorem 15.92; $\lambda_{\alpha_2} = 0$.) *It holds*

$$\int_a^x p(s) \Big(\sum_{j=1}^{M} |D_a^{\alpha_1} f_j(s)|^{\lambda_{\alpha_1}} \, |D_a^{\beta} f_j(s)|^{\lambda_\beta} \Big) ds \le$$

$$\frac{\|p(s)\|_\infty \, (x-a)^{(\beta \lambda_{\alpha_1} - \alpha_1 \lambda_{\alpha_1} + 1)}}{(\Gamma(\beta - \alpha_1 + 1))^{\lambda_{\alpha_1}} [\beta \lambda_{\alpha_1} - \alpha_1 \lambda_{\alpha_1} + 1]} \cdot \Big(\sum_{j=1}^{M} \|D_a^{\beta} f_j\|_\infty^{\lambda_{\alpha_1} + \lambda_\beta} \Big). \quad (15.225)$$

Proof. Based on Theorem 18 of [46]. $\quad\square$

It follows

Theorem 15.94. (As in Theorem 15.92; $\lambda_{\alpha_2} = \lambda_{\alpha_1} + \lambda_\beta$.) *It holds*

$$\int_a^x p(s) \Big\{ \Big\{ \sum_{j=1}^{M-1} [\, |D_a^{\alpha_1} f_j(s)|^{\lambda_{\alpha_1}} \, |D_a^{\alpha_2} f_{j+1}(s)|^{\lambda_{\alpha_1} + \lambda_\beta} \, |D_a^{\beta} f_j(s)|^{\lambda_\beta} +$$

$$|D_a^{\alpha_2} f_j(s)|^{\lambda_{\alpha_1} + \lambda_\beta} \, |D_a^{\alpha_1} f_{j+1}(s)|^{\lambda_{\alpha_1}} \, |D_a^{\beta} f_{j+1}(s)|^{\lambda_\beta} \,] \Big\} +$$

$$[\, |D_a^{\alpha_1} f_1(s)|^{\lambda_{\alpha_1}} \, |D_a^{\alpha_2} f_M(s)|^{\lambda_{\alpha_1} + \lambda_\beta} \, |D_a^{\beta} f_1(s)|^{\lambda_\beta} +$$

$$|D_a^{\alpha_2} f_1(s)|^{\lambda_{\alpha_1} + \lambda_\beta} \, |D_a^{\alpha_1} f_M(s)|^{\lambda_{\alpha_1}} \, |D_a^{\beta} f_M(s)|^{\lambda_\beta} \,] \Big\} ds \le$$

$$\le \Big\{ \frac{\|p(s)\|_\infty}{(\Gamma(\beta - \alpha_1 + 1))^{\lambda_{\alpha_1}} \, (\Gamma(\beta - \alpha_2 + 1))^{(\lambda_{\alpha_1} + \lambda_\beta)}}$$

$$\cdot \frac{2(x-a)^{(2\beta \lambda_{\alpha_1} - \alpha_1 \lambda_{\alpha_1} + \beta \lambda_\beta - \alpha_2 \lambda_{\alpha_1} - \alpha_2 \lambda_\beta + 1)}}{(2\beta \lambda_{\alpha_1} - \alpha_1 \lambda_{\alpha_1} + \beta \lambda_\beta - \alpha_2 \lambda_{\alpha_1} - \alpha_2 \lambda_\beta + 1)} \Big\} \cdot \Big(\sum_{j=1}^{M} \|D_a^{\beta} f_j\|_\infty^{2(\lambda_{\alpha_1} + \lambda_\beta)} \Big).$$

$$(15.226)$$

Proof. By Theorem 19 of [46]. $\quad\square$

We continue with

Theorem 15.95. (As in Theorem 15.92; $\lambda_\beta = 0$, $\lambda_{\alpha_1} = \lambda_{\alpha_2}$.) *It holds*

$$\int_a^x p(s) \Big\{ \Big\{ \sum_{j=1}^{M-1} [\, |D_a^{\alpha_1} f_j(s)|^{\lambda_{\alpha_1}} \, |D_a^{\alpha_2} f_{j+1}(s)|^{\lambda_{\alpha_1}} + |D_a^{\alpha_2} f_j(s)|^{\lambda_{\alpha_1}} \, |D_a^{\alpha_1} f_{j+1}(s)|^{\lambda_{\alpha_1}}] \Big\} +$$

$$[\, |D_a^{\alpha_1} f_1(s)|^{\lambda_{\alpha_1}} \, |D_a^{\alpha_2} f_M(s)|^{\lambda_{\alpha_1}} + |D_a^{\alpha_2} f_1(s)|^{\lambda_{\alpha_1}} \, |D_a^{\alpha_1} f_M(s)|^{\lambda_{\alpha_1}} \,] \Big\} ds$$

$$\le 2\rho^*(x-a) \Big[\sum_{j=1}^{M} \|D_a^{\beta} f_j\|_\infty^{2\lambda_{\alpha_1}} \Big]. \quad (15.227)$$

Here we have

$$\rho^*(x-a) :=$$

$$\frac{(x-a)^{(2\beta\lambda_{\alpha_1}-\alpha_1\lambda_{\alpha_1}-\alpha_2\lambda_{\alpha_1}+1)}\,\|p(s)\|_\infty}{(2\beta\lambda_{\alpha_1}-\alpha_1\lambda_{\alpha_1}-\alpha_2\lambda_{\alpha_1}+1)(\Gamma(\beta-\alpha_1+1))^{\lambda_{\alpha_1}}(\Gamma(\beta-\alpha_2+1))^{\lambda_{\alpha_1}}}. \tag{15.228}$$

Proof. Based on Theorem 20 of [46]. □

Next we give

Theorem 15.96. (As in Theorem 15.92; $\lambda_{\alpha_1}=0$, $\lambda_{\alpha_2}=\lambda_\beta$.) *It holds*

$$\int_a^x p(s)\Big\{\Big\{\sum_{j=1}^{M-1}[\,|D_a^{\alpha_2}f_{j+1}(s)|^{\lambda_{\alpha_2}}\,|D_a^\beta f_j(s)|^{\lambda_{\alpha_2}}+|D_a^{\alpha_2}f_j(s)|^{\lambda_{\alpha_2}}\,|D_a^\beta f_{j+1}(s)|^{\lambda_{\alpha_2}}]\Big\}+$$

$$[\,|D_a^{\alpha_2}f_M(s)|^{\lambda_{\alpha_2}}\,|D_a^\beta f_1(s)|^{\lambda_{\alpha_2}}+|D_a^{\alpha_2}f_1(s)|^{\lambda_{\alpha_2}}\,|D_a^\beta f_M(s)|^{\lambda_{\alpha_2}}\,]\Big\}ds\le$$

$$2\left(\frac{(x-a)^{(\beta\lambda_{\alpha_2}-\alpha_2\lambda_{\alpha_2}+1)}\,\|p(s)\|_\infty}{(\beta\lambda_{\alpha_2}-\alpha_2\lambda_{\alpha_2}+1)(\Gamma(\beta-\alpha_2+1))^{\lambda_{\alpha_2}}}\right)\cdot\left(\sum_{j=1}^M\|D_a^\beta f_j\|_\infty^{2\lambda_{\alpha_2}}\right). \tag{15.229}$$

Proof. Based on Theorem 21 of [46]. □

Some special cases follow.

Corollary 15.97. (To Theorem 15.95; all are as in Theorem 15.92; $\lambda_\beta=0$, $\lambda_{\alpha_1}=\lambda_{\alpha_2}$, $\alpha_2=\alpha_1+1$.) *It holds*

$$\int_a^x p(s)\Big\{\Big\{\sum_{j=1}^{M-1}[\,|D_a^{\alpha_1}f_j(s)|^{\lambda_{\alpha_1}}\,|D_a^{\alpha_1+1}f_{j+1}(s)|^{\lambda_{\alpha_1}}+$$

$$|D_a^{\alpha_1+1}f_j(s)|^{\lambda_{\alpha_1}}\,|D_a^{\alpha_1}f_{j+1}(s)|^{\lambda_{\alpha_1}}]\Big\}+$$

$$[\,|D_a^{\alpha_1}f_1(s)|^{\lambda_{\alpha_1}}\,|D_a^{\alpha_1+1}f_M(s)|^{\lambda_{\alpha_1}}+|D_a^{\alpha_1+1}f_1(s)|^{\lambda_{\alpha_1}}\,|D_a^{\alpha_1}f_M(s)|^{\lambda_{\alpha_1}}\,]\Big\}ds\le$$

$$2\left(\frac{(x-a)^{(2\beta\lambda_{\alpha_1}-2\alpha_1\lambda_{\alpha_1}-\lambda_{\alpha_1}+1)}\,\|p(s)\|_\infty}{(2\beta\lambda_{\alpha_1}-2\alpha_1\lambda_{\alpha_1}-\lambda_{\alpha_1}+1)(\beta-\alpha_1)^{\lambda_{\alpha_1}}(\Gamma(\beta-\alpha_1))^{2\lambda_{\alpha_1}}}\right)\cdot\left[\sum_{j=1}^M\|D_a^\beta f_j\|_\infty^{2\lambda_{\alpha_1}}\right]. \tag{15.230}$$

Proof. Based on Theorem 22 of [46]. □

Corollary 15.98. (to Corollary 15.97) *In detail; let $\alpha_1 \in \mathbb{R}_+$, $\beta > \alpha_1+1$, and let*

$$f_j \in L_1(a, x), \ j = 1, \ldots, M \in \mathbb{N}, \ a, \ x \in \mathbb{R}, \ a < x,$$

have, respectively, L_∞ fractional derivatives $D_a^\beta f_j$ in $[a, x]$, and let

$$D_a^{\beta-k} f_j(a) = 0, \ \text{for } k = 1, \ldots, [\beta] + 1; \ j = 1, \ldots, M.$$

Then

$$\int_a^x \left\{ \left\{ \sum_{j=1}^{M-1} [\, |D_a^{\alpha_1} f_j(s)| \, |D_a^{\alpha_1+1} f_{j+1}(s)| + |D_a^{\alpha_1} f_{j+1}(s)| \, |D_a^{\alpha_1+1} f_j(s)| \,] \right\} + \right.$$

$$\left. [\, |D_a^{\alpha_1} f_1(s)| \, |D_a^{\alpha_1+1} f_M(s)| + |D_a^{\alpha_1} f_M(s)| \, |D_a^{\alpha_1+1} f_1(s)| \,] \right\} ds \leq$$

$$\frac{(x-a)^{2(\beta-\alpha_1)}}{(\beta-\alpha_1)^2 \, (\Gamma(\beta-\alpha_1))^2} \left(\sum_{j=1}^M \|D_a^\beta f_j\|_\infty^2 \right). \tag{15.231}$$

Proof. Based on Corollary 23 of [46]. □

Corollary 15.99. (to Corollary 15.98) *It holds*

$$\int_a^x \left(\sum_{j=1}^M |D_a^{\alpha_1} f_j(s)| \, |D_a^{\alpha_1+1} f_j(s)| \right) ds \leq$$

$$\frac{(x-a)^{2(\beta-\alpha_1)}}{2(\beta-\alpha_1)^2 \, (\Gamma(\beta-\alpha_1))^2} \left(\sum_{j=1}^M \|D_a^\beta f_j\|_\infty^2 \right). \tag{15.232}$$

Proof. Based on inequality (15.155) of Proposition 15.64. □

Next we apply the previous results of this subsection to the spherical shell A. We make

Assumption 15.100. Let

$$\alpha_1, \alpha_2 \in \mathbb{R}_+, \ \beta > \alpha_1, \alpha_2, \ \beta - \alpha_i > (1/p), p > 1, i = 1, 2,$$

and for $j = 1, \ldots, M, \ M \in \mathbb{N}$, let $f_j \in L_1(A)$ with

$$\frac{\partial_{R_1}^\beta f_j(x)}{\partial r^\beta}, \ \in L_\infty(A), \ x \in A,$$

$$A := B(0, R_2) - \overline{B(0, R_1)} \subseteq \mathbb{R}^N, \ N \geq 2, \ 0 < R_1 < R_2.$$

Further assume that each $D_{R_1}^\beta f_j(rw) \in \mathbb{R}$ for almost all $r \in [R_1, R_2]$, for each $w \in S^{N-1}$, and for these $|D_{R_1}^\beta f_j(rw)| \le M_j$ for some $M_j > 0$; $j = 1, \ldots, M$. For each $w \in S^{N-1} - (\cup_{j=1}^M K(f_j))$, we assume that $f_j(\cdot\, w)$ has an L_∞ fractional derivative $D_{R_1}^\beta f_j(\cdot\, w)$ in $[R_1, R_2]$, and that

$$D_{R_1}^{\beta-k} f_j(R_1 w) = 0, \quad k = 1, \ldots, [\beta] + 1;$$

$j = 1, \ldots, M$. Let $\lambda_\beta > 0$ and $\lambda_{\alpha_1}, \lambda_{\alpha_2} \ge 0$, such that $\lambda_\beta < p$. If $\alpha_1 = 0$ we set $\lambda_{\alpha_1} = 1$, and if $\alpha_2 = 0$ we set $\lambda_{\alpha_2} = 1$.

We need

Notation 15.101. (on Assumption 15.100) We set

$$P_i(s) := \int_0^s (s-r)^{p(\beta-\alpha_i-1)/p-1} (r+R_1)^{(1-N/p-1)}\, dr,$$

$$i = 1, 2; \ 0 \le s \le R_2 - R_1, \tag{15.233}$$

$$A(s) := \frac{(s+R_1)^{(N-1)(1-(\lambda_\beta/p))} (P_1(s))^{\lambda_{\alpha_1}(p-1/p)} (P_2(s))^{\lambda_{\alpha_2}(p-1/p)}}{(\Gamma(\beta-\alpha_1))^{\lambda_{\alpha_1}} (\Gamma(\beta-\alpha_2))^{\lambda_{\alpha_2}}}, \tag{15.234}$$

and

$$A_0(R_2 - R_1) := \left(\int_0^{R_2-R_1} (A(s))^{p/(p-\lambda_\beta)}\, ds \right)^{(p-\lambda_\beta)/p}. \tag{15.235}$$

We present

Theorem 15.102. (All are as in Assumption 15.100 and Notation 15.101.) *Denote*

$$\varphi_1(R_2 - R_1) := \left(A_0(R_2-R_1)\Big|_{\lambda_{\alpha_2}=0} \right) \left(\frac{\lambda_\beta}{p} \right)^{(\lambda_\beta/p)}. \tag{15.236}$$

Let $\lambda_{\alpha_1} > 0$, $\lambda_{\alpha_2} = 0$, *and* $p = \lambda_{\alpha_1} + \lambda_\beta > 1$. *Then*

$$\int_A \left[\sum_{j=1}^M \left| \frac{\partial_{R_1}^{\alpha_1} f_j(x)}{\partial r^{\alpha_1}} \right|^{\lambda_{\alpha_1}} \left| \frac{\partial_{R_1}^\beta f_j(x)}{\partial r^\beta} \right|^{\lambda_\beta} \right] dx \le$$

$$\varphi_1(R_2 - R_1) \left[\int_A \left(\sum_{j=1}^M \left| \frac{\partial_{R_1}^\beta f_j(x)}{\partial r^\beta} \right|^p \right) dx \right]. \tag{15.237}$$

Proof. Based on Theorem 15.79 and similar to the proof of Theorem 15.67; notice here $\lambda_{S^{N-1}}(\cup_{j=1}^M K(f_j)) = 0$. □

Next we give

Theorem 15.103. (All are as in Assumption 15.100 and Notation 15.101.) *We denote*

$$\delta_3 := \begin{cases} 2^{\lambda_{\alpha_2}/\lambda_\beta} - 1, & \text{if } \lambda_{\alpha_2} \geq \lambda_\beta, \\ 1, & \text{if } \lambda_{\alpha_2} \leq \lambda_\beta, \end{cases} \tag{15.238}$$

and

$$\varphi_2(R_2 - R_1) := \left(A_0(R_2 - R_1)\Big|_{\lambda_{\alpha_1}=0} \right) 2^{(\lambda_{\alpha_2}/p)} \left(\frac{\lambda_\beta}{p} \right)^{(\lambda_\beta/p)} \delta_3^{(\lambda_\beta/p)}. \tag{15.239}$$

Here $\lambda_{\alpha_1} = 0$, $\lambda_{\alpha_2} > 0$ and $p = \lambda_\beta + \lambda_{\alpha_2} > 1$. Then

$$\int_A \Bigg\{ \Bigg\{ \sum_{j=1}^{M-1} \Bigg[\left| \frac{\partial_{R_1}^{\alpha_2} f_{j+1}(x)}{\partial r^{\alpha_2}} \right|^{\lambda_{\alpha_2}} \left| \frac{\partial_{R_1}^\beta f_j(x)}{\partial r^\beta} \right|^{\lambda_\beta} +$$

$$\left| \frac{\partial_{R_1}^{\alpha_2} f_j(x)}{\partial r^{\alpha_2}} \right|^{\lambda_{\alpha_2}} \left| \frac{\partial_{R_1}^\beta f_{j+1}(x)}{\partial r^\beta} \right|^{\lambda_\beta} \Bigg] \Bigg\} +$$

$$\Bigg[\left| \frac{\partial_{R_1}^{\alpha_2} f_M(x)}{\partial r^{\alpha_2}} \right|^{\lambda_{\alpha_2}} \left| \frac{\partial_{R_1}^\beta f_1(x)}{\partial r^\beta} \right|^{\lambda_\beta} + \left| \frac{\partial_{R_1}^{\alpha_2} f_1(x)}{\partial r^{\alpha_2}} \right|^{\lambda_{\alpha_2}} \left| \frac{\partial_{R_1}^\beta f_M(x)}{\partial r^\beta} \right|^{\lambda_\beta} \Bigg] dx \Bigg\} \leq$$

$$2\varphi_2(R_2 - R_1) \Bigg[\int_A \Bigg(\sum_{j=1}^M \left| \frac{\partial_{R_1}^\beta f_j(x)}{\partial r^\beta} \right|^p \Bigg) dx \Bigg]. \tag{15.240}$$

Proof. Based on Theorem 15.80 and similar to the proof of Theorem 15.67. □

The general case follows.

Theorem 15.104. (All are as in Assumption 15.100 and Notation 15.101.) *Here λ_{α_1}, $\lambda_{\alpha_2} > 0$, $p = \lambda_{\alpha_1} + \lambda_{\alpha_2} + \lambda_\beta > 1$. Denote*

$$\tilde{\gamma}_1 := \begin{cases} 2^{((\lambda_{\alpha_1}+\lambda_{\alpha_2})/\lambda_\beta)} - 1, & \text{if } \lambda_{\alpha_1} + \lambda_{\alpha_2} \geq \lambda_\beta, \\ 1, & \text{if } \lambda_{\alpha_1} + \lambda_{\alpha_2} \leq \lambda_\beta, \end{cases} \tag{15.241}$$

and

$$\varphi_3(R_2 - R_1) := A_0(R_2 - R_1) \left(\frac{\lambda_\beta}{(\lambda_{\alpha_1} + \lambda_{\alpha_2})p} \right)^{(\lambda_\beta/p)}$$

$$[\lambda_{\alpha_1}^{(\lambda_\beta/p)} + 2^{(\lambda_{\alpha_1}+\lambda_{\alpha_2})/p} (\tilde{\gamma}_1 \lambda_{\alpha_2})^{(\lambda_\beta/p)}]. \tag{15.242}$$

Then

$$\int_A \left\{ \sum_{j=1}^{M-1} \left[\left| \frac{\partial_{R_1}^{\alpha_1} f_j(x)}{\partial r^{\alpha_1}} \right|^{\lambda_{\alpha_1}} \left| \frac{\partial_{R_1}^{\alpha_2} f_{j+1}(x)}{\partial r^{\alpha_2}} \right|^{\lambda_{\alpha_2}} \left| \frac{\partial_{R_1}^{\beta} f_j(x)}{\partial r^{\beta}} \right|^{\lambda_{\beta}} + \right. \right.$$

$$\left| \frac{\partial_{R_1}^{\alpha_2} f_j(x)}{\partial r^{\alpha_2}} \right|^{\lambda_{\alpha_2}} \left| \frac{\partial_{R_1}^{\alpha_1} f_{j+1}(x)}{\partial r^{\alpha_1}} \right|^{\lambda_{\alpha_1}} \left| \frac{\partial_{R_1}^{\beta} f_{j+1}(x)}{\partial r^{\beta}} \right|^{\lambda_{\beta}} \right] +$$

$$\left[\left| \frac{\partial_{R_1}^{\alpha_1} f_1(x)}{\partial r^{\alpha_1}} \right|^{\lambda_{\alpha_1}} \left| \frac{\partial_{R_1}^{\alpha_2} f_M(x)}{\partial r^{\alpha_2}} \right|^{\lambda_{\alpha_2}} \left| \frac{\partial_{R_1}^{\beta} f_1(x)}{\partial r^{\beta}} \right|^{\lambda_{\beta}} + \right.$$

$$\left. \left. \left| \frac{\partial_{R_1}^{\alpha_2} f_1(x)}{\partial r^{\alpha_2}} \right|^{\lambda_{\alpha_2}} \left| \frac{\partial_{R_1}^{\alpha_1} f_M(x)}{\partial r^{\alpha_1}} \right|^{\lambda_{\alpha_1}} \left| \frac{\partial_{R_1}^{\beta} f_M(x)}{\partial r^{\beta}} \right|^{\lambda_{\beta}} \right] \right\} dx$$

$$\leq 2\varphi_3(R_2 - R_1) \left[\int_A \left(\sum_{j=1}^{M} \left| \frac{\partial_{R_1}^{\beta} f_j(x)}{\partial r^{\beta}} \right|^{p} \right) dx \right]. \tag{15.243}$$

Proof. Based on Theorem 15.81 and similar to the proof of Theorem 15.67. □

We continue with

Theorem 15.105. (All are as in Assumption 15.100.) *Here* $\alpha_2 = \alpha_1 + 1$, $\lambda_\alpha := \lambda_{\alpha_1} \geq 0$, $\lambda_{\alpha+1} := \lambda_{\alpha_2} \in (0, 1)$ *and* $p = \lambda_\alpha + \lambda_{\alpha_1} > 1$. *Denote*

$$\theta_3 := \begin{cases} 2^{(\lambda_\alpha / \lambda_{\alpha+1})} - 1, & \text{if } \lambda_\alpha \geq \lambda_{\alpha+1}, \\ 1, & \text{if } \lambda_\alpha \leq \lambda_{\alpha+1}, \end{cases} \tag{15.244}$$

$$L(R_2 - R_1) :=$$

$$\left[2 \frac{(1 - \lambda_{\alpha+1})}{(N - \lambda_{\alpha+1})} \left(R_2^{(N-\lambda_{\alpha+1})/(1-\lambda_{\alpha+1})} - R_1^{(N-\lambda_{\alpha+1})/(1-\lambda_{\alpha+1})} \right) \right]^{(1-\lambda_{\alpha+1})}$$

$$\left(\frac{\theta_3 \, \lambda_{\alpha+1}}{p} \right)^{\lambda_{\alpha+1}}, \tag{15.245}$$

and

$$P_1(R_2 - R_1) := \int_{R_1}^{R_2} (R_2 - r)^{(\beta-\alpha_1-1)p/(p-1)} \, r^{(1-N)/(p-1)} dr, \tag{15.246}$$

and

$$\Phi(R_2 - R_1) := L(R_2 - R_1) \left(\frac{(P_1(R_2 - R_1))^{(p-1)}}{(\Gamma(\beta - \alpha_1))^p} \right) 2^{(p-1)}. \tag{15.247}$$

Then

$$\int_A \left\{ \left\{ \sum_{j=1}^{M-1} \left[\left| \frac{\partial_{R_1}^{\alpha_1} f_j(x)}{\partial r^{\alpha_1}} \right|^{\lambda_\alpha} \left| \frac{\partial_{R_1}^{\alpha_1+1} f_{j+1}(x)}{\partial r^{\alpha_1+1}} \right|^{\lambda_{\alpha+1}} + \right. \right. \right.$$

$$\left| \frac{\partial_{R_1}^{\alpha_1} f_{j+1}(x)}{\partial r^{\alpha_1}} \right|^{\lambda_\alpha} \left| \frac{\partial_{R_1}^{\alpha_1+1} f_j(x)}{\partial r^{\alpha_1+1}} \right|^{\lambda_{\alpha+1}} \Big] \Big\}$$

$$+ \Big[\left| \frac{\partial_{R_1}^{\alpha_1} f_1(x)}{\partial r^{\alpha_1}} \right|^{\lambda_\alpha} \left| \frac{\partial_{R_1}^{\alpha_1+1} f_M(x)}{\partial r^{\alpha_1+1}} \right|^{\lambda_{\alpha+1}} + \left| \frac{\partial_{R_1}^{\alpha_1} f_M(x)}{\partial r^{\alpha_1}} \right|^{\lambda_\alpha} \left| \frac{\partial_{R_1}^{\alpha_1+1} f_1(x)}{\partial r^{\alpha_1+1}} \right|^{\lambda_{\alpha+1}} \Big] \Big\} dx$$

$$\leq 2\Phi(R_2 - R_1) \Big[\int_A \Big(\sum_{j=1}^{M} \left| \frac{\partial_{R_1}^{\beta} f_j(x)}{\partial r^{\beta}} \right|^{p} \Big) dx \Big]. \tag{15.248}$$

Proof. Based on Theorem 15.82 and similar to the proof of Theorem 15.67. □

We also give

Theorem 15.106. (All are as in Assumption 15.100 and Notation 15.101.) Here $\lambda_{\alpha_2} = \lambda_{\alpha_1} + \lambda_\beta$ and $p = 2(\lambda_{\alpha_1} + \lambda_\beta) > 1$. Denote

$$\tilde{T}(R_2 - R_1) := A_0(R_2 - R_1) \Big(\frac{2\lambda_\beta}{p} \Big)^{\lambda_\beta/p} 2^{-\lambda_\beta/p}. \tag{15.249}$$

Then

$$\int_A \Big\{ \Big\{ \sum_{j=1}^{M-1} \Big[\left| \frac{\partial_{R_1}^{\alpha_1} f_j(x)}{\partial r^{\alpha_1}} \right|^{\lambda_{\alpha_1}} \left| \frac{\partial_{R_1}^{\alpha_2} f_{j+1}(x)}{\partial r^{\alpha_2}} \right|^{\lambda_{\alpha_1}+\lambda_\beta} \left| \frac{\partial_{R_1}^{\beta} f_j(x)}{\partial r^{\beta}} \right|^{\lambda_\beta} +$$

$$\left| \frac{\partial_{R_1}^{\alpha_2} f_j(x)}{\partial r^{\alpha_2}} \right|^{\lambda_{\alpha_1}+\lambda_\beta} \left| \frac{\partial_{R_1}^{\alpha_1} f_{j+1}(x)}{\partial r^{\alpha_1}} \right|^{\lambda_{\alpha_1}} \left| \frac{\partial_{R_1}^{\beta} f_{j+1}(x)}{\partial r^{\beta}} \right|^{\lambda_\beta} \Big] \Big\} +$$

$$\Big[\left| \frac{\partial_{R_1}^{\alpha_1} f_1(x)}{\partial r^{\alpha_1}} \right|^{\lambda_{\alpha_1}} \left| \frac{\partial_{R_1}^{\alpha_2} f_M(x)}{\partial r^{\alpha_2}} \right|^{\lambda_{\alpha_1}+\lambda_\beta} \left| \frac{\partial_{R_1}^{\beta} f_1(x)}{\partial r^{\beta}} \right|^{\lambda_\beta} +$$

$$\left| \frac{\partial_{R_1}^{\alpha_2} f_1(x)}{\partial r^{\alpha_2}} \right|^{\lambda_{\alpha_1}+\lambda_\beta} \left| \frac{\partial_{R_1}^{\alpha_1} f_M(x)}{\partial r^{\alpha_1}} \right|^{\lambda_{\alpha_1}} \left| \frac{\partial_{R_1}^{\beta} f_M(x)}{\partial r^{\beta}} \right|^{\lambda_\beta} \Big] \Big\} dx$$

$$\leq 2\tilde{T}(R_2 - R_1) \Big[\int_A \Big(\sum_{j=1}^{M} \left| \frac{\partial_{R_1}^{\beta} f_j(x)}{\partial r^{\beta}} \right|^{p} \Big) dx \Big]. \tag{15.250}$$

Proof. Based on Theorem 15.83. □

Next we give L_∞ results on the shell A involving several functions. We make

Assumption 15.107. Let $\alpha_1, \alpha_2 \in \mathbb{R}_+$, $\beta > \alpha_1, \alpha_2$, and for $j = 1, \ldots, M$, $M \in \mathbb{N}$, let $f_j \in L_1(A)$ with

$$\frac{\partial_{R_1}^{\beta} f_j(x)}{\partial r^{\beta}}, \in L_\infty(A), \ x \in A,$$

$$A := B(0, R_2) - \overline{B(0, R_1)} \subseteq \mathbb{R}^N, \ N \geq 2, \ 0 < R_1 < R_2.$$

Further assume that each $D_{R_1}^\beta f_j(rw) \in \mathbb{R}$ for almost all $r \in [R_1, R_2]$, for each $w \in S^{N-1}$, and for these $|D_{R_1}^\beta f_j(rw)| \leq M_j$ for some $M_j > 0$; $j = 1, \ldots, M$. For each $w \in S^{N-1} - (\cup_{j=1}^M K(f_j))$, we assume that $f_j(\cdot \, w)$ has an L_∞ fractional derivative $D_{R_1}^\beta f_j(\cdot \, w)$ in $[R_1, R_2]$, and that

$$D_{R_1}^{\beta-k} f_j(R_1 w) = 0, \ k = 1, \ldots, [\beta] + 1;$$

$j = 1, \ldots, M$. Let $\lambda_{\alpha_1}, \lambda_{\alpha_2}, \lambda_\beta \geq 0$. If $\alpha_1 = 0$ we set $\lambda_{\alpha_1} = 1$, and if $\alpha_2 = 0$ we set $\lambda_{\alpha_2} = 1$.

We present

Theorem 15.108. (All here are as in Assumption 15.107.) *Set*

$$\rho(R_2 - R_1) =$$

$$\frac{R_2^{N-1}(R_2 - R_1)^{(\beta\lambda_{\alpha_1} - \alpha_1\lambda_{\alpha_1} + \beta\lambda_{\alpha_2} - \alpha_2\lambda_{\alpha_2} + 1)}}{(\Gamma(\beta - \alpha_1 + 1))^{\lambda_{\alpha_1}} (\Gamma(\beta - \alpha_2 + 1))^{\lambda_{\alpha_2}} (\beta\lambda_{\alpha_1} - \alpha_1\lambda_{\alpha_1} + \beta\lambda_{\alpha_2} - \alpha_2\lambda_{\alpha_2} + 1)}. \tag{15.251}$$

Then

$$\int_A \left\{ \left\{ \sum_{j=1}^{M-1} \left[\left| \frac{\partial_{R_1}^{\alpha_1} f_j(x)}{\partial r^{\alpha_1}} \right|^{\lambda_{\alpha_1}} \left| \frac{\partial_{R_1}^{\alpha_2} f_{j+1}(x)}{\partial r^{\alpha_2}} \right|^{\lambda_{\alpha_2}} \left| \frac{\partial_{R_1}^{\beta} f_j(x)}{\partial r^{\beta}} \right|^{\lambda_\beta} + \right. \right. \right.$$

$$\left. \left| \frac{\partial_{R_1}^{\alpha_2} f_j(x)}{\partial r^{\alpha_2}} \right|^{\lambda_{\alpha_2}} \left| \frac{\partial_{R_1}^{\alpha_1} f_{j+1}(x)}{\partial r^{\alpha_1}} \right|^{\lambda_{\alpha_1}} \left| \frac{\partial_{R_1}^{\beta} f_{j+1}(x)}{\partial r^{\beta}} \right|^{\lambda_\beta} \right] \right\} +$$

$$\left[\left| \frac{\partial_{R_1}^{\alpha_1} f_1(x)}{\partial r^{\alpha_1}} \right|^{\lambda_{\alpha_1}} \left| \frac{\partial_{R_1}^{\alpha_2} f_M(x)}{\partial r^{\alpha_2}} \right|^{\lambda_{\alpha_2}} \left| \frac{\partial_{R_1}^{\beta} f_1(x)}{\partial r^{\beta}} \right|^{\lambda_\beta} + \right.$$

$$\left. \left. \left| \frac{\partial_{R_1}^{\alpha_2} f_1(x)}{\partial r^{\alpha_2}} \right|^{\lambda_{\alpha_2}} \left| \frac{\partial_{R_1}^{\alpha_1} f_M(x)}{\partial r^{\alpha_1}} \right|^{\lambda_{\alpha_1}} \left| \frac{\partial_{R_1}^{\beta} f_M(x)}{\partial r^{\beta}} \right|^{\lambda_\beta} \right] \right\} dx$$

$$\leq \frac{2\pi^{N/2}}{\Gamma(N/2)} \rho(R_2 - R_1) \left\{ \sum_{j=1}^{M} [M_j^{2(\lambda_{\alpha_1} + \lambda_\beta)} + M_j^{2\lambda_{\alpha_2}}] \right\}. \tag{15.252}$$

Proof. Based on Theorem 15.92 and a similar proof to that of Theorem 15.73. □

Similarly we give

Theorem 15.109. (All are as in Assumption 15.107; $\lambda_{\alpha_2} = 0.$) *Then*

$$\int_A \left[\sum_{j=1}^{M} \left| \frac{\partial_{R_1}^{\alpha_1} f_j(x)}{\partial r^{\alpha_1}} \right|^{\lambda_{\alpha_1}} \left| \frac{\partial_{R_1}^{\beta} f_j(x)}{\partial r^{\beta}} \right|^{\lambda_{\beta}} \right] dx \le$$

$$\frac{R_2^{N-1}(R_2 - R_1)^{(\beta\lambda_{\alpha_1} - \alpha_1\lambda_{\alpha_1} + 1)}}{(\Gamma(\beta - \alpha_1 + 1))^{\lambda_{\alpha_1}} [\beta\lambda_{\alpha_1} - \alpha_1\lambda_{\alpha_1} + 1]} \left(\sum_{j=1}^{M} M_j^{(\lambda_{\alpha_1} + \lambda_{\beta})} \right) \frac{2\pi^{N/2}}{\Gamma(N/2)}.$$

$$(15.253)$$

Proof. Based on Theorem 15.93, similar to Theorem 15.73. $\square$

It follows

Theorem 15.110. (All are as in Assumption 15.107; $\lambda_{\alpha_2} = \lambda_{\alpha_1} + \lambda_{\beta}.$) *Then*

$$\int_A \Bigg\{ \Bigg\{ \sum_{j=1}^{M-1} \Bigg[\left| \frac{\partial_{R_1}^{\alpha_1} f_j(x)}{\partial r^{\alpha_1}} \right|^{\lambda_{\alpha_1}} \left| \frac{\partial_{R_1}^{\alpha_2} f_{j+1}(x)}{\partial r^{\alpha_2}} \right|^{\lambda_{\alpha_1} + \lambda_{\beta}} \left| \frac{\partial_{R_1}^{\beta} f_j(x)}{\partial r^{\beta}} \right|^{\lambda_{\beta}} +$$

$$\left| \frac{\partial_{R_1}^{\alpha_2} f_j(x)}{\partial r^{\alpha_2}} \right|^{\lambda_{\alpha_1} + \lambda_{\beta}} \left| \frac{\partial_{R_1}^{\alpha_1} f_{j+1}(x)}{\partial r^{\alpha_1}} \right|^{\lambda_{\alpha_1}} \left| \frac{\partial_{R_1}^{\beta} f_{j+1}(x)}{\partial r^{\beta}} \right|^{\lambda_{\beta}} \Bigg] \Bigg\}$$

$$+ \Bigg[\left| \frac{\partial_{R_1}^{\alpha_1} f_1(x)}{\partial r^{\alpha_1}} \right|^{\lambda_{\alpha_1}} \left| \frac{\partial_{R_1}^{\alpha_2} f_M(x)}{\partial r^{\alpha_2}} \right|^{\lambda_{\alpha_1} + \lambda_{\beta}} \left| \frac{\partial_{R_1}^{\beta} f_1(x)}{\partial r^{\beta}} \right|^{\lambda_{\beta}} +$$

$$\left| \frac{\partial_{R_1}^{\alpha_2} f_1(x)}{\partial r^{\alpha_2}} \right|^{\lambda_{\alpha_1} + \lambda_{\beta}} \left| \frac{\partial_{R_1}^{\alpha_1} f_M(x)}{\partial r^{\alpha_1}} \right|^{\lambda_{\alpha_1}} \left| \frac{\partial_{R_1}^{\beta} f_M(x)}{\partial r^{\beta}} \right|^{\lambda_{\beta}} \Bigg] \Bigg\} dx \le$$

$$\frac{R_2^{N-1}}{(\Gamma(\beta - \alpha_1 + 1))^{\lambda_{\alpha_1}} (\Gamma(\beta - \alpha_2 + 1))^{\lambda_{\alpha_1} + \lambda_{\beta}}} \cdot$$

$$\frac{(R_2 - R_1)^{(2\beta\lambda_{\alpha_1} - \alpha_1\lambda_{\alpha_1} + \beta\lambda_{\beta} - \alpha_2\lambda_{\alpha_1} - \alpha_2\lambda_{\beta} + 1)}}{[2\beta\lambda_{\alpha_1} - \alpha_1\lambda_{\alpha_1} + \beta\lambda_{\beta} - \alpha_2\lambda_{\alpha_1} - \alpha_2\lambda_{\beta} + 1]} \left[\sum_{j=1}^{M} M_j^{2(\lambda_{\alpha_1} + \lambda_{\beta})} \right] \frac{4\pi^{N/2}}{\Gamma(N/2)}.$$

$$(15.254)$$

Proof. Based on Theorem 15.94 and similar to the proof of Theorem 15.73. $\square$

We continue with

Theorem 15.111. (All are as in Assumption 15.107; here $\lambda_{\beta} = 0$, $\lambda_{\alpha_1} = \lambda_{\alpha_2}.$) *Then*

$$\int_A \Bigg\{ \Bigg\{ \sum_{j=1}^{M-1} \Bigg[\left| \frac{\partial_{R_1}^{\alpha_1} f_j(x)}{\partial r^{\alpha_1}} \right|^{\lambda_{\alpha_1}} \left| \frac{\partial_{R_1}^{\alpha_2} f_{j+1}(x)}{\partial r^{\alpha_2}} \right|^{\lambda_{\alpha_1}} +$$

$$\left| \frac{\partial_{R_1}^{\alpha_2} f_j(x)}{\partial r^{\alpha_2}} \right|^{\lambda_{\alpha_1}} \left| \frac{\partial_{R_1}^{\alpha_1} f_{j+1}(x)}{\partial r^{\alpha_1}} \right|^{\lambda_{\alpha_1}} \Big] \Big\} +$$

$$\Big[\left| \frac{\partial_{R_1}^{\alpha_1} f_1(x)}{\partial r^{\alpha_1}} \right|^{\lambda_{\alpha_1}} \left| \frac{\partial_{R_1}^{\alpha_2} f_M(x)}{\partial r^{\alpha_2}} \right|^{\lambda_{\alpha_1}} +$$

$$\left| \frac{\partial_{R_1}^{\alpha_2} f_1(x)}{\partial r^{\alpha_2}} \right|^{\lambda_{\alpha_1}} \left| \frac{\partial_{R_1}^{\alpha_1} f_M(x)}{\partial r^{\alpha_1}} \right|^{\lambda_{\alpha_1}} \Big] \Big\} dx \leq \frac{4\pi^{N/2}}{\Gamma(N/2)}$$

$$\left(\frac{R_2^{N-1}(R_2 - R_1)^{(2\beta\lambda_{\alpha_1} - \alpha_1\lambda_{\alpha_1} - \alpha_2\lambda_{\alpha_1} + 1)}}{(\Gamma(\beta - \alpha_1 + 1)\Gamma(\beta - \alpha_2 + 1))^{\lambda_{\alpha_1}} (2\beta\lambda_{\alpha_1} - \alpha_1\lambda_{\alpha_1} - \alpha_2\lambda_{\alpha_1} + 1)} \right)$$

$$\left(\sum_{j=1}^{M} M_j^{2\lambda_{\alpha_1}} \right). \tag{15.255}$$

Proof. Based on Theorem 15.95. $\square$

Next we give

Theorem 15.112. (All are as in Assumption 15.107; here $\lambda_{\alpha_1} = 0$, $\lambda_{\alpha_2} = \lambda_\beta$.) *Then*

$$\int_A \Big\{ \Big\{ \sum_{j=1}^{M-1} \Big[\left| \frac{\partial_{R_1}^{\alpha_2} f_{j+1}(x)}{\partial r^{\alpha_2}} \right|^{\lambda_{\alpha_2}} \left| \frac{\partial_{R_1}^{\beta} f_j(x)}{\partial r^{\beta}} \right|^{\lambda_{\alpha_2}} +$$

$$\left| \frac{\partial_{R_1}^{\alpha_2} f_j(x)}{\partial r^{\alpha_2}} \right|^{\lambda_{\alpha_2}} \left| \frac{\partial_{R_1}^{\beta} f_{j+1}(x)}{\partial r^{\beta}} \right|^{\lambda_{\alpha_2}} \Big] \Big\} +$$

$$\Big[\left| \frac{\partial_{R_1}^{\alpha_2} f_M(x)}{\partial r^{\alpha_2}} \right|^{\lambda_{\alpha_2}} \left| \frac{\partial_{R_1}^{\beta} f_1(x)}{\partial r^{\beta}} \right|^{\lambda_{\alpha_2}} +$$

$$\left| \frac{\partial_{R_1}^{\alpha_2} f_1(x)}{\partial r^{\alpha_2}} \right|^{\lambda_{\alpha_2}} \left| \frac{\partial_{R_1}^{\beta} f_M(x)}{\partial r^{\beta}} \right|^{\lambda_{\alpha_2}} \Big] \Big\} dx \leq \frac{4\pi^{N/2}}{\Gamma(N/2)}$$

$$\left(\frac{R_2^{N-1}(R_2 - R_1)^{(\beta\lambda_{\alpha_2} - \alpha_2\lambda_{\alpha_2} + 1)}}{(\beta\lambda_{\alpha_2} - \alpha_2\lambda_{\alpha_2} + 1)(\Gamma(\beta - \alpha_2 + 1))^{\lambda_{\alpha_2}}} \right) \left(\sum_{j=1}^{M} M_j^{2\lambda_{\alpha_2}} \right). \tag{15.256}$$

Proof. Based on Theorem 15.96. $\square$

We also mention a special case.

Corollary 15.113. (to Theorem 15.111) *All are as in Assumption 15.107; here $\lambda_\beta = 0$, $\lambda_{\alpha_1} = \lambda_{\alpha_2}$, $\alpha_2 = \alpha_1 + 1$. Then*

$$\int_A \left\{ \left\{ \sum_{j=1}^{M-1} \left[\left| \frac{\partial_{R_1}^{\alpha_1} f_j(x)}{\partial r^{\alpha_1}} \right|^{\lambda_{\alpha_1}} \left| \frac{\partial_{R_1}^{\alpha_1+1} f_{j+1}(x)}{\partial r^{\alpha_1+1}} \right|^{\lambda_{\alpha_1}} + \right. \right. \right.$$

$$\left. \left| \frac{\partial_{R_1}^{\alpha_1+1} f_j(x)}{\partial r^{\alpha_1+1}} \right|^{\lambda_{\alpha_1}} \left| \frac{\partial_{R_1}^{\alpha_1} f_{j+1}(x)}{\partial r^{\alpha_1}} \right|^{\lambda_{\alpha_1}} \right] \right\} +$$

$$\left[\left| \frac{\partial_{R_1}^{\alpha_1} f_1(x)}{\partial r^{\alpha_1}} \right|^{\lambda_{\alpha_1}} \left| \frac{\partial_{R_1}^{\alpha_1+1} f_M(x)}{\partial r^{\alpha_1+1}} \right|^{\lambda_{\alpha_1}} + \right.$$

$$\left. \left| \frac{\partial_{R_1}^{\alpha_1+1} f_1(x)}{\partial r^{\alpha_1+1}} \right|^{\lambda_{\alpha_1}} \left| \frac{\partial_{R_1}^{\alpha_1} f_M(x)}{\partial r^{\alpha_1}} \right|^{\lambda_{\alpha_1}} \right] \right\} dx \leq \frac{4\pi^{N/2}}{\Gamma(N/2)}$$

$$\left(\frac{R_2^{N-1}(R_2 - R_1)^{(2\beta\lambda_{\alpha_1} - 2\alpha_1\lambda_{\alpha_1} - \lambda_{\alpha_1} + 1)}}{(2\beta\lambda_{\alpha_1} - 2\alpha_1\lambda_{\alpha_1} - \lambda_{\alpha_1} + 1)(\beta - \alpha_1)^{\lambda_{\alpha_1}} (\Gamma(\beta - \alpha_1))^{2\lambda_{\alpha_1}}} \right) \left(\sum_{j=1}^{M} M_j^{2\lambda_{\alpha_1}} \right).$$

$$(15.257)$$

Proof. Based on Corollary 15.97.   □

We finish the chapter with the proof that $D^\alpha f$ of Lemma 15.7 (see (15.8), also other similar fractional derivatives here) are such that

$$D^\alpha f \in AC([0, x]) \text{ for } \beta - \alpha \geq 1 \text{ and } D^\alpha f \in C([0, x]), \text{ for } \beta - \alpha \in (0, 1).$$

The last derives from the next.

Proposition 15.114. *Let $r > 0$, $F \in L_\infty(a, b)$, and*

$$G(s) := \int_a^s (s - t)^{r-1} F(t) dt, \qquad (15.258)$$

all $s \in [a, b]$. Then $G \in AC([a, b])$ for $r \geq 1$ and $G \in C([a, b])$, only for $r \in (0, 1)$.

Proof. (1) Case $r \geq 1$. We use the definition of absolute continuity. So for every $\epsilon > 0$ we need $\delta > 0$: whenever (a_i, b_i), $i = 1, \ldots, n$, are disjoint open subintervals of $[a, b]$, then

$$\sum_{i=1}^{n} (b_i - a_i) < \delta \implies \sum_{i=1}^{n} |G(b_i) - G(a_i)| < \epsilon.$$

If $\|F\|_\infty = 0$, then $G(s) = 0$, for all $s \in [a, b]$, the trivial case and all fulfilled. So we assume $\|F\|_\infty \neq 0$. Hence we have

$$G(b_i) - G(a_i) = \int_a^{b_i} (b_i - t)^{r-1} F(t) dt - \int_a^{a_i} (a_i - t)^{r-1} F(t) dt =$$

$$\int_a^{a_i}(b_i-t)^{r-1}F(t)dt - \int_a^{a_i}(a_i-t)^{r-1}F(t)dt + \int_{a_i}^{b_i}(b_i-t)^{r-1}F(t)dt =$$

$$\int_a^{a_i}\left((b_i-t)^{r-1}-(a_i-t)^{r-1}\right)F(t)dt + \int_{a_i}^{b_i}(b_i-t)^{r-1}F(t)dt. \quad (15.259)$$

Call

$$I_i := \int_a^{a_i}|(b_i-t)^{r-1}-(a_i-t)^{r-1}|dt. \quad (15.260)$$

Thus

$$|G(b_i)-G(a_i)| \le \left[I_i + \frac{(b_i-a_i)^r}{r}\right]\|F\|_\infty := T_i. \quad (15.261)$$

If $r=1$, then $I_i=0$, and

$$|G(b_i)-G(a_i)| \le \|F\|_\infty(b_i-a_i), \quad (15.262)$$

for all $i := 1,\dots,n$.

If $r > 1$, then because $\left[(b_i-t)^{r-1}-(a_i-t)^{r-1}\right] \ge 0$, for all $t \in [a,a_i]$, we find

$$I_i = \int_a^{a_i}\left((b_i-t)^{r-1}-(a_i-t)^{r-1}\right)dt = \frac{(b_i-a)^r-(a_i-a)^r-(b_i-a_i)^r}{r}$$

$$= \frac{r(\xi-a)^{r-1}(b_i-a_i)-(b_i-a_i)^r}{r}, \quad (15.263)$$

for some $\xi \in (a_i,b_i)$. Therefore, it holds

$$I_i \le \frac{r(b-a)^{r-1}(b_i-a_i)-(b_i-a_i)^r}{r}, \quad (15.264)$$

and

$$\left(I_i + \frac{(b_i-a_i)^r}{r}\right) \le (b-a)^{r-1}(b_i-a_i). \quad (15.265)$$

That is,

$$T_i \le \|F\|_\infty(b-a)^{r-1}(b_i-a_i),$$

so that

$$|G(b_i)-G(a_i)| \le \|F\|_\infty(b-a)^{r-1}(b_i-a_i),\text{ for all } i=1,\dots,n. \quad (15.266)$$

So in the case of $r=1$, and by choosing $\delta := \epsilon/\|F\|_\infty$, we get

$$\sum_{i=1}^n |G(b_i)-G(a_i)| \le^{(15.262)} \|F\|_\infty\left(\sum_{i=1}^n(b_i-a_i)\right) \le \|F\|_\infty\delta = \epsilon,$$

$$(15.267)$$

proving for $r = 1$ that G is absolutely continuous. In the case of $r > 1$, and by choosing $\delta := \epsilon / \|F\|_\infty (b-a)^{r-1}$, we get

$$\sum_{i=1}^n |G(b_i) - G(a_i)| \leq^{(15.266)} \|F\|_\infty (b-a)^{r-1} \left(\sum_{i=1}^n (b_i - a_i) \right) \quad (15.268)$$

$$\leq \|F\|_\infty (b-a)^{r-1} \delta = \epsilon,$$

proving for $r > 1$ that G is absolutely continuous again.

(2) Case of $0 < r < 1$. Let $a_{i_*}, b_{i_*} \in [a, b] : a_{i_*} \leq b_{i_*}$. Then $(a_{i_*} - t)^{r-1} \geq (b_{i_*} - t)^{r-1}$, for all $t \in [a, a_{i_*}]$. Then

$$I_{i_*} = \int_a^{a_{i_*}} \left((a_{i_*} - t)^{r-1} - (b_{i_*} - t)^{r-1} \right) dt = \frac{(b_{i_*} - a_{i_*})^r}{r}$$

$$+ \left(\frac{(a_{i_*} - a)^r - (b_{i_*} - a)^r}{r} \right) \leq \frac{(b_{i_*} - a_{i_*})^r}{r}, \quad (15.269)$$

by $(a_{i_*} - a)^r - (b_{i_*} - a)^r < 0$. Therefore

$$I_{i_*} \leq \frac{(b_{i_*} - a_{i_*})^r}{r} \quad (15.270)$$

and

$$T_{i_*} \leq \frac{2(b_{i_*} - a_{i_*})^r}{r} \|F\|_\infty, \quad (15.271)$$

proving that

$$|G(b_{i_*}) - G(a_{i_*})| \leq \left(\frac{2\|F\|_\infty}{r} \right) (b_{i_*} - a_{i_*})^r, \quad (15.272)$$

which proves that G is continuous. Taking the special case of $a = 0$ and $F(t) = 1$, for all $t \in [0, b]$, we get that

$$G(s) = \frac{s^r}{r}, \text{ all } s \in [0, b], \text{ for } 0 < r < 1. \quad (15.273)$$

The last is a Lipschitz function of order $r \in (0, 1)$, which is not absolutely continuous. Consequently G for $r \in (0, 1)$ in general, cannot be absolutely continuous. That completes the proof. $\square$

16
Caputo Fractional Multivariate Opial-Type Inequalities over a Spherical Shell

Here is introduced the concept of the Caputo fractional radial derivative for a function defined on a spherical shell. Using polar coordinates we are able to derive multivariate Opial-type inequalities over a spherical shell of $\mathbb{R}^N$, $N \geq 2$, by studying the topic in all possibilities. Our results involve one, two, or more functions. We present many univariate Caputo fractional Opial-type inequalities, several of which are used to establish results on the shell. We give an application to prove the uniqueness of solution of a general partial differential equation on the shell. Also we apply our results for Riemann – Liouville fractional derivatives. This treatment relies on [58].

16.1   Introduction

This chapter is inspired by articles of Opial [315], Bessack [80], and Anastassiou, Koliha, and Pecaric [64, 65], and Anastassiou [46, 48].

Opial-type inequalities usually find applications in establishing the uniqueness of solution of initial value problems for differential equations and their systems; see Willett [406]. In this chapter we present a series of various Caputo fractional multivariate Opial type inequalities over spherical shells. To achieve our goal we use polar coordinates, and we introduce and use the Caputo fractional radial derivative. We work on the spherical shell, and not on the ball, because a radial derivative cannot be defined at zero. So, we reduce the problem to a univariate one.

G.A. Anastassiou, *Fractional Differentiation Inequalities*,   391
DOI 10.1007/978-0-387-98128-4_16, © Springer Science+Business Media, LLC 2009

Therefore we derive and use a large array of univariate Opial-type inequalities involving Caputo fractional derivatives; these are Caputo fractional derivatives defined at arbitrary anchor point $a \in \mathbb{R}$. In our results we involve one, two, or several functions. But first we need to develop an extensive background in two parts, then follow the main results.

At the end we give an application proving the uniqueness of solution for a general PDE initial value problem. Also we re-establish our results by involving Riemann – Liouville fractional derivatives defined at an arbitrary anchor point.

In this chapter to build our background regarding the Caputo fractional derivative we use the excellent monograph [134].

The Caputo derivative was introduced in 1967; see [102], and also see [112, 114].

It happens that the Riemann – Liouville fractional derivative has some disadvantages when modeling real-world phenomena with fractional differential equations. One reason is that the initial conditions there involve fractional derivatives that are difficult to connect with actual data, and so on. However, Caputo fractional derivative modeling involves initial conditions that are described by ordinary derivatives, much easier to write based on real-world data. So more and more in recent years the Caputo version is usually preferred when physical models are described, because the physical interpretation of the prescribed data is clear, and therefore it is in general possibly easier to gather these data, for example, by appropriate measurements. Also from the pure mathematics side there are reasons to prefer the Caputo fractional derivative.

16.2   Background—I

Here we follow [134].

We start with

Definition 16.1. Let $\nu \geq 0$; the operator J_a^ν, defined on $L_1[a, b]$ by

$$J_a^\nu f(x) := \frac{1}{\Gamma(\nu)} \int_a^x (x - t)^{\nu - 1} f(t) \, dt \qquad (16.1)$$

for $a \leq x \leq b$, is called the Riemann – Liouville fractional integral operator of order ν.

For $\nu = 0$, we set $J_a^0 := I$, the identity operator. Here Γ stands for the gamma function.

Theorem 16.2. [134] Let $f \in L_1[a, b]$, $\nu > 0$. Then, the integral $J_a^\nu f(x)$ exists for almost every $x \in [a, b]$.

Moreover, $J_a^\nu f \in L_1([a,b])$.

We need

Theorem 16.3. [134] *Let $m, n \geq 0$, $\Phi \in L_1([a,b])$.*
Then

$$J_a^m J_a^n \Phi = J_a^{m+n} \Phi \tag{16.2}$$

holds almost everywhere on $[a,b]$. If additionally $\Phi \in C([a,b])$ or $m+n \geq 1$, then the identity holds everywhere on $[a,b]$.

We give

Definition 16.4. [134] *Let $\nu \in \mathbb{R}_+$ and $m = \lceil \nu \rceil$; $\lceil \cdot \rceil$ is the ceiling of number. The operator D_a^ν, defined by*

$$D_a^\nu f := D^m J_a^{m-\nu} f, \quad D := \frac{d}{dx}, \tag{16.3}$$

is called the Riemann – Liouville fractional differential operator of order ν. For $\nu = 0$, we set $D_a^0 := I$, the identity operator. If $\nu \in \mathbb{N}$ then $D_a^\nu f = f^{(\nu)}$, the ordinary ν-order derivative.

Next we give

Definition 16.5. (p. 37, [134]) *Let $\nu \geq 0$ and $n := \lceil \nu \rceil$, $a \in \mathbb{R}$. Then, we define the operator*
$$\hat{D}_a^\nu f := J_a^{n-\nu} f^{(n)}, \tag{16.4}$$
whenever $f^{(n)} \in L_1([a,b])$.

Also we need

Theorem 16.6. (p. 37, [134]) *Let $\nu \geq 0$, $n := \lceil \nu \rceil$. Moreover assume that $f \in AC^n([a,b])$(the space of functions with absolutely continuous $(n-1)$st derivative). Then*

$$\hat{D}_a^\nu f = D_a^\nu(f - T_{n-1}(f;a)), \quad \text{a.e. on } [a,b], \tag{16.5}$$

where

$$T_{n-1}(f;a)(x) := \sum_{k=0}^{n-1} \frac{f^{(k)}(a)}{k!}(x-a)^k, \quad x \in [a,b], \tag{16.6}$$

is the Taylor polynomial of degree $n-1$ of f, centered at a.

Next we give the definition of Caputo fractional derivative [134].

Definition 16.7. (p. 38, [134]) Assume that f is such that $D_a^\nu (f - T_{n-1}(f;a))(x)$ exists for some $x \in [a,b]$. Then we define the Caputo fractional derivative by

$$D_{*a}^\nu f(x) := D_a^\nu (f - T_{n-1}(f;a))(x). \tag{16.7}$$

So the above definition applies to all points $x \in [a,b] : D_a^\nu (f - T_{n-1}(f;a))(x) \in \mathbb{R}$.

We have

Corollary 16.8. *Let* $\nu \geq 0$, $n := \lceil \nu \rceil$, $f \in AC^n([a,b])$. *Then the Caputo fractional derivative*

$$D_{*a}^\nu f(x) = \frac{1}{\Gamma(n-\nu)} \int_a^x (x-t)^{n-\nu-1} f^{(n)}(t)\, dt \tag{16.8}$$

exists almost everywhere for x *in* $[a,b]$.

We have

Corollary 16.9. *Let* $\nu \geq 0$, $n := \lceil \nu \rceil$, $f \in AC^n([a,b])$. *Then,* $D_{*a}^\nu f$ *exists iff* $D_a^\nu f$ *exists.*

Proof. By linearity of D_a^ν operator and assumption. □

We need

Lemma 16.10. [134] *Let* $\nu \geq 0$, $n = \lceil \nu \rceil$. *Assume that* f *is such that both* $D_{*a}^\nu f$ *and* $D_a^\nu f$ *exist.*
Then,

$$D_{*a}^\nu f(x) = D_a^\nu f(x) - \sum_{k=0}^{n-1} \frac{f^{(k)}(a)}{\Gamma(k-\nu+1)} (x-a)^{k-\nu}. \tag{16.9}$$

Lemma 16.11. [134] *All are as in Lemma 16.10.*
Additionally assume that $f^{(k)}(a) = 0$ *for* $k = 0, 1, \ldots, n-1$. *Then,*

$$D_{*a}^\nu f = D_a^\nu f. \tag{16.10}$$

In conclusion

Corollary 16.12. *Let* $\nu \geq 0$, $n := \lceil \nu \rceil$, $f \in AC^n([a,b])$, $D_{*a}^\nu f$ *exists or* $D_a^\nu f$ *exists, and* $f^{(k)}(a) = 0$, $k = 0, 1, \ldots, n-1$. *Then*

$$D_a^\nu f = D_{*a}^\nu f. \tag{16.11}$$

We need the following Taylor – Caputo formula.

Theorem 16.13. (p. 40, [134]) *Let* $\nu \geq 0$, $n := \lceil \nu \rceil$, $f \in AC^n([a,b])$. *Then*

$$f(x) = \sum_{k=0}^{n-1} \frac{f^{(k)}(a)}{k!} (x-a)^k + J_a^\nu D_{*a}^\nu f(x), \qquad (16.12)$$

$\forall x \in [a,b]$.

Clearly here $J_a^\nu D_{*a}^\nu f \in AC^n([a,b])$.

Corollary 16.14. *Let* $\nu \geq 0$, $n := \lceil \nu \rceil$, $f \in AC^n([a,b])$, *and* $f^{(k)}(a) = 0$, $k = 0,1,\ldots,n-1$. *Then*

$$f(x) = J_a^\nu D_{*a}^\nu f(x) = \frac{1}{\Gamma(\nu)} \int_a^x (x-t)^{\nu-1} D_{*a}^\nu f(t)\, dt. \qquad (16.13)$$

We need

Lemma 16.15. *Let* $\nu \geq \gamma + 1$, $\gamma \geq 0$. *Call* $n := \lceil \nu \rceil$, $m := \lceil \gamma \rceil$. *Then* $n - m \geq 1$; *that is,* $m \leq n - 1$.

Proof. Clearly $\nu \geq 1$ and $\nu > \gamma$, $\nu - \gamma \geq 1$. By $\gamma + 1 > m$ we get $\nu > m$, and $n > m$; that is, $\nu - m > 0$ and $n - m > 0$.

We see that $\nu \geq \gamma + 1 \geq [\gamma] + 1$, where $[\cdot]$ is the integral part. Thus $\nu \geq ([\gamma] + 1) \in \mathbb{N}$ and $\nu \geq [\nu] \geq [\gamma] + 1$.

Therefore

$$[\nu] - [\gamma] \geq 1, \qquad (16.14)$$

which is used next.

We distinguish the following cases.

(i) Let $\nu, \gamma \notin \mathbb{N}$; then $\lceil \nu \rceil = [\nu] + 1$, $\lceil \gamma \rceil = [\gamma] + 1$. By (16.14) we get $([\nu] + 1) - ([\gamma] + 1) \geq 1$. Hence $n - m \geq 1$.

(ii) Let $\nu, \gamma \in \mathbb{N}$; then $[\nu] = \lceil \nu \rceil = \nu$, $[\gamma] = \lceil \gamma \rceil = \gamma$. So by (16.14) $n - m \geq 1$.

(iii) Let $\nu \notin \mathbb{N}$, $\gamma \in \mathbb{N}$. Then $n = \lceil \nu \rceil = [\nu] + 1$, $m = \lceil \gamma \rceil = [\gamma] = \gamma$. Hence by (16.14) we have $(\lceil \nu \rceil - 1) - m \geq 1$, and $\lceil \nu \rceil - m \geq 2 > 1$. Hence $n - m > 1$.

(iv) Let $\nu \in \mathbb{N}$, $\gamma \notin \mathbb{N}$. Then $1 + \gamma < \lceil \gamma \rceil + 1 = \lceil \gamma + 1 \rceil$, and $1 + \gamma \leq \nu \in \mathbb{N}$ by assumption.

Therefore $\lceil \gamma \rceil + 1 \leq \nu$, and $\nu - \lceil \gamma \rceil \geq 1$. So that again $n - m \geq 1$.

The claim is proved in all cases. $\square$

We present the representation theorem.

Theorem 16.16. *Let* $\nu \geq \gamma + 1$, $\gamma \geq 0$. *Call* $n := \lceil \nu \rceil$, $m := \lceil \gamma \rceil$. *Assume* $f \in AC^n([a,b])$, *such that* $f^{(k)}(a) = 0$, $k = 0,1,\ldots,n-1$, *and* $D_{*a}^\nu f \in L_\infty(a,b)$.

Then

$$D_{*a}^{\gamma} f \in C\left([a,b]\right), \quad D_{*a}^{\gamma} f\left(x\right) = J_{a}^{m-\gamma} f^{(m)}\left(x\right), \qquad (16.15)$$

and

$$D_{*a}^{\gamma} f\left(x\right) = \frac{1}{\Gamma\left(\nu-\gamma\right)} \int_{a}^{x} \left(x-t\right)^{\nu-\gamma-1} D_{*a}^{\nu} f\left(t\right) dt, \qquad (16.16)$$

$\forall x \in [a,b]$.

Proof. If $\gamma = 0$ then (16.16) collapses to (16.13); also (16.15) is clear. So we assume $\gamma > 0$. By Lemma 16.15 we have $m \le n-1$. By the assumption $f \in AC^{n}\left([a,b]\right)$ we get that $f \in C^{n-1}\left([a,b]\right)$ and thus $f \in C^{m}\left([a,b]\right)$.

By Lemma 3.7, p. 41 of [134] we get that $D_{*a}^{\gamma} f = J_{a}^{m-\gamma} f^{(m)} \in C\left([a,b]\right)$ and $D_{*a}^{\gamma} f\left(a\right) = 0$, for $\gamma \notin \mathbb{N}$. Clearly the last statement is true also when $\gamma \in \mathbb{N}$, thus proving (16.15) and the first claim.

Recall that we have $\nu - m > 0$, and $\nu - 1 > 0$ by $\gamma > 0$. Using $\Gamma\left(p+1\right) = p\Gamma\left(p\right)$, $p > 0$, (16.13), and Theorem 7 of [48], we obtain

$$f^{(m)}\left(x\right) = J_{a}^{\nu-m} D_{*a}^{\nu} f\left(x\right), \quad \forall x \in [a,b]. \qquad (16.17)$$

Therefore we get

$$D_{*a}^{\gamma} f\left(x\right) = J_{a}^{m-\gamma} f^{(m)}\left(x\right) \stackrel{(16.17)}{=} J_{a}^{m-\gamma} J_{a}^{\nu-m} D_{*a}^{\nu} f\left(x\right)$$

(by Theorem 2.2, p. 14 of [134], and $\nu - \gamma \ge 1$)

$$= J_{a}^{\nu-\gamma} D_{*a}^{\nu} f\left(x\right), \quad \forall x \in [a,b],$$

thus proving (16.16). $\square$

We also give the representation theorem.

Theorem 16.17. *Let* $\nu \ge \gamma + 1$, $\gamma \ge 0$. *Call* $n := \lceil \nu \rceil$, $m := \lceil \gamma \rceil$. *Let* $f \in AC^{n}\left([a,b]\right)$, *such that* $f^{(k)}\left(a\right) = 0$, $k = 0,1,\ldots,n-1$. *Assume there exists* $D_{a}^{\nu} f\left(x\right) \in \mathbb{R}$, $\forall x \in [a,b]$, *and* $D_{a}^{\nu} f \in L_{\infty}\left(a,b\right)$. *Then*

$$D_{a}^{\gamma} f \in C\left([a,b]\right), \quad D_{a}^{\gamma} f\left(x\right) = J_{a}^{m-\gamma} f^{(m)}\left(x\right), \qquad (16.18)$$

$\forall x \in [a,b]$,

$$D_{a}^{\gamma} f\left(x\right) = \frac{1}{\Gamma\left(\nu-\gamma\right)} \int_{a}^{x} \left(x-t\right)^{\nu-\gamma-1} D_{a}^{\nu} f\left(t\right) dt, \qquad (16.19)$$

$\forall x \in [a,b]$.

Proof. By Corollaries 16.9 and 16.12 we get existing $D_{*a}^{\nu} f\left(x\right) \in \mathbb{R}$, and that $D_{*a}^{\nu} f\left(x\right) = D_{a}^{\nu} f\left(x\right)$, $\forall x \in [a,b]$. That is, $D_{*a}^{\nu} f \in L_{\infty}\left(a,b\right)$ and by (16.16) we have

$$D_{*a}^{\gamma} f\left(x\right) = \frac{1}{\Gamma\left(\nu-\gamma\right)} \int_{a}^{x} \left(x-t\right)^{\nu-\gamma-1} D_{a}^{\nu} f\left(t\right) dt, \quad \forall x \in [a,b]. \quad (16.20)$$

Because $D_{*a}^{\gamma}f \in C([a,b])$ we get $D_{*a}^{\gamma}f(x) \in \mathbb{R}$, $\forall x \in [a,b]$. And because $f \in C^m([a,b])$ then $f \in AC^m([a,b])$. By Corollary 16.9 $D_a^{\gamma}f$ exists. Also $f^{(k)}(a) = 0$, for $k = 0, 1, \ldots, m-1$. Thus by Corollary 16.12 we obtain $D_a^{\gamma}f(x) = D_{*a}^{\gamma}f(x)$, $\forall x \in [a,b]$. Now by (16.20) we have established (16.19). $\square$

16.3 Main Results

16.3.1 Results Involving One Function

We present the following theorem.

Theorem 16.18. *Let $\nu \geq \gamma + 1$, $\gamma \geq 0$. Call $n := \lceil \nu \rceil$ and assume $f \in AC^n([a,b])$ such that $f^{(k)}(a) = 0$, $k = 0, 1, \ldots, n-1$, and $D_{*a}^{\nu}f \in L_{\infty}(a,b)$. Let $p, q > 1$ such that $1/p + 1/q = 1$, $a \leq x \leq b$.
Then*

$$\int_a^x |D_{*a}^{\gamma}f(\omega)|\,|(D_{*a}^{\nu}f)(\omega)|\,d\omega \leq$$

$$\frac{(x-a)^{(p\nu - p\gamma - p + 2)/p}}{(\sqrt[q]{2})\,\Gamma(\nu - \gamma)\,((p\nu - p\gamma - p + 1)(p\nu - p\gamma - p + 2))^{1/p}}$$

$$\cdot \left(\int_a^x |D_{*a}^{\nu}f(\omega)|^q\,d\omega \right)^{2/q}. \tag{16.21}$$

Proof. Similar to Theorem 25.2, p. 545, [19], and Theorem 2.1 of [64]. $\square$

A related extreme case comes next.

Proposition 16.19. *All are as in Theorem 16.18, but with $p = 1$ and $q = \infty$, we find*

$$\int_a^x |D_{*a}^{\gamma}f(\omega)|\,|D_{*a}^{\nu}f(\omega)|\,d\omega \leq$$

$$\frac{(x-a)^{\nu - \gamma + 1}}{\Gamma(\nu - \gamma + 2)} \left(\|D_{*a}^{\nu}f\|_{\infty,(a,x)} \right)^2. \tag{16.22}$$

Proof. Similar to Proposition 25.1, p. 547, [19]. $\square$

The converse of (16.21) follows.

Theorem 16.20. *Let* $\nu \geq \gamma + 1$, $\gamma \geq 0$. *Call* $n := \lceil \nu \rceil$ *and assume* $f \in AC^n\left([a,b]\right)$ *such that* $f^{(k)}(a) = 0$, $k = 0,1,\ldots,n-1$, *and* $D_{*a}^\nu f$, $1/D_{*a}^\nu f \in L_\infty(a,b)$. *Suppose that* $D_{*a}^\nu f$ *is of fixed sign a.e. in* $[a,b]$. *Let* p,q *be such that* $0 < p < 1$, $q < 0$ *and* $1/p + 1/q = 1$, $a \leq x \leq b$. *Then*

$$\int_a^x \left|D_{*a}^\gamma f(\omega)\right| \left|D_{*a}^\nu f(\omega)\right| d\omega \geq$$

$$\frac{(x-a)^{(p\nu - p\gamma - p + 2)/p}}{\left(\sqrt[q]{2}\right) \Gamma(\nu - \gamma)\left((p\nu - p\gamma - p + 1)(p\nu - p\gamma - p + 2)\right)^{1/p}}$$

$$\cdot \left(\int_a^x \left|D_{*a}^\nu f(\omega)\right|^q d\omega\right)^{2/q}. \qquad (16.23)$$

Proof. Similar to Theorem 25.3, p. 547, [19], and Theorem 2.3 of [64]. □

We give

Theorem 16.21. *Let* $\nu \geq 2$, $k \geq 0$, $\nu \geq k + 2$. *Call* $n := \lceil \nu \rceil$ *and assume* $f \in AC^n\left([a,b]\right)$ *such that* $f^{(j)}(a) = 0$, $j = 0,1,\ldots,n-1$, *and* $D_{*a}^\nu f \in L_\infty(a,b)$. *Let* $p,q > 1$ *such that* $1/p + 1/q = 1$, $a \leq x \leq b$. *Then*

$$\int_a^x \left|D_{*a}^k f(\omega)\right| \left|D_{*a}^{k+1} f(\omega)\right| d\omega \leq$$

$$\frac{(x-a)^{2(p\nu - pk - p + 1)/p}}{2\left(\Gamma(\nu - k)\right)^2 (p\nu - pk - p + 1)^{2/p}}$$

$$\cdot \left(\int_a^x \left|D_{*a}^\nu f(\omega)\right|^q d\omega\right)^{2/q}. \qquad (16.24)$$

Proof. Similar to Theorem 25.4, p. 549, [19], and Theorem 2.4 of [64]. □

The extreme case follows.

Proposition 16.22. *Under the assumptions of Theorem 16.21 when* $p = 1$, $q = \infty$ *we find*

$$\int_a^x \left|D_{*a}^k f(\omega)\right| \left|D_{*a}^{k+1} f(\omega)\right| d\omega \leq$$

$$\frac{(x-a)^{2(\nu - k)}\left(\|D_{*a}^\nu f\|_{\infty,(a,x)}\right)^2}{2\left(\Gamma(\nu - k + 1)\right)^2}. \qquad (16.25)$$

Proof. Similar to Proposition 25.2, p. 551 of [19]. □

We give the related converse result.

Theorem 16.23. *Let* $\nu \geq 2, k \geq 0, \nu \geq k+2$. *Call* $n := \lceil \nu \rceil$. *Assume* $f \in AC^n([a,b])$ *such that* $f^{(j)}(a) = 0, j = 0, 1, \ldots, n-1$, *and* $D_{*a}^{\nu}f, 1/D_{*a}^{\nu}f \in L_{\infty}(a,b)$. *Suppose that* $D_{*a}^{\nu}f$ *is of fixed sign a.e. in* $[a,b]$. *Let* p, q *be such that* $0 < p < 1$, $q < 0$ *and* $1/p + 1/q = 1$, $a \leq x \leq b$. *Then*

$$\int_a^x \left| D_{*a}^k f(\omega) \right| \left| D_{*a}^{k+1} f(\omega) \right| d\omega \geq$$

$$\frac{(x-a)^{2(p\nu - pk - p + 1)/p}}{2\left(\Gamma(\nu - k)\right)^2 (p\nu - pk - p + 1)^{2/p}}$$

$$\cdot \left(\int_a^x |D_{*a}^{\nu}f(\omega)|^q \, d\omega \right)^{2/q}. \tag{16.26}$$

Proof. Similar to Theorem 25.5, p. 553 of [19]. □

Next we present

Theorem 16.24. *Let* $\gamma_i \geq 0, \nu \geq 1, \nu - \gamma_i \geq 1; i = 1, \ldots, l, n := \lceil \nu \rceil$, *and* $f \in AC^n([a,b])$ *such that* $f^{(k)}(a) = 0, k = 0, 1, \ldots, n-1$, *and* $D_{*a}^{\nu}f \in L_{\infty}(a,b)$. *Here* $a \leq x \leq b$; $q_1(x), q_2(x)$ *continuous functions on* $[a,b]$ *such that* $q_1(x) \geq 0, q_2(x) > 0$ *on* $[a,b]$, *and* $r_i > 0 : \sum_{i=1}^l r_i = r$. *Let* $s_1, s_1' > 1 : 1/s_1 + 1/s_1' = 1$ *and* $s_2, s_2' > 1 : 1/s_2 + 1/s_2' = 1$, *and* $p > s_2$.
Denote by

$$Q_1(x) := \left(\int_a^x (q_1(\omega))^{s_1'} \, d\omega \right)^{1/s_1'} \tag{16.27}$$

and

$$Q_2(x) := \left(\int_a^x (q_2(\omega))^{-s_2'/p} \, d\omega \right)^{r/s_2'}, \tag{16.28}$$

$$\sigma := \frac{p - s_2}{ps_2}. \tag{16.29}$$

Then

$$\int_a^x q_1(\omega) \prod_{i=1}^l \left| D_{*a}^{\gamma_i} f(\omega) \right|^{r_i} d\omega \leq Q_1(x) Q_2(x)$$

$$\prod_{i=1}^l \left\{ \frac{\sigma^{r_i \sigma}}{(\Gamma(\nu - \gamma_i))^{r_i} (\nu - \gamma_i - 1 + \sigma)^{r_i \sigma}} \right\}$$

$$\cdot \frac{(x-a)^{\left(\sum_{i=1}^{l}(\nu-\gamma_i-1)r_i+\sigma r+1/s_1\right)}}{\left(\left(\sum_{i=1}^{l}(\nu-\gamma_i-1)r_is_1\right)+rs_1\sigma+1\right)^{1/s_1}}$$

$$\cdot \left(\int_a^x q_2(\omega) |D_{*a}^\nu f(\omega)|^p \, d\omega\right)^{r/p}. \tag{16.30}$$

Proof. Similar to Theorem 26.1, p. 567 of [19], and Theorem 2.1 of [65]. □

The counterpart of the last theorem follows.

Theorem 16.25. Let $\gamma_i \geq 0$, $\nu \geq 1$, $\nu - \gamma_i \geq 1$; $i = 1, \ldots, l$, $n := \lceil \nu \rceil$, and $f \in AC^n([a,b])$ such that $f^{(k)}(a) = 0$, $k = 0, 1, \ldots, n-1$, and $D_{*a}^\nu f, 1/D_{*a}^\nu f \in L_\infty(a,b)$. Here $a \leq x \leq b$; $q_1(x), q_2(x) > 0$ are continuous functions on $[a,b]$ and $r_i > 0 : \sum_{i=1}^{l} r_i = r$. Let $0 < s_1$, $s_2 < 1$, and s_1', $s_2' < 0$ such that $1/s_1 + 1/s_1' = 1$, $1/s_2 + 1/s_2' = 1$. Assume that $D_{*a}^\nu f(t)$ is of fixed sign a.e. in $[a,b]$. Denote

$$Q_1(x) := \left(\int_a^x (q_1(\omega))^{s_1'} \, d\omega\right)^{1/s_1'}, \tag{16.31}$$

$$Q_2(x) := \left(\int_a^x (q_2(\omega))^{-s_2'} \, d\omega\right)^{r/s_2'}. \tag{16.32}$$

Set

$$\lambda := \frac{s_1 s_2}{s_1 s_2 - 1}. \tag{16.33}$$

Then

$$\int_a^x q_1(\omega) \left(\prod_{i=1}^{l} |D_{*a}^{\gamma_i} f(\omega)|^{r_i}\right) d\omega \geq$$

$$\frac{Q_1(x)\, Q_2(x)}{\prod_{i=1}^{l}\left\{(\Gamma(\nu-\gamma_i))^{r_i}\left((\nu-\gamma_i-1)s_2^2 s_1 + 1\right)^{(r_i/s_2^2 s_1)}\right\}}$$

$$\cdot \frac{(x-a)^{\left\{\left(\sum_{i=1}^{l} r_i((\nu-\gamma_i-1)s_1+s_2^{-2})\right)+1\right\}/s_1}}{\left\{\left(\sum_{i=1}^{l} r_i\left((\nu-\gamma_i-1)s_1+s_2^{-2}\right)\right)+1\right\}^{1/s_1}}$$

$$\cdot \left(\int_a^x q_2^{\lambda s_2}(\omega) |D_{*a}^\nu f(\omega)|^{\lambda s_2} \, d\omega\right)^{r/\lambda s_2}. \tag{16.34}$$

Proof. Similar to Theorem 26.2, p. 570 of [19], and Theorem 2.3 of [65]. □

A related extreme case comes next for $p = 1$ and $q = \infty$.

Theorem 16.26. *Let $\nu \geq \gamma_i + 1$, $\gamma_i \geq 0$; $i = 1, \ldots, l$. Call $n := \lceil \nu \rceil$ and assume $f \in AC^n([a, b])$ such that $f^{(k)}(a) = 0$, $k = 0, 1, \ldots, n - 1$, and $D_{*a}^{\nu} f \in L_{\infty}(a, b)$. Here $a \leq x \leq b$, with $0 \leq \tilde{q}(\omega) \in L_{\infty}(a, b)$ and $r_i > 0 : \sum_{i=1}^{l} r_i = r$. Then*

$$\int_a^x \tilde{q}(\omega) \prod_{i=1}^{l} |D_{*a}^{\gamma_i} f(\omega)|^{r_i} d\omega \leq$$

$$\left\{ \frac{\|\tilde{q}\|_{\infty, (a, x)} \left(\|D_{*a}^{\nu} f\|_{\infty, (a, x)} \right)^r}{\prod_{i=1}^{l} (\Gamma(\nu - \gamma_i + 1))^{r_i}} \right\} \cdot$$

$$\left\{ \frac{(x - a)^{r\nu - \sum_{i=1}^{l} r_i \gamma_i + 1}}{\left(r\nu - \sum_{i=1}^{l} r_i \gamma_i + 1 \right)} \right\}. \tag{16.35}$$

Proof. Similar to Proposition 26.1 of [19], p. 573, and Theorem 2.2 of [65]. □

We continue with the interesting

Theorem 16.27. *Let $k \geq 0$, $\gamma \geq 1$, $\nu \geq 2$, $n := \lceil \nu \rceil$, such that $\nu - \gamma \geq 1$, $\gamma - k \geq 1$, and $f \in AC^n([a, b])$ such that $f^{(j)}(a) = 0$, $j = 0, 1, \ldots, n - 1$, and $D_{*a}^{\nu} f \in L_{\infty}(a, b)$. Here $a \leq x \leq b$, $p, q > 1 : 1/p + 1/q = 1$. Then*

$$\int_a^x |D_{*a}^{\gamma} f(\omega)| |D_{*a}^k f(\omega)| d\omega \leq$$

$$\frac{2^{-1/p} (x - a)^{(2\nu - k - \gamma \ 1 + 2/q)}}{\Gamma(\nu - k) \Gamma(\nu - \gamma + 1) ((\nu - \gamma) q + 1)^{1/q}}$$

$$\cdot \frac{\left(\int_a^x |D_{*a}^{\nu} f(\omega)|^p d\omega \right)^{2/p}}{(2\nu q - kq - \gamma q - q + 2)^{1/q}}. \tag{16.36}$$

Proof. Similar to Theorem 26.3, p. 574 of [19], and Theorem 2.5 of [65]. □

We give

Theorem 16.28. *Let $\nu \geq \gamma_i + 1$, $\gamma_i \geq 0$, $i = 1, \ldots, k \in \mathbb{N} - \{1\}$, $n := \lceil \nu \rceil$. Assume $f \in AC^n([a, b])$ such that $f^{(j)}(a) = 0$, $j = 0, 1, \ldots, n - 1$, and $D_{*a}^{\nu} f \in L_{\infty}(a, b)$. Here $a \leq x \leq b$, $\gamma := \sum_{i=1}^{k} \gamma_i$. Let $p, q > 1$ such* that $1/p + 1/q = 1$. Furthermore, suppose that $|D_{*a}^\nu f(t)|$ is decreasing on $[a, x]$. Then

$$\int_a^x \prod_{i=1}^k |D_{*a}^{\gamma_i} f(\omega)| \, d\omega \le$$

$$\frac{p(x-a)^{(1+k\nu p-\gamma p/p)}}{\prod_{i=1}^k \Gamma(\nu - \gamma_i)(k\nu p - \gamma p - kp + 1)^{1/p}}$$

$$\cdot \frac{\left(\int_a^x |D_{*a}^\nu f(t)|^{kq} \, dt\right)^{1/q}}{(k\nu p - \gamma p - kp + p + 1)}. \tag{16.37}$$

Proof. Similar to Theorem 26.6, p. 581 of [19], and Theorem 2.6 of [65]. $\square$

The extreme case follows.

Theorem 16.29. *All are as in Theorem 16.28, but $p = 1$, $q = \infty$. Then*

$$\int_a^x \prod_{i=1}^k |D_{*a}^{\gamma_i} f(\omega)| \, d\omega \le$$

$$\cdot \frac{(x-a)^{k\nu - \gamma + 1} \left(\|D_{*a}^\nu f\|_{\infty,(a,x)}\right)^k}{\left(\prod_{i=1}^k \Gamma(\nu - \gamma_i)\right)(k\nu - \gamma - k + 1)(k\nu - \gamma + 1)}. \tag{16.38}$$

Proof. Similar to Proposition 26.4, p. 582 of [19], and Theorem 2.7 of [65]. $\square$

16.3.2 Results Involving Two Functions

We present the following theorem.

Theorem 16.30. *Let $\nu \ge \gamma_i + 1$, $\gamma_i \ge 0$, $i = 1, 2$, $n := \lceil \nu \rceil$, and f_1, $f_2 \in AC^n([a, b])$ such that $f_1^{(j)}(a) = f_2^{(j)}(a) = 0$, $j = 0, 1, \ldots, n-1$, $a \le x \le b$. Consider also $p(t) > 0$ and $q(t) \ge 0$, with all $p(t)$, $1/p(t)$, $q(t) \in L_\infty(a, b)$. Further assume $D_{*a}^\nu f_i \in L_\infty(a, b)$, $i = 1, 2$.*
Let $\lambda_\nu > 0$ and λ_α, $\lambda_\beta \ge 0$ such that $\lambda_\nu < p$, where $p > 1$. Set

$$P_k(\omega) := \int_a^\omega (\omega - t)^{(\nu - \gamma_k - 1)p/(p-1)}(p(t))^{-1/(p-1)} \, dt, \tag{16.39}$$

$k = 1, 2, \; a \le x \le b;$

$$A(\omega) := \frac{q(\omega)\,(P_1(\omega))^{\lambda_\alpha((p-1)/p)}\,(P_2(\omega))^{\lambda_\beta((p-1)/p)}\,(p(\omega))^{-\lambda_\nu/p}}{(\Gamma(\nu - \gamma_1))^{\lambda_\alpha}\,(\Gamma(\nu - \gamma_2))^{\lambda_\beta}},$$
$$(16.40)$$

$$A_0(x) := \left(\int_a^x (A(\omega))^{p/p - \lambda_\nu}\, d\omega \right)^{(p - \lambda_\nu)/p}, \qquad (16.41)$$

and

$$\delta_1 := \begin{cases} 2^{1 - (\lambda_\alpha + \lambda_\nu/p)}, & if \; \lambda_\alpha + \lambda_\nu \le p, \\ 1, & if \; \lambda_\alpha + \lambda_\nu \ge p. \end{cases} \qquad (16.42)$$

If $\lambda_\beta = 0$, we obtain that

$$\int_a^x q(\omega) \left[\left| D_{*a}^{\gamma_1} f_1(\omega) \right|^{\lambda_\alpha} \left| D_{*a}^{\nu} f_1(\omega) \right|^{\lambda_\nu} + \right.$$

$$\left. \left| D_{*a}^{\gamma_1} f_2(\omega) \right|^{\lambda_\alpha} \left| D_{*a}^{\nu} f_2(\omega) \right|^{\lambda_\nu} \right] d\omega \le \qquad (16.43)$$

$$\left(A_0(x) |_{\lambda_\beta = 0} \right) \left(\frac{\lambda_\nu}{\lambda_\alpha + \lambda_\nu} \right)^{\lambda_\nu/p} \delta_1$$

$$\left[\int_a^x p(\omega) \left[|D_{*a}^{\nu} f_1(\omega)|^p + |D_{*a}^{\nu} f_2(\omega)|^p \right] d\omega \right]^{((\lambda_\alpha + \lambda_\nu)/p)}.$$

Proof. Similar to Theorem 2 of [26] and Theorem 4 of [48]. □

The counterpart of the last theorem follows.

Theorem 16.31. *All here are as in Theorem 16.30.*
Denote
$$\delta_3 := \begin{cases} 2^{\lambda_\beta/\lambda_\nu} - 1, & if \; \lambda_\beta \ge \lambda_\nu, \\ 1, & if \; \lambda_\beta \le \lambda_\nu. \end{cases} \qquad (16.44)$$

If $\lambda_\alpha = 0$, then

$$\int_a^x q(\omega) \left[\left| D_{*a}^{\gamma_2} f_2(\omega) \right|^{\lambda_\beta} \left| D_{*a}^{\nu} f_1(\omega) \right|^{\lambda_\nu} + \right.$$

$$\left. \left| D_{*a}^{\gamma_2} f_1(\omega) \right|^{\lambda_\beta} \left| D_{*a}^{\nu} f_2(\omega) \right|^{\lambda_\nu} \right] d\omega \le \qquad (16.45)$$

$$\left(A_0(x) |_{\lambda_\alpha = 0} \right) 2^{p - \lambda_\nu/p} \left(\frac{\lambda_\nu}{\lambda_\beta + \lambda_\nu} \right)^{\lambda_\nu/p} \delta_3^{\lambda_\nu/p}$$

$$\left(\int_a^x p(\omega) \left[|D_{*a}^{\nu} f_1(\omega)|^p + |D_{*a}^{\nu} f_2(\omega)|^p \right] d\omega \right)^{((\lambda_\nu + \lambda_\beta)/p)},$$

all $a \le x \le b$.

Proof. Similar to Theorem 3 of [26] and Theorem 5 of [48]. $\square$

The complete case λ_α, $\lambda_\beta \neq 0$ follows.

Theorem 16.32. *All here are as in Theorem 16.30.*
Denote

$$\tilde{\gamma}_1 := \begin{cases} 2^{((\lambda_\alpha+\lambda_\beta)/\lambda_\nu)} - 1, & if \ \lambda_\alpha + \lambda_\beta \geq \lambda_\nu, \\ 1, & if \ \lambda_\alpha + \lambda_\beta \leq \lambda_\nu, \end{cases} \tag{16.46}$$

and

$$\tilde{\gamma}_2 := \begin{cases} 1, & if \ \lambda_\alpha + \lambda_\beta + \lambda_\nu \geq p, \\ 2^{1-(\lambda_\alpha+\lambda_\beta+\lambda_\nu/p)}, & if \ \lambda_\alpha + \lambda_\beta + \lambda_\nu \leq p. \end{cases} \tag{16.47}$$

Then

$$\int_a^x q(\omega) \left[|D_{*a}^{\gamma_1} f_1(\omega)|^{\lambda_\alpha} |D_{*a}^{\gamma_2} f_2(\omega)|^{\lambda_\beta} |D_{*a}^\nu f_1(\omega)|^{\lambda_\nu} + \right.$$

$$\left. |D_{*a}^{\gamma_2} f_1(\omega)|^{\lambda_\beta} |D_{*a}^{\gamma_1} f_2(\omega)|^{\lambda_\alpha} |D_{*a}^\nu f_2(\omega)|^{\lambda_\nu} \right] d\omega \leq \tag{16.48}$$

$$A_0(x) \left(\frac{\lambda_\nu}{(\lambda_\alpha + \lambda_\beta)(\lambda_\alpha + \lambda_\beta + \lambda_\nu)} \right)^{\lambda_\nu/p} \left[\lambda_\alpha^{\lambda_\nu/p} \tilde{\gamma}_2 + 2^{p-\lambda_\nu/p} (\tilde{\gamma}_1 \lambda_\beta)^{\lambda_\nu/p} \right]$$

$$\cdot \left(\int_a^x p(\omega) (|D_{*a}^\nu f_1(\omega)|^p + |D_{*a}^\nu f_2(\omega)|^p) d\omega \right)^{((\lambda_\alpha+\lambda_\beta+\lambda_\nu)/p)},$$

all $a \leq x \leq b$.

Proof. As for Theorem 4 of [26], and Theorem 6 of [48]. $\square$

We continue with a special important case.

Theorem 16.33. *Let* $\nu \geq \gamma_1 + 2$, $\gamma_1 \geq 0$, $n := \lceil \nu \rceil$, *and* $f_1, f_2 \in AC^n([a,b])$ *such that* $f_1^{(j)}(a) = f_2^{(j)}(a) = 0$, $j = 0, 1, \ldots, n-1$, $a \leq x \leq b$. *Consider also* $p(t) > 0$ *and* $q(t) \geq 0$, *with all* $p(t), 1/p(t), q(t) \in L_\infty(a,b)$. *Furthermore assume* $D_{*a}^\nu f_i \in L_\infty(a,b)$, $i = 1, 2$.
Let $\lambda_\alpha \geq 0$, $0 < \lambda_{\alpha+1} < 1$, *and* $p > 1$. *Denote*

$$\theta_3 := \begin{cases} 2^{\lambda_\alpha/\lambda_{\alpha+1}} - 1, & if \ \lambda_\alpha \geq \lambda_{\alpha+1}, \\ 1, & if \ \lambda_\alpha \leq \lambda_{\alpha+1}, \end{cases}, \tag{16.49}$$

$$L(x) := \left(2 \int_a^x (q(\omega))^{(1/1-\lambda_{\alpha+1})} d\omega \right)^{(1-\lambda_{\alpha+1})} \left(\frac{\theta_3 \lambda_{\alpha+1}}{\lambda_\alpha + \lambda_{\alpha+1}} \right)^{\lambda_{\alpha+1}}, \tag{16.50}$$

and

$$P_1(x) := \int_a^x (x-t)^{(\nu-\gamma_1-1)p/(p-1)} (p(t))^{-1/(p-1)} dt, \tag{16.51}$$

$$T(x) := L(x) \left(\frac{P_1(x)^{((p-1)/p)}}{\Gamma(\nu - \gamma_1)} \right)^{(\lambda_\alpha + \lambda_{\alpha+1})}, \tag{16.52}$$

$$\omega_1 := 2^{(p-1/p)(\lambda_\alpha + \lambda_{\alpha+1})}, \tag{16.53}$$

and

$$\Phi(x) := T(x)\,\omega_1. \tag{16.54}$$

Then

$$\int_a^x q(\omega) \left[\left| D_{*a}^{\gamma_1} f_1(\omega) \right|^{\lambda_\alpha} \left| D_{*a}^{\gamma_1+1} f_2(\omega) \right|^{\lambda_{\alpha+1}} + \right.$$

$$\left. \left| D_{*a}^{\gamma_1} f_2(\omega) \right|^{\lambda_\alpha} \left| D_{*a}^{\gamma_1+1} f_1(\omega) \right|^{\lambda_{\alpha+1}} \right] d\omega \leq$$

$$\Phi(x) \left[\int_a^x p(\omega) \left(\left| D_{*a}^{\nu} f_1(\omega) \right|^p + \right. \right.$$

$$\left. \left. \left| D_{*a}^{\nu} f_2(\omega) \right|^p \right) d\omega \right]^{((\lambda_\alpha + \lambda_{\alpha+1})/p)}, \tag{16.55}$$

all $a \leq x \leq b$.

Proof. As in Theorem 5 of [26], and Theorem 8 of [48]. $\square$

We give

Corollary 16.34. *All here are as in Theorem 16.30, with $\lambda_\beta = 0$, $p(t) = q(t) = 1$. Then*

$$\int_a^x \left[\left| D_{*a}^{\gamma_1} f_1(\omega) \right|^{\lambda_\alpha} \left| D_{*a}^{\nu} f_1(\omega) \right|^{\lambda_\nu} + \right.$$

$$\left. \left| D_{*a}^{\gamma_1} f_2(\omega) \right|^{\lambda_\alpha} \left| D_{*a}^{\nu} f_2(\omega) \right|^{\lambda_\nu} \right] d\omega \leq \tag{16.56}$$

$$C_1(x) \left(\int_a^x \left(\left| D_{*a}^{\nu} f_1(\omega) \right|^p + \left| D_{*a}^{\nu} f_2(\omega) \right|^p \right) d\omega \right)^{((\lambda_\alpha + \lambda_\nu)/p)},$$

all $a \leq x \leq b$, where

$$C_1(x) := \left(A_0(x) \mid_{\lambda_\beta = 0} \right) \left(\frac{\lambda_\nu}{\lambda_\alpha + \lambda_\nu} \right)^{\lambda_\nu/p} \delta_1, \tag{16.57}$$

$$\delta_1 := \begin{cases} 2^{1-((\lambda_\alpha + \lambda_\nu)/p)}, & \text{if } \lambda_\alpha + \lambda_\nu \leq p, \\ 1, & \text{if } \lambda_\alpha + \lambda_\nu \geq p. \end{cases} \tag{16.58}$$

We find that

$$\left(A_0(x) \mid_{\lambda_\beta = 0} \right) = \left\{ \left(\frac{(p-1)^{((\lambda_\alpha p - \lambda_\alpha)/p)}}{(\Gamma(\nu - \gamma_1))^{\lambda_\alpha} (\nu p - \gamma_1 p - 1)^{((\lambda_\alpha p - \lambda_\alpha)/p)}} \right) \right. .$$

$$\left(\frac{(p - \lambda_\nu)^{((p-\lambda_\nu)/p)}}{(\lambda_\alpha \nu p - \lambda_\alpha \gamma_1 p - \lambda_\alpha + p - \lambda_\nu)^{((p-\lambda_\nu)/p)}} \right) \right\} \cdot$$

$$(x - a)^{((\lambda_\alpha \nu p - \lambda_\alpha \gamma_1 p - \lambda_\alpha + p - \lambda_\nu)/p)} . \qquad (16.59)$$

Proof. As for Corollary 1 of [26], and Corollary 10 of [48]. □

Corollary 16.35. (All are as in Theorem 16.30; $\lambda_\beta = 0$, $p(t) = q(t) = 1$, $\lambda_\alpha = \lambda_\nu = 1$, $p = 2$.) In detail: let $\nu \geq \gamma_1 + 1$, $\gamma_1 \geq 0$, $n := \lceil \nu \rceil$, f_1, $f_2 \in AC^n([a,b]) : f_1^{(j)}(a) = f_2^{(j)}(a) = 0$, $j = 0, 1, \ldots, n - 1$, $a \leq x \leq b$, $D_{*a}^\nu f_i \in L_\infty(a,b)$, $i = 1, 2$.   Then

$$\int_a^x \left[\left| \left(D_{*a}^{\gamma_1} f_1 \right)(\omega) \right| \left| \left(D_{*a}^\nu f_1 \right)(\omega) \right| + \right.$$

$$\left. \left| \left(D_{*a}^{\gamma_1} f_2 \right)(\omega) \right| \left| \left(D_{*a}^\nu f_2 \right)(\omega) \right| \right] d\omega \leq \qquad (16.60)$$

$$\left(\frac{(x - a)^{(\nu - \gamma_1)}}{2\Gamma(\nu - \gamma_1) \sqrt{\nu - \gamma_1} \sqrt{2\nu - 2\gamma_1 - 1}} \right)$$

$$\left(\int_a^x \left[\left(\left(D_{*a}^\nu f_1 \right)(\omega) \right)^2 + \left(\left(D_{*a}^\nu f_2 \right)(\omega) \right)^2 \right] d\omega \right),$$

all $a \leq x \leq b$.

Corollary 16.36. (To Theorem 16.31; $\lambda_\alpha = 0$, $p(t) = q(t) = 1$.)
It holds

$$\int_a^x \left[\left| D_{*a}^{\gamma_2} f_2(\omega) \right|^{\lambda_\beta} \left| D_{*a}^\nu f_1(\omega) \right|^{\lambda_\nu} + \right.$$

$$\left. \left| D_{*a}^{\gamma_2} f_1(\omega) \right|^{\lambda_\beta} \left| D_{*a}^\nu f_2(\omega) \right|^{\lambda_\nu} \right] d\omega \leq \qquad (16.61)$$

$$C_2(x) \left(\int_a^x \left[\left| D_{*a}^\nu f_1(\omega) \right|^p + \left| D_{*a}^\nu f_2(\omega) \right|^p \right] d\omega \right)^{((\lambda_\nu + \lambda_\beta)/p)},$$

all $a \leq x \leq b$, where

$$C_2(x) := (A_0(x)|_{\lambda_\alpha = 0}) \, 2^{p - \lambda_\nu/p} \left(\frac{\lambda_\nu}{\lambda_\beta + \lambda_\nu} \right)^{\lambda_\nu/p} \delta_3^{\lambda_\nu/p}, \qquad (16.62)$$

$$\delta_3 := \left\{ \begin{array}{ll} 2^{\lambda_\beta/\lambda_\nu} - 1, & \text{if } \lambda_\beta \geq \lambda_\nu, \\ 1, & \text{if } \lambda_\beta \leq \lambda_\nu \end{array} \right\}. \qquad (16.63)$$

We find that

$$(A_0(x)|_{\lambda_\alpha = 0}) = \left\{ \left(\frac{(p - 1)^{((\lambda_\beta p - \lambda_\beta)/p)}}{(\Gamma(\nu - \gamma_2))^{\lambda_\beta} (\nu p - \gamma_2 p - 1)^{((\lambda_\beta p - \lambda_\beta)/p)}} \right) \right. \cdot$$

$$\left(\frac{(p - \lambda_\nu)^{((p-\lambda_\nu)/p)}}{(\lambda_\beta \nu p - \lambda_\beta \gamma_2 p - \lambda_\beta + p - \lambda_\nu)^{((p-\lambda_\nu)/p)}} \right) \right\}.$$

$$(x - a)^{((\lambda_\beta \nu p - \lambda_\beta \gamma_2 p - \lambda_\beta + p - \lambda_\nu)/p)}. \tag{16.64}$$

Corollary 16.37. (To Theorem 16.31; $\lambda_\alpha = 0$, $p(t) = q(t) = 1$, $\lambda_\beta = \lambda_\nu = 1$, $p = 2$.) *In detail: let* $\nu \geq \gamma_2 + 1$, $\gamma_2 \geq 0$, $n := \lceil \nu \rceil$, f_1, $f_2 \in AC^n([a,b]) : f_1^{(j)}(a) = f_2^{(j)}(a) = 0$, $j = 0,1,\ldots,n-1$, $a \leq x \leq b$, $D_{*a}^\nu f_i \in L_\infty(a,b)$, $i = 1,2$.
Then

$$\int_a^x \left[\left| D_{*a}^{\gamma_2} f_2(\omega) \right| \left| D_{*a}^\nu f_1(\omega) \right| + \right.$$

$$\left| D_{*a}^{\gamma_2} f_1(\omega) \right| \left| D_{*a}^\nu f_2(\omega) \right| \right] d\omega \leq$$

$$\left(\frac{(x-a)^{(\nu - \gamma_2)}}{\sqrt{2} \Gamma(\nu - \gamma_2) \sqrt{\nu - \gamma_2} \sqrt{2\nu - 2\gamma_2 - 1}} \right). \tag{16.65}$$

$$\left(\int_a^x \left((D_{*a}^\nu f_1(\omega))^2 + (D_{*a}^\nu f_2(\omega))^2 \right) d\omega \right),$$

all $a \leq x \leq b$.

We continue with related results regarding $\|\cdot\|_\infty$.

Theorem 16.38. *Let* $\nu \geq \gamma_i + 1$, $\gamma_i \geq 0$, $i = 1,2$, $n := \lceil \nu \rceil$, *and* f_1, $f_2 \in AC^n([a,b])$ *such that* $f_1^{(j)}(a) = f_2^{(j)}(a) = 0$, $j = 0,1,\ldots,n-1$; $a \leq x \leq b$. *Consider* $p(x) \geq 0$ *and* $p(x) \in L_\infty(a,b)$, *and assume* $D_{*a}^\nu f_i \in L_\infty(a,b)$, $i = 1,2$. *Let* $\lambda_\alpha, \lambda_\beta, \lambda_\nu \geq 0$. *Set*

$$T(x) := \frac{(x-a)^{(\nu\lambda_\alpha - \gamma_1\lambda_\alpha + \nu\lambda_\beta - \gamma_2\lambda_\beta + 1)}}{(\nu\lambda_\alpha - \gamma_1\lambda_\alpha + \nu\lambda_\beta - \gamma_2\lambda_\beta + 1)}$$

$$\cdot \frac{\|p(s)\|_{\infty,(a,x)}}{(\Gamma(\nu - \gamma_1 + 1))^{\lambda_\alpha} (\Gamma(\nu - \gamma_2 + 1))^{\lambda_\beta}}. \tag{16.66}$$

Then

$$\int_a^x p(\omega) \left[\left| D_{*a}^{\gamma_1} f_1(\omega) \right|^{\lambda_\alpha} \left| D_{*a}^{\gamma_2} f_2(\omega) \right|^{\lambda_\beta} \left| D_{*a}^\nu f_1(\omega) \right|^{\lambda_\nu} + \right.$$

$$\left| D_{*a}^{\gamma_2} f_1(\omega) \right|^{\lambda_\beta} \left| D_{*a}^{\gamma_1} f_2(\omega) \right|^{\lambda_\alpha} \left| D_{*a}^\nu f_2(\omega) \right|^{\lambda_\nu} \right] d\omega \leq$$

$$\frac{T(x)}{2} \left[\|D_{*a}^\nu f_1\|_{\infty,(a,x)}^{2(\lambda_\alpha + \lambda_\nu)} + \|D_{*a}^\nu f_1\|_{\infty,(a,x)}^{2\lambda_\beta} + \right.$$

$$\|D_{*a}^{\nu}f_2\|_{\infty,(a,x)}^{2\lambda_\beta} + \|D_{*a}^{\nu}f_2\|_{\infty,(a,x)}^{2(\lambda_\alpha+\lambda_\nu)}\Big],\qquad(16.67)$$

all $a \leq x \leq b$.

Proof. Similar to Theorem 7 of [26], and Theorem 18 of [48]. $\square$

We give

Corollary 16.39. (to Theorem 16.38) *Let* $\nu \geq \gamma_1+2$, $\gamma_1 \geq 0$, $n := \lceil \nu \rceil$, *and* f_1, $f_2 \in AC^n([a,b])$ *such that* $f_1^{(j)}(a) = f_2^{(j)}(a) = 0$, $j = 0,1,\ldots,n-1$; $D_{*a}^{\nu}f_i \in L_\infty(a,b)$, $i = 1,2$. *Then*

$$\int_a^x \Big[\big|D_{*a}^{\gamma_1}f_1(\omega)\big| \big|D_{*a}^{\gamma_1+1}f_2(\omega)\big| + $$

$$\big|D_{*a}^{\gamma_1+1}f_1(\omega)\big| \big|D_{*a}^{\gamma_1}f_2(\omega)\big| \Big]\, d\omega$$

$$\leq \frac{(x-a)^{2(\nu-\gamma_1)}}{2\,(\Gamma(\nu-\gamma_1+1))^2}$$

$$\Big[\|D_{*a}^{\nu}f_1\|_{\infty,(a,x)}^2 + \|D_{*a}^{\nu}f_2\|_{\infty,(a,x)}^2\Big],\qquad(16.68)$$

all $a \leq x \leq b$.

Next we give converse results involving two functions.

Theorem 16.40. *Let* $\gamma_j \geq 0, 1 \leq \nu - \gamma_j < 1/p$, $0 < p < 1$, $j = 1,2$; $n := \lceil \nu \rceil$, *and* f_1, $f_2 \in AC^n([a,b])$ *such that* $f_1^{(l)}(a) = f_2^{(l)}(a) = 0$, $l = 0,1,\ldots,n-1$, $a \leq x \leq b$. *Consider also* $p(t) > 0$ *and* $q(t) \geq 0$, *with all* $p(t)$, $1/p(t)$, $q(t)$, $1/q(t) \in L_\infty(a,b)$. *Further assume* $D_{*a}^{\nu}f_i \in L_\infty(a,b)$, $i = 1,2$, *each of which is of fixed sign a.e. on* $[a,b]$. *Let* $\lambda_\nu > 0$ *and* λ_α, $\lambda_\beta \geq 0$ *such that* $\lambda_\nu > p$.

Here $P_k(\omega)$, $A(\omega)$, $A_0(x)$ are as in (16.39), (16.40), and (16.41), respectively.

Set

$$\delta_1 := 2^{1-((\lambda_\alpha+\lambda_\nu)/p)}.\qquad(16.69)$$

If $\lambda_\beta = 0$, then

$$\int_a^x q(\omega)\Big[\big|D_{*a}^{\gamma_1}f_1(\omega)\big|^{\lambda_\alpha}\big|D_{*a}^{\nu}f_1(\omega)\big|^{\lambda_\nu} + $$

$$\big|D_{*a}^{\gamma_1}f_2(\omega)\big|^{\lambda_\alpha}\big|D_{*a}^{\nu}f_2(\omega)\big|^{\lambda_\nu}\Big]\,d\omega \geq$$

$$\big(A_0(x)\,|_{\lambda_\beta=0}\big)\left(\frac{\lambda_\nu}{\lambda_\alpha+\lambda_\nu}\right)^{\lambda_\nu/p}\delta_1$$

$$\left[\int_a^x p\left(\omega\right)\left[|D_{*a}^{\nu}f_1\left(\omega\right)|^p+|D_{*a}^{\nu}f_2\left(\omega\right)|^p\right]d\omega\right]^{(\lambda_\alpha+\lambda_\nu/p)}.\tag{16.70}$$

Proof. Similar to Theorem 5 of [51], and Theorem 4 of [47]. □

We continue with

Theorem 16.41. *All here are as in Theorem 16.40. Further assume* $\lambda_\beta \geq \lambda_\nu$. *Denote*

$$\delta_2 := 2^{1-(\lambda_\beta/\lambda_\nu)},$$

$$\delta_3 := (\delta_2 - 1)\,2^{-(\lambda_\beta/\lambda_\nu)}.\tag{16.71}$$

If $\lambda_\alpha = 0$, *then*

$$\int_a^x q\left(\omega\right)\left[\left|D_{*a}^{\gamma_2}f_2\left(\omega\right)\right|^{\lambda_\beta}|D_{*a}^{\nu}f_1\left(\omega\right)|^{\lambda_\nu}+\right.$$

$$\left.\left|D_{*a}^{\gamma_2}f_1\left(\omega\right)\right|^{\lambda_\beta}|D_{*a}^{\nu}f_2\left(\omega\right)|^{\lambda_\nu}\right]d\omega \geq$$

$$(A_0\left(x\right)|_{\lambda_\alpha=0})\,2^{p-\lambda_\nu/p}\left(\frac{\lambda_\nu}{\lambda_\beta+\lambda_\nu}\right)^{\lambda_\nu/p}\delta_3^{\lambda_\nu/p}$$

$$\cdot\left(\int_a^x p\left(\omega\right)\left[|D_{*a}^{\nu}f_1\left(\omega\right)|^p+|D_{*a}^{\nu}f_2\left(\omega\right)|^p\right]d\omega\right)^{((\lambda_\nu+\lambda_\beta)/p)},\tag{16.72}$$

all $a \leq x \leq b$.

Proof. Similar to Theorem 6 of [51], and Theorem 5 of [47]. □

We give

Theorem 16.42. *Let* $\nu \geq 2$ *and* $\gamma_1 \geq 0$ *such that* $2 \leq \nu - \gamma_1 < 1/p$, $0 < p < 1$, $n := \lceil \nu \rceil$. *Consider* $f_1, f_2 \in AC^n\left([a,b]\right)$ *such that* $f_1^{(l)}\left(a\right) = f_2^{(l)}\left(a\right) = 0$, $l = 0,1,\ldots,n-1$, $a \leq x \leq b$. *Assume that* $D_{*a}^{\nu}f_i \in L_\infty\left(a,b\right)$, $i = 1,2$, *each of which is of fixed sign a.e. on* $[a,b]$. *Consider also* $p\left(t\right) > 0$ *and* $q\left(t\right) \geq 0$, *with all* $p\left(t\right), 1/p\left(t\right), q\left(t\right), 1/q\left(t\right) \in L_\infty\left(a,b\right)$. *Let* $\lambda_\alpha \geq \lambda_{\alpha+1} > 1$. *Denote*

$$\theta_3 := \left(2^{1-(\lambda_\alpha/\lambda_{\alpha+1})}-1\right)2^{-\lambda_\alpha/\lambda_{\alpha+1}},\tag{16.73}$$

$L\left(x\right)$ *is as in* (16.50), $P_1\left(x\right)$ *as in* (16.51), $T\left(x\right)$ *as in* (16.52), ω_1 *as in* (16.53), *and* Φ *as in* (16.54). *Then*

$$\int_a^x q\left(\omega\right)\left[\left|D_{*a}^{\gamma_1}f_1\left(\omega\right)\right|^{\lambda_\alpha}\left|D_{*a}^{\gamma_1+1}f_2\left(\omega\right)\right|^{\lambda_{\alpha+1}}+\right.$$

$$\left|D_{*a}^{\gamma_1}f_2(\omega)\right|^{\lambda_\alpha}\left|D_{*a}^{\gamma_1+1}f_1(\omega)\right|^{\lambda_{\alpha+1}}\right]d\omega \geq$$

$$\Phi(x)\left[\int_a^x p(\omega)\left(|D_{*a}^\nu f_1(\omega)|^p + |D_{*a}^\nu f_2(\omega)|^p\right)d\omega\right]^{((\lambda_\alpha+\lambda_{\alpha+1})/p)}, \quad (16.74)$$

all $a \leq x \leq b$.

Proof. Similar to Theorem 7 of [51], and Theorem 7 of [47]. □

We have

Corollary 16.43. (To Theorem 16.40; $\lambda_\beta = 0$, $p(t) = q(t) = 1$.) *Then*

$$\int_a^x \left[|D_{*a}^{\gamma_1}f_1(\omega)|^{\lambda_\alpha}|D_{*a}^\nu f_1(\omega)|^{\lambda_\nu} + \right.$$

$$\left.|D_{*a}^{\gamma_1}f_2(\omega)|^{\lambda_\alpha}|D_{*a}^\nu f_2(\omega)|^{\lambda_\nu}\right]d\omega \geq \quad (16.75)$$

$$C_1(x)\left(\int_a^x (|D_{*a}^\nu f_1(\omega)|^p + |D_{*a}^\nu f_2(\omega)|^p)d\omega\right)^{((\lambda_\alpha+\lambda_\nu)/p)},$$

all $a \leq x \leq b$, *where*

$$C_1(x) := \left(A_0(x)|_{\lambda_\beta=0}\right)\left(\frac{\lambda_\nu}{\lambda_\alpha+\lambda_\nu}\right)^{\lambda_\nu/p}\delta_1, \quad (16.76)$$

$$\delta_1 := 2^{1-(\lambda_\alpha+\lambda_\nu/p)}. \quad (16.77)$$

Here $\left(A_0(x)|_{\lambda_\beta=0}\right)$ *is given by* (16.59).

Corollary 16.44. (To Theorem 16.41; $\lambda_\alpha = 0$, $p(t) = q(t) = 1$, $\lambda_\beta \geq \lambda_\nu$.) *Then*

$$\int_a^x \left[|D_{*a}^{\gamma_2}f_2(\omega)|^{\lambda_\beta}|D_{*a}^\nu f_1(\omega)|^{\lambda_\nu} + \right.$$

$$\left.|D_{*a}^{\gamma_2}f_1(\omega)|^{\lambda_\beta}|D_{*a}^\nu f_2(\omega)|^{\lambda_\nu}\right]d\omega \geq \quad (16.78)$$

$$C_2(x)\left(\int_a^x [|D_{*a}^\nu f_1(\omega)|^p + |D_{*a}^\nu f_2(\omega)|^p]d\omega\right)^{((\lambda_\nu+\lambda_\beta)/p)},$$

all $a \leq x \leq b$, *where*

$$C_2(x) := (A_0(x)|_{\lambda_\alpha=0})\, 2^{p-\lambda_\nu/p}\left(\frac{\lambda_\nu}{\lambda_\beta+\lambda_\nu}\right)^{\lambda_\nu/p}\delta_3^{\lambda_\nu/p}.$$

Here $(A_0(x)|_{\lambda_\alpha=0})$ *is given by* (16.64).

16.3.3   Results Involving Several Functions

We present

Theorem 16.45. *Here all notations, terms, and assumptions are as in Theorem 16.30, but for $f_j \in AC^n\left([a,b]\right)$, with $j = 1,\ldots,M \in \mathbb{N}$. Instead of δ_1 there, we define here*

$$\delta_1^* := \begin{cases} M^{1-(\lambda_\alpha + \lambda_\nu/p)}, & if \ \lambda_\alpha + \lambda_\nu \le p, \\ 2^{(\lambda_\alpha + \lambda_\nu/p)-1}, & if \ \lambda_\alpha + \lambda_\nu \ge p. \end{cases} \tag{16.79}$$

Call

$$\varphi_1\left(x\right) := \left(A_0\left(x\right)|_{\lambda_\beta = 0}\right)\left(\frac{\lambda_\nu}{\lambda_\alpha + \lambda_\nu}\right)^{\lambda_\nu/p}. \tag{16.80}$$

If $\lambda_\beta = 0$, then

$$\int_a^x q\left(\omega\right)\left(\sum_{j=1}^M \left|D_{*a}^{\gamma_1}f_j\left(\omega\right)\right|^{\lambda_\alpha}\left|D_{*a}^\nu f_j\left(\omega\right)\right|^{\lambda_\nu}\right)d\omega$$

$$\le \delta_1^*\varphi_1\left(x\right)\left[\int_a^x p\left(\omega\right)\left(\sum_{j=1}^M \left|D_{*a}^\nu f_j\left(\omega\right)\right|^p\right)d\omega\right]^{((\lambda_\alpha + \lambda_\nu)/p)}, \tag{16.81}$$

all $a \le x \le b$.

Proof. As in Theorem 2 of [27], and Theorem 4 of [46]. $\square$

We continue with

Theorem 16.46. *All here are as in Theorem 16.45.*
Denote

$$\delta_3 := \begin{cases} 2^{\lambda_\beta/\lambda_\nu} - 1, & if \ \lambda_\beta \ge \lambda_\nu, \\ 1, & if \ \lambda_\beta \le \lambda_\nu, \end{cases} \tag{16.82}$$

$$\varepsilon_2 := \begin{cases} 1, & if \ \lambda_\nu + \lambda_\beta \ge p, \\ M^{1-(\lambda_\nu + \lambda_\beta/p)} & if \ \lambda_\nu + \lambda_\beta \le p, \end{cases} \tag{16.83}$$

and

$$\varphi_2\left(x\right) := \left(A_0\left(x\right)|_{\lambda_\alpha = 0}\right)2^{(p-\lambda_\nu/p)}\left(\frac{\lambda_\nu}{\lambda_\beta + \lambda_\nu}\right)^{\lambda_\nu/p}\delta_3^{\lambda_\nu/p}. \tag{16.84}$$

If $\lambda_\alpha = 0$, then

$$\int_a^x q\left(\omega\right)\left\{\left\{\sum_{j=1}^{M-1}\left[\left|D_{*a}^{\gamma_2}f_{j+1}\left(\omega\right)\right|^{\lambda_\beta}\left|D_{*a}^\nu f_j\left(\omega\right)\right|^{\lambda_\nu} + \right.\right.\right.$$

$$\left. |D_{*a}^{\gamma_2} f_j\left(\omega\right)|^{\lambda_\beta} \, |D_{*a}^{\nu} f_{j+1}\left(\omega\right)|^{\lambda_\nu} \right] \right\} +$$

$$\left[\left| D_{*a}^{\gamma_2} f_M\left(\omega\right) \right|^{\lambda_\beta} |D_{*a}^{\nu} f_1\left(\omega\right)|^{\lambda_\nu} + \right.$$

$$\left. |D_{*a}^{\gamma_2} f_1\left(\omega\right)|^{\lambda_\beta} \, |D_{*a}^{\nu} f_M\left(\omega\right)|^{\lambda_\nu} \right] \right\} d\omega$$

$$\leq 2^{(\lambda_\nu + \lambda_\beta / p)} \varepsilon_2 \varphi_2\left(x\right) \cdot \left\{ \int_a^x p\left(\omega\right) \right.$$

$$\left. \cdot \left[\sum_{j=1}^{M} |D_{*a}^{\nu} f_j\left(\omega\right)|^p \right] d\omega \right\}^{((\lambda_\nu + \lambda_\beta) / p)} , \tag{16.85}$$

all $a \leq x \leq b$.

Proof. As for Theorem 3 of [27], and Theorem 5 of [46]. $\square$

We give the general case.

Theorem 16.47. *All are as in Theorem* 16.45.
Denote

$$\tilde{\gamma}_1 := \begin{cases} 2^{(\lambda_\alpha + \lambda_\beta / \lambda_\nu)} - 1, & \text{if } \lambda_\alpha + \lambda_\beta \geq \lambda_\nu, \\ 1, & \text{if } \lambda_\alpha + \lambda_\beta \leq \lambda_\nu, \end{cases} \tag{16.86}$$

and

$$\tilde{\gamma}_2 := \begin{cases} 1, & \text{if } \lambda_\alpha + \lambda_\beta + \lambda_\nu \geq p, \\ 2^{1-(\lambda_\alpha + \lambda_\beta + \lambda_\nu / p)}, & \text{if } \lambda_\alpha + \lambda_\beta + \lambda_\nu \leq p. \end{cases} \tag{16.87}$$

Set

$$\varphi_3\left(x\right) := A_0\left(x\right) \cdot \left(\frac{\lambda_\nu}{(\lambda_\alpha + \lambda_\beta)(\lambda_\alpha + \lambda_\beta + \lambda_\nu)} \right)^{\lambda_\nu / p}$$

$$\cdot \left[\lambda_\alpha^{\lambda_\nu / p} \tilde{\gamma}_2 + 2^{(p - \lambda_\nu / p)} (\tilde{\gamma}_1 \lambda_\beta)^{\lambda_\nu / p} \right], \tag{16.88}$$

and

$$\varepsilon_3 := \begin{cases} 1, & \text{if } \lambda_\alpha + \lambda_\beta + \lambda_\nu \geq p, \\ M^{1-(\lambda_\alpha + \lambda_\beta + \lambda_\nu / p)} & \text{if } \lambda_\alpha + \lambda_\beta + \lambda_\nu \leq p. \end{cases} \tag{16.89}$$

Then

$$\int_a^x q\left(\omega\right) \left[\sum_{j=1}^{M-1} \left[|(D_{*a}^{\gamma_1} f_j)\left(\omega\right)|^{\lambda_\alpha} \, |(D_{*a}^{\gamma_2} f_{j+1})\left(\omega\right)|^{\lambda_\beta} \, |(D_{*a}^{\nu} f_j)\left(\omega\right)|^{\lambda_\nu} \right. \right.$$

$$\left. + |(D_{*a}^{\gamma_2} f_j)\left(\omega\right)|^{\lambda_\beta} \, |(D_{*a}^{\gamma_1} f_{j+1})\left(\omega\right)|^{\lambda_\alpha} \, |(D_{*a}^{\nu} f_{j+1})\left(\omega\right)|^{\lambda_\nu} \right]$$

$$+ \left[\left| \left(D_{*a}^{\gamma_1} f_1 \right) (\omega) \right|^{\lambda_\alpha} \left| \left(D_{*a}^{\gamma_2} f_M \right) (\omega) \right|^{\lambda_\beta} \left| \left(D_{*a}^{\nu} f_1 \right) (\omega) \right|^{\lambda_\nu} \right.$$

$$\left. + \left| \left(D_{*a}^{\gamma_2} f_1 \right) (\omega) \right|^{\lambda_\beta} \left| \left(D_{*a}^{\gamma_1} f_M \right) (\omega) \right|^{\lambda_\alpha} \left| \left(D_{*a}^{\nu} f_M \right) (\omega) \right|^{\lambda_\nu} \right] \right] d\omega$$

$$\leq 2^{(\lambda_\alpha + \lambda_\beta + \lambda_\nu / p)} \varepsilon_3 \varphi_3 (x) \cdot \left\{ \int_a^x p(\omega) \right.$$

$$\left. \cdot \left[\sum_{j=1}^M \left| \left(D_{*a}^{\nu} f_j \right) (\omega) \right|^p \right] d\omega \right\}^{((\lambda_\alpha + \lambda_\beta + \lambda_\nu) / p)}, \qquad (16.90)$$

all $a \leq x \leq b$.

Proof. As for Theorem 4 of [27], and Theorem 6 of [46]. $\square$

We give

Theorem 16.48. *All here are as in Theorem 16.33, but for $f_j \in AC^n$* *$([a, b]), j = 1, \ldots, M \in \mathbb{N}$.*
Also put

$$\varepsilon_4 := \left\{ \begin{array}{ll} 1, & \text{if } \lambda_\alpha + \lambda_{\alpha+1} \geq p, \\ M^{1 - (\lambda_\alpha + \lambda_{\alpha+1} / p)} & \text{if } \lambda_\alpha + \lambda_{\alpha+1} \leq p. \end{array} \right\}. \qquad (16.91)$$

Then

$$\int_a^x q(\omega) \left\{ \left\{ \sum_{j=1}^{M-1} \left[\left| \left(D_{*a}^{\gamma_1} f_j \right) (\omega) \right|^{\lambda_\alpha} \left| \left(D_{*a}^{\gamma_1 + 1} f_{j+1} \right) (\omega) \right|^{\lambda_{\alpha+1}} \right. \right. \right.$$

$$\left. \left. + \left| \left(D_{*a}^{\gamma_1} f_{j+1} \right) (\omega) \right|^{\lambda_\alpha} \left| \left(D_{*a}^{\gamma_1 + 1} f_j \right) (\omega) \right|^{\lambda_{\alpha+1}} \right] \right\}$$

$$+ \left[\left| \left(D_{*a}^{\gamma_1} f_1 \right) (\omega) \right|^{\lambda_\alpha} \left| \left(D_{*a}^{\gamma_1 + 1} f_M \right) (\omega) \right|^{\lambda_{\alpha+1}} \right.$$

$$\left. \left. + \left| \left(D_{*a}^{\gamma_1} f_M \right) (\omega) \right|^{\lambda_\alpha} \left| \left(D_{*a}^{\gamma_1 + 1} f_1 \right) (\omega) \right|^{\lambda_{\alpha+1}} \right] \right\} d\omega$$

$$\leq 2^{(\lambda_\alpha + \lambda_{\alpha+1} / p)} \varepsilon_4 \Phi(x) \cdot \left[\int_a^x p(\omega) \right.$$

$$\left. \cdot \left(\sum_{j=1}^M \left| \left(D_{*a}^{\nu} f_j \right) (\omega) \right|^p \right) d\omega \right]^{((\lambda_\alpha + \lambda_{\alpha+1}) / p)}, \qquad (16.92)$$

all $a \leq x \leq b$.

Proof. As for Theorem 5 of [27], and Theorem 7 of [46]. $\square$

We continue with

Corollary 16.49. (To Theorem 16.45; $\lambda_\beta = 0$, $p(t) = q(t) = 1$, $\lambda_\alpha = \lambda_\nu = 1$, $p = 2$.) *In detail: let* $\nu \geq \gamma_1 + 1$, $\gamma_1 \geq 0$, $n := \lceil \nu \rceil$, $f_j \in AC^n([a,b])$, $j = 1, \ldots, M \in \mathbb{N}$; $a \leq x \leq b$, *and* $D_{*a}^\nu f_j \in L_\infty(a,b)$, $j = 1, \ldots, M$.
Here $f_j^{(l)}(a) = 0$, $l = 0, 1, \ldots, n-1$; $j = 1, \ldots, M$.
Then

$$\int_a^x \left(\sum_{j=1}^M |D_{*a}^{\gamma_1} f_j(\omega)| \, |D_{*a}^\nu f_j(\omega)| \right) d\omega \leq$$

$$\left(\frac{(x-a)^{\nu-\gamma_1}}{2\Gamma(\nu - \gamma_1)\sqrt{\nu - \gamma_1}\sqrt{2\nu - 2\gamma_1 - 1}} \right). \tag{16.93}$$

$$\left\{ \int_a^x \left[\sum_{j=1}^M (D_{*a}^\nu f_j(\omega))^2 \right] d\omega \right\},$$

all $a \leq x \leq b$.

Corollary 16.50. (To Theorem 16.46; $\lambda_\alpha = 0$, $p(t) = q(t) = 1$, $\lambda_\beta = \lambda_\nu = 1$, $p = 2$.) *In detail: let* $\nu \geq \gamma_2 + 1$, $\gamma_2 \geq 0$, $n := \lceil \nu \rceil$, $f_j \in AC^n([a,b])$, $D_{*a}^\nu f_j \in L_\infty(a,b)$, $j = 1, \ldots, M \in \mathbb{N}$; $a \leq x \leq b$.
Here $f_j^{(l)}(a) = 0$, $l = 0, 1, \ldots, n-1$; $j = 1, \ldots, M$.
Then

$$\int_a^x \left\{ \left\{ \sum_{j=1}^{M-1} \left[|(D_{*a}^{\gamma_2} f_{j+1})(\omega)| \, |(D_{*a}^\nu f_j)(\omega)| \right. \right. \right.$$

$$+ \left. |(D_{*a}^{\gamma_2} f_j)(\omega)| \, |(D_{*a}^\nu f_{j+1})(\omega)| \right] \right\}$$

$$+ \left[|(D_{*a}^{\gamma_2} f_M)(\omega)| \, |(D_{*a}^\nu f_1)(\omega)| \right.$$

$$+ \left. \left. |(D_{*a}^{\gamma_2} f_1)(\omega)| \, |(D_{*a}^\nu f_M)(\omega)| \right] \right\} d\omega$$

$$\leq \left(\frac{\sqrt{2}(x-a)^{(\nu-\gamma_2)}}{\Gamma(\nu - \gamma_2)\sqrt{\nu - \gamma_2}\sqrt{2\nu - 2\gamma_2 - 1}} \right) \tag{16.94}$$

$$\left\{ \int_a^x \left[\sum_{j=1}^M ((D_{*a}^\nu f_j)(\omega))^2 \right] d\omega \right\},$$

all $a \leq x \leq b$.

Corollary 16.51. (To Theorem 16.48; $\lambda_\alpha = 1$, $\lambda_{\alpha+1} = 1/2$, $p = 3/2$, $p(t) = q(t) = 1$.) *In detail: let* $\nu \geq \gamma_1 + 2$, $\gamma_1 \geq 0$, $n := \lceil \nu \rceil$, *and* $f_j \in$

$AC^n([a,b])$, $j = 1,\ldots,M \in \mathbb{N}$, such that $f_j^{(l)}(a) = 0$, $l = 0, 1, \ldots, n-1$, $a \le x \le b$. Assume also $D_{*a}^\nu f_j \in L_\infty(a,b)$, $j = 1,\ldots,M$.

Set

$$\Phi^*(x) := \left(\frac{2}{\sqrt{3\nu - 3\gamma_1 - 2}}\right) \frac{(x-a)^{(3\nu - 3\gamma_1 - 1/2)}}{(\Gamma(\nu - \gamma_1))^{3/2}}, \qquad (16.95)$$

all $a \le x \le b$. Then

$$\int_a^x \left\{ \left\{ \sum_{j=1}^{M-1} \left[|(D_{*a}^{\gamma_1} f_j)(\omega)| \sqrt{\left|\left(D_{*a}^{\gamma_1+1} f_{j+1}\right)(\omega)\right|} \right.\right.\right.$$

$$+ |(D_{*a}^{\gamma_1} f_{j+1})(\omega)| \left.\sqrt{\left|\left(D_{*a}^{\gamma_1+1} f_j\right)(\omega)\right|}\right]\right\}$$

$$+ \left[|(D_{*a}^{\gamma_1} f_1)(\omega)| \sqrt{\left|\left(D_{*a}^{\gamma_1+1} f_M\right)(\omega)\right|} \right.$$

$$\left.\left. + |(D_{*a}^{\gamma_1} f_M)(\omega)| \sqrt{\left|\left(D_{*a}^{\gamma_1+1} f_1\right)(\omega)\right|} \right]\right\} d\omega$$

$$\le 2\Phi^*(x) \cdot \left[\int_a^x \left(\sum_{j=1}^M |(D_{*a}^\nu f_j)(\omega)|^{3/2} \right) d\omega \right], \qquad (16.96)$$

all $a \le x \le b$.

We continue with results regarding $\|\cdot\|_\infty$.

Theorem 16.52. *All are as in Theorem 16.38 but for* $f_j \in AC^n([a,b])$, $j = 1,\ldots,M \in \mathbb{N}$. *Then*

$$\int_a^x p(\omega) \left\{ \left\{ \sum_{j=1}^{M-1} \left[|(D_{*a}^{\gamma_1} f_j)(\omega)|^{\lambda_\alpha} |(D_{*a}^{\gamma_2} f_{j+1})(\omega)|^{\lambda_\beta} |(D_{*a}^\nu f_j)(\omega)|^{\lambda_\nu} \right.\right.\right.$$

$$+ |(D_{*a}^{\gamma_2} f_j)(\omega)|^{\lambda_\beta} |(D_{*a}^{\gamma_1} f_{j+1})(\omega)|^{\lambda_\alpha} |(D_{*a}^\nu f_{j+1})(\omega)|^{\lambda_\nu} \Big]\Big\}$$

$$+ \left[|(D_{*a}^{\gamma_1} f_1)(\omega)|^{\lambda_\alpha} |(D_{*a}^{\gamma_2} f_M)(\omega)|^{\lambda_\beta} |(D_{*a}^\nu f_1)(\omega)|^{\lambda_\nu} \right.$$

$$\left.\left. + |(D_{*a}^{\gamma_2} f_1)(\omega)|^{\lambda_\beta} |(D_{*a}^{\gamma_1} f_M)(\omega)|^{\lambda_\alpha} |(D_{*a}^\nu f_M)(\omega)|^{\lambda_\nu} \right]\right\} d\omega$$

$$\le T(x) \left\{ \sum_{j=1}^M \left\{ \|D_{*a}^\nu f_j\|_{\infty,(a,x)}^{2(\lambda_\alpha + \lambda_\nu)} + \|D_{*a}^\nu f_j\|_{\infty,(a,x)}^{2\lambda_\beta} \right\} \right\}, \qquad (16.97)$$

all $a \le x \le b$.

Proof. Based on Theorem 16.38.   □

We give

Corollary 16.53. (to Theorem 16.52) *In detail: let* $\nu \geq \gamma_1 + 2$, $\gamma_1 \geq 0$, $n := \lceil \nu \rceil$ *and* $f_j \in AC^n([a,b])$, $j = 1,\ldots,M \in \mathbb{N}$, *such that* $f_j^{(l)}(a) = 0$, $l = 0,1,\ldots,n-1$; $j = 1,\ldots,M$; $a \leq x \leq b$. *Further suppose that* $D_{*a}^{\nu} f_j \in L_\infty(a,b)$, $j = 1,\ldots,M$. *Then*

$$\int_a^x \left\{ \left\{ \sum_{j=1}^{M-1} \left[\left| (D_{*a}^{\gamma_1} f_j)(\omega) \right| \left| \left(D_{*a}^{\gamma_1+1} f_{j+1} \right)(\omega) \right| \right. \right. \right.$$

$$+ \left. \left| \left(D_{*a}^{\gamma_1+1} f_j \right)(\omega) \right| \left| (D_{*a}^{\gamma_1} f_{j+1})(\omega) \right| \right] \right\}$$

$$+ \left[\left| (D_{*a}^{\gamma_1} f_1)(\omega) \right| \left| \left(D_{*a}^{\gamma_1+1} f_M \right)(\omega) \right| \right.$$

$$+ \left. \left. \left| \left(D_{*a}^{\gamma_1+1} f_1 \right)(\omega) \right| \left| (D_{*a}^{\gamma_1} f_M)(\omega) \right| \right] \right\} d\omega$$

$$\leq \left(\frac{(x-a)^{2(\nu - \gamma_1)}}{(\Gamma(\nu - \gamma_1 + 1))^2} \right) \left(\sum_{j=1}^{M} \| D_{*a}^{\nu} f_j \|_\infty^2 \right), \qquad (16.98)$$

all $a \leq x \leq b$.

We continue with the converse results.

Theorem 16.54. *Let* $\gamma_j \geq 0$, $1 \leq \nu - \gamma_j < 1/p$, $0 < p < 1$, $j = 1,2$; $n := \lceil \nu \rceil$, *and* $f_i \in AC^n([a,b])$, $i = 1,\ldots,M \in \mathbb{N}$, *such that* $f_i^{(l)}(a) = 0$, $l = 0,1,\ldots,n-1$; $i = 1,\ldots,M$; $a \leq x \leq b$. *Consider also* $p(t) > 0$ *and* $q(t) \geq 0$, *with all* $p(t)$, $1/p(t)$, $q(t)$, $1/q(t) \in L_\infty(a,b)$. *Further assume* $D_{*a}^{\nu} f_i \in L_\infty(a,b)$, $i = 1,\ldots,M$, *each of which is of fixed sign a.e. on* $[a,b]$. *Let* $\lambda_\nu > 0$ *and* $\lambda_\alpha, \lambda_\beta \geq 0$ *such that* $\lambda_\nu > p$.

Here $P_k(\omega)$, $k = 1,2$, $A(\omega)$, $A_0(x)$ *are as in* (16.39), (16.40), *and* (16.41), *respectively. Call*

$$\varphi_1(x) := \left(A_0(x) |_{\lambda_\beta=0} \right) \left(\frac{\lambda_\nu}{\lambda_\alpha + \lambda_\nu} \right)^{\lambda_\nu/p}, \qquad (16.99)$$

$$\delta_1^* := M^{1-(\lambda_\alpha + \lambda_\nu/p)}. \qquad (16.100)$$

If $\lambda_\beta = 0$, *then*

$$\int_a^x q(\omega) \left(\sum_{j=1}^{M} \left| D_{*a}^{\gamma_1} f_j(\omega) \right|^{\lambda_\alpha} \left| D_{*a}^{\nu} f_j(\omega) \right|^{\lambda_\nu} \right) d\omega \qquad (16.101)$$

$$\geq \delta_1^* \varphi_1 (x) \left[\int_a^x p(\omega) \left(\sum_{j=1}^{M} |D_{*a}^\nu f_j (\omega)|^p \right) d\omega \right]^{(\lambda_\alpha + \lambda_\nu / p)},$$

all $a \leq x \leq b$.

Proof. As for Theorem 11 of [47], and Theorem 11 of [51]. □

We continue with

Theorem 16.55. *All are as in Theorem 16.54. Assume* $\lambda_\beta \geq \lambda_\nu$. *Denote*

$$\varphi_2 (x) := (A_0 (x) |_{\lambda_\alpha = 0}) \, 2^{(p - \lambda_\nu / p)} \left(\frac{\lambda_\nu}{\lambda_\beta + \lambda_\nu} \right)^{\lambda_\nu / p} \delta_3^{\lambda_\nu / p}, \qquad (16.102)$$

where δ_3 *is as in* (16.71). *If* $\lambda_\alpha = 0$, *then*

$$\int_a^x q(\omega) \left\{ \left\{ \sum_{j=1}^{M-1} \left[\left| (D_{*a}^{\gamma_2} f_{j+1}) (\omega) \right|^{\lambda_\beta} \left| (D_{*a}^\nu f_j) (\omega) \right|^{\lambda_\nu} \right. \right. \right.$$

$$\left. + \left| (D_{*a}^{\gamma_2} f_j) (\omega) \right|^{\lambda_\beta} \left| (D_{*a}^\nu f_{j+1}) (\omega) \right|^{\lambda_\nu} \right] \right\}$$

$$+ \left[\left| (D_{*a}^{\gamma_2} f_M) (\omega) \right|^{\lambda_\beta} \left| (D_{*a}^\nu f_1) (\omega) \right|^{\lambda_\nu} \right.$$

$$\left. \left. + \left| (D_{*a}^{\gamma_2} f_1) (\omega) \right|^{\lambda_\beta} \left| (D_{*a}^\nu f_M) (\omega) \right|^{\lambda_\nu} \right] \right\} d\omega \geq$$

$$M^{1 - (\lambda_\nu + \lambda_\beta / p)} 2^{(\lambda_\nu + \lambda_\beta / p)} \varphi_2 (x) \cdot \left\{ \int_a^x p(\omega) \right.$$

$$\left. \left[\sum_{j=1}^{M} |(D_{*a}^\nu f_j) (\omega)|^p \right] d\omega \right\}^{((\lambda_\nu + \lambda_\beta)/p)}, \qquad (16.103)$$

all $a \leq x \leq b$.

Proof. As for Theorem 12 of [47], and Theorem 12 of [51]. □

We give

Theorem 16.56. *Let* $\nu \geq 2$ *and* $\gamma_1 \geq 0$ *such that* $2 \leq \nu - \gamma_1 < 1/p$, $0 < p < 1$, $n := \lceil \nu \rceil$. *Consider* $f_i \in AC^n ([a, b])$, $i = 1, \ldots, M \in \mathbb{N}$, *such that* $f_i^{(l)} (a) = 0$, $l = 0, 1, \ldots, n - 1$; $i = 1, \ldots, M$; $a \leq x \leq b$. *Assume that* $D_{*a}^\nu f_i \in L_\infty (a, b)$, $i = 1, \ldots, M$, *each of which is of fixed sign a.e. on* $[a, b]$. *Consider also* $p(t) > 0$ *and* $q(t) \geq 0$, *with all* $p(t)$, $1/p(t)$, $q(t)$, $1/q(t) \in L_\infty (a, b)$. *Let* $\lambda_\alpha \geq \lambda_{\alpha+1} > 1$.

Here θ_3 is as in (16.73), $L(x)$ as in (16.50), $P_1(x)$ as in (16.51), $T(x)$ as in (16.52), ω_1 as in (16.53), and Φ as in (16.54).

$$\int_a^x q(\omega) \left\{ \left\{ \sum_{j=1}^{M-1} \left[\left| (D_{*a}^{\gamma_1} f_j)(\omega) \right|^{\lambda_\alpha} \left| D_{*a}^{\gamma_1+1} f_{j+1}(\omega) \right|^{\lambda_{\alpha+1}} \right. \right. \right.$$

$$\left. + \left| (D_{*a}^{\gamma_1} f_{j+1})(\omega) \right|^{\lambda_\alpha} \left| D_{*a}^{\gamma_1+1} f_j(\omega) \right|^{\lambda_{\alpha+1}} \right] \right\}$$

$$+ \left[\left| (D_{*a}^{\gamma_1} f_1)(\omega) \right|^{\lambda_\alpha} \left| D_{*a}^{\gamma_1+1} f_M(\omega) \right|^{\lambda_{\alpha+1}} \right.$$

$$\left. \left. + \left| (D_{*a}^{\gamma_1} f_M)(\omega) \right|^{\lambda_\alpha} \left| D_{*a}^{\gamma_1+1} f_1(\omega) \right|^{\lambda_{\alpha+1}} \right] \right\} d\omega \geq$$

$$M^{1-(\lambda_\alpha+\lambda_{\alpha+1}/p)} 2^{(\lambda_\alpha+\lambda_{\alpha+1}/p)} \Phi(x) \cdot$$

$$\left[\int_a^x p(\omega) \left(\sum_{j=1}^M \left| (D_{*a}^\nu f_j)(\omega) \right|^p \right) d\omega \right]^{((\lambda_\alpha+\lambda_{\alpha+1})/p)} , \qquad (16.104)$$

all $a \leq x \leq b$.

Proof. As for Theorem 13 of [47] and Theorem 13 of [51]. □

We have the special cases.

Corollary 16.57. (To Theorem 16.54; $\lambda_\beta = 0$, $p(t) = q(t) = 1$.)
Then

$$\int_a^x \left(\sum_{j=1}^M \left| D_{*a}^{\gamma_1} f_j(\omega) \right|^{\lambda_\alpha} \left| D_{*a}^\nu f_j(\omega) \right|^{\lambda_\nu} \right) d\omega$$

$$\geq \delta_1^* \varphi_1(x) \left[\int_a^x \left[\sum_{j=1}^M \left| D_{*a}^\nu f_j(\omega) \right|^p \right] d\omega \right]^{((\lambda_\alpha+\lambda_\nu)/p)} , \qquad (16.105)$$

all $a \leq x \leq b$.
In (16.105), $\left(A_0(x) |_{\lambda_\beta=0} \right)$ of $\varphi_1(x)$ is given by (16.59).

Corollary 16.58. (To Theorem 16.55; $\lambda_\alpha = 0$, $p(t) = q(t) = 1$.) *It holds*

$$\int_a^x \left\{ \left\{ \sum_{j=1}^{M-1} \left[\left| (D_{*a}^{\gamma_2} f_{j+1})(\omega) \right|^{\lambda_\beta} \left| (D_{*a}^\nu f_j)(\omega) \right|^{\lambda_\nu} \right. \right. \right.$$

$$\left. \left. + \left| (D_{*a}^{\gamma_2} f_j)(\omega) \right|^{\lambda_\beta} \left| (D_{*a}^\nu f_{j+1})(\omega) \right|^{\lambda_\nu} \right] \right\}$$

$$+ \left[\left| \left(D_{*a}^{\gamma_2} f_M \right) (\omega) \right|^{\lambda_\beta} \left| \left(D_{*a}^{\nu} f_1 \right) (\omega) \right|^{\lambda_\nu} \right.$$

$$+ \left. \left| \left(D_{*a}^{\gamma_2} f_1 \right) (\omega) \right|^{\lambda_\beta} \left| \left(D_{*a}^{\nu} f_M \right) (\omega) \right|^{\lambda_\nu} \right] \right\} d\omega \ge$$

$$\left(M^{1 - (\lambda_\nu + \lambda_\beta / p)} \right) 2^{(\lambda_\nu + \lambda_\beta / p)} \varphi_2 (x) \cdot$$

$$\left\{ \int_a^x \left[\sum_{j=1}^{M} \left| \left(D_{*a}^{\nu} f_j \right) (\omega) \right|^p \right] d\omega \right\}^{((\lambda_\nu + \lambda_\beta) / p)}, \qquad (16.106)$$

all $a \le x \le b$.

In (16.106), $(A_0 (x) |_{\lambda_\alpha = 0})$ of $\varphi_2 (x)$ *is given by* (16.64).
Next we apply the above results on the spherical shell.

16.4   Background—II

Here we initially follow [356, pp. 149–150] and [383, pp. 87–88]. Let us
denote by $dx \equiv \lambda_{\mathbb{R}^N} (dx)$, $N \in \mathbb{N}$, the Lebesgue measure on $\mathbb{R}^N$, and
$S^{N-1} := \{ x \in \mathbb{R}^N : |x| = 1 \}$ the unit sphere on $\mathbb{R}^N$, where $|\cdot|$ stands for
the Euclidean norm in $\mathbb{R}^N$. Also denote the ball

$$B(0, R) := \{ x \in \mathbb{R}^N : |x| < R \} \subseteq \mathbb{R}^N, \ R > 0,$$

and the spherical shell

$$A := B(0, R_2) - \overline{B(0, R_1)}, \ 0 < R_1 < R_2.$$

For $x \in \mathbb{R}^N - \{0\}$ we can write uniquely $x = r\omega$, where $r = |x| > 0$, and
$\omega = x/r \in S^{N-1}$, $|\omega| = 1$. Clearly here

$$\mathbb{R}^N - \{0\} = (0, \infty) \times S^{N-1},$$

and the map

$$\Phi : \mathbb{R}^N - \{0\} \to S^{N-1} : \Phi (x) = \frac{x}{|x|}$$

is continuous.

Also $\overline{A} = [R_1, R_2] \times S^{N-1}$. Let us denote by $d\omega \equiv \lambda_{S^{N-1}} (\omega)$ the surface
measure on S^{N-1} to be defined as the image under Φ of $N \cdot \lambda_{\mathbb{R}^N}$ restricted
to the Borel class of $B(0, 1) - \{0\}$. More precisely the last definition is as
follows. Let $A \subset S^{N-1}$ be a Borel set, and let

$$\widetilde{A} := \{ ru : 0 < r < 1, \ u \in A \} \subset \mathbb{R}^N.$$

We define

$$\lambda_{S^{N-1}} (A) = N \cdot \lambda_{\mathbb{R}^N} \left(\widetilde{A} \right).$$

B_X here stands for the Borel class on space X.

We denote by

$$\omega_N \equiv \lambda_{S^{N-1}}\left(S^{N-1}\right) = \int_{S^{N-1}} d\omega = \frac{2\pi^{N/2}}{\Gamma\left(N/2\right)}$$

the surface area of S^{N-1} and we get the volume

$$|B\left(0,r\right)| = \frac{\omega_N r^N}{N} = \frac{2\pi^{N/2} r^N}{N\Gamma\left(N/2\right)},$$

so that

$$|B\left(0,1\right)| = \frac{2\pi^{N/2}}{N\Gamma\left(N/2\right)}.$$

Clearly here

$$Vol\left(A\right) = |A| = \frac{\omega_N\left(R_2^N - R_1^N\right)}{N} = \frac{2\pi^{N/2}\left(R_2^N - R_1^N\right)}{N\Gamma\left(N/2\right)}.$$

Next, define

$$\psi : (0,\infty) \times S^{N-1} \rightarrow \mathbb{R}^N - \{0\}$$

by $\psi\left(r,\omega\right) := r\omega$, ψ is a one-to-one and onto function; thus

$$\left(r,\omega\right) \equiv \psi^{-1}\left(x\right) = \left(|x|, \Phi\left(x\right)\right)$$

are called the polar coordinates of $x \in \mathbb{R}^N - \{0\}$.

Finally, define the measure R_N on $\left((0,\infty), \mathcal{B}_{(0,\infty)}\right)$ by

$$R_N\left(\Gamma\right) = \int_\Gamma r^{N-1} dr, \text{ any } \Gamma \in \mathcal{B}_{(0,\infty)}.$$

We mention the following very important theorem.

Theorem 16.59. (See Exercise 6, pp. 149–150 in [356] and Theorem 5.2.2 pp. 87–88 of [383].) *We have that* $\lambda_{\mathbb{R}^N} = \left(R_N \times \lambda_{S^{N-1}}\right) \circ \psi^{-1}$ *on* $\mathcal{B}_{\mathbb{R}^N - \{0\}}$.

In particular, if f is a nonnegative Borel measurable function on $\left(\mathbb{R}^N, \mathcal{B}_{\mathbb{R}^N}\right)$, then the Lebesgue integral

$$\int_{\mathbb{R}^N} f\left(x\right) dx = \int_{(0,\infty)} r^{N-1} \left(\int_{S^{N-1}} f\left(r\omega\right) \lambda_{S^{N-1}}\left(d\omega\right)\right) dr$$

$$= \int_{S^{N-1}} \left(\int_{(0,\infty)} f\left(r\omega\right) r^{N-1} dr\right) \lambda_{S^{N-1}}\left(d\omega\right). \tag{16.107}$$

Clearly (16.107) *is true for f a Borel integrable function taking values in* $\mathbb{R}$.

Based on Theorem 16.59 in [45] we prove the next result which is the main tool of this section.

Proposition 16.60. *Let*

$$f : A \to \mathbb{R},$$

be a Lebesgue integrable function, where

$$A := B(0, R_2) - \overline{B(0, R_1)}, \ 0 < R_1 < R_2. \tag{16.108}$$

Then

$$\int_A f(x)\, dx = \int_{S^{N-1}} \left(\int_{R_1}^{R_2} f(r\omega)\, r^{N-1} dr \right) d\omega.$$

We make

Remark 16.61. Let $F : \overline{A} = [R_1, R_2] \times S^{N-1} \to \mathbb{R}$ and for each $\omega \in S^{N-1}$ define

$$g_\omega(r) := F(r\omega) = F(x),$$

where $x \in \overline{A}$, with $A := B(0, R_2) - \overline{B(0, R_1)}$; $0 < R_1 \le r \le R_2$, $r = |x|$, $\omega = x/r \in S^{N-1}$.

For each $\omega \in S^{N-1}$ we assume that $g_\omega \in AC^n([R_1, R_2])$, where $n := \lceil \nu \rceil$, $\nu \ge 0$.

Thus, by Corollary 16.8, for almost all $r \in [R_1, R_2]$, there exists the Caputo fractional derivative

$$D^\nu_{*R_1} g_\omega(r) = \frac{1}{\Gamma(n-\nu)} \int_{R_1}^r (r-t)^{n-\nu-1} g_\omega^{(n)}(t)\, dt, \tag{16.109}$$

for all $\omega \in S^{N-1}$.

Now we are ready to give

Definition 16.62. Let $F : \overline{A} \to \mathbb{R}$, $\nu \ge 0$, $n := \lceil \nu \rceil$ such that $F(\cdot\omega) \in AC^n([R_1, R_2])$, for all $\omega \in S^{N-1}$.

We call the Caputo radial fractional derivative the following function

$$\frac{\partial^\nu_{*R_1} F(x)}{\partial r^\nu} := \frac{1}{\Gamma(n-\nu)} \int_{R_1}^r (r-t)^{n-\nu-1} \frac{\partial^n F(t\omega)}{\partial r^n}\, dt, \tag{16.110}$$

where $x \in \overline{A}$; that is, $x = r\omega$, $r \in [R_1, R_2]$, $\omega \in S^{N-1}$.

Clearly

$$\frac{\partial^0_{*R_1} F(x)}{\partial r^0} = F(x),$$

$$\frac{\partial^\nu_{*R_1} F(x)}{\partial r^\nu} = \frac{\partial^\nu F(x)}{\partial r^\nu}, \text{ if } \nu \in \mathbb{N}.$$

The above function (16.110) exists almost everywhere for $x \in \overline{A}$.

We justify this next.

Note 16.63. Call

$$\Lambda_1 := \left\{ r \in [R_1, R_2] : \frac{\partial^\nu_{*R_1} F(x)}{\partial r^\nu} \text{ does not exist} \right\}. \tag{16.111}$$

We have that Lebesgue measure $\lambda_{\mathbb{R}}(\Lambda_1) = 0$.

Call $\Lambda_N := \Lambda_1 \times S^{N-1}$. So there exists a Borel set $\Lambda_1^* \subset [R_1, R_2]$, such that $\Lambda_1 \subset \Lambda_1^*$, $\lambda_{\mathbb{R}}(\Lambda_1^*) = \lambda_{\mathbb{R}}(\Lambda_1) = 0$; thus $R_N(\Lambda_1^*) = 0$.

Consider now $\Lambda_N^* := \Lambda_1^* \times S^{N-1} \subset \overline{A}$, which is a Borel set of $\mathbb{R}^N - \{0\}$. Clearly then by Theorem 16.59, $\lambda_{\mathbb{R}^N}(\Lambda_N^*) = 0$, but $\Lambda_N \subset \Lambda_N^*$, implying $\lambda_{\mathbb{R}^N}(\Lambda_N) = 0$.

Consequently (16.110) exists a.e. in x with respect to $\lambda_{\mathbb{R}^N}$ on $\overline{A}$.

We give the following fundamental representation result.

Theorem 16.64. *Let $\nu \geq \gamma + 1$, $\gamma \geq 0$, $n := \lceil \nu \rceil$, $F : \overline{A} \to \mathbb{R}$ with $F \in L_1(A)$. Assume that $F(\cdot\omega) \in AC^n([R_1, R_2])$ for all $\omega \in S^{N-1}$, and that $\partial^\nu_{*R_1} F(\cdot\omega)/\partial r^\nu \in L_\infty(R_1, R_2)$ for all $\omega \in S^{N-1}$.*

Further assume that $\partial^\nu_{*R_1} F(x)/\partial r^\nu \in L_\infty(A)$. More precisely, for these $r \in [R_1, R_2]$, and for each $\omega \in S^{N-1}$, for which $D^\nu_{*R_1} F(r\omega)$ takes real values, there exists $M_1 > 0$ such that $\left| D^\nu_{*R_1} F(r\omega) \right| \leq M_1$.

We suppose that $\partial^i F(R_1\omega)/\partial r^j = 0$, $j = 0, 1, \ldots, n - 1$, for every $\omega \in S^{N-1}$. Then

$$\frac{\partial^\gamma_{*R_1} F(x)}{\partial r^\gamma} = D^\gamma_{*R_1} F(r\omega) =$$

$$\frac{1}{\Gamma(\nu - \gamma)} \int_{R_1}^r (r - t)^{\nu - \gamma - 1} \left(D^\nu_{*R_1} F \right)(t\omega) \, dt, \tag{16.112}$$

true $\forall x \in \overline{A}$; that is, true $\forall r \in [R_1, R_2]$ and $\forall \omega \in S^{N-1}$, $\gamma > 0$.
Here

$$D^\gamma_{*R_1} F(\cdot\omega) \in AC([R_1, R_2]), \tag{16.113}$$

$\forall \omega \in S^{N-1}$, $\gamma > 0$.
Furthermore

$$\frac{\partial^\gamma_{*R_1} F(x)}{\partial r^\gamma} \in L_\infty(A), \ \gamma > 0. \tag{16.114}$$

In particular, it holds

$$F(x) = F(r\omega) = \frac{1}{\Gamma(\nu)} \int_{R_1}^{r} (r-t)^{\nu-1} \left(D_{*R_1}^{\nu} F\right)(t\omega)\, dt, \qquad (16.115)$$

true $\forall x \in \overline{A}$; *that is, true* $\forall r \in [R_1, R_2]$ *and* $\forall \omega \in S^{N-1}$, *and*

$$F(\cdot\omega) \in AC([R_1, R_2]), \quad \forall \omega \in S^{N-1}. \qquad (16.116)$$

Proof. By our assumptions and Theorem 16.16, Corollary 16.14, we have valid (16.112) and (16.115). Also (16.113) is clear; see [45]. Property (16.116) is easy to prove.

Fixing $r \in [R_1, R_2]$, the function

$$\delta_r(t, \omega) := (r-t)^{\nu-\gamma-1} D_{*R_1}^{\nu} F(t\omega)$$

is measurable on

$$\left([R_1, r] \times S^{N-1}, \overline{\mathcal{B}_{[R_1, r]} \times \overline{\mathcal{B}}_{S^{N-1}}}\right).$$

Here $\overline{\mathcal{B}_{[R_1, r]} \times \overline{\mathcal{B}}_{S^{N-1}}}$ stands for the complete σ-algebra generated by $\mathcal{B}_{[R_1, r]} \times \overline{\mathcal{B}}_{S^{N-1}}$, where $\overline{\mathcal{B}}_X$ stands for the completion of $\mathcal{B}_X$.

Then we get that

$$\int_{S^{N-1}} \left(\int_{R_1}^{r} |\delta_r(t, \omega)|\, dt \right) d\omega =$$

$$\int_{S^{N-1}} \left(\int_{R_1}^{r} (r-t)^{\nu-\gamma-1} \left|D_{*R_1}^{\nu} F(t\omega)\right| dt \right) d\omega \leq \qquad (16.117)$$

$$\left\| \frac{\partial_{*R_1}^{\nu} F(x)}{\partial r^{\nu}} \right\|_{\infty, ([R_1, r] \times S^{N-1})} \left(\int_{S^{N-1}} \left(\int_{R_1}^{r} (r-t)^{\nu-\gamma-1} dt \right) d\omega \right)$$

$$\qquad (16.118)$$

$$= \left\| \frac{\partial_{*R_1}^{\nu} F(x)}{\partial r^{\nu}} \right\|_{\infty, ([R_1, r] \times S^{N-1})} \left(\frac{2\pi^{N/2}}{\Gamma(N/2)} \right) \frac{(r-R_1)^{\nu-\gamma}}{(\nu-\gamma)} \leq$$

$$\left\| \frac{\partial_{*R_1}^{\nu} F(x)}{\partial r^{\nu}} \right\|_{\infty, A} \left(\frac{2\pi^{N/2}}{\Gamma(N/2)} \right) \frac{(R_2-R_1)^{\nu-\gamma}}{(\nu-\gamma)} < \infty. \qquad (16.119)$$

Hence $\delta_r(t, \omega)$ is integrable on

$$\left([R_1, r] \times S^{N-1}, \overline{\mathcal{B}_{[R_1, r]} \times \overline{\mathcal{B}}_{S^{N-1}}}\right).$$

Consequently, by Fubini's theorem and (16.112), we obtain that $D_{*R_1}^{\gamma} F(r\omega)$, $\nu \geq \gamma + 1$, $\gamma > 0$ is integrable in ω over $\left(S^{N-1}, \overline{\mathcal{B}}_{S^{N-1}}\right)$. So we have that $D_{*R_1}^{\gamma} F(r\omega)$ is continuous in $r \in [R_1, R_2]$, $\forall \omega \in S^{N-1}$, and measurable in $\omega \in S^{N-1}$, $\forall r \in [R_1, R_2]$. So, it is a Carathéodory function. Here $[R_1, R_2]$ is a separable metric space and S^{N-1} is a measurable space, and the function takes values in $\mathbb{R}^* := \mathbb{R} \cup \{\pm\infty\}$, which is a metric space. Therefore by Theorem 20.15, p. 156 of [10], $(D_{*R_1}^{\gamma} F)(r\omega)$ is jointly $(\mathcal{B}_{[R_1,R_2]} \times \overline{\mathcal{B}}_{S^{N-1}})$-measurable on $[R_1, R_2] \times S^{N-1} = A$, that is, Lebesgue measurable on A. Indeed then we have that

$$\left| D_{*R_1}^{\gamma} F(r\omega) \right| \leq \frac{1}{\Gamma(\nu - \gamma)}$$

$$\int_{R_1}^{r} (r - t)^{\nu - \gamma - 1} \left| D_{*R_1}^{\nu} F(t\omega) \right| dt \qquad (16.120)$$

$$\leq \frac{\left\| D_{*R_1}^{\nu} F(\cdot\omega) \right\|_{\infty,[R_1,R_2]}}{\Gamma(\nu - \gamma)} \left(\int_{R_1}^{r} (r - t)^{\nu - \gamma - 1} dt \right) \leq \qquad (16.121)$$

$$\frac{M_1}{\Gamma(\nu - \gamma)} \frac{(r - R_1)^{\nu - \gamma}}{(\nu - \gamma)} \leq \frac{M_1}{\Gamma(\nu - \gamma - 1)} (R_2 - R_1)^{\nu - \gamma} := \tau < \infty,$$

for all $\omega \in S^{N-1}$ and for all $r \in [R_1, R_2]$.

We proved that

$$\left| D_{*R_1}^{\gamma} F(r\omega) \right| \leq \tau < \infty, \forall \omega \in S^{N-1}, \text{and } \forall r \in [R_1, R_2], \qquad (16.122)$$

hence proving $\partial_{*R_1}^{\gamma} F(x)/\partial r^{\gamma} \in L_{\infty}(A)$, $\gamma > 0$. We have completed our proof. □

16.5   Main Results on a Spherical Shell

16.5.1   Results Involving One Function

We give

Theorem 16.65. Let $\nu \geq \gamma_i + 1$, $\gamma_i \geq 0$, $i = 1, \ldots, l \in \mathbb{N}$, $n := \lceil \nu \rceil$; and $0 \leq \gamma_1 < \gamma_2 \leq \gamma_3 \leq \ldots \leq \gamma_l$. Here $f : \overline{A} \to \mathbb{R}$ is as in Theorem 16.64. Let $r_i > 0 : \sum_{i=1}^{l} r_i = p$. If $\gamma_1 = 0$ we set $r_1 = 1$. Let $s_1, s_1' > 1 :$ $1/s_1 + 1/s_1' = 1$, and $s_2, s_2' > 1 : 1/s_2 + 1/s_2' = 1$, and $p > s_2$, $N \geq 2$. Denote

$$Q_1(R_2) := \left(\frac{R_2^{(N-1)s_1'+1} - R_1^{(N-1)s_1'+1}}{(N-1)s_1' + 1} \right)^{1/s_1'}, \qquad (16.123)$$

$$Q_2(R_2) := \left(\frac{R_2^{(1-N)(s_2'/p)+1} - R_1^{(1-N)(s_2'/p)+1}}{(1-N)(s_2'/p) + 1} \right)^{p/s_2'}, \qquad (16.124)$$

and

$$\sigma := \frac{p - s_2}{p s_2}. \qquad (16.125)$$

Also call

$$C := Q_1 (R_2) Q_2 (R_2)$$

$$\prod_{i=1}^{l} \left\{ \frac{\sigma^{r_i \sigma}}{(\Gamma(\nu - \gamma_i))^{r_i} (\nu - \gamma_i - 1 + \sigma)^{r_i \sigma}} \right\}$$

$$\frac{(R_2 - R_1)^{\left(\sum_{i=1}^{l}(\nu - \gamma_i - 1)r_i + p/s_2 + 1/s_1 - 1\right)}}{\left(\left(\sum_{i=1}^{l}(\nu - \gamma_i - 1) r_i s_1\right) + s_1 \left(\frac{p}{s_2} - 1\right) + 1\right)^{1/s_1}}. \qquad (16.126)$$

Then

$$\int_A \prod_{i=1}^{l} \left| \frac{\partial_{*R_1}^{\gamma_i} f(x)}{\partial r^{\gamma_i}} \right|^{r_i} dx \le$$

$$C \int_A \left| \frac{\partial_{*R_1}^{\nu} f(x)}{\partial r^{\nu}} \right|^{p} dx. \qquad (16.127)$$

Proof. Clearly here $f(\cdot \omega)$ fulfills all the assumptions of Theorem 16.24, $\forall \omega \in S^{N-1}$.

We set there $q_1(r) = q_2(r) := r^{N-1}$, $r \in [R_1, R_2]$.

Hence by (16.30) we have

$$\int_{R_1}^{R_2} r^{N-1} \prod_{i=1}^{l} \left| \frac{\partial_{*R_1}^{\gamma_i} f(r\omega)}{\partial r^{\gamma_i}} \right|^{r_i} dr \le$$

$$C \int_{R_1}^{R_2} r^{N-1} \left| \frac{\partial_{*R_1}^{\nu} f(r\omega)}{\partial r^{\nu}} \right|^{p} dr, \forall \omega \in S^{N-1}. \qquad (16.128)$$

Therefore it holds

$$\int_{S^{N-1}} \left(\int_{R_1}^{R_2} r^{N-1} \prod_{i=1}^{l} \left| \frac{\partial_{*R_1}^{\gamma_i} f(r\omega)}{\partial r^{\gamma_i}} \right|^{r_i} dr \right) d\omega \le$$

$$C \int_{S^{N-1}} \left(\int_{R_1}^{R_2} r^{N-1} \left| \frac{\partial_{*R_1}^{\nu} f(r\omega)}{\partial r^{\nu}} \right|^{p} dr \right) d\omega. \qquad (16.129)$$

Using the conclusion of Theorem 16.64 and Proposition 16.60 we derive (16.127). □

We continue with the following extreme case.

Theorem 16.66. *Let $\nu \geq \gamma_i + 1$, $\gamma_i \geq 0$, $i = 1, \ldots, l \in \mathbb{N}$, $n := \lceil \nu \rceil$, and $0 \leq \gamma_1 < \gamma_2 \leq \gamma_3 \leq \ldots \leq \gamma_l$. Here $f : \overline{A} \to \mathbb{R}$ as in Theorem 16.64. Let $r_i > 0 : \sum_{i=1}^{l} r_i = r$. If $\gamma_1 = 0$ we set $r_1 = 1$, $N \geq 2$.*
Call

$$\widetilde{M} := \frac{R_2^{N-1} (R_2 - R_1)^{r\nu - \sum_{i=1}^{l} r_i \gamma_i + 1}}{\prod_{i=1}^{l} (\Gamma(\nu - \gamma_i + 1))^{r_i} \left(r\nu - \sum_{i=1}^{l} r_i \gamma_i + 1 \right)} > 0. \tag{16.130}$$

Then

$$\int_A \left(\prod_{i=1}^{l} \left| \frac{\partial_{*R_1}^{\gamma_i} f(x)}{\partial r^{\gamma_i}} \right|^{r_i} \right) dx \leq$$

$$\widetilde{M} M_1^r \frac{2\pi^{N/2}}{\Gamma(N/2)}. \tag{16.131}$$

Proof. Clearly here $f(\cdot\omega)$ fulfills all the assumptions of Theorem 16.26, $\forall \omega \in S^{N-1}$.
We set $\widetilde{q}(r) = r^{N-1}$, $r \in [R_1, R_2]$.
Hence by (16.35) we have

$$\int_{R_1}^{R_2} r^{N-1} \prod_{i=1}^{l} \left| D_{*R_1}^{\gamma_i} f(r\omega) \right|^{r_i} dr \leq$$

$$\left(\frac{R_2^{N-1} (R_2 - R_1)^{r\nu - \sum_{i=1}^{l} r_i \gamma_i + 1}}{\prod_{i=1}^{l} (\Gamma(\nu - \gamma_i + 1))^{r_i} \left(r\nu - \sum_{i=1}^{l} r_i \gamma_i + 1 \right)} \right) \left\| \frac{\partial_{*R_1}^{\nu} f(\cdot\omega)}{\partial r^{\nu}} \right\|_{\infty, [R_1, R_2]}$$

$$\leq \widetilde{M} M_1^r, \ \forall \omega \in S^{N-1}. \tag{16.132}$$

Hence

$$\int_{S^{N-1}} \left(\int_{R_1}^{R_2} r^{N-1} \prod_{i=1}^{l} \left| D_{*R_1}^{\gamma_i} f(r\omega) \right|^{r_i} dr \right) d\omega \leq$$

$$\widetilde{M} M_1^r \frac{2\pi^{N/2}}{\Gamma(N/2)}. \tag{16.133}$$

Using the conclusion of Theorem 16.64 and Proposition 16.60 we derive (16.131). $\square$

16.5.2 Results Involving Two Functions

We need to make the following assumption.

Assumption 16.67. Let $\nu \geq \gamma_i + 1$, $\gamma_i \geq 0$, $i = 1, 2$, $n := \lceil \nu \rceil$, f_1, $f_2 : \overline{A} \to \mathbb{R}$ with f_1, $f_2 \in L_1(A)$, where $A := B(0, R_2) - \overline{B(0, R_1)}$, $0 < R_1 < R_2$. Assume that $f_1(\cdot\omega)$, $f_2(\cdot\omega) \in AC^n([R_1, R_2])$ for all $\omega \in S^{N-1}$, and that $\partial^\nu_{*R_1} f_i(\cdot\omega)/\partial r^\nu \in L_\infty(R_1, R_2)$, for all $\omega \in S^{N-1}$; $i = 1, 2$. Further assume that $\partial^\nu_{*R_1} f_i(x)/\partial r^\nu \in L_\infty(A)$, $i = 1, 2$. More precisely, for these $r \in [R_1, R_2]$, $\forall \omega \in S^{N-1}$, for which $D^\nu_{*R_1} f_i(r\omega)$ takes real values, there exists $M_i > 0$ such that

$$\left| D^\nu_{*R_1} f_i(r\omega) \right| \leq M_i, \text{ for } i = 1, 2. \tag{16.134}$$

We suppose that

$$\frac{\partial^j f_i(R_1 \omega)}{\partial r^j} = 0, j = 0, 1, \ldots, n - 1,$$

$\forall \omega \in S^{N-1}$; $i = 1, 2$.

Let $\lambda_\nu > 0$, and λ_α, $\lambda_\beta \geq 0$, such that $\lambda_\nu < p$, where $p > 1$. If $\gamma_1 = 0$ we set $\lambda_\alpha = 1$ and if $\gamma_2 = 0$ we set $\lambda_\beta = 1$, here $N \geq 2$.

Assumption 16.67*. (continuation of Assumption 16.67)
Set

$$P_k(w) := \int_{R_1}^w (w - t)^{(\nu - \gamma_k - 1)p/(p-1)} t^{(1-N/p-1)} dt, \tag{16.135}$$

$k = 1, 2$, $R_1 \leq w \leq R_2$,

$$A(w) := \frac{w^{(N-1)(1-\lambda_\nu/p)} (P_1(w))^{\lambda_\alpha(p-1/p)} (P_2(w))^{\lambda_\beta((p-1)/p)}}{(\Gamma(\nu - \gamma_1))^{\lambda_\alpha} (\Gamma(\nu - \gamma_2))^{\lambda_\beta}}, \tag{16.136}$$

$$A_0(R_2) := \left(\int_{R_1}^{R_2} (A(w))^{p/(p-\lambda_\nu)} dw \right)^{(p-\lambda_\nu)/p}. \tag{16.137}$$

We present

Theorem 16.68. *All here are as in Assumptions* 16.67 *and* 16.67*; *especially assume* $\lambda_\alpha > 0$, $\lambda_\beta = 0$, *and* $p = \lambda_\alpha + \lambda_\nu > 1$. *Then*

$$\int_A \left[\left| \frac{\partial^{\gamma_1}_{*R_1} f_1(x)}{\partial r^{\gamma_1}} \right|^{\lambda_\alpha} \left| \frac{\partial^\nu_{*R_1} f_1(x)}{\partial r^\nu} \right|^{\lambda_\nu} + \right.$$

$$\left. \left| \frac{\partial_{*R_1}^{\gamma_1} f_2 (x)}{\partial r^{\gamma_1}} \right|^{\lambda_\alpha} \left| \frac{\partial_{*R_1}^{\nu} f_2 (x)}{\partial r^{\nu}} \right|^{\lambda_\nu} \right] dx \le \qquad (16.138)$$

$$(A_0 (R_2) |_{\lambda_\beta=0}) \left(\frac{\lambda_\nu}{p} \right)^{(\lambda_\nu/p)}$$

$$\int_A \left[\left| \frac{\partial_{*R_1}^{\nu} f_1 (x)}{\partial r^{\nu}} \right|^{p} + \left| \frac{\partial_{*R_1}^{\nu} f_2 (x)}{\partial r^{\nu}} \right|^{p} \right] dx.$$

Proof. We apply Theorem 16.30 here for every $\omega \in S^{N-1}$; here $p(r) = q(r) = r^{N-1}$, $r \in [R_1, R_2]$. We use Theorem 16.64 and Proposition 16.60. Thus the proof is similar to the proof of Theorem 16.65. $\square$

The counterpart of the last theorem follows.

Theorem 16.69. *All here are as in Assumptions* 16.67 *and* 16.67*; especially suppose* $\lambda_\alpha = 0$, $\lambda_\beta > 0$, $p = \lambda_\nu + \lambda_\beta > 1$.
Denote

$$\delta_3 := \begin{cases} 2^{\lambda_\beta/\lambda_\nu} - 1, & if \ \lambda_\beta \ge \lambda_\nu, \\ 1, & if \ \lambda_\beta \le \lambda_\nu. \end{cases} \qquad (16.139)$$

Then

$$\int_A \left[\left| \frac{\partial_{*R_1}^{\gamma_2} f_2 (x)}{\partial r^{\gamma_2}} \right|^{\lambda_\beta} \left| \frac{\partial_{*R_1}^{\nu} f_1 (x)}{\partial r^{\nu}} \right|^{\lambda_\nu} + \right.$$

$$\left. \left| \frac{\partial_{*R_1}^{\gamma_2} f_1 (x)}{\partial r^{\gamma_2}} \right|^{\lambda_\beta} \left| \frac{\partial_{*R_1}^{\nu} f_2 (x)}{\partial r^{\nu}} \right|^{\lambda_\nu} \right] dx \le \qquad (16.140)$$

$$(A_0 (R_2) |_{\lambda_\alpha=0}) \, 2^{\lambda_\beta/p} \left(\frac{\lambda_\nu}{p} \right)^{(\lambda_\nu/p)} \delta_3^{\lambda_\nu/p}$$

$$\int_A \left[\left| \frac{\partial_{*R_1}^{\nu} f_1 (x)}{\partial r^{\nu}} \right|^{p} + \left| \frac{\partial_{*R_1}^{\nu} f_2 (x)}{\partial r^{\nu}} \right|^{p} \right] dx.$$

Proof. Based on Theorem 16.31, similar to the proof of Theorem 16.68. $\square$

Theorem 16.70. *All here are as in Assumptions* 16.67 *and* 16.67*; especially suppose* λ_ν, λ_α, $\lambda_\beta > 0$, $p = \lambda_\alpha + \lambda_\beta + \lambda_\nu > 1$.
Denote

$$\tilde{\gamma}_1 := \begin{cases} 2^{(\lambda_\alpha+\lambda_\beta/\lambda_\nu)} - 1, & if \ \lambda_\alpha + \lambda_\beta \ge \lambda_\nu, \\ 1, & if \ \lambda_\alpha + \lambda_\beta \le \lambda_\nu. \end{cases} \qquad (16.141)$$

Then

$$\int_A \left[\left| \frac{\partial_{*R_1}^{\gamma_1} f_1(x)}{\partial r^{\gamma_1}} \right|^{\lambda_\alpha} \left| \frac{\partial_{*R_1}^{\gamma_2} f_2(x)}{\partial r^{\gamma_2}} \right|^{\lambda_\beta} \left| \frac{\partial_{*R_1}^{\nu} f_1(x)}{\partial r^{\nu}} \right|^{\lambda_\nu} + \right.$$

$$\left. \left| \frac{\partial_{*R_1}^{\gamma_2} f_1(x)}{\partial r^{\gamma_2}} \right|^{\lambda_\beta} \left| \frac{\partial_{*R_1}^{\gamma_1} f_2(x)}{\partial r^{\gamma_1}} \right|^{\lambda_\alpha} \left| \frac{\partial_{*R_1}^{\nu} f_2(x)}{\partial r^{\nu}} \right|^{\lambda_\nu} \right] dx \le$$

$$A_0(R_2) \left(\frac{\lambda_\nu}{(\lambda_\alpha + \lambda_\beta) p} \right)^{(\lambda_\nu/p)} \left[\lambda_\alpha^{\lambda_\nu/p} + 2^{(\lambda_\alpha+\lambda_\beta)/p} (\widetilde{\gamma}_1 \lambda_\beta)^{\lambda_\nu/p} \right]$$

$$\int_A \left(\left| \frac{\partial_{*R_1}^{\nu} f_1(x)}{\partial r^{\nu}} \right|^p + \left| \frac{\partial_{*R_1}^{\nu} f_2(x)}{\partial r^{\nu}} \right|^p \right) dx. \tag{16.142}$$

Proof. Based on Theorem 16.32, similar to the proof of Theorem 16.68. $\square$

We give the next special important case.

Theorem 16.71. *All are as in Assumption* 16.67 *without* λ_ν *there. Here* $\gamma_2 = \gamma_1 + 1$, $\lambda_\alpha \ge 0$, $\lambda_\beta := \lambda_{\alpha+1} \in (0,1)$, *and* $p = \lambda_\alpha + \lambda_{\alpha+1} > 1$.
Denote

$$\theta_3 := \begin{cases} 2^{(\lambda_\alpha/\lambda_{\alpha+1})} - 1 & \text{if } \lambda_\alpha \ge \lambda_{\alpha+1} \\ 1, & \text{if } \lambda_\alpha \le \lambda_{\alpha+1}, \end{cases} \tag{16.143}$$

$$L(R_2) := \left[2 \frac{(1 - \lambda_{\alpha+1})}{(N - \lambda_{\alpha+1})} \right.$$

$$\left. \left(R_2^{(N-\lambda_{\alpha+1})/(1-\lambda_{\alpha+1})} - R_1^{(N-\lambda_{\alpha+1})/(1-\lambda_{\alpha+1})} \right) \right]^{(1-\lambda_{\alpha+1})} \left(\frac{\theta_3 \lambda_{\alpha+1}}{p} \right)^{\lambda_{\alpha+1}}, \tag{16.144}$$

and

$$P_1(R_2) := \int_{R_1}^{R_2} (R_2 - t)^{(\nu-\gamma_1-1)p/(p-1)} t^{(1-N)/(p-1)} dt, \tag{16.145}$$

$$\Phi(R_2) := L(R_2) \left(\frac{P_1(R_2)^{(p-1)}}{(\Gamma(\nu - \gamma_1))^p} \right) 2^{p-1}. \tag{16.146}$$

Then

$$\int_A \left[\left| \frac{\partial_{*R_1}^{\gamma_1} f_1(x)}{\partial r^{\gamma_1}} \right|^{\lambda_\alpha} \left| \frac{\partial_{*R_1}^{\gamma_1+1} f_2(x)}{\partial r^{\gamma_1+1}} \right|^{\lambda_{\alpha+1}} + \right.$$

$$\left. \left| \frac{\partial_{*R_1}^{\gamma_1} f_2(x)}{\partial r^{\gamma_1}} \right|^{\lambda_\alpha} \left| \frac{\partial_{*R_1}^{\gamma_1+1} f_1(x)}{\partial r^{\gamma_1+1}} \right|^{\lambda_{\alpha+1}} \right] dx$$

$$\leq \Phi(R_2) \int_A \left(\left| \frac{\partial^\nu_{*R_1} f_1(x)}{\partial r^\nu} \right|^p + \left| \frac{\partial^\nu_{*R_1} f_2(x)}{\partial r^\nu} \right|^p \right) dx. \tag{16.147}$$

Proof. Based on Theorem 16.33, similar to the proof of Theorem 16.68.
□

We give an L_∞ result on the shell.
We need to make

Assumption 16.72. Let $\nu \geq \gamma_i + 1$, $\gamma_i \geq 0$, $i = 1, 2$, $n := [\nu]$, f_1, $f_2 : \overline{A} \to \mathbb{R}$ with $f_1, f_2 \in L_1(A)$, where $\mathbb{R}^N \supseteq A := B(0, R_2) - \overline{B}(0, R_1)$, $0 < R_1 < R_2$, $N \geq 2$. Assume that $f_1(\cdot\omega)$, $f_2(\cdot\omega) \in AC^n([R_1, R_2])$ for all $\omega \in S^{N-1}$, and that $\partial^\nu_{*R_1} f_i(\cdot\omega)/\partial r^\nu \in L_\infty(R_1, R_2)$, for all $\omega \in S^{N-1}$; $i = 1, 2$. Further assume that $\partial^\nu_{*R_1} f_i(x)/\partial r^\nu \in L_\infty(A)$, $i = 1, 2$. More precisely, for these $r \in [R_1, R_2]$, $\forall \omega \in S^{N-1}$, for which $D^\nu_{*R_1} f_i(r\omega)$ takes real values, there exists $M_i > 0$ such that

$$\left| D^\nu_{*R_1} f_i(r\omega) \right| \leq M_i, \text{ for } i = 1, 2. \tag{16.148}$$

We suppose that

$$\frac{\partial^j f_i(R_1\omega)}{\partial r^j} = 0, j = 0, 1, \ldots, n - 1,$$

$\forall \omega \in S^{N-1}$; $i = 1, 2$.
Let λ_ν, λ_α, $\lambda_\beta \geq 0$. If $\gamma_1 = 0$ we set $\lambda_\alpha = 1$ and if $\gamma_2 = 0$ we set $\lambda_\beta = 1$.

We present

Theorem 16.73. *All are as in Assumption 16.72.*
Set

$$T(R_2 - R_1) := \frac{R_2^{N-1}}{(\nu\lambda_\alpha - \gamma_1\lambda_\alpha + \nu\lambda_\beta - \gamma_2\lambda_\beta + 1)} \cdot$$

$$\frac{(R_2 - R_1)^{(\nu\lambda_\alpha - \gamma_1\lambda_\alpha + \nu\lambda_\beta - \gamma_2\lambda_\beta + 1)}}{(\Gamma(\nu - \gamma_1 + 1))^{\lambda_\alpha} (\Gamma(\nu - \gamma_2 + 1))^{\lambda_\beta}}. \tag{16.149}$$

Then

$$\int_A \left[\left| \frac{\partial^{\gamma_1}_{*R_1} f_1(x)}{\partial r^{\gamma_1}} \right|^{\lambda_\alpha} \left| \frac{\partial^{\gamma_2}_{*R_1} f_2(x)}{\partial r^{\gamma_2}} \right|^{\lambda_\beta} \left| \frac{\partial^\nu_{*R_1} f_1(x)}{\partial r^\nu} \right|^{\lambda_\nu} + \right.$$

$$\left. \left| \frac{\partial^{\gamma_2}_{*R_1} f_1(x)}{\partial r^{\gamma_2}} \right|^{\lambda_\beta} \left| \frac{\partial^{\gamma_1}_{*R_1} f_2(x)}{\partial r^{\gamma_1}} \right|^{\lambda_\alpha} \left| \frac{\partial^\nu_{*R_1} f_2(x)}{\partial r^\nu} \right|^{\lambda_\nu} \right] dx$$

$$\leq T\left(R_2 - R_1\right)\frac{\pi^{N/2}}{\Gamma\left(N/2\right)}$$

$$\left[M_1^{2(\lambda_\alpha + \lambda_\nu)} + M_1^{2\lambda_\beta} + M_2^{2\lambda_\beta} + M_2^{2(\lambda_\alpha + \lambda_\nu)}\right]. \tag{16.150}$$

Proof. Apply Theorem 16.38 for every $\omega \in S^{N-1}$; here $p(r) = r^{N-1}$, $r \in [R_1, R_2]$. It follows the proof of Theorem 16.66. Finally use Theorem 16.64 and Proposition 16.60. $\square$

16.5.3 Results Involving Several Functions

We need to make the following assumption.

Assumption 16.74. Let $\nu \geq \gamma_i + 1$, $\gamma_i \geq 0$, $i = 1, 2$, $n := \lceil \nu \rceil$, $f_j : \overline{A} \to \mathbb{R}$ with $f_j \in L_1(A)$, $j = 1, \ldots, M$, $M \in \mathbb{N}$, where $\mathbb{R}^N \supseteq A := B(0, R_2) - \overline{B(0, R_1)}$, $0 < R_1 < R_2$, $N \geq 2$. Assume that $f_j(\cdot\omega) \in AC^n([R_1, R_2])$ for all $\omega \in S^{N-1}$, and that $\partial_{*R_1}^\nu f_j(\cdot\omega)/\partial r^\nu \in L_\infty(R_1, R_2)$, for all $\omega \in S^{N-1}$; $j = 1, \ldots, M$. Further assume that $\partial_{*R_1}^\nu f_j(x)/\partial r^\nu \in L_\infty(A)$, $j = 1, \ldots, M$. More precisely, for these $r \in [R_1, R_2]$, $\forall \omega \in S^{N-1}$, for which $D_{*R_1}^\nu f_j(r\omega)$ takes real values, there exists $M_j > 0$ such that

$$\left|D_{*R_1}^\nu f_j(r\omega)\right| \leq M_j, \text{ for } j = 1, \ldots, M. \tag{16.151}$$

We suppose that

$$\frac{\partial^j f_j(R_1\omega)}{\partial r^k} = 0, k = 0, 1, \ldots, n - 1,$$

$\forall \omega \in S^{N-1}$; $j = 1, \ldots, M$.

Let $\lambda_\nu > 0$, and λ_α, $\lambda_\beta \geq 0$, such that $\lambda_\nu < p$, where $p > 1$. If $\gamma_1 = 0$ we set $\lambda_\alpha = 1$ and if $\gamma_2 = 0$ we set $\lambda_\beta = 1$.

We give

Theorem 16.75. Let $f_j, j = 1, \ldots, M$, as in Assumption 16.74. Let $\lambda_\nu > 0$, and $\lambda_\alpha > 0$; $\lambda_\beta \geq 0$, $p := \lambda_\alpha + \lambda_\nu > 1$. Set

$$P_k(w) := \int_{R_1}^w (w - t)^{(\nu - \gamma_k - 1)p/(p-1)} t^{(1 - N/p - 1)} dt, \tag{16.152}$$

$k = 1, 2$, $R_1 \leq w \leq R_2$,

$$A(w) := \frac{w^{(N-1)(1-(\lambda_\nu/p))} \left(P_1(w)\right)^{\lambda_\alpha((p-1)/p)} \left(P_2(w)\right)^{\lambda_\beta((p-1)/p)}}{\left(\Gamma\left(\nu - \gamma_1\right)\right)^{\lambda_\alpha} \left(\Gamma\left(\nu - \gamma_2\right)\right)^{\lambda_\beta}},$$

$$\tag{16.153}$$

$$A_0\left(R_2\right):=\left(\int_{R_1}^{R_2}\left(A\left(w\right)\right)^{p/\lambda_\alpha}dw\right)^{\lambda_\alpha/p}. \qquad (16.154)$$

Take the case of $\lambda_\beta = 0$. Then

$$\sum_{j=1}^{M}\int_A\left|\frac{\partial_{*R_1}^{\gamma_1}f_j\left(x\right)}{\partial r^{\gamma_1}}\right|^{\lambda_\alpha}\left|\frac{\partial_{*R_1}^{\nu}f_j\left(x\right)}{\partial r^{\nu}}\right|^{\lambda_\nu}dx$$

$$\le\left(A_0\left(R_2\right)|_{\lambda_\beta=0}\right)\left(\frac{\lambda_\nu}{p}\right)^{\left(\lambda_\nu/p\right)}$$

$$\left[\sum_{j=1}^{M}\left(\int_A\left|\frac{\partial_{*R_1}^{\nu}f_j\left(x\right)}{\partial r^{\nu}}\right|^{p}dx\right)\right]. \qquad (16.155)$$

Proof. As in Theorem 16.68, based on Theorem 16.45. □

We continue with

Theorem 16.76. *All basic assumptions are as in Theorem 16.75. Let $\lambda_\nu > 0$, $\lambda_\alpha = 0$; $\lambda_\beta > 0$, $p := \lambda_\nu + \lambda_\beta > 1$, P_2 defined by (16.152).*
Now it is

$$A\left(w\right):=\frac{w^{(N-1)(1-(\lambda_\nu/p))}\left(P_2\left(w\right)\right)^{\lambda_\beta((p-1)/p)}}{\left(\Gamma\left(\nu-\gamma_2\right)\right)^{\lambda_\beta}}, \qquad (16.156)$$

$$A_0\left(R_2\right):=\left(\int_{R_1}^{R_2}\left(A\left(w\right)\right)^{p/\lambda_\beta}dw\right)^{\lambda_\beta/p}. \qquad (16.157)$$

Denote

$$\delta_3:=\begin{cases} 2^{\lambda_\beta/\lambda_\nu}-1, & if\ \lambda_\beta\ge\lambda_\nu, \\ 1, & if\ \lambda_\beta\le\lambda_\nu. \end{cases} \qquad (16.158)$$

Call

$$\varphi_2\left(R_2\right):=A_0\left(R_2\right)2^{\lambda_\beta/p}\left(\frac{\lambda_\nu}{p}\right)^{\lambda_\nu/p}\delta_3^{\lambda_\nu/p}. \qquad (16.159)$$

Then

$$\int_A\left\{\left\{\sum_{j=1}^{M-1}\left[\left|\frac{\partial_{*R_1}^{\gamma_2}f_{j+1}\left(x\right)}{\partial r^{\gamma_2}}\right|^{\lambda_\beta}\left|\frac{\partial_{*R_1}^{\nu}f_j\left(x\right)}{\partial r^{\nu}}\right|^{\lambda_\nu}+\right.\right.\right.$$

$$\left.\left.\left.\left|\frac{\partial_{*R_1}^{\gamma_2}f_j\left(x\right)}{\partial r^{\gamma_2}}\right|^{\lambda_\beta}\left|\frac{\partial_{*R_1}^{\nu}f_{j+1}\left(x\right)}{\partial r^{\nu}}\right|^{\lambda_\nu}\right]\right\}\right\}+$$

$$\left[\left|\frac{\partial_{*R_1}^{\gamma_2} f_M(x)}{\partial r^{\gamma_2}}\right|^{\lambda_\beta}\left|\frac{\partial_{*R_1}^{\nu} f_1(x)}{\partial r^{\nu}}\right|^{\lambda_\nu}+\right.$$

$$\left.\left|\frac{\partial_{*R_1}^{\gamma_2} f_1(x)}{\partial r^{\gamma_2}}\right|^{\lambda_\beta}\left|\frac{\partial_{*R_1}^{\nu} f_M(x)}{\partial r^{\nu}}\right|^{\lambda_\nu}\right]\right\}dx \leq$$

$$2\varphi_2(R_2)\left[\sum_{j=1}^{M}\left(\int_A\left|\frac{\partial_{*R_1}^{\nu} f_j(x)}{\partial r^{\nu}}\right|^p dx\right)\right]. \tag{16.160}$$

Proof. As in Theorem 16.68, based on Theorem 16.46. □

We present the general case.

Theorem 16.77. *All basic assumptions are as in Theorem 16.75. Here* λ_ν, λ_α, $\lambda_\beta > 0$, $p := \lambda_\alpha + \lambda_\beta + \lambda_\nu > 1$, P_k *as in* (16.152)*, and* A *is as in* (16.153)*. Here*

$$A_0(R_2) := \left(\int_{R_1}^{R_2}(A(w))^{p/(\lambda_\alpha+\lambda_\beta)}dw\right)^{(\lambda_\alpha+\lambda_\beta)/p}, \tag{16.161}$$

$$\widetilde{\gamma}_1 := \begin{cases} 2^{(\lambda_\alpha+\lambda_\beta/\lambda_\nu)} - 1, & \text{if } \lambda_\alpha + \lambda_\beta \geq \lambda_\nu, \\ 1, & \text{if } \lambda_\alpha + \lambda_\beta \leq \lambda_\nu. \end{cases} \tag{16.162}$$

Put

$$\varphi_3(R_2) := A_0(R_2)\left(\frac{\lambda_\nu}{(\lambda_\alpha+\lambda_\beta)p}\right)^{(\lambda_\nu/p)}$$

$$\left[\lambda_\alpha^{(\lambda_\nu/p)} + 2^{(\lambda_\alpha+\lambda_\beta/p)}(\widetilde{\gamma}_1\lambda_\beta)^{(\lambda_\nu/p)}\right]. \tag{16.163}$$

Then

$$\int_A\left[\sum_{j=1}^{M-1}\left[\left|\frac{\partial_{*R_1}^{\gamma_1} f_j(x)}{\partial r^{\gamma_1}}\right|^{\lambda_\alpha}\left|\frac{\partial_{*R_1}^{\gamma_2} f_{j+1}(x)}{\partial r^{\gamma_2}}\right|^{\lambda_\beta}\left|\frac{\partial_{*R_1}^{\nu} f_j(x)}{\partial r^{\nu}}\right|^{\lambda_\nu}+\right.\right.$$

$$\left.\left|\frac{\partial_{*R_1}^{\gamma_2} f_j(x)}{\partial r^{\gamma_2}}\right|^{\lambda_\beta}\left|\frac{\partial_{*R_1}^{\gamma_1} f_{j+1}(x)}{\partial r^{\gamma_1}}\right|^{\lambda_\alpha}\left|\frac{\partial_{*R_1}^{\nu} f_{j+1}(x)}{\partial r^{\nu}}\right|^{\lambda_\nu}\right]+$$

$$\left[\left|\frac{\partial_{*R_1}^{\gamma_1} f_1(x)}{\partial r^{\gamma_1}}\right|^{\lambda_\alpha}\left|\frac{\partial_{*R_1}^{\gamma_2} f_M(x)}{\partial r^{\gamma_2}}\right|^{\lambda_\beta}\left|\frac{\partial_{*R_1}^{\nu} f_1(x)}{\partial r^{\nu}}\right|^{\lambda_\nu}+\right.$$

$$\left.\left.\left|\frac{\partial_{*R_1}^{\gamma_2} f_1(x)}{\partial r^{\gamma_2}}\right|^{\lambda_\beta}\left|\frac{\partial_{*R_1}^{\gamma_1} f_M(x)}{\partial r^{\gamma_1}}\right|^{\lambda_\alpha}\left|\frac{\partial_{*R_1}^{\nu} f_M(x)}{\partial r^{\nu}}\right|^{\lambda_\nu}\right]\right]dx \leq$$

$$2\varphi_3\left(R_2\right)\left[\sum_{j=1}^{M}\left(\int_A\left|\frac{\partial^{\nu}_{*R_1}f_j\left(x\right)}{\partial r^{\nu}}\right|^p dx\right)\right]. \tag{16.164}$$

Proof. As in Theorem 16.68, based on Theorem 16.47. □

We show the special important case next.

Theorem 16.78. *Let all be as in Assumption 16.74 without λ_ν there. Here $\gamma_2 = \gamma_1 + 1$, and let $\lambda_\alpha > 0$, $\lambda_\beta := \lambda_{\alpha+1}$, $0 < \lambda_{\alpha+1} < 1$, such that $p := \lambda_\alpha + \lambda_{\alpha+1} > 1$. Denote*

$$\theta_3 := \begin{cases} 2^{(\lambda_\alpha/\lambda_{\alpha+1})} - 1, & if\ \lambda_\alpha \geq \lambda_{\alpha+1}, \\ 1, & if\ \lambda_\alpha \leq \lambda_{\alpha+1}, \end{cases} \tag{16.165}$$

$$L\left(R_2\right) := \left[2\frac{(1-\lambda_{\alpha+1})}{(N-\lambda_{\alpha+1})}\right.$$

$$\left.\left(R_2^{(N-\lambda_{\alpha+1})/(1-\lambda_{\alpha+1})} - R_1^{(N-\lambda_{\alpha+1})/(1-\lambda_{\alpha+1})}\right)\right]^{(1-\lambda_{\alpha+1})}\left(\frac{\theta_3\lambda_{\alpha+1}}{\lambda_\alpha+\lambda_{\alpha+1}}\right)^{\lambda_{\alpha+1}}, \tag{16.166}$$

and

$$P\left(R_2\right) := \int_{R_1}^{R_2}\left(R_2 - t\right)^{(\nu-\gamma_1-1)(p/p-1)}t^{(1-N/p-1)}dt, \tag{16.167}$$

$$\Phi\left(R_2\right) := L\left(R_2\right)\left(\frac{P_1\left(R_2\right)^{(p-1)}}{(\Gamma\left(\nu-\gamma_1\right))^p}\right)2^{(p-1)}. \tag{16.168}$$

Then

$$\int_A\left\{\left\{\sum_{j=1}^{M-1}\left[\left|\frac{\partial^{\gamma_1}_{*R_1}f_j\left(x\right)}{\partial r^{\gamma_1}}\right|^{\lambda_\alpha}\left|\frac{\partial^{\gamma_1+1}_{*R_1}f_{j+1}\left(x\right)}{\partial r^{\gamma_1+1}}\right|^{\lambda_{\alpha+1}}+\right.\right.\right.$$

$$\left.\left.\left|\frac{\partial^{\gamma_1}_{*R_1}f_{j+1}\left(x\right)}{\partial r^{\gamma_1}}\right|^{\lambda_\alpha}\left|\frac{\partial^{\gamma_1+1}_{*R_1}f_j\left(x\right)}{\partial r^{\gamma_1+1}}\right|^{\lambda_{\alpha+1}}\right]\right\}+$$

$$\left[\left|\frac{\partial^{\gamma_1}_{*R_1}f_1\left(x\right)}{\partial r^{\gamma_1}}\right|^{\lambda_\alpha}\left|\frac{\partial^{\gamma_1+1}_{*R_1}f_M\left(x\right)}{\partial r^{\gamma_1+1}}\right|^{\lambda_{\alpha+1}}+\right.$$

$$\left.\left.\left|\frac{\partial^{\gamma_1}_{*R_1}f_M\left(x\right)}{\partial r^{\gamma_1}}\right|^{\lambda_\alpha}\left|\frac{\partial^{\gamma_1+1}_{*R_1}f_1\left(x\right)}{\partial r^{\gamma_1+1}}\right|^{\lambda_{\alpha+1}}\right]\right\}dx \leq$$

$$2\Phi\left(R_2\right)\left[\sum_{j=1}^{M}\left(\int_A\left|\frac{\partial^{\nu}_{*R_1}f_j\left(x\right)}{\partial r^{\nu}}\right|^p dx\right)\right].\tag{16.169}$$

Proof. As in Theorem 16.68, based on Theorem 16.48. □

We study the L_∞ case next.
We need to make

Assumption 16.79. Let $\nu \geq \gamma_i + 1$, $\gamma_i \geq 0$, $i = 1, 2$, $n := \lceil \nu \rceil$, $f_j : \overline{A} \to$
$\mathbb{R}$ with $f_j \in L_1\left(A\right)$, $j = 1, \ldots, M$, $M \in \mathbb{N}$, where $\mathbb{R}^N \supseteq A := B\left(0, R_2\right) -$
$\overline{B\left(0, R_1\right)}$, $0 < R_1 < R_2$, $N \geq 2$. Assume that $f_j\left(\cdot\omega\right) \in AC^n\left(\left[R_1, R_2\right]\right)$ for
all $\omega \in S^{N-1}$, and that $\partial^{\nu}_{*R_1}f_j\left(\cdot\omega\right)/\partial r^{\nu} \in L_\infty\left(R_1, R_2\right)$, for all $\omega \in S^{N-1}$;
$j = 1, \ldots, M$. Further assume that $\partial^{\nu}_{*R_1}f_j\left(x\right)/\partial r^{\nu} \in L_\infty\left(A\right)$, $j = 1, \ldots, M$.
More precisely, for these $r \in \left[R_1, R_2\right]$, $\forall \omega \in S^{N-1}$, for which $D^{\nu}_{*R_1}f_j\left(r\omega\right)$
takes real values, there exists $M_j > 0$ such that

$$\left|D^{\nu}_{*R_1}f_j\left(r\omega\right)\right| \leq M_j, \text{ for } j = 1, \ldots, M.\tag{16.170}$$

We suppose that

$$\frac{\partial^k f_j\left(R_1\omega\right)}{\partial r^k} = 0, k = 0, 1, \ldots, n-1,$$

$\forall \omega \in S^{N-1}$; $j = 1, \ldots, M$.
Let $\lambda_\nu, \lambda_\alpha, \lambda_\beta \geq 0$. If $\gamma_1 = 0$ we set $\lambda_\alpha = 1$ and if $\gamma_2 = 0$ we set $\lambda_\beta = 1$.

The last main result follows.

Theorem 16.80. *All are as in Assumption* 16.79.
Set

$$T\left(R_2\right) := \frac{R_2^{N-1}}{\left(\nu\lambda_\alpha - \gamma_1\lambda_\alpha + \nu\lambda_\beta - \gamma_2\lambda_\beta + 1\right)}\cdot$$

$$\frac{\left(R_2 - R_1\right)^{\left(\nu\lambda_\alpha - \gamma_1\lambda_\alpha + \nu\lambda_\beta - \gamma_2\lambda_\beta + 1\right)}}{\left(\Gamma\left(\nu - \gamma_1 + 1\right)\right)^{\lambda_\alpha}\left(\Gamma\left(\nu - \gamma_2 + 1\right)\right)^{\lambda_\beta}}.\tag{16.171}$$

Then

$$\int_A\left\{\left\{\sum_{j=1}^{M-1}\left[\left|\frac{\partial^{\gamma_1}_{*R_1}f_j\left(x\right)}{\partial r^{\gamma_1}}\right|^{\lambda_\alpha}\left|\frac{\partial^{\gamma_2}_{*R_1}f_{j+1}\left(x\right)}{\partial r^{\gamma_2}}\right|^{\lambda_\beta}\left|\frac{\partial^{\nu}_{*R_1}f_j\left(x\right)}{\partial r^{\nu}}\right|^{\lambda_\nu}+\right.\right.\right.$$

$$\left.\left.\left.\left|\frac{\partial^{\gamma_2}_{*R_1}f_j\left(x\right)}{\partial r^{\gamma_2}}\right|^{\lambda_\beta}\left|\frac{\partial^{\gamma_1}_{*R_1}f_{j+1}\left(x\right)}{\partial r^{\gamma_1}}\right|^{\lambda_\alpha}\left|\frac{\partial^{\nu}_{*R_1}f_{j+1}\left(x\right)}{\partial r^{\nu}}\right|^{\lambda_\nu}\right]\right]\right\}+$$

$$\left[\left[\left|\frac{\partial^{\gamma_1}_{*R_1} f_1(x)}{\partial r^{\gamma_1}}\right|^{\lambda_\alpha} \left|\frac{\partial^{\gamma_2}_{*R_1} f_M(x)}{\partial r^{\gamma_2}}\right|^{\lambda_\beta} \left|\frac{\partial^{\nu}_{*R_1} f_1(x)}{\partial r^{\nu}}\right|^{\lambda_\nu} + \right.\right.$$

$$\left.\left.\left|\frac{\partial^{\gamma_2}_{*R_1} f_1(x)}{\partial r^{\gamma_2}}\right|^{\lambda_\beta} \left|\frac{\partial^{\gamma_1}_{*R_1} f_M(x)}{\partial r^{\gamma_1}}\right|^{\lambda_\alpha} \left|\frac{\partial^{\nu}_{*R_1} f_M(x)}{\partial r^{\nu}}\right|^{\lambda_\nu}\right]\right\} dx \le$$

$$\frac{2\pi^{N/2}}{\Gamma(N/2)} T(R_2) \left\{\sum_{j=1}^{M-1}\left\{M_j^{2(\lambda_\alpha+\lambda_\nu)} + M_j^{2\lambda_\beta}\right\}\right\}. \tag{16.172}$$

Proof. Based on Theorem 16.52; here $p(r) = r^{N-1}$, $r \in [R_1, R_2]$. Apply (16.97) $\forall \omega \in S^{N-1}$. It follows the proof of Theorem 16.66. Finally use Theorem 16.64 and Proposition 16.60. $\square$

16.6 Applications

We need the following.

Corollary 16.81. (To Theorem 16.68; $f_1 = f_2$.) *All are as in Theorem 16.68. It holds*

$$\int_A \left|\frac{\partial^{\gamma_1}_{*R_1} f_1(x)}{\partial r^{\gamma_1}}\right|^{\lambda_\alpha} \left|\frac{\partial^{\nu}_{*R_1} f_1(x)}{\partial r^{\nu}}\right|^{\lambda_\nu} dx \le$$

$$(A_0(R_2)|_{\lambda_\beta=0}) \left(\frac{\lambda_\nu}{p}\right)^{(\lambda_\nu/p)} \left(\int_A \left|\frac{\partial^{\nu}_{*R_1} f_1(x)}{\partial r^{\nu}}\right|^p dx\right). \tag{16.173}$$

So setting $\lambda_\alpha = \lambda_\nu = 1$, $p = 2$ in (16.173), we obtain in detail

Proposition 16.82. *Let $\nu \ge \gamma + 1$, $\gamma \ge 0$, $n := \lceil \nu \rceil$, $f : \overline{A} \to \mathbb{R}$ with $f \in L_1(A)$, where $A := B(0, R_2) - \overline{B(0, R_1)} \subseteq \mathbb{R}^N$, $N \ge 2$, $0 < R_1 < R_2$. Assume that $f(\cdot\omega) \in AC^n([R_1, R_2])$, $\forall \omega \in S^{N-1}$, and that $\partial^{\nu}_{*R_1} f(\cdot\omega)/\partial r^{\nu} \in L_\infty(R_1, R_2)$, $\forall \omega \in S^{N-1}$. Further assume that $\partial^{\nu}_{*R_1} f(x)/\partial r^{\nu} \in L_\infty(A)$. More precisely, for these $r \in [R_1, R_2]$, $\forall \omega \in S^{N-1}$, for which $D^{\nu}_{*R_1} f(r\omega)$ takes real values, $\exists M_1 > 0$ such that*

$$\left|D^{\nu}_{*R_1} f(r\omega)\right| \le M_1.$$

Suppose that

$$\frac{\partial^j f(R_1\omega)}{\partial r^j} = 0, j = 0, 1, \dots, n-1, \forall \omega \in S^{N-1}.$$

Set

$$P(r) := \int_{R_1}^{r} (r - t)^{2(\nu - \gamma - 1)} t^{(1-N)} dt, \ R_1 \le r \le R_2, \qquad (16.174)$$

$$A(r) := \frac{r^{(N-1/2)} \sqrt{P(r)}}{\Gamma(\nu - \gamma)}, \qquad (16.175)$$

$$\widetilde{A}_0(R_2) := \left(\int_{R_1}^{R_2} (A(r))^2 \, dr \right)^{1/2}. \qquad (16.176)$$

Then

$$\int_A \left| \frac{\partial_{*R_1}^{\gamma} f(x)}{\partial r^{\gamma}} \right| \left| \frac{\partial_{*R_1}^{\nu} f(x)}{\partial r^{\nu}} \right| dx \le$$

$$\widetilde{A}_0(R_2) \, 2^{-1/2} \left(\int_A \left(\frac{\partial_{*R_1}^{\nu} f(x)}{\partial r^{\nu}} \right)^2 dx \right). \qquad (16.177)$$

When $\gamma = 0$ we get the following in detail.

Proposition 16.83. *Let* $\nu \ge 1$, $n := \lceil \nu \rceil$, $f : \overline{A} \to \mathbb{R}$ *with* $f \in L_1(A)$, *where* $A := B(0, R_2) - \overline{B(0, R_1)} \subseteq \mathbb{R}^N$, $N \ge 2$, $0 < R_1 < R_2$. *Assume that* $f(\cdot\omega) \in AC^n([R_1, R_2])$, $\forall \omega \in S^{N-1}$, *and that* $\partial_{*R_1}^{\nu} f(\cdot\omega)/\partial r^{\nu} \in L_{\infty}(R_1, R_2)$, $\forall \omega \in S^{N-1}$. *Further assume that* $\partial_{*R_1}^{\nu} f(x)/\partial r^{\nu} \in L_{\infty}(A)$. *More precisely, for these* $r \in [R_1, R_2]$, $\forall \omega \in S^{N-1}$, *for which* $D_{*R_1}^{\nu} f(r\omega)$ *takes real values,* $\exists M_1 > 0$ *such that*

$$|D_{*R_1}^{\nu} f(r\omega)| \le M_1.$$

Suppose that

$$\frac{\partial^j f(R_1 \omega)}{\partial r^j} = 0, j = 0, 1, \ldots, n - 1, \ \forall \omega \in S^{N-1}.$$

Set

$$P_0(r) := \int_{R_1}^{r} (r - t)^{2(\nu - 1)} t^{(1-N)} dt, \ R_1 \le r \le R_2, \qquad (16.178)$$

$$A_*(r) := \frac{r^{(N-1/2)} \sqrt{P_0(r)}}{\Gamma(\nu)}, \qquad (16.179)$$

$$\widetilde{A}_0(R_2) := \left(\int_{R_1}^{R_2} (A_*(r))^2 \, dr \right)^{1/2}. \qquad (16.180)$$

Then

$$\int_A |f(x)| \left| \frac{\partial_{*R_1}^\nu f(x)}{\partial r^\nu} \right| dx \le$$

$$\widetilde{\widetilde{A}}_0(R_2) \, 2^{-1/2} \left(\int_A \left(\frac{\partial_{*R_1}^\nu f(x)}{\partial r^\nu} \right)^2 dx \right). \tag{16.181}$$

Based on Corollary 16.35 we give

Proposition 16.84. *Let $\nu \ge \gamma + 1$, $\gamma \ge 0$, $n := \lceil \nu \rceil$, $f : \overline{A} \to \mathbb{R}$ with $f \in L_1(A)$, where $A := B(0, R_2) - \overline{B(0, R_1)} \subseteq \mathbb{R}^N$, $N \ge 2$, $0 < R_1 < R_2$. Assume that $f(\cdot \omega) \in AC^n([R_1, R_2])$, $\forall \omega \in S^{N-1}$, and that $\partial_{*R_1}^\nu f(\cdot \omega)/\partial r^\nu \in L_\infty(R_1, R_2)$, $\forall \omega \in S^{N-1}$.*
Suppose that

$$\frac{\partial^j f(R_1 \omega)}{\partial r^j} = 0, \; j = 0, 1, \ldots, n-1, \; \forall \omega \in S^{N-1}.$$

Then

$$(1) \int_{R_1}^r \left| D_{*R_1}^\gamma f(t\omega) \right| \left| D_{*R_1}^\nu f(t\omega) \right| dt \le$$

$$\left(\frac{(r - R_1)^{(\nu - \gamma)}}{2\Gamma(\nu - \gamma) \sqrt{\nu - \gamma} \sqrt{2\nu - 2\gamma - 1}} \right)$$

$$\left(\int_{R_1}^r \left(D_{*R_1}^\nu f(t\omega) \right)^2 dt \right), \; all \; R_1 \le r \le R_2, \; \forall \omega \in S^{N-1}. \tag{16.182}$$

$$(2) \; When \; \gamma = 0 \; we \; get$$

$$\int_{R_1}^r |f(t\omega)| \left| D_{*R_1}^\nu f(t\omega) \right| dt \le$$

$$\left(\frac{(r - R_1)^\nu}{2\Gamma(\nu) \sqrt{\nu} \sqrt{2\nu - 1}} \right) \left(\int_{R_1}^r \left(D_{*R_1}^\nu f(t\omega) \right)^2 dt \right), \tag{16.183}$$

all $R_1 \le r \le R_2$, $\forall \omega \in S^{N-1}$.
In particular we have

$$(3) \int_{R_1}^{R_2} |f(r\omega)| \left| D_{*R_1}^\nu f(r\omega) \right| dr \le$$

$$\left(\frac{(R_2 - R_1)^\nu}{2\Gamma(\nu) \sqrt{\nu} \sqrt{2\nu - 1}} \right) \left(\int_{R_1}^{R_2} \left(D_{*R_1}^\nu f(r\omega) \right)^2 dr \right), \; \forall \omega \in S^{N-1}. \tag{16.184}$$

Next we apply Proposition 16.84; see (16.183) for proving the uniqueness of solution in a PDE initial value problem on A.

Theorem 16.85. *Let $\nu > 1$, $\nu \notin \mathbb{N}$, $n := [\nu]$, $f : \overline{A} \to \mathbb{R}$ with $f \in L_1(A)$, where $A := B(0, R_2) - \overline{B(0, R_1)} \subseteq \mathbb{R}^N$, $N \geq 2$, $0 < R_1 < R_2$. Assume that $f(\cdot\omega) \in AC^n([R_1, R_2])$, $\forall \omega \in S^{N-1}$, and that $\partial^\nu_{*R_1} f(\cdot\omega)/\partial r^\nu \in AC([R_1, R_2])$, $\forall \omega \in S^{N-1}$. Further assume $D^\nu_{*R_1} f(x)/\partial r^\nu \in L_\infty(A)$, such that there exists $M_1 > 0$ with $\left| D^\nu_{*R_1} f(r\omega) \right| \leq M_1$, $\forall r \in [R_1, R_2]$, $\forall \omega \in S^{N-1}$. Suppose that*

$$\frac{\partial^j f(R_1\omega)}{\partial r^j} = 0, j = 0, 1, \ldots, n-1, \ \forall \omega \in S^{N-1}.$$

Consider the PDE

$$\frac{\partial}{\partial r}\left(\frac{\partial^\nu_{*R_1} f(x)}{\partial r^\nu}\right) = \theta(x) f(x), \tag{16.185}$$

$\forall x \in \overline{A}$, *where* $0 \neq \theta : \overline{A} \to \mathbb{R}$ *is continuous. If* (16.185) *has a solution then it is unique.*

Proof. We rewrite (16.185) as

$$\frac{\partial}{\partial r}\left(\frac{\partial^\nu_{*R_1} f(r\omega)}{\partial r^\nu}\right) = \theta(r\omega) f(r\omega), \tag{16.186}$$

valid $\forall r \in [R_1, R_2]$, $\forall \omega \in S^{N-1}$, and $0 \neq \theta : ([R_1, R_2] \times S^{N-1}) \to \mathbb{R}$ is continuous.

Assume f_1 and f_2 are the solution to (16.185); then

$$\frac{\partial}{\partial r}\left(\frac{\partial^\nu_{*R_1} f_1(r\omega)}{\partial r^\nu}\right) = \theta(r\omega) f_1(r\omega), \tag{16.187}$$

and

$$\frac{\partial}{\partial r}\left(\frac{\partial^\nu_{*R_1} f_2(r\omega)}{\partial r^\nu}\right) = \theta(r\omega) f_2(r\omega), \tag{16.188}$$

$\forall r \in [R_1, R_2]$, $\forall \omega \in S^{N-1}$.

Call $g := f_1 - f_2$; thus by subtraction in (16.187) we get

$$\frac{\partial}{\partial r}\left(\frac{\partial^\nu_{*R_1} g(r\omega)}{\partial r^\nu}\right) = \theta(r\omega) g(r\omega), \tag{16.189}$$

$\forall r \in [R_1, R_2]$, $\forall \omega \in S^{N-1}$. Of course

$$\frac{\partial^j g(R_1\omega)}{\partial r^j} = 0, \ j = 0, 1, \ldots, n-1, \ \forall \omega \in S^{N-1}.$$

Consequently we have

$$\left(\frac{\partial^\nu_{*R_1} g(r\omega)}{\partial r^\nu}\right) \frac{\partial}{\partial r}\left(\frac{\partial^\nu_{*R_1} g(r\omega)}{\partial r^\nu}\right)$$

$$= \theta\left(r\omega\right) g\left(r\omega\right) \left(\frac{\partial^{\nu}_{*R_1} g\left(r\omega\right)}{\partial r^{\nu}}\right), \tag{16.190}$$

$\forall r \in [R_1, R_2], \forall \omega \in S^{N-1}$.

Hence

$$\int_{R_1}^{r} \left(\frac{\partial^{\nu}_{*R_1} g\left(t\omega\right)}{\partial r^{\nu}}\right) \frac{\partial}{\partial r} \left(\frac{\partial^{\nu}_{*R_1} g\left(t\omega\right)}{\partial r^{\nu}}\right) dt$$

$$= \int_{R_1}^{r} \theta\left(t\omega\right) g\left(t\omega\right) \left(\frac{\partial^{\nu}_{*R_1} g\left(t\omega\right)}{\partial r^{\nu}}\right) dt, \tag{16.191}$$

$\forall r \in [R_1, R_2], \forall \omega \in S^{N-1}$.

Therefore we find

$$\left.\frac{\left(\frac{\partial^{\nu}_{*R_1} g(t\omega)}{\partial r^{\nu}}\right)^2}{2}\right|_{R_1}^{r} = \int_{R_1}^{r} \theta\left(t\omega\right) g\left(t\omega\right) \left(\frac{\partial^{\nu}_{*R_1} g\left(t\omega\right)}{\partial r^{\nu}}\right) dt, \tag{16.192}$$

$\forall r \in [R_1, R_2], \forall \omega \in S^{N-1}$.

Notice that $\partial^{\nu}_{*R_1} g\left(R_1\omega\right)/\partial r^{\nu} = 0, \forall \omega \in S^{N-1}$; see (16.110).

Consequently we find

$$\left(\frac{\partial^{\nu}_{*R_1} g\left(r\omega\right)}{\partial r^{\nu}}\right)^2 = 2 \left|\int_{R_1}^{r} \theta\left(t\omega\right) g\left(t\omega\right) \left(\frac{\partial^{\nu}_{*R_1} g\left(t\omega\right)}{\partial r^{\nu}}\right) dt\right|$$

$$\leq 2 \left\|g\right\|_{\infty} \int_{R_1}^{r} \left|g\left(t\omega\right)\right| \left|\frac{\partial^{\nu}_{*R_1} g\left(t\omega\right)}{\partial r^{\nu}}\right| dt \tag{16.193}$$

$$\overset{(16.183)}{\leq} \left(\frac{\left\|\theta\right\|_{\infty} \left(r - R_1\right)^{\nu}}{\Gamma\left(\nu\right) \sqrt{\nu}\sqrt{2\nu - 1}}\right) \tag{16.194}$$

$$\left(\int_{R_1}^{r} \left(D^{\nu}_{*R_1} g\left(t\omega\right)\right)^2 dt\right) \leq$$

$$\left(\frac{\left\|\theta\right\|_{\infty} \left(R_2 - R_1\right)^{\nu}}{\Gamma\left(\nu\right) \sqrt{\nu}\sqrt{2\nu - 1}}\right) \left(\int_{R_1}^{r} \left(D^{\nu}_{*R_1} g\left(t\omega\right)\right)^2 dt\right), \tag{16.195}$$

$\forall r \in [R_1, R_2], \forall \omega \in S^{N-1}$.

Call

$$K := \frac{\left\|\theta\right\|_{\infty} \left(R_2 - R_1\right)^{\nu}}{\Gamma\left(\nu\right) \sqrt{\nu}\sqrt{2\nu - 1}} > 0. \tag{16.196}$$

So we have proved that

$$\left(\frac{\partial^{\nu}_{*R_1} g\left(r\omega\right)}{\partial r^{\nu}}\right)^2 \leq K \left(\int_{R_1}^{r} \left(D^{\nu}_{*R_1} g\left(t\omega\right)\right)^2 dt\right), \tag{16.197}$$

$\forall r \in [R_1, R_2], \forall \omega \in S^{N-1}$. Here $D^{\nu}_{*R_1} g\left(\cdot\omega\right) \in C\left([R_1, R_2]\right), \forall \omega \in S^{N-1}$.

Hence by Grönwall's inequality we get $\left(D^\nu_{*R_1} g\left(r\omega\right)\right)^2 \equiv 0$, so that $D^\nu_{*R_1} g\left(r\omega\right) \equiv 0, \forall r \in [R_1, R_2]$, $\forall \omega \in S^{N-1}$. Thus $\partial/\partial r\left(\partial^\nu_{*R_1} g\left(r\omega\right)/\partial r^\nu\right) \equiv 0, \forall r \in [R_1, R_2]$, $\forall \omega \in S^{N-1}$. And by (16.188) we have $\theta\left(r\omega\right) g\left(r\omega\right) \equiv 0, \forall r \in [R_1, R_2]$, $\forall \omega \in S^{N-1}$, implying $g\left(r\omega\right) \equiv 0, \forall r \in [R_1, R_2]$, $\forall \omega \in S^{N-1}$.

Hence we proved $f_1\left(r\omega\right) = f_2\left(r\omega\right), \forall r \in [R_1, R_2]$, $\forall \omega \in S^{N-1}$. Thus

$$f_1\left(x\right) = f_2\left(x\right), \forall x \in \overline{A},$$

there by proving the claim. $\square$

We give the very important

Remark 16.86. From Corollary 16.12 we saw that: for $\nu \geq 0$, $n := \lceil\nu\rceil$, $f \in AC^n\left([a,b]\right)$, given that $D^\nu_a f\left(x\right)$ exists in $\mathbb{R}$, $\forall x \in [a,b]$, and $f^{(k)}\left(a\right) = 0$, $k = 0, 1, \ldots, n-1$, imply that $D^\nu_{*a} f = D^\nu_a f$. Also we saw in Theorems 16.16 and 16.17, that by adding to the assumptions of Theorem 16.16 that "there exists $D^\nu_a f\left(x\right) \in \mathbb{R}$, $\forall x \in [a,b]$," we can rewrite the conclusions of Theorem 16.16; that is, we can frame the conclusions of Theorem 16.17 in the language of Riemann – Liouville fractional derivatives. Notice there, under the above additional assumption, that $D^\gamma_a f\left(x\right) = D^\gamma_{*a} f\left(x\right)$, $\forall x \in [a,b]$ also holds.

This entire chapter is based on Theorem 16.16. So by adding to the assumptions of all of our results here for all functions involved that "there exists $D^\nu_a f\left(x\right) \in \mathbb{R}$, $\forall x \in [a,b]$," we can rewrite them all in terms of Riemann – Liouville fractional derivatives. Accordingly for the case of the spherical shell we need to add "there exists $D^\nu_{R_1} f\left(r\omega\right) \in \mathbb{R}$, $\forall r \in [R_1, R_2]$, for each $\omega \in S^{N-1}$," and all can be rewritten in terms of Riemann – Liouville radial fractional derivatives.

As examples we next present only a few of all those results that can be rewritten.

We present

Theorem 16.87. Let $\nu \geq \gamma + 1$, $\gamma \geq 0$. Call $n := \lceil\nu\rceil$ and assume $f \in AC^n\left([a,b]\right)$ such that $f^{(k)}\left(a\right) = 0$, $k = 0, 1, \ldots, n-1$, and $\exists D^\nu_a f\left(x\right) \in \mathbb{R}$, $\forall x \in [a,b]$ with $D^\nu_a f \in L_\infty\left(a,b\right)$. Let $p, q > 1$ such that $1/p + 1/q = 1$, $a \leq x \leq b$.
Then

$$\int_a^x \left|D^\gamma_a f\left(\omega\right)\right| \left|\left(D^\nu_a f\right)\left(\omega\right)\right| d\omega \leq$$

$$\frac{\left(x - a\right)^{(p\nu - p\gamma - p + 2)/p}}{\left(\sqrt[q]{2}\right) \Gamma\left(\nu - \gamma\right)\left(\left(p\nu - p\gamma - p + 1\right)\left(p\nu - p\gamma - p + 2\right)\right)^{1/p}}$$

$$\cdot \left(\int_a^x |D_a^\nu f(\omega)|^q \, d\omega \right)^{2/q}. \qquad (16.198)$$

Proof. Similar to Theorem 16.18. □

The converse result follows.

Theorem 16.88. Let $\nu \geq \gamma + 1$, $\gamma \geq 0$. Call $n := \lceil \nu \rceil$ and assume $f \in AC^n([a,b])$ such that $f^{(k)}(a) = 0$, $k = 0, 1, \ldots, n-1$, and $\exists D_a^\nu f(x) \in \mathbb{R}$, $\forall x \in [a,b]$ with $D_a^\nu f, 1/D_a^\nu f \in L_\infty(a,b)$. Suppose that $D_a^\nu f$ is of fixed sign a.e in $[a,b]$. Let p, q such that $0 < p < 1$, $q < 0$ and $1/p + 1/q = 1$, $a \leq x \leq b$.
  Then

$$\int_a^x |D_a^\gamma f(\omega)| \, |D_a^\nu f(\omega)| \, d\omega \geq$$

$$\frac{(x-a)^{(p\nu - p\gamma - p + 2)/p}}{\left(\sqrt[q]{2}\right) \Gamma(\nu - \gamma) \left((p\nu - p\gamma - p + 1)(p\nu - p\gamma - p + 2)\right)^{1/p}}$$

$$\cdot \left(\int_a^x |D_a^\nu f(\omega)|^q \, d\omega \right)^{2/q}. \qquad (16.199)$$

Proof. As in Theorem 16.20. □

We present

Theorem 16.89. Let $\nu \geq \gamma_i + 1$, $\gamma_i \geq 0$, $i = 1, 2$, $n := \lceil \nu \rceil$, and f_1, $f_2 \in AC^n([a,b])$ such that $f_1^{(j)}(a) = f_2^{(j)}(a) = 0$, $j = 0, 1, \ldots, n-1$, $a \leq x \leq b$. Consider also $p(t) > 0$ and $q(t) \geq 0$, with all $p(t)$, $1/p(t)$, $q(t) \in L_\infty(a,b)$. Further assume $\exists D_a^\nu f_i(x) \in \mathbb{R}$, $\forall x \in [a,b]$, and $D_a^\nu f_i \in L_\infty(a,b)$, $i = 1, 2$. Let $\lambda_\nu > 0$ and λ_α, $\lambda_\beta \geq 0$ such that $\lambda_\nu < p$, where $p > 1$.
  Here P_k is as in (16.39), $A(\omega)$ is as in (16.40), $A_0(x)$ as in (16.41), and δ_1 as in (16.42).
  If $\lambda_\beta = 0$, we obtain that

$$\int_a^x q(\omega) \left[|D_a^{\gamma_1} f_1(\omega)|^{\lambda_\alpha} |D_a^\nu f_1(\omega)|^{\lambda_\nu} + \right.$$

$$\left. |D_a^{\gamma_1} f_2(\omega)|^{\lambda_\alpha} |D_a^\nu f_2(\omega)|^{\lambda_\nu} \right] d\omega \leq \qquad (16.200)$$

$$\left(A_0(x)|_{\lambda_\beta = 0} \right) \left(\frac{\lambda_\nu}{\lambda_\alpha + \lambda_\nu} \right)^{\lambda_\nu/p} \delta_1$$

$$\left[\int_a^x p(\omega) \left[|D_a^\nu f_1(\omega)|^p + |D_a^\nu f_2(\omega)|^p \right] d\omega \right]^{((\lambda_\alpha + \lambda_\nu)/p)}.$$

Proof. As in Theorem 16.30. □

Corollary 16.90. (All are as in Theorem 16.89, $\lambda_\beta = 0$, $p(t) = q(t) = 1$, $\lambda_\alpha = \lambda_\nu = 1$, $p = 2$.) *In detail: let* $\nu \geq \gamma_1 + 1$, $\gamma_1 \geq 0$, $n := \lceil \nu \rceil$, f_1, $f_2 \in AC^n([a,b]) : f_1^{(j)}(a) = f_2^{(j)}(a) = 0$, $j = 0, 1, \ldots, n-1$, $a \leq x \leq b$; $\exists D_a^\nu f_i(x) \in \mathbb{R}$, $\forall x \in [a,b]$ *with* $D_a^\nu f_i \in L_\infty(a,b)$, $i = 1, 2$. *Then*

$$\int_a^x \left[|(D_a^{\gamma_1} f_1)(\omega)| \, |(D_a^\nu f_1)(\omega)| + \right.$$

$$\left. |(D_a^{\gamma_1} f_2)(\omega)| \, |(D_a^\nu f_2)(\omega)| \right] d\omega \leq \qquad (16.201)$$

$$\left(\frac{(x-a)^{(\nu-\gamma_1)}}{2\Gamma(\nu-\gamma_1)\sqrt{\nu-\gamma_1}\sqrt{2\nu-2\gamma_1-1}} \right)$$

$$\left(\int_a^x \left[((D_a^\nu f_1)(\omega))^2 + ((D_a^\nu f_2)(\omega))^2 \right] d\omega \right),$$

all $a \leq x \leq b$.

We need

Definition 16.91. Let $F : \overline{A} \to \mathbb{R}$, $\nu \geq 0$, $n := \lceil \nu \rceil$ such that $F(\cdot\omega) \in AC^n([R_1, R_2])$, for all $\omega \in S^{N-1}$. We call the Riemann – Liouville radial fractional derivative the following function

$$\frac{\partial_{R_1}^\nu F(x)}{\partial r^\nu} := \frac{1}{\Gamma(n-\nu)} \frac{\partial^n}{\partial r^n} \int_{R_1}^r (r-t)^{n-\nu-1} F(t\omega) \, dt, \qquad (16.202)$$

where $x \in \overline{A}$; that is, $x = r\omega$, $r \in [R_1, R_2]$, $\omega \in S^{N-1}$.

Clearly

$$\frac{\partial_{*R_1}^0 F(x)}{\partial r^0} = F(x),$$

and

$$\frac{\partial_{R_1}^\nu F(x)}{\partial r^\nu} = \frac{\partial^\nu F(x)}{\partial r^\nu}, \text{ if } \nu \in \mathbb{N}.$$

We give

Proposition 16.92. *Let* $\nu \geq \gamma + 1$, $\gamma \geq 0$, $n := \lceil \nu \rceil$, $f : \overline{A} \to \mathbb{R}$ *with* $f \in L_1(A)$, *where* $A := B(0, R_2) - \overline{B(0, R_1)} \subseteq \mathbb{R}^N$, $N \geq 2$, $0 < R_1 < R_2$. *Assume that* $f(\cdot\omega) \in AC^n([R_1, R_2])$, $\forall \omega \in S^{N-1}$, *and that* $\partial_{R_1}^\nu f(\cdot\omega)/\partial r^\nu \in L_\infty(R_1, R_2)$, $\forall \omega \in S^{N-1}$. *Further assume that* $\exists D_{R_1}^\nu f(r\omega) \in \mathbb{R}$, $\forall r \in$

$[R_1, R_2]$, for each $\omega \in S^{N-1}$, with $\partial_{R_1}^\nu f(x)/\partial r^\nu \in L_\infty(A)$. We suppose $\forall r \in [R_1, R_2]$ and $\forall \omega \in S^{N-1}$ that $\exists M_1 > 0$ such that $\left| D_{R_1}^\nu f(r\omega) \right| \leq M_1$.

Suppose that

$$\frac{\partial^j f(R_1\omega)}{\partial r^j} = 0, j = 0, 1, \ldots n-1, \forall \omega \in S^{N-1}.$$

Set

$$P(r) := \int_{R_1}^r (r-t)^{2(\nu-\gamma-1)} t^{(1-N)} dt, \quad R_1 \leq r \leq R_2, \tag{16.203}$$

also

$$A(r) := \frac{r^{(N-1/2)} \sqrt{P(r)}}{\Gamma(\nu-\gamma)}, \tag{16.204}$$

$$\widetilde{A}_0(R_2) := \left(\int_{R_1}^{R_2} (A(r))^2 \, dr \right)^{1/2}. \tag{16.205}$$

Then

$$\int_A \left| \frac{\partial_{R_1}^\gamma f(x)}{\partial r^\gamma} \right| \left| \frac{\partial_{R_1}^\nu f(x)}{\partial r^\nu} \right| dx \leq$$

$$\widetilde{A}_0(R_2) \, 2^{-1/2} \left(\int_A \left(\frac{\partial_{R_1}^\nu f(x)}{\partial r^\nu} \right)^2 dx \right). \tag{16.206}$$

Proof. Similarly as in Proposition 16.82. $\square$

17
Poincaré-Type Fractional Inequalities

Here we present Poincaré-type fractional inequalities involving fractional derivatives of Canavati, Riemann–Liouville, and Caputo types. The results are general L_p inequalities forward and reverse, univariate and multivariate, on a spherical shell. We give applications to ODEs and PDEs. We present also mean Poincaré-type fractional inequalities. This treatment relies on [57].

17.1 Introduction

This chapter is motivated by the famous Poincaré inequality; see [2]. Given a bounded domain $\Omega \subset \mathbb{R}^n$, $n \in \mathbb{N}$, it holds that

$$\|u\|_{L^p(\Omega)} \leq C \|\nabla u\|_{L^p(\Omega)}$$

for functions u with vanishing mean value over Ω, $1 \leq p \leq \infty$, under very general assumptions on Ω, where $\|\nabla u\|_{L^p(\Omega)}$ is defined as the L^p-norm of the Euclidean norm of ∇u.

Especially in [2] it is proven for a convex domain $\Omega \subset \mathbb{R}^n$ with diameter d that

$$\|u\|_{L^1(\Omega)} \leq \frac{d}{2} \|\nabla u\|_{L^1(\Omega)}$$

for any u with zero mean value on Ω, with the constant $1/2$ also being optimal.

G.A. Anastassiou, *Fractional Differentiation Inequalities*,
DOI 10.1007/978-0-387-98128-4_17, © Springer Science+Business Media, LLC 2009

We are also motivated by [126], where the authors prove the following. Let $1 < p < \infty$, $-\infty < a < b < \infty$. The best constant C (independent of a, b) for which the one-dimensional Poincaré inequality

$$\left\| f - \frac{\int_a^b f(t)\,dt}{b-a} \right\|_{L^1([a,b])} \leq C\,(b-a)^{2-1/p}\,\|f'\|_{L^p([a,b])}$$

holds for all Lipschitz continuous functions f, is $C = 1/2\,(1+p')^{-1/p'}$, where $p' > 1 : 1/p + 1/p' = 1$.

Here we present Poincaré-type inequalities with respect to fractional derivatives of Canavati, Riemann–Liouville, and Caputo types. We give univariate results on a closed interval of $\mathbb{R}$, and multivariate ones on a spherical shell. The surprising fact here is that we require only vanishing initial conditions of the derivatives up to a certain order for the involved function.

17.2   Fractional Poincaré Inequalities Results

Let $[a, b] \subseteq \mathbb{R}$; here see [19, 101]. Let x, $x_0 \in [a, b]$ such that $x \geq x_0$, x_0 is fixed, $f \in C([a, b])$, and define

$$(J_\nu^{x_0} f)(x) := \frac{1}{\Gamma(\nu)} \int_{x_0}^{x} (x-t)^{\nu-1} f(t)\,dt,$$

all $x_0 \leq x \leq b$, where $\nu > 0$, $n := [\nu]$ (integral part), $\alpha := \nu - n$, the generalized Riemann–Liouville integral, with Γ the gamma function.

We consider

$$C_{x_0}^\nu([a, b]) := \left\{ f \in C^n([a, b]) : J_{1-\alpha}^{x_0} f^{(n)} \in C^1([x_0, b]) \right\}.$$

Let $f \in C_{x_0}^\nu([a, b])$; we define the generalized ν-fractional derivative of f over $[x_0, b]$ as

$$D_{x_0}^\nu f := \left(J_{1-\alpha}^{x_0} f^{(n)} \right)',$$

the derivative with respect to x.

Clearly $D_{x_0}^\nu f \in C([x_0, b])$.

We need

Lemma 17.1. ([19, pp. 544–545]) *Let* $\nu \geq \gamma + 1$, $\gamma \geq 0$, $n := [\nu]$, *and* $f \in C_{x_0}^\nu([a, b])$, $x_0 \in [a, b]$. *Assume* $f^{(i)}(x_0) = 0$, $i = 0, 1, \dots, n-1$. *Then*

$$(D_{x_0}^\gamma f)(x) = \frac{1}{\Gamma(\nu-\gamma)} \int_{x_0}^{x} (x-t)^{\nu-\gamma-1} (D_{x_0}^\nu f)(t)\,dt, \qquad (17.1)$$

all $x \in [x_0, b]$.

We give Poincaré inequalities with respect to the above-defined Canavati-type fractional derivative.

Theorem 17.2. *Assumptions are as in Lemma 17.1. Let* $p, q > 1$:
$1/p + 1/q = 1$.
It holds

$$\int_{x_0}^{b} \left| D_{x_0}^{\gamma} f(x) \right|^q dx \le$$

$$\left[\frac{(b - x_0)^{q(\nu-\gamma)}}{\left(\Gamma(\nu - \gamma)\right)^q \left(p(\nu - \gamma - 1) + 1\right)^{(q/p)} q(\nu - \gamma)} \right] \left(\int_{x_0}^{b} \left| D_{x_0}^{\nu} f(t) \right|^q dt \right).$$
(17.2)

Proof. We have

$$\left| D_{x_0}^{\gamma} f(x) \right| \le \frac{1}{\Gamma(\nu - \gamma)} \int_{x_0}^{x} (x - t)^{\nu-\gamma-1} \left| D_{x_0}^{\nu} f(t) \right| dt \qquad (17.3)$$

(by the Hölder inequality)

$$\le \frac{1}{\Gamma(\nu - \gamma)} \left(\int_{x_0}^{x} (x - t)^{p(\nu-\gamma-1)} dt \right)^{1/p}$$

$$\left(\int_{x_0}^{x} \left| D_{x_0}^{\nu} f(t) \right|^q dt \right)^{1/q} = \qquad (17.4)$$

$$\frac{1}{\Gamma(\nu - \gamma)} \frac{(x - x_0)^{(\nu-\gamma-1)+1/p}}{(p(\nu - \gamma - 1) + 1)^{1/p}} \left(\int_{x_0}^{x} \left| D_{x_0}^{\nu} f(t) \right|^q dt \right)^{1/q} \qquad (17.5)$$

$$\le \frac{1}{\Gamma(\nu - \gamma)} \frac{(x - x_0)^{(\nu-\gamma-1)+1/p}}{(p(\nu - \gamma - 1) + 1)^{1/p}} \left(\int_{x_0}^{b} \left| D_{x_0}^{\nu} f(t) \right|^q dt \right)^{1/q}.$$

That is, we have

$$\left| D_{x_0}^{\gamma} f(x) \right| \le \left(\frac{(x - x_0)^{(\nu-\gamma-1)+1/p}}{\Gamma(\nu - \gamma)(p(\nu - \gamma - 1) + 1)^{1/p}} \right) \left(\int_{x_0}^{b} \left| D_{x_0}^{\nu} f(t) \right|^q dt \right)^{1/q},$$
(17.6)

$\forall x \in [x_0, b]$.
Consequently we find

$$\left| D_{x_0}^{\gamma} f(x) \right|^q \le \left(\frac{(x - x_0)^{q(\nu-\gamma-1)+q/p}}{\left(\Gamma(\nu - \gamma)\right)^q (p(\nu - \gamma - 1) + 1)^{q/p}} \right)$$

$$\left(\int_{x_0}^{b} \left| D_{x_0}^{\nu} f \left(t \right) \right|^{q} dt \right),\tag{17.7}$$

$\forall x \in [x_0, b]$.

Hence we obtain

$$\int_{x_0}^{b} \left| D_{x_0}^{\gamma} f \left(x \right) \right|^{q} dx \leq \left[\frac{\left(b - x_0 \right)^{q(\nu - \gamma - 1) + (q/p) + 1}}{\left(\Gamma \left(\nu - \gamma \right) \right)^{q} \left(p \left(\nu - \gamma - 1 \right) + 1 \right)^{(q/p)}} \cdot \right.$$

$$\left. \frac{1}{\left(q \left(\nu - \gamma - 1 \right) + \frac{q}{p} + 1 \right)} \right] \left(\int_{x_0}^{b} \left| D_{x_0}^{\nu} f \left(t \right) \right|^{q} dt \right).\tag{17.8}$$

□

Corollary 17.3. (to Theorem 17.2) *When* $\gamma = 0$ *we find*

$$\int_{x_0}^{b} \left| f \left(x \right) \right|^{q} dx \leq \left(\frac{\left(b - x_0 \right)^{q\nu}}{\left(\Gamma \left(\nu \right) \right)^{q} \left(p \left(\nu - 1 \right) + 1 \right)^{(q/p)} q\nu} \right)$$

$$\left(\int_{x_0}^{b} \left| D_{x_0}^{\nu} f \left(t \right) \right|^{q} dt \right).\tag{17.9}$$

Let $\alpha > 0$, $f \in L_1 \left(a, x \right)$, a, $x \in \mathbb{R}$, see [45, 64, 134].

We define the generalized Riemann–Liouville fractional derivative of f of order α by

$$D_a^{\alpha} f \left(s \right) := \frac{1}{\Gamma \left(m - \alpha \right)} \left(\frac{d}{ds} \right)^{m} \int_{a}^{s} \left(s - t \right)^{m - \alpha - 1} f \left(t \right) dt,$$

where $m := [\alpha] + 1$, $s \in [a, x]$, see also later Remark 17.46.

In addition, we set

$$D_a^0 f := f := J_0^a f,$$

$$J_{-\alpha}^a f := D_a^{\alpha} f, \text{ if } \alpha > 0,$$

$$D_a^{-\alpha} f := J_{\alpha}^a f, \text{ if } 0 < \alpha \leq 1,$$

$$D_a^n f = f^{(n)}, \text{ for } n \in \mathbb{N}.$$

We need

Definition 17.4. [45] We say that $f \in L_1 \left(a, x \right)$ has an L_∞ fractional derivative $D_a^{\alpha} f$ $\left(\alpha > 0 \right)$ in $[a, x]$, a, $x \in \mathbb{R}$, iff $D_a^{\alpha - k} f \in C \left([a, x] \right)$, $k = 1, \ldots, m := [\alpha] + 1$, and $D_a^{\alpha - 1} f \in AC \left([a, x] \right)$ (absolutely continuous functions) and $D_a^{\alpha} f \in L_\infty \left(a, x \right)$.

Lemma 17.5. [45] *Let* $\beta > \alpha \geq 0$, $f \in L_1(a,x)$, a, $x \in \mathbb{R}$, *have an* L_∞ *fractional derivative* $D_a^\beta f$ *in* $[a,x]$; *let* $D_a^{\beta-k} f(a) = 0$ *for* $k = 1, \ldots, [\beta]+1$. *Then*

$$D_a^\alpha f(s) = \frac{1}{\Gamma(\beta-\alpha)} \int_a^s (s-t)^{\beta-\alpha-1} D_a^\beta f(t)\,dt, \qquad (17.10)$$

$\forall s \in [a,x]$.

Here $D_a^\alpha f \in AC([a,x])$ for $\beta - \alpha \geq 1$, and $D_a^\alpha f \in C([a,x])$ for $\beta - \alpha \in (0,1)$.

We present Poincaré inequalities with respect to the above-defined generalized Riemann–Liouville fractional derivative.

Theorem 17.6. *Let all the assumptions be the same as in Lemma 17.5;* p, $q > 1 : 1/p + 1/q = 1$ *with* $p(\beta - \alpha - 1) + 1 > 0$. *Then*

$$\int_a^x |D_a^\alpha f(t)|^q\,dt \leq \left[\frac{(x-a)^{q(\beta-\alpha)}}{(\Gamma(\beta-\alpha))^q (p(\beta-\alpha-1)+1)^{(q/p)} q(\beta-\alpha)} \right]$$

$$\left(\int_a^x |D_a^\beta f(t)|^q\,dt \right). \qquad (17.11)$$

Corollary 17.7. *All are as in Theorem 17.6. Let* $\alpha = 0$; *then it holds that*

$$\int_a^x |f(t)|^q\,dt \leq \left[\frac{(x-a)^{q\beta}}{(\Gamma(\beta))^q (p(\beta-1)+1)^{(q/p)} q\beta} \right]$$

$$\left(\int_a^x |D_a^\beta f(t)|^q\,dt \right). \qquad (17.12)$$

Next we produce Poincaré inequalities with respect to the Caputo fractional derivative.

We mention

Definition 17.8. [134] *Let* $\nu \geq 0$, $n := \lceil \nu \rceil$; $\lceil \cdot \rceil$ *is the ceiling of the number* $f \in AC^n([a,b])$. *The Caputo fractional derivative is given by*

$$D_{*a}^\nu f(x) := \frac{1}{\Gamma(n-\nu)} \int_a^x (x-t)^{n-\nu-1} f^{(n)}(t)\,dt, \qquad (17.13)$$

$\forall x \in [a,b]$.

The above function $D_{*a}^\nu f(x)$ exists almost everywhere for $x \in [a,b]$.

We need

Proposition 17.9. [134] *Let $\nu \geq 0$, $n := \lceil \nu \rceil$, $f \in AC^n\left([a,b]\right)$. Then $D_{*a}^{\nu}f$ exists iff the generalized Riemann–Liouville fractional derivative $D_a^{\nu}f$ exists.*

Proposition 17.10. [134] *Let $\nu \geq 0$, $n = \lceil \nu \rceil$. Assume that f is such that both $D_{*a}^{\nu}f$ and $D_a^{\nu}f$ exist. Suppose that $f^{(k)}\left(a\right) = 0$ for $k = 0, 1, \ldots, n-1$. Then*

$$D_{*a}^{\nu}f = D_a^{\nu}f.$$

In conclusion

Corollary 17.11. [58] *Let $\nu \geq 0$, $n := \lceil \nu \rceil$, $f \in AC^n\left([a,b]\right)$, $D_{*a}^{\nu}f$ exists or $D_a^{\nu}f$ exists, and $f^{(k)}\left(a\right) = 0$, $k = 0, 1, \ldots, n-1$. Then*

$$D_a^{\nu}f = D_{*a}^{\nu}f.$$

We need

Proposition 17.12. [58] *Let $\nu \geq 0$, $n := \lceil \nu \rceil$, $f \in AC^n\left([a,b]\right)$, and $f^{(k)}\left(a\right) = 0$, $k = 0, 1, \ldots, n-1$. Then*

$$f\left(x\right) = \frac{1}{\Gamma\left(\nu\right)} \int_a^x \left(x-t\right)^{\nu-1} D_{*a}^{\nu}f\left(t\right)dt. \qquad (17.14)$$

We also need

Theorem 17.13. [58] *Let $\nu \geq \gamma + 1$, $\gamma \geq 0$. Call $n := \lceil \nu \rceil$. Assume $f \in AC^n\left([a,b]\right)$ such that $f^{(k)}\left(a\right) = 0$, $k = 0, 1, \ldots, n-1$, and $D_{*a}^{\nu}f \in L_{\infty}\left(a,b\right)$. Then $D_{*a}^{\gamma}f \in C\left([a,b]\right)$, and*

$$D_{*a}^{\gamma}f\left(x\right) = \frac{1}{\Gamma\left(\nu-\gamma\right)} \int_a^x \left(x-t\right)^{\nu-\gamma-1} D_{*a}^{\nu}f\left(t\right)dt, \qquad (17.15)$$

$\forall x \in [a,b]$.

Theorem 17.14. [58] *Let $\nu \geq \gamma+1$, $\gamma \geq 0$, $n := \lceil \nu \rceil$. Let $f \in AC^n\left([a,b]\right)$ such that $f^{(k)}\left(a\right) = 0$, $k = 0, 1, \ldots, n-1$. Assume $\exists D_a^{\nu}f\left(x\right) \in \mathbb{R}$, $\forall x \in [a,b]$, and $D_a^{\nu}f \in L_{\infty}\left(a,b\right)$. Then $D_a^{\gamma}f \in C\left([a,b]\right)$, and*

$$D_a^{\gamma}f\left(x\right) = \frac{1}{\Gamma\left(\nu-\gamma\right)} \int_a^x \left(x-t\right)^{\nu-\gamma-1} D_a^{\nu}f\left(t\right)dt, \qquad (17.16)$$

$\forall x \in [a,b]$.

Now we are ready to give

Theorem 17.15. *Let $\nu \geq \gamma+1$, $\gamma \geq 0$, $n := \lceil \nu \rceil$. Assume $f \in AC^n([a,b])$ such that $f^{(k)}(a) = 0$, $k = 0, 1, \ldots, n-1$, and $D_{*a}^\nu f \in L_\infty(a,b)$. Let p, $q > 1 : 1/p + 1/q = 1$.*
Then

$$\int_a^b |D_{*a}^\gamma f(x)|^q \, dx \leq$$

$$\left[\frac{(b-a)^{q(\nu-\gamma)}}{(\Gamma(\nu-\gamma))^q \, (p(\nu-\gamma-1)+1)^{(q/p)} \, q(\nu-\gamma)} \right] \left(\int_a^b |D_{*a}^\nu f(x)|^q \, dx \right).$$

$$(17.17)$$

Proof. Similar to Theorem 17.2. $\square$

We also give

Theorem 17.16. *Let $\nu \geq \gamma+1$, $\gamma \geq 0$, $n := \lceil \nu \rceil$. Let $f \in AC^n([a,b])$ such that $f^{(k)}(a) = 0$, $k = 0, 1, \ldots, n-1$. Assume $\exists D_a^\nu f(x) \in \mathbb{R}$, $\forall x \in [a,b]$, and $D_a^\nu f \in L_\infty(a,b)$. Let p, $q > 1 : 1/p + 1/q = 1$.*
Then

$$\int_a^b |D_a^\gamma f(x)|^q \, dx \leq$$

$$\left[\frac{(b-a)^{q(\nu-\gamma)}}{(\Gamma(\nu-\gamma))^q \, (p(\nu-\gamma-1)+1)^{(q/p)} \, q(\nu-\gamma)} \right] \left(\int_a^b |D_a^\nu f(x)|^q \, dx \right).$$

$$(17.18)$$

Proof. Similar to Theorem 17.2. $\square$

When $\gamma = 0$ we get

Proposition 17.17. *Same assumptions as in Theorem 17.15; $\gamma = 0$.*
Then

$$\int_a^b |f(x)|^q \, dx \leq$$

$$\left(\frac{(b-a)^{q\nu}}{(\Gamma(\nu))^q \, (p(\nu-1)+1)^{(q/p)} \, q\nu} \right) \left(\int_a^b |D_{*a}^\nu f(x)|^q \, dx \right). \qquad (17.19)$$

Similarly we have

Proposition 17.18. *Same assumptions as in Theorem 17.16; $\gamma = 0$.*
Then

$$\int_a^b |f(x)|^q \, dx \leq$$

$$\left(\frac{(b-a)^{q\nu}}{(\Gamma(\nu))^q \, (p(\nu-1)+1)^{(q/p)} \, q\nu}\right) \left(\int_a^b |D_a^\nu f(x)|^q \, dx\right). \qquad (17.20)$$

Special cases of the above results follow.

Corollary 17.19. (To Theorem 17.2; $p=q=2$.) *It holds*

$$\int_{x_0}^b \left(D_{x_0}^\gamma f(x)\right)^2 dx \le$$

$$\left[\frac{(b-x_0)^{2(\nu-\gamma)}}{(\Gamma(\nu-\gamma))^2 \, (2(\nu-\gamma)-1) \, 2(\nu-\gamma)}\right] \left(\int_{x_0}^b \left(D_{x_0}^\nu f(x)\right)^2 dx\right). \qquad (17.21)$$

Corollary 17.20. (To Theorem 17.6; $p=q=2$.) *Assume $\beta-\alpha>1/2$. Then*

$$\int_a^x \left(D_a^\alpha f(t)\right)^2 dt \le$$

$$\left[\frac{(x-a)^{2(\beta-\alpha)}}{(\Gamma(\beta-\alpha))^2 \, (2(\beta-\alpha)-1) \, 2(\beta-\alpha)}\right] \left(\int_a^x \left(D_a^\beta f(t)\right)^2 dt\right). \qquad (17.22)$$

Corollary 17.21. (To Theorem 17.15; $p=q=2$.) *It holds*

$$\int_a^b \left(D_{*a}^\gamma f(x)\right)^2 dx \le$$

$$\left[\frac{(b-a)^{2(\nu-\gamma)}}{(\Gamma(\nu-\gamma))^2 \, (2(\nu-\gamma)-1) \, 2(\nu-\gamma)}\right] \left(\int_a^b \left(D_{*a}^\nu f(x)\right)^2 dx\right). \qquad (17.23)$$

Corollary 17.22. (To Theorem 17.16; $p=q=2$.) *It holds*

$$\int_a^b \left(D_a^\gamma f(x)\right)^2 dx \le$$

$$\left[\frac{(b-a)^{2(\nu-\gamma)}}{(\Gamma(\nu-\gamma))^2 \, (2(\nu-\gamma)-1) \, 2(\nu-\gamma)}\right] \left(\int_a^b \left(D_a^\nu f(x)\right)^2 dx\right). \qquad (17.24)$$

Next we give converse results.

Theorem 17.23. *Let $\nu \ge \gamma+1$, $\gamma \ge 0$, $n := [\nu]$, $f \in C_{x_0}^\nu([a,b])$, $x_0 \in [a,b]$. Assume $f^{(i)}(x_0)=0$, $i=0,1,\ldots,n-1$. Assume that $D_{x_0}^\nu f$ is of fixed strict sign on $[x_0,b]$. Let $0<p<1$, $q<0$, such that $1/p+1/q=1$.*

Then

$$\int_{x_0}^{b} \left| D_{x_0}^{\gamma} f(t) \right|^{-q} dt \geq \left(\frac{1}{\left(\Gamma(\nu - \gamma) \right)^{-q}} \right)$$

$$\frac{(b - x_0)^{(-q(\nu-\gamma)+2)}}{(p(\nu - \gamma - 1) + 1)^{(1-q)} (-q(\nu - \gamma) + 2)} \left(\int_{x_0}^{b} \left| D_{x_0}^{\nu} f(t) \right|^q d \right)^{-1}. \quad (17.25)$$

Proof. Here $\left(D_{x_0}^{\gamma} f \right)(x_0) = 0$ by (17.1), but by assumption $\left(D_{x_0}^{\gamma} f \right)(x) \neq 0$, $\forall x \in (x_0, b]$.

By Lemma 17.1 and assumptions of the theorem we have

$$\left| D_{x_0}^{\gamma} f(x) \right| = \frac{1}{\Gamma(\nu - \gamma)} \int_{x_0}^{x} (x - t)^{\nu-\gamma-1} \left| D_{x_0}^{\nu} f(t) \right| dt \quad (17.26)$$

(by reverse Hölder's inequality)

$$\geq \frac{1}{\Gamma(\nu - \gamma)} \left(\int_{x_0}^{x} (x - t)^{p(\nu-\gamma-1)} dt \right)^{1/p} \left(\int_{x_0}^{x} \left| D_{x_0}^{\nu} f(t) \right|^q dt \right)^{1/q} \quad (17.27)$$

$$\geq \frac{1}{\Gamma(\nu - \gamma)} \frac{(x - x_0)^{(\nu-\gamma-1)+1/p}}{(p(\nu - \gamma - 1) + 1)^{1/p}} \left(\int_{x_0}^{b} \left| D_{x_0}^{\nu} f(t) \right|^q dt \right)^{1/q}. \quad (17.28)$$

So that

$$\left| D_{x_0}^{\gamma} f(x) \right| \geq \frac{1}{\Gamma(\nu - \gamma)} \frac{(x - x_0)^{(\nu-\gamma-1)+1/p}}{(p(\nu - \gamma - 1) + 1)^{1/p}} \left(\int_{x_0}^{b} \left| D_{x_0}^{\nu} f(t) \right|^q dt \right)^{1/q},$$
$$(17.29)$$

$\forall x \in [x_0, b]$.

Therefore

$$\left| D_{x_0}^{\gamma} f(x) \right|^{-q} \geq \left(\frac{1}{\left(\Gamma(\nu - \gamma) \right)^{-q}} \right)$$

$$\frac{(x - x_0)^{-q(\nu-\gamma-1)-q/p}}{(p(\nu - \gamma - 1) + 1)^{-q/p}} \left(\int_{x_0}^{b} \left| D_{x_0}^{\nu} f(t) \right|^q dt \right)^{-1}, \quad (17.30)$$

$\forall x \in [x_0, b]$.

Finally we obtain

$$\int_{x_0}^{b} \left| D_{x_0}^{\gamma} f(x) \right|^{-q} dx \geq \left(\frac{1}{\left(\Gamma(\nu - \gamma) \right)^{-q}} \right)$$

$$\frac{(b - x_0)^{-q(\nu-\gamma-1)-q/p+1}}{(p(\nu - \gamma - 1) + 1)^{-q/p} (-q(\nu - \gamma - 1) - q/p + 1)} \left(\int_{x_0}^{b} \left| D_{x_0}^{\nu} f(t) \right|^q dt \right)^{-1},$$
$$(17.31)$$

proving the claim. $\square$

Corollary 17.24. (To Theorem 17.23; $\gamma = 0$.) *It holds*

$$\int_{x_0}^{b} |f(t)|^{-q} \, dt \geq \left(\frac{1}{(\Gamma(\nu))^{-q}} \right)$$

$$\frac{(b-x_0)^{(-q\nu+2)}}{(p(\nu-1)+1)^{(1-q)} (-q\nu+2)} \left(\int_{x_0}^{b} |D_{x_0}^{\nu} f(t)|^q \, dt \right)^{-1}. \qquad (17.32)$$

We continue with

Theorem 17.25. *Let* $\beta > \alpha \geq 0$, $f \in L_1(a,x)$, a, $x \in \mathbb{R}$, *have an* L_∞ *fractional derivative* $D_a^\beta f$ *in* $[a,x]$, *and let* $D_a^{\beta-k} f(a) = 0$ *for* $k = 1, \ldots, [\beta] + 1$. *Assume that* $D_a^\beta f$ *has fixed sign a.e. on* $[a,x]$, *and* $1/D_a^\beta f \in L_\infty(a,x)$. *Let* $0 < p < 1$, $q < 0$, *such that* $1/p + 1/q = 1$, *with* $p(\beta - \alpha - 1) + 1 > 0$. *Then*

$$\int_{a}^{x} |D_a^\alpha f(t)|^{-q} \, dt \geq \left(\frac{1}{(\Gamma(\beta-\alpha))^{-q}} \right)$$

$$\frac{(x-a)^{(-q(\beta-\alpha)+2)}}{(p(\beta-\alpha-1)+1)^{(1-q)} (-q(\beta-\alpha)+2)} \left(\int_{a}^{x} |D_a^\beta f(t)|^q \, dt \right)^{-1}. \qquad (17.33)$$

Proof. Similar to Theorem 17.23; use (17.10). □

Corollary 17.26. (To Theorem 17.25; $\alpha = 0$.) *It holds*

$$\int_{a}^{x} |f(t)|^{-q} \, dt \geq \left(\frac{1}{(\Gamma(\beta))^{-q}} \right)$$

$$\frac{(x-a)^{(-q\beta+2)}}{(p(\beta-1)+1)^{(1-q)} (-q\beta+2)} \left(\int_{a}^{x} |D_a^\beta f(t)|^q \, dt \right)^{-1}. \qquad (17.34)$$

We present

Theorem 17.27. *Let* $\nu \geq \gamma + 1$, $\gamma \geq 0$, $n := \lceil \nu \rceil$, $f \in AC^n([a,b])$ *and* $f^{(k)}(a) = 0$, $k = 0, 1, \ldots, n-1$. *Assume that* $D_{*a}^\nu f$ *is of fixed sign a.e on* $[a,b]$, *and* $D_{*a}^\nu f$, $1/D_{*a}^\nu f \in L_\infty(a,b)$. *Let* $0 < p < 1$, $q < 0$, *such that* $1/p + 1/q = 1$.
Then

$$\int_{a}^{b} |D_{*a}^\gamma f(t)|^{-q} \, dt \geq \left(\frac{1}{(\Gamma(\nu-\gamma))^{-q}} \right)$$

$$\frac{(b-a)^{(-q(\nu-\gamma)+2)}}{(p(\nu-\gamma-1)+1)^{(1-q)} (-q(\nu-\gamma)+2)} \left(\int_{a}^{b} |D_{*a}^\nu f(t)|^q \, dt \right)^{-1}. \qquad (17.35)$$

Proof. Similar to Theorem 17.23; use (17.15).   □

Corollary 17.28. (To Theorem 17.27; $\gamma = 0$.) *It holds*

$$\int_a^b |f(t)|^{-q} \, dt \geq \left(\frac{1}{(\Gamma(\nu))^{-q}} \right)$$

$$\frac{(b-a)^{(-q\nu+2)}}{(p(\nu-1)+1)^{(1-q)}(-q\nu+2)} \left(\int_a^b |D_{*a}^{\nu} f(t)|^q \, dt \right)^{-1}. \qquad (17.36)$$

We give

Theorem 17.29. *Let* $\nu \geq \gamma + 1$, $\gamma \geq 0$, $n := \lceil \nu \rceil$, $f \in AC^n([a,b])$, *and* $f^{(k)}(a) = 0$, $k = 0, 1, \ldots, n-1$. *Suppose* $\exists D_a^{\nu} f(x) \in \mathbb{R}$, $\forall x \in [a,b]$, *and* $D_a^{\nu} f$, $1/D_a^{\nu} f \in L_{\infty}(a,b)$. *Here* $D_a^{\nu} f$ *is of fixed sign a.e on* $[a,b]$. *Let* $0 < p < 1$, $q < 0$, *such that* $1/p + 1/q = 1$.
Then

$$\int_a^b |D_a^{\gamma} f(t)|^{-q} \, dt \geq \left(\frac{1}{(\Gamma(\nu-\gamma))^{-q}} \right)$$

$$\frac{(b-a)^{(-q(\nu-\gamma)+2)}}{(p(\nu-\gamma-1)+1)^{(1-q)}(-q(\nu-\gamma)+2)} \left(\int_a^b |D_a^{\nu} f(t)|^q \, dt \right)^{-1}. \qquad (17.37)$$

Proof. As in Theorem 17.23; use (17.16).   □

Corollary 17.30. (To Theorem 17.29; $\gamma = 0$.) *It holds*

$$\int_a^b |f(t)|^{-q} \, dt \geq \left(\frac{1}{(\Gamma(\nu))^{-q}} \right)$$

$$\frac{(b-a)^{(-q\nu+2)}}{(p(\nu-1)+1)^{(1-q)}(-q\nu+2)} \left(\int_a^b |D_a^{\nu} f(t)|^q \, dt \right)^{-1}. \qquad (17.38)$$

Next we treat the easy but important case of $p = 1$.
We present for Canavati-type fractional derivatives

Theorem 17.31. *Let* $\nu \geq \gamma + 1$, $\gamma \geq 0$, $n := \lceil \nu \rceil$, *and* $f \in C_{x_0}^{\nu}([a,b])$, $x_0 \in [a,b]$. *Assume* $f^{(i)}(x_0) = 0$, $i = 0, 1, \ldots, n-1$.
Then

$$(1) \quad \int_{x_0}^b |D_{x_0}^{\gamma} f(t)| \, dt \leq \frac{(b-x_0)^{\nu-\gamma}}{\Gamma(\nu-\gamma+1)} \int_{x_0}^b |D_{x_0}^{\nu} f(t)| \, dt. \qquad (17.39)$$

Setting $\gamma = 0$ we get

$$(2) \quad \int_{x_0}^b |f(t)| \, dt \le \frac{(b - x_0)^{\nu}}{\Gamma(\nu + 1)} \int_{x_0}^b |D_{x_0}^{\nu} f(t)| \, dt. \qquad (17.40)$$

Proof. By (17.1) we have

$$|D_{x_0}^{\gamma} f(x)| \le \frac{1}{\Gamma(\nu - \gamma)} \int_{x_0}^x (x - t)^{\nu - \gamma - 1} |D_{x_0}^{\nu} f(t)| \, dt$$

$$\le \frac{(x - x_0)^{\nu - \gamma - 1}}{\Gamma(\nu - \gamma)} \int_{x_0}^b |D_{x_0}^{\nu} f(t)| \, dt, \qquad (17.41)$$

$\forall x \in [x_0, b]$.

Thus by integrating (17.41) over $[x_0, b]$ we find (17.39); setting there $\gamma = 0$ we obtain (17.40). $\square$

Similarly we find for Riemann–Liouville fractional derivatives

Theorem 17.32. *Let $\beta > \alpha + 1$, $\alpha \ge 0$, $f \in L_1(a, x)$, a, $x \in \mathbb{R}$, have an L_∞ fractional derivative $D_a^{\beta} f$ in $[a, x]$, and let $D_a^{\beta - k} f(a) = 0$ for $k = 1, \ldots, [\beta] + 1$. Then*

$$(1) \quad \int_a^x |D_a^{\alpha} f(t)| \, dt \le \frac{(x - a)^{\beta - \alpha}}{\Gamma(\beta - \alpha + 1)} \int_a^x |D_a^{\beta} f(t)| \, dt. \qquad (17.42)$$

When $\alpha = 0$ we find

$$(2) \quad \int_a^x |f(t)| \, dt \le \frac{(x - a)^{\beta}}{\Gamma(\beta + 1)} \int_a^x |D_a^{\beta} f(t)| \, dt. \qquad (17.43)$$

Proof. As in Theorem 17.31; based on Lemma 17.5. $\square$

Also we have for Caputo derivatives

Theorem 17.33. *Let $\nu \ge \gamma + 1$, $\gamma \ge 0$, $n := [\nu]$. Assume $f \in AC^n([a, b])$ such that $f^{(k)}(a) = 0$, $k = 0, 1, \ldots, n - 1$, and $D_{*a}^{\nu} f \in L_\infty(a, b)$. Then*

$$1) \quad \int_a^b |D_{*a}^{\gamma} f(t)| \, dt \le \frac{(b - a)^{\nu - \gamma}}{\Gamma(\nu - \gamma + 1)} \int_a^b |D_{*a}^{\nu} f(t)| \, dt. \qquad (17.44)$$

When $\gamma = 0$ we get

$$2) \quad \int_a^b |f(t)| \, dt \le \frac{(b - a)^{\nu}}{\Gamma(\nu + 1)} \int_a^b |D_{*a}^{\nu} f(t)| \, dt. \qquad (17.45)$$

Proof. As in Theorem 17.31; based on Theorem 17.13. $\square$

Coming back to Riemann–Liouville derivatives we get

Theorem 17.34. *Let* $\nu \geq \gamma + 1$, $\gamma \geq 0$, $n := \lceil \nu \rceil$. *Let* $f \in AC^n([a,b])$ *such that* $f^{(k)}(a) = 0$, $k = 0, 1, \ldots, n-1$. *Assume* $\exists D_a^\nu f(x) \in \mathbb{R}$, $\forall x \in [a,b]$, *and* $D_a^\nu f \in L_\infty(a,b)$. *Then*

(1)
$$\int_a^b |D_a^\gamma f(t)|\, dt \leq \frac{(b-a)^{\nu-\gamma}}{\Gamma(\nu-\gamma+1)} \int_a^b |D_a^\nu f(t)|\, dt. \qquad (17.46)$$

When $\gamma = 0$ *we get*

(2)
$$\int_a^b |f(t)|\, dt \leq \frac{(b-a)^{\nu}}{\Gamma(\nu+1)} \int_a^b |D_a^\nu f(t)|\, dt. \qquad (17.47)$$

Proof. As in Theorem 17.31; based on Theorem 17.14. $\square$

17.3 Applications of Fractional Poincaré Inequalities

Next we apply some of the above results to the multivariate case of the spherical shell.

Let $N \geq 2$, $S^{N-1} := \{x \in \mathbb{R}^N : |x| = 1\}$ the unit sphere on $\mathbb{R}^N$, where $|\cdot|$ stands for the Euclidean norm in $\mathbb{R}^N$. Also denote the ball

$$B(0,R) := \{x \in \mathbb{R}^N : |x| < R\} \subseteq \mathbb{R}^N, \; R > 0,$$

and the spherical shell

$$A := B(0, R_2) - \overline{B(0, R_1)}, \; 0 < R_1 < R_2.$$

For the following see [356, pp. 149–150] and [383, pp. 87–88].

For $x \in \mathbb{R}^N - \{0\}$ we can write uniquely $x = r\omega$, where $r = |x| > 0$, and $\omega = x/r \in S^{N-1}$, $|\omega| = 1$.

Clearly here

$$\mathbb{R}^N - \{0\} = (0, \infty) \times S^{N-1},$$

also

$$\overline{A} = [R_1, R_2] \times S^{N-1}.$$

In the sequel the following theorem is used frequently.

Theorem 17.35. *Let* $f : A \to \mathbb{R}$ *be a Lebesgue integrable function. Then*

$$\int_A f(x)\, dx = \int_{S^{N-1}} \left(\int_{R_1}^{R_2} f(r\omega)\, r^{N-1} dr \right) d\omega. \qquad (17.48)$$

So we are able to write an integral in polar form using the polar coordinates (r, ω).

Regarding the Canavati-type fractional derivative we need

Definition 17.36. (see [44]) Let $\nu > 0$, $n := [\nu]$, $\alpha := \nu - n$, $f \in C^n(\overline{A})$, and A is a spherical shell. Assume that there exists $\partial_{R_1}^\nu f(x)/\partial r^\nu \in C(\overline{A})$, given by

$$\frac{\partial_{R_1}^\nu f(x)}{\partial r^\nu} := \frac{1}{\Gamma(1-\alpha)} \frac{\partial}{\partial r} \left(\int_{R_1}^r (r-t)^{-\alpha} \frac{\partial^n f(t\omega)}{\partial r^n} dt \right), \qquad (17.49)$$

where $x \in \overline{A}$; that is, $x = r\omega$, $r \in [R_1, R_2]$, $\omega \in S^{N-1}$.

We call $\partial_{R_1}^\nu f/\partial r^\nu$ the radial Canavati-type fractional derivative of f of order ν. If $\nu = 0$, then set $\partial_{R_1}^\nu f(x)/\partial r^\nu := f(x)$.

We also need

Lemma 17.37. (see [44]) *Let* $\gamma \geq 0$, $\nu \geq 1$ *such that* $\nu - \gamma \geq 1$. *Let* $f \in C^n(\overline{A})$ *and there exists* $\partial_{R_1}^\nu f(x)/\partial r^\nu \in C(\overline{A})$, $x \in \overline{A}$, *and* A *a spherical shell. Further assume that* $\partial^j f(R_1\omega)/\partial r^j = 0$, $j = 0, 1, \ldots, n-1$, $n := [\nu]$, *and* $\forall \omega \in S^{N-1}$. *Then there exists* $\partial_{R_1}^\gamma f(x)/\partial r^\gamma \in C(\overline{A})$.

We present the following Canavati-type fractional Poincaré inequalities on the spherical shell.

Theorem 17.38. *Let* $\nu \geq \gamma + 1$, $\gamma \geq 0$, $n := [\nu]$, *and* $f \in C^n(\overline{A})$ *and there exists* $\partial_{R_1}^\nu f(x)/\partial r^\nu \in C(\overline{A})$, $x \in \overline{A}$, *and* A *a spherical shell. Further assume that* $\partial^j f(R_1\omega)/\partial r^j = 0$, $j = 0, 1, \ldots, n-1$; $\forall \omega \in S^{N-1}$. *Let* $p, q > 1$: $1/p + 1/q = 1$.

Then

$$1) \int_A \left| \frac{\partial_{R_1}^\gamma f(x)}{\partial r^\gamma} \right|^q dx \leq \left(\frac{R_2}{R_1} \right)^{N-1}$$

$$\left[\frac{(R_2 - R_1)^{q(\nu-\gamma)}}{q(\nu-\gamma)(\Gamma(\nu-\gamma))^q (p(\nu-\gamma-1)+1)^{q/p}} \right] \int_A \left| \frac{\partial_{R_1}^\nu f(x)}{\partial r^\nu} \right|^q dx. \qquad (17.50)$$

When $\gamma = 0$ *we get*

$$2) \int_A |f(x)|^q\, dx \leq \left(\frac{R_2}{R_1} \right)^{N-1}$$

$$\left[\frac{(R_2 - R_1)^{q\nu}}{q\nu \left(\Gamma \left(\nu \right) \right)^q \left(p \left(\nu - 1 \right) + 1 \right)^{q/p}} \right] \int_A \left| \frac{\partial_{R_1}^{\nu} f \left(x \right)}{\partial r^{\nu}} \right|^q dx. \qquad (17.51)$$

Proof. Because $R_1 \le r \le R_2$, $N \ge 2$, we have that $R_1^{N-1} \le r^{N-1} \le R_2^{N-1}$ and $R_2^{1-N} \le r^{1-N} \le R_1^{1-N}$. Hence

$$R_2^{1-N} \int_{R_1}^{R_2} r^{N-1} \left| D_{R_1}^{\gamma} f \left(r\omega \right) \right|^q dr \le$$

$$\int_{R_1}^{R_2} r^{1-N} r^{N-1} \left| D_{R_1}^{\gamma} f \left(r\omega \right) \right|^q dr =$$

$$\int_{R_1}^{R_2} \left| D_{R_1}^{\gamma} f \left(r\omega \right) \right|^q dr$$

$$\overset{(17.2)}{\le} \left(\frac{(R_2 - R_1)^{q(\nu - \gamma)}}{\left(\Gamma \left(\nu - \gamma \right) \right)^q \left(p \left(\nu - \gamma - 1 \right) + 1 \right)^{q/p} q \left(\nu - \gamma \right)} \right)$$

$$\left(\int_{R_1}^{R_2} r^{1-N} r^{N-1} \left| D_{R_1}^{\nu} f \left(r\omega \right) \right|^q dr \right) \le$$

$$\left(\frac{(R_2 - R_1)^{q(\nu - \gamma)} R_1^{1-N}}{\left(\Gamma \left(\nu - \gamma \right) \right)^q \left(p \left(\nu - \gamma - 1 \right) + 1 \right)^{q/p} q \left(\nu - \gamma \right)} \right)$$

$$\left(\int_{R_1}^{R_2} r^{N-1} \left| D_{R_1}^{\nu} f \left(r\omega \right) \right|^q dr \right). \qquad (17.52)$$

So we have established that

$$\int_{R_1}^{R_2} r^{N-1} \left| D_{R_1}^{\gamma} f \left(r\omega \right) \right|^q dr \le \left(\frac{R_2}{R_1} \right)^{N-1}$$

$$\left(\frac{(R_2 - R_1)^{q(\nu - \gamma)}}{\left(\Gamma \left(\nu - \gamma \right) \right)^q \left(p \left(\nu - \gamma - 1 \right) + 1 \right)^{q/p} q \left(\nu - \gamma \right)} \right)$$

$$\left(\int_{R_1}^{R_2} r^{N-1} \left| D_{R_1}^{\nu} f \left(r\omega \right) \right|^q dr \right), \; \forall \omega \in S^{N-1}. \qquad (17.53)$$

Then by integration of (17.53) on S^{N-1} we obtain

$$\int_{S^{N-1}} \left(\int_{R_1}^{R_2} \left| D_{R_1}^{\gamma} f \left(r\omega \right) \right|^q r^{N-1} dr \right) d\omega \le \left(\frac{R_2}{R_1} \right)^{N-1}$$

$$\left[\frac{(R_2 - R_1)^{q(\nu - \gamma)}}{q \left(\nu - \gamma \right) \left(\Gamma \left(\nu - \gamma \right) \right)^q \left(p \left(\nu - \gamma - 1 \right) + 1 \right)^{q/p}} \right]$$

$$\int_{S^{N-1}} \left(\int_{R_1}^{R_2} \left| D_{R_1}^{\nu} f\left(r\omega\right) \right|^q r^{N-1} dr \right) d\omega. \tag{17.54}$$

Using (17.48) in (17.54) we have proven the claim. □

Regarding the Riemann–Liouville fractional derivative we need

Remark 17.39. Here we follow [383] and denote by $\lambda_{\mathbb{R}^N}(x) \equiv dx$ the Lebesgue measure on $\mathbb{R}^N$, $N \geq 2$, and by $\lambda_{S^{N-1}}(\omega) \equiv d\omega$ the surface measure on S^{N-1}, where $\mathcal{B}_X$ stands for the Borel class on space X.
Define the measure R_N on $\left((0, \infty), \mathcal{B}_{(0,\infty)} \right)$ by

$$R_N\left(\Gamma\right) = \int_{\Gamma} r^{N-1} dr, \text{ any } \Gamma \in \mathcal{B}_{(0,\infty)}.$$

Now let

$$F \in L_1\left(A\right) = L_1\left([R_1, R_2] \times S^{N-1}\right).$$

For a fixed $\omega \in S^{N-1}$, define

$$g_\omega\left(r\right) := F\left(r\omega\right) = F\left(x\right),$$

where

$$x \in A := B\left(0, R_2\right) - \overline{B\left(0, R_1\right)},$$

$$0 < R_1 \leq r \leq R_2, \ r = |x|, \ \omega = \frac{x}{r} \in S^{N-1}.$$

By Fubini's theorem and Theorem 5.2.2, pp. 87 and 88 of [383], we have

$$g_\omega \in L_1\left([R_1, R_2], \mathcal{B}_{[R_1, R_2]}, R_N\right),$$

for $\lambda_{S^{N-1}} -$ almost every $\omega \in S^{N-1}$.
Call

$$K\left(F\right) := \left\{ \omega \in S^{N-1} : \ g_\omega \notin L_1\left([R_1, R_2], \mathcal{B}_{[R_1, R_2]}, R_N\right) \right\}$$

$$= \left\{ \omega \in S^{N-1} : \ F\left(\cdot\omega\right) \notin L_1\left([R_1, R_2], \mathcal{B}_{[R_1, R_2]}, R_N\right) \right\}. \tag{17.55}$$

That is,

$$\lambda_{S^{N-1}}\left(K\left(F\right)\right) = 0.$$

Of course,

$$\Theta\left(F\right) := [R_1, R_2] \times K\left(F\right) \subset A$$

and

$$\lambda_{\mathbb{R}^N}\left(\Theta\left(F\right)\right) = 0.$$

By the definition of the generalized Riemann–Liouville fractional derivative we then have (see also Remark 17.46 later)

$$\left(D_{R_1}^{\beta} g_\omega \right)(r) = \frac{1}{\Gamma\left(m - \beta\right)} \left(\frac{d}{dr}\right)^m$$

$$\int_0^{r-R_1} (r - R_1 - t)^{m-\beta-1} \, g_\omega \, (R_1 + t) \, dt, \qquad (17.56)$$

where

$$\beta > 0, \ m := [\beta] + 1, \ r \in [R_1, R_2].$$

If $\beta = 0$, then

$$\left(D_{R_1}^\beta g_\omega \right)(r) = g_\omega \, (r).$$

Formula (17.56) is written for all $\omega \in S^{N-1} - K\,(F)$. We set

$$\left(D_{R_1}^\beta g_\omega \right)(r) \equiv 0, \ \forall \omega \in K\,(F), \ \forall r \in [R_1, R_2], \ \text{any } \beta > 0.$$

The above lead to the following definition (see [45]).

Definition 17.40. Let $\beta > 0$, $m := [\beta] + 1$, $F \in L_1\,(A)$, and A is the spherical shell. We define

$$\frac{\partial_{R_1}^\beta F\,(x)}{\partial r^\beta} := \begin{cases} \frac{1}{\Gamma(m-\beta)} \left(\frac{d}{dr}\right)^m \int_0^{r-R_1} (r - R_1 - t)^{m-\beta-1} F\,((R_1 + t)\,\omega) \, dt, \\ \qquad\qquad \text{for } \omega \in S^{N-1} - K\,(F), \\ \qquad\qquad 0, \ \text{for } \omega \in K\,(F), \end{cases}$$

$$(17.57)$$

where

$$x = r\omega \in A, \ r \in [R_1, R_2], \ \omega \in S^{N-1}.$$

If $\beta = 0$, define

$$\frac{\partial_{R_1}^\beta F\,(x)}{\partial r^\beta} := F\,(x).$$

We call $\partial_{R_1}^\beta F\,(x)/\partial r^\beta$ the Riemann–Liouville radial fractional derivative of F of order β.

We need

Theorem 17.41. (see [45]) *Let $\beta > \alpha > 0$ and $F \in L_1\,(A)$. Assume that*

$$\frac{\partial_{R_1}^\beta F\,(x)}{\partial r^\beta} \in L_\infty\,(A).$$

Further suppose that $D_{R_1}^\beta F\,(r\omega)$ takes real values for almost all $r \in [R_1, R_2]$, for each $\omega \in S^{N-1}$, and for these

$$\left| D_{R_1}^\beta F\,(r\omega) \right| \le M_1$$

for some $M_1 > 0$.

For each $\omega \in S^{N-1} - K(F)$, *we assume that* $F(\cdot\omega)$ *has an* L_∞ *fractional derivative* $D_{R_1}^\beta F(\cdot\omega)$ *in* $[R_1, R_2]$, *and that*

$$D_{R_1}^{\beta-k} F(R_1\omega) = 0, \ k = 1, \ldots, [\beta] + 1.$$

Then

$$\frac{\partial_{R_1}^\alpha F(x)}{\partial r^\alpha} = \left(D_{R_1}^\alpha F\right)(r\omega) =$$

$$\frac{1}{\Gamma(\beta - \alpha)} \int_{R_1}^r (r - t)^{\beta-\alpha-1} \left(D_{R_1}^\beta F\right)(t\omega) \, dt, \qquad (17.58)$$

true $\forall x \in A$; *that is, true* $\forall r \in [R_1, R_2]$ *and* $\forall \omega \in S^{N-1}$.

  Here

$$\left(D_{R_1}^\alpha F\right)(\cdot\omega) \ \text{is in} \ AC([R_1, R_2]) \ \text{for} \ \beta - \alpha \geq 1$$

and

$$\left(D_{R_1}^\alpha F\right)(\cdot\omega) \ \text{is in} \ C([R_1, R_2]) \ \text{for} \ \beta - \alpha \in (0, 1),$$

$\forall \omega \in S^{N-1}$.

  Furthermore

$$\frac{\partial_{R_1}^\alpha F(x)}{\partial r^\alpha} \in L_\infty(A).$$

In particular, it holds

$$F(x) = F(r\omega) = \frac{1}{\Gamma(\beta)} \int_{R_1}^r (r - t)^{\beta-1} \left(D_{R_1}^\beta F\right)(t\omega) \, dt, \qquad (17.59)$$

$\forall r \in [R_1, R_2], \forall \omega \in S^{N-1} - K(F); \ x = r\omega,$ *and*

$$F(\cdot\omega) \ \text{is in} \ AC([R_1, R_2]) \ \text{for} \ \beta \geq 1$$

and

$$F(\cdot\omega) \ \text{is in} \ C([R_1, R_2]) \ \text{for} \ \beta \in (0, 1),$$

$\forall \omega \in S^{N-1} - K(F).$

We present Poincaré inequalities on the shell involving Riemann – Liouville radial fractional derivatives.

Theorem 17.42. *Let all terms and assumptions be as in Theorem 17.41. Let* $p, q > 1$: $1/p + 1/q = 1$ *with* $p(\beta - \alpha - 1) + 1 > 0$.
Then

$$(1) \ \int_A \left|\frac{\partial_{R_1}^\alpha F(x)}{\partial r^\alpha}\right|^q dx \leq \left(\frac{R_2}{R_1}\right)^{N-1}$$

$$\left[\frac{(R_2 - R_1)^{q(\beta-\alpha)}}{q(\beta - \alpha)(\Gamma(\beta - \alpha))^q (p(\beta - \alpha - 1) + 1)^{q/p}}\right] \int_A \left|\frac{\partial_{R_1}^\beta F(x)}{\partial r^\beta}\right|^q dx.$$

$$(17.60)$$

When $\alpha = 0$ we get

$$(2) \quad \int_A |F(x)|^q \, dx \le \left(\frac{R_2}{R_1} \right)^{N-1}$$

$$\left[\frac{(R_2 - R_1)^{q\beta}}{q\beta \, (\Gamma(\beta))^q \, (p(\beta - 1) + 1)^{q/p}} \right] \int_A \left| \frac{\partial_{R_1}^\beta F(x)}{\partial r^\beta} \right|^q \, dx. \qquad (17.61)$$

Proof. Similar to Theorem 17.38; based on Theorem 17.41. □

Regarding the Caputo fractional derivative we mention

Definition 17.43. (see [58]) Let $F : \overline{A} \to \mathbb{R}$, $\nu \ge 0$, $n := \lceil \nu \rceil$ such that $F(\cdot\omega) \in AC^n \left([R_1, R_2] \right)$, for all $\omega \in S^{N-1}$. We call the Caputo radial fractional derivative the following function

$$\frac{\partial_{*R_1}^\nu F(x)}{\partial r^\nu} := \frac{1}{\Gamma(n - \nu)} \int_{R_1}^r (r - t)^{n - \nu - 1} \frac{\partial^n F(t\omega)}{\partial r^n} \, dt, \qquad (17.62)$$

where $x \in \overline{A}$; that is, $x = r\omega$, $r \in [R_1, R_2]$, $\omega \in S^{N-1}$.
  Clearly

$$\frac{\partial_{*R_1}^0 F(x)}{\partial r^0} = F(x),$$

and

$$\frac{\partial_{*R_1}^\nu F(x)}{\partial r^\nu} = \frac{\partial^\nu F(x)}{\partial r^\nu}, \text{ if } \nu \in \mathbb{N}, \text{ the usual radial derivative.}$$

The above function (17.62) exists almost everywhere for $x \in \overline{A}$.

We mention the following fundamental representation result.

Theorem 17.44. (see [58]) *Let $\nu \ge \gamma + 1$, $\gamma \ge 0$, $n := \lceil \nu \rceil$, $F : \overline{A} \to \mathbb{R}$ with $F \in L_1(A)$. Assume that $F(\cdot\omega) \in AC^n \left([R_1, R_2] \right)$ for all $\omega \in S^{N-1}$, and that $\partial_{*R_1}^\nu F(\cdot\omega)/\partial r^\nu \in L_\infty(R_1, R_2)$ for all $\omega \in S^{N-1}$. Further assume that $\partial_{*R_1}^\nu F(x)/\partial r^\nu \in L_\infty(A)$.*
  More precisely, for these $r \in [R_1, R_2]$, for each $\omega \in S^{N-1}$, for which $D_{*R_1}^\nu F(r\omega)$ takes real values, there exists $M_1 > 0$ such that

$$\left| D_{*R_1}^\nu F(r\omega) \right| \le M_1.$$

We suppose that $\partial^j F(R_1\omega)/\partial r^j = 0$, $j = 0, 1, \ldots, n - 1$, for every $\omega \in S^{N-1}$. Then

$$\frac{\partial_{*R_1}^\gamma F(x)}{\partial r^\gamma} = D_{*R_1}^\gamma F(r\omega) =$$

$$\frac{1}{\Gamma(\nu-\gamma)}\int_{R_1}^{r}(r-t)^{\nu-\gamma-1}\left(D_{*R_1}^{\nu}F\right)(t\omega)\,dt, \qquad (17.63)$$

true $\forall x \in \overline{A}$; *that is, true* $\forall r \in [R_1, R_2]$ *and* $\forall \omega \in S^{N-1}$, $\gamma > 0$.
    Here

$$D_{*R_1}^{\gamma}F(\cdot\omega)\in AC\left([R_1,R_2]\right), \qquad (17.64)$$

$\forall \omega \in S^{N-1}$, $\gamma > 0$.
    Furthermore

$$\frac{\partial_{*R_1}^{\gamma}F(x)}{\partial r^{\gamma}}\in L_{\infty}(A),\ \gamma>0 \qquad (17.65)$$

In particular, it holds

$$F(x)=F(r\omega)=\frac{1}{\Gamma(\nu)}\int_{R_1}^{r}(r-t)^{\nu-1}\left(D_{*R_1}^{\nu}F\right)(t\omega)\,dt, \qquad (17.66)$$

true $\forall x \in A$; *that is, true* $\forall r \in [R_1, R_2]$ *and* $\forall \omega \in S^{N-1}$, *and*

$$F(\cdot\omega)\in AC\left([R_1,R_2]\right),\ \forall\omega\in S^{N-1}. \qquad (17.67)$$

We present Poincaré inequalities on the shell involving Caputo radial fractional derivatives.

Theorem 17.45. *All terms and assumptions for* f *are as in Theorem 17.44. Let* $p,q>1:\ 1/p+1/q=1$.
    Then

$$1)\ \int_A\left|\frac{\partial_{*R_1}^{\gamma}f(x)}{\partial r^{\gamma}}\right|^q dx\le\left(\frac{R_2}{R_1}\right)^{N-1}$$

$$\left[\frac{(R_2-R_1)^{q(\nu-\gamma)}}{q(\nu-\gamma)\left(\Gamma(\nu-\gamma)\right)^q\left(p(\nu-\gamma-1)+1\right)^{q/p}}\right]\int_A\left|\frac{\partial_{*R_1}^{\nu}f(x)}{\partial r^{\nu}}\right|^q dx. \qquad (17.68)$$

When $\gamma = 0$ *we get*

$$2)\ \int_A|f(x)|^q dx\le\left(\frac{R_2}{R_1}\right)^{N-1}$$

$$\left[\frac{(R_2-R_1)^{q\nu}}{q\nu\left(\Gamma(\nu)\right)^q\left(p(\nu-1)+1\right)^{q/p}}\right]\int_A\left|\frac{\partial_{*R_1}^{\nu}f(x)}{\partial r^{\nu}}\right|^q dx. \qquad (17.69)$$

We make

Remark 17.46. Regarding the two identical forms of the definition of generalized Riemann–Liouville fractional derivative, here $f\in L_1(a,x)$, $a<x$; $a,x\in\mathbb{R}$, iff $f_a\in L_1(0,x-a)$, where $f_a(t):=f(a+t)$.

The basic Riemann–Liouville derivative of f_a anchored at zero is defined and given by

$$(D^\nu f_a)(s') = \frac{1}{\Gamma(m-\nu)} \left(\frac{d}{ds'}\right)^m \int_0^{s'} (s'-t)^{m-\nu-1} f(a+t)\, dt, \quad (17.70)$$

$s' \in [0, x-a]$, $m := [\nu]+1$, $\nu > 0$.

Another way to define the generalized Riemann–Liouville fractional derivative (see [45, 64]) is

$$(D_a^\nu f)(s) := (D^\nu f_a)(s-a), \text{ for } s \in [a, x]. \tag{17.71}$$

But $0 \leq s-a \leq x-a$, calling $s' := s-a$, which is a one-to-one and onto map from $[a, x]$ onto $[0, x-a]$; we have $0 \leq s' \leq x-a$.

Consequently we get

$$(D^\nu f_a)(s-a) = \frac{1}{\Gamma(m-\nu)} \left(\frac{d}{d(s-a)}\right)^m$$

$$\left(\int_0^{s-a} ((s-a)-t)^{m-\nu-1} f(a+t)\, dt\right) = \tag{17.72}$$

$$\frac{1}{\Gamma(m-n)} \left(\frac{d}{ds}\right)^m \left(\int_0^{s-a} (s-(a+t))^{m-\nu-1} f(a+t)\, dt\right) =: (*). \tag{17.73}$$

In [65, 134], it is proved that

$$\int_a^s (s-t)^{m-\nu-1} |f(t)|\, dt < \infty \text{ for a.e. } s \in [a, x];$$

that is, there exists

$$\int_a^s (s-t)^{m-\nu-1} f(t)\, dt \text{ for a.e. } s \in [a, x].$$

Therefore from (17.73) we have

$$(*) = \frac{1}{\Gamma(m-\nu)} \left(\frac{d}{ds}\right)^m \int_0^s (s-t)^{m-\nu-1} f(t)\, dt \tag{17.74}$$

(change of variable) true a.e. for $s \in [a, x]$.

Because the derivative is defined only for real-valued functions (i.e., here only on existing integrals in (17.74)), we obviously get the identity of the two definitions. That is, $D_a^\nu f$ of (17.71) is such that

$$(D_a^\nu f)(s) = \frac{1}{\Gamma(m-\nu)} \left(\frac{d}{ds}\right)^m \int_a^s (s-t)^{m-\nu-1} f(t)\, dt, \tag{17.75}$$

$m := [\nu]+1$, $\nu > 0$, $s \in [a, x]$, as defined earlier in this chapter.

So based on the above Remark 17.46 we can redefine the Riemann–Liouville radial fractional derivative as follows.

Definition 17.47. Let $\beta > 0$, $m := [\beta] + 1$, $F \in L_1(A)$, and A is the spherical shell. We define

$$\frac{\partial_{R_1}^\beta F(x)}{\partial r^\beta} := \begin{cases} \frac{1}{\Gamma(m-\beta)} \left(\frac{\partial}{\partial r}\right)^m \int_{R_1}^r (r - t)^{m-\beta-1} F(t\omega)\, dt, \\ \qquad \text{for } \omega \in S^{N-1} - K(F), \\ \qquad 0, \text{ for } \omega \in K(F), \end{cases} \tag{17.76}$$

where

$$x = r\omega \in A, \ r \in [R_1, R_2], \ \omega \in S^{N-1}; \ K(F) \text{ as in (17.55)}.$$

If $\beta = 0$, define

$$\frac{\partial_{R_1}^\beta F(x)}{\partial r^\beta} := F(x).$$

So functions (17.57) and (17.76) are identical.

We need the following important representation result for Riemann–Liouville radial fractional derivatives.

Theorem 17.48. Let $\nu \geq \gamma + 1$, $\gamma \geq 0$, $n := [\nu]$, $F : \overline{A} \to \mathbb{R}$ with $F \in L_1(A)$. Assume that $F(\cdot\omega) \in AC^n([R_1, R_2])$, $\forall \omega \in S^{N-1}$, and that $\partial_{R_1}^\nu F(\cdot\omega)/\partial r^\nu$ is measurable on $[R_1, R_2]$, $\forall \omega \in S^{N-1}$. Also assume $\exists \partial_{R_1}^\nu F(r\omega)/\partial r^\nu \in \mathbb{R}$, $\forall r \in [R_1, R_2]$ and $\forall \omega \in S^{N-1}$, and $\partial_{R_1}^\nu F(x)/\partial r^\nu$ is measurable on $\overline{A}$. Suppose $\exists M_1 > 0$:

$$\left| \frac{\partial_{R_1}^\nu F(r\omega)}{\partial r^\nu} \right| \leq M_1, \ \forall (r, \omega) \in [R_1, R_2] \times S^{N-1}.$$

We suppose that $\partial^j F(R_1\omega)/\partial r^j = 0$, $j = 0, 1, \ldots, n - 1$; $\forall \omega \in S^{N-1}$. Then

$$\frac{\partial_{R_1}^\gamma F(x)}{\partial r^\gamma} = D_{R_1}^\gamma F(r\omega) =$$

$$\frac{1}{\Gamma(\nu - \gamma)} \int_{R_1}^r (r - t)^{\nu-\gamma-1} \left(D_{R_1}^\nu F\right)(t\omega)\, dt, \tag{17.77}$$

valid $\forall x \in \overline{A}$; that is, true $\forall r \in [R_1, R_2]$ and $\forall \omega \in S^{N-1}$; $\gamma > 0$.
  Here

$$D_{R_1}^\gamma F(\cdot\omega) \in AC([R_1, R_2]), \tag{17.78}$$

$\forall \omega \in S^{N-1}$, $\gamma > 0$.
  Furthermore

$$\frac{\partial_{R_1}^\gamma F(x)}{\partial r^\gamma} \in L_\infty(A), \ \gamma > 0. \tag{17.79}$$

In particular, it holds

$$F\left(x\right) = F\left(r\omega\right) = \frac{1}{\Gamma\left(\nu\right)} \int_{R_1}^{r} \left(r - t\right)^{\nu - 1} \left(D_{R_1}^{\nu} F\right)\left(t\omega\right) dt, \qquad (17.80)$$

true $\forall x \in A$; *that is, true* $\forall r \in [R_1, R_2]$ *and* $\forall \omega \in S^{N-1}$, *and*

$$F\left(\cdot\omega\right) \in AC\left(\left[R_1, R_2\right]\right), \ \forall \omega \in S^{N-1}. \qquad (17.81)$$

Proof. By our assumptions and Theorem 17.17, Corollary 14 of [58], we have (17.77) and (17.80). Also (17.78) is clear; see [45]. Property (17.81) is easy to prove.

Fixing $r \in [R_1, R_2]$, the function

$$\delta_r\left(t, \omega\right) := \left(r - t\right)^{\nu - \gamma - 1} D_{R_1}^{\nu} F\left(t\omega\right)$$

is measurable on

$$\left(\left[R_1, r\right] \times S^{N-1}, \overline{\mathcal{B}_{[R_1, r]} \times \overline{\mathcal{B}}_{S^{N-1}}}\right).$$

Here $\overline{\mathcal{B}_{[R_1, r]} \times \overline{\mathcal{B}}_{S^{N-1}}}$ stands for the complete σ-algebra generated by $\overline{\mathcal{B}}_{[R_1, r]} \times \overline{\mathcal{B}}_{S^{N-1}}$, where $\overline{\mathcal{B}}_X$ stands for the completion of $\mathcal{B}_X$.

Then we get that

$$\int_{S^{N-1}} \left(\int_{R_1}^{r} \left|\delta_r\left(t, \omega\right)\right| dt\right) d\omega =$$

$$\int_{S^{N-1}} \left(\int_{R_1}^{r} \left(r - t\right)^{\nu - \gamma - 1} \left|D_{R_1}^{\nu} F\left(t\omega\right)\right| dt\right) d\omega \leq \qquad (17.82)$$

$$\left\|\frac{\partial_{R_1}^{\nu} F\left(x\right)}{\partial r^{\nu}}\right\|_{\infty, \left(\left[R_1, r\right] \times S^{N-1}\right)} \left(\int_{S^{N-1}} \left(\int_{R_1}^{r} \left(r - t\right)^{\nu - \gamma - 1} dt\right) d\omega\right)$$

$$= \left\|\frac{\partial_{R_1}^{\nu} F\left(x\right)}{\partial r^{\nu}}\right\|_{\infty, \left(\left[R_1, r\right] \times S^{N-1}\right)} \left(\frac{2\pi^{N/2}}{\Gamma\left(N/2\right)}\right) \frac{\left(r - R_1\right)^{\nu - \gamma}}{\left(\nu - \gamma\right)} \leq \qquad (17.83)$$

$$\left\|\frac{\partial_{R_1}^{\nu} F\left(x\right)}{\partial r^{\nu}}\right\|_{\infty, A} \left(\frac{2\pi^{N/2}}{\Gamma\left(N/2\right)}\right) \frac{\left(R_2 - R_1\right)^{\nu - \gamma}}{\left(\nu - \gamma\right)} < \infty \qquad (17.84)$$

Hence $\delta_r\left(t, \omega\right)$ is integrable on

$$\left(\left[R_1, r\right] \times S^{N-1}, \overline{\mathcal{B}_{[R_1, r]} \times \overline{\mathcal{B}}_{S^{N-1}}}\right).$$

Consequently, by Fubini's theorem and (17.77), we obtain that $D_{R_1}^{\gamma} F\left(r\omega\right)$, $\nu \geq \gamma + 1$, $\gamma > 0$, is integrable in ω over $\left(S^{N-1}, \overline{\mathcal{B}}_{S^{N-1}}\right)$.

So we have that $D_{R_1}^\gamma F (r\omega)$ is continuous in $r \in [R_1, R_2]$, $\forall \omega \in S^{N-1}$, and measurable in $\omega \in S^{N-1}$, $\forall r \in [R_1, R_2]$. So, it is a Carathéodory function. Here $[R_1, R_2]$ is a separable metric space and S^{N-1} is a measurable space, and the function takes values in $\mathbb{R}^* := \mathbb{R} \cup \{\pm\infty\}$, which is a metric space. Therefore by Theorem 20.15, p. 156 of [10], $\left(D_{R_1}^\gamma F\right)(r\omega)$ is jointly $\left(\mathcal{B}_{[R_1,R_2]} \times \overline{\mathcal{B}}_{S^{N-1}}\right)$-measurable on $[R_1, R_2] \times S^{N-1} = A$, that is, Lebesgue measurable on A. Then we have that

$$\left| D_{R_1}^\gamma F (r\omega) \right| \le \frac{1}{\Gamma(\nu - \gamma)} \int_{R_1}^r (r-t)^{\nu-\gamma-1} \left| D_{R_1}^\nu F (t\omega) \right| dt \qquad (17.85)$$

$$\le \frac{\left\| D_{R_1}^\nu F (\cdot \omega) \right\|_{\infty, [R_1, R_2]}}{\Gamma(\nu - \gamma)} \left(\int_{R_1}^r (r-t)^{\nu-\gamma-1} dt \right) \le \qquad (17.86)$$

$$\frac{M_1}{\Gamma(\nu-\gamma)} \frac{(r-R_1)^{\nu-\gamma}}{(\nu-\gamma)} \le \frac{M_1}{\Gamma(\nu-\gamma+1)} (R_2 - R_1)^{\nu-\gamma} := \tau < \infty,$$

for all $\omega \in S^{N-1}$ and for all $r \in [R_1, R_2]$.

Hence we have shown that

$$\left| D_{R_1}^\gamma F (r\omega) \right| \le \tau < \infty, \quad \forall \omega \in S^{N-1} \text{ and } \forall r \in [R_1, R_2] \qquad (17.87)$$

and

$$\frac{\partial_{R_1}^\gamma F (x)}{\partial r^\gamma} \in L_\infty (A), \ \gamma > 0.$$

We have completed our proof. $\square$

Next we give Poincaré inequalities on the shell involving Riemann–Liouville radial fractional derivatives.

Theorem 17.49. *Let all terms and assumptions for f be as in Theorem 17.48, and $p, q > 1$: $1/p + 1/q = 1$.*

Then

$$(1) \quad \int_A \left| \frac{\partial_{R_1}^\gamma f (x)}{\partial r^\gamma} \right|^q dx \le \left(\frac{R_2}{R_1} \right)^{N-1}$$

$$\left[\frac{(R_2 - R_1)^{q(\nu-\gamma)}}{q(\nu-\gamma) (\Gamma(\nu-\gamma))^q (p(\nu-\gamma-1)+1)^{q/p}} \right] \int_A \left| \frac{\partial_{R_1}^\nu f (x)}{\partial r^\nu} \right|^q dx.$$

$$(17.88)$$

When $\gamma = 0$ we get

$$(2) \quad \int_A |f (x)|^q dx \le \left(\frac{R_2}{R_1} \right)^{N-1}$$

$$\left[\frac{(R_2 - R_1)^{q\nu}}{q\nu (\Gamma(\nu))^q (p(\nu-1)+1)^{q/p}} \right] \int_A \left| \frac{\partial_{R_1}^\nu f (x)}{\partial r^\nu} \right|^q dx. \qquad (17.89)$$

Next we apply (17.69) to a Caputo fractional PDE.

Theorem 17.50. *Let* $1 \leq \nu_1 < \nu_2 < \cdots < \nu_k$, $n_k := \lceil \nu_k \rceil$, $k \in \mathbb{N}$, $F : \overline{A} \to \mathbb{R}$ *with* $F \in L_1(A)$. *Assume that* $F(\cdot\omega) \in AC^{n_k}([R_1, R_2])$, $\forall \omega \in S^{N-1}$, *and that* $\partial_{*R_1}^{\nu_l} F(\cdot\omega)/\partial r^{\nu_l} \in L_\infty (R_1, R_2)$, $\forall \omega \in S^{N-1}$; $l = 1, \ldots, k$. *Suppose* $\partial_{*R_1}^{\nu_l} F(x)/\partial r^{\nu_l} \in L_\infty(A)$, $l = 1, \ldots, k$, *such that* $\exists M_l > 0$: $\left| D_{*R_1}^{\nu_l} F(r\omega) \right| \leq M_l$, $l = 1, \ldots, k$, *a.e.* $r \in [R_1, R_2]$, $\forall \omega \in S^{N-1}$. *Also suppose* $\partial^j F(R_1\omega)/\partial r^j = 0$, $j = 0, 1, \ldots, n_k - 1$, $\forall \omega \in S^{N-1}$. *Let also* B, C_l, $1/C_l \in L_\infty(A)$, $l = 1, \ldots, k$, *with all* $C_l \geq 0$.
The above-described F *fulfills*

$$\sum_{l=1}^{k} C_l(x) \left(\frac{\partial_{*R_1}^{\nu_l} F(x)}{\partial r^{\nu_l}} \right)^2 = B(x), \ \forall x \in \overline{A}. \tag{17.90}$$

Define

$$\delta := \frac{1}{2} \left(\frac{R_2}{R_1} \right)^{N-1} \max_{1 \leq l \leq k} \left(\frac{(R_2 - R_1)^{2\nu_l}}{\nu_l (\Gamma(\nu_l))^2 (2\nu_l - 1)} \right), \tag{17.91}$$

$$\rho := \max_{1 \leq l \leq k} \left\| \frac{1}{C_l} \right\|_{L_\infty(A)}. \tag{17.92}$$

Then

$$\|F\|_{L_2(A)} \leq \sqrt{\frac{\delta\rho}{k}} \left(\int_A B(x) \, dx \right)^{1/2}. \tag{17.93}$$

Proof. Set $n_l := \lceil \nu_l \rceil$, $l = 1, \ldots, k-1$. Clearly, $F(\cdot\omega) \in AC^{n_l}([R_1, R_2])$, $\forall \omega \in S^{N-1}$; $l = 1, \ldots, k-1$. Also

$$\frac{\partial^j F(R_1\omega)}{\partial r^j} = 0, \ j = 0, 1, \ldots, n_l - 1, \forall \omega \in S^{N-1},$$

as $n_l \leq n_k$, for all $l = 1, \ldots, k-1$.
So all assumptions of Theorem 17.44 are fulfilled for F and fractional orders ν_l, $l = 1, \ldots, k$. Thus by choosing $p = q = 2$ we apply (17.69), and we obtain for $l = 1, \ldots, k$ that

$$\int_A (F(x))^2 \, dx \leq \left(\frac{R_2}{R_1} \right)^{N-1}$$

$$\left[\frac{(R_2 - R_1)^{2\nu_l}}{2\nu_l (\Gamma(\nu_l))^2 (2\nu_l - 1)} \right] \int_A \left(\frac{\partial_{*R_1}^{\nu_l} F(x)}{\partial r^{\nu_l}} \right)^2 \, dx \tag{17.94}$$

$$\leq \delta \int_A (C_l(x))^{-1} (C_l(x)) \left(\frac{\partial_{*R_1}^{\nu_l} F(x)}{\partial r^{\nu_l}} \right)^2 \, dx$$

$$\leq \delta \left\| \frac{1}{C_l} \right\|_{L_\infty (A)} \int_A C_l(x) \left(\frac{\partial_{*R_1}^{\nu_l} F(x)}{\partial r^{\nu_l}} \right)^2 dx \qquad (17.95)$$

$$\leq \delta \rho \int_A C_l(x) \left(\frac{\partial_{*R_1}^{\nu_l} F(x)}{\partial r^{\nu_l}} \right)^2 dx. \qquad (17.96)$$

That is,

$$\int_A (F(x))^2 \, dx \leq \delta \rho \int_A C_l(x) \left(\frac{\partial_{*R_1}^{\nu_l} F(x)}{\partial r^{\nu_l}} \right)^2 dx, \qquad (17.97)$$

for all $l = 1, \ldots, k$.

Consequently, summing (17.97) over all possible l values gives

$$\int_A (F(x))^2 \, dx \leq \frac{\delta \rho}{k} \int_A \left(\sum_{l=1}^k C_l(x) \left(\frac{\partial_{*R_1}^{\nu_l} F(x)}{\partial r^{\nu_l}} \right)^2 \right) dx$$

$$= \frac{\delta \rho}{k} \int_A B(x) \, dx, \qquad (17.98)$$

proving the claim. $\square$

In the simpler case of an ordinary differential equation we apply (17.19) next.

Theorem 17.51. *Let* $1 \leq \nu_1 < \nu_2 < \ldots < \nu_k$, $n_k := \lceil \nu_k \rceil$, $k \in \mathbb{N}$. *Assume* $f \in AC^{n_k}([a,b])$ *such that* $f^{(j)}(a) = 0$, $j = 0, 1, \ldots, n_k - 1$, *and* $D_{*a}^{\nu_l} f \in L_\infty(a,b)$, *for all* $l = 1, \ldots, k$. *Let also* B, C_l, $1/C_l \in L_\infty(a,b)$, $l = 1, \ldots, k$, *with all* $C_l \geq 0$. *The above described* f *satisfies*

$$\sum_{l=1}^k C_l(x) \left(D_{*a}^{\nu_l} f(x) \right)^2 = B(x), \ \forall x \in [a, b]. \qquad (17.99)$$

Call

$$\delta^* := \frac{1}{2} \max_{1 \leq l \leq k} \left(\frac{(b-a)^{2\nu_l}}{\nu_l \left(\Gamma(\nu_l) \right)^2 (2\nu_l - 1)} \right), \qquad (17.100)$$

$$\rho^* := \max_{1 \leq l \leq k} \left\| \frac{1}{C_l} \right\|_{L_\infty(a,b)}. \qquad (17.101)$$

Then

$$\|f\|_{L_2(a,b)} \leq \sqrt{\frac{\delta^* \rho^*}{k}} \left(\int_a^b B(x) \, dx \right)^{1/2}. \qquad (17.102)$$

Proof. Set $n_l := \lceil \nu_l \rceil$, $l = 1, \ldots, k - 1$. Clearly here $f \in AC^{n_l}([a,b])$, $l = 1, \ldots, k - 1$. Also $f^{(j)}(a) = 0$, $j = 0, 1, \ldots, n_l - 1$, $l = 1, \ldots, k - 1$. So all assumptions of Theorem 17.15 are fulfilled for f and fractional orders ν_l, $l = 1, \ldots, k - 1$. Thus by choosing $p = q = 2$ we apply (17.19), for $l = 1, \ldots, k$, to find

$$\int_a^b (f(x))^2 \, dx \leq \left(\frac{(b-a)^{2\nu_l}}{2\nu_l \left(\Gamma(\nu_l) \right)^2 (2\nu_l - 1)} \right) \left(\int_a^b \left(D_{*a}^{\nu_l} f(x) \right)^2 \, dx \right) \tag{17.103}$$

$$\leq \delta^* \int_a^b (C_l(x))^{-1} (C_l(x)) \left(D_{*a}^{\nu_l} f(x) \right)^2 \, dx$$

$$\leq \delta^* \rho^* \int_a^b C_l(x) \left(D_{*a}^{\nu_l} f(x) \right)^2 \, dx. \tag{17.104}$$

That is,

$$\int_a^b (f(x))^2 \, dx \leq \delta^* \rho^* \int_a^b C_l(x) \left(D_{*a}^{\nu_l} f(x) \right)^2 \, dx, \tag{17.105}$$

for all $l = 1, \ldots, k$.

The last imply

$$\int_a^b (f(x))^2 \, dx \leq \frac{\delta^* \rho^*}{k} \int_a^b \left(\sum_{l=1}^{k} C_l(x) \left(D_{*a}^{\nu_l} f(x) \right)^2 \right) dx$$

$$= \frac{\delta^* \rho^*}{k} \int_a^b B(x) \, dx, \tag{17.106}$$

proving the claim. $\square$

One can give similar applications to ODEs and PDEs involving fractional derivatives of Canavati-type and Riemann–Liouville-type.

We finish this section with L_1 results on the shell.

We present for Canavati-type radial fractional derivatives

Theorem 17.52. *Let $\nu \geq \gamma + 1$, $\gamma \geq 0$, $n := [\nu]$, and $f \in C^n \left(\overline{A} \right)$ and there exists $\partial_{R_1}^{\nu} f(x) / \partial r^{\nu} \in C \left(\overline{A} \right)$, $x \in \overline{A}$, and A a spherical shell. Further assume $\partial^j f (R_1 \omega) / \partial r^j = 0$, $j = 0, 1, \ldots, n - 1$; $\forall \omega \in S^{N-1}$.*
Then

$$(1) \quad \int_A \left| \frac{\partial_{R_1}^{\gamma} f(x)}{\partial r^{\gamma}} \right| dx \leq \left(\frac{R_2}{R_1} \right)^{N-1}$$

$$\frac{(R_2 - R_1)^{\nu - \gamma}}{\Gamma(\nu - \gamma + 1)} \int_A \left| \frac{\partial_{R_1}^{\nu} f(x)}{\partial r^{\nu}} \right| dx. \tag{17.107}$$

When $\gamma = 0$ we get

$$(2) \quad \int_A |f(x)| \, dx \leq \left(\frac{R_2}{R_1} \right)^{N-1} \frac{(R_2 - R_1)^{\nu}}{\Gamma(\nu + 1)} \int_A \left| \frac{\partial_{R_1}^{\nu} f(x)}{\partial r^{\nu}} \right| dx. \tag{17.108}$$

Proof. Because $R_1 \leq r \leq R_2$, $N \geq 2$, we have that $R_2^{1-N} \leq r^{1-N} \leq R_1^{1-N}$. Hence

$$R_2^{1-N} \int_{R_1}^{R_2} r^{N-1} \left| D_{R_1}^\gamma f (r\omega) \right| dr \leq$$

$$\int_{R_1}^{R_2} r^{1-N} r^{N-1} \left| D_{R_1}^\gamma f (r\omega) \right| dr = \int_{R_1}^{R_2} \left| D_{R_1}^\gamma f (r\omega) \right| dr \qquad (17.109)$$

$$\overset{(17.39)}{\leq} \frac{(R_2 - R_1)^{\nu - \gamma}}{\Gamma (\nu - \gamma + 1)} \int_{R_1}^{R_2} r^{1-N} r^{N-1} \left| D_{R_1}^\nu f (r\omega) \right| dr \leq$$

$$\frac{R_1^{1-N} (R_2 - R_1)^{\nu - \gamma}}{\Gamma (\nu - \gamma + 1)} \int_{R_1}^{R_2} r^{N-1} \left| D_{R_1}^\nu f (r\omega) \right| dr, \qquad (17.110)$$

$\forall \omega \in S^{N-1}$. We have proven so far

$$\int_{R_1}^{R_2} \left| D_{R_1}^\gamma f (r\omega) \right| r^{N-1} dr \leq \left(\frac{R_2}{R_1} \right)^{N-1} \frac{(R_2 - R_1)^{\nu - \gamma}}{\Gamma (\nu - \gamma + 1)}$$

$$\int_{R_1}^{R_2} \left| D_{R_1}^\nu f (r\omega) \right| r^{N-1} dr, \; \forall \omega \in S^{N-1}. \qquad (17.111)$$

Thus by integration of (17.111) on S^{N-1} we find

$$\int_{S^{N-1}} \left(\int_{R_1}^{R_2} \left| D_{R_1}^\gamma f (r\omega) \right| r^{N-1} dr \right) d\omega \leq$$

$$\left(\left(\frac{R_2}{R_1} \right)^{N-1} \frac{(R_2 - R_1)^{\nu - \gamma}}{\Gamma (\nu - \gamma + 1)} \right) \left(\int_{S^{N-1}} \left(\int_{R_1}^{R_2} \left| D_{R_1}^\nu f (r\omega) \right| r^{N-1} dr \right) d\omega \right).$$
$$(17.112)$$

Using (17.48) on (17.112) we have proven the claim. $\square$

We present for Riemann–Liouville radial fractional derivatives

Theorem 17.53. *Let all terms and assumptions for f be as in Theorem 17.41; here $\beta > \alpha + 1$, $\alpha \geq 0$.*
Then

(1) $\displaystyle \int_A \left| \frac{\partial_{R_1}^\alpha f (x)}{\partial r^\alpha} \right| dx \leq \left(\frac{R_2}{R_1} \right)^{N-1} \frac{(R_2 - R_1)^{\beta - \alpha}}{\Gamma (\beta - \alpha + 1)} \int_A \left| \frac{\partial_{R_1}^\beta f (x)}{\partial r^\beta} \right| dx.$
$$(17.113)$$

When $\alpha = 0$ we get

(2) $\displaystyle \int_A \left| f (x) \right| dx \leq \left(\frac{R_2}{R_1} \right)^{N-1} \frac{(R_2 - R_1)^\beta}{\Gamma (\beta + 1)} \int_A \left| \frac{\partial_{R_1}^\beta f (x)}{\partial r^\beta} \right| dx. \qquad (17.114)$

Proof. Similar to Theorem 17.52, using (17.42). $\square$

We give for Caputo radial fractional derivatives

Theorem 17.54. *Let all terms and assumptions for f be as in Theorem 17.44.*
Then

(1)
$$\int_A \left| \frac{\partial^\gamma_{*R_1} f(x)}{\partial r^\gamma} \right| dx \leq \left(\frac{R_2}{R_1} \right)^{N-1} \frac{(R_2 - R_1)^{\nu-\gamma}}{\Gamma(\nu - \gamma + 1)} \int_A \left| \frac{\partial^\nu_{*R_1} f(x)}{\partial r^\nu} \right| dx.$$
$$(17.115)$$

When $\gamma = 0$ we obtain

(2)
$$\int_A |f(x)| \, dx \leq \left(\frac{R_2}{R_1} \right)^{N-1} \frac{(R_2 - R_1)^{\nu}}{\Gamma(\nu + 1)} \int_A \left| \frac{\partial^\nu_{*R_1} f(x)}{\partial r^\nu} \right| dx. \quad (17.116)$$

Proof. Similar to Theorem 17.52, using (17.44). $\square$

Going back to Riemann–Liouville radial fractional derivatives we have

Theorem 17.55. *Let all terms and assumptions for f be as in Theorem 17.48.*
Then

(1)
$$\int_A \left| \frac{\partial^\gamma_{R_1} f(x)}{\partial r^\gamma} \right| dx \leq \left(\frac{R_2}{R_1} \right)^{N-1} \frac{(R_2 - R_1)^{\nu-\gamma}}{\Gamma(\nu - \gamma + 1)} \int_A \left| \frac{\partial^\nu_{R_1} f(x)}{\partial r^\nu} \right| dx.$$
$$(17.117)$$

When $\gamma = 0$ we obtain

(2)
$$\int_A |f(x)| \, dx \leq \left(\frac{R_2}{R_1} \right)^{N-1} \frac{(R_2 - R_1)^{\nu}}{\Gamma(\nu + 1)} \int_A \left| \frac{\partial^\nu_{R_1} f(x)}{\partial r^\nu} \right| dx. \quad (17.118)$$

Proof. Similar to Theorem 17.52, using (17.46). $\square$

17.4   Fractional Mean Poincaré Inequalities

First we give Canavati-type fractional derivatives related results.

Theorem 17.56. *Let $\gamma \geq 0$ and $1+\gamma \leq \nu_1 < \nu_2 < \ldots < \nu_k$, $n_k := [\nu_k]$, $k \in \mathbb{N}$. Assume $f \in \bigcap_{l=1}^{k} C^{\nu_l}_{x_0}([a,b])$, $x_0 \in [a,b]$. Assume $f^{(i)}(x_0) = 0$,*

$i = 0, 1, \ldots, n_k - 1$; $p, q > 1$: $1/p + 1/q = 1$. *Call*

$$\Lambda_\gamma := \max_{1 \le l \le k} \left[\frac{(b - x_0)^{(\nu_l - \gamma)}}{(\Gamma(\nu_l - \gamma))(p(\nu_l - \gamma - 1) + 1)^{1/p}(q(\nu_l - \gamma))^{1/q}} \right].$$

(17.119)

Then

$$(1) \ \left\| D_{x_0}^\gamma f \right\|_{L_q(x_0, b)} \le \left(\frac{\Lambda_\gamma}{k} \right) \left(\sum_{l=1}^k \left\| D_{x_0}^{\nu_l} f \right\|_{L_q(x_0, b)} \right).$$

(17.120)

When $\gamma = 0$ we get

$$(2) \ \left\| f \right\|_{L_q(x_0, b)} \le \left(\frac{\Lambda_0}{k} \right) \left(\sum_{l=1}^k \left\| D_{x_0}^{\nu_l} f \right\|_{L_q(x_0, b)} \right),$$

(17.121)

where

$$\Lambda_0 := \max_{1 \le l \le k} \left[\frac{(b - x_0)^{\nu_l}}{(\Gamma(\nu_l))(p(\nu_l - 1) + 1)^{1/p}(q\nu_l)^{1/q}} \right].$$

(17.122)

Proof. Here we apply Theorem 17.2.
By (17.2) we obtain

$$\left\| D_{x_0}^\gamma f \right\|_{L_q(x_0, b)} \le \left[\frac{(b - x_0)^{(\nu_l - \gamma)}}{(\Gamma(\nu_l - \gamma))(p(\nu_l - \gamma - 1) + 1)^{1/p}(q(\nu_l - \gamma))^{1/q}} \right]$$

$$\left\| D_{x_0}^{\nu_l} f \right\|_{L_q(x_0, b)} \le \Lambda_\gamma \left\| D_{x_0}^{\nu_l} f \right\|_{L_q(x_0, b)}.$$

(17.123)

That is, for $l = 1, \ldots, k$ holds

$$\left\| D_{x_0}^\gamma f \right\|_{L_q(x_0, b)} \le \Lambda_\gamma \left\| D_{x_0}^{\nu_l} f \right\|_{L_q(x_0, b)}.$$

(17.124)

Consequently by addition of (17.124) we have

$$k \left\| D_{x_0}^\gamma f \right\|_{L_q(x_0, b)} \le \Lambda_\gamma \left(\sum_{l=1}^k \left\| D_{x_0}^{\nu_l} f \right\|_{L_q(x_0, b)} \right),$$

(17.125)

proving (17.120). $\quad \square$

We continue with Riemann–Liouville fractional derivatives related results.

Theorem 17.57. *Let $0 \le \alpha < \beta_1 < \beta_2 < \ldots < \beta_k$, $k \in \mathbb{N}$; $f \in L_1(a, x)$, a, $x \in \mathbb{R}$, have L_∞ fractional derivatives $D_a^{\beta_l} f$ in $[a, x]$, with*

$D_a^{\beta_l - k} f(a) = 0$ for $k = 1, \ldots, [\beta_l] + 1$; $l = 1, \ldots, k$. Let $p, q > 1$: $1/p + 1/q = 1$ with $p(\beta_l - \alpha - 1) + 1 > 0$, all $l = 1, \ldots, k$. Put

$$K_\alpha := \max_{1 \le l \le k} \left[\frac{(x-a)^{(\beta_l - \alpha)}}{(\Gamma(\beta_l - \alpha))(p(\beta_l - \alpha - 1) + 1)^{1/p}(q(\beta_l - \alpha))^{1/q}} \right],$$

(17.126)

and

$$K_0 := \max_{1 \le l \le k} \left[\frac{(x-a)^{\beta_l}}{(\Gamma(\beta_l))(p(\beta_l - 1) + 1)^{1/p}(q\beta_l)^{1/q}} \right].$$

(17.127)

Then

(1) $\quad \|D_a^\alpha f\|_{L_q(a,x)} \le \left(\dfrac{K_\alpha}{k} \right) \left(\displaystyle\sum_{l=1}^{k} \|D_a^{\beta_l} f\|_{L_q(a,x)} \right).$

(17.128)

When $\alpha = 0$ we obtain

(2) $\quad \|f\|_{L_q(a,x)} \le \dfrac{K_0}{k} \left(\displaystyle\sum_{l=1}^{k} \|D_a^{\beta_l} f\|_{L_q(a,x)} \right).$

(17.129)

Proof. Similar to Theorem 17.56, now using Lemma 17.5 and Theorem 17.6; see (17.11) there. $\square$

We continue with Caputo fractional derivative related results.

Theorem 17.58. *Let* $\gamma \ge 0$ *and* $1 + \gamma \le \nu_1 < \nu_2 < \ldots < \nu_k$, $n_k := \lceil \nu_k \rceil$, $k \in \mathbb{N}$. *Assume* $f \in AC^{n_k}([a,b])$ *such that* $f^{(k)}(a) = 0$, $k = 0, 1, \ldots, n_k - 1$, *and* $D_{*a}^{\nu_l} f \in L_\infty(a,b)$, $l = 1, \ldots, k$. *Let* $p, q > 1$: $1/p + 1/q = 1$. *Set*

$$M_\gamma := \max_{1 \le l \le k} \left\{ \frac{(b-a)^{(\nu_l - \gamma)}}{(\Gamma(\nu_l - \gamma))(p(\nu_l - \gamma - 1) + 1)^{(1/p)}(q(\nu_l - \gamma))^{1/q}} \right\},$$

(17.130)

and

$$M_0 := \max_{1 \le l \le k} \left\{ \frac{(b-a)^{\nu_l}}{(\Gamma(\nu_l))(p(\nu_l - 1) + 1)^{(1/p)}(q\nu_l)^{1/q}} \right\}.$$

(17.131)

Then

(1) $\quad \|D_{*a}^\gamma f\|_{L_q(a,b)} \le \left(\dfrac{M_\gamma}{k} \right) \left(\displaystyle\sum_{l=1}^{k} \|D_{*a}^{\nu_l} f\|_{L_q(a,b)} \right).$

(17.132)

When $\gamma = 0$ we obtain

$$(2) \quad \|f\|_{L_q(a,b)} \leq \left(\frac{M_0}{k}\right) \left(\sum_{l=1}^{k} \|D_{*a}^{\nu_l} f\|_{L_q(a,b)}\right). \qquad (17.133)$$

Proof. Similar to Theorem 17.56; based on (17.17). □

We also give a related average result again regarding Riemann–Liouville fractional derivatives.

Theorem 17.59. *Let $\gamma \geq 0$ and $1+\gamma \leq \nu_1 < \nu_2 < \ldots < \nu_k$, $n_k := \lceil \nu_k \rceil$, $k \in \mathbb{N}$. Assume $f \in AC^{n_k}([a,b])$ such that $f^{(k)}(a) = 0$, $k = 0, 1, \ldots, n_k - 1$. Assume $\exists D_a^{\nu_l} f(x) \in \mathbb{R}$, $\forall x \in [a,b]$, and $D_a^{\nu_l} f \in L_\infty(a,b)$, for all $l = 1, \ldots, k$. Let $p, q > 1 : 1/p + 1/q = 1$. Here M_γ, M_0 as in (17.130), (17.131), respectively. Then*

$$(1) \quad \|D_a^\gamma f\|_{L_q(a,b)} \leq \left(\frac{M_\gamma}{k}\right) \left(\sum_{l=1}^{k} \|D_a^{\nu_l} f\|_{L_q(a,b)}\right). \qquad (17.134)$$

When $\gamma = 0$ we have

$$(2) \quad \|f\|_{L_q(a,b)} \leq \left(\frac{M_0}{k}\right) \left(\sum_{l=1}^{k} \|D_a^{\nu_l} f\|_{L_q(a,b)}\right). \qquad (17.135)$$

Proof. Similar to Theorem 17.56; based on (17.18). □

Next we present converse fractional mean Poincaré inequalities. First we give Canavati-type fractional derivative related results.

Theorem 17.60. *Let $\gamma \geq 0$ and $1+\gamma \leq \nu_1 < \nu_2 < \ldots < \nu_k$, $n_k := [\nu_k]$, $k \in \mathbb{N}$. Assume $f \in \bigcap_{l=1}^{k} C_{x_0}^{\nu_l}([a,b])$, $x_0 \in [a,b]$. Assume $f^{(i)}(x_0) = 0$, $i = 0, 1, \ldots, n_k - 1$; $0 < p < 1$, $q < 0 : 1/p + 1/q = 1$. Further suppose each $D_{x_0}^{\nu_l} f$ is of fixed strict sign on $[x_0, b]$; all $l = 1, \ldots, k$. Call*

$$\theta_\gamma := \min_{1 \leq l \leq k} \left\{ \frac{(b-x_0)^{((\nu_l-\gamma)-2/q)}}{\Gamma(\nu_l - \gamma)(p(\nu_l - \gamma - 1) + 1)^{(1-1/q)}(-q(\nu_l - \gamma) + 2)^{(-1/q)}} \right\}, \qquad (17.136)$$

and

$$\theta_0 := \min_{1 \leq l \leq k} \left\{ \frac{(b-x_0)^{(\nu_l-2/q)}}{\Gamma(\nu_l)(p(\nu_l - 1) + 1)^{(1-1/q)}(-q\nu_l + 2)^{(-1/q)}} \right\}. \qquad (17.137)$$

Then

$$(1) \quad \left\| D_{x_0}^{\gamma} f \right\|_{L_{(-q)}(x_0,b)} \geq \frac{\theta_{\gamma}}{k} \left(\sum_{l=1}^{k} \left\| D_{x_0}^{\nu_l} f \right\|_{L_q(x_0,b)} \right). \tag{17.138}$$

When $\gamma = 0$ we get

$$(2) \quad \left\| f \right\|_{L_{(-q)}(x_0,b)} \geq \frac{\theta_0}{k} \left(\sum_{l=1}^{k} \left\| D_{x_0}^{\nu_l} f \right\|_{L_q(x_0,b)} \right). \tag{17.139}$$

Proof. By (17.25) we obtain

$$\left\| D_{x_0}^{\gamma} f \right\|_{L_{(-q)}(x_0,b)} \geq \frac{(b-x_0)^{((\nu_l-\gamma)-2/q)} \left\| D_{x_0}^{\nu_l} f \right\|_{L_q(x_0,b)}}{\Gamma(\nu_l - \gamma)(p(\nu_l - \gamma - 1) + 1)^{(1-1/q)} (-q(\nu_l - \gamma) + 2)^{(-1/q)}}$$

$$\geq \theta_{\gamma} \left\| D_{x_0}^{\nu_l} f \right\|_{L_q(x_0,b)}. \tag{17.140}$$

That is,

$$\left\| D_{x_0}^{\gamma} f \right\|_{L_{(-q)}(x_0,b)} \geq \theta_{\gamma} \left\| D_{x_0}^{\nu_l} f \right\|_{L_q(x_0,b)}; \tag{17.141}$$

for all $l = 1, \ldots, k$.

Consequently by addition of (17.141) we have

$$k \left\| D_{x_0}^{\gamma} f \right\|_{L_{(-q)}(x_0,b)} \geq \theta_{\gamma} \left(\sum_{l=1}^{k} \left\| D_{x_0}^{\nu_l} f \right\|_{L_q(x_0,b)} \right), \tag{17.142}$$

proving the claim. $\square$

We continue with Riemann–Liouville fractional derivatives related results.

Theorem 17.61. *Let $0 \leq \alpha < \beta_1 < \beta_2 < \ldots < \beta_k$, $k \in \mathbb{N}$; $f \in L_1(a,x)$, a, $x \in \mathbb{R}$, have L_{∞} fractional derivatives $D_a^{\beta_l} f$ in $[a,x]$, with $D_a^{\beta_l - k} f(a) = 0$ for $k = 1, \ldots, [\beta_l] + 1$; $l = 1, \ldots, k$. Let $0 < p < 1, q < 0$, such that $1/p + 1/q = 1$, with $p(\beta_l - \alpha - 1) + 1 > 0$, all $l = 1, \ldots, k$. Further assume that each $D_a^{\beta_l} f$ has fixed sign a.e. on $[a,x]$, and $1/D_a^{\beta_l} f \in L_{\infty}(a,x)$, for all $l = 1, \ldots, k$. Set*

$$\theta_{\alpha}^* := \min_{1 \leq l \leq k} \left\{ \frac{(x-a)^{((\beta_l - \alpha) - 2/q)}}{\Gamma(\beta_l - \alpha)(p(\beta_l - \alpha - 1) + 1)^{(1-1/q)} (-q(\beta_l - \alpha) + 2)^{(-1/q)}} \right\} \tag{17.143}$$

and

$$\theta_0^* := \min_{1 \leq l \leq k} \left\{ \frac{(x-a)^{(\beta_l - 2/q)}}{\Gamma(\beta_l)(p(\beta_l - 1) + 1)^{(1-1/q)} (-q\beta_l + 2)^{(-1/q)}} \right\}. \tag{17.144}$$

Then

$$(1) \ \|D_a^\alpha f\|_{L_{(-q)}(a,x)} \geq \frac{\theta_\alpha^*}{k} \left(\sum_{l=1}^{k} \|D_a^{\beta_l} f\|_{L_q(a,x)} \right). \tag{17.145}$$

When $\alpha = 0$ *we obtain*

$$(2) \ \|f\|_{L_{(-q)}(a,x)} \geq \frac{\theta_0^*}{k} \left(\sum_{l=1}^{k} \|D_a^{\beta_l} f\|_{L_q(a,x)} \right). \tag{17.146}$$

Proof. Similar to Theorem 17.60; based on (17.33). □

We continue with Caputo fractional derivatives related results.

Theorem 17.62. *Let* $\gamma \geq 0$ *and* $1+\gamma \leq \nu_1 < \nu_2 < \ldots < \nu_k$, $n_k := \lceil \nu_k \rceil$, $k \in \mathbb{N}$. *Assume* $f \in AC^{n_k}([a,b])$ *such that* $f^{(k)}(a) = 0$, $k = 0,1,\ldots,n_k-1$, *and* $D_{*a}^{\nu_l} f, 1/D_{*a}^{\nu_l} f \in L_\infty(a,b)$, $l = 1,\ldots,k$. *Suppose each* $D_{*a}^{\nu_l} f$ *is of fixed sign a.e. on* $[a,b]$; *all* $l = 1,\ldots,k$. *Let* $0 < p < 1$, $q < 0$, *such that* $1/p + 1/q = 1$. *Set*

$$A_\gamma := \min_{1 \leq l \leq k} \left\{ \frac{(b-a)^{((\nu_l-\gamma)-2/q)}}{\Gamma(\nu_l - \gamma)(p(\nu_l - \gamma - 1) + 1)^{(1-1/q)}(-q(\nu_l - \gamma) + 2)^{(-1/q)}} \right\}, \tag{17.147}$$

and

$$A_0 := \min_{1 \leq l \leq k} \left\{ \frac{(b-a)^{(\nu_l-2/q)}}{\Gamma(\nu_l)(p(\nu_l - 1) + 1)^{(1-1/q)}(-q\nu_l + 2)^{(-1/q)}} \right\}. \tag{17.148}$$

Then

$$(1) \ \|D_{*a}^\gamma f\|_{L_{(-q)}(a,b)} \geq \frac{A_\gamma}{k} \left(\sum_{l=1}^{k} \|D_{*a}^{\nu_l} f\|_{L_q(a,b)} \right). \tag{17.149}$$

When $\gamma = 0$ *we get*

$$(2) \ \|f\|_{L_{(-q)}(a,b)} \geq \frac{A_0}{k} \left(\sum_{l=1}^{k} \|D_{*a}^{\nu_l} f\|_{L_q(a,b)} \right). \tag{17.150}$$

Proof. Similar to Theorem 17.60; based on (17.35). □

We come back to Riemann–Liouville fractional derivatives next.

Theorem 17.63. *Let* $\gamma \geq 0$ *and* $1+\gamma \leq \nu_1 < \nu_2 < \ldots < \nu_k$, $n_k := \lceil \nu_k \rceil$, $k \in \mathbb{N}$. *Assume* $f \in AC^{n_k}([a,b])$ *such that* $f^{(k)}(a) = 0$, $k = 0,1,\ldots,n_k-1$.

Suppose $\exists D_a^{\nu_l} f(x) \in \mathbb{R}$, $\forall x \in [a,b]$, and $D_a^{\nu_l} f$, $1/D_a^{\nu_l} f \in L_\infty(a,b)$; $l = 1,\ldots,k$. Assume each $D_a^{\nu_l} f$ is of fixed sign a.e. on $[a,b]$, all $l = 1,\ldots,k$. Let $0 < p < 1$, $q < 0 : 1/p + 1/q = 1$. Here A_γ is as in (17.147), and A_0 is as in (17.148).
Then

$$(1) \quad \|D_a^\gamma f\|_{L_{(-q)}(a,b)} \geq \frac{A_\gamma}{k} \left(\sum_{l=1}^{k} \|D_a^{\nu_l} f\|_{L_q(a,b)} \right). \qquad (17.151)$$

When $\gamma = 0$ we get

$$(2) \quad \|f\|_{L_{(-q)}(a,b)} \geq \frac{A_0}{k} \left(\sum_{l=1}^{k} \|D_a^{\nu_l} f\|_{L_q(a,b)} \right). \qquad (17.152)$$

Proof. Similar to Theorem 17.60; based on (17.37). $\square$

17.5 Applications of Fractional Mean Poincaré Inequalities

Next we apply some of the results of Section 17.4 to the multivariate case of the spherical shell. All terminology and symbols are as in Section 17.3.

We present the following Canavati-type fractional radial mean Poincaré inequalities on the spherical shell.

Theorem 17.64. *Let $\gamma \geq 0$ and $1+\gamma \leq \nu_1 < \nu_2 < \ldots < \nu_k$, $n_k := \lceil \nu_k \rceil$, $k \in \mathbb{N}$. Let $f \in C^n(\overline{A})$; there exist $\partial_{R_1}^{\nu_l} f(x)/\partial r^{\nu_l} \in C(\overline{A})$, $x \in \overline{A}$, A and a spherical shell, for all $l = 1,\ldots,k$. Further assume that $\partial^j f(R_1\omega)/\partial r^j = 0$, $j = 0,1,\ldots,n_k - 1$; $\forall \omega \in S^{N-1}$. Let $p, q > 1 : 1/p + 1/q = 1$. Set*

$$\tilde{\Lambda}_\gamma := \left[\max_{1 \leq l \leq k} \left\{ \frac{(R_2 - R_1)^{(\nu_l - \gamma)}}{(q(\nu_l - \gamma))^{1/q}(\Gamma(\nu_l - \gamma))(p(\nu_l - \gamma - 1) + 1)^{1/p}} \right\} \right] \left(\frac{R_2}{R_1} \right)^{(N-1)/q}, \qquad (17.153)$$

and

$$\tilde{\Lambda}_0 := \left[\max_{1 \leq l \leq k} \left\{ \frac{(R_2 - R_1)^{\nu_l}}{(q\nu_l)^{1/q}(\Gamma(\nu_l))(p(\nu_l - 1) + 1)^{1/p}} \right\} \right] \left(\frac{R_2}{R_1} \right)^{((N-1)/q)}. \qquad (17.154)$$

Then

$$1) \quad \left\| \frac{\partial_{R_1}^\gamma f}{\partial r^\gamma} \right\|_{L_q(A)} \leq \frac{\tilde{\Lambda}_\gamma}{k} \left(\sum_{l=1}^{k} \left\| \frac{\partial_{R_1}^{\nu_l} f}{\partial r^{\nu_l}} \right\|_{L_q(A)} \right). \qquad (17.155)$$

When $\gamma = 0$ we get

$$2) \quad \|f\|_{L_q(A)} \leq \frac{\tilde{\Lambda}_0}{k} \left(\sum_{l=1}^{k} \left\| \frac{\partial_{R_1}^{\nu_l} f}{\partial r^{\nu_l}} \right\|_{L_q(A)} \right). \tag{17.156}$$

Proof. By (17.50) we get

$$\left\| \frac{\partial_{R_1}^{\gamma} f}{\partial r^{\gamma}} \right\|_{L_q(A)} \leq \tilde{\Lambda}_{\gamma} \left\| \frac{\partial_{R_1}^{\nu_l} f}{\partial r^{\nu_l}} \right\|_{L_q(A)}, \tag{17.157}$$

for all $l = 1, \ldots, k$.

Adding all (17.157) we derive (17.155).

Setting $\gamma = 0$ in (17.155) we obtain (17.156). $\square$

Next follow Riemann–Liouville fractional radial mean Poincaré inequalities on the spherical shell.

Theorem 17.65. *Let $0 \leq \alpha < \beta_1 < \beta_2 < \ldots < \beta_k$, $k \in \mathbb{N}$; $f \in L_1(A)$, $\partial_{R_1}^{\beta_k} f(x)/\partial r^{\beta_k} \in L_\infty(A)$. Further assume that $D_{R_1}^{\beta_l} f(r\omega)$ takes real values for almost all $r \in [R_1, R_2]$, for each $\omega \in S^{N-1}$, and for these $\left| D_{R_1}^{\beta_l} f(r\omega) \right| \leq M_l$, where $M_l > 0$, for all $l = 1, \ldots, k$. For each $\omega \in S^{N-1} - K(f)$, we assume that $f(\cdot\omega)$ have L_∞ fractional derivatives $D_{R_1}^{\beta_l} f(\cdot\omega)$ in $[R_1, R_2]$, and that*

$$D_{R_1}^{\beta_l - \rho} f(R_1\omega) = 0, \quad \rho = 1, \ldots, [\beta_l] + 1, \tag{17.158}$$

for all $l = 1, \ldots, k$. Let $p, q > 1 : 1/p + 1/q = 1$ with $p(\beta_l - \alpha - 1) + 1 > 0$, all $l = 1, \ldots, k$. Set

$$\tilde{K}_{\gamma} := \left(\frac{R_2}{R_1} \right)^{(N-1/q)}$$

$$\left\{ \max_{1 \leq l \leq k} \left\{ \frac{(R_2 - R_1)^{(\beta_l - \alpha)}}{(q(\beta_l - \alpha))^{1/q} (\Gamma(\beta_l - \alpha)) (p(\beta_l - \alpha - 1) + 1)^{1/p}} \right\} \right\}, \tag{17.159}$$

and

$$\tilde{K}_0 := \left(\frac{R_2}{R_1} \right)^{(N-1/q)} \left\{ \max_{1 \leq l \leq k} \left\{ \frac{(R_2 - R_1)^{\beta_l}}{(q\beta_l)^{1/q} (\Gamma(\beta_l)) (p(\beta_l - 1) + 1)^{1/p}} \right\} \right\}. \tag{17.160}$$

Then

$$(1) \quad \left\| \frac{\partial_{R_1}^{\alpha} f}{\partial r^{\alpha}} \right\|_{L_q(A)} \leq \frac{\tilde{K}_{\gamma}}{k} \left(\sum_{l=1}^{k} \left\| \frac{\partial_{R_1}^{\beta_l} f}{\partial r^{\beta_l}} \right\|_{L_q(A)} \right). \tag{17.161}$$

When $\alpha = 0$ we get

$$(2) \quad \|f\|_{L_q(A)} \leq \frac{\widetilde{K}_0}{k} \left(\sum_{l=1}^{k} \left\| \frac{\partial_{R_1}^{\beta_l} f}{\partial r^{\beta_l}} \right\|_{L_q(A)} \right). \qquad (17.162)$$

Proof. As in Theorem 17.64; based on Theorems 17.41 and 17.42. □

Next are Caputo fractional radial mean Poincaré inequalities on the spherical shell.

Theorem 17.66. *Let $\gamma \geq 0$ and $1+\gamma \leq \nu_1 < \nu_2 < \ldots < \nu_k$, $n_k := \lceil \nu_k \rceil$, $k \in \mathbb{N}$; $f : \overline{A} \to \mathbb{R}$ with $f \in L_1(A)$. Assume that $f(\cdot\omega) \in AC^{n_k}([R_1, R_2])$ for all $\omega \in S^{N-1}$, and that $\partial_{*R_1}^{\nu_l} f(\cdot\omega)/\partial r^{\nu_l} \in L_\infty(R_1, R_2)$ for all $\omega \in S^{N-1}$, $l = 1, \ldots, k$. Further assume that $\partial_{*R_1}^{\nu_l} F(x)/\partial r^{\nu_l} \in L_\infty(A)$, for all $l = 1, \ldots, k$. More precisely, for these $r \in [R_1, R_2]$, for each $\omega \in S^{N-1}$, for which $D_{*R_1}^{\nu_l} f(r\omega)$ takes real values, there exist $M_l > 0$ such that $\left| D_{*R_1}^{\nu_l} f(r\omega) \right| \leq M_l$, for all $l = 1, \ldots, k$. We suppose that $\partial^j f(R_1\omega)/\partial r^j = 0$, $j = 0, 1, \ldots, n_k - 1$, for every $\omega \in S^{N-1}$. Let $p, q > 1$ such that $1/p + 1/q = 1$.*
Here $\widetilde{\Lambda}_\gamma$ as in (17.153), and $\widetilde{\Lambda}_0$ as in (17.154). Then

$$(1) \quad \left\| \frac{\partial_{*R_1}^{\gamma} f}{\partial r^{\gamma}} \right\|_{L_q(A)} \leq \frac{\widetilde{\Lambda}_\gamma}{k} \left(\sum_{l=1}^{k} \left\| \frac{\partial_{*R_1}^{\nu_l} f}{\partial r^{\nu_l}} \right\|_{L_q(A)} \right). \qquad (17.163)$$

When $\gamma = 0$ we get

$$(2) \quad \|f\|_{L_q(A)} \leq \frac{\widetilde{\Lambda}_0}{k} \left(\sum_{l=1}^{k} \left\| \frac{\partial_{*R_1}^{\nu_l} f}{\partial r^{\nu_l}} \right\|_{L_q(A)} \right). \qquad (17.164)$$

Proof. Similar to Theorem 17.64, using Theorems 17.44 and 17.45. □

We finish this chapter by returning to Riemann–Liouville fractional radial mean Poincaré inequalities on the spherical shell.

Theorem 17.67. *Let $\gamma \geq 0$ and $1+\gamma \leq \nu_1 < \nu_2 < \ldots < \nu_k$, $n_k := \lceil \nu_k \rceil$, $k \in \mathbb{N}$; $f : \overline{A} \to \mathbb{R}$ with $f \in L_1(A)$. Assume that $f(\cdot\omega) \in AC^{n_k}([R_1, R_2])$ $\forall \omega \in S^{N-1}$, and that $\partial_{R_1}^{\nu_l} f(\cdot\omega)/\partial r^{\nu_l}$ is measurable on $[R_1, R_2]$, $\forall \omega \in S^{N-1}$; $l = 1, \ldots, k$. Also assume $\exists \partial_{R_1}^{\nu_l} f(r\omega)/\partial r^{\nu_l} \in \mathbb{R}$, $\forall r \in [R_1, R_2]$ and $\forall \omega \in S^{N-1}$, and $\partial_{R_1}^{\nu_l} f(x)/\partial r^{\nu_l}$ is measurable on $\overline{A}$; all $l = 1, \ldots, k$. Suppose $\exists M_l > 0 : \left| \partial_{R_1}^{\nu_l} f(r\omega)/\partial r^{\nu_l} \right| \leq M_l$, $\forall (r, \omega) \in [R_1, R_2] \times S^{N-1}$; all $l = 1, \ldots, k$. We suppose that $\partial^j f(R_1\omega)/\partial r^j = 0$, $j = 0, 1, \ldots, n_k - 1$,*

$\forall \omega \in S^{N-1}$. *Let* $p, q > 1 : 1/p + 1/q = 1$, $\widetilde{\Lambda}_\gamma$ *as in* (17.153), *and* $\widetilde{\Lambda}_0$ *as in* (17.154).

Then

$$(1) \quad \left\| \frac{\partial_{R_1}^\gamma f}{\partial r^\gamma} \right\|_{L_q(A)} \leq \frac{\widetilde{\Lambda}_\gamma}{k} \left(\sum_{l=1}^k \left\| \frac{\partial_{R_1}^{\nu_l} f}{\partial r^{\nu_l}} \right\|_{L_q(A)} \right). \qquad (17.165)$$

When $\gamma = 0$ *we obtain*

$$(2) \quad \|f\|_{L_q(A)} \leq \frac{\widetilde{\Lambda}_0}{k} \left(\sum_{l=1}^k \left\| \frac{\partial_{R_1}^{\nu_l} f}{\partial r^{\nu_l}} \right\|_{L_q(A)} \right). \qquad (17.166)$$

Proof. Similar to Theorem 17.64, using Theorems 17.48 and 17.49. $\square$

18

Various Sobolev-Type Fractional Inequalities

Here we present various univariate Sobolev-type fractional inequalities involving fractional derivatives of Canavati, Riemann–Liouville, and Caputo types. The results are general L_p inequalities forward and converse on a closed interval. We give an application to a fractional ODE. We present also the mean Sobolev-type fractional inequalities. This treatment relies on [56].

18.1 Introduction

This chapter is motivated by the famous Sobolev-type inequality (see [159, p. 263]), the Gagliardo–Nirenberg–Sobolev inequality: *Assume $1 \leq p < n$. Then there exists a constant C, depending only on p and n, such that*

$$\|u\|_{L^{p^*}(\mathbb{R}^n)} \leq C \|Du\|_{L^p(\mathbb{R}^n)},$$

for all $u \in C_c^1(\mathbb{R}^n)$.

Here $p^* := np/n - p$, $p^* > p$, and Du is the gradient of u.

Also we are motivated by the following result (p. 265, [159]): estimates for Sobolev space $W^{1,p}$, $1 \leq p < n$: *let U be a bounded open subset of $\mathbb{R}^n$, and suppose the boundary ∂U is C^1. Assume $1 \leq p < n$, and $u \in W^{1,p}(U)$. Then $u \in L^{p^*}(U)$, with the estimate*

$$\|u\|_{L^{p^*}(U)} \leq C \|u\|_{W^{1,p}(U)},$$

the constant C depending only on p, n, and U.

So here we derive general fractional Sobolev-type inequalities on a closed interval of the real line.

As a typical case we mention our results that involve the Caputo fractional derivative. Please see the following results of Theorem 18.21, Corollary 18.22, Theorem 18.23, and Theorem 18.24, the converse inequality in Theorem 18.33, and the application in Theorem 18.39. We also produce corresponding results for the Canavati-type and Riemann–Liouville-type fractional derivatives.

18.2   Various Univariate Sobolev-Type Fractional Inequalities

Let $[a, b] \subseteq \mathbb{R}$ here; see [19, 101]. Let x, $x_0 \in [a, b]$ such that $x \geq x_0$, x_0 is fixed, $f \in C([a, b])$, and define

$$(J_\nu^{x_0} f)(x) := \frac{1}{\Gamma(\nu)} \int_{x_0}^x (x - t)^{\nu - 1} f(t) \, dt,$$

all $x_0 \leq x \leq b$, where $\nu > 0$, $n := [\nu]$ (integral part), $\alpha := \nu - n$, the generalized Riemann–Liouville integral, with Γ the gamma function.

We consider

$$C_{x_0}^\nu([a, b]) := \left\{ f \in C^n([a, b]) : J_{1-\alpha}^{x_0} f^{(n)} \in C^1([x_0, b]) \right\}.$$

Let $f \in C_{x_0}^\nu([a, b])$; we define the generalized ν-fractional derivative of f over $[x_0, b]$ as

$$D_{x_0}^\nu f := \left(J_{1-\alpha}^{x_0} f^{(n)} \right)'.$$

Clearly $D_{x_0}^\nu f \in C([x_0, b])$.

We need

Lemma 18.1. ([19, pp. 544–545]) *Let* $\nu \geq \gamma + 1$, $\gamma \geq 0$, $n := [\nu]$, *and* $f \in C_{x_0}^\nu([a, b])$, $x_0 \in [a, b]$. *Assume* $f^{(i)}(x_0) = 0$, $i = 0, 1, \ldots, n - 1$. *Then*

$$(D_{x_0}^\gamma f)(x) = \frac{1}{\Gamma(\nu - \gamma)} \int_{x_0}^x (x - t)^{\nu - \gamma - 1} (D_{x_0}^\nu f)(t) \, dt, \qquad (18.1)$$

all $x \in [x_0, b]$.

We give Sobolev-type inequalities with respect to the above-defined Canavati ([19, 101]) type fractional derivative.

Theorem 18.2. *Let* $\nu \geq \gamma + 1$, $\gamma \geq 0$, $n := [\nu]$, *and* $f \in C^{\nu}_{x_0}([a,b])$, $x_0 \in [a,b]$. *Assume* $f^{(i)}(x_0) = 0$, $i = 0,1,\ldots,n-1$. *Let* $r \geq 1$; $p, q > 1$: $1/p + 1/q = 1$. *Then*

$$\left\| D^{\gamma}_{x_0} f \right\|_{L_r(x_0,b)} \leq$$

$$\frac{(b-x_0)^{\nu-\gamma+1/r-1/q} \left\| D^{\nu}_{x_0} f \right\|_{L_q(x_0,b)}}{(\Gamma(\nu-\gamma))(p(\nu-\gamma-1)+1)^{1/p} \left[r\left(\nu-\gamma-\frac{1}{q}\right)+1 \right]^{1/r}}. \tag{18.2}$$

Proof. We have

$$\left| D^{\gamma}_{x_0} f(x) \right| \leq \frac{1}{\Gamma(\nu-\gamma)} \int_{x_0}^{x} (x-t)^{\nu-\gamma-1} \left| D^{\nu}_{x_0} f(t) \right| dt \tag{18.3}$$

(by the Hölder inequality)

$$\leq \frac{1}{\Gamma(\nu-\gamma)} \left(\int_{x_0}^{x} (x-t)^{p(\nu-\gamma-1)} dt \right)^{1/p}$$

$$\left(\int_{x_0}^{x} \left| D^{\nu}_{x_0} f(t) \right|^q dt \right)^{1/q} = \tag{18.4}$$

$$\frac{1}{\Gamma(\nu-\gamma)} \frac{(x-x_0)^{(\nu-\gamma-1)+1/p}}{(p(\nu-\gamma-1)+1)^{1/p}}$$

$$\left(\int_{x_0}^{x} \left| D^{\nu}_{x_0} f(t) \right|^q dt \right)^{1/q}$$

$$\leq \frac{1}{\Gamma(\nu-\gamma)} \frac{(x-x_0)^{(\nu-\gamma-1)+1/p}}{(p(\nu-\gamma-1)+1)^{1/p}}$$

$$\left(\int_{x_0}^{b} \left| D^{\nu}_{x_0} f(t) \right|^q dt \right)^{1/q}. \tag{18.5}$$

That is,

$$\left| D^{\gamma}_{x_0} f(x) \right| \leq \left[\frac{(x-x_0)^{(\nu-\gamma-1)+1/p}}{\Gamma(\nu-\gamma)(p(\nu-\gamma-1)+1)^{1/p}} \right]$$

$$\left(\int_{x_0}^{b} \left| D^{\nu}_{x_0} f(t) \right|^q dt \right)^{1/q}, \tag{18.6}$$

$\forall x \in [x_0, b]$.

That is,

$$\left|D_{x_0}^{\gamma} f(x)\right| \le \frac{(x-x_0)^{(\nu-\gamma-1/q)}}{\Gamma(\nu-\gamma)(p(\nu-\gamma-1)+1)^{1/p}}\left\|D_{x_0}^{\nu} f\right\|_{L_q(x_0,b)}, \quad (18.7)$$

$\forall x \in [x_0, b]$.
Hence, by $r \ge 1$, we obtain

$$\left|D_{x_0}^{\gamma} f(x)\right|^r \le \frac{(x-x_0)^{r(\nu-\gamma-1/q)}}{(\Gamma(\nu-\gamma))^r (p(\nu-\gamma-1)+1)^{r/p}}\left\|D_{x_0}^{\nu} f\right\|_{L_q(x_0,b)}^r, \quad (18.8)$$

$\forall x \in [x_0, b]$.
Consequently it holds

$$\int_{x_0}^{b}\left|D_{x_0}^{\gamma} f(x)\right|^r dx \le$$

$$\frac{(b-x_0)^{r(\nu-\gamma-1/q)+1}\left\|D_{x_0}^{\nu} f\right\|_{L_q(x_0,b)}^r}{(\Gamma(\nu-\gamma))^r (p(\nu-\gamma-1)+1)^{r/p}\left[r\left(\nu-\gamma-\frac{1}{q}\right)+1\right]}. \quad (18.9)$$

So that

$$\left\|D_{x_0}^{\gamma} f\right\|_{L_r(x_0,b)} \le$$

$$\frac{(b-x_0)^{\nu-\gamma-1/q+1/r}\left\|D_{x_0}^{\nu} f\right\|_{L_q(x_0,b)}}{(\Gamma(\nu-\gamma))(p(\nu-\gamma-1)+1)^{1/p}\left[r\left(\nu-\gamma-\frac{1}{q}\right)+1\right]^{1/r}}, \quad (18.10)$$

proving the claim. □

Corollary 18.3. (to Theorem 18.2) *When* $\gamma = 0$ *it holds*

$$\|f\|_{L_r(x_0,b)} \le \frac{(b-x_0)^{\nu+1/r-1/q}\left\|D_{x_0}^{\nu} f\right\|_{L_q(x_0,b)}}{(\Gamma(\nu))(p(\nu-1)+1)^{1/p}\left[r\left(\nu-\frac{1}{q}\right)+1\right]^{1/r}}. \quad (18.11)$$

Next we give the corresponding L_1 result.

Theorem 18.4. *Let* $\nu \ge \gamma+1$, $\gamma \ge 0$, $n := [\nu]$, *and* $f \in C_{x_0}^{\nu}([a,b])$, $x_0 \in [a,b]$. *Assume* $f^{(i)}(x_0) = 0$, $i = 0, 1, \ldots, n-1$; $r \ge 1$. *Then*

(1) $\left\|D_{x_0}^{\gamma} f\right\|_{L_r(x_0,b)} \le \dfrac{(b-x_0)^{\nu-\gamma-1+1/r}\left\|D_{x_0}^{\nu} f\right\|_{L_1(x_0,b)}}{(\Gamma(\nu-\gamma))[r(\nu-\gamma-1)+1]^{1/r}}. \quad (18.12)$

When $\gamma = 0$ *we get*

(2) $\|f\|_{L_r(x_0,b)} \le \dfrac{(b-x_0)^{\nu-1+1/r}\left\|D_{x_0}^{\nu} f\right\|_{L_1(x_0,b)}}{(\Gamma(\nu))[r(\nu-1)+1]^{1/r}}. \quad (18.13)$

Proof. By (18.1) we have

$$\left| D_{x_0}^{\gamma} f\left(x\right) \right| \leq \frac{1}{\Gamma\left(\nu - \gamma\right)} \int_{x_0}^{x} \left(x - t\right)^{\nu - \gamma - 1} \left| D_{x_0}^{\nu} f\left(t\right) \right| dt$$

$$\leq \frac{\left(x - x_0\right)^{\nu - \gamma - 1}}{\Gamma\left(\nu - \gamma\right)} \int_{x_0}^{b} \left| D_{x_0}^{\nu} f\left(t\right) \right| dt, \qquad (18.14)$$

$\forall x \in [x_0, b]$.

That is,

$$\left| D_{x_0}^{\gamma} f\left(x\right) \right| \leq \frac{\left(x - x_0\right)^{\nu - \gamma - 1}}{\Gamma\left(\nu - \gamma\right)} \left\| D_{x_0}^{\nu} f \right\|_{L_1(x_0, b)}, \qquad (18.15)$$

$\forall x \in [x_0, b]$.

Consequently

$$\left| D_{x_0}^{\gamma} f\left(x\right) \right|^r \leq \frac{\left(x - x_0\right)^{r(\nu - \gamma - 1)}}{\left(\Gamma\left(\nu - \gamma\right)\right)^r} \left\| D_{x_0}^{\nu} f \right\|_{L_1(x_0, b]}^r, \qquad (18.16)$$

$\forall x \in [x_0, b]$ and

$$\int_{x_0}^{b} \left| D_{x_0}^{\gamma} f\left(x\right) \right|^r dx \leq \frac{\left(b - x_0\right)^{r(\nu - \gamma - 1) + 1} \left\| D_{x_0}^{\nu} f \right\|_{L_1(x_0, b)}^r}{\left(\Gamma\left(\nu - \gamma\right)\right)^r \left(r\left(\nu - \gamma - 1\right) + 1\right)}, \qquad (18.17)$$

proving the claim. □

We continue with the following mean Sobolev-type inequalities result.

Theorem 18.5. *Let* $\gamma \geq 0$ *and* $1 + \gamma \leq \nu_1 < \nu_2 < \ldots < \nu_k$, $n_k := [\nu_k]$, $k \in \mathbb{N}$. *Assume* $f \in \bigcap_{l=1}^{k} C_{x_0}^{\nu_l}\left([a, b]\right)$, $x_0 \in [a, b]$. *Suppose* $f^{(i)}\left(x_0\right) = 0$, $i = 0, 1, \ldots, n_k - 1$. *Let* $r \geq 1$; $p, q > 1 : 1/p + 1/q = 1$.
Call

$$\Lambda_{\gamma} := \max_{1 \leq l \leq k}$$

$$\left[\frac{\left(b - x_0\right)^{\nu_l - \gamma + 1/r - 1/q}}{\left(\Gamma\left(\nu_l - \gamma\right)\right) \left[p\left(\nu_l - \gamma - 1\right) + 1\right]^{1/p} \left[r\left(\nu_l - \gamma - \frac{1}{q}\right) + 1\right]^{1/r}} \right]. \qquad (18.18)$$

Then

(1) $\left\| D_{x_0}^{\gamma} f \right\|_{L_r(x_0, b)} \leq \left(\dfrac{\Lambda_{\gamma}}{k}\right) \left(\displaystyle\sum_{l=1}^{k} \left\| D_{x_0}^{\nu_l} f \right\|_{L_q(x_0, b)}\right). \qquad (18.19)$

When $\gamma = 0$ *we find*

(2) $\left\| f \right\|_{L_r(x_0, b)} \leq \left(\dfrac{\Lambda_0}{k}\right) \left(\displaystyle\sum_{l=1}^{k} \left\| D_{x_0}^{\nu_l} f \right\|_{L_q(x_0, b)}\right), \qquad (18.20)$

where

$$\Lambda_0 := \max_{1 \le l \le k}$$

$$\left[\frac{(b - x_0)^{\nu_l + 1/r - 1/q}}{(\Gamma(\nu_l)) [p(\nu_l - 1) + 1]^{1/p} \left[r\left(\nu_l - \frac{1}{q}\right) + 1 \right]^{1/r}} \right]. \tag{18.21}$$

Proof. Easy; based on Lemma 18.1, Theorem 18.2, and Corollary 18.3. $\square$

The corresponding L_1 result follows.

Theorem 18.6. *Let* $\gamma \ge 0$ *and* $1 + \gamma \le \nu_1 < \nu_2 < \ldots < \nu_k$, $n_k := [\nu_k]$, $k \in \mathbb{N}$. *Assume* $f \in \bigcap_{l=1}^{k} C_{x_0}^{\nu_l}([a, b])$ *and* $x_0 \in [a, b]$. *Suppose* $f^{(i)}(x_0) = 0$, $i = 0, 1, \ldots, n_k - 1$; $r \ge 1$.
Call

$$\Lambda_\gamma^* := \max_{1 \le l \le k} \left[\frac{(b - x_0)^{(\nu_l - \gamma - 1 + 1/r)}}{(\Gamma(\nu_l - \gamma)) [r(\nu_l - \gamma - 1) + 1]^{1/r}} \right], \tag{18.22}$$

and

$$\Lambda_0^* := \max_{1 \le l \le k} \left[\frac{(b - x_0)^{(\nu_l - 1 + 1/r)}}{(\Gamma(\nu_l)) [r(\nu_l - 1) + 1]^{1/r}} \right]. \tag{18.23}$$

Then

$$(1) \ \left\| D_{x_0}^\gamma f \right\|_{L_r(x_0, b)} \le \left(\frac{\Lambda_\gamma^*}{k} \right) \left(\sum_{l=1}^{k} \left\| D_{x_0}^{\nu_l} f \right\|_{L_1(x_0, b)} \right). \tag{18.24}$$

When $\gamma = 0$ *we obtain*

$$(2) \ \|f\|_{L_r(x_0, b)} \le \left(\frac{\Lambda_0^*}{k} \right) \left(\sum_{l=1}^{k} \left\| D_{x_0}^{\nu_l} f \right\|_{L_1(x_0, b)} \right). \tag{18.25}$$

Proof. Easy; based on Lemma 18.1 and Theorem 18.4. $\square$

Let $\alpha > 0$, $f \in L_1(a, x)$, $a, x \in \mathbb{R}$; see [45, 64, 134].
We define the generalized Riemann–Liouville fractional derivative of f of order α by

$$D_a^\alpha f(s) := \frac{1}{\Gamma(m - \alpha)} \left(\frac{d}{ds} \right)^m \int_a^s (s - t)^{m - \alpha - 1} f(t) \, dt,$$

where $m := [\alpha] + 1$, $s \in [a, x]$; see also [57, Remark 46].

In addition, we set

$$D_a^0 f := f := J_0^a f,$$

$$J_{-\alpha}^a f := D_a^\alpha f, \text{ if } \alpha > 0,$$

$$D_a^{-\alpha} f := J_\alpha^a f, \text{ if } 0 < \alpha \le 1,$$

$$D_a^n f = f^{(n)}, \text{ for } n \in \mathbb{N}.$$

We need

Definition 18.7. [45] We say that $f \in L_1(a,x)$ has an L_∞ fractional derivative $D_a^\alpha f$ ($\alpha > 0$) in $[a,x]$, a, $x \in \mathbb{R}$, iff $D_a^{\alpha-k} f \in C([a,x])$, $k = 1, \ldots, m := [\alpha] + 1$, and $D_a^{\alpha-1} f \in AC([a,x])$ (absolutely continuous functions) and $D_a^\alpha f \in L_\infty(a,x)$.

Lemma 18.8. [45] *Let* $\beta > \alpha \ge 0$, $f \in L_1(a,x)$, a, $x \in \mathbb{R}$, *have an* L_∞ *fractional derivative* $D_a^\beta f$ *in* $[a,x]$, *and let* $D_a^{\beta-k} f(a) = 0$ *for* $k = 1, \ldots, [\beta] + 1$. *Then*

$$D_a^\alpha f(s) = \frac{1}{\Gamma(\beta - \alpha)} \int_a^s (s-t)^{\beta-\alpha-1} D_a^\beta f(t)\, dt, \qquad (18.26)$$

$\forall s \in [a,x]$.
   Here $D_a^\alpha f \in AC([a,x])$ *for* $\beta - \alpha \ge 1$, *and* $D_a^\alpha f \in C([a,x])$ *for* $\beta - \alpha \in (0,1)$.

We present Sobolev-type inequalities with respect to the above-defined generalized Riemann–Liouville fractional derivative.

Theorem 18.9. *Let* $\beta > \alpha \ge 0$, $f \in L_1(a,x)$, a, $x \in \mathbb{R}$, *have an* L_∞ *fractional derivative* $D_a^\beta f$ *in* $[a,x]$, *and let* $D_a^{\beta-k} f(a) = 0$, *for* $k = 1, \ldots, [\beta]+1$. *Let* $r \ge 1$; $p, q > 1 : 1/p + 1/q = 1$ *with* $p(\beta - \alpha - 1) + 1 > 0$. *Then*

$$\|D_a^\alpha f\|_{L_r(a,x)} \le \frac{(x-a)^{\beta-\alpha+1/r-1/q}}{(\Gamma(\beta-\alpha))(p(\beta-\alpha-1)+1)^{1/p}}$$

$$\frac{\|D_a^\beta f\|_{L_q(a,x)}}{\left[r\left(\beta - \alpha - \frac{1}{q}\right) + 1 \right]^{1/r}}. \qquad (18.27)$$

Proof. Similar to Theorem 18.2; now based on Lemma 18.8. □

Corollary 18.10. (to Theorem 18.9) *When* $\alpha = 0$ *it holds*

$$\|f\|_{L_r(a,x)} \le \frac{(x-a)^{\beta+1/r-1/q} \|D_a^\beta f\|_{L_q(a,x)}}{(\Gamma(\beta))(p(\beta-1)+1)^{1/p} \left[r\left(\beta - \frac{1}{q}\right) + 1 \right]^{1/r}}. \qquad (18.28)$$

Next we give the corresponding L_1 result.

Theorem 18.11. *Let $\beta > \alpha + 1$, $\alpha \geq 0$, $f \in L_1(a,x)$, a, $x \in \mathbb{R}$, have an L_∞ fractional derivative $D_a^\beta f$ in $[a,x]$ and let $D_a^{\beta-k} f(a) = 0$, for $k = 1, \ldots, [\beta] + 1$; $r \geq 1$. Then*

(1) $\|D_a^\alpha f\|_{L_r(a,x)} \leq \dfrac{(x-a)^{(\beta-\alpha-1+1/r)} \|D_a^\beta f\|_{L_1(a,x)}}{(\Gamma(\beta-\alpha)) [r(\beta-\alpha-1)+1]^{1/r}}.$ (18.29)

When $\alpha = 0$ we find

(2) $\|f\|_{L_r(a,x)} \leq \dfrac{(x-a)^{(\beta-1+1/r)} \|D_a^\beta f\|_{L_1(a,x)}}{(\Gamma(\beta)) [r(\beta-1)+1]^{1/r}}.$ (18.30)

Proof. As in Theorem 18.4; now based on Lemma 18.8. $\square$

We continue with mean Sobolev-type inequalities.

Theorem 18.12. *Let $0 \leq \alpha < \beta_1 < \beta_2 < \ldots < \beta_k$, $k \in \mathbb{N}$; $f \in L_1(a,x)$, a, $x \in \mathbb{R}$, have L_∞ fractional derivatives $D_a^{\beta_l} f$ in $[a,x]$, with $D_a^{\beta_l-k} f(a) = 0$ for $k = 1, \ldots, [\beta_l] + 1$; $l = 1, \ldots, k$. Let $p, q > 1 : 1/p + 1/q = 1$ with $p(\beta_l - \alpha - 1) + 1 > 0$, all $l = 1, \ldots, k$; $r \geq 1$. Put*

$$K_\alpha := \max_{1 \leq l \leq k}$$

$$\left[\frac{(x-a)^{\beta_l - \alpha + 1/r - 1/q}}{(\Gamma(\beta_l - \alpha)) (p(\beta_l - \alpha - 1)+1)^{1/p} \left[r\left(\beta_l - \alpha - \frac{1}{q}\right)+1\right]^{1/r}} \right],$$ (18.31)

and

$$K_0 := \max_{1 \leq l \leq k} \left[\frac{(x-a)^{\beta_l + 1/r - 1/q}}{(\Gamma(\beta_l)) (p(\beta_l - 1)+1)^{1/p} \left(r\left(\beta_l - \frac{1}{q}\right)+1\right)^{1/r}} \right].$$ (18.32)

Then

(1) $\|D_a^\alpha f\|_{L_r(a,x)} \leq \left(\dfrac{K_\alpha}{k}\right) \left(\displaystyle\sum_{l=1}^{k} \|D_a^{\beta_l} f\|_{L_q(a,x)}\right).$ (18.33)

When $\alpha = 0$ we get

(2) $\|f\|_{L_r(a,x)} \leq \left(\dfrac{K_0}{k}\right) \left(\displaystyle\sum_{l=1}^{k} \|D_a^{\beta_l} f\|_{L_q(a,x)}\right).$ (18.34)

Proof. Based on Theorem 18.9 and Corollary 18.10. □

The corresponding average L_1 result follows.

Theorem 18.13. *Let* $\alpha \geq 0$ *and* $\alpha + 1 < \beta_1 < \beta_2 < \ldots < \beta_k$, $k \in \mathbb{N}$; $f \in L_1(a, x)$, a, $x \in \mathbb{R}$, *have* L_∞ *fractional derivatives* $D_a^{\beta_l} f$ *in* $[a, x]$, *with* $D_a^{\beta_l - k} f(a) = 0$ *for* $k = 1, \ldots, [\beta_l] + 1$; $l = 1, \ldots, k$, *and* $r \geq 1$.
Set

$$K_\alpha^* := \max_{1 \leq l \leq k} \left[\frac{(x-a)^{(\beta_l - \alpha - 1 + 1/r)}}{(\Gamma(\beta_l - \alpha))(r(\beta_l - \alpha - 1) + 1)^{1/r}} \right], \tag{18.35}$$

$$K_0^* := \max_{1 \leq l \leq k} \left[\frac{(x-a)^{(\beta_l - 1 + 1/r)}}{(\Gamma(\beta_l))(r(\beta_l - 1) + 1)^{1/r}} \right]. \tag{18.36}$$

Then

(1) $\quad \|D_a^\alpha f\|_{L_r(a,x)} \leq \left(\dfrac{K_\alpha^*}{k} \right) \left(\displaystyle\sum_{l=1}^{k} \|D_a^{\beta_l} f\|_{L_1(a,x)} \right).$ $\tag{18.37}$

When $\alpha = 0$ *we get*

(2) $\quad \|f\|_{L_r(a,x)} \leq \left(\dfrac{K_0^*}{k} \right) \left(\displaystyle\sum_{l=1}^{k} \|D_a^{\beta_l} f\|_{L_1(a,x)} \right).$ $\tag{18.38}$

Proof. Based on Theorem 18.11. □

Next we produce Sobolev-type inequalities with respect to the Caputo fractional derivative.
We mention

Definition 18.14. [134] Let $\nu \geq 0$, $n := \lceil \nu \rceil$; $\lceil \cdot \rceil$ is the ceiling of the number $f \in AC^n([a,b])$. We call the Caputo fractional derivative

$$D_{*a}^\nu f(x) := \frac{1}{\Gamma(n-\nu)} \int_a^x (x-t)^{n-\nu-1} f^{(n)}(t)\, dt, \tag{18.39}$$

$\forall x \in [a, b]$.
The above function $D_{*a}^\nu f(x)$ exists almost everywhere for $x \in [a, b]$.

We need

Proposition 18.15. [134] *Let* $\nu \geq 0$, $n := \lceil \nu \rceil$ *and* $f \in AC^n([a,b])$. *Then* $D_{*a}^\nu f$ *exists iff the generalized Riemann–Liouville fractional derivative* $D_a^\nu f$ *exists.*

Proposition 18.16. [134] *Let $\nu \geq 0$, $n := \lceil \nu \rceil$. Assume that f is such that both $D_{*a}^\nu f$ and $D_a^\nu f$ exist. Suppose that $f^{(k)}(a) = 0$ for $k = 0, 1, \ldots, n - 1$.*

Then

$$D_{*a}^\nu f = D_a^\nu f.$$

In conclusion

Corollary 18.17. [58] *Let $\nu \geq 0$, $n := \lceil \nu \rceil$, $f \in AC^n([a,b])$, $D_{*a}^\nu f$ exists or $D_a^\nu f$ exists, and $f^{(k)}(a) = 0$, $k = 0, 1, \ldots, n - 1$. Then*

$$D_a^\nu f = D_{*a}^\nu f.$$

We need

Theorem 18.18. [58] *Let $\nu \geq 0$, $n := \lceil \nu \rceil$, $f \in AC^n([a,b])$, and $f^{(k)}(a) = 0$, $k = 0, 1, \ldots, n - 1$. Then*

$$f(x) = \frac{1}{\Gamma(\nu)} \int_a^x (x - t)^{\nu - 1} D_{*a}^\nu f(t) \, dt. \tag{18.40}$$

We also need

Theorem 18.19. [58] *Let $\nu \geq \gamma + 1$, $\gamma \geq 0$. Call $n := \lceil \nu \rceil$. Assume $f \in AC^n([a,b])$ such that $f^{(k)}(a) = 0$, $k = 0, 1, \ldots, n - 1$, and $D_{*a}^\nu f \in L_\infty(a,b)$. Then $D_{*a}^\gamma f \in C([a,b])$, and*

$$D_{*a}^\gamma f(x) = \frac{1}{\Gamma(\nu - \gamma)} \int_a^x (x - t)^{\nu - \gamma - 1} D_{*a}^\nu f(t) \, dt, \tag{18.41}$$

$\forall x \in [a, b]$.

Theorem 18.20. [58] *Let $\nu \geq \gamma + 1$, $\gamma \geq 0$, $n := \lceil \nu \rceil$. Let $f \in AC^n([a,b])$ such that $f^{(k)}(a) = 0$, $k = 0, 1, \ldots, n-1$. Assume $\exists D_a^\nu f(x) \in \mathbb{R}$, $\forall x \in [a,b]$, and $D_a^\nu f \in L_\infty(a,b)$. Then $D_a^\gamma f \in C([a,b])$, and*

$$D_a^\gamma f(x) = \frac{1}{\Gamma(\nu - \gamma)} \int_a^x (x - t)^{\nu - \gamma - 1} D_a^\nu f(t) \, dt, \tag{18.42}$$

$\forall x \in [a, b]$.

Now we are ready to give

Theorem 18.21. *Let* $\nu \geq \gamma + 1$, $\gamma \geq 0$, $n := \lceil \nu \rceil$. *Assume* $f \in AC^n([a,b])$ *such that* $f^{(k)}(a) = 0$, $k = 0, 1, \ldots, n-1$, *and* $D_{*a}^{\nu} f \in L_{\infty}(a,b)$. *Let* $p, q > 1 : 1/p + 1/q = 1$, $r \geq 1$. *Then*

$$\|D_{*a}^{\gamma} f\|_{L_r(a,b)} \leq \frac{(b-a)^{\nu - \gamma + 1/r - 1/q}}{(\Gamma(\nu - \gamma))(p(\nu - \gamma - 1) + 1)^{1/p}}$$

$$\frac{\|D_{*a}^{\nu} f\|_{L_q(a,b)}}{\left[r\left(\nu - \gamma - \frac{1}{q}\right) + 1\right]^{1/r}}. \tag{18.43}$$

Proof. As in Theorem 18.2; based on Theorems 18.18 and 18.19.   □

Corollary 18.22. (to Theorem 18.21) *When* $\gamma = 0$ *it holds*

$$\|f\|_{L_r(a,b)} \leq \frac{(b-a)^{\nu + 1/r - 1/q} \|D_{*a}^{\nu} f\|_{L_q(a,b)}}{(\Gamma(\nu))(p(\nu - 1) + 1)^{1/p} \left(r\left(\nu - \frac{1}{q}\right) + 1\right)^{1/r}}. \tag{18.44}$$

Next we give the corresponding L_1 result.

Theorem 18.23. *Let* $\nu \geq \gamma + 1$, $\gamma \geq 0$, $n := \lceil \nu \rceil$. *Assume* $f \in AC^n([a,b])$ *such that* $f^{(k)}(a) = 0$, $k = 0, 1, \ldots, n-1$, *and* $D_{*a}^{\nu} f \in L_{\infty}(a,b)$, $r \geq 1$. *Then*

(1) $\|D_{*a}^{\gamma} f\|_{L_r(a,b)} \leq \dfrac{(b-a)^{(\nu - \gamma - 1 + 1/r)} \|D_{*a}^{\nu} f\|_{L_1(a,b)}}{(\Gamma(\nu - \gamma))(r(\nu - \gamma - 1) + 1)^{1/r}}.$ \hfill (18.45)

When $\gamma = 0$ *we get*

(2) $\|f\|_{L_r(a,b)} \leq \dfrac{(b-a)^{(\nu - 1 + 1/r)} \|D_{*a}^{\nu} f\|_{L_1(a,b)}}{(\Gamma(\nu))(r(\nu - 1) + 1)^{1/r}}.$ \hfill (18.46)

Proof. Similar to Theorem 18.4; based on Theorems 18.18 and 18.19. □

We continue with mean Sobolev-type inequalities.

Theorem 18.24. *Let* $\gamma \geq 0$ *and* $1 + \gamma \leq \nu_1 < \nu_2 < \ldots < \nu_k$, $n_k := \lceil \nu_k \rceil$, $k \in \mathbb{N}$. *Assume* $f \in AC^{n_k}([a,b])$ *such that* $f^{(k)}(a) = 0$, $k = 0, 1, \ldots, n_k - 1$, *and* $D_{*a}^{\nu_l} f \in L_{\infty}(a,b)$, $l = 1, \ldots, k$. *Let* $p, q > 1 : 1/p + 1/q = 1$, $r \geq 1$. *Call*

$$M_{\gamma} := \max_{1 \leq l \leq k}$$

$$\left[\frac{(b-a)^{\nu_l - \gamma + 1/r - 1/q}}{(\Gamma(\nu_l - \gamma)) \left[p(\nu_l - \gamma - 1) + 1 \right]^{1/p} \left[r\left(\nu_l - \gamma - \frac{1}{q}\right) + 1 \right]^{1/r}} \right], \qquad (18.47)$$

and

$$M_0 := \max_{1 \le l \le k} \left[\frac{(b-a)^{\nu_l + 1/r - 1/q}}{(\Gamma(\nu_l)) \left[p(\nu_l - 1) + 1 \right]^{1/p} \left(r\left(\nu_l - \frac{1}{q}\right) + 1 \right)^{1/r}} \right]. \qquad (18.48)$$

Then

$$(1) \ \|D_{*a}^{\gamma} f\|_{L_r(a,b)} \le \left(\frac{M_\gamma}{k} \right) \left(\sum_{l=1}^{k} \|D_{*a}^{\nu_l} f\|_{L_q(a,b)} \right). \qquad (18.49)$$

When $\gamma = 0$ we obtain

$$(2) \ \|f\|_{L_r(a,b)} \le \left(\frac{M_0}{k} \right) \left(\sum_{l=1}^{k} \|D_{*a}^{\nu_l} f\|_{L_q(a,b)} \right). \qquad (18.50)$$

Proof. Easy; based on Theorems 18.18, 18.19, 18.21 and Corollary 18.22. □

The corresponding average L_1 result follows.

Theorem 18.25. *Let $\gamma \ge 0$ and $1+\gamma \le \nu_1 < \nu_2 < \ldots < \nu_k$, $n_k := \lceil \nu_k \rceil$, $k \in \mathbb{N}$. Assume $f \in AC^{n_k}([a,b])$ such that $f^{(k)}(a) = 0$, $k = 0, 1, \ldots, n_k - 1$, and $D_{*a}^{\nu_l} f \in L_\infty(a,b)$, $l = 1, \ldots, k$, $r \ge 1$. Set*

$$M_\gamma^* := \max_{1 \le l \le k} \left[\frac{(b-a)^{(\nu_l - \gamma - 1 + 1/r)}}{(\Gamma(\nu_l - \gamma))(r(\nu_l - \gamma - 1) + 1)^{1/r}} \right], \qquad (18.51)$$

and

$$M_0^* := \max_{1 \le l \le k} \left[\frac{(b-a)^{(\nu_l - 1 + 1/r)}}{(\Gamma(\nu_l))(r(\nu_l - 1) + 1)^{1/r}} \right]. \qquad (18.52)$$

Then

$$1) \ \|D_{*a}^{\gamma} f\|_{L_r(a,b)} \le \left(\frac{M_\gamma^*}{k} \right) \left(\sum_{l=1}^{k} \|D_{*a}^{\nu_l} f\|_{L_1(a,b)} \right). \qquad (18.53)$$

When $\gamma = 0$ we get

$$2) \ \|f\|_{L_r(a,b)} \le \left(\frac{M_0^*}{k} \right) \left(\sum_{l=1}^{k} \|D_{*a}^{\nu_l} f\|_{L_1(a,b)} \right). \qquad (18.54)$$

Proof. Easy; based on Theorem 18.23.   □

Next we return to Riemann–Liouville fractional derivatives related results from another perspective.

Theorem 18.26. *Let* $\nu \geq \gamma + 1$, $\gamma \geq 0$, $n := \lceil \nu \rceil$. *Assume* $f \in AC^n([a,b])$ *such that* $f^{(k)}(a) = 0$, $k = 0,1,\ldots,n-1$, *and* $D_a^\nu f \in L_\infty(a,b)$. *Assume* $\exists D_a^\nu f(x) \in \mathbb{R}$, $\forall x \in [a,b]$. *Let* $p, q > 1 : 1/p + 1/q = 1$, $r \geq 1$. *Then*

$$\|D_a^\gamma f\|_{L_r(a,b)} \leq$$

$$\frac{(b-a)^{\nu-\gamma+1/r-1/q}\,\|D_a^\nu f\|_{L_q(a,b)}}{(\Gamma(\nu-\gamma))\,(p\,(\nu-\gamma-1)+1)^{1/p}\left[r\left(\nu-\gamma-\tfrac{1}{q}\right)+1\right]^{1/r}}, \tag{18.55}$$

Proof. As in Theorem 18.2; based on Theorem 18.20.   □

Corollary 18.27. (to Theorem 18.26) *When* $\gamma = 0$ *it holds*

$$\|f\|_{L_r(a,b)} \leq \frac{(b-a)^{\nu+1/r-1/q}\,\|D_a^\nu f\|_{L_q(a,b)}}{(\Gamma(\nu))\,(p\,(\nu-1)+1)^{1/p}\left(r\left(\nu-\tfrac{1}{q}\right)+1\right)^{1/r}}. \tag{18.56}$$

Next we give the corresponding L_1 result.

Theorem 18.28. *Let* $\nu \geq \gamma + 1$, $\gamma \geq 0$, $n := \lceil \nu \rceil$. *Assume* $f \in AC^n([a,b])$ *such that* $f^{(k)}(a) = 0$, $k = 0,1,\ldots,n-1$, *and* $D_a^\nu f \in L_\infty(a,b)$, $r \geq 1$. *Assume* $\exists D_a^\nu f(x) \in \mathbb{R}$, $\forall x \in [a,b]$. *Then*

$$(1)\ \|D_a^\gamma f\|_{L_r(a,b)} \leq \frac{(b-a)^{(\nu-\gamma-1+1/r)}\,\|D_a^\nu f\|_{L_1(a,b)}}{(\Gamma(\nu-\gamma))\,(r\,(\nu-\gamma-1)+1)^{1/r}} \tag{18.57}$$

When $\gamma = 0$ *we obtain*

$$(2)\ \|f\|_{L_r(a,b)} \leq \frac{(b-a)^{(\nu-1+1/r)}\,\|D_a^\nu f\|_{L_1(a,b)}}{(\Gamma(\nu))\,(r\,(\nu-1)+1)^{1/r}}. \tag{18.58}$$

Proof. Similar to Theorem 18.4; based on Theorem 18.20.   □

We continue with mean Sobolev-type inequalities.

Theorem 18.29. *Let* $\gamma \geq 0$ *and* $1 + \gamma \leq \nu_1 < \nu_2 < \ldots < \nu_k$, $n_k := \lceil \nu_k \rceil$, *and* $k \in \mathbb{N}$. *Assume* $f \in AC^{n_k}([a,b])$ *such that* $f^{(k)}(a) = 0$, $k = 0,1,\ldots,$

$n_k - 1$, and $D_a^{\nu_l} f \in L_\infty(a,b)$, $l = 1, \ldots, k$. Let $p, q > 1 : 1/p + 1/q = 1$, $r \geq 1$. Further assume $\exists D_a^\nu f(x) \in \mathbb{R}$, $\forall x \in [a,b]$. Here M_γ is as in (18.47) and M_0 as in (18.48). Then

$$(1) \quad \|D_a^\gamma f\|_{L_r(a,b)} \leq \left(\frac{M_\gamma}{k}\right) \left(\sum_{l=1}^{k} \|D_a^{\nu_l} f\|_{L_q(a,b)}\right). \qquad (18.59)$$

When $\gamma = 0$ we obtain

$$(2) \quad \|f\|_{L_r(a,b)} \leq \left(\frac{M_0}{k}\right) \left(\sum_{l=1}^{k} \|D_a^{\nu_l} f\|_{L_q(a,b)}\right). \qquad (18.60)$$

Proof. Easy; based on Theorem 18.26 and Corollary 18.27. □

The corresponding average L_1 result follows.

Theorem 18.30. Let $\gamma \geq 0$ and $1 + \gamma \leq \nu_1 < \nu_2 < \ldots < \nu_k$, $n_k := \lceil \nu_k \rceil$, and $k \in \mathbb{N}$. Assume $f \in AC^{n_k}([a,b])$ such that $f^{(k)}(a) = 0$, $k = 0, 1, \ldots, n_k - 1$, and $D_a^{\nu_l} f \in L_\infty(a,b)$, $l = 1, \ldots, k$, $r \geq 1$. Further assume $\exists D_a^\nu f(x) \in \mathbb{R}$, $\forall x \in [a,b]$. Here M_γ^* is as in (18.51) and M_0^* as in (18.52). Then

$$(1) \quad \|D_a^\gamma f\|_{L_r(a,b)} \leq \left(\frac{M_\gamma^*}{k}\right) \left(\sum_{l=1}^{k} \|D_a^{\nu_l} f\|_{L_1(a,b)}\right). \qquad (18.61)$$

When $\gamma = 0$ we get

$$(2) \quad \|f\|_{L_r(a,b)} \leq \left(\frac{M_0^*}{k}\right) \left(\sum_{l=1}^{k} \|D_a^{\nu_l} f\|_{L_1(a,b)}\right). \qquad (18.62)$$

Proof. Easy; based on Theorem 18.28. □

Next we give converse Sobolev-type results, starting with the Canavati-type fractional derivative.

Theorem 18.31. Let $\nu \geq \gamma + 1$, $\gamma \geq 0$, $n := [\nu]$, $f \in C_{x_0}^\nu([a,b])$, $x_0 \in [a,b]$. Assume $f^{(i)}(x_0) = 0$, $i = 0, 1, \ldots, n - 1$. Assume that $D_{x_0}^\nu f$ is of fixed strict sign on $[x_0, b]$. Let $0 < p < 1$, $q < 0$, such that $1/p + 1/q = 1$, $r \geq 1$. Then

$$(1) \quad \left\|D_{x_0}^\gamma f\right\|_{L_r(x_0,b)} \geq$$

$$\frac{(b - x_0)^{(\nu - \gamma - 1/q + 1/r)} \left\|D_{x_0}^\nu f\right\|_{L_q(x_0,b)}}{\Gamma(\nu - \gamma)(p(\nu - \gamma - 1) + 1)^{1/p} \left(r\left(\nu - \gamma - \frac{1}{q}\right) + 1\right)^{1/r}}. \qquad (18.63)$$

When $\gamma = 0$ we obtain

$$(2)\ \|f\|_{L_r(x_0,b)} \geq \frac{(b-x_0)^{(\nu-1/q+1/r)} \left\|D_{x_0}^\nu f\right\|_{L_q(x_0,b)}}{\Gamma(\nu)\,(p(\nu-1)+1)^{1/p}\,\left(r\left(\nu-\frac{1}{q}\right)+1\right)^{1/r}}. \qquad (18.64)$$

Proof. Here $\left(D_{x_0}^\gamma f\right)(x_0) = 0$ by (18.1), but by assumption $\left(D_{x_0}^\gamma f\right)(x) \neq 0$, $\forall x \in (x_0, b]$.

By Lemma 18.1 and assumptions of the theorem we have

$$\left|D_{x_0}^\gamma f(x)\right| = \frac{1}{\Gamma(\nu-\gamma)} \int_{x_0}^x (x-t)^{\nu-\gamma-1} \left|D_{x_0}^\nu f(t)\right| dt$$

(by the reverse Hölder's inequality)

$$\geq \frac{1}{\Gamma(\nu-\gamma)} \left(\int_{x_0}^x (x-t)^{p(\nu-\gamma-1)}\, dt\right)^{1/p}$$

$$\left(\int_{x_0}^x \left|D_{x_0}^\nu f(t)\right|^q dt\right)^{1/q}$$

$$\geq \frac{1}{\Gamma(\nu-\gamma)} \frac{(x-x_0)^{(\nu-\gamma-1)+1/p}}{(p(\nu-\gamma-1)+1)^{1/p}}$$

$$\left(\int_{x_0}^b \left|D_{x_0}^\nu f(t)\right|^q dt\right)^{1/q}. \qquad (18.65)$$

So that

$$\left|D_{x_0}^\gamma f(x)\right| \geq \frac{1}{\Gamma(\nu-\gamma)}$$

$$\frac{(x-x_0)^{(\nu-\gamma-1)+1/p}}{(p(\nu-\gamma-1)+1)^{1/p}} \left(\int_{x_0}^b \left|D_{x_0}^\nu f(t)\right|^q dt\right)^{1/q}, \qquad (18.66)$$

$\forall x \in [x_0, b]$. Thus

$$\left|D_{x_0}^\gamma f(x)\right|^r \geq \frac{1}{(\Gamma(\nu-\gamma))^r}$$

$$\frac{(x-x_0)^{r(\nu-\gamma-1)+r/p}}{(p(\nu-\gamma-1)+1)^{r/p}} \left\|D_{x_0}^\nu f\right\|_{L_q(x_0,b)}^r, \qquad (18.67)$$

$\forall x \in [x_0, b]$. Hence

$$\int_{x_0}^b \left|D_{x_0}^\gamma f(x)\right|^r dx \geq \frac{\left\|D_{x_0}^\nu f\right\|_{L_q(x_0,b)}^r}{(\Gamma(\nu-\gamma))^r}$$

$$\frac{1}{(p(\nu - \gamma - 1) + 1)^{r/p}} \int_{x_0}^{b} (x - x_0)^{r(\nu - \gamma - 1) + r/p} \, dx \qquad (18.68)$$

$$= \frac{\left\| D_{x_0}^{\nu} f \right\|_{L_q(x_0, b)}^r}{(\Gamma(\nu - \gamma))^r} \frac{1}{(p(\nu - \gamma - 1) + 1)^{r/p}}$$

$$\frac{(b - x_0)^{r(\nu - \gamma - 1) + r/p + 1}}{\left(r(\nu - \gamma - 1) + \frac{r}{p} + 1 \right)}.$$

So we have established

$$\left\| D_{x_0}^{\gamma} f \right\|_{L_r(x_0, b)} \geq \frac{\left\| D_{x_0}^{\nu} f \right\|_{L_q(x_0, b)}}{\Gamma(\nu - \gamma)} \frac{1}{(p(\nu - \gamma - 1) + 1)^{1/p}}$$

$$\frac{(b - x_0)^{(\nu - \gamma - 1) + 1/p + 1/r}}{\left(r(\nu - \gamma - 1) + \frac{r}{p} + 1 \right)^{1/r}}, \qquad (18.69)$$

proving the claim. $\quad \square$

Converse results follow related to the Riemann–Liouville fractional derivative.

Theorem 18.32. *Let $\beta > \alpha \geq 0$, $f \in L_1(a, x)$, a, $x \in \mathbb{R}$, have L_∞ fractional derivative $D_a^{\beta} f$ in $[a, x]$, and let $D_a^{\beta - k} f(a) = 0$ for $k = 1, \ldots, [\beta] + 1$. Assume that $D_a^{\beta} f$ has fixed sign a.e. on $[a, x]$, and $1/D_a^{\beta} f \in L_\infty(a, x)$. Let $0 < p < 1$, $q < 0$, such that $1/p + 1/q = 1$, with $p(\beta - \alpha - 1) + 1 > 0$; $r \geq 1$. Then*

$$(1) \quad \left\| D_a^{\alpha} f \right\|_{L_r(a, x)} \geq$$

$$\frac{(x - a)^{(\beta - \alpha - 1/q + 1/r)} \left\| D_a^{\beta} f \right\|_{L_q(a, x)}}{\Gamma(\beta - \alpha) (p(\beta - \alpha - 1) + 1)^{1/p} \left(r\left(\beta - \alpha - \frac{1}{q} \right) + 1 \right)^{1/r}}. \qquad (18.70)$$

When $\alpha = 0$ we get

$$(2) \quad \left\| f \right\|_{L_r(a, x)} \geq \frac{(x - a)^{(\beta - 1/q + 1/r)} \left\| D_a^{\beta} f \right\|_{L_q(a, x)}}{\Gamma(\beta) (p(\beta - 1) + 1)^{1/p} \left(r\left(\beta - \frac{1}{q} \right) + 1 \right)^{1/r}}. \qquad (18.71)$$

Proof. Similar to Theorem 18.31; based on (18.26). $\quad \square$

We continue with converse results related to the Caputo fractional derivative.

Theorem 18.33. *Let $\nu \geq \gamma + 1$, $\gamma \geq 0$, $n := \lceil \nu \rceil$, $f \in AC^n([a,b])$, and $f^{(k)}(a) = 0$, $k = 0, 1, \ldots, n - 1$. Assume that $D_{*a}^\nu f$ is of fixed sign a.e. on $[a,b]$, and $D_{*a}^\nu f$, $1/D_{*a}^\nu f \in L_\infty(a,b)$. Let $0 < p < 1$, $q < 0$, such that $1/p + 1/q = 1$; $r \geq 1$. Then*

$$(1) \quad \|D_{*a}^\gamma f\|_{L_r(a,b)} \geq$$

$$\frac{(b-a)^{(\nu-\gamma-1/q+1/r)} \|D_{*a}^\nu f\|_{L_q(a,b)}}{\Gamma(\nu-\gamma)(p(\nu-\gamma-1)+1)^{1/p}\left(r\left(\nu-\gamma-\frac{1}{q}\right)+1\right)^{1/r}}. \tag{18.72}$$

When $\gamma = 0$ we get

$$(2) \quad \|f\|_{L_r(a,b)} \geq \frac{(b-a)^{(\nu-1/q+1/r)} \|D_{*a}^\nu f\|_{L_q(a,b)}}{\Gamma(\nu)(p(\nu-1)+1)^{1/p}\left(r\left(\nu-\frac{1}{q}\right)+1\right)^{1/r}}. \tag{18.73}$$

Proof. Similar to Theorem 18.31; based on (18.40), (18.41). □

Next we give converse results again for the Riemann–Liouville fractional derivative.

Theorem 18.34. *Let $\nu \geq \gamma + 1$, $\gamma \geq 0$, $n := \lceil \nu \rceil$, $f \in AC^n([a,b])$, and $f^{(k)}(a) = 0$, $k = 0, 1, \ldots, n - 1$. Suppose $\exists D_a^\nu f(x) \in \mathbb{R}, \forall x \in [a,b]$, and $D_a^\nu f$, $1/D_a^\nu f \in L_\infty(a,b)$. Here $D_a^\nu f$ is of fixed sign a.e. on $[a,b]$. Let $0 < p < 1$, $q < 0$, such that $1/p + 1/q = 1$; $r \geq 1$. Then*

$$(1) \quad \|D_a^\gamma f\|_{L_r(a,b)} \geq$$

$$\frac{(b-a)^{(\nu-\gamma-1/q+1/r)} \|D_a^\nu f\|_{L_q(a,b)}}{\Gamma(\nu-\gamma)(p(\nu-\gamma-1)+1)^{1/p}\left(r\left(\nu-\gamma-\frac{1}{q}\right)+1\right)^{1/r}}. \tag{18.74}$$

When $\gamma = 0$ we get

$$(2) \quad \|f\|_{L_r(a,b)} \geq \frac{(b-a)^{(\nu-1/q+1/r)} \|D_a^\nu f\|_{L_q(a,b)}}{\Gamma(\nu)(p(\nu-1)+1)^{1/p}\left(r\left(\nu-\frac{1}{q}\right)+1\right)^{1/r}}. \tag{18.75}$$

Proof. Similar to Theorem 18.31; based on (18.42). □

Next we give converse mean Sobolev type inequalities.

Theorem 18.35. *Let $\gamma \geq 0$ and $1 + \gamma \leq \nu_1 < \nu_2 < \ldots < \nu_k$, $n_k := \lceil \nu_k \rceil$, and $k \in \mathbb{N}$. Assume $f \in \bigcap_{l=1}^k C_{x_0}^{\nu_l}([a,b])$, $x_0 \in [a,b]$. Assume $f^{(i)}(x_0) = 0$,*

$i = 0, 1, \ldots, n_k - 1; \ 0 < p < 1, \ q < 0 : 1/p + 1/q = 1; \ r \geq 1$. *Further suppose each* $D_{x_0}^{\nu_l} f$ *is of fixed strict sign on* $[x_0, b]$, *all* $l = 1, \ldots, k$. *Call*

$$\theta_\gamma := \min_{1 \leq l \leq k}$$

$$\left\{ \frac{(b - x_0)^{(\nu_l - \gamma - 1/q + 1/r)}}{\Gamma(\nu_l - \gamma)\left(p(\nu_l - \gamma - 1) + 1\right)^{1/p} \left(r\left(\nu_l - \gamma - \frac{1}{q}\right) + 1\right)^{1/r}} \right\}, \quad (18.76)$$

and

$$\theta_0 := \min_{1 \leq l \leq k} \left\{ \frac{(b - x_0)^{(\nu_l - 1/q + 1/r)}}{\Gamma(\nu_l)\left(p(\nu_l - 1) + 1\right)^{1/p} \left(r\left(\nu_l - \frac{1}{q}\right) + 1\right)^{1/r}} \right\}. \quad (18.77)$$

Then

$$(1) \ \left\| D_{x_0}^\gamma f \right\|_{L_r(x_0, b)} \geq \frac{\theta_\gamma}{k} \left(\sum_{l=1}^k \left\| D_{x_0}^{\nu_l} f \right\|_{L_q(x_0, b)} \right). \quad (18.78)$$

When $\gamma = 0$ *we obtain*

$$(2) \ \left\| f \right\|_{L_r(x_0, b)} \geq \frac{\theta_0}{k} \left(\sum_{l=1}^k \left\| D_{x_0}^{\nu_l} f \right\|_{L_q(x_0, b)} \right). \quad (18.79)$$

Proof. By (18.63) we have

$$\left\| D_{x_0}^\gamma f \right\|_{L_r(x_0, b)} \geq$$

$$\frac{(b - x_0)^{(\nu_l - \gamma - 1/q + 1/r)} \left\| D_{x_0}^{\nu_l} f \right\|_{L_q(x_0, b)}}{\Gamma(\nu_l - \gamma)\left(p(\nu_l - \gamma - 1) + 1\right)^{1/p} \left(r\left(\nu_l - \gamma - \frac{1}{q}\right) + 1\right)^{1/r}}$$

$$\geq \theta_\gamma \left\| D_{x_0}^{\nu_l} f \right\|_{L_q(x_0, b)}, \quad (18.80)$$

for all $l = 1, \ldots, k$. That is,

$$\left\| D_{x_0}^\gamma f \right\|_{L_r(x_0, b)} \geq \theta_\gamma \left\| D_{x_0}^{\nu_l} f \right\|_{L_q(x_0, b)}, \quad (18.81)$$

for all $l = 1, \ldots, k$.

Adding the above (18.81) we obtain

$$k \left\| D_{x_0}^\gamma f \right\|_{L_r(x_0, b)} \geq \theta_\gamma \left(\sum_{l=1}^k \left\| D_{x_0}^{\nu_l} f \right\|_{L_q(x_0, b)} \right), \quad (18.82)$$

establishing our claim. $\quad \square$

We continue with Riemann–Liouville fractional derivatives.

Theorem 18.36. *Let $0 \leq \alpha < \beta_1 < \beta_2 < \ldots < \beta_k$, $k \in \mathbb{N}$; $f \in L_1(a, x)$, $a, x \in \mathbb{R}$, have L_∞ fractional derivatives $D_a^{\beta_l} f$ in $[a, x]$, with $D_a^{\beta_l - k} f(a) = 0$ for $k = 1, \ldots, [\beta_l] + 1$; $l = 1, \ldots, k$. Let $0 < p < 1$, $q < 0$, such that $1/p + 1/q = 1$, with $p(\beta_l - \alpha - 1) > 0$, all $l = 1, \ldots, k$; $r \geq 1$. Further assume that each $D_a^{\beta_l} f$ has fixed sign a.e. on $[a, x]$, and $1/D_a^{\beta_l} f \in L_\infty(a, x)$, for all $l = 1, \ldots, k$.*
Put

$$\theta_\alpha^* := \min_{1 \leq l \leq k}$$

$$\left\{ \frac{(x - a)^{(\beta_l - \alpha - 1/q + 1/r)}}{\Gamma(\beta_l - \alpha)(p(\beta_l - \alpha - 1) + 1)^{1/p} \left(r \left(\beta_l - \alpha - \frac{1}{q} \right) + 1 \right)^{1/r}} \right\}, \quad (18.83)$$

and

$$\theta_0^* := \min_{1 \leq l \leq k} \left\{ \frac{(x - a)^{(\beta_l - 1/q + 1/r)}}{\Gamma(\beta_l)(p(\beta_l - 1) + 1)^{1/p} \left(r \left(\beta_l - \frac{1}{q} \right) + 1 \right)^{1/r}} \right\}. \quad (18.84)$$

Then

$$(1) \quad \|D_a^\alpha f\|_{L_r(a,x)} \geq \frac{\theta_\alpha^*}{k} \left(\sum_{l=1}^{k} \|D_a^{\beta_l} f\|_{L_q(a,x)} \right). \quad (18.85)$$

When $\alpha = 0$ we get

$$(2) \quad \|f\|_{L_r(a,x)} \geq \frac{\theta_0^*}{k} \left(\sum_{l=1}^{k} \|D_a^{\beta_l} f\|_{L_q(a,x)} \right). \quad (18.86)$$

Proof. As in Theorem 18.35; based on Theorem 18.32. □

We continue with Caputo fractional derivatives.

Theorem 18.37. *Let $\gamma \geq 0$ and $1 + \gamma \leq \nu_1 < \nu_2 < \ldots < \nu_k$, $n_k := \lceil \nu_k \rceil$, $k \in \mathbb{N}$. Assume $f \in AC^{n_k}([a, b])$ such that $f^{(k)}(a) = 0$, $k = 0, 1, \ldots, n_k - 1$, and $D_{*a}^{\nu_l} f$, $1/D_{*a}^{\nu_l} f \in L_\infty(a, b)$, $l = 1, \ldots, k$. Suppose each $D_{*a}^{\nu_l} f$ is of fixed sign a.e. on $[a, b]$, all $l = 1, \ldots, k$. Let $0 < p < 1$, $q < 0$, such that $1/p + 1/q = 1$; $r \geq 1$.*
Set

$$A_\gamma := \min_{1 \leq l \leq k}$$

$$\left\{ \frac{(b-a)^{(\nu_l - \gamma - 1/q + 1/r)}}{\Gamma(\nu_l - \gamma)\left(p(\nu_l - \gamma - 1) + 1\right)^{1/p}\left(r\left(\nu_l - \gamma - \frac{1}{q}\right) + 1\right)^{1/r}} \right\}, \quad (18.87)$$

and

$$A_0 := \min_{1 \le l \le k} \left\{ \frac{(b-a)^{(\nu_l - 1/q + 1/r)}}{\Gamma(\nu_l)\left(p(\nu_l - 1) + 1\right)^{1/p}\left(r\left(\nu_l - \frac{1}{q}\right) + 1\right)^{1/r}} \right\}. \quad (18.88)$$

Then

$$(1)\ \|D^{\gamma}_{*a} f\|_{L_r(a,b)} \ge \frac{A_{\gamma}}{k}\left(\sum_{l=1}^{k}\|D^{\nu_l}_{*a} f\|_{L_q(a,b)}\right). \quad (18.89)$$

When $\gamma = 0$ we get

$$(2)\ \|f\|_{L_r(a,b)} \ge \frac{A_0}{k}\left(\sum_{l=1}^{k}\|D^{\nu_l}_{*a} f\|_{L_q(a,b)}\right). \quad (18.90)$$

Proof. As in Theorem 18.35; based on Theorem 18.33. $\square$

Next we come back to Riemann–Liouville fractional derivatives.

Theorem 18.38. *Let $\gamma \ge 0$ and $1 + \gamma \le \nu_1 < \nu_2 < \ldots < \nu_k$, $n_k := \lceil \nu_k \rceil$, and $k \in \mathbb{N}$. Assume $f \in AC^{n_k}([a,b])$ such that $f^{(k)}(a) = 0$, $k = 0, 1, \ldots, n_k - 1$; $\exists D^{\nu_l}_a f(x) \in \mathbb{R}$, $\forall x \in [a,b]$, and $D^{\nu_l}_a f$, $1/D^{\nu_l}_a f \in L_{\infty}(a,b)$, $l = 1, \ldots, k$. Suppose each $D^{\nu_l}_a f$ is of fixed sign a.e. on $[a,b]$, all $l = 1, \ldots, k$. Let $0 < p < 1$, $q < 0$, such that $1/p + 1/q = 1$; $r \ge 1$. Here A_{γ} is as in (18.87), and A_0 as in (18.88). Then*

$$(1)\ \|D^{\gamma}_a f\|_{L_r(a,b)} \ge \frac{A_{\gamma}}{k}\left(\sum_{l=1}^{k}\|D^{\nu_l}_a f\|_{L_q(a,b)}\right). \quad (18.91)$$

When $\gamma = 0$ we get

$$(2)\ \|f\|_{L_r(a,b)} \ge \frac{A_0}{k}\left(\sum_{l=1}^{k}\|D^{\nu_l}_a f\|_{L_q(a,b)}\right). \quad (18.92)$$

Proof. As in Theorem 18.35; based on Theorem 18.34. $\square$

18.3   Applications

Next we give an application to ordinary fractional differential equations involving Caputo fractional derivatives.

Theorem 18.39. *Let* $1 \leq \nu_1 < \nu_2 < \ldots < \nu_k$, $n_k := \lceil \nu_k \rceil$, *and* $k \in \mathbb{N}$. *Assume* $f \in AC^{n_k}([a,b])$ *such that* $f^{(j)}(a) = 0$, $j = 0, 1, \ldots, n_k - 1$, *and* $D_{*a}^{\nu_l} f \in L_\infty(a,b)$, *all* $l = 1, \ldots, k$; $r \geq 1$. *Let also* B, C_l, $1/C_l \in L_\infty(a,b)$, $l = 1, \ldots, k$, *with all* $C_l \geq 0$. *The above-described* f *satisfies*

$$\sum_{l=1}^{k} C_l(x) \left(D_{*a}^{\nu_l} f(x) \right)^2 = B(x), \ \forall x \in [a,b]. \tag{18.93}$$

Call

$$\delta^* := \max_{1 \leq l \leq k} \left\{ \frac{(b-a)^{(2\nu_l + 2/r - 1)}}{(\Gamma(\nu_l))^2 (2\nu_l - 1) \left(r\left(\nu_l - \frac{1}{2}\right) + 1 \right)^{2/r}} \right\}, \tag{18.94}$$

$$\rho^* := \max_{1 \leq l \leq k} \left\| \frac{1}{C_l} \right\|_{L_\infty(a,b)}. \tag{18.95}$$

Then

$$\|f\|_{L_r(a,b)} \leq \sqrt{\frac{\delta^* \rho^*}{k}} \left(\int_a^b B(x)\, dx \right)^{1/2}. \tag{18.96}$$

Proof. Set $n_l := \lceil \nu_l \rceil$, $l = 1, \ldots, k-1$. Clearly here $f \in AC^{n_l}([a,b])$, $l = 1, \ldots, k-1$. Also $f^{(j)}(a) = 0$, $j = 0, 1, \ldots, n_l - 1$, $l = 1, \ldots, k-1$. So all assumptions of Theorem 18.21 are fulfilled for f and fractional orders ν_l, $l = 1, \ldots, k-1$. Thus by choosing $p = q = 2$ we apply (18.44), for $l = 1, \ldots, k$, to obtain

$$\|f\|_{L_r(a,b)} \leq \frac{(b-a)^{\nu_l + 1/r - 1/2} \left(\int_a^b \left(D_{*a}^{\nu_l} f(x) \right)^2 dx \right)^{1/2}}{\Gamma(\nu_l)(2\nu_l - 1)^{1/2} \left(r\left(\nu_l - \frac{1}{2}\right) + 1 \right)^{1/r}}. \tag{18.97}$$

Hence it holds

$$\|f\|_{L_r(a,b)}^2 \leq \frac{(b-a)^{(2\nu_l + 2/r - 1)} \left(\int_a^b \left(D_{*a}^{\nu_l} f(x) \right)^2 dx \right)}{(\Gamma(\nu_l))^2 (2\nu_l - 1) \left(r\left(\nu_l - \frac{1}{2}\right) + 1 \right)^{2/r}} \leq$$

$$\delta^* \int_a^b (C_l(x))^{-1} (C_l(x)) \left(D_{*a}^{\nu_l} f(x) \right)^2 dx \leq$$

$$\delta^* \rho^* \int_a^b C_l(x) \left(D_{*a}^{\nu_l} f(x) \right)^2 dx. \tag{18.98}$$

That is,

$$\|f\|_{L_r(a,b)}^2 \le \delta^* \rho^* \int_a^b C_l(x) \left(D_{*a}^{\nu_l} f(x)\right)^2 dx, \tag{18.99}$$

for all $l = 1, \ldots, k$.

The last imply by addition that

$$\|f\|_{L_r(a,b)}^2 \le \frac{\delta^* \rho^*}{k} \left(\int_a^b \left(\sum_{l=1}^k C_l(x) \left(D_{*a}^{\nu_l} f(x)\right)^2 \right) dx \right)$$

$$= \frac{\delta^* \rho^*}{k} \int_a^b B(x) \, dx, \tag{18.100}$$

proving the claim. $\square$

One can give similar applications to fractional ODEs involving fractional derivatives of Canavati-type and Riemann–Liouville-type.

19

General Hilbert–Pachpatte-Type Integral Inequalities

In this chapter we present very general weighted Hilbert–Pachpatte-type integral inequalities. These are with regard to ordinary derivatives and fractional derivatives of Riemann–Liouville and Canavati types, and also in regard to general derivatives of Widder-type and linear differential operators. These results apply to continuous functions and some to integrable functions. This treatment relies on [41].

19.1 Introduction

In this chapter we present a series of weighted very general Hilbert–Pachpatte-type integral inequalities regarding ordinary and fractional derivatives, derivatives of Widder-type, and linear differential operators. First we derive a very general result in Theorem 19.3, for which the rest of the results are applications.

The results here are motivated by the original Hilbert double integral inequality.

Theorem 19.1. [191, Theorem 316]. *If $p > 1$, $q = p/(p-1)$ and*

$$\int_0^\infty f^p(x)dx \leq F, \qquad \int_0^\infty g^q(y)dy \leq G,$$

then

$$\int_0^\infty \int_0^\infty \frac{f(x)g(y)}{x+y}dxdy < \frac{\pi}{\sin(\pi/p)}F^{1/p}G^{1/q} \qquad (19.1)$$

G.A. Anastassiou, *Fractional Differentiation Inequalities*,
DOI 10.1007/978-0-387-98128-4_19, © Springer Science+Business Media, LLC 2009

where f, g are nonnegative measurable functions, unless

$$f \equiv 0 \quad or \quad g \equiv 0.$$

The constant $\pi \operatorname{cosec}(\pi/p)$ is the best possible in (19.1).

Also the results here are motivated by

Theorem 19.2. (Pachpatte [320, Theorem 1]). *Let $n \geq 1$ and $0 \leq k \leq n-1$ be integers. Let $u \in C^n([0, x])$ and $v \in C^n([0, y])$, where $x > 0$, $y > 0$, and let $u^{(j)}(0) = v^{(j)}(0) = 0$ for $j \in \{0, \dots, n-1\}$. Then*

$$\int_0^x \int_0^y \frac{|u^{(k)}(s)| \, |v^{(k)}(t)|}{s^{2n-2k-1} + t^{2n-2k-1}} ds \, dt \leq M(n, k, x, y)$$

$$\left(\int_0^x (x-s) \left(u^{(n)}(s) \right)^2 ds \right)^{1/2} \left(\int_0^y (y-t) \left(v^{(n)}(t) \right)^2 dt \right)^{1/2}, \qquad (19.2)$$

where

$$M(n, k, x, y) := \frac{1}{2} \frac{\sqrt{xy}}{[(n-k-1)!]^2 (2n-2k-1)}.$$

Also of great motivation to this chapter are the articles [147, 187, 188].

19.2   Main Results

We present the following general result.

Theorem 19.3. *Here for $i \in \{1, \dots, n\}$, take $x_i > 0$, and assume*

$$u_i \in L_1(0, x_i), \quad g_i \in L_\infty((0, x_i)^2), \quad \Phi_i \in L_\infty(0, x_i),$$

with $g_i, \Phi_i \geq 0$. Take $r_i \geq 0$, $p_i, q_i > 1 \colon 1/p_i + 1/q_i = 1$, and $w_i > 0$ such that $\sum_{i=1}^n w_i = \Omega_n$, and $a_i, b_i \in [0, 1]$ such that $a_i + b_i = 1$. Call

$$\varphi_i(s_i) := \int_0^{s_i} \left(g_i(s_i, \tau_i) \right)^{(a_i + b_i q_i) r_i} d\tau_i. \qquad (19.3)$$

Suppose $\varphi_i(s_i) > 0$, with the exception $\varphi_i(0) = 0$. If

$$|u_i(s_i)| \leq \int_0^{s_i} \left(g_i(s_i, \tau_i) \right)^{r_i} \Phi_i(\tau_i) d\tau_i, \quad s_i \in [0, x_i], \quad i = 1, \dots, n, \qquad (19.4)$$

then

$$I_1 \quad : \quad = \int_0^{x_1} \cdots \int_0^{x_n} \frac{\prod_{i=1}^n |u_i(s_i)|}{\left(\frac{1}{\Omega_n} \sum_{i=1}^n w_i (\varphi_i(s_i))^{1/q_i w_i} \right)^{\Omega_n}} ds_1 \cdots ds_n$$

$$\leq \prod_{i=1}^n x_i^{1/q_i} \left[\int_0^{x_i} (\Phi_i(\tau_i))^{p_i} \left(\int_{\tau_i}^{x_i} (g_i(s_i, \tau_i))^{a_i r_i} ds_i \right) d\tau_i \right]^{1/p_i} \qquad (19.5)$$

Proof. We write

$$\left(g_i(s_i, \tau_i)\right)^{r_i} \Phi_i(\tau_i) = \left(g_i(s_i, \tau_i)\right)^{\left(a_i/q_i + b_i\right)r_i} \times \left(g_i(s_i, \tau_i)\right)^{\left(a_i/p_i\right)r_i} \Phi_i(\tau_i).$$

Notice here that the two sections of g_i belong to $L_\infty(0, x_i)$, a.e.
  Using Hölder's inequality we find

$$\begin{aligned}
|u_i(s_i)| \;\leq\; & \left(\int_0^{s_i} \left(g_i(s_i, \tau_i)\right)^{\left(a_i + b_i q_i\right)r_i} d\tau_i\right)^{1/q_i} \\
& \times \left(\int_0^{s_i} \left(g_i(s_i, \tau_i)\right)^{a_i r_i} \left(\Phi_i(\tau_i)\right)^{p_i} d\tau_i\right)^{1/p_i}.
\end{aligned}$$

That is, we have that

$$|u_i(s_i)| \leq \left(\varphi_i(s_i)\right)^{1/q_i} \left(\int_0^{s_i} \left(g_i(s_i, \tau_i)\right)^{a_i r_i} \left(\Phi_i(\tau_i)\right)^{p_i} d\tau_i\right)^{1/p_i}, \quad i = 1, \ldots, n.$$

Set $r_i^* = 1/q_i w_i$.
  By the means inequality ([297, p. 15]) we have

$$\left(\prod_{i=1}^n \left(\varphi_i(s_i)\right)^{w_i r_i^*}\right) \leq \left(\frac{1}{\Omega_n} \sum_{i=1}^n w_i \varphi_i^{r_i^*}(s_i)\right)^{\Omega_n}.$$

Therefore we find

$$\begin{aligned}
\prod_{i=1}^n |u_i(s_i)| \;\leq\; & \left(\frac{1}{\Omega_n} \sum_{i=1}^n w_i \varphi_i^{r_i^*}(s_i)\right)^{\Omega_n} \\
& \times \prod_{i=1}^n \left(\int_0^{s_i} \left(g_i(s_i, \tau_i)\right)^{a_i r_i} \left(\Phi_i(\tau_i)\right)^{p_i} d\tau_i\right)^{1/p_i}.
\end{aligned}$$

That is,

$$\frac{\prod\limits_{i=1}^n |u_i(s_i)|}{\left(\frac{1}{\Omega_n} \sum\limits_{i=1}^n w_i \varphi_i^{r_i^*}(s_i)\right)^{\Omega_n}} \leq \prod_{i=1}^n \left(\int_0^{s_i} \left(g_i(s_i, \tau_i)\right)^{a_i r_i} \left(\Phi_i(\tau_i)\right)^{p_i} d\tau_i\right)^{1/p_i},$$

whenever at least one $s_i > 0$, $i \in \{1, \ldots, n\}$, where $s_i \in [0, x_i]$. Thus we obtain

$$\int_0^{x_1} \cdots \int_0^{x_n} \frac{\prod\limits_{i=1}^{n} |u_i(s_i)|}{\left(\frac{1}{\Omega_n} \sum\limits_{i=1}^{n} w_i \varphi_i^{r_i^*}(s_i)\right)^{\Omega_n}} ds_1 \cdots ds_n$$

$$\leq \int_0^{x_1} \cdots \int_0^{x_n} \prod_{i=1}^{n} \left(\int_0^{s_i} (g_i(s_i, \tau_i))^{a_i r_i} (\Phi_i(\tau_i))^{p_i} d\tau_i\right)^{1/p_i} ds_1 \cdots ds_n$$

$$= \prod_{i=1}^{n} \left(\int_0^{x_i} \left(\int_0^{s_i} (g_i(s_i, \tau_i))^{a_i r_i} (\Phi_i(\tau_i))^{p_i} d\tau_i\right)^{1/p_i} ds_i\right)$$

(using again Hölder's inequality)

$$\leq \prod_{i=1}^{n} x_i^{1/q_i} \left[\int_0^{x_i} \left(\int_0^{s_i} (g_i(s_i, \tau_i))^{a_i r_i} (\Phi_i(\tau_i))^{p_i} d\tau_i\right) ds_i\right]^{1/p_i}$$

(by changing the order of integration)

$$= \prod_{i=1}^{n} x_i^{1/q_i} \left[\int_0^{x_i} \left(\int_{\tau_i}^{x_i} (g_i(s_i, \tau_i))^{a_i r_i} (\Phi_i(\tau_i))^{p_i} ds_i\right) d\tau_i\right]^{1/p_i}$$

$$= \prod_{i=1}^{n} x_i^{1/q_i} \left[\int_0^{x_i} (\Phi_i(\tau_i))^{p_i} \left(\int_{\tau_i}^{x_i} (g_i(s_i, \tau_i))^{a_i r_i} ds_i\right) d\tau_i\right]^{1/p_i},$$

proving the claim.   □

We give

Corollary 19.4. *All are as in Theorem 19.3 and* $1/p := \sum_{i=1}^{n} 1/p_i$, $r > 0$. *Then*

$$I_1 \leq p^{1/rp} \sum_{i=1}^{n} x_i^{1/q_i} \left[\sum_{i=1}^{n} \frac{1}{p_i} \left\{\int_0^{x_i} (\Phi_i(\tau_i))^{p_i} \right.\right.$$

$$\left.\left. \times \left(\int_{\tau_i}^{x_i} (g_i(s_i, \tau_i))^{a_i r_i} ds_i\right) d\tau_i\right\}^r\right]^{1/rp}.$$

Proof. By the inequality of means, for any $A_i \geq 0$ and $r > 0$, we derive

$$\prod_{i=1}^{n} A_i^{1/p_i} \leq \left(p \sum_{i=1}^{n} \frac{1}{p_i} A_i^r\right)^{1/rp}.$$

We use here (19.5).   □

Next we apply Theorem 19.3.
We give

Theorem 19.5. *Let $u_i \in C^{m_i}([0, x_i])$, $x_i > 0$, $m_i \in \mathbb{N}$, be such that $u_i^{(j)}(0) = 0$ for $j \in \{0, \ldots, m_i - 1\}$, $i = 1, \ldots, n$. Let $k_i \in \{0, \ldots, m_i - 1\}$; p_i, $q_i > 1$: $1/p_i + 1/q_i = 1$; $a_i, b_i \in [0, 1]$: $a_i + b_i = 1$. Let also $w_i > 0$ such that $\sum_{i=1}^{n} w_i = \Omega_n$. Then*

$$
I_2 \quad : \quad = \int_0^{x_1} \cdots \int_0^{x_n} \frac{\prod_{i=1}^{n} |u_i^{(k_i)}(s_i)|}{\left(\frac{1}{\Omega_n} \sum_{i=1}^{n} w_i \frac{s_i^{[(a_i + b_i q_i)(m_i - k_i - 1) + 1]/q_i w_i}}{[(a_i + b_i q_i)(m_i - k_i - 1) + 1]^{1/q_i w_i}} \right)^{\Omega_n}} ds_1 \cdots ds_n
$$

$$
\leq \quad A \prod_{i=1}^{n} x_i^{1/q_i} \left[\int_0^{x_i} |u_i^{(m_i)}(\tau_i)|^{p_i} (x_i - \tau_i)^{[a_i(m_i - k_i - 1) + 1]} d\tau_i \right]^{1/p_i}, \quad (19.6)
$$

where

$$
A := \prod_{i=1}^{n} \left(\frac{1}{(m_i - k_i - 1)![a_i(m_i - k_i - 1) + 1]^{1/p_i}} \right).
$$

Proof. By [320, Equation (7)] we have

$$
u_i^{(k_i)}(s_i) = \frac{1}{(m_i - k_i - 1)!} \int_0^{s_i} (s_i - \tau_i)^{m_i - k_i - 1} u_i^{(m_i)}(\tau_i) d\tau_i,
$$

where $s_i \in [0, x_i]$, $i = 1, \ldots, n$. We use Theorem 19.3.
We set

$$
\Phi_i(\tau_i) := \frac{|u_i^{(m_i)}(\tau_i)|}{(m_i - k_i - 1)!}, \quad \text{all } 0 \leq \tau_i \leq x_i, \quad i = 1, \ldots, n.
$$

We also set

$$
g_i(s_i, \tau_i) := |s_i - \tau_i|, \quad \text{all } s_i, \tau_i \in [0, x_i],
$$

and $r_i := m_i - k_i - 1$. Clearly then

$$
\varphi_i(s_i) \quad = \quad \int_0^{s_i} (s_i - \tau_i)^{(a_i + b_i q_i)(m_i - k_i - 1)} d\tau_i
$$

$$
= \quad \frac{s_i^{(a_i + b_i q_i)(m_i - k_i - 1) + 1}}{(a_i + b_i q_i)(m_i - k_i - 1) + 1}, \quad s_i \in [0, x_i], \quad i = 1, \ldots, n.
$$

We see here that

$$
|u_i^{(k_i)}(s_i)| \leq \int_0^{s_i} (s_i - \tau_i)^{m_i - k_i - 1} \frac{|u_i^{(m_i)}(\tau_i)|}{(m_i - k_i - 1)!} d\tau_i,
$$

where $s_i \in [0, x_i]$, $i = 1, \ldots, n$. We also find

$$\int_{\tau_i}^{x_i} (s_i - \tau_i)^{a_i(m_i-k_i-1)} ds_i = \frac{(x_i - \tau_i)^{a_i(m_i-k_i-1)+1}}{a_i(m_i - k_i - 1) + 1}.$$

Now the claim is clear. $\square$

We give

Corollary 19.6. *All are as in Theorem 19.5 and* $1/p := \sum_{i=1}^n 1/p_i$, $r > 0$. *Then*

$$I_2 \leq p^{1/rp} A \prod_{i=1}^n x_i^{1/q_i} \left[\sum_{i=1}^n \frac{1}{p_i} \right.$$

$$\left. \times \left\{ \int_0^{x_i} |u_i^{(m_i)}(\tau_i)|^{p_i} (x_i - \tau_i)^{[a_i(m_i-k_i-1)+1]} d\tau_i \right\}^r \right]^{1/rp}.$$

Proof. By using inequality of means and (19.6). $\square$

We present

Theorem 19.7. *Let* $u_i \in C^{m_i+1}([0, x_i])$ *be such that* $u_i^{(j)}(0) = 0$ *for* $j \in \{0, \ldots, m_i\}$, *and let* $\rho \in C^1([0, \infty))$, *where* $\rho > 0$; $x_i > 0$, $m_i \in \mathbb{N}$, $j \in \{0, \ldots, m_i-1\}$, $i = 1, \ldots, n$. *Let* $k_i \in \{0, \ldots, m_i-1\}$; $p_i, q_i > 1 : 1/p_i + 1/q_i = 1$; $a_i, b_i \in [0, 1] : a_i + b_i = 1$. *Also Let* $w_i > 0 : \sum_{i=1}^n w_i = \Omega_n$. *Then*

$$I_3 \quad : \quad = \int_0^{x_1} \cdots \int_0^{x_n} \frac{\prod_{i=1}^n |u_i^{(k_i)}(s_i)|}{\left(\frac{1}{\Omega_n} \sum_{i=1}^n w_i \frac{s_i^{[(a_i+b_iq_i)(m_i-k_i-1)+1]/q_iw_i}}{[(a_i+b_iq_i)(m_i-k_i-1)+1]^{1/q_iw_i}} \right)^{\Omega_n}} ds_1 \cdots ds_n$$

$$\leq A \prod_{i=1}^n x_i^{1/q_i} \left[\int_0^{x_i} \frac{\tau_i^{p_i-1}}{(\rho(\tau_i))^{p_i}} \left(\int_0^{\tau_i} |(\rho(\sigma_i)u_i^{(m_i)}(\sigma_i))'|^{p_i} d\sigma_i \right) \right.$$

$$\left. \times (x_i - \tau_i)^{[a_i(m_i-k_i-1)+1]} d\tau_i \right]^{1/p_i}, \tag{19.7}$$

where

$$A := \prod_{i=1}^n \left(\frac{1}{(m_i - k_i - 1)![a_i(m_i - k_i - 1) + 1]^{1/p_i}} \right).$$

Proof. By [320, Equation (14)] we have

$$u_i^{(k_i)}(s_i) \quad = \quad \frac{1}{(m_i - k_i - 1)!} \int_0^{s_i} (s_i - \tau_i)^{m_i-k_i-1}$$

$$\times \left(\frac{1}{\rho(\tau_i)} \int_0^{\tau_i} (\rho(\sigma_i)u_i^{(m_i)}(\sigma_i))' d\sigma_i \right) d\tau_i.$$

By Hölder's inequality we obtain

$$\int_0^{\tau_i} \left|\left(\rho(\sigma_i)u_i^{(m_i)}(\sigma_i)\right)'\right| d\sigma_i \leq \tau_i^{1/q_i} \left(\int_0^{\tau_i} \left|\left(\rho(\sigma_i)u_i^{(m_i)}(\sigma_i)\right)'\right|^{p_i} d\sigma_i\right)^{1/p_i}.$$

We set here

$$\Phi_i(\tau_i) := \frac{1}{(m_i - k_i - 1)!} \frac{\tau_i^{1/q_i}}{\rho(\tau_i)} \left(\int_0^{\tau_i} \left|\left(\rho(\sigma_i)u_i^{(m_i)}(\sigma_i)\right)'\right|^{p_i} d\sigma_i\right)^{1/p_i}.$$

The claim now is clear. $\quad\square$

We give

Corollary 19.8. *All are as in Theorem 19.7 and* $1/p := \sum_{i=1}^n 1/p_i$, $r > 0$. *Then*

$$I_3 \leq p^{1/rp} A \prod_{i=1}^n x_i^{1/q_i} \left[\sum_{i=1}^n \frac{1}{p_i} \left\{\int_0^{x_i} \frac{\tau_i^{p_i-1}}{(\rho(\tau_i))^{p_i}} \left(\int_0^{\tau_i} \left|\left(\rho(\sigma_i)u_i^{(m_i)}(\sigma_i)\right)'\right|^{p_i} d\sigma_i\right)\right.\right.$$
$$\left.\left.\times (x_i - \tau_i)^{[a_i(m_i-k_i-1)+1]} d\tau_i\right\}\right]^{1/rp}.$$

Proof. By using inequality of means and (19.7). $\quad\square$

We give

Theorem 19.9. *Let* $u_i \in C^{2m_i}([0, x_i])$, $\rho \in C^m([0, \infty))$, $\rho > 0$, *with* $m = \max m_i$, *be such that* $u_i^{(j)}(0) = 0$, *and* $\left(\rho(s_i)u_i^{(m_i)}(s_i)\right)^{(j)} = 0$ *at* $s_i = 0$, *for* $j \in \{0, \ldots, m_i - 1\}$; $j \in \{1, \ldots, n\}$, $x_i > 0$, $m_i \in \mathbb{N}$. *Let* $k_i \in \{0, \ldots, m_i - 1\}$; $p_i, q_i > 1: 1/p_i + 1/q_i = 1$; $a_i, b_i \in [0, 1]: a_i + b_i = 1$. *Also let* $w_i > 0: \sum_{i=1}^n w_i = \Omega_n$. *Then*

$$I_4 \quad := \int_0^{x_1} \cdots \int_0^{x_n} \frac{\prod_{i=1}^n |u_i^{(k_i)}(s_i)|}{\left(\frac{1}{\Omega_n} \sum_{i=1}^n w_i \frac{s_i^{[(a_i+b_iq_i)(m_i-k_i-1)+1]/q_iw_i}}{[(a_i+b_iq_i)(m_i-k_i-1)+1]^{1/q_iw_i}}\right)^{\Omega_n}} ds_1 \cdots ds_n$$

$$\leq B \prod_{i=1}^n x_i^{1/q_i} \left[\int_0^{x_i} \frac{\tau_i^{(q_i(m_i-1)+1)(p_i-1)}}{(\rho(\tau_i))^{p_i}} \left(\int_0^{\tau_i} \left|\left(\rho(\sigma_i)u_i^{(m_i)}(\sigma_i)\right)^{(m_i)}\right|^{p_i} d\sigma_i\right)\right.$$

$$\left.\times (x_i - \tau_i)^{[a_i(m_i-k_i-1)+1]} d\tau_i\right]^{1/p_i}, \tag{19.8}$$

where

$$B := \prod_{i=1}^n$$

$$\left(\frac{1}{[(m_i - 1)!(m_i - k_i - 1)!(q_i(m_i - 1) + 1)^{1/q_i}] \left[a_i(m_i - k_i - 1) + 1 \right]^{1/p_i}} \right).$$

Proof. By [320, Equation (21)], we get that

$$u_i^{(k_i)}(s_i) = \frac{1}{(m_i - 1)!(m_i - k_i - 1)!} \int_0^{s_i} (s_i - \tau_i)^{m_i - k_i - 1}$$
$$\times \left(\frac{1}{\rho(\tau_i)} \int_0^{\tau_i} (\tau_i - \sigma_i)^{m_i - 1} \left(\rho(\sigma_i) u_i^{(m_i)}(\sigma_i) \right)^{(m_i)} d\sigma_i \right) d\tau_i.$$

Put

$$F_i(\sigma_i) := \left| \left(\rho(\sigma_i) u_i^{(m_i)}(\sigma_i) \right)^{(m_i)} \right|.$$

Using Hölder's inequality we derive

$$\int_0^{\tau_i} (\tau_i - \sigma_i)^{m_i - 1} F_i(\sigma_i) d\sigma_i$$
$$\leq \left(\int_0^{\tau_i} (\tau_i - \sigma_i)^{q_i(m_i - 1)} d\sigma_i \right)^{1/q_i} \left(\int_0^{\tau_i} F_i(\sigma_i)^{p_i} d\sigma_i \right)^{1/p_i}$$
$$= \frac{\tau_i^{(q_i(m_i - 1) + 1)/q_i}}{(q_i(m_i - 1) + 1)^{1/q_i}} \left(\int_0^{\tau_i} F_i(\sigma_i)^{p_i} d\sigma_i \right)^{1/p_i}.$$

Set now

$$\Phi_i(\tau_i) := W_i \frac{\tau_i^{(q_i(m_i - 1) + 1)/q_i}}{\rho(\tau_i)} \left(\int_0^{\tau_i} \left| \left(\rho(\sigma_i) u_i^{(m_i)}(\sigma_i) \right)^{(m_i)} \right|^{p_i} d\sigma_i \right)^{1/p_i},$$

where

$$W_i := \frac{1}{(m_i - 1)!(m_i - k_i - 1)!(q_i(m_i - 1) + 1)^{1/q_i}}.$$

Now apply Theorem 19.3. The claim is now clear. $\square$

We give

Corollary 19.10. *All are as in Theorem 19.9 and $1/p := \sum_{i=1}^n 1/p_i$, $r > 0$. Then*

$$I_4 \leq p^{1/rp} B \prod_{i=1}^n x_i^{1/q_i} \left[\sum_{i=1}^n \frac{1}{p_i} \left\{ \int_0^{x_i} \frac{\tau_i^{(q_i(m_i - 1) + 1)(p_i - 1)}}{(\rho(\tau_i))^{p_i}} \right. \right.$$
$$\left. \left. \times \left(\int_0^{\tau_i} \left| \left(\rho(\sigma_i) u_i^{(m_i)}(\sigma_i) \right)^{(m_i)} \right|^{p_i} d\sigma_i \right) (x_i - \tau_i)^{[a_i(m_i - k_i - 1) + 1]} d\tau_i \right\}^r \right]^{1/rp}.$$

Proof. By using inequality of means and (19.8). $\square$

Note. Similar theorems to our Theorems 19.5–19.9 appeared earlier in [188].

We need (see [187])

Definition 19.11. Let $\alpha > 0$. For any $f \in L_1(0,x)$; $x > 0$, the *Riemann–Liouville fractional integral* of f of order α is defined by

$$(J_\alpha f)(s) := \frac{1}{\Gamma(\alpha)} \int_0^s (s-t)^{\alpha-1} f(t)dt, \quad s \in [0,x], \qquad (19.9)$$

and the *Riemann–Liouville fractional derivative* of f of order α by

$$\Delta^\alpha f(s) := \frac{1}{\Gamma(m-\alpha)} \left(\frac{d}{ds}\right)^m \int_0^s (s-t)^{m-\alpha-1} f(t)dt, \qquad (19.10)$$

where $m := [\alpha] + 1$, $[\cdot]$ is the integral part. In addition, we set

$$\Delta^0 f := f =: J_0 f, \quad J_{-\alpha} f := \Delta^\alpha f \text{ if } \alpha > 0, \quad \Delta^{-\alpha} f := J_\alpha f \text{ if } 0 < \alpha \le 1.$$

If $\alpha \in \mathbb{N}$, then $\Delta^\alpha f = f^{(\alpha)}$.

Definition 19.12. [187]. We say that $f \in L_1(0,x)$ has an L_∞ *fractional derivative* $\Delta^\alpha f$ in $[0,x]$ iff

$$\Delta^{\alpha-k} f \in C([0,x]), \quad k = 1, \ldots, m := [\alpha] + 1; \quad \alpha > 0,$$

and $\Delta^{\alpha-1} f \in AC([0,x])$ (absolutely continuous functions) and $\Delta^\alpha f \in L_\infty(0,x)$.

We also need

Lemma 19.13. [187]. *Let $\alpha \ge 0$, $\beta > \alpha$, let $f \in L_1(0,x)$ have an L_∞ fractional derivative $\Delta^\beta f$ in $[0,x]$, and let $\Delta^{\beta-k} f(0) = 0$ for $k = 1, \ldots, [\beta] + 1$. Then*

$$\Delta^\alpha f(s) = \frac{1}{\Gamma(\beta-\alpha)} \int_0^s (s-t)^{\beta-\alpha-1} \Delta^\beta f(t)dt, \quad s \in [0,x]. \qquad (19.11)$$

Clearly here $\Delta^\alpha f \in L_\infty(0,x)$, thus $\Delta^\alpha f \in L_1(0,x)$.

We give

Theorem 19.14. *Let $i \in \{1, \ldots, n\}$, $x_i > 0$, $\alpha_i \ge 0$, $\beta_i \ge \alpha_i + 1$. Let $f_i \in L_1(0,x_i)$ have L_∞ fractional derivatives $\Delta^{\beta_i} f_i$ in $[0,x_i]$, and let*

$$\Delta^{\beta_i-k_i} f_i(0) = 0 \quad \text{for } k_i = 1, \ldots, [\beta_i] + 1.$$

Let $p_i, q_i > 1$: $1/p_i + 1/q_i = 1$, and $w_i > 0$ such that $\sum_{i=1}^n w_i = \Omega_n$, and $a_i, b_i \in [0,1]$ such that $a_i + b_i = 1$. Then

$$I_5 \quad : \quad = \int_0^{x_1} \cdots \int_0^{x_n} \frac{\prod_{i=1}^n |\Delta^{\alpha_i} f_i(s_i)|}{\left(\frac{1}{\Omega_n} \sum_{i=1}^n w_i \frac{s_i^{[(a_i+b_iq_i)(\beta_i-\alpha_i-1)+1]/q_iw_i}}{((a_i+b_iq_i)(\beta_i-\alpha_i-1)+1)^{1/q_iw_i}}\right)^{\Omega_n}} ds_1 \cdots ds_n$$

$$\leq \quad C \prod_{i=1}^n x_i^{1/q_i} \left[\int_0^{x_i} |\Delta^{\beta_i} f_i(t_i)|^{p_i} (x_i - t_i)^{[a_i(\beta_i-\alpha_i-1)+1]} dt_i\right]^{1/p_i}, \quad (19.12)$$

where

$$C := \prod_{i=1}^n \left(\frac{1}{\Gamma(\beta_i - \alpha_i)(a_i(\beta_i - \alpha_i - 1) + 1)^{1/p_i}}\right).$$

Proof. Here we again apply Theorem 19.3. By Lemma 19.13 we get that $\Delta^{\alpha_i} f_i \in L_1(0, x_i)$, also $\Delta^{\beta_i} f_i \in L_\infty(0, x_i)$, and

$$|\Delta^{\alpha_i} f_i(s_i)| \leq \frac{1}{\Gamma(\beta_i - \alpha_i)} \int_0^{s_i} (s_i - t_i)^{\beta_i - \alpha_i - 1} |\Delta^{\beta_i} f_i(t_i)| dt_i, \quad \forall s_i \in [0, x_i].$$

We put $u_i := \Delta^{\alpha_i} f_i$, $g_i(s_i, t_i) := |s_i - t_i|$,

$$\Phi_i(t_i) := \frac{|\Delta^{\beta_i} f_i(t_i)|}{\Gamma(\beta_i - \alpha_i)} \quad \text{and} \quad r_i := \beta_i - \alpha_i - 1, \quad i = 1, \ldots, n.$$

Here

$$\varphi_i(s_i) := \int_0^{s_i} (s_i - t_i)^{(a_i+b_iq_i)(\beta_i-\alpha_i-1)} dt_i$$

$$= \frac{s_i^{(a_i+b_iq_i)(\beta_i-\alpha_i-1)+1}}{(a_i + b_iq_i)(\beta_i - \alpha_i - 1) + 1} > 0, \quad \forall s \in (0, x_i],$$

with $\varphi_i(0) = 0$. Now the proof is clear. $\square$

Note. Theorem 19.14 is related and is similar to Theorem 3.2 of [187], but in terms of weights is more general.

We give

Corollary 19.15. *All are as in Theorem 19.14 and* $1/p := \sum_{i=1}^n 1/p_i$, $r > 0$. *Then*

$$I_5 \leq p^{1/rp} C \prod_{i=1}^n x_i^{1/q_i} \left[\sum_{i=1}^n \frac{1}{p_i} \left\{\int_0^{x_i} |\Delta^{\beta_i} f_i(t_i)|^{p_i} (x_i - t_i)^{(a_i(\beta_i-\alpha_i-1)+1)} dt_i\right\}^r\right]^{1/rp}.$$

Proof. By using inequality of means and (19.12). □

We need

Definition 19.16. [17, 19, 101, pp. 539–540]. Let $\nu > 0$, $n := [\nu]$ and $\alpha := \nu - n$ $(0 < \alpha < 1)$. Let $f \in C([0,x])$, $s \in [0,x]$. We consider the *Riemann–Liouville fractional integral*

$$(J_\nu f)(s) = \frac{1}{\Gamma(\nu)} \int_0^s (s-t)^{\nu-1} f(t) dt.$$

We consider the subspace $C^\nu([0,x])$ of $C^n([0,x])$:

$$C^\nu([0,x]) := \{f \in C^n([0,x]) : J_{1-\alpha} f^{(n)} \in C^1([0,x])\}.$$

Hence, letting $f \in C^\nu([0,x])$, we define the *generalized ν-fractional derivative* of f over $[0,x]$ as

$$f^{(\nu)} := \left(J_{1-\alpha} f^{(n)}\right)'.$$

Notice that

$$\left(J_{1-\alpha} f^{(n)}\right)(s) = \frac{1}{\Gamma(1-\alpha)} \int_0^s (s-t)^{-\alpha} f^{(n)}(t) dt$$

exists for $f \in C^\nu([0,x])$.

We also need

Lemma 19.17. [17, 19, pp. 544–545]. Let $\nu_i \geq 1$, $\gamma_i \geq 0$, $\nu_i - \gamma_i - 1 \geq 0$, $i = 1, \ldots, n$. Call $n_i := [\nu_i]$. Let $f_i \in C^{\nu_i}([0,x_i])$, $x_i > 0$, such that $f_i^{(j)}(0) = 0$, $j = 0, 1, \ldots, n_i - 1$. It follows that $f_i \in C^{\gamma_i}([0,x_i])$ and

$$f_i^{(\gamma_i)}(s_i) = \frac{1}{\Gamma(\nu_i - \gamma_i)} \int_0^{s_i} (s_i - t_i)^{\nu_i - \gamma_i - 1} f_i^{(\nu_i)}(t_i) dt_i, \ \forall s_i \subset [0,x_i], \ i = 1, \ldots, n.$$

$$(19.13)$$

We make

Remark 19.18. (on Lemma 19.17). That is, by (19.13) we have

$$\left|f_i^{(\gamma_i)}(s_i)\right| \leq \int_0^{s_i} (s_i - t_i)^{\nu_i - \gamma_i - 1} \frac{|f_i^{(\nu_i)}(t_i)|}{\Gamma(\nu_i - \gamma_i)} dt_i, \ \forall s_i \in [0,x_i]; \ i = 1, \ldots, n.$$

$$(19.14)$$

We set here

$$u_i := f_i^{(\gamma_i)}, \ \ g_i(s_i, t_i) := |s_i - t_i|, \ \ r_i := \nu_i - \gamma_i - 1 \geq 0,$$

$$\Phi_i(t_i) := \frac{|f_i^{(\nu_i)}(t_i)|}{\Gamma(\nu_i - \gamma_i)}, \quad \text{all } s_i, t_i \in [0, x_i]; \ i = 1, \ldots, n.$$

So that (19.4) is fulfilled.

By applying Theorem 19.3 now we obtain

Theorem 19.19. *Let $i \in \{1, \ldots, n\}$, $x_i > 0$, $\nu_i \geq 1$, $\gamma_i \geq 0$, $\nu_i - \gamma_i - 1 \geq 0$, $n_i := [\nu_i]$. Let $f_i \in C^{\nu_i}([0, x_i])$, such that $f_i^{(j)}(0) = 0$, $j = 0, 1, \ldots, n_i - 1$. Let $p_i, q_i > 1 : 1/p_i + 1/q_i = 1$, and $w_i > 0$ such that $\sum_{i=1}^{n} w_i = \Omega_n$, and $a_i, b_i \in [0, 1]$ such that $a_i + b_i = 1$. Then*

$$I_6 \ : \ = \int_0^{x_1} \cdots \int_0^{x_n} \frac{\prod\limits_{i=1}^{n} |f_i^{(\gamma_i)}(s_i)|}{\left(\frac{1}{\Omega_n} \sum\limits_{i=1}^{n} w_i \frac{s_i^{((a_i + b_i q_i)(\nu_i - \gamma_i - 1)+1)/q_i w_i}}{((a_i + b_i q_i)(\nu_i - \gamma_i - 1)+1)^{1/q_i w_i}} \right)^{\Omega_n}} ds_1 \cdots ds_n$$

$$\leq \ D \prod_{i=1}^{n} x_i^{1/q_i} \left[\int_0^{x_i} |f_i^{(\nu_i)}(t_i)|^{p_i} (x_i - t_i)^{(a_i(\nu_i - \gamma_i - 1)+1)} dt_i \right]^{1/p_i}, \quad (19.15)$$

where

$$D := \prod_{i=1}^{n} \left(\frac{1}{\Gamma(\nu_i - \gamma_i)(a_i(\nu_i - \gamma_i - 1) + 1)^{1/p_i}} \right).$$

We give

Corollary 19.20. *All are as in Theorem 19.19 and $1/p := \sum_{i=1}^{n} 1/p_i$, $r > 0$. Then*

$$I_6 \ \leq \ p^{1/rp} D \prod_{i=1}^{n} x_i^{1/q_i} \left[\sum_{i=1}^{n} \frac{1}{p_i} \left\{ \int_0^{x_i} |f_i^{(\nu_i)}(t_i)|^{p_i} \right. \right.$$

$$\left. \left. \times (x_i - t_i)^{(a_i(\nu_i - \gamma_i - 1)+1)} dt_i \right\}^{r} \right]^{1/rp}.$$

Proof. By using inequality of means and (19.15). □

We make

Remark 19.21. The following are taken from [405]. Let $f, u_0, u_1, \ldots,$ $u_n \in C^{n+1}([a, b])$, $n \geq 0$, and the Wronskians

$$W_i(x) := W[u_0(x), u_1(x), \ldots, u_i(x)] :=$$

$$\begin{vmatrix} u_0(x) & u_1(x) & \cdots & u_i(x) \\ u_0'(x) & u_1'(x) & \cdots & u_i'(x) \\ \vdots & & & \\ u_0^{(i)}(x) & u_1^{(i)}(x) & \cdots & u_i^{(i)}(x) \end{vmatrix}, \quad i = 0, 1, \ldots, n. \quad (19.16)$$

Assume $W_i(x) > 0$ over $[a, b]$.

For $i \geq 0$, the differential operator of order i (Widder derivative):

$$L_i f(x) := \frac{W[u_0(x), u_1(x), \ldots, u_{i-1}(x), f(x)]}{W_{i-1}(x)}, \quad (19.17)$$

$i = 1, \ldots, n+1$; $L_0 f(x) := f(x)$, $\forall x \in [a, b]$. Consider also

$$g_i(x, t) := \frac{1}{W_i(t)} \begin{vmatrix} u_0(t) & u_1(t) & \cdots & u_i(t) \\ u_0'(t) & u_1'(t) & \cdots & u_i'(t) \\ \vdots & & & \\ u_0^{(i-1)}(t) & u_1^{(i-1)}(t) & \cdots & u_i^{(i-1)}(t) \\ u_0(x) & u_1(x) & \cdots & u_i(x) \end{vmatrix}, \quad (19.18)$$

$i = 1, 2, \ldots, n$; $g_0(x, t) := u_0(x)/u_0(t)$, $\forall x, t \in [a, b]$.

Example [405] Sets $\{u_0, u_1, \ldots, u_n\}$ are $\{1, x, x^2, \ldots, x^n\}$,

$$\{1, \sin x, -\cos x, -\sin 2x, \cos 2x, \ldots, (-1)^{n-1} \sin nx, (-1)^n \cos nx\},$$

and so on.

We also mention the generalized Widder–Taylor formula; see [405].

Theorem 19.22. *Let the functions $f, u_0, u_1, \ldots, u_n \in C^{n+1}([a, b])$, and the Wronskians $W_0(x), W_1(x), \ldots, W_n(x) > 0$ on $[a, b]$, $x \in [a, b]$. Then for $t \in [a, b]$ we have*

$$f(x) = f(t)\frac{u_0(x)}{u_0(t)} + L_1 f(t) g_1(x, t) + \cdots + L_n f(t) g_n(x, t) + R_n(x), \quad (19.19)$$

where

$$R_n(x) := \int_t^x g_n(x, s) L_{n+1} f(s) ds. \quad (19.20)$$

For example, one could take $u_0(x) = c > 0$. If $u_i(x) = x^i$, $i = 0, 1, \ldots, n$, defined on $[a, b]$, then $L_i f(t) = f^{(i)}(t)$, and

$$g_i(x, t) = \frac{(x-t)^i}{i!}, \quad t \in [a, b].$$

We need

Corollary 19.23. (on Theorem 19.22). *By additionally assuming for fixed $t \in [a, b]$ that $L_i f(t) = 0$, $i = 0, 1, \ldots, n$, we get that*

$$f(x) = \int_x^t g_n(x, s) L_{n+1} f(s) ds, \quad \forall x \in [a, b]. \qquad (19.21)$$

We make

Remark 19.24. Here we make use of (19.21). For $i = 1, \ldots, n$ let the function sets

$$\{ f_i, u_{i0}, u_{i1}, \ldots, u_{in_i} \} \subseteq C^{n_i+1}([0, x_i]), \quad x_i > 0, \quad n_i \in \mathbb{N}.$$

Assume that the Wronskians $W_{i0}, W_{i1}, \ldots, W_{in_i} > 0$ on $[0, x_i]$, $i = 1, \ldots, n$. Also assume that

$$L_{ij} f_i(0) = 0; \quad \text{all } j = 0, 1, \ldots, n_i, \quad \text{for all } i = 1, \ldots, n.$$

By (19.21) we obtain that

$$f_i(s_i) = \int_0^{s_i} g_{in_i}(s_i, t_i) L_{i, n_i+1} f_i(t_i) dt_i, \quad \text{all } i = 1, \ldots, n, \ \forall s_i \in [0, x_i].$$
$$(19.22)$$

Thus

$$|f_i(s_i)| \leq \int_0^{s_i} g_{in_i}(s_i, t_i) \big| L_{i, n_i+1} f_i(t_i) \big| dt_i, \quad i = 1, \ldots, n, \ \forall s_i \in [0, x_i].$$
$$(19.23)$$

From [405] we derive that $g_{in_i}(s_i, t_i) \geq 0$ when $s_i \geq t_i \geq 0$, $i = 1, \ldots, n$. When $s_i > t_i$ then

$$g_{in_i}(s_i, t_i) > 0.$$

Using Theorem 19.3 we derive

Theorem 19.25. *Here for $i \in \{1, \ldots, n\}$, $x_i > 0$, assume $\{ f_i, u_{i0}, u_{i1}, \ldots, u_{in_i} \} \subseteq C^{n_i+1}([0, x_i])$, $n_i \in \mathbb{N}$. Suppose that the Wronskians $W_{i0}, W_{i1}, \ldots, W_{in_i} > 0$ on $[0, x_i]$. Further assume that*

$$L_{ij} f_i(0) = 0; \quad \text{all } j = 0, 1, \ldots, n_i, \quad \text{for all } i = 1, \ldots, n.$$

Take $p_i, q_i > 1 \colon 1/p_i + 1/q_i = 1$, and $w_i > 0$ such that $\sum_{i=1}^n w_i = \Omega_n$, and $a_i, b_i \in [0, 1]$ such that $a_i + b_i = 1$. Call

$$\psi_i(s_i) := \int_0^{s_i} \big(g_{in_i}(s_i, t_i) \big)^{(a_i + b_i q_i)} dt_i. \qquad (19.24)$$

Then

$$I_7 := \int_0^{x_1} \cdots \int_0^{x_n} \frac{\prod_{i=1}^n |f_i(s_i)|}{\left(\frac{1}{\Omega_n} \sum_{i=1}^n w_i(\psi_i(s_i))^{1/q_i w_i}\right)^{\Omega_n}} ds_1 \cdots ds_n$$

$$\leq \prod_{i=1}^n x_i^{1/q_i} \left[\int_0^{x_i} \left|L_{i,n_i+1} f_i(t_i)\right|^{p_i} \right.$$

$$\left. \left(\int_{t_i}^{x_i} (g_{in_i}(s_i, t_i))^{a_i} ds_i \right) dt_i \right]^{1/p_i} . \tag{19.25}$$

Proof. In Theorem 19.3 put $u_i := f_i$, $g_i := |g_{in_i}|$, $\Phi_i := |L_{i,n_i+1}(f_i)|$, $\varphi_i := \psi_i$, $i = 1, \ldots, n$. Notice here that $f_i \in C([0, x_i])$,

$$|g_{in_i}| \in C([0, x_i]^2), \quad |L_{i,n_i+1}(f_i)| \in C([0, x_i]), \quad r_i = 1; \text{ all } i = 1, \ldots, n.$$

So (19.4) is fulfilled by (19.23). Also for $s_i > 0$ we get that $\psi_i(s_i) > 0$ with $\psi_i(0) = 0$. Finally we apply (19.5) to get (19.25). $\square$

We give

Corollary 19.26. *All are as in Theorem 19.25 and $1/p := \sum_{i=1}^n 1/p_i$, $r > 0$. Then*

$$I_7 \leq p^{1/rp} \sum_{i=1}^n x_i^{1/q_i} \left[\sum_{i=1}^n \frac{1}{p_i} \left\{ \int_0^{x_i} \left|L_{i,n_i+1} f_i(t_i)\right|^{p_i} \right.\right.$$

$$\left.\left. \times \left(\int_{t_i}^{x_i} (g_{in_i}(s_i, t_i))^{a_i} ds_i \right) dt_i \right\}^r \right]^{1/rp} .$$

Proof. By using inequality of means and (19.25). $\square$

We need

Setting 19.27. Here we use the notation from [239, pp. 145–154]. Let I be a closed interval of $\mathbb{R}$. Let $\alpha_j(x)$, $j = 0, 1, \ldots, m-1$ ($m \in \mathbb{N}$), $h(x)$ be continuous functions on I, and let

$$L = D^m + \alpha_{m-1}(x)D^{m-1} + \cdots + \alpha_0(x)$$

be a fixed linear differential operator on $C^m(I)$. Let $y_1(x), \ldots, y_m(x)$ be a set of linear independent solutions to $Ly = 0$. Here the associated *Green's function* for L is given by

$$H(x,t) := \begin{vmatrix} y_1(t) & \cdots & y_m(t) \\ y_1'(t) & \cdots & y_m'(t) \\ \vdots & & \\ y_1^{(m-2)}(t) & \cdots & y_m^{(m-2)}(t) \\ y_1(x) & \cdots & y_m(x) \end{vmatrix} \Bigg/ \begin{vmatrix} y_1(t) & \cdots & y_m(t) \\ y_1'(t) & \cdots & y_m'(t) \\ \vdots & & \\ y_1^{(m-2)}(t) & \cdots & y_m^{(m-2)}(t) \\ y_1^{(m-1)}(t) & \cdots & y_m^{(m-1)}(t) \end{vmatrix}, \qquad (19.26)$$

which is a continuous function on I^2. Take a fixed $x_0 \in I$; then

$$y(x) = \int_{x_0}^{x} H(x,t)h(t)dt, \quad \text{all } x \in I \qquad (19.27)$$

is the unique solution of the initial value problem

$$Ly = h; \quad y^{(j)}(x_0) = 0, \quad j = 0, 1, \ldots, m-1. \qquad (19.28)$$

Based on Setting 19.27 we make

Remark 19.28. Let $i \in \{1, \ldots, n\}$, $[0, x_i] \subseteq \mathbb{R}$, $x_i > 0$. Let $\alpha_{i,j}(s_i)$, $j = 0, 1, \ldots, m_i - 1$ $(m_i \in \mathbb{N})$, $h_i(s_i)$ be continuous functions on $[0, x_i]$, and let

$$L_i = D^{m_i} + \alpha_{i,m_i-1}(s_i)D^{m_i-1} + \cdots + \alpha_{i,0}(s_i)$$

be a fixed linear differential operator on $C^{m_i}([0, x_i])$. Let $y_{i1}(s_i), \ldots, y_{im_i}(s_i)$ be a set of linear independent solutions to $L_i y = 0$. Here the associated *Green's function* for L_i is denoted by $H_i(s_i, t_i)$, which is a continuous function on $[0, x_i]^2$. Then

$$y_i(s_i) = \int_0^{s_i} H_i(s_i, t_i)h_i(t_i)dt_i, \quad \text{all } s_i \in [0, x_i] \qquad (19.29)$$

is the unique solution of the initial value problem

$$L_i y = h_i; \quad y^{(j)}(0) = 0, \quad j = 0, 1, \ldots, m_i - 1; \quad \text{all } i = 1, \ldots, n. \qquad (19.30)$$

Let $\varepsilon_i > 0$, then

$$|y_i(s_i)| \leq \int_0^{s_i} \left(|H_i(s_i, t_i)| + \varepsilon_i \right) |h_i(t_i)|dt_i, \quad i = 1, \ldots, n. \qquad (19.31)$$

We give

Theorem 19.29. *All elements are as in Remark* 19.28. *Take* $p_i, q_i >$
$1: 1/p_i + 1/q_i = 1$, *and* $w_i > 0$ *such that* $\sum_{i=1}^{n} w_i = \Omega_n$, *and* $a_i, b_i \in$
$[0,1]: a_i + b_i = 1$. *Call*

$$\psi_i(s_i) := \int_0^{s_i} \left(|H_i(s_i,t_i)| + \varepsilon_i\right)^{(a_i+b_iq_i)} dt_i, \quad i=1,\dots,n. \tag{19.32}$$

Then

$$I_8 := \int_0^{x_1} \cdots \int_0^{x_n} \frac{\prod_{i=1}^{n} |y_i(s_i)|}{\left(\frac{1}{\Omega_n} \sum_{i=1}^{n} w_i(\psi_i(s_i))^{1/q_iw_i}\right)^{\Omega_n}} ds_1 \cdots ds_n$$

$$\leq \prod_{i=1}^{n} x_i^{1/q_i} \left[\int_0^{x_i} |h_i(t_i)|^{p_i}\right.$$

$$\left.\left(\int_{t_i}^{x_i} \left(|H_i(s_i,t_i)| + \varepsilon_i\right)^{a_i} ds_i\right) dt_i\right]^{1/p_i}. \tag{19.33}$$

Proof. In Theorem 19.3 put $u_i := y_i$, $g_i := |H_i(\cdot,\cdot)| + \varepsilon_i$, $\Phi_i := |h_i|$,
$\varphi_i := \psi_i$, $i=1,\dots,n$. Notice here that $y_i \in C([0,x_i])$, $(|H_i(\cdot,\cdot)| + \varepsilon_i) \in$
$C([0,x_i]^2)$, $|h_i| \in C([0,x_i])$, $r_i = 1$; all $i=1,\dots,n$. So (19.4) is fulfilled by
(19.31). Also for $s_i > 0$ we get $\psi_i(s_i) > 0$ with $\psi_i(0) = 0$. Finally we apply
(19.5) to get (19.33). $\quad\square$

We give

Corollary 19.30. *All are as in Remark* 19.28 *and Theorem* 19.29, *and*
$1/p = \sum_{i=1}^{n} 1/p_i$, $r > 0$. *Then*

$$I_8 \leq p^{1/rp} \prod_{i=1}^{n} x_i^{1/q_i} \left[\sum_{i=1}^{n} \frac{1}{p_i}\left\{\int_0^{x_i} |h_i(t_i)|^{p_i}\right.\right.$$

$$\left.\left.\times \left(\int_{t_i}^{x_i} \left(|H_i(s_i,t_i)| + \varepsilon_i\right)^{a_i} ds_i\right) dt_i\right\}^r\right]^{1/rp}.$$

Proof. By using inequality of means and (19.33). $\quad\square$

Note 19.31. One can give several examples based on all the above results
by assuming for instance, that all $a_i = 1$, $b_i = 0$ or $a_i = 0$ and $b_i = 1$, with
or without

$$q_i = n, \quad w_i = \frac{1}{n}, \quad p_i = \frac{n}{n-1}, \quad r_i = \bar{r}; \quad \text{all } i=1,\dots,n,$$

also for $p_i = q_i = n = 2$. Especially and additionally in Theorems 19.5–19.9 one can assume $m_i = m$, $k_i = k$. Also in Theorem 19.14 we can set $\alpha_i = \alpha$, $\beta_i = \beta$; in Theorem 19.19 we can set $\nu_i = \nu$, $\gamma_i = \gamma$, all $i = 1, \ldots, n$, and so on.

20
General Multivariate Hilbert–Pachpatte-Type Integral Inequalities

In this chapter we present general weighted Hilbert–Pachpatte-type multivariate integral inequalities. These are with regard to ordinary partial derivatives and fractional partial derivatives of Canavati and Riemann–Liouville types. These results are applications of a general theorem we present for integrable multivariate functions. This treatment relies on [42].

20.1 Introduction

In this chapter we present several weighted general Hilbert–Pachpatte-type multidimensional integral inequalities for ordinary partial derivatives and fractional partial derivatives of Canavati and Riemann–Liouville kinds. First we establish a very general multivariate result regarding integrable functions (see Theorem 20.3) to which the rest of the results are applications. Along the way many other interesting results are given. Our results are mainly inspired by [189].

To stimulate the reader's interest, we give an interesting result from [322]. In this theorem, $H(I \times J)$ denotes the class of functions $u \in C^{(n-1, m-1)}(I \times J)$ such that $D_1^i u(0, t) = 0$, $0 \leq i \leq n-1$, $t \in J$, $D_2^j u(s, 0) = 0$, $0 \leq j \leq m-1$, $s \in I$, and $D_1^n D_2^{m-1} u(s, t)$ and $D_1^{n-1} D_2^m u(s, t)$ are absolutely continuous on $I \times J$. Here I, J are intervals of the type $I_\xi = [0, \xi)$, $\xi \in \mathbb{R}$.

Theorem 20.1. [322, Theorem 1] Let $u(s, t) \in H(I_x \times I_y)$ and $v(k, r) \in H(I_z \times I_w)$. Then, for $0 \leq i \leq n-1$, $0 \leq j \leq m-1$, the following inequality holds.

G.A. Anastassiou, *Fractional Differentiation Inequalities*,   523
DOI 10.1007/978-0-387-98128-4_20, © Springer Science+Business Media, LLC 2009

$$\int_0^x \int_0^y \left(\int_0^z \int_0^w \frac{|D_1^i D_2^j u(s,t) D_1^i D_2^j v(k,r)| dk dr}{s^{2n-2i-1} t^{2m-2j-1} + k^{2n-2i-1} r^{2m-2j-1}} \right) ds dt$$

$$\leq \frac{1}{2} (A_{i,j} B_{i,j})^2 \sqrt{xyzw} \left(\int_0^x \int_0^y (x-s)(y-t) \left(D_1^n D_2^m u(s,t) \right)^2 ds dt \right)^{1/2}$$

$$\times \left(\int_0^z \int_0^w (z-k)(w-r) \left(D_1^n D_2^m u(k,r) \right)^2 dk dr \right)^{1/2}, \qquad (20.1)$$

where

$$A_{ij} \quad : \quad = \frac{1}{(n-i-1)!(m-j-1)!},$$

$$B_{ij} \quad = \quad \frac{1}{(2n-2i-1)(2m-2j-1)}. \qquad (20.2)$$

This chapter's results generalize the results of [189, 322], and then mainly treat fractional partial derivatives.

20.2   Symbols and Basics

By $\mathbb{Z}(\mathbb{Z}_+)$ and $\mathbb{R}(\mathbb{R}_+)$ we denote the sets of all (nonnegative) integers and (nonnegative) real numbers. We work with functions of d variables, where d is a fixed positive integer, writing the variable as a vector $s = (s^1, \ldots, s^d) \in \mathbb{R}^d$. A multi-index m is an element $m = (m^1, \ldots, m^d)$ of $\mathbb{Z}_+^d$. As usual, the factorial of a multi-index m is defined by $m! = \cdots m^d!$. An integer j may be regarded as the multi-index $(j, \ldots, j)$ depending on the context. For vectors in $\mathbb{R}^d$ and multi-indices we use the usual operations of vector addition and multiplication of vectors by scalars. We write $s \leq \tau$ ($s < \tau$) if $s^j \leq \tau^j$ ($s^j < \tau^j$) for $1 \leq j \leq d$. The same convention applies to multi-indices. In particular, $s \geq 0$ ($s > 0$) means $s^j \geq 0$ ($s^j > 0$) if $1 \leq j \leq d$.

If $s = (s^1, \ldots, s^d) \in \mathbb{R}^d$ and $s > 0$, we define the *cell*

$$Q(s) = [0, s^1] \times \cdots \times [0, s^j] \times \cdots \times [0, s^d];$$

replacing the factor $[0, s^j]$ by $\{0\}$ in this product, we have the *face* $\partial_j Q(s)$ of $Q(s)$.

Let $s = (s^1, \ldots, s^d)$, $\tau = (\tau^1, \ldots, \tau^d) \in \mathbb{R}^d$, $s, \tau > 0$, let $k = (k^1, \ldots, k^d)$ be a multi-index, and let $u : Q(s) \to \mathbb{R}$. Write $D_j = \partial/\partial s^j$. We use the following notation:

$$s^\tau \quad = \quad (s^1)^{\tau^1} \cdots (s^d)^{\tau^d},$$

$$D^k u(s) \quad = \quad D_1^{k^1} \cdots D_d^{k^d} u(s), \qquad (20.3)$$

$$\int_0^s u(\tau) d\tau \quad = \quad \int_0^{s^1} \cdots \int_0^{s^d} u(\tau) d\tau^1 \cdots d\tau^d.$$

An exponent $\alpha \in \mathbb{R}$ in the expression s^α, where $s \in \mathbb{R}^d$, is regarded as a multiexponent; that is, $s^\alpha = s^{(\alpha,\dots,\alpha)}$. Let $N \in \mathbb{N}$ be fixed.

The following notation and hypotheses are used throughout the chapter.

$$I = \{1,\dots,N\}$$

$m_i,\ i \in I$	$m_i = (m_i^1,\dots,m_i^d) \in \mathbb{Z}_+^d$
$x_i,\ i \in I$	$x_i = (x_i^1,\dots,x_i^d) \in \mathbb{R}^d,\ x_i > 0$
$p_i, q_i,\ i \in I$	$p_i, q_i > 1,\ \frac{1}{p_i} + \frac{1}{q_i} = 1$
$p,\ q$	$\frac{1}{p} = \sum_{i=1}^{N}\frac{1}{p_i},\ \frac{1}{q} = \sum_{i=1}^{N}\frac{1}{q_i}$
$a_i, b_i,\ i \in I$	$a,b_i \in \mathbb{R}_+,\ a_i + b_i = 1$
$w_i,\ i \in I$	$w_i \in \mathbb{R},\ w_i > 0,\ \sum_{i=1}^{N} w_i = \Omega_N.$

Throughout this chapter, u_i, v_i, Φ_i denote functions from $Q(x_i)$ to $\mathbb{R}$ that are measurable with further L_p-type assumptions. If m is a multi-index and $x \in \mathbb{R}^d$, $x > 0$, then $C^m(Q(x))$ denotes the set of all functions $u : Q(x) \to \mathbb{R}$ that possess continuous derivatives $D^k u$, where $0 \le k \le m$.

We further denote $\alpha_i := (a_i + b_i q_i)c_i \in \mathbb{R}^d$, $\beta_i := a_i c_i \in \mathbb{R}^d$, for $c_i \in \mathbb{R}^d$, $i \in I$. In particular we observe that

$$x_i^{1/q_i} = (x_i^1)^{1/q_i} \cdots (x_i^d)^{1/q_i}, \quad \prod_{i=1}^{N}(\alpha_i + 1)^{1/q_i} = \prod_{i=1}^{N}\prod_{j=1}^{d}(\alpha_i^j + 1)^{1/q_i}.$$

We give and need the next basic result.

Proposition 20.2. *In the space $C^m(Q(x))$, m is a multi-index; the order of differentiation does not matter. That is, $D^k u(s)$ (see (20.3)) remains the same under any permutation of $(D_1^{k^1},\dots,D_d^{k^d})$.*

Proof. It is enough to prove it for $d = 2$, which we prove by mathematical induction. So let $f \in C^{(\tilde{m},n)}(Q(x))$; $\tilde{m}, n \in \mathbb{N}$. We prove in a neighborhood of $(x_0, y_0) \in Q(x)$ that

$$\frac{\partial^j}{\partial x^j}\frac{\partial^k f}{\partial y^k} = P\left(\frac{\partial^j}{\partial x^j},\frac{\partial^k}{\partial y^k}\right) f, \quad 0 \le j \le \tilde{m},\ 0 \le k \le n,$$

P stands for any permutation of the involved differentials. $\square$

Claim: It holds

$$\frac{\partial}{\partial x}\frac{\partial^k}{\partial y^k}f = \frac{\partial^k}{\partial y^k}\frac{\partial}{\partial x}f, \quad 0 \le k \le n. \tag{20.4}$$

Proof of Claim: For $n = 0$ we have (20.4) is true trivially.

Using Theorem 6.20, p. 121 of [68] (i.e., if f_x, f_y, f_{yx} are continuous, then f_{xy} exists and $f_{yx} = f_{xy}$), we have that (20.4) is true when $n = 1$.

Next we use mathematical induction. Assume the claim true for n; we prove it for $n + 1$. So let $f \in C^{(\tilde{m}, n+1)}$. We prove that

$$\frac{\partial}{\partial x} \frac{\partial^{n+1}}{\partial y^{n+1}} f = \frac{\partial^{n+1}}{\partial y^{n+1}} \frac{\partial}{\partial x} f.$$

Notice that $C^{(\tilde{m}, n+1)}(Q(x)) \subset C^{(\tilde{m}, n)}(Q(x))$, and for $0 \le k \le n$ by the induction hypothesis it holds

$$\frac{\partial}{\partial x} \frac{\partial^k}{\partial y^k} f = \frac{\partial^k}{\partial y^k} \frac{\partial}{\partial x} f.$$

But $g := \partial / \partial y f \in C^{(\tilde{m}, n)}$. Hence by the induction hypothesis we have

$$\frac{\partial}{\partial x} \frac{\partial^n}{\partial y^n} g = \frac{\partial^n}{\partial y^n} \frac{\partial}{\partial x} g;$$

That is,

$$\frac{\partial}{\partial x} \frac{\partial^{n+1} f}{\partial y^{n+1}} = \frac{\partial^n}{\partial y^n} \frac{\partial}{\partial x} \frac{\partial f}{\partial y} = \frac{\partial^{n+1}}{\partial y^{n+1}} \frac{\partial f}{\partial x},$$

proving the claim.

Let now $0 \le \ell \le \tilde{m}$, $0 \le k \le n$. Using the above claim we obtain

$$\frac{\partial^\ell}{\partial x^\ell} \frac{\partial^k}{\partial y^k} f = \frac{\partial^{\ell-1}}{\partial x^{\ell-1}} \left(\frac{\partial}{\partial x} \frac{\partial^k}{\partial y^k} f \right) = \frac{\partial^{\ell-1}}{\partial x^{\ell-1}} \left(\frac{\partial^k}{\partial y^k} \frac{\partial}{\partial x} f \right)$$

$$\left(\text{set } h := \frac{\partial}{\partial x} f \in C^{(\tilde{m}-1, n)} \right)$$

$$= \frac{\partial^{\ell-1}}{\partial x^{\ell-1}} \left(\frac{\partial^k}{\partial y^k} h \right) = \frac{\partial^{\ell-2}}{\partial x^{\ell-2}} \left(\frac{\partial}{\partial x} \frac{\partial^k}{\partial y^k} h \right)$$

$$= \frac{\partial^{\ell-2}}{\partial x^{\ell-2}} \left(\frac{\partial^k}{\partial y^k} \frac{\partial}{\partial x} h \right)$$

$$\left(\text{set } \rho := \frac{\partial h}{\partial x} \in C^{(\tilde{m}-2, n)} \right)$$

$$= \frac{\partial^{\ell-2}}{\partial x^{\ell-2}} \left(\frac{\partial^k \rho}{\partial x^k} \right) = \frac{\partial^{\ell-3}}{\partial x^{\ell-3}} \left(\frac{\partial}{\partial x} \frac{\partial^k \rho}{\partial y^k} \right)$$

$$= \frac{\partial^{\ell-3}}{\partial x^{\ell-3}} \left(\frac{\partial^k}{\partial y^k} \frac{\partial}{\partial x} \rho \right) \cdots$$

$$= P \left(\frac{\partial^\ell}{\partial x^\ell}, \frac{\partial^k}{\partial y^k} \right) f, \quad \text{true.} \quad \square$$

20.3   Main Results

We give our first very general result on which the next results rely.

Theorem 20.3. *Let* $v_i \in L_1(Q(x_i))$, $\Phi_i \in L_\infty(Q(x_i))$, $\Phi_i \geq 0$, $c_i > -1$; $c_i \in \mathbb{R}^d$, $i = 1, \ldots, N$. Set

$$U := \frac{1}{\prod_{i=1}^{N} [(\alpha_i + 1)^{1/q_i}(\beta_i + 1)^{1/p_i}]}.$$

Let $J_i := \{j \in \{1, \ldots, d\} \mid -1 < c_i^j < 0\}$. *If* $J_i \neq \emptyset$ *we take*

$$b_i > \max_{j \in J_i} \left(\frac{(1 + c_i^j)}{c_i^j (1 - q_i)} \right).$$

Suppose

$$|v_i(s_i)| \leq \int_0^{s_i} (s_i - \tau_i)^{c_i} \Phi_i(\tau_i) d\tau_i, \quad \forall s_i \in Q(x_i), \ i = 1, \ldots, N. \quad (20.5)$$

Then

$$M \quad : \quad = \int_0^{x_1} \cdots \int_0^{x_N} \frac{\prod_{i=1}^{N} |v_i(s_i)| ds_1 \cdots ds_N}{\left(\frac{1}{\Omega_N} \sum_{i=1}^{N} w_i s_i^{(\alpha_i+1)/q_i w_i} \right)^{\Omega_N}}$$

$$\leq \quad U \prod_{i=1}^{N} x_i^{1/q_i} \prod_{i=1}^{N} \left(\int_0^{x_i} (x_i - s_i)^{\beta_i+1} \Phi_i(s_i)^{p_i} ds_i \right)^{1/p_i}. \quad (20.6)$$

Proof. One can write

$$(s_i - \tau_i)^{c_i} \Phi_i(\tau_i) = (s_i - \tau_i)^{(a_i/q_i+b_i)c_i}(s_i - \tau_i)^{(a_i/p_i)c_i} \Phi_i(\tau_i).$$

Next we rewrite (20.5), and apply Hölder's inequality and Fubini's theorem to get

$$|v_i(s_i)| \leq \int_0^{s_i} (s_i - \tau_i)^{(a_i/q_i+b_i)c_i}(s_i - \tau_i)^{(a_i/p_i)c_i}\Phi_i(\tau_i)d\tau_i$$

$$\leq \left(\int_0^{s_i} (s_i - \tau_i)^{(a_i+b_i q_i)c_i} d\tau_i \right)^{1/q_i} \times \left(\int_0^{s_i} (s_i - \tau_i)^{a_i c_i}\Phi_i(\tau_i)^{p_i} d\tau_i \right)^{1/p_i}$$

$$= \frac{s_i^{(\alpha_i+1)/q_i}}{(\alpha_i+1)^{1/q_i}} \left(\int_0^{s_i} (s_i - \tau_i)^{\beta_i}\Phi_i(\tau_i)^{p_i} d\tau_i \right)^{1/p_i}.$$

That is, we found that

$$|v_i(s_i)| \leq \frac{s_i^{(\alpha_i+1)/q_i}}{(\alpha_i+1)^{1/q_i}} \left(\int_0^{s_i} (s_i - \tau_i)^{\beta_i}\Phi_i(\tau_i)^{p_i} d\tau_i \right)^{1/p_i}. \quad (20.7)$$

Using the inequality of general means [297, p. 15] we have

$$\prod_{i=1}^{N} s_i^{(\alpha_i+1)/q_i} \leq \left(\frac{1}{\Omega_N} \sum_{i=1}^{N} w_i s_i^{(\alpha_i+1)/q_i w_i}\right)^{\Omega_N}. \tag{20.8}$$

We also have by (20.7) that

$$\prod_{i=1}^{N} |v_i(s_i)| \leq \prod_{i=1}^{N} \frac{s_i^{(\alpha_i+1)/q_i}}{(\alpha_i+1)^{1/q_i}} \prod_{i=1}^{N} \left(\int_0^{s_i} (s_i-\tau_i)^{\beta_i} \Phi_i(\tau_i)^{p_i} d\tau_i\right)^{1/p_i}. \tag{20.9}$$

Put

$$W := \frac{1}{\prod\limits_{i=1}^{N} (\alpha_i+1)^{1/q_i}}.$$

Using (20.9) and (20.8) we find

$$\prod_{i=1}^{N} |v_i(s_i)| \leq W \left(\frac{1}{\Omega_N} \sum_{i=1}^{N} w_i s_i^{(\alpha_i+1)/q_i w_i}\right)^{\Omega_N}$$

$$\prod_{i=1}^{N} \left(\int_0^{s_i} (s_i-\tau_i)^{\beta_i} \Phi_i(\tau_i)^{p_i} d\tau_i\right)^{1/p_i}. \tag{20.10}$$

Thus

$$\frac{\prod\limits_{i=1}^{N} |v_i(s_i)|}{\left(\frac{1}{\Omega_N} \sum\limits_{i=1}^{N} w_i s_i^{(\alpha_i+1)/q_i w_i}\right)^{\Omega_N}} \leq W \prod_{i=1}^{N} \left(\int_0^{s_i} (s_i-\tau_i)^{\beta_i} \Phi_i(\tau_i)^{p_i} d\tau_i\right)^{1/p_i},$$

$$\tag{20.11}$$

whenever at least one $s_i > 0$, $i \in \{1, \ldots, N\}$, where $s_i \in Q(x_i)$.

Next we integrate (20.11), use Hölder's inequality and Fubini's theorem, and at the end we change the order of integration. So we have

$$\int_0^{x_1} \cdots \int_0^{x_N} \frac{\prod\limits_{i=1}^{N} |v_i(s_i)| ds_1 \cdots ds_N}{\left(\frac{1}{\Omega_N} \sum\limits_{i=1}^{N} w_i s_i^{(\alpha_i+1)/q_i w_i}\right)^{\Omega_N}}$$

$$\leq W \prod_{i=1}^{N} \left[\int_0^{x_i} \left(\int_0^{s_i} (s_i-\tau_i)^{\beta_i} \Phi_i(\tau_i)^{p_i} d\tau_i\right)^{1/p_i} ds_i\right]$$

$$\leq W \prod_{i=1}^{N} x_i^{1/q_i} \left(\int_0^{x_i} \left(\int_0^{s_i} (s_i - \tau_i)^{\beta_i} \Phi_i(\tau_i)^{p_i} d\tau_i \right) ds_i \right)^{1/p_i}$$

$$= \frac{W}{\prod_{i=1}^{N} (\beta_i + 1)^{1/p_i}} \sum_{i=1}^{N} x_i^{1/q_i} \prod_{i=1}^{N} \left(\int_0^{x_i} (x_i - \tau_i)^{\beta_i+1} \Phi_i(\tau_i)^{p_i} d\tau_i \right)^{1/p_i},$$

proving (20.6). $\square$

Note. Theorem 20.3 and the proof generalize a lot and resemble Theorem 3.1 and its proof in [189].

Corollary 20.4. *All are as in Theorem 20.3. Then*

$$M \leq p^{1/rp} U \prod_{i=1}^{N} x_i^{1/q_i} \left[\sum_{i=1}^{N} \frac{1}{p_i} \left(\int_0^{x_i} (x_i - s_i)^{\beta_i+1} \Phi_i(s_i)^{p_i} ds_i \right)^r \right]^{1/rp},$$

$$\tag{20.12}$$

where $r > 0$.

Proof. By the inequality of means, for any $A_i \geq 0$, $r > 0$, we obtain

$$\prod_{i=1}^{N} A_i^{1/p_i} \leq \left(p \left(\sum_{i=1}^{N} \frac{1}{p_i} A_i^r \right) \right)^{1/rp}.$$

Plug into it

$$A_i := \int_0^{x_i} (x_i - s_i)^{\beta_i+1} \phi_i(s_i)^{p_i} ds_i.$$

Thus inequality (20.12) is now established. $\square$

Next we apply Theorem 20.3 to ordinary partial derivatives.

Convention 20.5. Here we assume that m_i, k_i are multi-indices satisfying $0 \leq k_i < m_i - 1$, and write

$$\alpha_i = (a_i + b_i q_i)(m_i - k_i - 1), \quad \beta_i = a_i(m_i - k_i - 1). \tag{20.13}$$

Recall that $m_i - k_i - 1 = (m_i^1 - k_i^1 - 1, \ldots, m_1^d - k_i^d - 1)$.

We give

Theorem 20.6. *Under Convention 20.5, let* $u_i \in C^{m_i}(Q(x_i))$ *be such that* $D_j^r u_i(s_i) = 0$ *for* $s_i \in \partial_j Q(x_i)$, $0 \leq r \leq m_i^j - 1$, $1 \leq j \leq d$, $i \in I$. *Then*

$$M_1 := \int_0^{x_1} \cdots \int_0^{x_N} \frac{\prod_{i=1}^{N} |D^{k_i} u_i(s_i)| ds_1 \cdots ds_N}{\left(\frac{1}{\Omega_N} \sum_{i=1}^{N} w_i s_i^{(\alpha_i+1)/q_i w_i} \right)^{\Omega_N}}$$

$$\leq U_1 \prod_{i=1}^{N} x_i^{1/q_i} \prod_{i=1}^{N} \left(\int_0^{x_i} (x_i - s_i)^{\beta_i + 1} |D^{m_i} u_i(s_i)|^{p_i} ds_i \right)^{1/p_i}, \qquad (20.14)$$

where

$$U_1 := \frac{1}{\displaystyle\prod_{i=1}^{N} \left[(m_i - k_i - 1)!(\alpha_i + 1)^{1/q_i}(\beta_i + 1)^{1/p_i} \right]}. \qquad (20.15)$$

Proof. Under the assumptions of the theorem we have from [319] that

$$D^{k_i} u_i(s_i) = \frac{1}{(m_i - k_i - 1)!} \int_0^{s_i} (s_i - \tau_i)^{m_i - k_i - 1} D^{m_i} u_i(\tau_i) d\tau_i, \quad i \in I.$$
$$(20.16)$$

So using (20.6) with $v_i(s_i) = D^{k_i} u_i(s_i)$, $c_i = m_i - k_i - 1$, and $\Phi_i(s_i) = |D^{m_i} u_i(s_i)|/(m_i - k_1 - 1)!$ we get (20.14). $\square$

We give

Corollary 20.7. *All are as in Theorem 20.6. Then*

$$M_1 \leq p^{1/rp} U_1 \prod_{i=1}^{N} x_i^{1/q_i} \left[\sum_{i=1}^{N} \frac{1}{p_i} \left(\int_0^{x_i} (x_i - s_i)^{\beta_i + 1} |D^{m_i} u_i(s_i)|^{p_i} ds_i \right)^r \right]^{1/rp},$$
$$(20.17)$$

where $r > 0$.

Proof. Apply (20.12). $\square$

Note. Theorem 20.6 and Corollary 20.7 are similar but generalize Theorem 4.1 and Corollary 4.2 of [189].

We proceed with

Theorem 20.8. *Let $u_i \in C^{\vec{1}}(Q(x_i))$, $\vec{1} = (1, \ldots, 1) \in \mathbb{N}^d$, be such that $u_i(s_i) = 0$ for $s_i \in \partial_j(Q(s_i))$, $1 \leq j \leq d$, $i \in I$. Then*

$$M_2 : \; = \; \int_0^{x_1} \cdots \int_0^{x_N} \frac{\displaystyle\prod_{i=1}^{N} |u_i(s_i)| ds_1 \cdots ds_N}{\left(\frac{1}{\Omega_N} \displaystyle\sum_{i=1}^{N} w_i s_i^{1/q_i w_i} \right)^{\Omega_N}}$$

$$\leq \; \prod_{i=1}^{N} x_i^{1/q_i} \prod_{i=1}^{N} \left(\int_0^{x_i} (x_i - s_i) |D^{\vec{1}} u_i(s_i)|^{p_i} ds_i \right)^{1/p_i}. \quad (20.18)$$

Proof. By [319] we have that

$$u_i(s_i) = \int_0^{s_i} D^{\vec{1}} u_i(\tau_i) d\tau_i, \quad i \in I.$$

In Theorem 20.3 set $v_i(s_i) = u_i(s_i)$, $c_i = 0$, $\Phi_i(s_i) = |D^{\vec{1}} u_i(s_i)|$ (see here $\alpha_i = \beta_i = 0$, $U = 1$), and the result follows. $\square$

We have

Corollary 20.9. *All are as in Theorem 20.8. Then*

$$M_2 \le p^{1/rp} \prod_{i=1}^{N} x_i^{1/q_i} \left[\sum_{i=1}^{N} \frac{1}{p_i} \left(\int_0^{x_i} (x_i - s_i) |D^{\vec{1}} u_i(s_i)|^{p_i} ds_i \right)^r \right]^{1/rp}.$$

$$(20.19)$$

Proof. Apply (20.12). $\square$

Note 20.10. Theorem 20.8 and Corollary 20.9 are similar and generalize Theorem 4.3 and Corollary 4.4 of [189].

Next we apply Theorem 20.3 to fractional partial derivatives. To keep things simple we work at $d = 2$.

We start with those in the Canavati [101] sense. But first we need to build some necessary background. We need the following univariate concept.

Definition 20.11. [101, 17, 19, pp. 539–540] Let $\nu > 0$, $\tilde{n} := [\nu]$, $[\cdot]$ the integral part, and $\alpha := \nu - \tilde{n}$ $(0 < \alpha < 1)$. Let $f \in C([0, x])$, $x \in \mathbb{R}_+ - \{0\}$, $s \in [0, x]$. We consider the *Riemann–Liouville fractional integral*

$$(J_\nu f)(s) = \frac{1}{\Gamma(\nu)} \int_0^s (s - t)^{\nu-1} f(t) dt. \qquad (20.20)$$

We consider the subspace $C^\nu([0, x])$ of $C^{\tilde{n}}([0, x])$:

$$C^\nu([0, x]) := \{ f \in C^{\tilde{n}}([0, x]) : J_{1-\alpha} f^{(\tilde{n})} \in C^1([0, x]) \}.$$

Hence, let $f \in C^\nu([0, x])$; we define the *generalized ν -fractional derivative* of f over $[0, x]$ as

$$f^{(\nu)} := \left(J_{1-\alpha} f^{(\tilde{n})} \right)'. \qquad (20.21)$$

Notice that

$$\left(J_{1-\alpha} f^{(\tilde{n})} \right) = \frac{1}{\Gamma(1 - \alpha)} \int_0^s (s - t)^{-\alpha} f^{(\tilde{n})}(t) dt$$

exists for $f \in C^\nu([0, x])$, $\forall s \in [0, x]$; $x \in \mathbb{R}_+ - \{0\}$.

We need the following univariate result.

Lemma 20.12. [17, 19, pp. 544–545] *Let $\nu \geq 1, \gamma \geq 0, \nu - \gamma - 1 \geq 0$, $\tilde{n} := [\nu]$. Let $f \in C^\nu([0, x])$, $x \in \mathbb{R}_+ - \{0\}$, such that $f^{(j)}(0) = 0$, $j = 0, 1, \ldots, \tilde{n} - 1$. It follows that $f \in C^\gamma([0, x])$ and*

$$f^{(\gamma)}(s) = \frac{1}{\Gamma(\nu - \gamma)} \int_0^s (s - t)^{\nu - \gamma - 1} f^{(\nu)}(t) dt, \quad \forall s \in [0, x], \ x \in \mathbb{R}_+ - \{0\}.$$

(20.22)

We define fractional partial derivatives in the Canavati sense.

Definition 20.13. (i) Let $\nu > 0$, $\tilde{n} := [\nu]$, $\alpha := \nu - \tilde{n}$; assume here

$$f(\cdot, t) \in C^\nu([0, x]); \ x \in \mathbb{R}_+ - \{0\}, \ \forall t \in [0, y]; \ y \in \mathbb{R}_+ - \{0\}.$$

We define the *ν-partial fractional derivative* of f with respect to s: $\partial^\nu f(\cdot, t)/\partial s^\nu$ as

$$\frac{\partial^\nu f(s, t)}{\partial s^\nu} := \frac{\partial}{\partial s}\left(J_{1-\alpha} \frac{\partial^{\tilde{n}} f(s, t)}{\partial s^{\tilde{n}}}\right), \quad \forall (s, t) \in [0, x] \times [0, y].$$

(20.23)

(ii) Let $\mu > 0$, $m := [\mu]$, $\beta := \mu - m$; assume here

$$f(s, \cdot) \in C^\mu([0, y]); \ y \in \mathbb{R}_+ - \{0\}, \ \forall s \in [0, x]; \ x \in \mathbb{R}_+ - \{0\}.$$

We define the *μ-partial fractional derivative* of f with respect to t: $\partial^\mu f(s, \cdot)/\partial t^\mu$ as

$$\frac{\partial^\mu f(s, t)}{\partial t^\mu} := \frac{\partial}{\partial t}\left(J_{1-\beta} \frac{\partial^m f(s, t)}{\partial t^m}\right), \quad \forall (s, t) \in [0, x] \times [0, y].$$

(20.24)

We present

Theorem 20.14. *Let $f: [0, x] \times [0, y] \to \mathbb{R}$, $x, y \in \mathbb{R}_+ - \{0\}$. Let $\nu > 0$, $n := [\nu]$, $\alpha := \nu - n$; $\rho > 0$, $\tilde{r} := [\rho]$, $\delta := \rho - \tilde{r}$. Assume*

$$\frac{\partial^\rho f}{\partial s_2^\rho}(\cdot, s_2) \in C^\nu([0, x]), \quad \forall s_2 \in [0, y].$$

Suppose also that $\partial^\nu f/\partial s_1^\nu(s_1, \cdot) \in C^\rho([0, y])$, $\forall s_1 \in [0, x]$. That is, there exist the fractional mixed partial derivatives $\partial^{\nu+\rho} f/\partial s_1^\nu \partial s_2^\rho$ and $\partial^{\rho+\nu} f/\partial s_2^\rho \partial s_1^\nu$ on $[0, x] \times [0, y]$, that are meant the obvious iterated way. Suppose further:
(1) $f \in C^{n,\tilde{r}}([0, x] \times [0, y])$.
(2)

$$G_1(t, s_2) := \int_0^{s_2} (s_2 - s)^{-\delta} \frac{\partial^{\tilde{r}} f}{\partial s_2^{\tilde{r}}}(t, s) ds \in C^{n,1}([0, x] \times [0, y]).$$

(20.25)

(3)

$$M_1(t, s_2) := \int_0^{s_2} (s_2 - s)^{-\delta} \frac{\partial^{n+\tilde{r}}}{\partial s_1^n \partial s_2^{\tilde{r}}} f(t, s) ds \in C^1([0, x]) \times [0, y]). \quad (20.26)$$

(4) *Let*

$$F(s_1, s_2) := \int_0^{s_1} \int_0^{s_2} (s_1 - t)^{-\alpha} (s_2 - s)^{-\delta} \frac{\partial^{n+\tilde{r}}}{\partial s_1^n \partial s_2^{\tilde{r}}} f(t, s) ds dt. \quad (20.27)$$

Assume $F_{s_1}, F_{s_2}, F_{s_2 s_1} \in C([0, x] \times [0, y])$.
(5)

$$G_2(s_1, s) := \int_0^{s_1} (s_1 - t)^{-\alpha} \frac{\partial^n}{\partial s_1^n} f(t, s) dt \in C^{1, \tilde{r}}([0, x] \times [0, y]). \quad (20.28)$$

(6)

$$M_2(s_1, s) := \int_0^{s_1} (s_1 - t)^{-\alpha} \frac{\partial^{n+\tilde{r}}}{\partial s_1^n \partial s_2^{\tilde{r}}} f(t, s) dt \in C^1([0, x] \times [0, y]). \quad (20.29)$$

Then

$$\frac{\partial^{\nu + \rho} f}{\partial s_1^\nu \partial s_2^\rho} = \frac{\partial^{\rho + \nu} f}{\partial s_2^\rho \partial s_1^\nu} \quad (20.30)$$

and are continuous on $[0, x] \times [0, y]$.

Proof. We observe the following

$$\frac{\partial^\nu}{\partial s_1^\nu} \left(\frac{\partial^\rho}{\partial s_2^\rho} f(s_1, s_2) \right) = \frac{1}{\Gamma(1 - \alpha)} \frac{\partial}{\partial s_1} \left(\int_0^{s_1} (s_1 - t)^{-\alpha} \frac{\partial^n}{\partial s_1^n} \left(\frac{\partial^\rho}{\partial s_2^\rho} f(t, s_2) \right) dt \right)$$

$$= \frac{1}{\Gamma(1 - \alpha)\Gamma(1 - \delta)} \frac{\partial}{\partial s_1} \left[\int_0^{s_1} (s_1 - t)^{-\alpha} \frac{\partial^n}{\partial s_1^n} \left\{ \frac{\partial}{\partial s_2} \left(\int_0^{s_2} (s_2 - s)^{-\delta} \frac{\partial^{\tilde{r}}}{\partial s_2^{\tilde{r}}} f(t, s) ds \right) \right\} dt \right]$$

(by Proposition 20.2)

$$= \frac{1}{\Gamma(1 - \alpha)\Gamma(1 - \delta)} \frac{\partial}{\partial s_1} \left[\int_0^{s_1} (s_1 - t)^{-\alpha} \frac{\partial}{\partial s_2} \left\{ \frac{\partial^n}{\partial s_1^n} \left(\int_0^{s_2} (s_2 - s)^{-\delta} \frac{\partial^{\tilde{r}}}{\partial s_2^{\tilde{r}}} f(t, s) ds \right) \right\} dt \right]$$

(by Theorem 24.5, p. 193 of [10])

$$= \frac{1}{\Gamma(1 - \alpha)\Gamma(1 - \delta)} \frac{\partial}{\partial s_1} \left[\int_0^{s_1} (s_1 - t)^{-\alpha} \frac{\partial}{\partial s_2} \left(\int_0^{s_2} (s_2 - s)^{-\delta} \frac{\partial^{n+\tilde{r}}}{\partial s_1^n \partial s_2^{\tilde{r}}} f(t, s) ds \right) dt \right]$$

$$= \frac{1}{\Gamma(1 - \alpha)\Gamma(1 - \delta)} \frac{\partial^2}{\partial s_1 \partial s_2} \left(\int_0^{s_1} \int_0^{s_2} (s_1 - t)^{-\alpha} (s_2 - s)^{-\delta} \frac{\partial^{n+\tilde{r}}}{\partial s_1^n \partial s_2^{\tilde{r}}} f(t, s) ds dt \right) =: A.$$

Similarly we have

$$\frac{\partial^\rho}{\partial s_2^\rho}\left(\frac{\partial^\nu}{\partial s_1^\nu}f(s_1,s_2)\right) = \frac{1}{\Gamma(1-\delta)}\frac{\partial}{\partial s_2}\left(\int_0^{s_2}(s_2-s)^{-\delta}\frac{\partial^{\tilde{r}}}{\partial s_2^{\tilde{r}}}\left(\frac{\partial^\nu}{\partial s_1^\nu}f(s_1,s)\right)ds\right)$$

$$= \frac{1}{\Gamma(1-\alpha)\Gamma(1-\delta)}\frac{\partial}{\partial s_2}\left[\int_0^{s_2}(s_2-s)^{-\delta}\frac{\partial^{\tilde{r}}}{\partial s_2^{\tilde{r}}}\left\{\frac{\partial}{\partial s_1}\left(\int_0^{s_1}(s_1-t)^{-\alpha}\frac{\partial^n}{\partial s_1^n}f(t,s)dt\right)\right\}ds\right]$$

(by Proposition 20.2)

$$= \frac{1}{\Gamma(1-\alpha)\Gamma(1-\delta)}\frac{\partial}{\partial s_2}\left(\int_0^{s_2}(s_2-s)^{-\delta}\frac{\partial}{\partial s_1}\left(\frac{\partial^{\tilde{r}}}{\partial s_2^{\tilde{r}}}\left(\int_0^{s_1}(s_1-t)^{-\alpha}\frac{\partial^n}{\partial s_1^n}f(t,s)dt\right)\right)ds\right)$$

(by Theorem 24.5, p. 193 of [10])

$$= \frac{1}{\Gamma(1-\alpha)\Gamma(1-\delta)}\frac{\partial}{\partial s_2}\left(\int_0^{s_2}(s_2-s)^{-\delta}\frac{\partial}{\partial s_1}\left(\int_0^{s_1}(s_1-t)^{-\alpha}\frac{\partial^{n+\tilde{r}}}{\partial s_1^n\partial s_2^{\tilde{r}}}f(t,s)dt\right)ds\right)$$

$$= \frac{1}{\Gamma(1-\alpha)\Gamma(1-\delta)}\frac{\partial^2}{\partial s_2\partial s_1}\left(\int_0^{s_2}\int_0^{s_1}(s_2-s)^{-\delta}(s_1-t)^{-\alpha}\frac{\partial^{n+\tilde{r}}}{\partial s_1^n\partial s_2^{\tilde{r}}}f(t,s)dtds\right)$$

$$\frac{1}{\Gamma(1-\alpha)\Gamma(1-\delta)}\frac{\partial^2}{\partial s_1\partial s_2}\left(\int_0^{s_1}\int_0^{s_2}(s_1-t)^{-\alpha}(s_2-s)^{-\delta}\frac{\partial^{n+\tilde{r}}}{\partial s_1^n\partial s_2^{\tilde{r}}}f(t,s)dtds\right) = A,$$

proving the claim. □

We need

Lemma 20.15. *Let* $f \in L_\infty([0,A]\times[0,B])$, *where* $A,B \in \mathbb{R}_+ - \{0\}$. *Let*

$$F(x,y) := \int_0^x\int_0^y(x-t)^\alpha(y-s)^\beta f(t,s)dtds, \quad \forall x \in [0,A], \forall y \in [0,B]; \alpha,\beta \in \mathbb{R}_+.$$

Then

$$F \in C([0,A]\times[0,B]).$$

Proof. We notice that

$$F(x,y) = \int_0^x\int_0^y|x-t|^\alpha|y-s|^\beta f(t,s)dtds$$

$$= \int_0^A\int_0^B \chi_{[0,x]\times[0,y]}(t,s)|x-t|^\alpha|y-s|^\beta f(t,s)dtds,$$

$$\forall x \in [0,A], \forall y \in [0,B].$$

Let $x_n \in [0,A]$, $y_n \in [0,B]$, $n \in \mathbb{N}$: $(x_n,y_n) \to (x,y)$, as $n \to \infty$, $x \in A$, $y \in B$. We have, as $n \to \infty$, that

$$\chi_{[0,x_n]\times[0,y_n]}(t,s) \longrightarrow \chi_{[0,x]\times[0,y]}(t,s) \quad \text{a.e. in } (t,s) \in [0,A]\times[0,B].$$

Also, as $n \to \infty$, we have that

$$|x_n - t|^\alpha |y_n - s|^\beta f(t, s) \to |x - t|^\alpha |y - s|^\beta f(t, s), \quad \forall (t, s) \in [0, A] \times [0, B].$$

Consequently it holds

$$\chi_{[0, x_n] \times [0, y_n]}(t, s) |x_n - t|^\alpha |y_n - s|^\beta f(t, s) \longrightarrow \chi_{[0, x] \times [0, y]}(t, s) |x - t|^\alpha |y - s|^\beta f(t, s),$$

a.e. in $(t, s) \in [0, A] \times [0, B]$. Furthermore

$$\chi_{[0, x_n] \times [0, y_n]}(t, s) |x_n - t|^\alpha |y_n - s|^\beta |f(t, s)| \le A^\alpha B^\beta |f(t, s)|, \quad \forall (t, s) \in [0, A] \times [0, B],$$

and $A^\alpha B^\beta |f(t, s)|$ is integrable. Hence by the dominated convergence theorem we obtain, as $n \to \infty$, that

$$
\begin{aligned}
F(x_n, y_n) &= \int_0^A \int_0^B \chi_{[0, x_n] \times [0, y_n]}(t, s) |x_n - t|^\alpha |y_n - s|^\beta f(t, s) dt ds \\
&\longrightarrow \int_0^A \int_0^B \chi_{[0, x] \times [0, y]}(t, s) |x - t|^\alpha |y - s|^\beta f(t, s) dt ds = F(x, y),
\end{aligned}
$$

proving that F is jointly continuous in (x, y). $\square$

We present

Theorem 20.16. *Here $i = 1, \ldots, N$. Let $\nu_i \ge 1, \gamma_i \ge 0, \nu_i - \gamma_i - 1 \ge 0$, $n_i = [\nu_i]$, and $\mu_i, \rho_i \ge 1$: $\mu_i - \rho_i - 1 \ge 0$, $m_i := [\mu_i]$, $r_i := [\rho_i]$; $x_i, y_i \in \mathbb{R}_+ - \{0\}$. Let $f_i \in C^{n_i, r_i}([0, x_i] \times [0, y_i])$. Assume that*

$$\frac{\partial^{\rho_i} f_i}{\partial s_{i2}^{\rho_i}}(\cdot, s_{i2}) \in C^{\nu_i}([0, x_i]), \quad \forall s_{i2} \in [0, y_i],$$

and

$$\frac{\partial^j}{\partial s_{i1}^j}\left(\frac{\partial^{\rho_i} f_i}{\partial s_{i2}^{\rho_i}}\right)(0, s_{i2}) = 0, \quad all \ j = 0, 1, \ldots, n_i - 1; \ \forall s_{i2} \in [0, y_i].$$

Also suppose that

$$\frac{\partial^{\nu_i} f_i}{\partial s_{i1}^{\nu_i}}(s_{i1}, \cdot) \in C^{\mu_i}([0, y_i]), \quad \forall s_{i1} \in [0, x_i],$$

and

$$\frac{\partial^j}{\partial s_{i2}^j}\left(\frac{\partial^{\nu_i} f_i}{\partial s_{i1}^{\nu_i}}\right)(s_{i1}, 0) = 0, \quad all \ j = 0, 1, \ldots, m_i - 1; \ \forall s_{i1} \in [0, x_i].$$

We suppose $\partial^{\mu_i + \nu_i} f_i / \partial s_{i2}^{\mu_i} \partial s_{i1}^{\nu_i} \in L_\infty([0, x_i] \times [0, y_i])$. Further assume that f_i fulfills all assumptions of Theorem 20.14 (by indexing all there by i). Then we get the continuous function

$$\frac{\partial^{\gamma_i + \rho_i} f_i(s_{i1}, s_{i2})}{\partial s_{i1}^{\gamma_i} \partial s_{i2}^{\rho_i}} = \frac{1}{\Gamma(\nu_i - \gamma_i)\Gamma(\mu_i - \rho_i)} \int_0^{s_{i1}} \int_0^{s_{i2}} (s_{i1} - t_{i1})^{\nu_i - \gamma_i - 1}(s_{i2} - t_{i2})^{\mu_i - \rho_i - 1}$$

$$\times \frac{\partial^{\mu_i+\nu_i} f_i}{\partial s_{i2}^{\mu_i} \partial s_{i1}^{\nu_i}}(t_{i1}, t_{i2}) dt_{i2} dt_{i1}, \quad \forall (s_{i1}, s_{i2}) \in [0, x_i] \times [0, y_i], \qquad (20.31)$$

all $i = 1, \ldots, N$.

Proof. By using Lemma 20.12 we obtain that

$$\frac{\partial^{\rho_i} f_i}{\partial s_{i2}^{\rho_i}}(\cdot, s_{i2}) \in C^{\gamma_i}([0, x_i]), \quad \forall s_{i2} \in [0, y_i].$$

Furthermore we get

$$\frac{\partial^{\gamma_i+\rho_i} f_i}{\partial s_{i1}^{\gamma_i} \partial s_{i2}^{\rho_i}}(s_{i1}, s_{i2}) = \frac{\partial^{\gamma_i}}{\partial s_{i1}^{\gamma_i}} \left(\frac{\partial^{\rho_i} f_i}{\partial s_{i2}^{\rho_i}} \right)(s_{i1}, s_{i2})$$

$$= \frac{1}{\Gamma(\nu_i - \gamma_i)} \int_0^{s_{i1}} (s_{i1} - t_{i1})^{\nu_i-\gamma_i-1} \frac{\partial^{\nu_i}}{\partial s_{i1}^{\nu_i}} \left(\frac{\partial^{\rho_i} f_i}{\partial s_{i2}^{\rho_i}}(t_{i1}, s_{i2}) \right) dt_{i1}$$

$$= \frac{1}{\Gamma(\nu_i - \gamma_i)} \int_0^{s_{i1}} (s_{i1} - t_{i1})^{\nu_i-\gamma_i-1} \frac{\partial^{\nu_i+\rho_i}}{\partial s_{i1}^{\nu_i} \partial s_{i2}^{\rho_i}} f_i(t_{i1}, s_{i2}) dt_{i1},$$

$$\forall s_{i1} \in [0, x_i]; \quad \forall s_{i2} \in [0, y_i]. \qquad (20.32)$$

Again applying Lemma 20.12 we get

$$\frac{\partial^{\nu_i} f_i}{\partial s_{i1}^{\nu_i}}(t_{i1}, \cdot) \in C^{\rho_i}([0, y_i]), \quad \forall t_{i1} \in [0, x_i],$$

and

$$\frac{\partial^{\rho_i}}{\partial s_{i2}^{\rho_i}} \left(\frac{\partial^{\nu_i} f_i}{\partial s_{i1}^{\nu_i}} \right)(t_{i1}, s_{i2})$$

$$= \frac{1}{\Gamma(\mu_i - \rho_i)} \int_0^{s_{i2}} (s_{i2} - t_{i2})^{\mu_i-\rho_i-1} \frac{\partial^{\mu_i}}{\partial s_{i2}^{\mu_i}} \left(\frac{\partial^{\nu_i} f_i}{\partial s_{i1}^{\nu_i}} \right)(t_{i1}, t_{i2}) dt_{i2},$$

$$\forall s_{i2} \in [0, y_i]; \quad \forall t_{i1} \in [0, x_i].$$

Thus we have

$$\frac{\partial^{\rho_i+\nu_i} f_i}{\partial s_{i2}^{\rho_i} \partial s_{i1}^{\nu_i}}(t_{i1}, s_{i2}) = \frac{1}{\Gamma(\mu_i - \rho_i)} \int_0^{s_{i2}} (s_{i2} - t_{i2})^{\mu_i-\rho_i-1}$$

$$\times \frac{\partial^{\mu_i+\nu_i} f_i}{\partial s_{i2}^{\mu_i} \partial s_{i1}^{\nu_i}}(t_{i1}, t_{i2}) dt_{i2}, \forall s_{i2} \in [0, y_i]; \forall t_{i1} \in [0, x_i]. \qquad (20.33)$$

By Theorem 20.14 we obtain that

$$\frac{\partial^{\rho_i+\nu_i} f_i}{\partial s_{i2}^{\rho_i} \partial s_{i1}^{\nu_i}} = \frac{\partial^{\nu_i+\rho_i} f_i}{\partial s_{i1}^{\nu_i} \partial s_{i2}^{\rho_i}} \quad \text{over } [0, x_i] \times [0, y_i]. \qquad (20.34)$$

Using (20.34) and plugging (20.33) into (20.32) we finally find

$$
\frac{\partial^{\gamma_i+\rho_i} f_i(s_{i1}, s_{i2})}{\partial s_{i1}^{\gamma_i} \partial s_{i2}^{\rho_i}} = \frac{1}{\Gamma(\nu_i - \gamma_i)\Gamma(\mu_i - \rho_i)} \int_0^{s_{i1}} (s_{i1} - t_{i1})^{\nu_i - \gamma_i - 1}
$$

$$
\times \left(\int_0^{s_{i2}} (s_{i2} - t_{i2})^{\mu_i - \rho_i - 1} \right.
$$

$$
\left. \frac{\partial^{\mu_i+\nu_i}}{\partial s_{i2}^{\mu_i} \partial s_{i1}^{\nu_i}} f_i(t_{i1}, t_{i2}) dt_{i2} \right) dt_{i1}, \tag{20.35}
$$

all $i = 1, \ldots, N$; there by proving the claim. Continuity of $\partial^{\gamma_i+\rho_i} f_i(s_{i1}, s_{i2})/ \partial s_{i1}^{\gamma_i} \partial s_{i2}^{\rho_i}$ is arrived at via Lemma 20.15. $\square$

We continue with an application of Theorem 20.3.

Theorem 20.17. *All here are as in Theorem 20.16. Call* $\alpha_{i1} := (a_i + b_i q_i)(\nu_i - \gamma_i - 1)$, $\alpha_{i2} := (a_i + b_i q_i)(\mu_i - \rho_i - 1)$; $\beta_{i1} := a_i(\nu_i - \gamma_i - 1)$, $\beta_{i2} := a_i(\mu_i - \rho_i - 1)$; *all* $i = 1, \ldots, N$. *Then*

$$
M^* := \int_0^{x_1} \int_0^{y_1} \int_0^{x_2} \int_0^{y_2} \cdots \int_0^{x_N} \int_0^{y_N}
$$

$$
\frac{\prod_{i=1}^{N} \left| \frac{\partial^{\gamma_i+\rho_i} f_i(s_{i1}, s_{i2})}{\partial s_{i1}^{\gamma_i} \partial s_{i2}^{\rho_i}} \right| ds_{11} ds_{12} ds_{21} ds_{22} \cdots ds_{N1} ds_{N2}}{\left(\frac{1}{\Omega_N} \sum_{i=1}^{N} w_i s_{i1}^{(\alpha_{i1}+1)/q_i w_i} s_{i2}^{(\alpha_{i2}+1)/q_i w_i} \right)^{\Omega_N}}
$$

$$
\leq U^* \prod_{i=1}^{N} (x_i y_i)^{1/q_i} \sum_{i=1}^{N} \left(\int_0^{x_i} \int_0^{y_i} (x_i - s_{i1})^{\beta_{i1}+1} (y_i - s_{i2})^{\beta_{i2}+1} \right.
$$

$$
\left. \times \left(\left| \frac{\partial^{\mu_i+\nu_i} f_i}{\partial s_{i2}^{\mu_i} \partial s_{i1}^{\nu_i}} (s_{i1}, s_{i2}) \right| \right)^{p_i} ds_{i1} ds_{i2} \right)^{1/p_i}. \tag{20.36}
$$

Here

$$
U^* := \frac{1}{\prod_{i=1}^{N} \left[((\alpha_{i1}+1)(\alpha_{i2}+1))^{1/q_i} ((\beta_{i1}+1)(\beta_{i2}+1))^{1/p_i} \right]}. \tag{20.37}
$$

Proof. Here we use (20.31). Call

$$
v_i(s_{i1}, s_{i2}) := \frac{\partial^{\gamma_i+\rho_i} f_i(s_{i1}, s_{i2})}{\partial s_{i1}^{\gamma_i} \partial s_{i2}^{\rho_i}} \in C([0, x_i] \times [0, y_i])
$$

and by (20.31),

$$\Phi_i(s_{i1}, s_{i2}) := \left| \frac{\partial^{\mu_i + \nu_i} f_i(s_{i1}, s_{i2})}{\partial s_{i2}^{\mu_i} \partial s_{i1}^{\nu_i}} \right| \frac{1}{\Gamma(\nu_i - \gamma_i)\Gamma(\mu_i - \rho_i)} \in L_\infty([0, x_i] \times [0, y_i]),$$

and use Theorem 20.3. Then (20.36) is obvious. $\square$

We give

Corollary 20.18. *All are as in Theorem 20.17. Then*

$$M^* \leq p^{1/rp} U^* \prod_{i=1}^{N} (x_i y_i)^{1/q_i} \left[\sum_{i=1}^{N} \frac{1}{p_i} \left(\int_0^{x_i} \int_0^{y_i} (x_i - s_{i1})^{\beta_{i1}+1} (y_i - s_{i2})^{\beta_{i2}+1} \right. \right.$$

$$\times \left. \left. \left(\left| \frac{\partial^{\mu_i + \nu_i} f_i}{\partial s_{i2}^{\mu_i} \partial s_{i1}^{\nu_i}} (s_{i1}, s_{i2}) \right| \right)^{p_i} ds_{i1} ds_{i2} \right)^r \right]^{1/rp}, \tag{20.38}$$

where $r \in \mathbb{R}_+ - \{0\}$.

Proof. By (20.36) and (20.12). $\square$

Finally, we deal with the Riemann–Liouville fractional partial derivatives. Again we work at $d = 2$.
First we need to build the necessary background.

Definition 20.19. (See [187].) Let $\alpha \in \mathbb{R}_+ - \{0\}$. For any $f \in L_1(0, x)$; $x \in \mathbb{R}_+ - \{0\}$, the *Riemann–Liouville fractional integral* of f of order α is defined by

$$(J_\alpha f)(s) := \frac{1}{\Gamma(\alpha)} \int_0^s (s - t)^{\alpha - 1} f(t) dt, \quad \forall s \in [0, x], \tag{20.39}$$

and the *Riemann–Liouville fractional derivative* of f of *order* α by

$$\Delta^\alpha f(s) := \frac{1}{\Gamma(m - \alpha)} \left(\frac{d}{ds} \right)^m \int_0^s (s - t)^{m - \alpha - 1} f(t) dt, \tag{20.40}$$

where $m := [\alpha] + 1$, $[\cdot]$ is the integral part. In addition, we set $\Delta^0 f := f =: \mathcal{J}_0 f$, $J_{-\alpha} f := \Delta^\alpha f$ if $\alpha > 0$, and $\Delta^{-\alpha} f := J_\alpha f$, if $0 < \alpha \leq 1$. If $\alpha \in \mathbb{N}$, then $\Delta^\alpha f = f^{(\alpha)}$.

Definition 20.20. [187] We say that $f \in L_1(0, x)$ has an L_∞ fractional derivative $\Delta^\alpha f$ in $[0, x]$, $x \in \mathbb{R}_+ - \{0\}$, iff $\Delta^{\alpha - k} f \in C([0, x])$, $k = 1, \ldots, m := [\alpha] + 1$; $\alpha \in \mathbb{R}_+ - \{0\}$, and $\Delta^{\alpha - 1} f \in AC([0, x])$ (absolutely continuous functions) and $\Delta^\alpha f \in L_\infty(0, x)$.

We also need

Lemma 20.21. [187] *Let* $\alpha \in \mathbb{R}_+$, $\beta > \alpha$, *let* $f \in L_1(0,x)$, $x \in \mathbb{R}_+ - \{0\}$, *have an* L_∞ *fractional derivative* $\Delta^\beta f$ *in* $[0,x]$, *and let* $\Delta^{\beta-k} f(0) = 0$ *for* $k = 1, \ldots, [\beta] + 1$. *Then*

$$\Delta^\alpha f(s) = \frac{1}{\Gamma(\beta - \alpha)} \int_0^s (s-t)^{\beta-\alpha-1} \Delta^\beta f(t) dt, \quad \forall s \in [0,x]. \quad (20.41)$$

Clearly here $\Delta^\alpha f \in L_\infty(0,x)$; *thus* $\Delta^\alpha f \in L_1(0,x)$.

We need

Definition 20.22. Let $f : [0,x] \times [0,y] \to \mathbb{R}$, $x, y \in \mathbb{R}_+ - \{0\}$, such that $f(\cdot, s_2) \in L_1(0,x)$, $\forall s_2 \in [0,y]$; $f(s_1, \cdot) \in L_1(0,y)$, $\forall s_1 \in [0,x]$; $\alpha, \beta \in \mathbb{R}_+ - \{0\}$. We define the *Riemann–Liouville fractional partial derivatives* with respect to s_1 (resp., s_2) of f of order α (resp., order β) by

$$\Delta^\alpha_{s_1} f(s_1, s_2) := \frac{1}{\Gamma(m-\alpha)} \left(\frac{\partial}{\partial s_1}\right)^m \int_0^{s_1} (s_1 - t_1)^{m-\alpha-1} f(t_1, s_2) dt_1,$$
$$(20.42)$$

where $m := [\alpha] + 1$, $\forall s_1 \in [0,x]$, all $s_2 \in [0,y]$; and

$$\Delta^\beta_{s_2} f(s_1, s_2) := \frac{1}{\Gamma(r-\beta)} \left(\frac{\partial}{\partial s_2}\right)^r \int_0^{s_2} (s_2 - t_2)^{r-\beta-1} f(s_1, t_2) dt_2, \quad (20.43)$$

where $r := [\beta] + 1$, $\forall s_2 \in [0,y]$, all $s_1 \in [0,x]$.
We define the *Riemann–Liouville fractional mixed (iterated) partial derivatives* as

$$\Delta^\alpha_{s_1} \Delta^\beta_{s_2} f(s_1, s_2) \quad : \quad = \Delta^\alpha_{s_1} \left(\Delta^\beta_{s_2} f(s_1, s_2)\right) := \Delta^{\alpha+\beta}_{s_1 s_2} f(s_1, s_2),$$
$$\Delta^\beta_{s_2} \Delta^\alpha_{s_1} f(s_1, s_2) \quad : \quad = \Delta^\beta_{s_2} \left(\Delta^\alpha_{s_1} f(s_1, s_2)\right) := \Delta^{\beta+\alpha}_{s_2 s_1} f(s_1, s_2),$$
$$\vee(s_1, s_2) \quad \in \quad [0,x] \times [0,y],$$

defined the obvious way via (20.40).

We need and give a commutativity result for the above partials.

Theorem 20.23. *Let* $f \in L_\infty((0,x_1) \times (0,x_2))$, $x_1, x_2 \in \mathbb{R}_+ - \{0\}$. *Let* $\beta_1, \alpha_2 \in \mathbb{R}_+ - \{0\}$, $r_1 := [\beta_1] + 1$, $m_2 := [\alpha_2] + 1$. *Assume that* $f(s_1, \cdot) \in L_1(0,x_2)$ *has* $\Delta^{\alpha_2}_{s_2} f(s_1, \cdot) \in L_\infty(0,x_2)$, $\forall s_1 \in [0,x_1]$, *and be given* $\Delta^{\alpha_2}_{s_2} f(\cdot, s_2) \in L_1(0,x_1)$ *form* $\Delta^{\beta_1}_{s_1}(\Delta^{\alpha_2}_{s_2} f(\cdot, s_2))$ *over* $[0,x_1]$, $\forall s_2 \in [0,x_2]$. *Also assume that* $f(\cdot, s_2) \in L_1(0,x_1)$ *has* $\Delta^{\beta_1}_{s_1} f(\cdot, s_2) \in L_\infty(0,x_1)$, $\forall s_2 \in [0,x_2]$, *and be given* $\Delta^{\beta_1}_{s_1} f(s_1, \cdot) \in L_1(0,x_2)$ *form* $\Delta^{\alpha_2}_{s_2}(\Delta^{\beta_1}_{s_1} f(s_1, \cdot))$ *over* $[0,x_2]$, $\forall s_1 \in [0,x_1]$.
We suppose further:

(1)

$$M_1(t_1, s_2) := \int_0^{s_2} (s_2 - t_2)^{m_2 - \alpha_2 - 1} f(t_1, t_2) dt_2 \in C^{m_2}([0, x_1] \times [0, x_2]).$$
(20.44)

(2)

$$M_2(s_1, t_2) := \int_0^{s_1} (s_1 - t_1)^{r_1 - \beta_1 - 1} f(t_1, t_2) dt_1 \in C^{r_1}([0, x_1] \times [0, x_2]).$$
(20.45)

(3)

$$\lambda(s_1, s_2) := \int_0^{s_1} \int_0^{s_2} (s_1 - t_1)^{r_1 - \beta_1 - 1} (s_2 - t_2)^{m_2 - \alpha_2 - 1} f(t_1, t_2) dt_2 dt_1$$

$$\in C^{r_1, m_2}([0, x_1] \times [0, x_2]).$$
(20.46)

Then there exist $\Delta_{s_1 s_2}^{\beta_1 + \alpha_2} f$, $\Delta_{s_2 s_1}^{\alpha_2 + \beta_1} f$, *are continuous; and*

$$\Delta_{s_1 s_2}^{\beta_1 + \alpha_2} f = \Delta_{s_2 s_1}^{\alpha_2 + \beta_1} f$$
(20.47)

on $[0, x_1] \times [0, x_2]$.

Proof. We observe that

$$\Delta_{s_1 s_2}^{\beta_1 + \alpha_2} f(s_1, s_2) = \Delta_{s_1}^{\beta_1} \left(\Delta_{s_2}^{\alpha_2} f(s_1, s_2) \right)$$

$$= \frac{1}{\Gamma(r_1 - \beta_1)} \left(\frac{\partial}{\partial s_1} \right)^{r_1} \int_0^{s_1} (s_1 - t_1)^{r_1 - \beta_1 - 1} \Delta_{s_2}^{\alpha_2} f(t_1, s_2) dt_1$$

$$= \frac{1}{\Gamma(r_1 - \beta_1)\Gamma(m_2 - \alpha_2)} \left(\frac{\partial}{\partial s_1} \right)^{r_1} \int_0^{s_1} (s_1 - t_1)^{r_1 - \beta_1 - 1}$$

$$\times \left(\left(\frac{\partial}{\partial s_2} \right)^{m_2} \int_0^{s_2} (s_2 - t_2)^{m_2 - \alpha_2 - 1} f(t_1, t_2) dt_2 \right) dt_1$$

(by [10, Theorem 24.5, p. 193])

$$= \frac{1}{\Gamma(r_1 - \beta_1)\Gamma(m_2 - \alpha_2)} \left(\frac{\partial}{\partial s_1} \right)^{r_1} \left(\frac{\partial}{\partial s_2} \right)^{m_2}$$

$$\int_0^{s_1} \int_0^{s_2} (s_1 - t_1)^{r_1 - \beta_1 - 1} (s_2 - t_2)^{m_2 - \alpha_2 - 1} f(t_1, t_2) dt_2 dt_1$$

$$= \frac{1}{\Gamma(r_1 - \beta_1)\Gamma(m_2 - \alpha_2)} \frac{\partial^{r_1 + m_2}}{\partial s_1^{r_1} \partial s_2^{m_2}}$$

$$\int_0^{s_1} \int_0^{s_2} (s_1 - t_1)^{r_1 - \beta_1 - 1} (s_2 - t_2)^{m_2 - \alpha_2 - 1} f(t_1, t_2) dt_2 dt_1 =: B < +\infty.$$

Also we have

$$\Delta_{s_2 s_1}^{\alpha_2+\beta_1} f(s_1, s_2) = \Delta_{s_2}^{\alpha_2}\left(\Delta_{s_1}^{\beta_1} f(s_1, s_2)\right)$$

$$= \frac{1}{\Gamma(m_2-\alpha_2)}\left(\frac{\partial}{\partial s_2}\right)^{m_2} \int_0^{s_2} (s_2-t_2)^{m_2-\alpha_2-1} \Delta_{s_1}^{\beta_1} f(s_1, t_2) dt_2$$

$$= \frac{1}{\Gamma(r_1-\beta_1)\Gamma(m_2-\alpha_2)}\left(\frac{\partial}{\partial s_2}\right)^{m_2} \int_0^{s_2} (s_2-t_2)^{m_2-\alpha_2-1}$$

$$\left(\left(\frac{\partial}{\partial s_1}\right)^{r_1} \int_0^{s_1} (s_1-t_1)^{r_1-\beta_1-1} f(t_1, t_2) dt_1\right) dt_2$$

(by [10, Theorem 24.5, p. 193])

$$= \frac{1}{\Gamma(r_1-\beta_1)\Gamma(m_2-\alpha_2)}\left(\frac{\partial}{\partial s_2}\right)^{m_2}\left(\frac{\partial}{\partial s_1}\right)^{r_1}$$

$$\int_0^{s_2}\left(\int_0^{s_1} (s_2-t_2)^{m_2-\alpha_2-1} (s_1-t_1)^{r_1-\beta_1-1} f(t_1, t_2) dt_1\right) dt_2$$

$$= \frac{1}{\Gamma(r_1-\beta_1)\Gamma(m_2-\alpha_2)} \frac{\partial^{m_2+r_1}}{\partial s_2^{m_2} \partial s_1^{r_1}}$$

$$\int_0^{s_1}\int_0^{s_2} (s_1-t_1)^{r_1-\beta_1-1} (s_2-t_2)^{m_2-\alpha_2-1} f(t_1, t_2) dt_2 dt_1$$

(by Proposition 20.2)

$$= \frac{1}{\Gamma(r_1-\beta_1)\Gamma(m_2-\alpha_2)} \frac{\partial^{r_1+m_2}}{\partial s_1^{r_1} \partial s_2^{m_2}}$$

$$\int_0^{s_1}\int_0^{s_2} (s_1-t_1)^{r_1-\beta_1-1} (s_2-t_2)^{m_2-\alpha_2-1} f(t_1, t_2) dt_2 dt_1 = B,$$

proving the claim. □

We proceed with

Theorem 20.24. *Here $i = 1, \ldots, N$. Let $\alpha_{i1}, \alpha_{i2} \in \mathbb{R}_+$; $\beta_{i1}, \beta_{i2} \in \mathbb{R}_+ - \{0\}$: $\beta_{i1} \geq \alpha_{i1}+1$, $\beta_{i2} \geq \alpha_{i2}+1$; $f_i \in L_\infty((0, x_{i1}) \times (0, x_{i2}))$, $x_{i1}, x_{i2} \in \mathbb{R}_+ - \{0\}$. Assume that $f_i(s_{i1}, \cdot) \in L_1(0, x_{i2})$ has $\Delta_{s_{i2}}^{\alpha_{i2}} f_i(s_{i1}, \cdot) \in L_\infty(0, x_{i2})$, $\forall s_{i1} \in [0, x_{i1}]$. Assume that $\Delta_{s_{i2}}^{\alpha_{i2}} f_i(\cdot, s_{i2}) \in L_1(0, x_{i1})$ has an L_∞ partial fractional derivative $\Delta_{s_{i1}}^{\beta_{i1}}\left(\Delta_{s_{i2}}^{\alpha_{i2}} f_i(\cdot, s_{i2})\right)$ in $[0, x_{i1}]$, and*

$$\Delta_{s_{i1}}^{\beta_{i1}-k_{i1}}\left(\Delta_{s_{i2}}^{\alpha_{i2}} f_i(0, s_{i2})\right) = 0, \quad \textit{for } k_{i1} = 1, \ldots, [\beta_{i1}]+1, \ \forall s_{i2} \in [0, x_{i2}].$$

Suppose that $f_i(\cdot, s_{i2}) \in L_1(0, x_{i1})$ has $\Delta_{s_{i1}}^{\beta_{i1}} f_i(\cdot, s_{i2}) \in L_\infty(0, x_{i1})$, $\forall s_{i2} \in [0, x_{i2}]$. Suppose that $\Delta_{s_{i1}}^{\beta_{i1}} f_i(t_{i1}, \cdot) \in L_1(0, x_{i2})$ has an L_∞ fractional derivative $\Delta_{s_{i2}}^{\beta_{i2}}\left(\Delta_{s_{i1}}^{\beta_{i1}} f_i(t_{i1}, \cdot)\right)$ in $[0, x_{i2}]$, and let

$$\Delta_{s_{i2}}^{\beta_{i2}-k_{i2}}\left(\Delta_{s_{i1}}^{\beta_{i1}} f_i(t_{i1}, 0)\right) = 0,$$

for all $k_{i2} = 1, \ldots, [\beta_{i2}]+1$, $\forall t_{i1} \in [0, x_{i1}]$. Suppose $\Delta_{s_{i2}s_{i1}}^{\beta_{i2}+\beta_{i1}} f_i \in L_\infty((0, x_{i1})$ $\times (0, x_{i2}))$. Further assume that f_i fulfills all assumptions of Theorem 20.23 (by indexing all there by i). Then we get the continuous function

$$\Delta_{s_{i1}s_{i2}}^{\alpha_{i1}+\alpha_{i2}} f_i(s_{i1}, s_{i2}) = \frac{1}{\Gamma(\beta_{i1} - \alpha_{i1})\Gamma(\beta_{i2} - \alpha_{i2})}$$

$$\times \int_0^{s_{i1}} \int_0^{s_{i2}} (s_{i1}-t_{i1})^{\beta_{i1}-\alpha_{i1}-1}(s_{i2}-t_{i2})^{\beta_{i2}-\alpha_{i2}-1}\Delta_{s_{i2}s_{i1}}^{\beta_{i2}+\beta_{i1}} f_i(t_{i1}, t_{i2})dt_{i2}dt_{i1},$$

$$\forall (s_{i1}, s_{i2}) \in [0, x_{i1}] \times [0, x_{i2}], \quad \text{all } i = 1, \ldots, N. \tag{20.48}$$

Proof. By using Lemma 20.21 we get

$$\Delta_{s_{i1}}^{\alpha_{i1}} \left(\Delta_{s_{i2}}^{\alpha_{i2}} f_i(s_{i1}, s_{i2})\right)$$

$$= \frac{1}{\Gamma(\beta_{i1} - \alpha_{i1})} \int_0^{s_{i1}} (s_{i1} - t_{i1})^{\beta_{i1}-\alpha_{i1}-1}\Delta_{s_{i1}}^{\beta_{i1}} \left(\Delta_{s_{i2}}^{\alpha_{i2}} f_i(t_{i1}, s_{i2})\right)dt_{i1},$$

$$\tag{20.49}$$

$\forall s_{i1} \in [0, x_{i1}]$, all $s_{i2} \in [0, x_{i2}]$. Also by Lemma 20.21 we have

$$\Delta_{s_{i2}}^{\alpha_{i2}} \left(\Delta_{s_{i1}}^{\beta_{i1}} f_i(t_{i1}, s_{i2})\right)$$

$$= \frac{1}{\Gamma(\beta_{i2} - \alpha_{i2})} \int_0^{s_{i2}} (s_{i2}-t_{i2})^{\beta_{i2}-\alpha_{i2}-1}\Delta_{s_{i2}}^{\beta_{i2}} \left(\Delta_{s_{i1}}^{\beta_{i1}} f_i(t_{i1}, t_{i2})\right)dt_{i2}, \tag{20.50}$$

$\forall s_{i2} \in [0, x_{i2}]$, all $t_{i1} \in [0, x_{i1}]$. That is,

$$\Delta_{s_{i2}s_{i1}}^{\alpha_{i2}+\beta_{i1}} f_i(t_{i1}, s_{i2})$$

$$= \frac{1}{\Gamma(\beta_{i2} - \alpha_{i2})} \int_0^{s_{i2}} (s_{i2} - t_{i2})^{\beta_{i2}-\alpha_{i2}-1}\Delta_{s_{i2}s_{i1}}^{\beta_{i2}+\beta_{i1}} f_i(t_{i1}, t_{i2})dt_{i2}, \tag{20.51}$$

$\forall s_{i2} \in [0, x_{i2}]$, all $t_{i1} \in [0, x_{i1}]$.

By Theorem 20.23 we have existence and continuity of partials and

$$\Delta_{s_{i1}s_{i2}}^{\beta_{i1}+\alpha_{i2}} f_i = \Delta_{s_{i2}s_{i1}}^{\alpha_{i2}+\beta_{i1}} f_i, \tag{20.52}$$

on $[0, x_{i1}] \times [0, x_{i2}]$, all $i = 1, \ldots, N$. Using (20.52) and (20.51) into (20.49) we derive (20.48). Continuity of $\Delta_{s_{i1}s_{i2}}^{\alpha_{i1}+\alpha_{i2}} f_i(s_{i1}, s_{i2})$ is arrived at via Lemma 20.15. □

Next we give the final application of Theorem 20.3.

Theorem 20.25. *All here are as in Theorem 20.24. Call $\tilde{\alpha}_{i1} := (a_i + b_i q_i)(\beta_{i1} - \alpha_{i1} - 1)$, $\tilde{\alpha}_{i2} := (a_i + b_i q_i)(\beta_{i2} - \alpha_{i2} - 1)$; $\tilde{\beta}_{i1} := a_i(\beta_{i1} - \alpha_{i1} - 1)$,*

$\tilde{\beta}_{i2} := a_i(\beta_{i2} - \alpha_{i2} - 1)$; all $i = 1, \ldots, N$. Then

$$\Theta \quad : \quad = \int_0^{x_{11}} \int_0^{x_{12}} \int_0^{x_{21}} \int_0^{x_{22}} \cdots \int_0^{x_{N1}} \int_0^{x_{N2}}$$

$$\frac{\prod_{i=1}^N \left| \Delta_{s_{i1}s_{i2}}^{\alpha_{i1}+\alpha_{i2}} f_i(s_{i1}, s_{i2}) \right| ds_{11} ds_{12} ds_{21} ds_{22} \cdots ds_{N1} ds_{N2}}{\left(\frac{1}{\Omega_N} \sum_{i=1}^N w_i s_{i1}^{(\tilde{\alpha}_{i1}+1)/q_i w_i} s_{i2}^{(\tilde{\alpha}_{i2}+1)/q_i w_i} \right)^{\Omega_N}}$$

$$\leq \quad \tilde{U} \prod_{i=1}^N (x_{i1}x_{i2})^{1/q_i} \prod_{i=1}^N \left(\int_0^{x_{i1}} \int_0^{x_{i2}} (x_{i1} - s_{i1})^{\tilde{\beta}_{i1}+1}(x_{i2} - s_{i2})^{\tilde{\beta}_{i2}+1} \right.$$

$$\times \quad \left. \left| \Delta_{s_{i2}s_{i1}}^{\beta_{i2}+\beta_{i1}} f_i(s_{i1}, s_{i2}) \right|^{p_i} ds_{i1} ds_{i2} \right)^{1/p_i}. \tag{20.53}$$

Here

$$\tilde{U} := \frac{1}{\prod_{i=1}^N \left[((\tilde{\alpha}_{i1}+1)(\tilde{\alpha}_{i2}+1))^{1/q_i} ((\tilde{\beta}_{i1}+1)(\tilde{\beta}_{i2}+1))^{1/p_i} \right]}. \tag{20.54}$$

Proof. Here we use (20.48). Call

$$v_i(s_{i1}, s_{i2}) := \Delta_{s_{i1}s_{i2}}^{\alpha_{i1}+\alpha_{i2}} f_i(s_{i1}, s_{i2}) \in C([0, x_{i1}] \times [0, x_{i2}])$$

by (20.48), and

$$\Phi_i(s_{i1}, s_{i2}) := \left| \Delta_{s_{i2}s_{i1}}^{\beta_{i2}+\beta_{i1}} f_i(s_{i1}, s_{i2}) \right|$$

$$\frac{1}{\Gamma(\beta_{i1} - \alpha_{i1})\Gamma(\beta_{i2} - \alpha_{i2})} \in L_\infty([0, x_{i1}] \times [0, x_{i2}]),$$

and use Theorem 20.3. Then (20.53) is obvious. $\quad \square$

We give

Corollary 20.26. *All are as in Theorem* 20.25. *Then*

$$\theta \quad \leq \quad p^{1/rp} \tilde{U} \prod_{i=1}^N (x_{i1}x_{i2})^{1/q_i} \left[\sum_{i=1}^N \frac{1}{p_i} \left(\int_0^{x_{i1}} \right. \right.$$

$$\int_0^{x_{i2}} (x_{i1} - s_{i1})^{\tilde{\beta}_{i1}+1}(x_{i2} - s_{i2})^{\tilde{\beta}_{i2}+1}$$

$$\times \left. \left. \left| \Delta_{s_{i2}s_{i1}}^{\beta_{i2}+\beta_{i1}} f_i(s_{i1}, s_{i2}) \right|^{p_i} ds_{i1} ds_{i2} \right)^r \right]^{1/rp}, \tag{20.55}$$

where $r \in \mathbb{R}_+ - \{0\}$.

Proof. By (20.53) and (20.12). □

Note 20.27. One can give several examples based on all the above results by assuming, for instance, that all $a_i = 1$, $b_1 = 0$ or $a_i = 0$ and $b_i = 1$; with or without $q_i = N$, $w_i = 1/N$, $p_i = N/N - 1$, $c_i = c^*$; all $i = 1, \ldots, N$. Also for $p_i = q_i = N = 2$, all $i = 1, \ldots, N$. Especially and additionally in Theorem 20.6 one can assume $m_i = m^*$, $k_i = k^*$, all $i = 1, \ldots, N$. Also in Theorem 20.17 one can put $\mu_i = \mu^*$, $\rho_i = \rho^*$, $\nu_i = \nu^*$, $\gamma_i = \gamma^*$; all $i = 1, \ldots, N$.

In Theorem 20.25 one can put $\alpha_{i1} = \alpha^*$, $\beta_{i1} = \beta^*$; $\alpha_{i2} = \tilde{\alpha}$, $\beta_{i2} = \tilde{\beta}$, all $i = 1, \ldots, N$, and so on. Wishing to keep the length of the chapter short, we do not show these applications here.

21
Other Hilbert–Pachpatte-Type Fractional Integral Inequalities

This is a continuation of Chapters 19 and 20.

We present here very general weighted univariate and multivariate Hilbert–Pachpatte-type integral inequalities. These involve Caputo and Riemann–Liouville fractional derivatives and fractional partial derivatives of the mentioned types. This treatment is based on [54]. Of great motivation to write this chapter have been [191, Theorem 316], [320, Theorem 1], [322, Theorem 1], and [147, 187, 188].

21.1 Background

Here we follow [134].

We start with

Definition 21.1. Let $\nu \geq 0$ and the operator J_a^ν, defined on $L_1(a, b)$ by

$$J_a^\nu f(x) := \frac{1}{\Gamma(\nu)} \int_a^x (x - t)^{\nu - 1} f(t) \, dt \tag{21.1}$$

for $a \leq x \leq b$, be called the Riemann–Liouville fractional integral operator of order ν.

For $\nu = 0$, we set $J_a^0 := I$, the identity operator. Here Γ stands for the gamma function. Let $\alpha > 0$, $f \in L_1(a, b)$, $a, b \in \mathbb{R}$; see [134]. Here $[\cdot]$ stands for the integral part of the number.

G.A. Anastassiou, *Fractional Differentiation Inequalities*, 545
DOI 10.1007/978-0-387-98128-4_21, © Springer Science+Business Media, LLC 2009

We define the generalized Riemann–Liouville fractional derivative of f of order α by

$$D_a^\alpha f(s) := \frac{1}{\Gamma(m-\alpha)} \left(\frac{d}{ds}\right)^m \int_a^s (s-t)^{m-\alpha-1} f(t)\, dt, \qquad (21.2)$$

where $m := [\alpha] + 1$, $s \in [a, b]$; see also [57], Remark 46 there.

In addition, we set

$$D_a^0 f := f,$$

$$J_a^{-\alpha} f := D_a^\alpha f, \text{ if } \alpha > 0,$$

$$D_a^{-\alpha} f := J_a^\alpha f, \text{ if } 0 < \alpha \le 1,$$

$$D_a^n f = f^{(n)}, \text{ for } n \in \mathbb{N}. \qquad (21.3)$$

We need to mention

Definition 21.2. [134] Let $\nu \ge 0$, $n := \lceil \nu \rceil$ and $\lceil \cdot \rceil$ be the ceiling of the number, $f \in AC^n([a, b])$. We call the Caputo fractional derivative

$$D_{*a}^\nu f(x) := \frac{1}{\Gamma(n-\nu)} \int_a^x (x-t)^{n-\nu-1} f^{(n)}(t)\, dt, \qquad (21.4)$$

$\forall x \in [a, b]$.

The above function $D_{*a}^\nu f(x)$ exists almost everywhere for $x \in [a, b]$. Here $AC^n([a, b])$ is the space of functions with absolutely continuous $(n-1)$st derivative, $n \in \mathbb{N}$.

We need

Proposition 21.3. [134] Let $\nu \ge 0$, $n := \lceil \nu \rceil$ and $f \in AC^n([a, b])$. Then $D_{*a}^\nu f$ exists iff the generalized Riemann–Liouville fractional derivative $D_a^\nu f$ exists.

Proposition 21.4. [134] Let $\nu \ge 0$, $n := \lceil \nu \rceil$. Assume that f is such that both $D_{*a}^\nu f$ and $D_a^\nu f$ exist. Suppose that $f^{(k)}(a) = 0$ for $k = 0, 1, \ldots, n-1$.

Then

$$D_{*a}^\nu f = D_a^\nu f. \qquad (21.5)$$

In conclusion

Corollary 21.5. [58] Let $\nu \ge 0$, $n := \lceil \nu \rceil$, $f \in AC^n([a, b])$, $D_{*a}^\nu f$ exists or $D_a^\nu f$ exists, and $f^{(k)}(a) = 0$, $k = 0, 1, \ldots, n-1$.

Then

$$D_a^\nu f = D_{*a}^\nu f. \tag{21.6}$$

We need

Theorem 21.6. [58] *Let* $\nu \geq 0$, $n := \lceil \nu \rceil$, $f \in AC^n ([a,b])$, *and* $f^{(k)} (a) = 0$, $k = 0, 1, \ldots, n - 1$.
Then

$$f(x) = \frac{1}{\Gamma(\nu)} \int_a^x (x-t)^{\nu-1} D_{*a}^\nu f(t) \, dt. \tag{21.7}$$

We also need

Theorem 21.7. [58] *Let* $\nu \geq \gamma + 1$, $\gamma \geq 0$. *Call* $n := \lceil \nu \rceil$. *Assume* $f \in AC^n ([a,b])$ *such that* $f^{(k)} (a) = 0$, $k = 0, 1, \ldots, n - 1$, *and* $D_{*a}^\nu f \in L_\infty (a, b)$. *Then* $D_{*a}^\gamma f \in AC ([a,b])$, *and*

$$D_{*a}^\gamma f(x) = \frac{1}{\Gamma(\nu - \gamma)} \int_a^x (x-t)^{\nu-\gamma-1} D_{*a}^\nu f(t) \, dt, \tag{21.8}$$

$\forall x \in [a,b]$.

Theorem 21.8. [58] *Let* $\nu \geq \gamma + 1$, $\gamma \geq 0$, *and* $n := \lceil \nu \rceil$. *Let* $f \in AC^n ([a,b])$ *such that* $f^{(k)} (a) = 0$, $k = 0, 1, \ldots, n - 1$. *Assume* $\exists D_a^\nu f(x) \in \mathbb{R}$, $\forall x \in [a,b]$, *and* $D_a^\nu f \in L_\infty (a, b)$. *Then* $D_a^\gamma f \in AC ([a,b])$, *and*

$$D_a^\gamma f(x) = \frac{1}{\Gamma(\nu - \gamma)} \int_a^x (x-t)^{\nu-\gamma-1} D_a^\nu f(t) \, dt, \tag{21.9}$$

$\forall x \in [a,b]$.

In [41] we proved the following result which we use here.

Theorem 21.9. *Here for* $i \in \{1, \ldots, n\}$, *take* $x_i > 0$, *and assume*

$$u_i \in L_1 (0, x_i), \ g_i \in L_\infty \left((0, x_i)^2 \right), \ \Phi_i \in L_\infty (0, x_i),$$

with g_i, $\Phi_i \geq 0$. *Take* $\tau_i \geq 0$, p_i, $q_i > 1 : 1/p_i + 1/q_i = 1$, *and* $\omega_i > 0$ *such that* $\sum_{i=1}^n \omega_i = \Omega_n$, *and* a_i, $b_i \in [0,1]$ *such that* $a_i + b_i = 1$. *Call*

$$\varphi_i (s_i) := \int_0^{s_i} (g_i (s_i, \tau_i))^{(a_i + b_i q_i)\tau_i} \, d\tau_i. \tag{21.10}$$

Assume $\varphi_i (s_i) > 0$, *with the exception* $\varphi_i (0) = 0$. *If*

$$|u_i (s_i)| \leq \int_0^{s_i} (g_i (s_i, \tau_i))^{\tau_i} \Phi_i (\tau_i) \, d\tau_i, \ s_i \in [0, x_i], \ i = 1, \ldots, n, \tag{21.11}$$

then

$$I_1 := \int_0^{x_1} \cdots \int_0^{x_n} \frac{\prod\limits_{i=1}^n |u_i\,(s_i)|}{\left(\frac{1}{\Omega_n} \sum\limits_{i=1}^n \omega_i\,(\varphi_i\,(s_i))^{1/q_i\omega_i}\right)^{\Omega_n}} ds_1 \ldots ds_n$$

$$\leq \prod_{i=1}^n x_i^{1/q_i} \left[\int_0^{x_i} (\Phi_i\,(\tau_i))^{p_i} \left(\int_{\tau_i}^{x_i} (g_i\,(s_i,\tau_i))^{a_i\tau_i}\,ds_i\right) d\tau_i\right]^{1/p_i}. \quad (21.12)$$

Also from [41] we need

Corollary 21.10. *All are as in Theorem 21.9 and* $1/p := \sum_{i=1}^n 1/p_i$, $\tau > 0$. *Then*

$$I_1 \leq p^{1/\tau p} \sum_{i=1}^n x_i^{1/q_i} \left[\sum_{i=1}^n \frac{1}{p_i} \left\{\int_0^{x_i} (\Phi_i\,(\tau_i))^{p_i}\right.\right.$$

$$\left.\left.\times \left(\int_{\tau_i}^{x_i} (g_i\,(s_i,\tau_i))^{a_i\tau_i}\,ds_i\right) d\tau_i\right\}^\tau\right]^{1/\tau p}. \quad (21.13)$$

For the multivariate results, Section 21.3, we need

Symbols 21.11. By $\mathbb{Z}\,(\mathbb{Z}_+)$ and $\mathbb{R}\,(\mathbb{R}_+)$ we denote the sets of all (nonnegative) integers and (nonnegative) real numbers. In the following we work with functions of d variables, where d is a fixed positive integer, writing the variable as a vector $s = (s^1, \ldots, s^d) \in \mathbb{R}^d$. A multi-index m is an element $m = (m^1, \ldots, m^d)$ of $\mathbb{Z}_+^d$. An usual, the factorial of a multi-index m is defined by $m! = m^1! \ldots m^d!$. An integer j may be regarded as the multi-index $(j, \ldots, j)$ depending on the context. For vectors in $\mathbb{R}^d$ and multi-indices we use the usual operations of vector addition and multiplication of vectors by scalars. We write $s \leq \tau$ $(s < \tau)$ if $s^j \leq \tau^j$ $(s^j < \tau^j)$ for $1 \leq j \leq d$. The same convention applies to multi-indices. In particular, $s \geq 0$ $(s > 0)$ means $s^j \geq 0$ $(s^j > 0)$ if $1 \leq j \leq d$.

If $s = (s^1, \ldots, s^d) \in \mathbb{R}^d$ and $s > 0$, we define the cell

$$Q\,(s) = [0, s^1] \times \cdots \times [0, s^j] \times \cdots \times [0, s^d];$$

replacing the factor $[0, s^j]$ by $\{0\}$ in this product, we have the face $\partial_j Q\,(s)$ of $Q\,(s)$.

Let $s = (s^1, \ldots, s^d)$, $\tau = (\tau^1, \ldots, \tau^d) \in \mathbb{R}^d$, $s, \tau > 0$, let $k = (k^1, \ldots, k^d)$ be a multi-index, and let $u : Q\,(s) \to \mathbb{R}$. Write $D_j = \partial/\partial s^j$. We use the following notation.

$$s^\tau = (s^1)^{\tau^1} \ldots (s^d)^{\tau^d},$$

$$D^k u(s) = D_1^{k^1} \dots D_d^{k^d} u(s),$$

$$\int_0^s u(\tau)\, d\tau = \int_0^{s^1} \dots \int_0^{s^d} u(\tau)\, d\tau^1 \dots d\tau^d. \tag{21.14}$$

An exponent $\alpha \in \mathbb{R}$ in the expression s^α, where $s \in \mathbb{R}^d$, is regarded as a multiexponent; that is, $s^\alpha = s^{(\alpha, \dots, \alpha)}$. Let $N \in \mathbb{N}$ be fixed.

The following notation and hypotheses are used in the following.

$$
\begin{aligned}
&I = \{1, \dots, N\} \\
&m_i,\ i \in I && m_i = (m_i^1, \dots, m_i^d) \in \mathbb{Z}_+^d \\
&x_i,\ i \in I && x_i = (x_i^1, \dots, x_i^d) \in \mathbb{R}^d,\ x_i > 0 \\
&p_i,\ q_i,\ i \in I && p_i,\ q_i > 1,\ \tfrac{1}{p_i} + \tfrac{1}{q_i} = 1 \\
&p,\ q && \tfrac{1}{p} = \sum_{i=1}^N \tfrac{1}{p_i},\ \tfrac{1}{q} = \sum_{i=1}^N \tfrac{1}{q_i} \\
&a_i,\ b_i,\ i \in I && a,\ b_i \in \mathbb{R}_+,\ a_i + b_i = 1 \\
&\omega_i,\ i \in I && \omega_i \in \mathbb{R},\ \omega_i > 0,\ \sum_{i=1}^N \omega_i = \Omega_N.
\end{aligned}
\tag{21.15}
$$

In the sequel, u_i, v_i, Φ_i denote functions from $Q(x_i)$ to $\mathbb{R}$ that are measurable with further L_p-type assumptions. If m is a multi-index and $x \in \mathbb{R}^d$, $x > 0$, then $C^m(Q(x))$ denotes the set of all functions $u : Q(x) \to \mathbb{R}$ that possesses continuous derivatives $D^k u$, where $0 \le k \le m$. In [42] we proved that in $C^m(Q(x))$ the order of differentiation does not matter.

We further denote $\alpha_i := (a_i + b_i q_i) c_i \in \mathbb{R}^d$, $\beta_i := a_i c_i \in \mathbb{R}^d$, for $c_i \in \mathbb{R}^d$, $i \in I$. In particular we observe that

$$x_i^{1/q_i} = (x_i^1)^{1/q_i} \dots (x_i^d)^{1/q_i},$$

$$\prod_{i=1}^N (\alpha_i + 1)^{1/q_i} = \prod_{i=1}^N \prod_{j=1}^d (\alpha_i^j + 1)^{1/q_i}. \tag{21.16}$$

From [42] we use the following.

Theorem 21.12. Let $v_i \in L_1(Q(x_i))$, $\Phi_i \in L_\infty(Q(x_i))$, $\Phi_i \ge 0$, $c_i > -1$; $c_i \in \mathbb{R}^d$, $i = 1, \dots, N$.

Set

$$U := \frac{1}{\displaystyle\prod_{i=1}^N \left[(\alpha_i + 1)^{1/q_i} (\beta_i + 1)^{1/p_i}\right]}. \tag{21.17}$$

Let $J_i := \left\{ j \in \{1, \dots, d\} \mid -1 < c_i^j < 0 \right\}$. If $J_i \ne \emptyset$ we take

$$b_j > \max_{j \in J_i} \left(\frac{(1 + c_i^j)}{c_i^j (1 - q_i)} \right). \tag{21.18}$$

Assume

$$|v_i(s_i)| \le \int_0^{s_i} (s_i - \tau_i)^{c_i} \Phi_i(\tau_i) d\tau_i, \ \forall s_i \in Q(x_i), \ i = 1, \ldots, N. \quad (21.19)$$

Then

$$M_1 := \int_0^{x_1} \cdots \int_0^{x_N} \frac{\prod_{i=1}^N |v_i(s_i)| ds_1 \ldots ds_N}{\left(\frac{1}{\Omega_N} \sum_{i=1}^N \omega_i s_i^{(\alpha_i+1)/q_i\omega_i}\right)^{\Omega_N}}$$

$$\le U \prod_{i=1}^N x_i^{1/q_i} \prod_{i=1}^N \left(\int_0^{x_i} (x_i - s_i)^{\beta_i+1} \Phi_i(s_i)^{p_i} ds_i\right)^{1/p_i}. \quad (21.20)$$

Finally from [42] we use

Corollary 21.13. *All are as in Theorem 21.12. Then*

$$M_1 \le p^{1/\tau p} U \prod_{i=1}^N x_i^{1/q_i}$$

$$\left[\sum_{i=1}^N \frac{1}{p_i} \left(\int_0^{x_i} (x_i - s_i)^{\beta_i+1} \Phi_i(s_i)^{p_i} ds_i\right)^\tau\right]^{1/\tau p}, \quad (21.21)$$

where $\tau > 0$.

21.2 Univariate Results

We present the first main result of this chapter.

Theorem 21.14. *Let $i \in \{1, \ldots, n\}$, $x_i > 0$, $\gamma_i \ge 0$, $\nu_i \ge \gamma_i + 1$. Let $f_i \in AC^{n_i}([0, x_i])$, $n_i := \lceil \nu_i \rceil$, such that $f_i^{(k_i)}(0) = 0$, $k_i = 0, 1, \ldots, n_i - 1$, and $D_{*0}^{\nu_i} f_i \in L_\infty(0, x_i)$. Let $p_i, q_i > 1 : 1/p_i + 1/q_i = 1$, and $\omega_i > 0$ such that $\sum_{i=1}^n \omega_i = \Omega_n$, and $a_i, b_i \in [0, 1]$ such that $a_i + b_i = 1$. Then*

$$I_2 := \int_0^{x_1} \cdots \int_0^{x_n} \frac{\prod_{i=1}^n \left|D_{*0}^{\gamma_i} f_i(s_i)\right| ds_1 \ldots ds_n}{\left(\frac{1}{\Omega_n} \sum_{i=1}^n \omega_i \frac{s_i^{[(a_i+b_iq_i)(\nu_i-\gamma_i-1)+1]/q_i\omega_i}}{((a_i+b_iq_i)(\nu_i-\gamma_i-1)+1)^{1/q_i\omega_i}}\right)^{\Omega_n}}$$

$$\le C \prod_{i=1}^n x_i^{1/q_i} \left[\int_0^{x_i} |D_{*0}^{\nu_i} f_i(t_i)|^{p_i} (x_i - t_i)^{[a_i(\nu_i-\gamma_i-1)+1]} dt_i\right]^{1/p_i}, \quad (21.22)$$

where

$$C := \prod_{i=1}^{n} \left(\frac{1}{\Gamma\left(\nu_i - \gamma_i\right)\left(a_i\left(\nu_i - \gamma_i - 1\right)+1\right)^{1/p_i}} \right). \qquad (21.23)$$

Proof. By Theorem 21.7 we get

$$D_{*0}^{\gamma_i} f_i\left(s_i\right) = \frac{1}{\Gamma\left(\nu_i - \gamma_i\right)} \int_0^{s_i} \left(s_i - t_i\right)^{\nu_i - \gamma_i - 1} D_{*0}^{\nu_i} f_i\left(t_i\right) dt_i, \qquad (21.24)$$

$\forall s_i \in [0, x_i]$.

Here we apply Theorem 21.9. By Theorem 21.7 we have that $D_{*0}^{\gamma_i} f_i \in AC\left([0, x_i]\right)$; also by assumption $D_{*0}^{\nu_i} f_i \in L_\infty\left(0, x_i\right)$; and by (21.24) we have

$$\left|D_{*0}^{\gamma_i} f_i\left(s_i\right)\right| \leq \frac{1}{\Gamma\left(\nu_i - \gamma_i\right)} \int_0^{s_i} \left(s_i - t_i\right)^{\nu_i - \gamma_i - 1} \left|D_{*0}^{\nu_i} f_i\left(t_i\right)\right| dt_i, \qquad (21.25)$$

$\forall s_i \in [0, x_i]$.

We set

$$u_i := D_{*0}^{\gamma_i} f_i, g_i\left(s_i, t_i\right) := \left|s_i - t_i\right|,$$

$$\Phi_i\left(t_i\right) := \frac{\left|D_{*0}^{\nu_i} f_i\left(t_i\right)\right|}{\Gamma\left(\nu_i - \gamma_i\right)} \text{ and }$$

$$\tau_i := \nu_i - \gamma_i - 1, \ i = 1, \ldots, n.$$

Here

$$\varphi_i\left(s_i\right) := \int_0^{s_i} \left(s_i - t_i\right)^{(a_i + b_i q_i)(\nu_i - \gamma_i - 1)} dt_i =$$

$$\frac{s_i^{(a_i + b_i q_i)(\nu_i - \gamma_i - 1)+1}}{\left(a_i + b_i q_i\right)\left(\nu_i - \gamma_i - 1\right)+1} > 0, \ \forall s_i \in (0, x_i], \qquad (21.26)$$

with $\varphi_i\left(0\right) = 0$. Now the proof is clear. $\square$

We give

Corollary 21.15. *All are as in Theorem 21.14 and* $1/p := \sum_{i=1}^{n} 1/p_i$, $\tau > 0$. *Then*

$$I_2 \leq p^{1/\tau p} C \prod_{i=1}^{n} x_i^{1/q_i} \left[\sum_{i=1}^{n} \frac{1}{p_i} \right.$$

$$\left. \left\{ \int_0^{x_i} \left|D_{*0}^{\nu_i} f_i\left(t_i\right)\right|^{p_i} \left(x_i - t_i\right)^{(a_i(\nu_i - \gamma_i - 1)+1)} dt_i \right\}^\tau \right]^{1/\tau p}. \qquad (21.27)$$

Proof. By using (21.22) and the inequality of means: for any $A_i \geq 0$ and $\tau > 0$, we obtain

$$\prod_{i=1}^{n} A_i^{1/p_i} \leq \left(p \sum_{i=1}^{n} \frac{1}{p_i} A_i^{\tau} \right)^{1/\tau p}. \tag{21.28}$$

Here we put

$$A_i := \int_0^{x_i} |D_{*0}^{\nu_i} f_i(t_i)|^{p_i} (x_i - t_i)^{[a_i(\nu_i - \gamma_i - 1)+1]} \, dt_i, \tag{21.29}$$

$i = 1, \ldots, n$. □

We continue with

Theorem 21.16. *Let* $i \in \{1, \ldots, n\}$, $x_i > 0$, $\gamma_i \geq 0$, $\nu_i \geq \gamma_i + 1$. *Let* $f_i \in AC^{n_i}([0, x_i])$, $n_i := \lceil \nu_i \rceil$, *such that* $f_i^{(k_i)}(0) = 0$, $k_i = 0, 1, \ldots, n_i - 1$, *and there exists* $D_0^{\nu_i} f_i(s_i) \in \mathbb{R}$, *all* $s_i \in [0, x_i]$, *and* $D_0^{\nu_i} f_i \in L_\infty(0, x_i)$. *Let* $p_i, q_i > 1 : 1/p_i + 1/q_i = 1$, *and* $\omega_i > 0$ *such that* $\sum_{i=1}^{n} \omega_i = \Omega_n$, *and* $a_i, b_i \in [0, 1]$ *such that* $a_i + b_i = 1$. *Then*

$$I_3 := \int_0^{x_1} \cdots \int_0^{x_n} \frac{\prod_{i=1}^{n} |D_0^{\gamma_i} f_i(s_i)| \, ds_1 \ldots ds_n}{\left(\frac{1}{\Omega_n} \sum_{i=1}^{n} \omega_i \frac{s_i^{((a_i+b_i q_i)(\nu_i - \gamma_i - 1)+1)/q_i \omega_i}}{((a_i+b_i q_i)(\nu_i - \gamma_i - 1)+1)^{1/q_i \omega_i}} \right)^{\Omega_n}}$$

$$\leq C \prod_{i=1}^{n} x_i^{1/q_i} \left[\int_0^{x_i} |D_0^{\nu_i} f_i(t_i)|^{p_i} (x_i - t_i)^{[a_i(\nu_i - \gamma_i - 1)+1]} \, dt_i \right]^{1/p_i}, \tag{21.30}$$

where C *is as in* (21.23).

Proof. Very similar to the proof of Theorem 21.14, now using Theorem 21.8 and applying Theorem 21.9. □

We also give

Corollary 21.17. *All are as in Theorem 21.16 and* $1/p := \sum_{i=1}^{n} 1/p_i$, $\tau > 0$. *Then*

$$I_3 \leq p^{1/\tau p} C \prod_{i=1}^{n} x_i^{1/q_i} \left[\sum_{i=1}^{n} \frac{1}{p_i} \right.$$

$$\left. \left\{ \int_0^{x_i} |D_0^{\nu_i} f_i(t_i)|^{p_i} (x_i - t_i)^{(a_i(\nu_i - \gamma_i - 1)+1)} \, dt_i \right\}^{\tau} \right]^{1/\tau p}. \tag{21.31}$$

Proof. By using (21.30) and (21.28). □

One can give a variety of applications of the derived inequalities (21.22), (21.27), (21.30), and (21.31), for various specific values of the involved parameters n, p_i, q_i, ω_i, a_i, and b_i.

21.3   Multivariate Results

We need the following.

Definition 21.18. (See also [42].) Let $f : [0, x] \times [0, y] \to \mathbb{R}$, x, $y \in \mathbb{R}_+ - \{0\}$, such that $f(\cdot, s_2) \in L_1(0, x)$, $\forall s_2 \in [0, y]$; $f(s_1, \cdot) \in L_1(0, y)$, $\forall s_1 \in [0, x]$; α, $\beta \in \mathbb{R}_+ - \{0\}$.

We define the Riemann–Liouville fractional partial derivatives with respect to s_1 (resp., s_2) of f of order α (resp., order β) by

$$D_{s_1}^\alpha f(s_1, s_2) := \frac{1}{\Gamma(m - \alpha)} \left(\frac{\partial}{\partial s_1} \right)^m$$

$$\int_0^{s_1} (s_1 - t_1)^{m - \alpha - 1} f(t_1, s_2) \, dt_1, \tag{21.32}$$

where $m := [\alpha] + 1$, $\forall s_1 \in [0, x]$, all $s_2 \in [0, y]$; and

$$D_{s_2}^\beta f(s_1, s_2) := \frac{1}{\Gamma(\tau - \beta)} \left(\frac{\partial}{\partial s_2} \right)^\tau$$

$$\int_0^{s_2} (s_2 - t_2)^{\tau - \beta - 1} f(s_1, t_2) \, dt_2, \tag{21.33}$$

where $\tau := [\beta] + 1$, $\forall s_2 \in [0, y]$, all $s_1 \in [0, x]$.

We define the Riemann–Liouville fractional mixed (iterated) partial derivatives as

$$D_{s_1}^\alpha D_{s_2}^\beta f(s_1, s_2) := D_{s_1}^\alpha \left(D_{s_2}^\beta f(s_1, s_2) \right) := D_{s_1 s_2}^{\alpha + \beta} f(s_1, s_2),$$

$$D_{s_2}^\beta D_{s_1}^\alpha f(s_1, s_2) := D_{s_2}^\beta \left(D_{s_1}^\alpha f(s_1, s_2) \right) := D_{s_2 s_1}^{\beta + \alpha} f(s_1, s_2), \tag{21.34}$$

$\forall (s_1, s_2) \in [0, x] \times [0, y]$, defined the obvious iterated way via (21.2).

Clearly in the last (21.34) we assume that $D_{s_2}^\beta f(s_1, s_2) \in L_1(0, x)$, $\forall s_2 \in [0, y]$, and $D_{s_1}^\alpha f(s_1, s_2) \in L_1(0, y)$, $\forall s_1 \in [0, x]$.

We also need

Definition 21.19. Let $f : [0, x] \times [0, y] \to \mathbb{R}$, x, $y \in \mathbb{R}_+ - \{0\}$, such that $f(\cdot, s_2) \in AC^n([0, x])$, $\forall s_2 \in [0, y]$, where $n := \lceil \nu \rceil$, $\nu > 0$; $f(s_1, \cdot) \in AC^m([0, y])$, $\forall s_1 \in [0, x]$, where $m := \lceil \mu \rceil$, $\mu > 0$.

We define the Caputo fractional partial derivatives with respect to s_1 (resp., s_2) of f of order ν (resp., order μ) by

$$D_{*s_1}^\nu f(s_1, s_2) := \frac{1}{\Gamma(n - \nu)}$$

$$\int_0^{s_1} (s_1 - t_1)^{n - \nu - 1} \frac{\partial^n}{\partial s_1^n} f(t_1, s_2) \, dt_1, \tag{21.35}$$

$\forall s_1 \in [0, x]$, all $s_2 \in [0, y]$; and

$$D_{*s_2}^{\mu} f\left(s_1, s_2\right) := \frac{1}{\Gamma\left(m - \mu\right)}$$

$$\int_0^{s_2} \left(s_2 - t_2\right)^{m-\mu-1} \frac{\partial^m}{\partial s_2^m} f\left(s_1, t_2\right) dt_2, \tag{21.36}$$

$\forall s_2 \in [0, y]$, all $s_1 \in [0, x]$.

We define the Caputo fractional mixed (iterated) partial derivatives as

$$D_{*s_1}^{\nu} D_{*s_2}^{\mu} f\left(s_1, s_2\right) := D_{*s_1}^{\nu} \left(D_{*s_2}^{\mu} f\left(s_1, s_2\right)\right) := D_{*s_1 s_2}^{\nu+\mu} f\left(s_1, s_2\right),$$

$$D_{*s_2}^{\mu} D_{*s_1}^{\nu} f\left(s_1, s_2\right) := D_{*s_2}^{\mu} \left(D_{*s_1}^{\nu} f\left(s_1, s_2\right)\right) := D_{*s_2 s_1}^{\mu+\nu} f\left(s_1, s_2\right), \tag{21.37}$$

$\forall \left(s_1, s_2\right) \in [0, x] \times [0, y]$, defined the obvious iterated way via (21.4).

Clearly in the last (21.37) we assume that $D_{*s_2}^{\mu} f\left(s_1, s_2\right) \in AC^n\left([0, x]\right)$, all $s_2 \in [0, y]$, and that $D_{*s_1}^{\nu} f\left(s_1, s_2\right) \in AC^m\left([0, y]\right)$, all $s_1 \in [0, x]$.

We need

Theorem 21.20. *Let f be as in Definition 21.19. Additionally assume that $f \in C^{(n,m)}\left([0, x] \times [0, y]\right)$.*

Then

$$D_{*s_1 s_2}^{\nu+\mu} f\left(s_1, s_2\right) = D_{*s_2 s_1}^{\mu+\nu} f\left(s_1, s_2\right), \tag{21.38}$$

for all $\left(s_1, s_2\right) \in [0, x] \times [0, y]$.

Proof. We observe that

$$D_{*s_1 s_2}^{\nu+\mu} f\left(s_1, s_2\right) = D_{*s_1}^{\nu} \left(D_{*s_2}^{\mu} f\left(s_1, s_2\right)\right) =$$

$$\frac{1}{\Gamma\left(n - \nu\right)} \int_0^{s_1} \left(s_1 - t_1\right)^{n-\nu-1} \frac{\partial^n}{\partial s_1^n} \left(D_{*s_2}^{\mu} f\left(t_1, s_2\right)\right) dt_1 =$$

$$\frac{1}{\Gamma\left(n - \nu\right) \Gamma\left(m - \mu\right)} \int_0^{s_1} \left(s_1 - t_1\right)^{n-\nu-1} \frac{\partial^n}{\partial s_1^n}$$

$$\left(\int_0^{s_2} \left(s_2 - t_2\right)^{m-\mu-1} \frac{\partial^m}{\partial s_2^m} f\left(t_1, t_2\right) dt_2\right) dt_1$$

(by [10, Theorem 24.5, p. 193], repeated application)

$$= \frac{1}{\Gamma\left(n - \nu\right) \Gamma\left(m - \mu\right)} \int_0^{s_1}$$

$$\left(\int_0^{s_2} \left(s_1 - t_1\right)^{n-\nu-1} \left(s_2 - t_2\right)^{m-\mu-1} \frac{\partial^{n+m}}{\partial s_1^n \partial s_2^m} f\left(t_1, t_2\right) dt_2\right) dt_1$$

(by Fubini's theorem and commutativity of ordinary partials)

$$= \frac{1}{\Gamma\left(m-\mu\right)\Gamma\left(n-\nu\right)} \int_0^{s_2}$$

$$\left(\int_0^{s_1} (s_2 - t_2)^{m-\mu-1} (s_1 - t_1)^{n-\nu-1} \frac{\partial^{m+n}}{\partial s_2^m \partial s_1^n} f\left(t_1, t_2\right) dt_1\right) dt_2$$

(by [10, Theorem 24.5, p. 193], repeated application)

$$= \frac{1}{\Gamma\left(m-\mu\right)\Gamma\left(n-\nu\right)} \int_0^{s_2} (s_2 - t_2)^{m-\mu-1} \frac{\partial^m}{\partial s_2^m}$$

$$\left(\int_0^{s_1} (s_1 - t_1)^{n-\nu-1} \frac{\partial^n}{\partial s_1^n} f\left(t_1, t_2\right) dt_1\right) dt_2 =$$

$$\frac{1}{\Gamma\left(m-\mu\right)} \int_0^{s_2} (s_2 - t_2)^{m-\mu-1} \frac{\partial^m}{\partial s_2^m} \left(D_{*s_1}^\nu f\left(s_1, t_2\right)\right) dt_2$$

$$= D_{*s_2}^\mu \left(D_{*s_1}^\nu f\left(s_1, s_2\right)\right) = D_{*s_2 s_1}^{\mu+\nu} f\left(s_1, s_2\right),$$

proving the commutativity of Caputo fractional mixed partials under the assumptions of the theorem. □

We give the counterpart of the last theorem.

Theorem 21.21. *All are as in Theorem 21.20. Assume that* $\mathbb{R} \ni D_{s_2}^\mu f$ (s_1, s_2) *exists in* $s_2 \in [0, y]$, $\forall s_1 \in [0, x]$, *and* $\partial^l f/\partial s_2^l (s_1, 0) = 0$, $l = 0, 1, \ldots, m-1$, $\forall s_1 \in [0, x]$. *Assume that* $D_{s_2}^\mu f (s_1, s_2) \in AC^n ([0, x])$, $\forall s_2 \in [0, y]$, *and* $\mathbb{R} \ni D_{s_1}^\nu \left(D_{s_2}^\mu f (s_1, s_2)\right)$ *exists in* $s_1 \in [0, x]$, $\forall s_2 \in [0, y]$, *with* $\partial^k/\partial s_1^k \left(D_{s_2}^\mu f (0, s_2)\right) = 0$, *for* $k = 0, 1, \ldots, n-1$, $\forall s_2 \in [0, y]$. *Also suppose that* $\mathbb{R} \ni D_{s_1}^\nu f (s_1, s_2)$ *exists in* $s_1 \in [0, x]$, $\forall s_2 \in [0, y]$, *and* $\partial^k f/\partial s_1^k (0, s_2) = 0$, $k = 0, 1, \ldots, n-1$, $\forall s_2 \in [0, y]$. *And suppose that* $D_{s_1}^\nu f (s_1, s_2) \in AC^m ([0, y])$, $\forall s_1 \in [0, x]$, *and* $\mathbb{R} \ni D_{s_2}^\mu \left(D_{s_1}^\nu f (s_1, s_2)\right)$ *exists in* $s_2 \in [0, y]$, $\forall s_1 \in [0, x]$, *and* $\partial^l/\partial s_2^l \left(D_{s_1}^\nu f (s_1, 0)\right) = 0$, $l = 0, 1, \ldots, m-1$, $\forall s_1 \in [0, x]$.

Then

$$D_{s_1 s_2}^{\nu+\mu} f\left(s_1, s_2\right) = D_{s_2 s_1}^{\mu+\nu} f\left(s_1, s_2\right), \tag{21.39}$$

for all $(s_1, s_2) \in [0, x] \times [0, y]$.

Proof. We apply Theorem 21.20 and use Corollary 21.5 repeatedly. Namely we have the following.

We notice

$$D_{*s_1 s_2}^{\nu+\mu} f\left(s_1, s_2\right) = D_{*s_1}^\nu \left(D_{*s_2}^\mu f\left(s_1, s_2\right)\right) =$$

$$D_{*s_1}^\nu \left(D_{s_2}^\mu f\left(s_1, s_2\right)\right) = D_{s_1}^\nu \left(D_{s_2}^\mu f\left(s_1, s_2\right)\right) = D_{s_1 s_2}^{\nu+\mu} f\left(s_1, s_2\right).$$

Similarly we have

$$D_{*s_2 s_1}^{\mu+\nu} f(s_1, s_2) = D_{*s_2}^{\mu} \left(D_{*s_1}^{\nu} f(s_1, s_2) \right) =$$

$$D_{*s_2}^{\mu} \left(D_{s_1}^{\nu} f(s_1, s_2) \right) = D_{s_2}^{\mu} \left(D_{s_1}^{\nu} f(s_1, s_2) \right) = D_{s_2 s_1}^{\mu+\nu} f(s_1, s_2).$$

Now using (21.38) we prove the claim. □

We need

Lemma 21.22. [42] Let $f \in L_\infty([0,A] \times [0,B])$, where $A, B \in \mathbb{R}_+ - \{0\}$. Let

$$F(x,y) := \int_0^x \int_0^y (x-t)^\alpha (y-s)^\beta f(t,s) \, dt ds, \qquad (21.40)$$

$\forall x \in [0, A]$, $\forall y \in [0, B]$; $\alpha, \beta \in \mathbb{R}_+$.
Then $F \in C([0,A] \times [0,B])$.

We further present

Theorem 21.23. *Here* $i = 1, \ldots, N$. *Let* $\alpha_{i1}, \alpha_{i2} \in \mathbb{R}_+$; $\beta_{i1}, \beta_{i2} \in \mathbb{R}_+ - \{0\}$: $\beta_{i1} \geq \alpha_{i1} + 1$, $\beta_{i2} \geq \alpha_{i2} + 1$; $x_{i1}, x_{i2} \in \mathbb{R}_+ - \{0\}$, $f_i : [0, x_{i1}] \times [0, x_{i2}] \longmapsto \mathbb{R}$. *Call* $n_{i1} := \lceil \beta_{i1} \rceil$, $n_{i2} := \lceil \beta_{i2} \rceil$; $\lambda_{i1} := \lceil \alpha_{i1} \rceil$, $\lambda_{i2} := \lceil \alpha_{i2} \rceil$. *The mixed partials* $D_{*s_{i1} s_{i2}}^{\alpha_{i1}+\alpha_{i2}} f_i$, $D_{*s_{i2} s_{i1}}^{\beta_{i2}+\beta_{i1}} f_i$, $D_{*s_{i1} s_{i2}}^{\beta_{i1}+\alpha_{i2}} f_i$, *and* $D_{*s_{i2} s_{i1}}^{\alpha_{i2}+\beta_{i1}} f_i$ *are considered according to Definition 21.19.*
Assume that $f_i \in C^{(n_{i1},\lambda_{i2})}([0, x_{i1}] \times [0, x_{i2}])$ *and are as in Definition 21.19.*
Suppose $D_{*s_{i2}}^{\alpha_{i2}} f_i(\cdot, s_{i2}) \in AC^{n_{i1}}([0, x_{i1}])$, *such that*

$$\frac{\partial^{k_i}}{\partial s_{i1}^{k_i}} \left(D_{*s_{i2}}^{\alpha_{i2}} f_i(0, s_{i2}) \right) = 0, \ k_i = 0, 1, \ldots, n_{i1} - 1, \text{ and}$$

$$D_{*s_{i1} s_{i2}}^{\beta_{i1}+\alpha_{i2}} f_i(\cdot, s_{i2}) \in L_\infty(0, x_{i1}), \text{ all these } \forall s_{i2} \in [0, x_{i2}].$$

Suppose that $D_{*s_{i1}}^{\beta_{i1}} f_i(t_{i1}, \cdot) \in AC^{n_{i2}}([0, x_{i2}])$, *such that*

$$\frac{\partial^{l_i}}{\partial s_{i2}^{l_i}} \left(D_{*s_{i1}}^{\beta_{i1}} f_i(t_{i1}, 0) \right) = 0, \ l_i = 0, 1, \ldots, n_{i2} - 1, \text{ and}$$

$$D_{*s_{i2} s_{i1}}^{\beta_{i2}+\beta_{i1}} f_i(t_{i1}, \cdot) \in L_\infty(0, x_{i2}), \text{ all these } \forall t_{i1} \in [0, x_{i1}].$$

Further suppose

$$D_{*s_{i2} s_{i1}}^{\beta_{i2}+\beta_{i1}} f_i \in L_\infty((0, x_{i1}) \times (0, x_{i2})).$$

Then we obtain the continuous function

$$D_{*s_{i1}s_{i2}}^{\alpha_{i1}+\alpha_{i2}} f_i (s_{i1}, s_{i2}) = \frac{1}{\Gamma(\beta_{i1} - \alpha_{i1})\Gamma(\beta_{i2} - \alpha_{i2})}$$

$$\int_0^{s_{i1}} \int_0^{s_{i2}} (s_{i1} - t_{i1})^{\beta_{i1}-\alpha_{i1}-1} (s_{i2} - t_{i2})^{\beta_{i2}-\alpha_{i2}-1} D_{*s_{i2}s_{i1}}^{\beta_{i2}+\beta_{i1}} f_i (t_{i1}, t_{i2})$$

$$(21.41)$$

$$dt_{i2}dt_{i1}, \ \forall (s_{i1}, s_{i2}) \in [0, x_{i1}] \times [0, x_{i2}], \ \text{all } i = 1, \ldots, N.$$

Proof. By using Theorem 21.7 we have

$$D_{*s_{i1}}^{\alpha_{i1}} \left(D_{*s_{i2}}^{\alpha_{i2}} f_i (s_{i1}, s_{i2}) \right) = \frac{1}{\Gamma(\beta_{i1} - \alpha_{i1})}$$

$$\int_0^{s_{i1}} (s_{i1} - t_{i1})^{\beta_{i1}-\alpha_{i1}-1} D_{*s_{i1}}^{\beta_{i1}} \left(D_{*s_{i2}}^{\alpha_{i2}} f_i (t_{i1}, s_{i2}) \right) dt_{i1}, \qquad (21.42)$$

$$\forall s_{i1} \in [0, x_{i1}], \ \text{all } s_{i2} \in [0, x_{i2}].$$

Again by Theorem 21.7 we obtain

$$D_{*s_{i2}}^{\alpha_{i2}} \left(D_{*s_{i1}}^{\beta_{i1}} f_i (t_{i1}, s_{i2}) \right) = \frac{1}{\Gamma(\beta_{i2} - \alpha_{i2})}$$

$$\int_0^{s_{i2}} (s_{i2} - t_{i2})^{\beta_{i2}-\alpha_{i2}-1} D_{*s_{i2}}^{\beta_{i2}} \left(D_{*s_{i1}}^{\beta_{i1}} f_i (t_{i1}, t_{i2}) \right) dt_{i2}, \qquad (21.43)$$

$$\forall s_{i2} \in [0, x_{i2}], \ \text{all } t_{i1} \in [0, x_{i1}].$$

That is,

$$D_{*s_{i2}s_{i1}}^{\alpha_{i2}+\beta_{i1}} f_i (t_{i1}, s_{i2}) = \frac{1}{\Gamma(\beta_{i2} - \alpha_{i2})}$$

$$\int_0^{s_{i2}} (s_{i2} - t_{i2})^{\beta_{i2}-\alpha_{i2}-1} D_{*s_{i2}s_{i1}}^{\beta_{i2}+\beta_{i1}} f_i (t_{i1}, t_{i2}) dt_{i2}, \qquad (21.44)$$

$$\forall s_{i2} \in [0, x_{i2}], \ \text{all } t_{i1} \in [0, x_{i1}].$$

Using Theorem 21.20 we obtain

$$D_{*s_{i1}s_{i2}}^{\beta_{i1}+\alpha_{i2}} f_i = D_{*s_{i2}s_{i1}}^{\alpha_{i2}+\beta_{i1}} f_i, \qquad (21.45)$$

on $[0, x_{i1}] \times [0, x_{i2}]$, all $i = 1, \ldots, N$.

Using (21.45) and (21.44) into (21.42) we derive (21.41). Continuity of $D_{*s_{i1}s_{i2}}^{\alpha_{i1}+\alpha_{i2}} f_i (s_{i1}, s_{i2})$ is arrived at via Lemma 21.22. $\square$

Next we apply Theorem 21.12 for $d = 2$ at the fractional level.

Theorem 21.24. *All here are as in Theorem 21.23. Call* $\tilde{\alpha}_{i1} := (a_i + b_i q_i)$ $(\beta_{i1} - \alpha_{i1} - 1)$, $\tilde{\alpha}_{i2} := (a_i + b_i q_i)(\beta_{i2} - \alpha_{i2} - 1)$; $\tilde{\beta}_{i1} := a_i (\beta_{i1} - \alpha_{i1} - 1)$, $\tilde{\beta}_{i2} := a_i (\beta_{i2} - \alpha_{i2} - 1)$; *all* $i = 1, \ldots, N$.
Then

$$M_2 := \int_0^{x_{11}} \int_0^{x_{12}} \int_0^{x_{21}} \int_0^{x_{22}} \cdots \int_0^{x_{N1}} \int_0^{x_{N2}}$$

$$\frac{\prod_{i=1}^{N} \left| D_{*s_{i1}s_{i2}}^{\alpha_{i1}+\alpha_{i2}} f_i(s_{i1}, s_{i2}) \right| ds_{11} ds_{12} ds_{21} ds_{22} \ldots ds_{N1} ds_{N2}}{\left(\frac{1}{\Omega_N} \sum_{i=1}^{N} \omega_i s_{i1}^{(\tilde{\alpha}_{i1}+1)/q_i \omega_i} s_{i2}^{(\tilde{\alpha}_{i2}+1)/q_i \omega_i} \right)^{\Omega_N}}$$

$$\leq \tilde{U} \prod_{i=1}^{N} (x_{i1} x_{i2})^{1/q_i} \prod_{i=1}^{N} \left(\int_0^{x_{i1}} \int_0^{x_{i2}} (x_{i1} - s_{i1})^{\tilde{\beta}_{i1}+1} \right. \tag{21.46}$$

$$\left. (x_{i2} - s_{i2})^{\tilde{\beta}_{i2}+1} \left| D_{*s_{i2}s_{i1}}^{\beta_{i2}+\beta_{i1}} f_i(s_{i1}, s_{i2}) \right|^{p_i} ds_{i1} ds_{i2} \right)^{1/p_i}.$$

Here

$$\tilde{U} := \frac{1}{\prod_{i=1}^{N} \left[((\tilde{\alpha}_{i1}+1)(\tilde{\alpha}_{i2}+1))^{1/q_i} \left((\tilde{\beta}_{i1}+1)(\tilde{\beta}_{i2}+1) \right)^{1/p_i} \right]}. \tag{21.47}$$

Proof. Here we use (21.41). Call

$$v_i(s_{i1}, s_{i2}) := D_{*s_{i1}s_{i2}}^{\alpha_{i1}+\alpha_{i2}} f_i(s_{i1}, s_{i2}) \in C([0, x_{i1}] \times [0, x_{i2}])$$

by (21.41), and

$$\Phi_i(s_{i1}, s_{i2}) := \frac{\left| D_{*s_{i2}s_{i1}}^{\beta_{i2}+\beta_{i1}} f_i(s_{i1}, s_{i2}) \right|}{\Gamma(\beta_{i1} - \alpha_{i1})\Gamma(\beta_{i2} - \alpha_{i2})} \in L_\infty([0, x_{i1}] \times [0, x_{i2}])$$

by assumption; also set

$$c_{i1} := \beta_{i1} - \alpha_{i1} - 1, \ c_{i2} := \beta_{i2} - \alpha_{i2} - 1,$$

and use Theorem 21.12. Then (21.46) is obvious. □

We give

Corollary 21.25. *All here are as in Theorem 21.24. Then*

$$M_2 \leq p^{1/\tau p} \tilde{U} \prod_{i=1}^{N} (x_{i1} x_{i2})^{1/q_i} \left[\sum_{i=1}^{N} \frac{1}{p_i} \left(\int_0^{x_{i1}} \int_0^{x_{i2}} (x_{i1} - s_{i1})^{\tilde{\beta}_{i1}+1} \right. \right.$$

$$\left. (x_{i2} - s_{i2})^{\tilde{\beta}_{i2}+1} \left| D_{*s_{i2}s_{i1}}^{\beta_{i2}+\beta_{i1}} f_i (s_{i1}, s_{i2}) \right|^{p_i} ds_{i1}ds_{i2} \right)^{\tau} \right]^{1/\tau p}, \qquad (21.48)$$

where $\tau > 0$.

Proof. By (21.46) and (21.21). $\square$

We further give

Theorem 21.26. *Here $i = 1, \ldots, N$. Let $\alpha_{i1}, \alpha_{i2} \in \mathbb{R}_+$; $\beta_{i1}, \beta_{i2} \in \mathbb{R}_+ - \{0\}$: $\beta_{i1} \geq \alpha_{i1} + 1$, $\beta_{i2} \geq \alpha_{i2} + 1$; $x_{i1}, x_{i2} \in \mathbb{R}_+ - \{0\}$, $f_i : [0, x_{i1}] \times [0, x_{i2}] \longmapsto \mathbb{R}$. Call $n_{i1} := \lceil \beta_{i1} \rceil$, $n_{i2} := \lceil \beta_{i2} \rceil$; $\lambda_{i1} := \lceil \alpha_{i1} \rceil$, $\lambda_{i2} := \lceil \alpha_{i2} \rceil$. The mixed partials $D_{s_{i1}s_{i2}}^{\alpha_{i1}+\alpha_{i2}} f_i$, $D_{s_{i2}s_{i1}}^{\beta_{i2}+\beta_{i1}} f_i$, $D_{s_{i1}s_{i2}}^{\beta_{i1}+\alpha_{i2}} f_i$, and $D_{s_{i2}s_{i1}}^{\alpha_{i2}+\beta_{i1}} f_i$ are considered according to Definition 21.18.*
Assume that

$$D_{s_{i1}s_{i2}}^{\beta_{i1}+\alpha_{i2}} f_i = D_{s_{i2}s_{i1}}^{\alpha_{i2}+\beta_{i1}} f_i. \qquad (21.49)$$

(For sufficient conditions for (21.49) see Theorem 21.21.)
Suppose $D_{s_{i2}}^{\alpha_{i2}} f_i (\cdot, s_{i2}) \in AC^{n_{i1}} ([0, x_{i1}])$, such that

$$\frac{\partial^{k_i}}{\partial s_{i1}^{k_i}} \left(D_{s_{i2}}^{\alpha_{i2}} f_i (0, s_{i2}) \right) = 0, \; k_i = 0, 1, \ldots, n_{i1} - 1,$$

and there exists in $\mathbb{R}$ the next

$$D_{s_{i1}s_{i2}}^{\beta_{i1}+\alpha_{i2}} f_i (\cdot, s_{i2}) \in L_\infty (0, x_{i1}), \; \text{all these } \forall s_{i2} \in [0, x_{i2}].$$

Suppose that $D_{s_{i1}}^{\beta_{i1}} f_i (t_{i1}, \cdot) \in AC^{n_{i2}} ([0, x_{i2}])$, such that

$$\frac{\partial^{l_i}}{\partial s_{i2}^{l_i}} \left(D_{s_{i1}}^{\beta_{i1}} f_i (t_{i1}, 0) \right) = 0, \; l_i = 0, 1, \ldots, n_{i2} - 1,$$

and there exists in $\mathbb{R}$ the next

$$D_{s_{i2}s_{i1}}^{\beta_{i2}+\beta_{i1}} f_i (t_{i1}, \cdot) \in L_\infty (0, x_{i2}), \; \text{all these } \forall t_{i1} \in [0, x_{i1}].$$

Further assume

$$D_{s_{i2}s_{i1}}^{\beta_{i2}+\beta_{i1}} f_i \in L_\infty ((0, x_{i1}) \times (0, x_{i2})).$$

Then we obtain the continuous function

$$D_{s_{i1}s_{i2}}^{\alpha_{i1}+\alpha_{i2}} f_i (s_{i1}, s_{i2}) = \frac{1}{\Gamma (\beta_{i1} - \alpha_{i1}) \Gamma (\beta_{i2} - \alpha_{i2})}$$

$$\int_0^{s_{i1}} \int_0^{s_{i2}} (s_{i1} - t_{i1})^{\beta_{i1}-\alpha_{i1}-1} (s_{i2} - t_{i2})^{\beta_{i2}-\alpha_{i2}-1} D_{s_{i2}s_{i1}}^{\beta_{i2}+\beta_{i1}} f_i (t_{i1},t_{i2})$$

$$dt_{i2}dt_{i1}, \ \forall \, (s_{i1}, s_{i2}) \in [0, x_{i1}] \times [0, x_{i2}], \ \text{all } i = 1, \dots, N. \qquad (21.50)$$

Proof. By using Theorem 21.8 we have

$$D_{s_{i1}}^{\alpha_{i1}} \left(D_{s_{i2}}^{\alpha_{i2}} f_i \left(s_{i1}, s_{i2} \right) \right) = \frac{1}{\Gamma \left(\beta_{i1} - \alpha_{i1} \right)}$$

$$\int_0^{s_{i1}} (s_{i1} - t_{i1})^{\beta_{i1} - \alpha_{i1} - 1} \, D_{s_{i1}}^{\beta_{i1}} \left(D_{s_{i2}}^{\alpha_{i2}} f_i \left(t_{i1}, s_{i2} \right) \right) dt_{i1}, \qquad (21.51)$$

$$\forall s_{i1} \in [0, x_{i1}], \ \text{all } s_{i2} \in [0, x_{i2}].$$

Again by Theorem 21.8 we obtain

$$D_{s_{i2}}^{\alpha_{i2}} \left(D_{s_{i1}}^{\beta_{i1}} f_i \left(t_{i1}, s_{i2} \right) \right) = \frac{1}{\Gamma \left(\beta_{i2} - \alpha_{i2} \right)}$$

$$\int_0^{s_{i2}} (s_{i2} - t_{i2})^{\beta_{i2} - \alpha_{i2} - 1} \, D_{s_{i2}}^{\beta_{i2}} \left(D_{s_{i1}}^{\beta_{i1}} f_i \left(t_{i1}, t_{i2} \right) \right) dt_{i2}, \qquad (21.52)$$

$$\forall s_{i2} \in [0, x_{i2}], \ \text{all } t_{i1} \in [0, x_{i1}].$$

That is,

$$D_{s_{i2}s_{i1}}^{\alpha_{i2} + \beta_{i1}} f_i \left(t_{i1}, s_{i2} \right) = \frac{1}{\Gamma \left(\beta_{i2} - \alpha_{i2} \right)}$$

$$\int_0^{s_{i2}} (s_{i2} - t_{i2})^{\beta_{i2} - \alpha_{i2} - 1} \, D_{s_{i2}s_{i1}}^{\beta_{i2} + \beta_{i1}} f_i \left(t_{i1}, t_{i2} \right) dt_{i2}, \qquad (21.53)$$

$$\forall s_{i2} \in [0, x_{i2}], \ \text{all } t_{i1} \in [0, x_{i1}].$$

Using (21.49) and (21.53) into (21.51) we derive (21.50). Continuity of $D_{s_{i1}s_{i2}}^{\alpha_{i1} + \alpha_{i2}} f_i \left(s_{i1}, s_{i2} \right)$ is arrived at via Lemma 21.22. $\square$

Next we apply Theorem 21.12 for $d = 2$ at the fractional level again.

Theorem 21.27. *All here are as in Theorem 21.26. Call $\tilde{\alpha}_{i1} := (a_i + b_i q_i)$ $(\beta_{i1} - \alpha_{i1} - 1)$, $\tilde{\alpha}_{i2} := (a_i + b_i q_i)(\beta_{i2} - \alpha_{i2} - 1)$; $\tilde{\beta}_{i1} := a_i(\beta_{i1} - \alpha_{i1} - 1)$, $\tilde{\beta}_{i2} := a_i(\beta_{i2} - \alpha_{i2} - 1)$; all $i = 1, \dots, N$.*
Then

$$M_3 := \int_0^{x_{11}} \int_0^{x_{12}} \int_0^{x_{21}} \int_0^{x_{22}} \cdots \int_0^{x_{N1}} \int_0^{x_{N2}}$$

$$\frac{\displaystyle\prod_{i=1}^{N} \left| D_{s_{i1}s_{i2}}^{\alpha_{i1} + \alpha_{i2}} f_i \left(s_{i1}, s_{i2} \right) \right| \, ds_{11} ds_{12} ds_{21} ds_{22} \dots ds_{N1} ds_{N2}}{\left(\frac{1}{\Omega_N} \displaystyle\sum_{i=1}^{N} \omega_i s_{i1}^{(\tilde{\alpha}_{i1} + 1)/q_i \omega_i} s_{i2}^{(\tilde{\alpha}_{i2} + 1)/q_i \omega_i} \right)^{\Omega_N}}$$

$$\leq \tilde{U} \prod_{i=1}^{N} (x_{i1} x_{i2})^{1/q_i} \prod_{i=1}^{N} \left(\int_0^{x_{i1}} \int_0^{x_{i2}} (x_{i1} - s_{i1})^{\tilde{\beta}_{i1} + 1} \right. \qquad (21.54)$$

$$\left(x_{i2} - s_{i2}\right)^{\tilde{\beta}_{i2}+1} \left| D_{s_{i2}s_{i1}}^{\beta_{i2}+\beta_{i1}} f_i\left(s_{i1}, s_{i2}\right)\right|^{p_i} ds_{i1}ds_{i2}\right)^{1/p_i}.$$

Here

$$\tilde{U} := \frac{1}{\prod\limits_{i=1}^{N} \left[\left((\tilde{\alpha}_{i1}+1)(\tilde{\alpha}_{i2}+1)\right)^{1/q_i} \left(\left(\tilde{\beta}_{i1}+1\right)\left(\tilde{\beta}_{i2}+1\right)\right)^{1/p_i}\right]}. \qquad (21.55)$$

Proof. Here we use (21.50). Call

$$v_i\left(s_{i1}, s_{i2}\right) := D_{s_{i1}s_{i2}}^{\alpha_{i1}+\alpha_{i2}} f_i\left(s_{i1}, s_{i2}\right) \in C\left([0, x_{i1}] \times [0, x_{i2}]\right)$$

by (21.50), and

$$\Phi_i\left(s_{i1}, s_{i2}\right) := \frac{\left| D_{s_{i2}s_{i1}}^{\beta_{i2}+\beta_{i1}} f_i\left(s_{i1}, s_{i2}\right)\right|}{\Gamma\left(\beta_{i1} - \alpha_{i1}\right)\Gamma\left(\beta_{i2} - \alpha_{i2}\right)} \in L_\infty\left([0, x_{i1}] \times [0, x_{i2}]\right)$$

by assumption; also set

$$c_{i1} := \beta_{i1} - \alpha_{i1} - 1, \ c_{i2} := \beta_{i2} - \alpha_{i2} - 1,$$

and use Theorem 21.12. Then (21.54) is obvious. $\square$

We give

Corollary 21.28. *All here are as in Theorem 21.27. Then*

$$M_3 \le p^{1/\tau p} \tilde{U} \prod_{i=1}^{N} \left(x_{i1}x_{i2}\right)^{1/q_i} \left[\sum_{i=1}^{N} \frac{1}{p_i} \left(\int_0^{x_{i1}} \int_0^{x_{i2}} \left(x_{i1} - s_{i1}\right)^{\tilde{\beta}_{i1}+1}\right.\right.$$

$$\left.\left.\left(x_{i2} - s_{i2}\right)^{\tilde{\beta}_{i2}+1} \left| D_{s_{i2}s_{i1}}^{\beta_{i2}+\beta_{i1}} f_i\left(s_{i1}, s_{i2}\right)\right|^{p_i} ds_{i1}ds_{i2}\right)^{\tau}\right]^{1/\tau p}, \qquad (21.56)$$

where $\tau > 0$.

Proof. By (21.54) and (21.21). $\square$

One can give a variety of applications of the derived inequalities (21.46), (21.48), (21.54), and (21.56), for various specific values of the involved parameters, such as N, p_i, q_i, a_i, b_i, ω_i, α_{i1}, α_{i2}, β_{i1}, and β_{i2}.

22

Canavati Fractional and Other Approximation of Csiszar's f-Divergence

Here are presented various sharp and nearly optimal probabilistic inequalities that give best or nearly best estimates for the Csiszar's f-divergence. These involve Canavati fractional and ordinary derivatives of the directing function f. Also given are lower bounds for the Csiszar's distance. The Csiszar's discrimination is the most essential and general measure for the comparison between two probability measures. This treatment is based on [28].

22.1 Preliminaries

Throughout this chapter we use the following.

(I) Let f be a convex function from $(0, +\infty)$ into $\mathbb{R}$ that is strictly convex at 1 with $f(1) = 0$. Let $(X, \mathcal{A}, \lambda)$ be a measure space, where λ is a finite or a σ-finite measure on $(X, \mathcal{A})$. And let μ_1, μ_2 be two probability measures on $(X, \mathcal{A})$ such that $\mu_1 \ll \lambda$, $\mu_2 \ll \lambda$ (absolutely continuous); for example, $\lambda = \mu_1 + \mu_2$. Denote by $p = d\mu_1/d\lambda$, $q = d\mu_2/d\lambda$ the (densities) Radon–Nikodym derivatives of μ_1, μ_2 with respect to λ. Here we suppose that

$$a < a \leq \frac{p}{q} \leq b, \quad \text{a.e. on } X$$

and $a \leq 1 \leq b$.

The quantity

$$\Gamma_f(\mu_1, \mu_2) = \int_X q(x) f\left(\frac{p(x)}{q(x)}\right) d\lambda(x), \tag{22.1}$$

G.A. Anastassiou, *Fractional Differentiation Inequalities*,
DOI 10.1007/978-0-387-98128-4_22, © Springer Science+Business Media, LLC 2009

was introduced by I. Csiszar in 1967 (see [128]), and is called *f-divergence* of the probability measures μ_1 and μ_2. By Lemma 1.1 of [128], the integral (22.1) is well defined and $\Gamma_f(\mu_1, \mu_2) \geq 0$ with equality only when $\mu_1 = \mu_2$. In [128] the author without proof mentions that $\Gamma_f(\mu_1, \mu_2)$ does not depend on the choice of λ. We give a proof of that fact in our setting.

Lemma 22.1. *Call*

$$p = \frac{d\mu_1}{d(\mu_1 + \mu_2)}, \quad q = \frac{d\mu_2}{d(\mu_1 + \mu_2)}, \quad p^* = \frac{d\mu_1}{d\lambda}, \quad q^* = \frac{d\mu_2}{d\lambda}.$$

Then

$$\int_X q(x) f\left(\frac{p(x)}{q(x)}\right) d(\mu_1 + \mu_2) = \int_X q^*(x) f\left(\frac{p^*(x)}{q^*(x)}\right) d\lambda. \qquad (22.2)$$

Proof. Because $f: (0, +\infty) \to \mathbb{R}$ is a convex function it is continuous there and Borel measurable, making $f(p(x)/q(x))$ X-measurable.

In general let μ, ν be σ-finite measures on X such that $\nu \ll \mu$ and a g-measurable function on X; then it is known that

$$\int_X g_\pm \, d\nu = \int_X g_\pm \left(\frac{d\nu}{d\mu}\right) d\mu,$$

respectively, where $g = g_+ - g_-$, $g_\pm \geq 0$. Finally we obtain

$$\int_X g \, d\nu = \int_X g \left(\frac{d\nu}{d\mu}\right) d\mu. \qquad (22.3)$$

Next let $E \in \mathcal{A}$ be such that $\lambda(E) = 0$; then $\mu_1(E) = \mu_2(E) = 0$ and $(\mu_1 + \mu_2)(E) = 0$; that is, $\mu_1 + \mu_2 \ll \lambda$. Clearly here we have that

$$\mu_1 \ll \mu_1 + \mu_2 \ll \lambda, \qquad (22.4)$$

$$\mu_2 \ll \mu_1 + \mu_2 \ll \lambda. \qquad (22.5)$$

Then it is known that

$$\frac{d\mu_1}{d\lambda} = \frac{d\mu_1}{d(\mu_1 + \mu_2)} \frac{d(\mu_1 + \mu_2)}{d\lambda}, \qquad (22.6)$$

and

$$\frac{d\mu_2}{d\lambda} = \frac{d\mu_2}{d(\mu_1 + \mu_2)} \frac{d(\mu_1 + \mu_2)}{d\lambda}. \qquad (22.7)$$

That is,

$$p^* = p \frac{d(\mu_1 + \mu_2)}{d\lambda} \qquad (22.8)$$

and

$$q^* = q\frac{d(\mu_1 + \mu_2)}{d\lambda}. \qquad (22.9)$$

Here by assumption

$$0 < a \le \frac{p}{q}, \; \frac{p^*}{q^*} \le b, \quad \text{a.e. on } X;$$

that is, $q, q^* > 0$ a.e. on X. That is, $d(\mu_1 + \mu_2)/d\lambda > 0$ a.e. on X. Therefore from the above we get

$$\frac{p^*}{q^*} = \frac{p}{q} \quad \text{a.e. on } X. \qquad (22.10)$$

Consequently it holds

$$\int_X q(x)f\left(\frac{p(x)}{q(x)}\right) d(\mu_1+\mu_2) = \int_X f\left(\frac{p^*(x)}{q^*(x)}\right) d\mu_2 = \int_X q^*(x)f\left(\frac{p^*(x)}{q^*(x)}\right) d\lambda.$$

$\square$

The concept of f-divergence was introduced first in [127] as a generalization of Kullback's "information for discrimination" or I-divergence (generalized entropy) [240, 241] and of Rényi's "information gain" (I-divergence of order α) [345]. In fact the I-divergence of order 1 equals

$$\Gamma_{u \log_2 u}(\mu_1, \mu_2).$$

The choice $f(u) = (u-1)^2$ again produces a known measure of difference of distributions that is called χ^2-divergence; of course the total variation distance $|\mu_1 - \mu_2| = \int_X |p(x) - q(x)| \, d\lambda(x)$ equals $\Gamma_{|u-1|}(\mu_1, \mu_2)$.

Here by assuming $f(1) = 0$ we can consider $\Gamma_f(\mu_1, \mu_2)$ as a measure of the difference between the probability measures μ_1, μ_2. The f-divergence is in general asymmetric in μ_1 and μ_2. But because f is convex and strictly convex at 1 (see Lemma 22.2 next) so is

$$f^*(u) = uf\left(\frac{1}{u}\right) \qquad (22.11)$$

and as in [128] we find

$$\Gamma_f(\mu_2, \mu_1) = \Gamma_{f^*}(\mu_1, \mu_2). \qquad (22.12)$$

Lemma 22.2. Let $f: (0, +\infty) \to \mathbb{R}$ be a strictly convex function at 1. Then so is $f^*(u) = uf(1/u)$, $u \in (0, +\infty)$.

Proof. Let $0 < u_1 < 1 < u_2$; then $1/u_2 < 1 < 1/u_1$. Hence

$$f\left(\frac{2}{u_1 + u_2}\right) = f\left(\left(\frac{u_1}{u_1 + u_2}\right)\frac{1}{u_1} + \left(\frac{u_2}{u_1 + u_2}\right)\frac{1}{u_2}\right)$$

$$< \left(\frac{u_1}{u_1 + u_2}\right)f\left(\frac{1}{u_1}\right) + \left(\frac{u_2}{u_1 + u_2}\right)f\left(\frac{1}{u_2}\right).$$

Clearly we get

$$\left(\frac{u_1 + u_2}{2}\right) f\left(\frac{1}{\left(\frac{u_1+u_2}{2}\right)}\right) < \frac{u_1 f\left(\frac{1}{u_1}\right) + u_2 f\left(\frac{1}{u_2}\right)}{2}.$$

Thus we have

$$f^*\left(\frac{u_1 + u_2}{2}\right) < \frac{f^*(u_1) + f^*(u_2)}{2},$$

proving our claim. □

In information theory and statistics many other concrete divergences are used that are special cases of the above general Csiszar f-divergence, such as Hellinger distance D_H, α-divergence D_α, Bhattacharyya distance D_B, harmonic distance D_{Ha}, Jeffrey's distance D_J, and triangular discrimination D_Δ; for all these see, for example [78, 145]. The problem of finding and estimating the *proper distance* (*or difference or discrimination*) of two probability distributions is one of the major questions in probability theory.

The above f-divergence measures in their various forms have also been applied to anthropology, genetics, finance, economics, political science, biology, approximation of probability distributions, signal processing, and pattern recognition. A great inspiration for this chapter has been the very important monograph on the topic by S. Dragomir [145].

(II) In the sequel we follow [101]. Let $g \in C([0,1])$. Let ν be a positive number, $n := [\nu]$ and $\alpha := \nu - n$ $(0 < \alpha < 1)$. Define

$$(J_\nu g)(x) := \frac{1}{\Gamma(\nu)} \int_0^x (x - t)^{\nu-1} g(t)\, dt, \quad 0 \le x \le 1, \qquad (22.13)$$

the *Riemann–Liouville integral*, where Γ is the gamma function. We define the subspace $C^\nu([0,1])$ or $C^n([0,1])$ as follows.

$$C^\nu([0,1]) := \{g \in C^n([0,1]) : J_{1-\alpha} D^n g \in C^1([0,1])\},$$

where $D := d/dx$. So for $g \in C^\nu([0,1])$, we define the Canavati *ν-fractional derivative* of g as

$$D^\nu g := D J_{1-\alpha} D^n g. \qquad (22.14)$$

When $\nu \ge 1$ we have the Taylor formula

$$\begin{aligned} g(t) = {}& g(0) + g'(0)t + g''(0)\frac{t^2}{2!} + \cdots + g^{(n-1)}(0)\frac{t^{n-1}}{(n-1)!} \\ & + (J_\nu D^\nu g)(t), \text{ for all } t \in [0,1]. \end{aligned} \qquad (22.15)$$

When $0 < \nu < 1$ we find

$$g(t) = (J_\nu D^\nu g)(t), \text{ for all } t \in [0,1]. \qquad (22.16)$$

Next we transfer the above notions over to arbitrary $[a, b] \subseteq \mathbb{R}$ (see [17]). Let $x, x_0 \in [a, b]$ such that $x \geq x_0$, where x_0 is fixed. Let $f \in C([a, b])$ and define

$$(J_\nu^{x_0} f)(x) := \frac{1}{\Gamma(\nu)} \int_{x_0}^x (x - t)^{\nu-1} f(t) \, dt, \quad x_0 \leq x \leq b, \qquad (22.17)$$

the *generalized Riemann–Liouville integral.* We define the subspace $C_{x_0}^\nu([a, b])$ of $C^n([a, b])$:

$$C_{x_0}^\nu([a, b]) := \{ f \in C^n([a, b]) \colon J_{1-\alpha}^{x_0} D^n f \in C^1([x_0, b]) \}.$$

For $f \in C_{x_0}^\nu([a, b])$, we define the *generalized ν -fractional derivative of f over $[x_0, b]$* as

$$D_{x_0}^\nu f := D J_{1-\alpha}^{x_0} f^{(n)} \quad (f^{(n)} := D^n f). \qquad (22.18)$$

Observe that

$$(J_{1-\alpha}^{x_0} f^{(n)})(x) = \frac{1}{\Gamma(1-\alpha)} \int_{x_0}^x (x - t)^{-\alpha} f^{(n)}(t) \, dt$$

exists for $f \in C_{x_0}^\nu([a, b])$.

We recall the following generalization of Taylor's formula (see [17, 101]).

Theorem 22.3. *Let $f \in C_{x_0}^\nu([a, b])$, $x_0 \in [a, b]$, fixed.*

(i) *If $\nu \geq 1$ then*

$$f(x) = f(x_0) + f'(x_0)(x - x_0) + f''(x_0)\frac{(x - x_0)^2}{2}$$

$$+ \cdots + f^{(n-1)}(x_0)\frac{(x - x_0)^{n-1}}{(n-1)!}$$

$$+ (J_\nu^{x_0} D_{x_0}^\nu f)(x), \quad \text{for all } x \in [a, b] \colon x \geq x_0. \qquad (22.19)$$

(ii) *If $0 < \nu < 1$ then*

$$f(x) = (J_\nu^{x_0} D_{x_0}^\nu f)(x), \quad \text{for all } x \in [a, b] \colon x \geq x_0. \qquad (22.20)$$

We make

Remark 22.4. (1) $(D_{x_0}^n f) = f^{(n)}$, $n \in \mathbb{N}$.

(2) Let $f \in C_{x_0}^\nu([a, b])$, $\nu \geq 1$ and $f^{(i)}(x_0) = 0$, $i = 0, 1, \ldots, n - 1$; $n := [\nu]$. Then by (22.19)

$$f(x) = (J_\nu^{x_0} D_{x_0}^\nu f)(x).$$

That is,

$$f(x) = \frac{1}{\Gamma(\nu)} \int_{x_0}^{x} (x-t)^{\nu-1} (D_{x_0}^{\nu} f)(t)\, dt, \qquad (22.21)$$

for all $x \in [a, b]$ with $x \geq x_0$. Notice that (22.21) is also true when $0 < \nu < 1$.

We also make

Remark 22.5. Let $\nu \geq 1$, $\gamma \geq 0$, such that $\nu - \gamma \geq 1$, so that $\gamma < \nu$. Call $n := [\nu]$, $\alpha := \nu - n$; $m := [\gamma]$, $\rho := \gamma - m$. Note that $\nu - m \geq 1$ and $n - m \geq 1$. Let $f \in C_{x_0}^{\nu}([a, b])$ be such that $f^{(i)}(x_0) = 0$, $i = 0, 1, \dots, n-1$. Hence by (22.19)

$$f(x) = (J_{\nu}^{x_0} D_{x_0}^{\nu} f)(x), \quad \text{for all } x \in [a, b]: x \geq x_0.$$

Therefore by Leibnitz's formula and $\Gamma(p+1) = p\Gamma(p)$, $p > 0$, we get that

$$f^{(m)}(x) = (J_{\nu-m}^{x_0} D_{x_0}^{\nu} f)(x), \quad \text{for all } x \geq x_0.$$

It follows that $f \in C_{x_0}^{\gamma}([a, b])$ and thus $(D_{x_0}^{\gamma} f)(x) := (DJ_{1-\rho}^{x_0} f^{(m)})(x)$ exists for all $x \geq x_0$.

We easily obtain

$$(D_{x_0}^{\gamma} f)(x) = D((J_{1-\rho}^{x_0} f^{(m)})(x)) = \frac{1}{\Gamma(\nu-\gamma)} \int_{x_0}^{x} (x-t)^{(\nu-\gamma)-1} (D_{x_0}^{\nu} f)(t)\, dt,$$

$$(22.22)$$

and thus

$$(D_{x_0}^{\gamma} f)(x) = (J_{\nu-\gamma}^{x_0} (D_{x_0}^{\nu} f))(x)$$

and is continuous in x on $[x_0, b]$.

22.2   Main Results

Again f and the whole setting is as in Section 22.1(I). Other notations are as in Section 22.1(II). We present the following.

Theorem 22.6. *Let $a < b$, $1 \leq \nu < 2$, $f \in C_a^{\nu}([a, b])$, and $a \leq p(x)/q(x) \leq b$, a.e. on X. Then*

$$\Gamma_f(\mu_1, \mu_2) \leq \frac{\|D_a^{\nu} f\|_{\infty, [a,b]}}{\Gamma(\nu+1)} \int_X (q(x))^{1-\nu} (p(x) - aq(x))^{\nu}\, d\lambda(x). \quad (22.23)$$

Proof. Here $n := [\nu] = 1$ and from (22.21) we have

$$f(x) = \frac{1}{\Gamma(\nu)} \int_a^x (x-w)^{\nu-1} (D_a^{\nu} f)(w)\, dw, \quad \text{all } a \leq x \leq b. \quad (22.24)$$

Then

$$|f(x)| \leq \frac{1}{\Gamma(\nu)} \int_a^x (x-w)^{\nu-1} |(D_a^\nu f)(w)| \, dw$$

$$\leq \frac{\|D_a^\nu f\|_{\infty,[a,b]}}{\Gamma(\nu)} \int_a^x (x-w)^{\nu-1} \, dw$$

$$= \frac{\|D_a^\nu f\|_{\infty,[a,b]}}{\Gamma(\nu+1)} (x-a)^\nu. \tag{22.25}$$

That is, we get

$$|f(x)| \leq \frac{\|D_a^\nu f\|_{\infty,[a,b]}}{\Gamma(\nu+1)} (x-a)^\nu, \quad \text{for all } a \leq x \leq b. \tag{22.26}$$

Consequently we obtain

$$\Gamma_f(\mu_1,\mu_2) = \int_X q(x) f\left(\frac{p(x)}{q(x)}\right) d\lambda \leq \int_X q(x) \left|f\left(\frac{p(x)}{q(x)}\right)\right| d\lambda$$

$$\leq \left(\frac{\|D_a^\nu f\|_{\infty,[a,b]}}{\Gamma(\nu+1)}\right) \int_X q(x) \left(\frac{p(x)}{q(x)} - a\right)^\nu d\lambda$$

$$= \frac{\|D_a^\nu f\|_{\infty,[a,b]}}{\Gamma(\nu+1)} \int_X (q(x))^{1-\nu} (p(x) - aq(x))^\nu \, d\lambda(x), \tag{22.27}$$

thereby proving (22.23). □

The counterpart of the previous result follows.

Theorem 22.7. Let $a < b$, $\nu \geq 2$, $n := [\nu]$, $f \in C_a^\nu([a,b])$, $f^{(i)}(a) = 0$, $i = 0, 1, \ldots, n-1$, and $a \leq p(x)/q(x) \leq b$, a.e. on X. Then

$$\Gamma_f(\mu_1,\mu_2) \leq \frac{\|D_a^\nu f\|_{\infty,[a,b]}}{\Gamma(\nu+1)} \int_X (q(x))^{1-\nu} (p(x) - aq(x))^\nu \, d\lambda(x). \tag{22.28}$$

Proof. As in the proof of Theorem 22.6. □

Next we give an L_α estimate.

Theorem 22.8. Let $a < b$, $\nu \geq 1$, $n := [\nu]$, $f \in C_a^\nu([a,b])$, $f^{(i)}(a) = 0$, $i = 0, 1, \ldots, n-1$, and $a \leq p(x)/q(x) \leq b$, a.e. on X. Let $\alpha, \beta > 1$: $1/\alpha + 1/\beta = 1$. Then

$$\Gamma_f(\mu_1,\mu_2) \leq \frac{\|D_a^\nu f\|_{\alpha,[a,b]}}{\Gamma(\nu)(\beta(\nu-1)+1)^{1/\beta}} \int_X (q(x))^{2-\nu-1/\beta} (p(x) - aq(x))^{\nu-1+1/\beta} \, d\lambda(x). \tag{22.29}$$

Proof. From (22.21) we have

$$f(x) = \frac{1}{\Gamma(\nu)} \int_a^x (x-w)^{\nu-1}(D_a^\nu f)(w)\, dw, \quad \text{for all } a \le x \le b. \quad (22.30)$$

Also $(D_a^\nu f)(w) \in C([a, b])$. Hence

$$
\begin{aligned}
|f(x)| &\le \frac{1}{\Gamma(\nu)} \int_a^x (x-w)^{\nu-1}|(D_a^\nu f)(w)|\, dw \\
&\le \frac{1}{\Gamma(\nu)} \left(\int_a^x (x-w)^{\beta(\nu-1)}\, dw \right)^{1/\beta} \left(\int_a^x |(D_a^\nu f)(w)|^\alpha\, dw \right)^{1/\alpha} \\
&\le \frac{\|D_a^\nu f\|_{\alpha,[a,b]}}{\Gamma(\nu)} \left(\int_a^x (x-w)^{\beta(\nu-1)}\, dw \right)^{1/\beta} \\
&= \frac{\|D_a^\nu f\|_{\alpha,[a,b]}}{\Gamma(\nu)} \left(\frac{(x-a)^{\beta(\nu-1)+1}}{\beta(\nu-1)+1} \right)^{1/\beta}. \quad (22.31)
\end{aligned}
$$

That is, we find that

$$|f(x)| \le \frac{\|D_a^\nu f\|_{\alpha,[a,b]}}{\Gamma(\nu)(\beta(\nu-1)+1)^{1/\beta}}(x-a)^{(\nu-1+1/\beta)}. \quad (22.32)$$

Therefore

$$
\begin{aligned}
\Gamma_f(\mu_1, \mu_2) &\le \int_X q\left|f\left(\frac{p}{q}\right)\right| d\lambda \le \frac{\|D_a^\nu f\|_{\alpha,[a,b]}}{\Gamma(\nu)(\beta(\nu-1)+1)^{1/\beta}} \int_X q\left(\frac{p}{q}-a\right)^{\nu-1+1/\beta} d\lambda \\
&= \frac{\|D_a^\nu f\|_{\alpha,[a,b]}}{\Gamma(\nu)(\beta(\nu-1)+1)^{1/\beta}} \int_X q^{2-\nu-1/\beta}(p-aq)^{\nu-1+1/\beta}\, d\lambda, \quad (22.33)
\end{aligned}
$$

proving (22.29). $\square$

An L_1 estimate follows.

Theorem 22.9. *Let* $a < b$, $\nu \ge 1$, $n := [\nu]$, $f \in C_a^\nu([a, b])$, $f^{(i)}(a) = 0$, $i = 0, 1, \ldots, n-1$, *and* $a \le p(x)/q(x) \le b$, *a.e. on* X. *Then*

$$\Gamma_f(\mu_1, \mu_2) \le \frac{\|D_a^\nu f\|_{1,[a,b]}}{\Gamma(\nu)} \int_X q(x)^{2-\nu}(p(x) - aq(x))^{\nu-1}\, d\lambda(x). \quad (22.34)$$

Proof. Again from (22.21) and (22.30) we have

$$
\begin{aligned}
|f(x)| &\le \frac{1}{\Gamma(\nu)} \int_a^x (x-w)^{\nu-1}|D_a^\nu f(w)|\, dw \le \frac{1}{\Gamma(\nu)}(x-a)^{\nu-1} \int_a^x |D_a^\nu f(w)|\, dw \\
&\le \frac{1}{\Gamma(\nu)}(x-a)^{\nu-1} \int_a^b |D_a^\nu f(w)|\, dw = \frac{1}{\Gamma(\nu)}(x-a)^{\nu-1}\|D_a^\nu f\|_{1,[a,b]}. \quad (22.35)
\end{aligned}
$$

That is,

$$|f(x)| \leq \frac{\|D_a^\nu f\|_{1,[a,b]}}{\Gamma(\nu)} (x-a)^{\nu-1}, \qquad (22.36)$$

for all x in $[a,b]$. Thus

$$\Gamma_f(\mu_1,\mu_2) \leq \int_X q \left| f\left(\frac{p}{q}\right) \right| d\lambda \leq \left(\int_X q \left(\frac{p}{q}-a\right)^{\nu-1} d\lambda \right) \frac{\|D_a^\nu f\|_{1,[a,b]}}{\Gamma(\nu)}$$

$$= \frac{\|D_a^\nu f\|_{1,[a,b]}}{\Gamma(\nu)} \left(\int_X q^{2-\nu}(p-aq)^{\nu-1} d\lambda \right), \qquad (22.37)$$

proving (22.34). $\square$

Remark 22.10. Let $f: [a,b] \to \mathbb{R}$ convex,

$$f \in C^1([a,b]), \quad x_0 \in [a,b], \quad a \neq b.$$

Then

$$f(x) \geq f(x_0) + f'(x_0)(x-x_0), \quad \text{for all } x \in [a,b]. \qquad (22.38)$$

Then one can easily prove that

$$\Delta(x_0) := \frac{1}{b-a} \int_a^b f(x)\, dx - f(x_0) \geq f'(x_0)\left[\frac{a+b}{2} - x_0\right]; \qquad (22.39)$$

see also S. Dragomir and C. E. M. Pearce [146, p. 9, Theorem 18]. In this chapter's setting

$$0 < a \leq \frac{p(x)}{q(x)} \leq b, \quad \text{a.e. on } X,$$

and by calling

$$M := \frac{1}{b-a} \int_a^b f(x)\, dx \qquad (22.40)$$

we get

$$M - f\left(\frac{p(x)}{q(x)}\right) \geq f'\left(\frac{p(x)}{q(x)}\right)\left[\frac{a+b}{2} - \frac{p(x)}{q(x)}\right], \quad \text{a.e. on } X. \qquad (22.41)$$

Then

$$q(x)M - q(x)f\left(\frac{p(x)}{q(x)}\right) \geq f'\left(\frac{p(x)}{q(x)}\right)\left[q(x)\frac{(a+b)}{2} - p(x)\right], \quad \text{a.e. on } X. \qquad (22.42)$$

Here $q(x) > 0$ a.e. on X. Therefore by integrating (22.42) against λ we obtain

$$M - \int_X q(x)f\left(\frac{p(x)}{q(x)}\right) d\lambda(x) \geq \int_X f'\left(\frac{p(x)}{q(x)}\right)\left(\frac{q(x)(a+b)}{2} - p(x)\right) d\lambda(x). \qquad (22.43)$$

We have established

Theorem 22.11. *Let all elements be as in Section 22.1(I), hold, and additionally assume that $f \in C^1([a,b])$, $a \neq b$. Then*

$$\Gamma_f(\mu_1, \mu_2) \leq \frac{1}{b-a} \int_a^b f(x)\, dx - \int_X f'\left(\frac{p(x)}{q(x)}\right)\left(\frac{q(x)(a+b)}{2} - p(x)\right) d\lambda.$$

$$(22.44)$$

Remark 22.12. Let all elements be as in Section 22.1(I), hold, and $f \in C^1([a,b])$. Then by (22.38) we get

$$f(x) \geq f'(1)(x-1) \qquad (22.45)$$

and

$$f\left(\frac{p(x)}{q(x)}\right) \geq f'(1)\left(\frac{p(x)}{q(x)} - 1\right), \quad \text{a.e. on } X. \qquad (22.46)$$

Thus

$$q(x) f\left(\frac{p(x)}{q(x)}\right) \geq f'(1)(p(x) - q(x)), \quad \text{a.e. on } X. \qquad (22.47)$$

Integrating (22.47) against λ over X we find

$$\Gamma_f(\mu_1, \mu_2) \geq 0 \qquad (22.48)$$

which is known and mentioned in [128].

Remark 22.13. Let $n \in \mathbb{N}$, $f \in C^{n+1}([a,b])$, $[a,b] \subset \mathbb{R}$, such that $f^{(n+1)} \geq 0$ (≤ 0); then by Taylor's formula we get that

$$f(x) \geq (\leq) \sum_{i=0}^n f^{(i)}(x_0) \frac{(x-x_0)^i}{i!}, \qquad (22.49)$$

respectively, for any $x, x_0 \in [a,b]$: $x \geq x_0$. Inequalities (22.49) are valid also when n is odd and $x \leq x_0$. Take $0 < a \leq 1 \leq b$, $a \leq p/q \leq b$, a.e. on X and $f(1) = 0$, along with $f \in C^{n+1}([a,b])$, n odd, such that $f^{(n+1)} \geq 0$ (≤ 0); then

$$f(x) \geq (\leq) \sum_{i=1}^n f^{(i)}(1) \frac{(x-1)^i}{i!}, \qquad (22.50)$$

for all x in $[a,b]$. Hence

$$q f\left(\frac{p}{q}\right) \geq (\leq) \sum_{i=1}^n f^{(i)}(1) \frac{q\left(\frac{p}{q} - 1\right)^i}{i!}. \qquad (22.51)$$

And finally we obtain

$$\Gamma_f(\mu_1, \mu_2) \geq (\leq) \sum_{i=2}^{n} \frac{f^{(i)}(1)}{i!} \int_X q^{1-i}(p-q)^i \, d\lambda, \quad n \text{ odd.} \qquad (22.52)$$

Remark 22.14. Let n be odd, $f \in C^{n+1}([a,b])$, $[a,b] \subset \mathbb{R}$, such that $f^{(n+1)} \geq 0 \ (\leq 0)$, $x, x_0 \in [a,b]$; then as before we have

$$f(x) - f(x_0) \geq (\leq) \sum_{i=1}^{n} f^{(i)}(x_0) \frac{(x-x_0)^i}{i!}. \qquad (22.53)$$

Then

$$\frac{1}{b-a} \int_a^b f(x) \, dx - f(x_0) = \frac{1}{b-a} \int_a^b (f(x) - f(x_0)) \, dx$$

$$\geq (\leq) \frac{1}{b-a} \int_a^b \sum_{i=1}^{n} \frac{f^{(i)}(x_0)}{i!} (x-x_0)^i \, dx$$

$$= \frac{1}{b-a} \sum_{i=1}^{n} \frac{f^{(i)}(x_0)}{i!} \int_a^b (x-x_0)^i \, dx$$

$$= \frac{1}{b-a} \sum_{i=1}^{n} \frac{f^{(i)}(x_0)}{(i+1)!} \left((x-x_0)^{i+1} \big|_a^b \right)$$

$$= \frac{1}{b-a} \sum_{i=1}^{n} \frac{f^{(i)}(x_0)}{(i+1)!} \left[(b-x_0)^{i+1} \right. \qquad (22.54)$$

$$\left. - (a-x_0)^{i+1} \right].$$

We obtained

$$\frac{1}{b-a} \int_a^b f(x) \, dx - f(x_0) \geq (\leq) \sum_{i=1}^{n} \frac{f^{(i)}(x_0)}{(i+1)!} \left[\sum_{k=0}^{i} (b-x_0)^{i-k}(a-x_0)^k \right], \qquad (22.55)$$

where n is odd.

Next let $0 < a \leq p/q \leq b$, a.e. on X. Then from (22.55) we find

$$\frac{q}{b-a} \int_a^b f(x) dx - qf\left(\frac{p}{q}\right)$$

$$\geq (\leq) \sum_{i=1}^{n} \frac{f^{(i)}\left(\frac{p}{q}\right)}{(i+1)!} \left[\sum_{k=0}^{i} q \left(b - \frac{p}{q}\right)^{i-k} \left(a - \frac{p}{q}\right)^k \right]$$

$$= \sum_{i=1}^{n} \frac{f^{(i)}\left(\frac{p}{q}\right)}{(i+1)!} \left[\sum_{k=0}^{i} q^{1-i}(bq-p)^{i-k}(aq-p)^k \right]. \qquad (22.56)$$

So when n is odd we get

$$\frac{1}{b-a} \int_a^b f(x)\,dx - \Gamma_f(\mu_1, \mu_2)$$

$$\geq (\leq) \sum_{i=1}^n \frac{1}{(i+1)!} \left(\sum_{k=0}^i \int_X f^{(i)} \left(\frac{p}{q} \right) q^{1-i}(bq-p)^{(i-k)}(aq-p)^k d\lambda \right).$$
$$(22.57)$$

We have proved

Theorem 22.15. *Let n be odd and $f \in C^{n+1}([a,b])$, such that $f^{(n+1)} \geq 0$ (≤ 0), $0 < a \leq p/q \leq b$, a.e. on X. Then*

$$\Gamma_f(\mu_1, \mu_2) \leq (\geq) \frac{1}{b-a} \int_a^b f(x)\,dx$$

$$-\sum_{i=1}^n \frac{1}{(i+1)!} \left(\sum_{k=0}^i \int_X f^{(i)} \left(\frac{p}{q} \right) q^{1-i}(bq-p)^{(i-k)}(aq-p)^k d\lambda \right). \quad (22.58)$$

Remark 22.16. Let n be odd, $f \in C^{n+1}([a,b])$, $a,b \in \mathbb{R}$, such that $f^{(n+1)} \geq 0$ (≤ 0), $x, x_0 \in [a,b]$, and μ be any probability measure on $[a,b]$; then by (22.53) we get

$$\int_{[a,b]} f(x)\,d\mu - f(x_0) \geq (\leq) \sum_{i=1}^n \frac{f^{(i)}(x_0)}{i!} \int_{[a,b]} (x-x_0)^i d\mu. \quad (22.59)$$

Comment 22.17. Assume all singletons of X belong to $\mathcal{A}$. Let the Dirac measure $\delta_{x_0} \ll \lambda$, $x_0 \in X$; then there exists density $\hat{f} \geq 0$ such that

$$\delta_{x_0}(E) = \int_E \hat{f}\,d\lambda, \quad \text{for all } E \in \mathcal{A},$$

(here $\hat{f}$ is the Radon–Nikodym derivative of δ_{x_0} with respect to λ). Clearly

$$1 = \delta_{x_0}(\{x_0\}) = \int_{\{x_0\}} \hat{f}\,d\lambda = \hat{f}(x_0)\lambda(\{x_0\}). \quad (22.60)$$

Because $\hat{f}$ is real-valued we must have $\lambda(\{x_0\}) \neq 0$ and

$$\hat{f}(x_0) = \frac{1}{\lambda(\{x_0\})} > 0. \quad (22.61)$$

Also

$$0 = \delta_{x_0}(X - \{x_0\}) = \int_{X-\{x_0\}} \hat{f}\,d\lambda.$$

Therefore

$$\hat{f}\big|_{X-\{x_0\}} = 0, \quad \text{a.e. on } X. \tag{22.62}$$

So let $\mu_2 = \delta_{x_0}$; then $q = \hat{f}$ and

$$\Gamma_f(\mu_1, \delta_{x_0}) = \int_X \hat{f} f\left(\frac{p}{\hat{f}}\right) d\lambda = f(\lambda(\{x_0\})p(x_0)), \tag{22.63}$$

is trivial and of no interest for further study.

Finally we give

Comment 22.18. Let $f \in C_{x_0}^{\nu}([a,b])$, $x_0 \in [a,b] \subset \mathbb{R}$, and $\nu \geq 1$, the remainder of the fractional Taylor formula (see Theorem 22.3 and (22.19)) is

$$\left(J_{\nu}^{x_0}(D_{x_0}^{\nu}f)\right)(x) = \frac{1}{\Gamma(\nu)} \int_{x_0}^{x} (x-t)^{\nu-1}(D_{x_0}^{\nu}f)(t)\, dt, \tag{22.64}$$

for all $x_0 \leq x \leq b$. If

$$(D_{x_0}^{\nu}f)(t) \; \begin{matrix} \geq 0 \\ (\leq 0) \end{matrix}$$

over $[x_0, b]$ then

$$\left(J_{\nu}^{x_0}(D_{x_0}^{\nu}f)\right)(x) \; \begin{matrix} \geq 0 \\ (\leq 0) \end{matrix}$$

over $[x_0, b]$. The last implies (by (22.19)) that

$$f(x) \geq (\leq) \sum_{i=0}^{n-1} f^{(i)}(x_0)\frac{(x-x_0)^i}{i!}, \tag{22.65}$$

where $n := [\nu]$, for all $x_0 \leq x \leq b$. According to Section 22.1(I) we take here f to be convex from $(0, +\infty)$ into $\mathbb{R}$ which is strictly convex at 1 with $f(1) = 0$. Also we take $0 < a \leq p/q \leq b$ a.e. on X, with $a \leq 1 \leq b$.

Let $\nu \geq 1$ and additionally assume that $f \in C_a^{\nu}([a,b])$ such that $(D_a^{\nu}f)(t) \geq 0 \ (\leq 0)$ over $[a,b]$; then we find

$$f(x) \geq (\leq) \sum_{i=0}^{n-1} f^{(i)}(a)\frac{(x-a)^i}{i!}, \tag{22.66}$$

for all $a \leq x \leq b$, $n := [\nu]$. Then

$$qf\left(\frac{p}{q}\right) \geq (\leq) \sum_{i=0}^{n-1} f^{(i)}(a)\frac{q\left(\frac{p}{q}-a\right)^i}{i!}, \tag{22.67}$$

a.e. on X. Consequently we obtain

$$\Gamma_f(\mu_1, \mu_2) \geq (\leq) \sum_{i=0}^{n-1} \frac{f^{(i)}(a)}{i!} \int_X q^{1-i}(p - qa)^i \, d\lambda, \qquad (22.68)$$

where $n := [\nu]$, $\nu \geq 1$.

23

Caputo and Riemann–Liouville Fractional Approximation of Csiszar's f-Divergence

Here are presented various tight probabilistic inequalities that give nearly best estimates for the Csiszar's f-divergence. These involve Riemann–Liouville and Caputo fractional derivatives of the directing function f. Also is given a lower bound for the Csiszar's distance. The Csiszar's discrimination is the most essential and general measure for the comparison between two probability measures. This is a continuation of Chapter 22 and is based on [55].

23.1 Preliminaries

Throughout this chapter we use the following.

(I) Let f be a convex function from $(0, +\infty)$ into $\mathbb{R}$ that is strictly convex at 1 with $f(1) = 0$. Let $(X, \mathcal{A}, \lambda)$ be a measure space, where λ is a finite or a σ-finite measure on $(X, \mathcal{A})$. And let μ_1, μ_2 be two probability measures on $(X, \mathcal{A})$ such that $\mu_1 \ll \lambda, \mu_2 \ll \lambda$ (absolutely continuous); for example, $\lambda = \mu_1 + \mu_2$. Denote by $p = d\mu_1/d\lambda$, $q = d\mu_2/d\lambda$ the (densities) Radon–Nikodym derivatives of μ_1, μ_2 with respect to λ. Here we assume that

$$0 < a \le \frac{p}{q} \le b, \quad \text{a.e. on } X$$

and $a \le 1 \le b$.

G.A. Anastassiou, *Fractional Differentiation Inequalities*, 577
DOI 10.1007/978-0-387-98128-4_23, © Springer Science+Business Media, LLC 2009

The quantity

$$\Gamma_f \left(\mu_1, \mu_2 \right) = \int_X q\left(x \right) f \left(\frac{p\left(x \right)}{q\left(x \right)} \right) d\lambda\left(x \right), \qquad (23.1)$$

was introduced by I. Csiszar in 1967 (see [128]), and is called f-divergence of the probability measures μ_1 and μ_2. By Lemma 1.1 of [128], the integral (23.1) is well defined and $\Gamma_f \left(\mu_1, \mu_2 \right) \geq 0$ with equality only when $\mu_1 = \mu_2$. In [128] the author without proof mentions that $\Gamma_f \left(\mu_1, \mu_2 \right)$ does not depend on the choice of λ.

For a proof of the last see [28], Lemma 1.1; see here Lemma 22.1.

The concept of f-divergence was introduced first in [127] as a generalization of Kullback's "information for discrimination" or I-divergence (generalized entropy) [240, 241] and of Rényi's "information gain" (I-divergence of order α) [345]. In fact the I-divergence of order 1 equals

$$\Gamma_{u \log_2 u} \left(\mu_1, \mu_2 \right).$$

The choice $f\left(u \right) = \left(u - 1 \right)^2$ again produces a known measure of difference of distributions called $\varkappa^2$- divergence; of course the total variation distance $\left| \mu_1 - \mu_2 \right| = \int_X \left| p\left(x \right) - q\left(x \right) \right| d\lambda\left(x \right)$ equals $\Gamma_{\left| u-1 \right|} \left(\mu_1, \mu_2 \right)$.

Here by assuming $f\left(1 \right) = 0$ we can consider $\Gamma_f \left(\mu_1, \mu_2 \right)$ as a measure of the difference between the probability measures μ_1, μ_2. The f-divergence is in general asymmetric in μ_1 and μ_2. But because f is convex and strictly convex at 1 (see Lemma 2, [28]) so is

$$f^* \left(u \right) = u f \left(\frac{1}{u} \right) \qquad (23.2)$$

and as in [128] we get

$$\Gamma_f \left(\mu_2, \mu_1 \right) = \Gamma_{f^*} \left(\mu_1, \mu_2 \right). \qquad (23.3)$$

In information theory and statistics many other concrete divergences are used that are special cases of the above general Csiszar f-divergence, such as Hellinger distance D_H, α-divergence D_α, Bhattacharyya distance D_B, harmonic distance D_{H_α}, Jeffrey's distance D_J, and triangular discrimination D_Δ; for all these, see, for example [78, 145]. The problem of finding and estimating the proper distance (or difference or discrimination) of two probability distributions is one of the major questions in probability theory.

The above f-divergence measures in their various forms have also been applied to anthropology, genetics, finance, economics, political science, biology, approximation of probability distributions, signal processing, and pattern recognition. A great inspiration for this chapter has been the very important monograph on the topic by S. Dragomir [145].

(II) Here we follow [134].

We start with

Definition 23.1. Let $\nu \geq 0$; the operator J_a^ν, defined on $L_1(a, b)$ by

$$J_a^\nu f(x) := \frac{1}{\Gamma(\nu)} \int_a^x (x - t)^{\nu - 1} f(t) \, dt \qquad (23.4)$$

for $a \leq x \leq b$, is called the Riemann–Liouville fractional integral operator of order ν.

For $\nu = 0$, we set $J_a^0 := I$, the identity operator. Here Γ stands for the gamma function.

Let $\alpha > 0$, $f \in L_1(a, b)$, $a, b \in \mathbb{R}$; see [134]. Here $[\cdot]$ stands for the integral part of the number.

We define the generalized Riemann–Liouville fractional derivative of f of order α by

$$D_a^\alpha f(s) := \frac{1}{\Gamma(m - \alpha)} \left(\frac{d}{ds} \right)^m \int_a^s (s - t)^{m - \alpha - 1} f(t) \, dt,$$

where $m := [\alpha] + 1$, $s \in [a, b]$; see also [57], Remark 46 there.

In addition, we set

$$D_a^0 f := f,$$

$$J_a^{-\alpha} f := D_a^\alpha f, \quad \text{if } \alpha > 0,$$

$$D_a^{-\alpha} f := J_a^\alpha f, \quad \text{if } 0 < \alpha \leq 1,$$

$$D_a^n f = f^{(n)}, \quad \text{for } n \in \mathbb{N}. \qquad (23.5)$$

We need

Definition 23.2. [45] We say that $f \in L_1(a, b)$ has an L_∞ fractional derivative $D_a^\alpha f$ $(\alpha > 0)$ in $[a, b]$, $a, b \in \mathbb{R}$, iff $D_a^{\alpha - k} f \in C([a, b])$, $k = 1, \ldots, m := [\alpha] + 1$, and $D_a^{\alpha - 1} f \in AC([a, b])$ (absolutely continuous functions) and $D_a^\alpha f \in L_\infty(a, b)$.

Lemma 23.3. [45] *Let* $\beta > \alpha \geq 0$, $f \in L_1(a, b)$, $a, b \in \mathbb{R}$, *have an* L_∞ *fractional derivative* $D_a^\beta f$ *in* $[a, b]$, *and let* $D_a^{\beta - k} f(a) = 0$ *for* $k = 1, \ldots, [\beta] + 1$. *Then*

$$D_a^\alpha f(s) = \frac{1}{\Gamma(\beta - \alpha)} \int_a^s (s - t)^{\beta - \alpha - 1} D_a^\beta f(t) \, dt, \qquad (23.6)$$

$\forall s \in [a, b]$.

Here $D_a^\alpha f \in AC([a, b])$ *for* $\beta - \alpha \geq 1$, *and* $D_a^\alpha f \in C([a, b])$ *for* $\beta - \alpha \in (0, 1)$.

Here $AC^n([a, b])$ *is the space of functions with absolutely continuous* $(n - 1)st$ *derivative.*

We need to mention

Definition 23.4. [134] Let $\nu \geq 0$, $n := \lceil \nu \rceil$; $\lceil \cdot \rceil$ is the ceiling of the number, $f \in AC^n([a,b])$. We call the Caputo fractional derivative

$$D_{*a}^{\nu}f(x) := \frac{1}{\Gamma(n-\nu)} \int_a^x (x-t)^{n-\nu-1} f^{(n)}(t)\, dt, \qquad (23.7)$$

$\forall x \in [a,b]$.

The above function $D_{*a}^{\nu}f(x)$ exists almost everywhere for $x \in [a,b]$.

We need

Proposition 23.5. [134] Let $\nu \geq 0$, $n := \lceil \nu \rceil$, $f \in AC^n([a,b])$. Then $D_{*a}^{\nu}f$ exists iff the generalized Riemann–Liouville fractional derivative $D_a^{\nu}f$ exists.

Proposition 23.6. [134] Let $\nu \geq 0$, $n := \lceil \nu \rceil$. Assume that f is such that both $D_{*a}^{\nu}f$ and $D_a^{\nu}f$ exist. Suppose that $f^{(k)}(a) = 0$ for $k = 0, 1, \ldots, n-1$. Then

$$D_{*a}^{\nu}f = D_a^{\nu}f. \qquad (23.8)$$

In conclusion

Corollary 23.7. [58] Let $\nu \geq 0$, $n := \lceil \nu \rceil$, $f \in AC^n([a,b])$, $D_{*a}^{\nu}f$ exists or $D_a^{\nu}f$ exists, and $f^{(k)}(a) = 0$, $k = 0, 1, \ldots, n-1$.
Then

$$D_a^{\nu}f = D_{*a}^{\nu}f. \qquad (23.9)$$

We need

Theorem 23.8. [58] Let $\nu \geq 0$, $n := \lceil \nu \rceil$, $f \in AC^n([a,b])$, and $f^{(k)}(a) = 0$, $k = 0, 1, \ldots, n-1$.
Then

$$f(x) = \frac{1}{\Gamma(\nu)} \int_a^x (x-t)^{\nu-1} D_{*a}^{\nu}f(t)\, dt. \qquad (23.10)$$

We also need

Theorem 23.9. [58] Let $\nu \geq \gamma + 1$, $\gamma \geq 0$. Call $n := \lceil \nu \rceil$. Assume $f \in AC^n([a,b])$ such that $f^{(k)}(a) = 0$, $k = 0, 1, \ldots, n-1$, and $D_{*a}^{\nu}f \in L_\infty(a,b)$. Then $D_{*a}^{\gamma}f \in AC([a,b])$, and

$$D_{*a}^{\gamma}f(x) = \frac{1}{\Gamma(\nu-\gamma)} \int_a^x (x-t)^{\nu-\gamma-1} D_{*a}^{\nu}f(t)\, dt, \qquad (23.11)$$

$\forall x \in [a,b]$.

Theorem 23.10. [58] *Let $\nu \geq \gamma+1$, $\gamma \geq 0$, $n := \lceil \nu \rceil$. Let $f \in AC^n\left([a,b]\right)$ such that $f^{(k)}\left(a\right) = 0$, $k = 0,1,\ldots,n-1$. Assume $\exists D_a^\nu f\left(x\right) \in \mathbb{R}$, $\forall x \in [a,b]$, and $D_a^\nu f \in L_\infty\left(a,b\right)$. Then $D_a^\gamma f \in AC\left([a,b]\right)$, and*

$$D_a^\gamma f\left(x\right) = \frac{1}{\Gamma\left(\nu-\gamma\right)} \int_a^x \left(x-t\right)^{\nu-\gamma-1} D_a^\nu f\left(t\right) dt, \qquad (23.12)$$

$\forall x \in [a,b]$.

23.2 Results

Here f and the whole setting are as in section 23.1, Preliminaries (I). We present first results regarding the Riemann–Liouville fractional derivative.

Theorem 23.11. *Let $\beta > 0$, $f \in L_1\left(a,b\right)$, have an L_∞ fractional derivative $D_a^\beta f$ in $[a,b]$, and let $D_a^{\beta-k} f\left(a\right) = 0$ for $k = 1,\ldots,[\beta]+1$. Also it holds $0 < a \leq p\left(x\right)/q\left(x\right) \leq b$, a.e. on X, $a < b$. Then*

$$\Gamma_f\left(\mu_1,\mu_2\right) \leq \frac{\left\|D_a^\beta f\right\|_{\infty,[a,b]}}{\Gamma\left(\beta+1\right)}$$

$$\int_X q\left(x\right)^{1-\beta}\left(p\left(x\right)-aq\left(x\right)\right)^\beta d\lambda\left(x\right). \qquad (23.13)$$

Proof. By (23.6), $\alpha = 0$, we obtain

$$f\left(s\right) = \frac{1}{\Gamma\left(\beta\right)} \int_a^s \left(s-t\right)^{\beta-1} D_a^\beta f\left(t\right) dt, \qquad (23.14)$$

all $a \leq s \leq b$.
Then

$$\left|f\left(s\right)\right| \leq \frac{1}{\Gamma\left(\beta\right)} \int_a^s \left(s-t\right)^{\beta-1} \left|D_a^\beta f\left(t\right)\right| dt$$

$$\leq \frac{\left\|D_a^\beta f\right\|_{\infty,[a,b]}}{\Gamma\left(\beta\right)} \int_a^s \left(s-t\right)^{\beta-1} dt$$

$$= \frac{\left\|D_a^\beta f\right\|_{\infty,[a,b]}}{\Gamma\left(\beta\right)} \frac{\left(s-a\right)^\beta}{\beta}$$

$$= \frac{\left\| D_a^\beta f \right\|_{\infty,[a,b]}}{\Gamma(\beta+1)} (s-a)^\beta, \quad \text{all } a \le s \le b. \tag{23.15}$$

That is, we have that

$$|f(s)| \le \frac{\left\| D_a^\beta f \right\|_{\infty,[a,b]}}{\Gamma(\beta+1)} (s-a)^\beta, \quad \text{all } a \le s \le b. \tag{23.16}$$

Consequently we find

$$\Gamma_f(\mu_1,\mu_2) = \int_X q(x) f\left(\frac{p(x)}{q(x)}\right) d\lambda(x)$$

$$\le \frac{\left\| D_a^\beta f \right\|_{\infty,[a,b]}}{\Gamma(\beta+1)} \int_X q(x) \left(\frac{p(x)}{q(x)} - a\right)^\beta d\lambda(x)$$

$$= \frac{\left\| D_a^\beta f \right\|_{\infty,[a,b]}}{\Gamma(\beta+1)} \int_X q(x)^{1-\beta} (p(x) - aq(x))^\beta d\lambda(x), \tag{23.17}$$

proving the claim. □

Next we give an L_δ result.

Theorem 23.12. *Same assumptions as in Theorem 23.11. Let $\gamma, \delta > 1$:*
$1/\gamma + 1/\delta = 1$ *and* $\gamma(\beta-1)+1 > 0$.
Then

$$\Gamma_f(\mu_1,\mu_2) \le \frac{\left\| D_a^\beta f \right\|_{\delta,[a,b]}}{\Gamma(\beta)(\gamma(\beta-1)+1)^{1/\gamma}}$$

$$\int_X q(x)^{2-\beta-1/\gamma} (p(x) - aq(x))^{\beta-1+1/\gamma} d\lambda(x). \tag{23.18}$$

Proof. By (23.6), $\alpha = 0$, we get again

$$f(s) = \frac{1}{\Gamma(\beta)} \int_a^s (s-t)^{\beta-1} D_a^\beta f(t)\, dt, \tag{23.19}$$

all $a \le s \le b$.
Hence

$$|f(s)| \le \frac{1}{\Gamma(\beta)} \int_a^s (s-t)^{\beta-1} \left| D_a^\beta f(t) \right| dt$$

$$\le \frac{1}{\Gamma(\beta)} \left(\int_a^s (s-t)^{\gamma(\beta-1)}\, dt \right)^{1/\gamma} \left(\int_a^s \left| D_a^\beta f(t) \right|^\delta dt \right)^{1/\delta}$$

$$\le \frac{\left\| D_a^\beta f \right\|_{\delta,[a,b]}}{\Gamma(\beta)} \frac{(s-a)^{\beta-1+1/\gamma}}{(\gamma(\beta-1)+1)^{1/\gamma}}, \tag{23.20}$$

all $a \le s \le b$.

That is,

$$|f(s)| \leq \frac{\|D_a^\beta f\|_{\delta,[a,b]}}{\Gamma(\beta)} \frac{(s-a)^{\beta-1+1/\gamma}}{(\gamma(\beta-1)+1)^{1/\gamma}}, \qquad (23.21)$$

all $a \leq s \leq b$.

Consequently we find

$$\Gamma_f(\mu_1, \mu_2) \leq \int_X q \left| f\left(\frac{p}{q}\right) \right| d\lambda$$

$$\leq \frac{\|D_a^\beta f\|_{\delta,[a,b]}}{\Gamma(\beta)(\gamma(\beta-1)+1)^{1/\gamma}} \int_X q \left(\frac{p}{q}-a\right)^{\beta-1+1/\gamma} d\lambda$$

$$= \frac{\|D_a^\beta f\|_{\delta,[a,b]}}{\Gamma(\beta)(\gamma(\beta-1)+1)^{1/\gamma}} \int_X q^{2-\beta-1/\gamma} (p-aq)^{\beta-1+1/\gamma} d\lambda, \qquad (23.22)$$

proving the claim. $\square$

It follows an L_1 estimate.

Theorem 23.13. *Same assumptions as in Theorem 23.11. Let $\beta \geq 1$. Then*

$$\Gamma_f(\mu_1, \mu_2) \leq \frac{\|D_a^\beta f\|_{1,[a,b]}}{\Gamma(\beta)}$$

$$\left(\int_X (q(x))^{2-\beta} (p(x) - aq(x))^{\beta-1} d\lambda(x) \right). \qquad (23.23)$$

Proof. By (23.19) we have

$$|f(s)| \leq \frac{1}{\Gamma(\beta)} \int_a^s (s-t)^{\beta-1} \left| D_a^\beta f(t) \right| dt$$

$$\leq \frac{(s-a)^{\beta-1}}{\Gamma(\beta)} \int_a^b \left| D_a^\beta f(t) \right| dt$$

$$= \frac{(s-a)^{\beta-1}}{\Gamma(\beta)} \|D_a^\beta f\|_{1,[a,b]}. \qquad (23.24)$$

That is,

$$|f(s)| \leq \frac{(s-a)^{\beta-1}}{\Gamma(\beta)} \|D_a^\beta f\|_{1,[a,b]}, \qquad (23.25)$$

for all s in $[a,b]$. Thus

$$\Gamma_f(\mu_1, \mu_2) \leq \int_X q \left| f\left(\frac{p}{q}\right) \right| d\lambda$$

584   23. Fractional Approximation of Csiszar's f-Divergence

$$\leq \frac{\|D_a^\beta f\|_{1,[a,b]}}{\Gamma(\beta)} \int_X q \left(\frac{p}{q} - a\right)^{\beta-1} d\lambda$$

$$= \frac{\|D_a^\beta f\|_{1,[a,b]}}{\Gamma(\beta)} \left(\int_X q^{2-\beta} (p - aq)^{\beta-1} d\lambda\right), \qquad (23.26)$$

proving the claim. □

We continue with results regarding the Caputo fractional derivative.

Theorem 23.14. *Let* $\nu > 0$, $n := \lceil \nu \rceil$, $f \in AC^n([a,b])$, *and* $f^{(k)}(a) = 0$, $k = 0, 1, \ldots, n-1$. *Assume* $D_{*a}^\nu f \in L_\infty(a,b)$, $0 < a \leq p(x)/q(x) \leq b$, *a.e. on* X, $a < b$.
Then

$$\Gamma_f(\mu_1, \mu_2) \leq \frac{\|D_{*a}^\nu f\|_{\infty,[a,b]}}{\Gamma(\nu+1)}$$

$$\int_X q(x)^{1-\nu} (p(x) - aq(x))^\nu d\lambda(x). \qquad (23.27)$$

Proof. Similar to Theorem 23.11, using Theorem 23.8. □

Next we give an L_δ result.

Theorem 23.15. *Assume all are as in Theorem 23.14. Let* γ, $\delta > 1$: $1/\gamma + 1/\delta = 1$, *and* $\gamma(\nu - 1) + 1 > 0$.
Then

$$\Gamma_f(\mu_1, \mu_2) \leq \frac{\|D_{*a}^\nu f\|_{\delta,[a,b]}}{\Gamma(\nu)(\gamma(\nu-1)+1)^{1/\gamma}}$$

$$\int_X q(x)^{2-\nu-1/\gamma} (p(x) - aq(x))^{\nu-1+1/\gamma} d\lambda(x). \qquad (23.28)$$

Proof. Similar to Theorem 23.12, using Theorem 23.8. □

It follows an L_1 estimate.

Theorem 23.16. *Assume all are as in Theorem 23.14. Let* $\nu \geq 1$. *Then*

$$\Gamma_f(\mu_1, \mu_2) \leq \frac{\|D_{*a}^\nu f\|_{1,[a,b]}}{\Gamma(\nu)}$$

$$\left(\int_X (q(x))^{2-\nu} (p(x) - aq(x))^{\nu-1} d\lambda(x)\right). \qquad (23.29)$$

Proof. Similar to Theorem 23.13, using Theorem 23.8. □

Regarding again the Riemann–Liouville fractional derivative we need:

Corollary 23.17. *Let* $\nu \geq 0$, $n := \lceil \nu \rceil$, $f \in AC^n\left([a,b]\right)$, $\exists D_a^\nu f\left(x\right) \in \mathbb{R}$, $\forall x \in [a,b]$, *and* $f^{(k)}\left(a\right) = 0$, $k = 0, 1, \ldots, n-1$. *Then*

$$f\left(x\right) = \frac{1}{\Gamma\left(\nu\right)} \int_a^x \left(x - t\right)^{\nu-1} D_a^\nu f\left(t\right) dt. \tag{23.30}$$

Proof. By Corollary 23.7 and Theorem 23.8. □

We continue with results again regarding the Riemann–Liouville fractional derivative.

Theorem 23.18. *Let* $\nu > 0$, $n := \lceil \nu \rceil$, $f \in AC^n\left([a,b]\right)$, $\exists D_a^\nu f\left(x\right) \in \mathbb{R}$, $\forall x \in [a,b]$, *and* $f^{(k)}\left(a\right) = 0$, $k = 0, 1, \ldots, n-1$. *Assume* $D_a^\nu f \in L_\infty\left(a,b\right)$, $0 < a \leq p\left(x\right)/q\left(x\right) \leq b$, *a.e. on* X, $a < b$.
Then

$$\Gamma_f\left(\mu_1, \mu_2\right) \leq \frac{\|D_a^\nu f\|_{\infty,[a,b]}}{\Gamma\left(\nu+1\right)}$$

$$\int_X q\left(x\right)^{1-\nu}\left(p\left(x\right) - aq\left(x\right)\right)^\nu d\lambda\left(x\right). \tag{23.31}$$

Proof. Similar to Theorem 23.11, using Corollary 23.17. □

Next we give the corresponding L_δ result.

Theorem 23.19. *Assume all are as in Theorem 23.18. Let* $\gamma, \delta > 1$: $1/\gamma + 1/\delta = 1$, *and* $\gamma\left(\nu - 1\right) + 1 > 0$.
Then

$$\Gamma_f\left(\mu_1, \mu_2\right) \leq \frac{\|D_a^\nu f\|_{\delta,[a,b]}}{\Gamma\left(\nu\right)\left(\gamma\left(\nu - 1\right) + 1\right)^{1/\gamma}}$$

$$\int_X q\left(x\right)^{2-\nu-1/\gamma}\left(p\left(x\right) - aq\left(x\right)\right)^{\nu-1+1/\gamma} d\lambda\left(x\right). \tag{23.32}$$

Proof. Similar to Theorem 23.12, using Corollary 23.17. □

The L_1 estimate follows.

Theorem 23.20. *Assume all are as in Theorem 23.18. Let* $\nu \geq 1$. *Then*

$$\Gamma_f\left(\mu_1, \mu_2\right) \leq \frac{\|D_a^\nu f\|_{1,[a,b]}}{\Gamma\left(\nu\right)}$$

$$\left(\int_X (q(x))^{2-\nu} (p(x) - aq(x))^{\nu-1} \, d\lambda(x) \right). \qquad (23.33)$$

Proof. Similar to Theorem 23.13, using Corollary 23.17. □

We need

Theorem 23.21. (Taylor expansion for Caputo derivatives, [134, p. 40])
Assume $\nu \geq 0$, $n = \lceil \nu \rceil$, *and* $f \in AC^n([a,b])$.
Then

$$f(x) = \sum_{k=0}^{n-1} \frac{f^{(k)}(a)}{k!} (x-a)^k + \frac{1}{\Gamma(\nu)} \int_a^x (x-t)^{\nu-1} D_{*a}^\nu f(t) \, dt, \quad (23.34)$$

$\forall x \in [a,b]$.

We make

Remark 23.22. Let $\nu > 0$, $n = \lceil \nu \rceil$, and $f \in AC^n([a,b])$.
If $D_{*a}^\nu f \underset{(\leq 0)}{\geq} 0$ over $[a,b]$, then

$$\int_a^x (x-t)^{\nu-1} D_{*a}^\nu f(t) \, dt \underset{(\leq 0)}{\geq} 0 \quad \text{on } [a,b].$$

By (23.34) then we obtain

$$f(x) \geq \ (\leq) \sum_{k=0}^{n-1} \frac{f^{(k)}(a)}{k!} (x-a)^k, \qquad (23.35)$$

$\forall x \in [a,b]$. Hence

$$qf\left(\frac{p}{q}\right) \geq \ (\leq) \sum_{k=0}^{n-1} \frac{f^{(k)}(a)}{k!} q \left(\frac{p}{q} - a\right)^k, \qquad (23.36)$$

a.e. on X.
Consequently we derive

$$\Gamma_f(\mu_1, \mu_2) \geq \ (\leq) \sum_{k=0}^{n-1} \frac{f^{(k)}(a)}{k!} \int_X q^{1-k} (p - aq)^k \, d\lambda. \qquad (23.37)$$

We have established

Theorem 23.23. *Let* $\nu > 0$, $n = \lceil \nu \rceil$, *and* $f \in AC^n([a,b])$.

If $D^{\nu}_{*a}f \underset{(\leq 0)}{\geq} 0$ on $[a, b]$, then

$$\Gamma_f(\mu_1, \mu_2) \geq \ (\leq) \sum_{k=0}^{n-1} \frac{f^{(k)}(a)}{k!}$$

$$\left(\int_X (q(x))^{1-k} (p(x) - aq(x))^k \, d\lambda(x) \right). \tag{23.38}$$

We finish with

Remark 23.24. Using Lemma 23.3, Theorem 23.9, and Theorem 23.10 and in their settings, for g any of $D^{\alpha}_a f$, $D^{\gamma}_{*a} f$, $D^{\gamma}_a f$, that fulfill the conditions and assumptions of section 23.1, Preliminaries (I), we can find similar estimates as above for $\Gamma_g(\mu_1, \mu_2)$.

24
Canavati Fractional Ostrowski-Type Inequalities

Optimal upper bounds are given to the deviation of an initial value of a function $f \in C_{x_0}^{\nu}([a,b])$, $x_0 \in [a,b]$, $\nu > 0$ from the corresponding average of f. These bounds are of type $A \cdot \|D_{x_0}^{\nu} f\|_{\infty,[x_0,b]}$, where A is the smallest universal constant; that is, the produced inequalities are sharp and attained. Here $D_{x_0}^{\nu} f$ is the ν-order Canavati-type fractional derivative of f. This chapter was inspired by the work of Ostrowski [318], 1938, and of the author's [13], 1995. This treatment relies on [24].

24.1 Background

In the sequel we follow [101]. Let $g \in C([0,1])$, $n := [\nu]$, $\nu > 0$, and $\alpha := \nu - n$ $(0 < \alpha < 1)$. Define

$$(J_{\nu} g)(x) := \frac{1}{\Gamma(\nu)} \int_0^x (x-t)^{\nu-1} g(t) dt, \quad 0 \le x \le 1, \qquad (24.1)$$

the *Riemann–Liouville integral*, where Γ is the gamma function $\Gamma(\nu) := \int_0^{\infty} e^{-t} t^{\nu-1} dt$. We define the subspace $C^{\nu}([0,1])$ of $C^n([0,1])$:

$$C^{\nu}([0,1]) := \{g \in C^n([0,1]) : J_{1-\alpha} g^{(n)} \in C^1([0,1])\}.$$

Thus letting $g \in C^{\nu}([0,1])$, we define the Canavati ν- *fractional derivative* of g as

$$D^{\nu} g := D J_{1-\alpha} g^{(n)}, \quad D := \frac{d}{dx}. \qquad (24.2)$$

G.A. Anastassiou, *Fractional Differentiation Inequalities*,
DOI 10.1007/978-0-387-98128-4_24, © Springer Science+Business Media, LLC 2009

When $\nu \geq 1$ we have the Taylor formula

$$
\begin{aligned}
g(t) \quad = \quad & g(0) + g'(0)t + g''(0)\frac{t^2}{2!} + \cdots + g^{(n-1)}(0)\frac{t^{n-1}}{(n-1)!} \\
& + (J_\nu D^\nu g)(t), \quad \forall t \in [0,1].
\end{aligned}
\tag{24.3}
$$

When $0 < \nu < 1$ we find

$$
g(t) = (J_\nu D^\nu g)(t), \quad \forall t \in [0,1].
\tag{24.4}
$$

Next we carry the above notions over to an arbitrary $[a,b] \subseteq \mathbf{R}$ (see [17]). Let $x, x_0 \in [a,b]$ such that $x \geq x_0$, x_0 is fixed. Let $f \in C([a,b])$ and define

$$
(J_\nu^{x_0} f)(x) := \frac{1}{\Gamma(\nu)} \int_{x_0}^x (x-t)^{\nu-1} f(t)dt, \quad x_0 \leq x \leq b,
\tag{24.5}
$$

the *generalized Riemann–Liouville integral*. We define the subspace $C_{x_0}^\nu([a,b])$ of $C^n([a,b])$:

$$
C_{x_0}^\nu([a,b]) := \{ f \in C^n([a,b]) : J_{1-\alpha}^{x_0} f^{(n)} \in C^1([x_0,b]) \}.
$$

For $f \in C_{x_0}^\nu([a,b])$, we define the *generalized ν -fractional derivative of f over $[x_0,b]$* as

$$
D_{x_0}^\nu f := D J_{1-\alpha}^{x_0} f^{(n)}.
\tag{24.6}
$$

We observe that $D_{x_0}^n f = f^{(n)}$, $n \in \mathbf{N}$. Notice that

$$
(J_{1-\alpha}^{x_0} f^{(n)})(x) = \frac{1}{\Gamma(1-\alpha)} \int_{x_0}^x (x-t)^{-\alpha} f^{(n)}(t)dt
$$

exists for $f \in C_{x_0}^\nu([a,b])$.

We recall the following fractional generalization of Taylor's formula (see [17, 101]).

Theorem 24.1. *Let $f \in C_{x_0}^\nu([a,b])$, $x_0 \in [a,b]$ fixed.*
(i) *If $\nu \geq 1$ then*

$$
\begin{aligned}
f(x) \quad = \quad & f(x_0) + f'(x_0)(x-x_0) + f''(x_0)\frac{(x-x_0)^2}{2} + \\
& \cdots + f^{(n-1)}(x_0)\frac{(x-x_0)^{n-1}}{(n-1)!} \\
& + (J_\nu^{x_0} D_{x_0}^\nu f)(x), \quad \text{all } x \in [a,b] : x \geq x_0.
\end{aligned}
\tag{24.7}
$$

(ii) *If $0 < \nu < 1$ we get*

$$
f(x) = (J_\nu^{x_0} D_{x_0}^\nu f)(x), \quad \text{all } x \in [a,b] : x \geq x_0.
\tag{24.8}
$$

Here we use (24.7).

24.2   Results

We present the first fractional Ostrowski-type inequality.

Theorem 24.2. *Let* $1 \le \nu < 2$ *and* $f \in C_{x_0}^{\nu}([a, b])$, $a \le x_0 < b$, x_0 *fixed.*
Then

$$\left| \frac{1}{b - x_0} \int_{x_0}^{b} f(y) dy - f(x_0) \right| \le \frac{\|D_{x_0}^{\nu} f\|_{\infty, [x_0, b]}}{\Gamma(\nu + 2)} (b - x_0)^{\nu}. \qquad (24.9)$$

Proof. Here $n := [\nu] = 1$. From (24.7) we have

$$f(y) - f(x_0) = \frac{1}{\Gamma(\nu)} \int_{x_0}^{y} (y - w)^{\nu - 1} (D_{x_0}^{\nu} f)(w) dw, \quad \forall y \ge x_0.$$

Thus

$$\left| \frac{1}{b - x_0} \int_{x_0}^{b} f(y) dy - f(x_0) \right| = \left| \frac{1}{b - x_0} \int_{x_0}^{b} (f(y) - f(x_0)) dy \right|$$

$$\le \frac{1}{b - x_0} \int_{x_0}^{b} |f(y) - f(x_0)| dy$$

$$\le \frac{1}{b - x_0} \int_{x_0}^{b} \left(\frac{1}{\Gamma(\nu)} \int_{x_0}^{y} (y - w)^{\nu - 1} |D_{x_0}^{\nu} f(w)| dw \right) dy$$

$$\le \frac{1}{b - x_0} \int_{x_0}^{b} \frac{1}{\Gamma(\nu)} \|D_{x_0}^{\nu} f\|_{\infty, [x_0, b]} \left(\int_{x_0}^{y} (y - w)^{\nu - 1} dw \right) dy$$

$$= \frac{\|D_{x_0}^{\nu} f\|_{\infty, [x_0, b]}}{(b - x_0) \Gamma(\nu)} \cdot \int_{x_0}^{b} \frac{(y - x_0)^{\nu}}{\nu} dy$$

(by $\Gamma(\nu + 1) = \nu \Gamma(\nu), \nu > 0$)

$$= \frac{\|D_{x_0}^{\nu} f\|_{\infty, [x_0, b]}}{(b - x_0) \Gamma(\nu + 1)} \cdot \frac{(b - x_0)^{\nu + 1}}{\nu + 1}$$

$$= \frac{\|D_{x_0}^{\nu} f\|_{\infty, [x_0, b]} \cdot (b - x_0)^{\nu}}{\Gamma(\nu + 2)}.$$

We have proved (24.9).   □

Next we give another more general fractional Ostrowski-type inequality.

Theorem 24.3. *Let* $a \le x_0 < b$ *fixed. Let* $f \in C_{x_0}^{\nu}([a, b])$, $\nu \ge 2$, $n := [\nu]$.
Assume $f^{(i)}(x_0) = 0$, $i = 1, \ldots, n - 1$. *Then*

$$\left| \frac{1}{b - x_0} \int_{x_0}^{b} f(y) dy - f(x_0) \right| \le \frac{\|D_{x_0}^{\nu} f\|_{\infty, [x_0, b]}}{\Gamma(\nu + 2)} \cdot (b - x_0)^{\nu}. \qquad (24.10)$$

Proof. Again from (24.7) we have

$$f(y) - f(x_0) = \frac{1}{\Gamma(\nu)} \int_{x_0}^{y} (y - w)^{\nu-1} (D_{x_0}^{\nu} f)(w) dw, \quad \forall y \geq x_0.$$

We observe that

$$|f(y) - f(x_0)| \leq \frac{1}{\Gamma(\nu)} \| D_{x_0}^{\nu} f \|_{\infty, [x_0, b]} \int_{x_0}^{y} (y - w)^{\nu-1} dw$$

$$= \frac{1}{\Gamma(\nu)} \| D_{x_0}^{\nu} f \|_{\infty, [x_0, b]} \frac{(y - x_0)^{\nu}}{\nu} = \frac{\| D_{x_0}^{\nu} f \|_{\infty, [x_0, b]}}{\Gamma(\nu + 1)} \cdot (y - x_0)^{\nu}.$$

That is,

$$|f(y) - f(x_0)| \leq \frac{\| D_{x_0}^{\nu} f \|_{\infty, [x_0, b]}}{\Gamma(\nu + 1)} \cdot (y - x_0)^{\nu}, \quad \forall y \geq x_0. \tag{24.11}$$

Therefore we get

$$\left| \frac{1}{b - x_0} \int_{x_0}^{b} f(y) dy - f(x_0) \right| = \left| \frac{1}{b - x_0} \int_{x_0}^{b} (f(y) - f(x_0)) dy \right|$$

$$\leq \frac{1}{b - x_0} \int_{x_0}^{b} |f(y) - f(x_0)| dy$$

$$\overset{(24.11)}{\leq} \frac{1}{b - x_0} \int_{x_0}^{b} \frac{\| D_{x_0}^{\nu} f \|_{\infty, [x_0, b]}}{\Gamma(\nu + 1)} \cdot (y - x_0)^{\nu} dy$$

$$= \frac{1}{b - x_0} \cdot \frac{\| D_{x_0}^{\nu} f \|_{\infty, [x_0, b]}}{\Gamma(\nu + 1)} \cdot \frac{(b - x_0)^{\nu+1}}{\nu + 1}$$

$$= \frac{\| D_{x_0}^{\nu} f \|_{\infty, [x_0, b]}}{\Gamma(\nu + 2)} \cdot (b - x_0)^{\nu}.$$

That proves (24.10). $\square$

Remark 24.4. Let $\mu, \nu > 0$ such that $\nu \leq \mu$, and $n := [\nu]$, $\alpha := \nu - n$. Consider

$$\phi_{\mu}(x) := \frac{(x - x_0)^{\mu}}{\Gamma(\mu + 1)}, \quad x_0 \leq x \leq b. \tag{24.12}$$

Then

$$\phi_{\mu}^{(n)}(x) = \frac{\mu(\mu - 1)(\mu - 2) \cdots (\mu - n + 2)(\mu - n + 1)(x - x_0)^{\mu-n}}{\Gamma(\mu + 1)},$$

and

$$\phi_{\mu}^{(n)}(x) = \frac{(x - x_0)^{\mu-n}}{\Gamma(\mu - n + 1)} = \phi_{\mu-n}(x).$$

Next we find that

$$(J^{x_0}_{1-\alpha}\phi_{\mu-n})(x) = \frac{1}{\Gamma(1-\alpha)} \int_{x_0}^{x} (x-t)^{-\alpha} \frac{(t-x_0)^{\mu-n}}{\Gamma(\mu-n+1)} dt$$

$$= \frac{1}{\Gamma(1-\alpha)} \int_{x_0}^{x} (x-t)^{(1-\alpha)-1} \frac{(t-x_0)^{(\mu-n+1)-1}}{\Gamma(\mu-n+1)} dt$$

(by [404, p. 256]; notice that $1-\alpha, \mu-n+1 > 0$)

$$= \frac{1}{\Gamma(1-\alpha)\Gamma(\mu-n+1)} \cdot$$

$$\frac{\Gamma(1-\alpha)\Gamma(\mu-n+1)}{\Gamma(1-\alpha+\mu-n+1)}(x-x_0)^{(1-\alpha+\mu-n)}$$

$$= \frac{(x-x_0)^{(\mu-\nu+1)}}{\Gamma(\mu-\nu+2)} \cdot$$

That is,

$$(J^{x_0}_{1-\alpha}\phi_{\mu-n})(x) = \frac{(x-x_0)^{(\mu-\nu+1)}}{\Gamma(\mu-\nu+2)}. \tag{24.13}$$

Hence

$$(D^{\nu}_{x_0}\phi_{\mu})(x) = (DJ^{x_0}_{1-\alpha}\phi^{(n)})(x) = (DJ^{x_0}_{1-\alpha}\phi_{\mu-n})(x)$$

$$= \left(\frac{(x-x_0)^{\mu-\nu+1}}{\Gamma(\mu-\nu+2)}\right)' = \frac{(x-x_0)^{\mu-\nu}}{\Gamma(\mu-\nu+1)},$$

so that

$$(D^{\nu}_{x_0}\phi_{\mu})(x) = \frac{(x-x_0)^{\mu-\nu}}{\Gamma(\mu-\nu+1)} = \phi_{\mu-\nu}(x), \tag{24.14}$$

and

$$D^{\nu}_{x_0}(x-x_0)^{\mu} = \frac{\Gamma(\mu+1)}{\Gamma(\mu-\nu+1)}(x-x_0)^{\mu-\nu}, \quad 0 < \nu \le \mu, \ x_0 \le x \le b. \tag{24.15}$$

In particular we get

$$D^{\nu}_{x_0}(x-x_0)^{\nu} = \Gamma(\nu+1), \quad (\text{by } \Gamma(1) = 1).$$

Consequently it holds

$$\|D^{\nu}_{x_0}(x-x_0)^{\nu}\|_{\infty,[x_0,b]} = \Gamma(\nu+1). \tag{24.16}$$

Proposition 24.5. *Inequality* (24.9) *is sharp; namely it is attained by*

$$f(x) := (x-x_0)^{\nu}, \quad 1 \le \nu < 2, \ x \in [a,b].$$

Proof. Observe that

$$\text{R.H.S.}(24.9) = \frac{1}{b - x_0} \int_{x_0}^{b} (y - x_0)^{\nu} dy = \frac{(b - x_0)^{\nu}}{\nu + 1}.$$

Also we see

$$\text{L.H.S.}(24.9) \overset{(24.16)}{=} \frac{\Gamma(\nu + 1)}{\Gamma(\nu + 2)} (b - x_0)^{\nu} = \frac{\Gamma(\nu + 1)}{(\nu + 1)\Gamma(\nu + 1)} (b - x_0)^{\nu} = \frac{(b - x_0)^{\nu}}{\nu + 1}.$$

That is, both sides of (24.9) are equal. □

Proposition 24.6. *Inequality* (24.10) *is sharp; namely it is attained by*

$$f(x) := (x - x_0)^{\nu}, \quad \nu \geq 2, \; x \in [a, b].$$

Proof. Observe that

$$f^{(i)}(x_0) = ((x - x_0)^{\nu})^{(i)} \Big|_{x = x_0} = 0, \quad i = 1, \ldots, n - 1.$$

Again we have

$$\text{R.H.S.}(24.10) = \frac{(b - x_0)^{\nu}}{\nu + 1},$$

and

$$\text{L.H.S.}(24.10) = \frac{(b - x_0)^{\nu}}{\nu + 1}.$$

That is, both sides of (24.10) are equal. □

Comment 24.7. In the fractional Ostrowski-type inequalities, under the same initial conditions—assumption, as in the integer ordinary derivative case—one can derive results for higher-order (fractional) derivatives, appearing in the R.H.S.s of the corresponding inequalities. Thus, compare Theorem 24.2, $1 \leq \nu < 2$, with an ordinary case of order $n = 1$; see [13]. Also compare Theorem 24.3, $\nu \geq 2$, where $n + 1 \leq \nu < n + 2$, $n + 1 = [\nu]$, with an ordinary case of order $n + 1$; see [13].

25

Multivariate Canavati Fractional Ostrowski-Type Inequalities

Optimal upper bounds are given to the deviation of a value of a multivariate function of a fractional space from its average, over convex and compact subsets of $\mathbb{R}^N$, $N \geq 2$. In particular we work over rectangles, balls, and spherical shells. These bounds involve the supremum and L_∞ norms of related multivariate Canavati fractional derivatives of the involved function. The presented inequalities are sharp; namely they are attained. This chapter has been motivated by the works of Ostrowski [318], 1938, and Anasstasiou [24], 2003, and the chapter is based on [43].

25.1 Background

In the sequel we follow Canavati [101]. Let $g \in C([0,1])$, $n := [\nu]$, $\nu > 0$, and $\alpha := \nu - n$ $(0 < \alpha < 1)$. Define

$$(\mathcal{J}_\nu g)(x) := \frac{1}{\Gamma(\nu)} \int_0^x (x-t)^{\nu-1} g(t)dt, \quad 0 \leq x \leq 1, \tag{25.1}$$

the *Riemann – Liouville fractional integral*, where Γ is the gamma function $\Gamma(\nu) := \int_0^\infty e^{-t}t^{\nu-1}dt$. We define the subspace $C^\nu([0,1])$ of $C^n([0,1])$:

$$C^\nu([0,1]) := \{g \in C^n([0,1]) : \mathcal{J}_{1-\alpha}g^{(n)} \in C^1([0,1])\}. \tag{25.2}$$

G.A. Anastassiou, *Fractional Differentiation Inequalities*,
DOI 10.1007/978-0-387-98128-4_25, © Springer Science+Business Media, LLC 2009

So let $g \in C^{\nu}([0,1])$; we define the Canavati ν *-fractional derivative* of g as

$$g^{(\nu)} := (\mathcal{J}_{1-\alpha}\, g^{(n)})'. \tag{25.3}$$

When $\nu \geq 1$ we have the fractional Taylor formula ([101])

$$g(t) = g(0) + g'(0)t + g''(0)\frac{t^2}{2!} + \cdots + g^{(n-1)}(0)\frac{t^{n-1}}{(n-1)!} + (\mathcal{J}_{\nu}g^{(\nu)})(t), \ \forall\, t \in [0,1], \tag{25.4}$$

and when $0 < \nu < 1$ we find

$$g(t) = (\mathcal{J}_{\nu}g^{(\nu)})(t), \ \forall\, t \in [0,1]. \tag{25.5}$$

Next we carry the above notions over to an arbitrary interval $[a,b] \subseteq \mathbb{R}$ (see Anastassiou [17]). Let $x, x_0 \in [a,b]$ such that $x \geq x_0$, x_0 is fixed. Let $f \in C([a,b])$ and define

$$(\mathcal{J}_{\nu}^{x_0} f)(x) := \frac{1}{\Gamma(\nu)} \int_{x_0}^{x} (x-t)^{\nu-1} f(t)dt, \ x_0 \leq x \leq b, \tag{25.6}$$

the *generalized Riemann – Liouville integral*. We define the subspace $C_{x_0}^{\nu}([a,b])$ of $C^n([a,b])$:

$$C_{x_0}^{\nu}([a,b]) := \{f \in C^n([a,b]) : \mathcal{J}_{1-\alpha}^{x_0}\, f^{(n)} \in C^1([x_0,b])\}. \tag{25.7}$$

For $f \in C_{x_0}^{\nu}([a,b])$, we define the *generalized ν-fractional derivative of f over* $[x_0,b]$, as

$$D_{x_0}^{\nu}\, f := \left(\mathcal{J}_{1-\alpha}^{x_0}\, f^{(n)}\right)'. \tag{25.8}$$

We observe that $D_{x_0}^{n}\, f = f^{(n)}$, $n \in \mathbb{N}$. Notice that

$$\mathcal{J}_{1-\alpha}^{x_0}\, f^{(n)}(x) = \frac{1}{\Gamma(1-\alpha)} \int_{x_0}^{x} (x-t)^{-\alpha}\, f^{(n)}(t)dt \tag{25.9}$$

exists for $f \in C_{x_0}^{\nu}([a,b])$.

We mention the following generalization of the fractional Taylor formula (see Anastassiou [17] and Canavati [101]).

Theorem 25.1. *Let* $f \in C_{x_0}^{\nu}([a,b])$, $x_0 \in [a,b]$ *fixed.*
(i) If $\nu \geq 1$, *then*

$$f(x) = f(x_0) + f'(x_0)(x-x_0) + f''(x_0)\frac{(x-x_0)^2}{2!} + \cdots +$$

$$f^{(n-1)}(x_0)\frac{(x-x_0)^{n-1}}{(n-1)!} + (\mathcal{J}_{\nu}^{x_0} D_{x_0}^{\nu} f)(x), \quad \text{all} \quad x \in [a,b] : x \geq x_0. \tag{25.10}$$

(ii) If $0 < \nu < 1$, we get

$$f(x) = (\mathcal{J}_\nu^{x_0} D_{x_0}^\nu f)(x), \quad \text{all} \quad x \in [a,b] : x \geq x_0. \tag{25.11}$$

We also mention from Anastassiou [40], the basic multivariate fractional Taylor formula.

Theorem 25.2. *Let $f \in C^1(Q)$, where Q is convex and compact $\subseteq \mathbb{R}^N$, $N \geq 2$. For fixed $x_0, z \in Q$, assume that as a function of $t \in [0,1]$: $f_{x_i}(x_0 + t(z - x_0)) \in C^{\nu-1}([0,1])$, all $i = 1, \ldots, N$, where $\nu \in [1,2)$. Then*

$$f(z) = f(x_0) + \sum_{i=1}^{N} \frac{(z_i - x_{0i})}{\Gamma(\nu)} \int_0^1 (1-t)^{\nu-1} (f_{x_i}(x_0 + t(z - x_0)))^{(\nu-1)} dt,$$
$$\tag{25.12}$$

where $z = (z_1, \ldots, z_N)$, $x = (x_{01}, \ldots, x_{0N})$.

The following general multivariate fractional Taylor formula also comes from Anastassiou [40].

Theorem 25.3. *Let $f \in C^n(Q)$, Q compact and convex $\subseteq \mathbb{R}^N$, $N \geq 2$; here $\nu \geq 1$ such that $n = [\nu]$. For fixed $x_0, z \in Q$ assume that as functions of $t \in [0,1]$: $f_\alpha (x_0 + t(z - x_0)) \in C^{(\nu-n)}([0,1])$, for all $\alpha := (\alpha_1, \ldots, \alpha_N)$, $\alpha_i \in \mathbb{Z}^+$, $i = 1, \ldots, N$; $|\alpha| := \sum_{i=1}^N \alpha_i = n$. Then*
(i)

$$f(z) = f(x_0) + \sum_{i=1}^{N}(z_i - x_{0i})\frac{\partial f}{\partial x_i}(x_0)+$$

$$\frac{\left[\left(\sum_{i=1}^N (z_i - x_{0i})\frac{\partial}{\partial x_i}\right)^2 f\right](x_0)}{2!} + \cdots + \frac{\left[\left(\sum_{i=1}^N (z_i - x_{0i})\frac{\partial}{\partial x_i}\right)^{n-1} f\right](x_0)}{(n-1)!} +$$

$$\frac{1}{\Gamma(\nu)} \int_0^1 (1-t)^{\nu-1} \left\{\left[\left(\sum_{i=1}^N (z_i - x_{0i})\frac{\partial}{\partial x_i}\right)^n f\right]^{(\nu-n)} (x_0 + t(z - x_0))\right\} dt.$$
$$\tag{25.13}$$

(ii) If all $f_\alpha(x_0) = 0$, $\alpha := (\alpha_1, \ldots, \alpha_N)$, $\alpha_i \in \mathbb{Z}^+$, $i = 1, \ldots, N$, $|\alpha| := \sum_{i=1}^N \alpha_i = l$, $l = 1, \ldots, n-1$, then

$$f(z) - f(x_0) =$$

$$\frac{1}{\Gamma(\nu)} \int_0^1 (1-t)^{\nu-1} \left\{\left[\left(\sum_{i=1}^N (z_i - x_{0i})\frac{\partial}{\partial x_i}\right)^n f\right]^{(\nu-n)} (x_0 + t(z - x_0))\right\} dt.$$
$$\tag{25.14}$$

In Anastassiou [24] we proved the following Ostrowski-type results.

Theorem 25.4. Let $1 \le \nu < 2$ and $f \in C_{x_0}^{\nu}([a,b])$, $a \le x_0 < b$, x_0 fixed. Then

$$\left| \frac{1}{b - x_0} \int_{x_0}^{b} f(y)dy - f(x_0) \right| \le \frac{\|D_{x_0}^{\nu}f\|_{\infty,[x_0,b]}}{\Gamma(\nu + 2)}(b - x_0)^{\nu}. \qquad (25.15)$$

Inequality (25.15) is sharp; namely it is attained by $f(x) := (x - x_0)^{\nu}$, $1 \le \nu < 2$, $x \in [a,b]$.

Also in [24] we gave

Theorem 25.5. Let $a \le x_0 < b$ be fixed. Let $f \in C_{x_0}^{\nu}([a,b])$, $\nu \ge 2$, $n := [\nu]$. Assume $f^{(i)}(x_0) = 0$, $i = 1, \ldots, n-1$. Then

$$\left| \frac{1}{b - x_0} \int_{x_0}^{b} f(y)dy - f(x_0) \right| \le \frac{\|D_{x_0}^{\nu}f\|_{\infty,[x_0,b]}}{\Gamma(\nu + 2)}(b - x_0)^{\nu}. \qquad (25.16)$$

Inequality (25.16) is sharp; namely it is attained by

$$f(x) := (x - x_0)^{\nu}, \quad \nu \ge 2, \quad x \in [a,b].$$

Establishing sharpness in (25.15) and (25.16), we proved first that [24]

$$\|D_{x_0}^{\nu}(x - x_0)^{\nu}\|_{\infty,[x_0,b]} = \Gamma(\nu + 1). \qquad (25.17)$$

In this chapter, motivated by (25.15) and (25.16), we present various multivariate fractional Ostrowski-type inequalities.

25.2   Results

We present the first main result of the chapter.

Theorem 25.6. Let $f \in C^1(Q)$, where Q is convex and compact $\subseteq \mathbb{R}^N$, $N \ge 2$. For fixed $x_0 \in Q$ and any $z \in Q$ assume that as a function of $t \in [0,1] : f_{x_i}(x_0 + t(z - x_0)) \in C^{\nu-1}([0,1])$, all $i = 1, \ldots, N$, where $\nu \in [1,2)$. Then

$$\left| f(x_0) - \frac{\int_Q f(z)dz}{Vol(Q)} \right| \le$$

$$\frac{\max\limits_{1 \le i \le N} \|(f_{x_i}(x_0 + t(z - x_0)))^{(\nu-1)}\|_{\infty,(t,z)\in[0,1]\times Q}}{\Gamma(\nu + 1)Vol(Q)} \int_Q \|z - x_0\|_{l_1} dz. \qquad (25.18)$$

Proof. From (25.12) we obtain

$$f(z) - f(x_0) = \sum_{i=1}^{N} \frac{(z_i - x_{0i})}{\Gamma(\nu)} \int_0^1 (1-t)^{\nu-1} (f_{x_i}(x_0 + t(z - x_0)))^{(\nu-1)} dt,$$

(25.19)

and

$$|f(z) - f(x_0)| \leq \sum_{i=1}^{N} \frac{|z_i - x_{0i}|}{\Gamma(\nu)} \int_0^1 (1-t)^{\nu-1} |(f_{x_i}(x_0 + t(z - x_0)))^{(\nu-1)}| dt \leq$$

$$\frac{1}{\Gamma(\nu+1)} \sum_{i=1}^{N} |x_i - x_{0i}| \; \| (f_{x_i}(x_0 + t(z - x_0)))^{(\nu-1)} \|_{\infty, t \in [0,1]}. \quad (25.20)$$

That is,

$$|f(z) - f(x_0)| \leq \frac{1}{\Gamma(\nu+1)} \| z - x_0 \|_{l_1} \max_{1 \leq i \leq N} \|(f_{x_i}(x_0 + t(z - x_0))^{(\nu-1)}\|_{\infty, (t,z) \in [0,1] \times Q},$$

(25.21)

$\forall z \in Q, \quad x_0 \in Q$ fixed.

Hence we have

$$\left| \frac{\int_Q f(z) dz}{Vol(Q)} - f(x_0) \right| = \left| \frac{\int_Q (f(z) - f(x_0)) dz}{Vol(Q)} \right| \leq \frac{1}{Vol(Q)} \int_Q |f(z) - f(x_0)| dz$$

$$\overset{(25.21)}{\leq} \frac{\max_{1 \leq i \leq N} \|(f_{x_i}(x_0 + t(z - x_0)))^{(\nu-1)}\|_{\infty, (t,z) \in [0,1] \times Q}}{\Gamma(\nu+1) \, Vol(Q)} \int_Q \|z - x_0\|_{l_1} dz,$$

(25.22)

proving the claim. $\quad\square$

Next we give

Theorem 25.7. *Let* $f \in C^n(Q)$, Q *compact and convex* $\subseteq \mathbb{R}^N$, $N \geq 2$; *here* $\nu \geq 1$ *such that* $n - [\nu]$. *For fixed* $x_0 \in Q$ *and any* $z \in Q$ *assume that as functions of* $t \in [0,1]: f_\alpha(x_0 + t(z - x_0)) \in C^{\nu-n}([0,1])$, *for all* $\alpha : (\alpha_1, ... \alpha_N)$, $\alpha_i \in \mathbb{Z}^+$, $i = 1, ..., N$; $|\alpha| := \sum_{i=1}^N \alpha_i = n$. *Assume* $f_\alpha(x_0) = 0$, *all* $\alpha := (\alpha_1, ..., \alpha_N)$, $\alpha_i \in \mathbb{Z}^+$, $i = 1, ..., N$, $|\alpha| = l$, $l = 1, ..., n-1$. *Call*

$$\|D^{\nu-n} f(x_0 + t(z - x_0))\|_{\infty, (t,z) \in [0,1] \times Q} =$$

$$\max_{|\alpha|=n} \|f_\alpha^{(\nu-n)}(x_0 + t(z - x_0))\|_{\infty, (t,z) \in [0,1] \times Q}. \quad (25.23)$$

Then

$$\left| f(x_0) - \frac{\int_Q f(z) dz}{Vol(Q)} \right| \leq \frac{\|D^{\nu-n} f(x_0 + t(z - x_0))\|_{\infty, (t,z) \in [0,1] \times Q}}{\Gamma(\nu+1) \, Vol(Q)} \int_Q \|z - x_0\|_{l_1}^n dz.$$

(25.24)

Proof. From (25.14) we have

$$|f(z) - f(x_0)| \leq$$

$$\frac{1}{\Gamma(\nu)} \int_0^1 (1-t)^{\nu-1} \left| \left\{ \left[\left(\sum_{i=1}^N (z_i - x_{0i}) \frac{\partial}{\partial x_i} \right)^n f \right]^{(\nu-n)} (x_0 + t(z - x_0)) \right\} \right| dt$$

$$\leq \frac{1}{\Gamma(\nu+1)} \left\| \left\{ \left[\left(\sum_{i=1}^N (z_i - x_{0i}) \frac{\partial}{\partial x_i} \right)^n f \right]^{(\nu-n)} (x_0 + t(z - x_0)) \right\} \right\|_{\infty, t \in [0,1]} \leq$$

$$\frac{1}{\Gamma(\nu-1)} (\|z - x_0\|_{l_1})^n \|D^{\nu-n} f(x_0 + t(z - x_0))\|_{\infty, (t,z) \in [0,1] \times Q}. \quad (25.25)$$

That is, we find

$$|f(z) - f(x_0)| \leq \frac{(\|z - x_0\|_{l_1})^n}{\Gamma(\nu+1)} \|D^{\nu-n} f(x_0 + t(z - x_0))\|_{\infty, (t,z) \in [0,1] \times Q}, \quad (25.26)$$

$\forall z \in Q, \quad x_0 \in Q$ fixed.

Therefore as before in (25.22) we have that

$$\left| \frac{\int_Q f(z) dz}{Vol(Q)} - f(x_0) \right| \leq \frac{1}{Vol(Q)} \int_Q |f(z) - f(x_0)| dz$$

$$\overset{(25.26)}{\leq} \frac{\|D^{\nu-n} f(x_0 + t(z - x_0))\|_{\infty, (t,z) \in [0,1] \times Q}}{\Gamma(\nu+1) Vol(Q)} \int_Q (\|z - x_0\|_{l_1})^n dz, \quad (25.27)$$

proving the claim. $\square$

We continue with

Theorem 25.8. Let $Q := [x_0, b] \times [c, d]$, $x_0 \in [a, b)$, and $f \in C([a, b] \times [c, d])$. Let $1 \leq \nu < 2$ and $\partial_{x_0}^\nu f / \partial x^\nu \in C_{x_0}^\nu([a, b])$, $y_0 \in [a, b]$.
Then

$$\left| \frac{1}{(b - x_0)(d - c)} \int_Q f(x, y) dx \, dy - f(x_0, y_0) \right| \leq$$

$$\frac{1}{d - c} \int_c^d |f(x_0, y) - f(x_0, y_0)| dy + \frac{(b - x_0)^\nu}{\Gamma(\nu+2)} \left\| \frac{\partial_{x_0}^\nu f}{\partial x^\nu} \right\|_{\infty, Q}. \quad (25.28)$$

Proof. By (25.10) we have

$$f(x, y) - f(x_0, y) = \frac{1}{\Gamma(\nu)} \int_{x_0}^x (x - t)^{\nu-1} \frac{\partial_{x_0}^\nu f}{\partial x^\nu}(t, y) dt, \quad (25.29)$$

$x \geq x_0$, all $y \in [c, d]$.
That is,

$$|f(x, y) - f(x_0, y)| \leq \frac{1}{\Gamma(\nu)} \left\| \frac{\partial_{x_0}^\nu f}{\partial x^\nu} \right\|_{\infty, Q} \int_{x_0}^x (x - t)^{\nu - 1} dt, \qquad (25.30)$$

and

$$|f(x, y) - f(x_0, y)| \leq \frac{(x - x_0)^\nu}{\Gamma(\nu + 1)} \left\| \frac{\partial_{x_0}^\nu f}{\partial x^\nu} \right\|_{\infty, Q}, \qquad (25.31)$$

all $x \geq x_0$, all $y \in [c, d]$.
However, it holds

$$|f(x, y) - f(x_0, y_0)| \leq |f(x, y) - f(x_0, y)| + |f(x_0, y) - f(x_0, y_0)| \leq$$

$$|f(x_0, y) - f(x_0, y_0)| + \frac{1}{\Gamma(\nu + 1)} \left\| \frac{\partial_{x_0}^\nu f}{\partial x^\nu} \right\|_{\infty, Q} (x - x_0)^\nu, \qquad (25.32)$$

$\forall \, x \geq x_0, \quad \forall \, y \in [c, d]$.
Consequently we derive

$$\left| \frac{1}{(b - x_0)(d - c)} \int_{[x_0, b] \times [c, d]} f(x, y) dx \, dy - f(x_0, y_0) \right| =$$

$$\frac{1}{(b - x_0)(d - c)} \left| \int_{[x_0, b] \times [c, d]} (f(x, y) - f(x_0, y_0)) dx \, dy \right| \leq$$

$$\frac{1}{(b - x_0)(d - c)} \int_{[x_0, b] \times [c, d]} |f(x, y) - f(x_0, y_0)| dx \, dy \qquad (25.33)$$

$$\overset{(25.32)}{\leq} \frac{1}{(b - x_0)(d - c)} \left[(b - x_0) \int_c^d |f(x_0, y) - f(x_0, y_0)| dy + \right.$$

$$\left. \frac{(d - c)}{\Gamma(\nu + 1)} \left\| \frac{\partial_{x_0}^\nu f}{\partial x^\nu} \right\|_{\infty, Q} \int_{x_0}^b (x - x_0)^\nu dx \right]$$

$$= \frac{1}{(b - x_0)(d - c)} \left[(b - x_0) \int_c^d |f(x_0, y) - f(x_0, y_0)| dy + \right.$$

$$\left. \frac{(b - x_0)^{\nu + 1}(d - c)}{\Gamma(\nu + 2)} \left\| \frac{\partial_{x_0}^\nu f}{\partial x^\nu} \right\|_{\infty, Q} \right] \qquad (25.34)$$

$$= \frac{1}{d - c} \int_c^d |f(x_0, y) - f(x_0, y_0)| dy + \frac{(b - x_0)^\nu}{\Gamma(\nu + 2)} \left\| \frac{\partial_{x_0}^\nu f}{\partial x^\nu} \right\|_{\infty, Q}, \qquad (25.35)$$

proving the claim. $\square$

We further have

Theorem 25.9. *Let* $Q := [x_0, b] \times [c, d]$, $x_0 \in [a, b)$, *and* $f \in C^n([a, b] \times [c, d])$. *Let* $\nu \geq 2$ *such that* $n = [\nu]$ *and* $\partial_{x_0}^{\nu} f / \partial x^{\nu} \in C_{x_0}^{\nu}([a, b])$, $y_0 \in [a, b]$. *We further assume that* $\partial^j f(x_0, y) / \partial x^j = 0$, $j = 1, \ldots, n - 1$.
Then

$$\left| \frac{1}{(b - x_0)(d - c)} \int_Q f(x, y) dx\, dy - f(x_0, y_0) \right| \leq$$

$$\frac{1}{d - c} \int_c^d |f(x_0, y) - f(x_0, y_0)| dy + \frac{(b - x_0)^{\nu}}{\Gamma(\nu + 2)} \left\| \frac{\partial_{x_0}^{\nu} f}{\partial x^{\nu}} \right\|_{\infty, Q}. \qquad (25.36)$$

Proof. By (25.10) we get again

$$f(x, y) - f(x_0, y) = \frac{1}{\Gamma(\nu)} \int_{x_0}^x (x - t)^{\nu - 1} \frac{\partial_{x_0}^{\nu} f}{\partial x^{\nu}}(t, y) dt, \qquad (25.37)$$

$x \geq x_0$, $\forall\, y \in [c, d]$.
And again

$$|f(x, y) - f(x_0, y)| \leq \frac{(x - x_0)^{\nu}}{\Gamma(\nu + 1)} \left\| \frac{\partial_{x_0}^{\nu} f}{\partial x^{\nu}} \right\|_{\infty, Q}, \qquad (25.38)$$

$\forall\, x \geq x_0$, $\forall\, y \in [c, d]$.
Also, it holds again

$$|f(x, y) - f(x_0, y_0)| \leq |f(x_0, y) - f(x_0; y_0)| + \frac{(x - x_0)^{\nu}}{\Gamma(\nu + 1)} \left\| \frac{\partial_{x_0}^{\nu} f}{\partial x^{\nu}} \right\|_{\infty, Q},$$
$$(25.39)$$

$\forall\, x \geq x_0$, $\forall\, y \in [c, d]$.
Integrating (25.39) over Q we prove (25.36). □

Similar to (25.28) and (25.36) one can prove inequalities in more than two variables.

Next we study fractional Ostrowski-type inequalities over balls and spherical shells. For that we make

Remark 25.10. We define the ball $B(0, R) := \{x \in \mathbb{R}^N : |x| < R\} \subseteq \mathbb{R}^N$, $N \geq 2$, $R > 0$, and the sphere $S^{N-1} := \{x \in \mathbb{R}^N : |x| = 1\}$, where $|\cdot|$ is the Euclidean norm.
Let $d\omega$ be the element of surface measure on S^{N-1} and let $\omega_N = \int_{S^{N-1}} d\omega = 2\pi^{N/2}/\Gamma(N/2)$. For $x \in \mathbb{R}^N - \{0\}$ we can write uniquely $x = r\omega$, where $r = |x| > 0$ and $\omega = x/r \in S^{N-1}$, $|\omega| = 1$. Note that $\int_{B(0,R)} dy =$

$\omega_N R^N / N$ is the Lebesgue measure of the ball. For $F \in C(\overline{B(0,R)})$ we have $\int_{B(0,R)} F(x) dx = \int_{S^{N-1}} \left(\int_0^R F(r\omega) r^{N-1} dr \right) d\omega$; we use this formula frequently.

The function $f : \overline{B(0,R)} \to \mathbb{R}$ is *radial* if there exists a function g such that $f(x) = g(r)$, where $r = |x|$, $r \in [0,R]$, $\forall x \in \overline{B(0,R)}$. Here we suppose that $g \in C_0^\nu([0,R])$, $1 \leq \nu < 2$.

By (25.10) we have

$$g(s) - g(0) = \frac{1}{\Gamma(\nu)} \int_0^s (s-w)^{\nu-1} (D_0^\nu g)(w) dw, \qquad (25.40)$$

$\forall s \in [0,R]$.

Thus

$$|g(s) - g(0)| \leq \frac{s^\nu}{\Gamma(\nu+1)} \|D_0^\nu g\|_{\infty,[0,R]}, \forall s \in [0,R]. \qquad (25.41)$$

Next we observe that

$$\left| f(0) - \frac{\int_{B(0,R)} f(y) dy}{Vol(B(0,R))} \right| = \left| g(0) - \frac{\int_{S^{N-1}} (\int_0^R g(s) s^{N-1} ds) d\omega}{\int_{S^{N-1}} (\int_0^R s^{N-1} ds) d\omega} \right| = \qquad (25.42)$$

$$\left| g(0) - \frac{N}{R^N} \int_0^R g(s) s^{N-1} ds \right| = \frac{N}{R^N} \left| \int_0^R s^{N-1} (g(0) - g(s)) ds \right| \leq \qquad (25.43)$$

$$\frac{N}{R^N} \int_0^R s^{N-1} |g(s) - g(0)| ds \overset{(25.41)}{\leq} \frac{\|D_0^\nu g\|_\infty}{\Gamma(\nu+1)} \frac{N}{R^N} \int_0^R s^{\nu+N-1} ds =$$

$$\frac{\|D_0^\nu g\|_\infty N R^\nu}{\Gamma(\nu+1)(\nu+N)}. \qquad (25.44)$$

That is, we have proved that

$$\left| f(0) - \frac{\int_{B(0,R)} f(y) dy}{Vol(B(0,R))} \right| = \left| g(0) - \frac{N}{R^N} \int_0^R g(s) s^{N-1} ds \right| \leq \frac{\|D_0^\nu g\|_\infty N R^\nu}{\Gamma(\nu+1)(\nu+N)}. \qquad (25.45)$$

The last inequality (25.45) is sharp; namely it is attained by $g(r) = r^\nu$, $1 \leq \nu < 2$, $r \in [0,R]$. Indeed by (25.17) we get $\|D_0^\nu x^\nu\|_{\infty,[0,R]} = \Gamma(\nu+1)$.

Notice also that

$$\text{L.H.S.}(25.45) = \frac{N}{R^N} \int_0^R s^{\nu+N-1} ds = \frac{N R^\nu}{\nu+N} = \text{R.H.S.}(25.45), \qquad (25.46)$$

proving optimality.

We have established

Theorem 25.11. *Let* $f : \overline{B(0,R)} \to \mathbb{R}$ *that is radial; that is, there exists* g *such that* $f(x) = g(r)$, $r = |x|$, $\forall\, x \in \overline{B(0,R)}$. *Assume that* $g \in C_0^\nu([0,R])$, $1 \le \nu < 2$. *Then*

$$\left| f(0) - \frac{\int_{B(0,R)} f(y)dy}{Vol(B(0,R))} \right| = \left| g(0) - \frac{N}{R^N} \int_0^R g(s)s^{N-1}ds \right| \le \frac{\|D_0^\nu g\|_\infty N\, R^\nu}{\Gamma(\nu+1)(\nu+N)}.$$
$$(25.47)$$

Inequality (25.47) *is sharp; that is, it is attained by* $g(r) = r^\nu$.

We continue the previous remark.

Remark 25.12. We treat here the general, not necessarily radial, case of $f \in C(\overline{B(0,R)})$. For any fixed $\omega \in S^{N-1}$ the function $f(\cdot\omega)$ is radial on [0,R]. We suppose that $f(\cdot\omega) \in C_0^\nu([0,R])$, $1 \le \nu < 2$. That is, $\exists\, \partial_0^\nu f(r\omega)/\partial r^\nu$ and is continuous in $r \in [0,R]$, for any $\omega \in S^{N-1}$. Here we have

$$\frac{\partial_0^\nu f(r\omega)}{\partial r^\nu} = \frac{\partial}{\partial r}\left(\mathcal{J}_{2-\nu}\left(\frac{\partial f}{\partial r}(\cdot\omega) \right) \right)(r) =$$

$$\frac{1}{\Gamma(2-\nu)} \frac{\partial}{\partial r}\left(\int_0^r (r-t)^{1-\nu}\frac{\partial f}{\partial r}(t\omega)dt \right). \qquad (25.48)$$

For $x \ne 0$ (i.e. $x = r\omega$, $r > 0$, $\omega \in S^{N-1}$), the *fractional radial derivative* $\partial_0^\nu f(x)/\partial r^\nu$ is defined as in (25.48). Clearly

$$\left. \frac{\partial_0^\nu f(x)}{\partial r^\nu} \right|_{x=0}$$

is not defined.

We mention

Lemma 25.13. *All are as in Remark* 25.12. *The function* $\partial_0^\nu f(x)/\partial r^\nu$ *is measurable over* $\overline{B(0,R)} - \{0\}$.

Proof. For each $n \in \mathbb{N}$ define

$$g_n(r,\omega) := n\left[f\left(\left(r - \frac{1}{n}\right)\omega \right) - f(r\omega) \right] = \frac{f\left(\left(r - \frac{1}{n}\right)\omega \right) - f(r\omega)}{\frac{1}{n}}$$

and note that each g_n is jointly measurable in (r,ω) because it is jointly continuous in (r,ω) by $f \in C(\overline{B(0,R)})$; here $r \in (0,R]$ and $\omega \in S^{N-1}$. In view of $g_n(r,\omega) \to \partial f(r\omega)/\partial r$ as $n \to \infty$, we get that $\partial f(r\omega)/\partial r$ is jointly measurable in $(r,\omega) \in (0,R] \times s^{N-1} = \overline{B(0,R)} - \{0\}$.

Then $\partial f/\partial r(r\cdot)$ is measurable in $\omega \in S^{N-1}$, $\forall\, r \in (0,R]$. Thus the integral $I_\varepsilon(r,\omega) = \int_0^{r-\varepsilon}(r-t)^{1-\nu}\partial f/\partial r(t\omega)dt$, $r \in (0,R]$, $\omega \in S^{N-1}$, $\epsilon > 0$ small; because it is a limit of Riemann sums, it is measurable in $\omega \in$

S^{N-1}. Because $(r-t)^{1-\nu}\partial f/\partial r(t\omega)$ is integrable over $[0,r]$, we get that $I_\varepsilon(r,\omega)$ is continuous in $r-\varepsilon$, $\forall\varepsilon > 0$ small. Thus

$$\lim_{\varepsilon\to 0} I_\varepsilon(r,\omega) = I(r,\omega) := \int_0^r (r-t)^{1-\nu}\frac{\partial f}{\partial r}(t\omega)dt,$$

proving $I(r,\omega)$ measurable in $\omega\in s^{N-1}$, $\forall r\in(0,R]$.

But, by the assumption $f(\cdot\omega)\in C_0^\nu([0,R])$, we have that $I(r,\omega)$ is continuous in $r\in[0,R]$, $\forall\omega\in s^{N-1}$. Therefore by the Carathéodory theorem (see [10, p. 156]) we get that $I(r,\omega)$ is jointly measurable in $(r,\omega)\in(0,R]\times S^{N-1}$, as being a Carathéodory function. Because

$$\frac{\partial I(r,\omega)}{\partial r} = \lim_{n\to\infty} n\left[I(r-\frac{1}{n},\omega) - I(r,\omega)\right]$$

and also $I(r-1/n,\omega)$ is jointly measurable in $(r,\omega)\in(0,R]\times S^{n-1}$, we get that $\partial I(r,\omega)/\partial r$ is jointly measurable in $(r,\omega)\in\overline{B(0,R)}-\{0\}$, proving the claim. $\square$

We need

Lemma 25.14. *All are as in Remark 25.12. Additionally assume that $\partial_0^\nu f(x)/\partial r^\nu$ is continuous on $\overline{B(0,R)}-\{0\}$, and*

$$K := \left\|\frac{\partial_0^\nu f(x)}{\partial r^\nu}\right\|_{L_\infty(B(0,R))} = \operatorname{ess\,sup}\left|\frac{\partial_0^\nu f(x)}{\partial r^\nu}\right|_{B(0,R)} < \infty.$$

Then

$$\left\|\frac{\partial_0^\nu f(r\omega)}{\partial r^\nu}\right\|_{\infty,(r\in[0,R])} \leq K, \quad \forall\,\omega\in s^{N-1}. \tag{25.49}$$

Proof. In the radial case (25.49) is obvious. Also it is obvious if

$$\left\|\frac{\partial_0^\nu f(r\omega)}{\partial r^\nu}\right\|_{\infty,[0,R]} = \left|\frac{\partial_0^\nu f(r_0\omega)}{\partial r^\nu}\right|,$$

for some $r_0\in(0,R]$. The only difficulty here is if for specific $\omega_0\in S^{N-1}$ we have that

$$\left\|\frac{\partial_0^\nu f(r\omega_0)}{\partial r^\nu}\right\|_{\infty,[0,R]} = \left|\frac{\partial_0^\nu f(0)}{\partial r^\nu}\right|.$$

Then it is evident, for very small $r^* > 0$, that by continuity of $\partial_0^\nu f(\cdot\,\omega_0)/\partial r^\nu$ we have

$$\left|\frac{\partial_0^\nu f(r^*\omega_0)}{\partial r^\nu}\right| \approx \left|\frac{\partial_0^\nu f(0)}{\partial r^\nu}\right|.$$

If

$$\left|\frac{\partial_0^\nu f(0)}{\partial r^\nu}\right| > \operatorname{ess\,sup}\left|\frac{\partial_0^\nu f(x)}{\partial r^\nu}\right|_{B(0,R)},$$

then

$$\left| \frac{\partial_0^\nu f(r^*\omega_0)}{\partial r^\nu} \right| > \text{ess sup} \left| \frac{\partial_0^\nu f(x)}{\partial r^\nu} \right|_{B(0,R)} = \left\| \frac{\partial_0^\nu f(x)}{\partial r^\nu} \right\|_{\infty, \overline{B(0,R)} - \{0\}},$$

a contradiction. □

Remark 25.15. (continuation) By (25.47) we obtain

$$\left| f(0) - \frac{N}{R^N} \int_0^R f(s\omega) s^{N-1} ds \right| \leq \frac{\left\| \frac{\partial_0^\nu f(r\omega)}{\partial r^\nu} \right\|_{\infty, (r \in [0,R])} N R^\nu}{\Gamma(\nu+1)(\nu+N)} \leq \frac{K N R^\nu}{\Gamma(\nu+1)(\nu+N)}.$$

Consequently we find

$$\left| f(0) - \frac{N}{\omega_N R^N} \int_{S^{N-1}} \left(\int_0^R f(s\omega) s^{N-1} ds \right) d\omega \right| \leq \frac{K N R^\nu}{\Gamma(\nu+1)(\nu+N)}.$$

That proves

$$\left| f(0) - \frac{\int_{B(0,R)} f(y) dy}{Vol(B(0,R))} \right| \leq \frac{K N R^\nu}{\Gamma(\nu+1)(\nu+N)}. \tag{25.50}$$

We have established

Theorem 25.16. Let $f \in C(\overline{B(0,R)})$ that is not necessarily radial, and assume that $f(\cdot \omega) \in C_0^\nu([0,R])$, $1 \leq \nu < 2$, for any $\omega \in S^{N-1}$. Suppose also $\partial_0^\nu f(x)/\partial r^\nu$ is continuous on $\overline{B(0,R)} - \{0\}$, and that

$$\left\| \frac{\partial_0^\nu f(x)}{\partial r^\nu} \right\|_{L_\infty(B(0,R))} < \infty.$$

Then

$$\left| f(0) - \frac{\int_{B(0,R)} f(y) dy}{Vol(B(0,R))} \right| \leq \frac{N R^\nu}{\Gamma(\nu+1)(\nu+N)} \left\| \frac{\partial_0^\nu f(x)}{\partial r^\nu} \right\|_{L_\infty(B(0,R))}. \tag{25.51}$$

We make

Remark 25.17. Let the spherical shell $A := B(0, R_2) - \overline{B(0, R_1)}$, $0 < R_1 < R_2$, $A \subseteq \mathbb{R}^N$, $N \geq 2$, $x \in \bar{A}$. Consider $f \in C^1(\bar{A})$ and assume that there exists $\partial_{R_1}^\nu f(x)/\partial r^\nu \in C(\bar{A})$, $1 \leq \nu < 2$; $x = r\omega, r \in [R_1, R_2], \omega \in S^{N-1}$; where $\partial_{R_1}^\nu f(x)/\partial r^\nu = 1/\Gamma(2-\nu) \partial/\partial r \left(\int_{R_1}^r (r-t)^{1-\nu} \partial f/\partial r(t\omega) dt \right)$.

Clearly here $f(r\omega) \in C^1([R_1, R_2])$ and $\partial^\nu_{R_1} f(r\omega)/\partial r^\nu \in C([R_1, R_2])$, $\forall \omega \in S^{N-1}$. For $F \in C(\bar{A})$ it holds $\int_A F(x)dx = \int_{S^{N-1}} \left(\int_{R_1}^{R_2} F(r\omega) r^{N-1} dr \right) d\omega$; we often exploit this formula here.

Initially we assume that f is *radial*; that is, there exists g such that $f(x) = g(r)$. Here $Vol(A) = \omega_N(R_2^N - R_1^N)/N$. Then we get via the polar method that

$$\left| f(R_1\omega) - \frac{\int_A f(y)dy}{Vol(A)} \right| = \left| g(R_1) - \left(\frac{N}{R_2^N - R_1^N} \right) \int_{R_1}^{R_2} g(s) s^{N-1} ds \right|$$

$$= \left(\frac{N}{R_2^N - R_1^N} \right) \left| \int_{R_1}^{R_2} (g(R_1) - g(s)) s^{N-1} ds \right| \leq$$

$$\left(\frac{N}{R_2^N - R_1^N} \right) \int_{R_1}^{R_2} |g(R_1) - g(s)| s^{N-1} ds =: (\star). \tag{25.52}$$

Here by (25.10) we get for $s \geq R_1$,

$$g(s) - g(R_1) = \frac{1}{\Gamma(\nu)} \int_{R_1}^{s} (s - w)^{\nu-1} (D^\nu_{R_1} g)(w) dw. \tag{25.53}$$

Thus

$$|g(s) - g(R_1)| \leq \frac{\|D^\nu_{R_1} g\|_{\infty, [R_1, R_2]}}{\Gamma(\nu + 1)} (s - R_1)^\nu, \tag{25.54}$$

$\forall s \geq R_1$.

Consequently it holds

$$(\star) \leq \left(\frac{N(\|D^\nu_{R_1} g\|_{\infty, [R_1, R_2]})}{(R_2^N - R_1^N) \Gamma(\nu + 1)} \right) \int_{R_1}^{R_2} (s - R_1)^\nu s^{N-1} ds =$$

$$= \frac{N \|D^\nu_{R_1} g\|_{\infty, [R_1, R_2]}}{(R_2^N - R_1^N) \Gamma(\nu + 1)} I =: (\star\star). \tag{25.55}$$

Here

$$I := \int_{R_1}^{R_2} (s - R_1)^\nu s^{N-1} ds = (-1)^{N-1} \int_{R_1}^{R_2} (-s)^{N-1} (s - R_1)^\nu ds$$

$$= (-1)^{N-1} \int_{R_1}^{R_2} (-R_2 + R_2 - s)^{N-1} (s - R_1)^\nu ds = \tag{25.56}$$

$$(-1)^{N-1} \int_{R_1}^{R_2} \left[\sum_{k=0}^{N-1} \binom{N-1}{k} (-R_2)^{N-1-k} (R_2 - s)^k \right] (s - R_1)^\nu ds = ((-1)^{N-1})^2$$

$$\left(\sum_{k=0}^{N-1} \binom{N-1}{k} (-1)^{-k} R_2^{N-k-1} \int_{R_1}^{R_2} (R_2 - s)^{(k+1)-1} (s - R_1)^{(\nu+1)-1} ds \right)$$

$$= \sum_{k=0}^{N-1} \binom{N-1}{k} (-1)^k \, R_2^{N-k-1} \frac{\Gamma(k+1)\,\Gamma(\nu+1)}{\Gamma(k+\nu+2)} (R_2 - R_1)^{k+\nu+1} =$$

$$\Gamma(\nu+1)(N-1)! \sum_{k=0}^{N-1} \frac{(-1)^k}{(n-k-1)!} R_2^{n-k-1} \frac{(R_2-R_1)^{k+\nu+1}}{\Gamma(k+\nu+2)}. \qquad (25.57)$$

That is,

$$I = \int_{R_1}^{R_2} (s-R_1)^\nu s^{N-1} ds =$$

$$\Gamma(\nu+1)\,(N-1)! \sum_{k=0}^{N-1} \frac{(-1)^k}{(N-k-1)!} R_2^{N-k-1} \frac{(R_2-R_1)^{k+\nu+1}}{\Gamma(k+\nu+2)}. \qquad (25.58)$$

Continuing with (25.55) via (25.58), we have

$$(\star\star) = \left(\frac{N!}{R_2^N - R_1^N} \|D_{R_1}^\nu g\|_{\infty,[R_1,R_2]} \right) \left(\sum_{k=0}^{N-1} \frac{(-1)^k}{(N-k-1)!} R_2^{N-k-1} \frac{(R_2-R_1)^{k+\nu+1}}{\Gamma(k+\nu+2)} \right). \qquad (25.59)$$

Hence in the radial case we established

$$\left| f(R_1\omega) - \frac{\int_A f(y)dy}{Vol(A)} \right| \le \left(\frac{N! \|D_{R_1}^\nu g\|_{\infty,[R_1,R_2]}}{R_2^N - R_1^N} \right)$$

$$\left(\sum_{k=0}^{N-1} \frac{(-1)^k}{(N-k-1)!} R_2^{N-k-1} \frac{(R_2-R_1)^{k+\nu+1}}{\Gamma(k+\nu+2)} \right). \qquad (25.60)$$

Inequality (25.60) is attained by $g(s) := (s-R_1)^\nu$, $1 \le \nu < 2$, $s \in [R_1, R_2]$. Indeed, we observe that

$$\text{L.H.S.}(25.60) = \left(\frac{N}{R_2^N - R_1^N} \right) \int_{R_1}^{R_2} (s-R_1)^\nu s^{N-1} ds =$$

$$\frac{\Gamma(\nu+1)\,N!}{(R_2^N - R_1^N)} \left(\sum_{k=0}^{N-1} \frac{(-1)^k}{(N-k-1)!} R_2^{N-k-1} \frac{(R_2-R_1)^{k+\nu+1}}{\Gamma(k+\nu+2)} \right) = \text{R.H.S.}(25.60);$$
$$\qquad (25.61)$$

by (25.17) that says

$$\|D_{R_1}^\nu (s-R_1)^\nu\|_{\infty,[R_1,R_2]} = \Gamma(\nu+1). \qquad (25.62)$$

We have established

Theorem 25.18. Let $A := B(0, R_2) - \overline{B(0, R_1)}$, $0 < R_1 < R_2$, $A \subseteq \mathbb{R}^N$, $N \ge 2$. Consider $f : \bar{A} \to \mathbb{R}$ that is radial; that is, there exists g such

that $f(x) = g(r)$, $x = r\omega, r \in [R_1, R_2]$, $\omega \in S^{N-1}$, $x \in \bar{A}$. Suppose that $g \in C^\nu_{R_1}([R_1, R_2])$, $1 \leq \nu < 2$.
Then

$$\left| f(R_1\omega) - \frac{\int_A f(y)dy}{Vol(A)} \right| = \left| g(R_1) - \left(\frac{N}{R_2^N - R_1^N} \right) \int_{R_1}^{R_2} g(s)s^{N-1}ds \right| \leq$$

$$\left(\frac{N! \, \|D^\nu_{R_1} g\|_{\infty, [R_1, R_2]}}{R_2^N - R_1^N} \right) \left(\sum_{k=0}^{N-1} \frac{(-1)^k}{(N-k-1)!} R_2^{N-k-1} \frac{(R_2 - R_1)^{k+\nu+1}}{\Gamma(k+\nu+2)} \right).$$

$$(25.63)$$

Inequality (25.63) is sharp; namely it is attained by $g(s) = (s - R_1)^\nu$, $s \in [R_1, R_2]$.

We continue the last remark.

Remark 25.19. We treat the nonradial case here. For fixed $\omega \in S^{N-1}$ the function $f(r\omega)$ is radial over $[R_1, R_2]$. We apply (25.63) for $g = f(\cdot \omega)$ to get:

$$\left| f(R_1\omega) - \left(\frac{N}{R_2^N - R_1^N} \right) \int_{R_1}^{R_2} f(s\omega)s^{N-1}ds \right| \leq$$

$$\left(\frac{N! \, \left\| \frac{\partial^\nu_{R_1} f}{\partial r^\nu} \right\|_{\infty, \bar{A}}}{R_2^N - R_1^N} \right) \left(\sum_{k=0}^{N-1} \frac{(-1)^k}{(N-k-1)!} R_2^{N-k-1} \frac{(R_2 - R_1)^{k+\nu+1}}{\Gamma(k+\nu+2)} \right).$$

$$(25.64)$$

Hence it holds

$$\left| \frac{\int_{S^{N-1}} f(R_1\omega)d\omega}{\omega_N} - \frac{N}{(R_2^N - R_1^N)\omega_N} \int_{S^{N-1}} \left(\int_{R_1}^{R_2} f(s\omega)s^{N-1}ds \right) d\omega \right|$$

$$\leq \left(\frac{N! \, \left\| \frac{\partial^\nu_{R_1} f}{\partial r^\nu} \right\|_{\infty, \bar{A}}}{R_2^N - R_1^N} \right) \left(\sum_{k=0}^{N-1} \frac{(-1)^k}{(N-k-1)!} R_2^{N-k-1} \frac{(R_2 - R_1)^{k+\nu+1}}{\Gamma(k+\nu+2)} \right)$$

$$=: C \left\| \frac{\partial^\nu_{R_1} f}{\partial r^\nu} \right\|_{\infty, \bar{A}}. \qquad (25.65)$$

That is, we proved that

$$\left| \frac{\Gamma(\frac{N}{2}) \int_{S^{N-1}} f(R_1\omega)d\omega}{2\pi^{N/2}} - \frac{\int_A f(y)dy}{Vol(A)} \right| \leq C \left\| \frac{\partial^\nu_{R_1} f}{\partial r^\nu} \right\|_{\infty, \bar{A}}. \qquad (25.66)$$

However, we have for $x \in \bar{A}$:

$$\left| f(x) - \frac{\int_A f(y)dy}{Vol(A)} \right| \leq \left| f(x) - \frac{\Gamma(\frac{N}{2}) \int_{S^{N-1}} f(R_1\omega)d\omega}{2\pi^{N/2}} \right| + C \left\| \frac{\partial^\nu_{R_1} f}{\partial r^\nu} \right\|_{\infty, \bar{A}}. \qquad (25.67)$$

We have established the following result.

Theorem 25.20. *Consider $f \in C^1(\bar{A})$ such that there exists $\partial_{R_1}^{\nu} f(x)/\partial r^{\nu} \in C(\bar{A})$, $1 \le \nu < 2$, $x \in \bar{A}$. Then*

$$\left| f(x) - \frac{\int_A f(y)dy}{Vol(A)} \right| \le \left| f(x) - \frac{\Gamma(\frac{N}{2}) \int_{S^{N-1}} f(R_1\omega)d\omega}{2\pi^{N/2}} \right| +$$

$$\left(\frac{N!}{R_2^N - R_1^N} \right) \left(\sum_{k=0}^{N-1} \frac{(-1)^k}{(N-k-1)!} R_2^{N-k-1} \frac{(R_2 - R_1)^{k+\nu+1}}{\Gamma(k+\nu+2)} \right) \left\| \frac{\partial_{R_1}^{\nu} f}{\partial r^{\nu}} \right\|_{\infty,\bar{A}}.$$

$$(25.68)$$

We make

Remark 25.21. This continues Remarks 25.17 and 25.19. Here we establish higher-order multivariate fractional Ostrowski-type inequalities over spherical shells.

Here $\nu \ge 2$, $n := [\nu] \ge 2$, $\alpha := \nu - n$. Consider $f \in C^n(\bar{A})$, which implies that $f(r\omega) \in C^n([R_1, R_2])$, $\forall \omega \in S^{N-1}$. Furthermore assume that there exists $\partial_{R_1}^{\nu} f(x)/\partial r^{\nu} \in C(\bar{A})$, $x \in \bar{A}$; $x = r\omega$, $r \in [R_1, R_2]$, $\omega \in S^{N-1}$, where $\partial_{R_1}^{\nu} f(x)/\partial r^{\nu} = 1/\Gamma(1-\alpha)\partial/\partial r \left(\int_{R_1}^{r} (r-t)^{-\alpha} \partial^n f(t,\omega)/\partial r^n dt \right)$. The last implies $\partial_{R_1}^{\nu} f(r\omega)/\partial r^{\nu} \in C([R_1, R_2])$, $\forall \omega \in S^{N-1}$. We start again with f being radial; that is, $\exists g : f(x) = g(r)$, $r \in [R_1, R_2]$, $x \in \bar{A}$.

We have

$$\left| f(R_1\omega) - \frac{\int_A f(y)dy}{Vol(A)} \right| = \left(\frac{N}{R_2^N - R_1^N} \right) \left| \int_{R_1}^{R_2} (g(s) - g(R_1))s^{N-1}ds \right| =: (\star).$$

$$(25.69)$$

By (25.10) we get

$$g(s) - g(R_1) = \sum_{k=1}^{n-1} g^{(k)}(R_1) \frac{(s-R_1)^k}{k!} + \frac{1}{\Gamma(\nu)} \int_{R_1}^{s} (s-w)^{\nu-1}(D_{R_1}^{\nu}g)(w)dw,$$

$$(25.70)$$

all $s \ge R_1$.

Consequently it holds

$$(\star) = \left(\frac{N}{R_2^N - R_1^N} \right) \left[\sum_{k=1}^{n-1} \frac{|g^{(k)}(R_1)|}{k!} \left| \int_{R_1}^{R_2} s^{N-1}(s-R_1)^k ds \right| + \right.$$

$$\left. \frac{1}{\Gamma(\nu)} \int_{R_1}^{R_2} s^{N-1} \left| \int_{R_1}^{s} (s-w)^{\nu-1}(D_{R_1}^{\nu}g)(w)dw \right| ds \right] \le (\text{ by } (25.58))$$

$$(25.71)$$

$$\left(\frac{N}{R_2^N - R_1^N}\right)\left[(N-1)!\sum_{k=1}^{n-1}|g^{(k)}(R_1)|\left|\sum_{\lambda=0}^{N-1}\frac{(-1)^\lambda}{(N-\lambda-1)!}R_2^{N-\lambda-1}\frac{(R_2-R_1)^{\lambda+k+1}}{(\lambda+k+1)!}\right|\right.$$

$$\left.+\frac{\|D_{R_1}^\nu g\|_{\infty,[R_1,R_2]}}{\Gamma(\nu+1)}\left(\int_{R_1}^{R_2}(s-R_1)^\nu s^{N-1}ds\right)\right] \stackrel{(25.58)}{=}$$

$$\left(\frac{N!}{R_2^N - R_1^N}\right)\left[\sum_{k=1}^{n-1}|g^{(k)}(R_1)|\left|\sum_{\lambda=0}^{N-1}\frac{(-1)^\lambda}{(N-\lambda-1)!}R_2^{N-\lambda-1}\frac{(R_2-R_1)^{\lambda+k+1}}{(\lambda+k+1)!}\right|+\right.$$

$$\left.(\|D_{R_1}^\nu g\|_{\infty,[R_1,R_2]})\left(\sum_{\lambda=0}^{N-1}\frac{(-1)^\lambda}{(N-\lambda-1)!}R_2^{N-\lambda-1}\frac{(R_2-R_1)^{\lambda+\nu+1}}{\Gamma(\lambda+\nu+2)}\right)\right]. \quad (25.72)$$

We have established the following result.

Theorem 25.22. *Here* $\nu \geq 2$, $n := [\nu]$. *Suppose that f is radial; that is,* $f(x) = g(r)$, $r \in [R_1, R_2]$, $\forall x \in \bar{A}$. *Assume that* $g \in C_{R_1}^\nu([R_1, R_2])$. *Then*

$$E := \left|f(R_1\omega) - \frac{\int_A f(y)dy}{Vol(A)}\right| = \left|g(R_1) - \left(\frac{N}{R_2^N - R_1^N}\right)\int_{R_1}^{R_2}g(s)s^{N-1}ds\right| \leq$$

$$(25.73)$$

$$\left(\frac{N!}{R_2^N - R_1^N}\right)\left[\sum_{k=1}^{n-1}|g^{(k)}(R_1)|\left|\sum_{\lambda=0}^{N-1}\frac{(-1)^\lambda}{(N-\lambda-1)!}R_2^{N-\lambda-1}\frac{(R_2-R_1)^{\lambda+k+1}}{(\lambda+k+1)!}\right|+\right.$$

$$\left.(\|D_{R_1}^\nu g\|_{\infty,[R_1,R_2]})\left(\sum_{\lambda=0}^{N-1}\frac{(-1)^\lambda}{(N-\lambda-1)!}R_2^{N-\lambda-1}\frac{(R_2-R_1)^{\lambda+\nu+1}}{\Gamma(\lambda+\nu+2)}\right)\right].$$

Inequality (25.73) *is sharp; namely it is attained by* $g^*(s) = (s-R_1)^\nu$, $s \in [R_1, R_2]$.

Proof of Sharpness. Again by (25.17) we get

$$\|D_{R_1}^\nu g^*\|_{\infty,[R_1,R_2]} = \Gamma(\nu+1). \quad (25.74)$$

Also it holds $g^{*(k)}(R_1) = 0$, $k = 1, \ldots, n-1$.
Thus

$$\text{L.H.S.}(25.73) = \left(\frac{N}{R_2^N - R_1^N}\right)\int_{R_1}^{R_2}(s-R_1)^\nu s^{N-1}ds$$

$$= \left(\frac{N!\,\Gamma(\nu+1)}{R_2^N - R_1^N}\right)\left[\sum_{\lambda=0}^{N-1}\frac{(-1)^\lambda}{(N-\lambda-1)!}R_2^{N-\lambda-1}\frac{(R_2-R_1)^{\lambda+\nu+1}}{\Gamma(\lambda+\nu+2)}\right] = \text{R.H.S.}(25.73).$$

$$(25.75)$$

$\square$

We give

Corollary 25.23. *In the terms and assumptions of Theorem 25.22, addition-ally suppose that* $g^{(k)}(R_1) = 0$, $k = 1, \ldots, n-1$.
Then

$$E \le \left(\frac{N!}{R_2^N - R_1^N}\right)\left(\sum_{\lambda=0}^{N-1}\frac{(-1)^\lambda}{(N-\lambda-1)!}R_2^{N-\lambda-1}\frac{(R_2-R_1)^{\lambda+\nu+1}}{\Gamma(\lambda+\nu+2)}\right).$$

$$\|D_{R_1}^\nu g\|_{\infty,[R_1,R_2]}. \tag{25.76}$$

We continue Remark 25.21 with

Remark 25.24. We treat here the general, not necessarily radial, case of f. We apply (25.73) to $f(r\omega)$, ω fixed, $r \in [R_1, R_2]$. We then have

$$\left|f(R_1\omega) - \left(\frac{N}{R_2^N - R_1^N}\right)\int_{R_1}^{R_2}f(s\omega)s^{N-1}ds\right| \le \left(\frac{N!}{R_2^N - R_1^N}\right)$$

$$\left[\sum_{k=1}^{n-1}\left|\frac{\partial^k f}{\partial r^k}(R_1\omega)\right|\left|\sum_{\lambda=0}^{N-1}\frac{(-1)^\lambda}{(N-\lambda-1)!}R_2^{N-\lambda-1}\frac{(R_2-R_1)^{\lambda+k+1}}{(\lambda+k+1)!}\right|\right.$$

$$\left.+\left\|\frac{\partial_{R_1}^\nu f}{\partial r^\nu}\right\|_{\infty,\bar{A}}\left(\sum_{\lambda=0}^{N-1}\frac{(-1)^\lambda}{(N-\lambda-1)!}R_2^{N-\lambda-1}\frac{(R_2-R_1)^{\lambda+\nu+1}}{\Gamma(\lambda+\nu+2)}\right)\right]. \tag{25.77}$$

Therefore

$$\left|\frac{\int_{S^{N-1}}f(R_1\omega)d\omega}{\omega_N} - \frac{N}{(R_2^N-R_1^N)\omega_N}\int_{S^{N-1}}\left(\int_{R_1}^{R_2}f(s\omega)s^{N-1}ds\right)d\omega\right|$$

$$\le \left(\frac{N!}{R_2^N - R_1^N}\right)\left[\sum_{k=1}^{n-1}\left(\frac{\int_{S^{N-1}}\left|\frac{\partial^k f}{\partial r^k}(R_1\omega)\right|d\omega}{\omega_N}\right)\right.$$

$$\left|\sum_{\lambda=0}^{N-1}\frac{(-1)^\lambda}{(N-\lambda-1)!}R_2^{N-\lambda-1}\frac{(R_2-R_1)^{\lambda+k+1}}{(\lambda+k+1)!}\right|$$

$$\left.+\left\|\frac{\partial_{R_1}^\nu f}{\partial r^\nu}\right\|_{\infty,\bar{A}}\left(\sum_{\lambda=0}^{N-1}\frac{(-1)^\lambda}{(N-\lambda-1)!}R_2^{N-\lambda-1}\frac{(R_2-R_1)^{\lambda+\nu+1}}{(\lambda+\nu+2)}\right)\right] =: \delta. \tag{25.78}$$

That is,

$$\left|\frac{\Gamma(\frac{N}{2})\int_{S^{N-1}}f(R_1\omega)d\omega}{2\pi^{N/2}} - \frac{\int_A f(y)dy}{Vol(A)}\right| \le \delta. \tag{25.79}$$

Consequently it holds for $x \in \bar{A}$ that

$$\left|f(x) - \frac{\int_A f(y)dy}{Vol(A)}\right| \le \left|f(x) - \frac{\Gamma(\frac{N}{2})\int_{S^{N-1}}f(R_1\omega)d\omega}{2\pi^{N/2}}\right| + \delta. \tag{25.80}$$

We have established the next result.

Theorem 25.25. *Here* $\nu \geq 2$, $n := [\nu]$. *Consider* $f \in C^n(\bar{A})$ *and assume there exists* $\partial_{R_1}^\nu f(x)/\partial r^\nu \in C(\bar{A})$, $x \in \bar{A}$. *Then*

$$M := \left| f(x) - \frac{\int_A f(y)dy}{Vol(A)} \right| \leq \left| f(x) - \frac{\Gamma(\frac{N}{2}) \int_{S^{N-1}} f(R_1\omega)d\omega}{2\pi^{N/2}} \right| +$$

$$\left(\frac{N!}{R_2^N - R_1^N} \right) \left[\frac{\Gamma(\frac{N}{2})}{2\pi^{N/2}} \sum_{k=1}^{n-1} \left(\int_{S^{N-1}} \left| \frac{\partial^k f}{\partial r^k}(R_1\omega) \right| d\omega \right) \right.$$

$$\left| \sum_{\lambda=0}^{N-1} \frac{(-1)^\lambda}{(N-\lambda-1)!} R_2^{N-\lambda-1} \frac{(R_2 - R_1)^{\lambda+k+1}}{(\lambda+k+1)!} \right| +$$

$$+ \left\| \frac{\partial_{R_1}^\nu f}{\partial r^\nu} \right\|_{\infty,\bar{A}} \left(\sum_{\lambda=0}^{N-1} \frac{(-1)^\lambda}{(N-\lambda-1)!} R_2^{N-\lambda-1} \frac{(R_2 - R_1)^{\lambda+\nu+1}}{(\lambda+\nu+2)} \right) \right]. \qquad (25.81)$$

We finish with

Corollary 25.26. *In the terms and assumptions of Theorem 25.25, additionally suppose that* $\partial^k f/\partial r^k, k = 1, \ldots, n-1$, *vanish on* $\partial B(0, R_1)$. *Then*

$$M \leq \left| f(x) - \frac{\Gamma(\frac{N}{2})}{2\pi^{N/2}} \int_{S^{N-1}} f(R_1\omega)d\omega \right| +$$

$$\left(\frac{N!}{R_2^N - R_1^N} \right) \left(\sum_{\lambda=0}^{N-1} \frac{(-1)^\lambda}{(N-\lambda-1)!} R_2^{N-\lambda-1} \frac{(R_2 - R_1)^{\lambda+\nu+1}}{\Gamma(\lambda+\nu+2)} \right) \left\| \frac{\partial_{R_1}^\nu f}{\partial r^\nu} \right\|_{\infty,\bar{A}}.$$

$$(25.82)$$

26
Caputo Fractional Ostrowski-Type Inequalities

Optimal upper bounds are given to the deviation of a value of a univariate or multivariate function of a Caputo fractional derivative related space from its average, over convex and compact subsets of $\mathbb{R}^N$, $N \geq 1$. In particular we work over closed intervals, rectangles, balls, and spherical shells. These bounds involve the supremum and L_∞ norms of related univariate or multivariate Caputo fractional derivatives of the involved functions. The derived inequalities are sharp; namely they are attained by simple functions. This chapter has been motivated by the works of Ostrowski [318], 1938, and of the author's [24], 2003 and [43], 2007, and the chapter also relies on [52].

26.1 Background

We start with

Definition 26.1. [134] Let $\nu \geq 0$; the operator J_a^ν, defined on $L_1(a, b)$ by

$$J_a^\nu f(x) := \frac{1}{\Gamma(\nu)} \int_a^x (x - t)^{\nu-1} f(t)\, dt \qquad (26.1)$$

for $a \leq x \leq b$, is called the Riemann–Liouville fractional integral operator of order ν. For $\nu = 0$, we set $J_a^0 := I$, the identity operator. Here Γ stands for the gamma function. By Theorem 2.1 of [134, p. 13], $J_a^\nu f(x)$, $\nu > 0$, exists for almost all $x \in [a, b]$ and $J_a^\nu f \in L_1(a, b)$, where $f \in L_1(a, b)$.

G.A. Anastassiou, *Fractional Differentiation Inequalities*,   615
DOI 10.1007/978-0-387-98128-4_26, © Springer Science+Business Media, LLC 2009

Here $AC^n\left(\left[a,b\right]\right)$ is the space of functions with absolutely continuous $(n-1)$st derivative.

We need to mention

Definition 26.2. [58, 134] Let $\nu \geq 0$, $n := \lceil \nu \rceil$; $\lceil \cdot \rceil$ is the ceiling of the number, $f \in AC^n\left(\left[a,b\right]\right)$. We call the Caputo fractional derivative

$$D_{*a}^{\nu} f\left(x\right) := \frac{1}{\Gamma\left(n-\nu\right)} \int_a^x \left(x-t\right)^{n-\nu-1} f^{(n)}\left(t\right) dt, \qquad (26.2)$$

$\forall x \in [a,b]$.

The above function $D_{*a}^{\nu} f\left(x\right)$ exists almost everywhere for $x \in [a,b]$. If $\nu \in \mathbb{N}$, then $D_{*a}^{\nu} f = f^{(\nu)}$ the ordinary derivative; it is also $D_{*a}^{0} f = f$.

We need

Theorem 26.3. (Taylor expansion for Caputo derivatives, [134, p. 40])
Assume $\nu \geq 0$, $n := \lceil \nu \rceil$, and $f \in AC^n\left(\left[a,b\right]\right)$. Then

$$f\left(x\right) = \sum_{k=0}^{n-1} \frac{f^{(k)}\left(a\right)}{k!} \left(x-a\right)^k + \frac{1}{\Gamma\left(\nu\right)} \int_a^x \left(x-t\right)^{\nu-1} D_{*a}^{\nu} f\left(t\right) dt, \qquad (26.3)$$

$\forall x \in [a,b]$.
Additionally assume

$$f^{(k)}\left(a\right) = 0, \ \ k = 1, \ldots, n-1;$$

then

$$f\left(x\right) - f\left(a\right) = \frac{1}{\Gamma\left(\nu\right)} \int_a^x \left(x-t\right)^{\nu-1} D_{*a}^{\nu} f\left(t\right) dt. \qquad (26.4)$$

Next we mention the multivariate analogue of (26.3) and (26.4) (see [53]).

Remark 26.4. Let Q be a compact and convex subset of $\mathbb{R}^k$, $k \geq 2$; $z := \left(z_1, \ldots, z_k\right)$, $x_0 := \left(x_{01}, \ldots, x_{0k}\right) \in Q$. Let $f \in C^n\left(Q\right)$, $n \in \mathbb{N}$.
Set

$$g_z\left(t\right) := f\left(x_0 + t\left(z - x_0\right)\right),$$

$$0 \leq t \leq 1; \ g_z\left(0\right) = f\left(x_0\right), \ g_z\left(1\right) = f\left(z\right). \qquad (26.5)$$

Then

$$g_z^{(j)}\left(t\right) = \left[\left(\sum_{i=1}^{k} \left(z_i - x_{0i}\right) \frac{\partial}{\partial x_i}\right)^j f\right]\left(x_0 + t\left(z - x_0\right)\right), \qquad (26.6)$$

$j = 0, 1, 2, \ldots, n$, and

$$g_z^{(n)}\left(0\right) = \left[\left(\sum_{i=1}^{k} \left(z_i - x_{0i}\right) \frac{\partial}{\partial x_i}\right)^n f\right]\left(x_0\right). \qquad (26.7)$$

If all

$$f_\alpha (x_0) := \frac{\partial^\alpha f}{\partial x^\alpha} (x_0) = 0, \ \alpha := (\alpha_1, \ldots, \alpha_k),$$

$$\alpha_i \in \mathbb{Z}^+, \ i = 1, \ldots, k; \ |\alpha| := \sum_{i=1}^k \alpha_i =: l,$$

then

$$g_z^{(l)} (0) = 0, \text{ where } l \in \{0, 1, \ldots, n\}.$$

We quote that

$$g_z' (t) = \sum_{i=1}^k (z_i - x_{0i}) \frac{\partial f}{\partial x_i} (x_0 + t(z - x_0)). \qquad (26.8)$$

When $f \in C^2 (Q)$, $Q \subseteq \mathbb{R}^2$, we have

$$g_z'' (t) = (z_1 - x_{01})^2 \frac{\partial^2 f}{\partial x_1^2} (x_0 + t(z - x_0)) + 2(z_1 - x_{01})$$

$$(z_2 - x_{02}) \frac{\partial^2 f}{\partial x_1 \partial x_2} (x_0 + t(z - x_0))$$

$$+ (z_2 - x_{02})^2 \frac{\partial^2 f}{\partial x_2^2} (x_0 + t(z - x_0)), \qquad (26.9)$$

and so on.

Clearly here $g_z \in C^n ([0,1])$, hence $g_z \in AC^n ([0,1])$.

In [53] we proved the following general multivariate fractional Taylor formula.

Theorem 26.5. [53] *Let $\nu > 0$, $n = \lceil \nu \rceil$, $f \in C^n (Q)$, where Q is a compact und convex subset of $\mathbb{R}^k$, $k \geq 2$; $z := (z_1, \ldots, z_k)$, $x_0 := (x_{01}, \ldots, x_{0k}) \in Q$. Then*

$$(1) \ f(z) - f(x_0) + \sum_{i=1}^k (z_i - x_{0i}) \frac{\partial f(x_0)}{\partial x_i}$$

$$+ \sum_{l=2}^{n-1} \frac{\left[\left(\sum_{i=1}^k (z_i - x_{0i}) \frac{\partial}{\partial x_i} \right)^l f \right] (x_0)}{l!} +$$

$$\frac{1}{\Gamma(\nu)} \int_0^1 (1-t)^{\nu-1} \left[J_0^{n-\nu} \left\{ \left[\left(\sum_{i=1}^k (z_i - x_{0i}) \frac{\partial}{\partial x_i} \right)^n \right.\right.\right.$$

$$\left.\left.\left. f \right] (x_0 + t(z - x_0)) \right\} \right] dt. \qquad (26.10)$$

Additionally assume that

$$f_\alpha (x_0) = 0, \ \alpha := (\alpha_1, \ldots, \alpha_k), \ \alpha_i \in \mathbb{Z}^+, \ i = 1, \ldots, k;$$

$$|\alpha| := \sum_{i=1}^{k} \alpha_i =: r, \ r = 1, \ldots, n - 1; \ \text{then}$$

$$(2) \ f(z) - f(x_0) = \frac{1}{\Gamma(\nu)} \int_0^1 (1-t)^{\nu-1} \left[J_0^{n-\nu} \right.$$

$$\left\{ \left[\left(\sum_{i=1}^{k} (z_i - x_{0i}) \frac{\partial}{\partial x_i} \right)^n f \right] \right.$$

$$(x_0 + t(z - x_0))\}] \, dt =: R_\nu. \tag{26.11}$$

Remark 26.6. [53] (on Theorem 26.5)
Set

$$G_\nu(t) := J_0^{n-\nu} \left\{ \left[\left(\sum_{i=1}^{k} (z_i - x_{0i}) \frac{\partial}{\partial x_i} \right)^n f \right] \right.$$

$$(x_0 + t(z - x_0))\}, \ t \in [0,1], \tag{26.12}$$

which shows up in R_ν and is continuous; see Proposition 114 of [45] and $R_\nu \in \mathbb{R}$. So we can rewrite

$$R_\nu = \frac{1}{\Gamma(\nu)} \int_0^1 (1-t)^{\nu-1} G_\nu(t) \, dt, \ \nu > 0. \tag{26.13}$$

We mention

Theorem 26.7. [53] *All are as in Theorem 26.5. Let R_ν be the remainder in* (26.10) *(same as in* (26.11)), *and $G_\nu(t)$, $t \in [0,1]$ as in* (26.12), *$\nu \geq 1$. Then*

$$|R_\nu| \leq \min \left\{ \frac{\|G_\nu\|_{L_1([0,1])}}{\Gamma(\nu)}, \frac{\|G_\nu\|_{L_q([0,1])}}{\Gamma(\nu)(p(\nu-1)+1)^{1/p}} \right.$$

$$\left. , \frac{\|G_\nu\|_{L_2([0,1])}}{\Gamma(\nu)\sqrt{2\nu-1}}, \frac{\|G_\nu\|_{\infty,[0,1]}}{\Gamma(\nu+1)} \right\} =: \Lambda_\nu, \tag{26.14}$$

where $p, q > 1 : 1/p + 1/q = 1$.

26.2   Univariate Results

We present here our first Ostrowski-type result.

Theorem 26.8. *Let $\nu \geq 0$, $n = \lceil \nu \rceil$, and $f \in AC^n([a,b])$. Assume that*

$$f^{(k)}(a) = 0, \ k = 1, \ldots, n-1, \ \text{and} \ D_{*a}^\nu f \in L_\infty([a,b]).$$

Then

$$\left| \frac{1}{b-a} \int_a^b f(x)\, dx - f(a) \right| \leq \frac{\|D_{*a}^\nu f\|_{\infty,[a,b]}}{\Gamma(\nu+2)} (b-a)^\nu . \tag{26.15}$$

Proof. By (26.4) we get

$$|f(x) - f(a)| \leq \frac{1}{\Gamma(\nu)} \int_a^x (x-t)^{\nu-1} \left| D_{*a}^\nu f(t) \right| dt$$

$$\leq \frac{1}{\Gamma(\nu)} \left(\int_a^x (x-t)^{\nu-1}\, dt \right) \|D_{*a}^\nu f\|_{\infty,[a,b]}$$

$$= \frac{(x-a)^\nu}{\Gamma(\nu+1)} \|D_{*a}^\nu f\|_{\infty,[a,b]} .$$

That is, we have

$$|f(x) - f(a)| \leq \frac{\|D_{*a}^\nu f\|_{\infty,[a,b]}}{\Gamma(\nu+1)} (x-a)^\nu , \tag{26.16}$$

$\forall x \in [a,b]$.

Therefore we get

$$\left| \frac{1}{b-a} \int_a^b f(x)\, dx - f(a) \right| = \left| \frac{1}{b-a} \int_a^b (f(x) - f(a))\, dx \right|$$

$$\leq \frac{1}{b-a} \int_a^b |f(x) - f(a)|\, dx \overset{(26.16)}{\leq} \frac{\|D_{*a}^\nu f\|_{\infty,[a,b]}}{\Gamma(\nu+1)(b-a)}$$

$$\int_a^b (x-a)^\nu\, dx = \frac{\|D_{*a}^\nu f\|_{\infty,[a,b]}}{\Gamma(\nu+2)} (b-a)^\nu .$$

This proves (26.15). □

We continue with

Theorem 26.9. *Let $\nu \geq 1$, $n = \lceil \nu \rceil$, and $f \in AC^n([a,b])$. Assume that*

$$f^{(k)}(a) = 0, \quad k = 1, \ldots, n-1, \text{ and } D_{*a}^\nu f \in L_1([a,b]).$$

Then

$$\left| \frac{1}{b-a} \int_a^b f(x)\, dx - f(a) \right| \leq \frac{\|D_{*a}^\nu f\|_{L_1([a,b])}}{\Gamma(\nu+1)} (b-a)^{\nu-1}. \tag{26.17}$$

Proof. Again by (26.4) we get

$$|f(x) - f(a)| \leq \frac{1}{\Gamma(\nu)} \int_a^x (x-t)^{\nu-1} \left| D_{*a}^\nu f(t) \right| dt$$

$$\leq \frac{1}{\Gamma(\nu)} (x-a)^{\nu-1} \int_a^x \left| D_{*a}^\nu f(t) \right| dt \leq$$

$$\frac{1}{\Gamma(\nu)} (x-a)^{\nu-1} \left\| D_{*a}^\nu f \right\|_{L_1([a,b])}.$$

That is, we have

$$\left| f(x) - f(a) \right| \leq \frac{\left\| D_{*a}^\nu f \right\|_{L_1([a,b])}}{\Gamma(\nu)} (x-a)^{\nu-1}, \tag{26.18}$$

$\forall x \in [a, b]$.

Therefore we get

$$\left| \frac{1}{b-a} \int_a^b f(x)\, dx - f(a) \right| \leq \frac{1}{b-a} \int_a^b \left| f(x) - f(a) \right| dx \overset{(26.18)}{\leq}$$

$$\frac{\left\| D_{*a}^\nu f \right\|_{L_1([a,b])}}{\Gamma(\nu)(b-a)} \int_a^b (x-a)^{\nu-1}\, dx =$$

$$\frac{\left\| D_{*a}^\nu f \right\|_{L_1([a,b])}}{\Gamma(\nu+1)} (b-a)^{\nu-1}.$$

This proves (26.17). $\square$

We also give

Theorem 26.10. Let $p, q > 1 : 1/p + 1/q = 1$, and $\nu > 1 - 1/p$, $n = \lceil \nu \rceil$, and $f \in AC^n([a,b])$. Assume that

$$f^{(k)}(a) = 0, \ k = 1, \ldots, n-1, \ \text{and} \ D_{*a}^\nu f \in L_q([a,b]).$$

Then

$$\left| \frac{1}{b-a} \int_a^b f(x)\, dx - f(a) \right| \leq$$

$$\frac{\left\| D_{*a}^\nu f \right\|_{L_q([a,b])}}{\Gamma(\nu)(p(\nu-1)+1)^{1/p}(\nu+1/p)} (b-a)^{\nu-1+1/p}. \tag{26.19}$$

Proof. Again by (26.4) we obtain

$$\left| f(x) - f(a) \right| \leq \frac{1}{\Gamma(\nu)} \int_a^x (x-t)^{\nu-1} \left| D_{*a}^\nu f(t) \right| dt$$

$$\leq \frac{1}{\Gamma(\nu)} \left(\int_a^x (x-t)^{p(\nu-1)}\, dt \right)^{1/p} \left(\int_a^x \left| D_{*a}^\nu f(t) \right|^q dt \right)^{1/q} \leq$$

$$\frac{1}{\Gamma(\nu)} \frac{(x-a)^{\nu-1+1/p}}{(p(\nu-1)+1)^{1/p}} \left\| D_{*a}^\nu f \right\|_{L_q([a,b])}.$$

That is, we have

$$|f(x) - f(a)| \le \frac{\|D_{*a}^{\nu} f\|_{L_q([a,b])}}{\Gamma(\nu)(p(\nu-1)+1)^{1/p}} (x-a)^{\nu-1+1/p}, \qquad (26.20)$$

$\forall x \in [a,b]$.

Consequently we get

$$\left| \frac{1}{b-a} \int_a^b f(x)\, dx - f(a) \right| \le \frac{1}{b-a} \int_a^b |f(x) - f(a)|\, dx \overset{(26.20)}{\le}$$

$$\frac{\|D_{*a}^{\nu} f\|_{L_q([a,b])}}{\Gamma(\nu)(p(\nu-1)+1)^{1/p}(b-a)} \int_a^b (x-a)^{\nu-1+1/p}\, dx$$

$$= \frac{\|D_{*a}^{\nu} f\|_{L_q([a,b])}}{\Gamma(\nu)(p(\nu-1)+1)^{1/p}(\nu+1/p)} (b-a)^{\nu-1+1/p}.$$

This proves (26.19). □

Corollary 26.11. (To Theorem 26.10; $p = q = 2$ case.) *Let* $\nu > 1/2$, $n = \lceil \nu \rceil$, *and* $f \in AC^n([a,b])$. *Assume that*

$$f^{(k)}(a) = 0, \ k = 1, \ldots, n-1, \ \text{and } D_{*a}^{\nu} f \in L_2([a,b]).$$

Then

$$\left| \frac{1}{b-a} \int_a^b f(x)\, dx - f(a) \right| \le$$

$$\frac{\|D_{*a}^{\nu} f\|_{L_2([a,b])}}{\Gamma(\nu)(\sqrt{2\nu-1})(\nu+\frac{1}{2})} (b-a)^{\nu-1/2}. \qquad (26.21)$$

Proof. Apply (26.19). □

We finish this section with

Proposition 26.12. *Inequality (26.15) is sharp; namely it is attained by*

$$f(x) = (x-a)^{\nu}, \ \nu > 0, \ \nu \notin \mathbb{N}, \ x \in [a,b].$$

Proof. Here the function

$$f(x) = (x-a)^{\nu}, \ \nu > 0, \ \nu \notin \mathbb{N}, \ x \in [a,b],$$

belongs to $AC^n([a,b])$, $n = \lceil \nu \rceil$. We observe that

$$f'(x) = \nu(x-a)^{\nu-1}, \ f''(x) = \nu(\nu-1)(x-a)^{\nu-2}, \ldots,$$

$$f^{(n-1)}(x) = \nu(\nu-1)(\nu-2)\ldots(\nu-n+2)(x-a)^{\nu-n+1}$$

$$f^{(n)}(x) = \nu(\nu-1)\ldots(\nu-n+1)(x-a)^{\nu-n}.$$

Thus

$$D_{*a}^{\nu}f(x) \overset{(26.2)}{=} \frac{1}{\Gamma(n-\nu)}\int_a^x (x-t)^{(n-\nu)-1}$$

$$\nu(\nu-1)\ldots(\nu-n+1)(t-a)^{\nu-n}\,dt$$

$$= \frac{\nu(\nu-1)\ldots(\nu-n+1)}{\Gamma(n-\nu)}\int_a^x (x-t)^{(n-\nu)-1}(t-a)^{(\nu-n+1)-1}\,dt$$

(by Whittaker and Watson [404, p. 256]; notice that $n-\nu$, $\nu-n+1>0$)

$$= \frac{\nu(\nu-1)\ldots(\nu-n+1)}{\Gamma(n-\nu)}\frac{\Gamma(n-\nu)\Gamma(\nu-n+1)}{\Gamma(1)}$$

$$= \nu(\nu-1)\ldots(\nu-n+1)\Gamma(\nu-n+1) = \Gamma(\nu+1).$$

That is,

$$D_{*a}^{\nu}f(x) = \Gamma(\nu+1),\ \forall x\in[a,b]. \tag{26.22}$$

Also we see that

$$f^{(k)}(a) = 0,\ k = 0,1,\ldots,n-1,\ \text{and}\ D_{*a}^{\nu}f \in L_{\infty}([a,b]).$$

So f fulfills all the assumptions of Theorem 26.8.
  Next we find

$$\text{R.H.S. } (26.15) = \frac{\Gamma(\nu+1)}{\Gamma(\nu+2)}(b-a)^{\nu} = \frac{(b-a)^{\nu}}{(\nu+1)}. \tag{26.23}$$

Furthermore we have

$$\text{L.H.S. } (26.15) = \left|\frac{1}{b-a}\int_a^b (x-a)^{\nu}\,dx\right|$$

$$= \frac{1}{(b-a)}\frac{(b-a)^{\nu+1}}{(\nu+1)} = \frac{(b-a)^{\nu}}{(\nu+1)}. \tag{26.24}$$

Clearly R.H.S. (26.15) = L.H.S. (26.15), proving the claim. $\square$

26.3   Multivariate Results

We present our first multivariate Ostrowski-type fractional inequality in the Caputo sense.

Theorem 26.13. *Let $\nu > 0$, $n = \lceil \nu \rceil$, $f \in C^n(Q)$, where Q is a nonempty compact and convex subset of $\mathbb{R}^k$, $k \geq 2$; $z := (z_1, \ldots, z_k)$, $x_0 := (x_{01}, \ldots, x_{0k}) \in Q$. Further assume that*

$$f_\alpha(x_0) = 0, \ \alpha := (\alpha_1, \ldots, \alpha_k), \ \alpha_i \in \mathbb{Z}^+, \ i = 1, \ldots, k;$$

$$|\alpha| := \sum_{i=1}^k \alpha_i =: r, \ r = 1, \ldots, n-1,$$

where $x_0 \in Q$ is fixed.
Then

$$\left| \frac{\int_Q f(z)\, dz}{Vol(Q)} - f(x_0) \right| \leq$$

$$\left(\frac{\max_{|\alpha|=n} \left\| J_0^{n-\nu} f_\alpha(x_0 + t(z - x_0)) \right\|_{\infty, (t,z) \in [0,1] \times Q}}{\Gamma(\nu+1)\, Vol(Q)} \right) \int_Q \|z - x_0\|_{l_1}^n \, dz. \quad (26.25)$$

Proof. Notice that (see (26.12))

$$|G_\nu(t)| = \left| \left[\left(\sum_{i=1}^k (z_i - x_{0i}) J_0^{n-\nu} \frac{\partial}{\partial x_i} \right)^n f \right] \right.$$

$$(x_0 + t(z - x_0))| \leq \left(\|z - x_0\|_{l_1} \right)^n \quad (26.26)$$

$$\max_{|\alpha|=n} \left\| J_0^{n-\nu} f_\alpha(x_0 + t(z - x_0)) \right\|_{\infty, (t,z) \in [0,1] \times Q},$$

$$\forall z \in Q, \ x_0 \in Q \text{ being fixed}, \ 0 \leq t \leq 1.$$

Hence

$$\|G_\nu\|_{\infty, t \in [0,1]} \leq \left(\|z - x_0\|_{l_1} \right)^n$$

$$\max_{|\alpha|=n} \left\| J_0^{n-\nu} f_\alpha(x_0 + t(z - x_0)) \right\|_{\infty, (t,z) \in [0,1] \times Q}, \quad (26.27)$$

and by (26.11) and (26.14) we get

$$|f(z) - f(x_0)| \leq \Lambda_\nu \leq \frac{\|G_\nu\|_{\infty, t \in [0,1]}}{\Gamma(\nu+1)} \overset{(26.27)}{\leq} \frac{\left(\|z - x_0\|_{l_1} \right)^n}{\Gamma(\nu+1)}$$

$$\max_{|\alpha|=n} \left\| J_0^{n-\nu} f_\alpha(x_0 + t(z - x_0)) \right\|_{\infty, (t,z) \in [0,1] \times Q}, \quad (26.28)$$

$\forall z \in Q$.
Consequently we obtain

$$\left| \frac{\int_Q f(z)\, dz}{Vol(Q)} - f(x_0) \right| = \left| \frac{\int_Q (f(z) - f(x_0))\, dz}{Vol(Q)} \right| \leq$$

$$\frac{1}{Vol\left(Q\right)}\int_Q \left|f\left(z\right)-f\left(x_0\right)\right|dz \le$$

$$\left(\frac{\max\limits_{|\alpha|=n}\left\|J_0^{n-\nu}f_\alpha\left(x_0+t\left(z-x_0\right)\right)\right\|_{\infty,(t,z)\in[0,1]\times Q}}{\Gamma\left(\nu+1\right)Vol\left(Q\right)}\right)\int_Q \left\|z-x_0\right\|_{l_1}^n dz, \quad (26.29)$$

proving the claim. $\square$

We further have

Theorem 26.14. *Let* $f:[a,b]\times[c,d]\to\mathbb{R}$ *be Lebesgue integrable, also* $f\left(a,\cdot\right)$ *integrable on* $[c,d]$ *and* $f\left(\cdot,y\right)\in AC^n\left([a,b]\right)$, $\forall y\in[c,d]$, *where* $n=\lceil\nu\rceil$, $\nu\ge 0$, *and* $\partial^k/\partial x^k f\left(a,y\right)=0$, $k=1,\ldots,n-1$, $\forall y\in[c,d]$. *Here* $\left\|\cdot\right\|_{\infty,[a,b]\times[c,d]}$ *is the supremum norm.*

Further suppose that $\partial_{*a}^\nu f\left(x,y\right)/\partial x^\nu\in B\left([a,b]\times[c,d]\right)$ *bounded functions,* $(x,y)\in[a,b]\times[c,d]$.

Then

$$\left|\frac{1}{\left(b-a\right)\left(d-c\right)}\int_{[a,b]\times[c,d]}f\left(x,y\right)dxdy-f\left(a,c\right)\right|\le$$

$$\frac{1}{\left(d-c\right)}\int_{[c,d]}\left|f\left(a,y\right)-f\left(a,c\right)\right|dy+\frac{\left(b-a\right)^\nu}{\Gamma\left(\nu+2\right)}\left\|\frac{\partial_{*a}^\nu f}{\partial x^\nu}\right\|_{\infty,[a,b]\times[c,d]}. \quad (26.30)$$

Proof. Because $f\left(\cdot,y\right)\in AC^n\left([a,b]\right)$, $\forall y\in[c,d]$, and $\partial^k/\partial x^k f\left(a,y\right)=0$, $k=1,\ldots,n-1$, $\forall y\in[c,d]$, by (26.4) we have

$$f\left(x,y\right)-f\left(a,y\right)=\frac{1}{\Gamma\left(\nu\right)}\int_a^x\left(x-t\right)^{\nu-1}\frac{\partial_{*a}^\nu f}{\partial x^\nu}\left(t,y\right)dt, \quad (26.31)$$

$\forall x\in[a,b]$, $\forall y\in[c,d]$.

Because

$$\frac{\partial_{*a}^\nu f}{\partial x^\nu}\left(x,y\right)\in B\left([a,b]\times[c,d]\right),$$

we obtain

$$\left|f\left(x,y\right)-f\left(a,y\right)\right|\le\frac{\left(x-a\right)^\nu}{\Gamma\left(\nu+1\right)}\left\|\frac{\partial_{*a}^\nu f}{\partial x^\nu}\right\|_{\infty,[a,b]\times[c,d]}, \quad (26.32)$$

$\forall x\in[a,b]$, $\forall y\in[c,d]$.

But it holds

$$\left|f\left(x,y\right)-f\left(a,c\right)\right|\le\left|f\left(x,y\right)-f\left(a,y\right)\right|$$

$$+\left|f\left(a,y\right)-f\left(a,c\right)\right|\le\left|f\left(a,y\right)-f\left(a,c\right)\right|+$$

$$\frac{\left(x-a\right)^\nu}{\Gamma\left(\nu+1\right)}\left\|\frac{\partial_{*a}^\nu f}{\partial x^\nu}\right\|_{\infty,[a,b]\times[c,d]}, \quad (26.33)$$

$\forall x\in[a,b]$, $\forall y\in[c,d]$.

Consequently we have

$$\left| \frac{1}{(b-a)(d-c)} \int_{[a,b]\times[c,d]} f(x,y)\,dxdy - f(a,c) \right| =$$

$$\frac{1}{(b-a)(d-c)} \left| \int_{[a,b]\times[c,d]} (f(x,y) - f(a,c))\,dxdy \right| \leq \qquad (26.34)$$

$$\frac{1}{(b-a)(d-c)} \int_{[a,b]\times[c,d]} |f(x,y) - f(a,c)|\,dxdy \quad \overset{(26.33)}{\leq}$$

$$\frac{1}{(b-a)(d-c)} \int_{[a,b]\times[c,d]} |f(a,y) - f(a,c)|\,dxdy + \qquad (26.35)$$

$$\frac{1}{(b-a)(d-c)} \frac{1}{\Gamma(\nu+1)} \left\| \frac{\partial_{*a}^{\nu} f}{\partial x^{\nu}} \right\|_{\infty,[a,b]\times[c,d]}$$

$$\int_{[a,b]\times[c,d]} (x-a)^{\nu}\,dxdy = \frac{1}{(d-c)} \int_{[c,d]} |f(a,y) - f(a,c)|\,dy$$

$$+ \frac{(b-a)^{\nu}}{\Gamma(\nu+2)} \left\| \frac{\partial_{*a}^{\nu} f}{\partial x^{\nu}} \right\|_{\infty,[a,b]\times[c,d]}, \qquad (26.36)$$

proving the claim. □

Similar to (26.30) one can prove inequalities in more than two variables.

Next we study fractional Ostrowski-type inequalities over balls and spherical shells. For that we need to make

Remark 26.15. We define the ball

$$B(0,R) := \left\{ x \in \mathbb{R}^N : |x| < R \right\} \subseteq \mathbb{R}^N, \ N \geq 2, \ R > 0,$$

and the sphere

$$S^{N-1} := \left\{ x \in \mathbb{R}^N : |x| = 1 \right\},$$

where $|\cdot|$ is the Euclidean norm.

Let $d\omega$ be the element of surface measure on S^{N-1} and let

$$\omega_N = \int_{S^{N-1}} d\omega = \frac{2\pi^{N/2}}{\Gamma(N/2)}.$$

For $x \in \mathbb{R}^N - \{0\}$ we can write uniquely $x = r\omega$, where $r = |x| > 0$ and $\omega = x/r \in S^{N-1}$, $|\omega| = 1$. Note that

$$\int_{B(0,R)} dy = \frac{\omega_N R^N}{N}$$

is the Lebesgue measure of the ball. Following [356, pp. 149–150, Exercise 6] and [383, pp. 87–88, Theorem 5.2.2] we can write for $F : \overline{B\,(0,R)} \to \mathbb{R}$ a Lebesgue integrable function that

$$\int_{B(0,R)} F\,(x)\,dx = \int_{S^{N-1}} \left(\int_0^R F\,(rw)\,r^{N-1}dr \right) dw; \qquad (26.37)$$

we use this formula often.

Initially the function $f : \overline{B\,(0,R)} \to \mathbb{R}$ is radial; that is, there exists a function g such that $f\,(x) = g\,(r)$, where $r = |x|$, $r \in [0,R]$, $\forall x \in \overline{B\,(0,R)}$. Here we assume that $g \in AC^n\,([0,R])$, $n = \lceil \nu \rceil$, $\nu \geq 0$, and $g^{(k)}\,(0) = 0$, $k = 1, \ldots, n-1$.

By (26.4) we get

$$g\,(s) - g\,(0) = \frac{1}{\Gamma\,(\nu)} \int_0^s (s-t)^{\nu-1}\,D_{*0}^{\nu}g\,(t)\,dt, \qquad (26.38)$$

$\forall s \in [0,R]$.

Further assume that $D_{*0}^{\nu}g \in L_\infty\,([0,R])$.

By (26.38) we obtain

$$|g\,(s) - g\,(0)| \leq \frac{s^{\nu}}{\Gamma\,(\nu+1)}\,\|D_{*0}^{\nu}g\|_{\infty,[0,R]}\,, \qquad (26.39)$$

$\forall s \in [0,R]$.

Next we observe that

$$\left| f\,(0) - \frac{\int_{B(0,R)} f\,(y)\,dy}{Vol\,(B\,(0,R))} \right| =$$

$$\left| g\,(0) - \frac{\int_{S^{N-1}} \left(\int_0^R g\,(s)\,s^{N-1}ds \right) dw}{\int_{S^{N-1}} \left(\int_0^R s^{N-1}ds \right) dw} \right| = \qquad (26.40)$$

$$\left| g\,(0) - \frac{N}{R^N} \int_0^R g\,(s)\,s^{N-1}ds \right| = \frac{N}{R^N}$$

$$\left| \int_0^R s^{N-1}\,(g\,(0) - g\,(s))\,ds \right| \leq \qquad (26.41)$$

$$\frac{N}{R^N} \int_0^R s^{N-1}\,|g\,(0) - g\,(s)|\,ds \overset{(26.39)}{\leq}$$

$$\frac{N}{R^N}\,\frac{\|D_{*0}^{\nu}g\|_{\infty,[0,R]}}{\Gamma\,(\nu+1)} \int_0^R s^{N-1}s^{\nu}ds = \qquad (26.42)$$

$$\frac{N}{R^N}\,\frac{\|D_{*0}^{\nu}g\|_{\infty,[0,R]}}{\Gamma\,(\nu+1)}\,\frac{R^{\nu+N}}{(\nu+N)} = \frac{\left(\|D_{*0}^{\nu}g\|_{\infty,[0,R]} \right) N R^{\nu}}{\Gamma\,(\nu+1)\,(\nu+N)}. \qquad (26.43)$$

That is, we have proved that

$$\left| f\left(0\right) - \frac{\int_{B(0,R)} f\left(y\right) dy}{Vol\left(B\left(0,R\right)\right)} \right| =$$

$$\left| g\left(0\right) - \frac{N}{R^N} \int_0^R g\left(s\right) s^{N-1} ds \right| \leq \frac{\left(\left\| D_{*0}^\nu g \right\|_{\infty,[0,R]} \right) NR^\nu}{\Gamma\left(\nu+1\right)\left(\nu+N\right)}. \tag{26.44}$$

The last inequality (26.44) is sharp; namely it is attained by $\bar{g}\left(r\right) = r^\nu$, $\nu > 0$, $r \in [0,R]$. Indeed by (26.2) we get

$$D_{*0}^\nu \bar{g}\left(r\right) = \Gamma\left(\nu+1\right),$$

and

$$\left\| D_{*0}^\nu \bar{g} \right\|_{\infty,[0,R]} = \Gamma\left(\nu+1\right).$$

So that in that case

$$R.H.S \; (26.44) = \frac{NR^\nu}{\nu+N}. \tag{26.45}$$

But

$$L.H.S \; (26.44) = \frac{N}{R^N} \int_0^R s^{\nu+N-1} ds$$

$$= \frac{N}{R^N} \frac{R^{\nu+N}}{\left(\nu+N\right)} = \frac{NR^\nu}{\nu+N}, \tag{26.46}$$

proving equality in (26.44).

We have established the following.

Theorem 26.16. *Let $f : \overline{B\left(0,R\right)} \to \mathbb{R}$ that is radial; that is, there exists g such that $f\left(x\right) = g\left(r\right)$, $r = |x|$, $\forall x \in \overline{B\left(0,R\right)}$. Assume that $g \in AC^n\left([0,R]\right)$, $n = \lceil \nu \rceil$, $\nu \geq 0$, and*

$$g^{(k)}\left(0\right) = 0, \; k = 1,\ldots,n-1, \; and \; D_{*0}^\nu g \in L_\infty\left([0,R]\right).$$

Then

$$\left| f\left(0\right) - \frac{\int_{B(0,R)} f\left(y\right) dy}{Vol\left(B\left(0,R\right)\right)} \right| =$$

$$\left| g\left(0\right) - \frac{N}{R^N} \int_0^R g\left(s\right) s^{N-1} ds \right| \leq \frac{\left(\left\| D_{*0}^\nu g \right\|_{\infty,[0,R]} \right) NR^\nu}{\Gamma\left(\nu+1\right)\left(\nu+N\right)}. \tag{26.47}$$

Inequality (26.47) is sharp; it is attained by $g\left(r\right) = r^\nu$, $\nu > 0$.

We make

Remark 26.17. Let the spherical shell $A := B\left(0,R_2\right) - \overline{B\left(0,R_1\right)}$, $0 < R_1 < R_2$, $A \subseteq \mathbb{R}^N$, $N \geq 2$, $x \in \overline{A}$. Consider again that $f : \overline{A} \to \mathbb{R}$ is radial; that is, there exists g such that $f(x) = g(r)$, $r = |x|$, $r \in [R_1, R_2]$, $\forall x \in \overline{A}$. Here again x can be written uniquely as $x = r\omega$, where $r = |x| > 0$, and $\omega = x/r \in S^{N-1}$, $|\omega| = 1$.

Following [356, pp. 149–150, Exercise 6] and [383, pp. 87–88, Theorem 5.2.2] we can write for $F : \overline{A} \to \mathbb{R}$ a Lebesgue integrable function that

$$\int_A F(x)\, dx = \int_{S^{N-1}} \left(\int_{R_1}^{R_2} F(r\omega)\, r^{N-1} dr \right) d\omega. \qquad (26.48)$$

Here

$$Vol(A) = \frac{\omega_N \left(R_2^N - R_1^N \right)}{N},$$

and we assume that $g \in AC^n([R_1, R_2])$, $n = \lceil \nu \rceil$, $\nu \geq 0$, and $g^{(k)}(R_1) = 0$, $k = 1, \ldots, n-1$.

By (26.4) we get

$$g(s) - g(R_1) = \frac{1}{\Gamma(\nu)} \int_{R_1}^s (s-t)^{\nu-1} D_{*R_1}^\nu g(t)\, dt, \qquad (26.49)$$

$\forall s \in [R_1, R_2]$.

Further assume that $D_{*R_1}^\nu g \in L_\infty([R_1, R_2])$. By (26.49) we obtain

$$|g(s) - g(R_1)| \leq \frac{(s-R_1)^\nu}{\Gamma(\nu+1)} \left\| D_{*R_1}^\nu g \right\|_{\infty, [R_1, R_2]}, \qquad (26.50)$$

$\forall s \in [R_1, R_2]$.

Next we observe that

$$\left| f(R_1\omega) - \frac{\int_A f(y)\, dy}{Vol(A)} \right| =$$

$$\left| g(R_1) - \left(\frac{N}{R_2^N - R_1^N} \right) \int_{R_1}^{R_2} g(s)\, s^{N-1} ds \right| = \qquad (26.51)$$

$$\left(\frac{N}{R_2^N - R_1^N} \right) \left| \int_{R_1}^{R_2} (g(R_1) - g(s))\, s^{N-1} ds \right| \leq \qquad (26.52)$$

$$\left(\frac{N}{R_2^N - R_1^N} \right) \int_{R_1}^{R_2} |g(R_1) - g(s)|\, s^{N-1} ds \overset{(26.50)}{\leq}$$

$$\left(\frac{N}{R_2^N - R_1^N} \right) \left(\frac{\left\| D_{*R_1}^\nu g \right\|_{\infty, [R_1, R_2]}}{\Gamma(\nu+1)} \right) \int_{R_1}^{R_2} (s-R_1)^\nu\, s^{N-1} ds \qquad (26.53)$$

(by [43], there (2.41))

$$= \left(\frac{N}{R_2^N - R_1^N} \right) \left(\frac{\left\| D_{*R_1}^\nu g \right\|_{\infty, [R_1, R_2]}}{\Gamma(\nu+1)} \right) (\Gamma(\nu+1)(N-1)!$$

$$\sum_{k=0}^{N-1} \frac{(-1)^k}{(N-k-1)!} R_2^{N-k-1} \frac{(R_2 - R_1)^{k+\nu+1}}{\Gamma(k+\nu+2)} \Bigg) \tag{26.54}$$

$$= \left(\frac{N! \left(\| D_{*R_1}^\nu g \|_{\infty, [R_1, R_2]} \right)}{R_2^N - R_1^N} \right)$$

$$\left(\sum_{k=0}^{N-1} \frac{(-1)^k}{(N-k-1)!} R_2^{N-k-1} \frac{(R_2 - R_1)^{k+\nu+1}}{\Gamma(k+\nu+2)} \right). \tag{26.55}$$

Hence in the radial case we proved

$$\left| f(R_1\omega) - \frac{\int_A f(y)\, dy}{Vol(A)} \right| =$$

$$\left| g(R_1) - \left(\frac{N}{R_2^N - R_1^N} \right) \int_{R_1}^{R_2} g(s)\, s^{N-1} ds \right| \le$$

$$\left(\frac{N! \left(\| D_{*R_1}^\nu g \|_{\infty, [R_1, R_2]} \right)}{R_2^N - R_1^N} \right)$$

$$\left(\sum_{k=0}^{N-1} \frac{(-1)^k}{(N-k-1)!} R_2^{N-k-1} \frac{(R_2 - R_1)^{k+\nu+1}}{\Gamma(k+\nu+2)} \right). \tag{26.56}$$

Inequality (26.56) is attained by

$$g(s) := (s - R_1)^\nu, \ \nu > 0, \ s \in [R_1, R_2].$$

Indeed we observe that

$$\text{L.H.S. } (26.56) = \left(\frac{N}{R_2^N - R_1^N} \right) \int_{R_1}^{R_2} (s - R_1)^\nu s^{N-1} ds =$$

$$\frac{\Gamma(\nu+1) N!}{(R_2^N - R_1^N)} \left(\sum_{k=0}^{N-1} \frac{(-1)^k}{(N-k-1)!} R_2^{N-k-1} \frac{(R_2 - R_1)^{k+\nu+1}}{\Gamma(k+\nu+2)} \right). \tag{26.57}$$

But by (26.22) we get

$$D_{*R_1}^\nu g(s) = \Gamma(\nu+1), \ \forall s \in [R_1, R_2]. \tag{26.58}$$

That is,

$$\| D_{*R_1}^\nu g \|_{\infty, [R_1, R_2]} = \Gamma(\nu+1). \tag{26.59}$$

The last implies that

$$\text{R.H.S. } (26.56) = \left(\frac{\Gamma(\nu+1) N!}{R_2^N - R_1^N} \right)$$

$$\left(\sum_{k=0}^{N-1} \frac{(-1)^k}{(N-k-1)!} R_2^{N-k-1} \frac{(R_2 - R_1)^{k+\nu+1}}{\Gamma(k+\nu+2)} \right). \tag{26.60}$$

So by (26.57) and (26.60) we get

$$\text{L.H.S. } (26.56) = \text{R.H.S. } (26.56), \tag{26.61}$$

proving sharpness of (26.56).

So putting the above together we derive

Theorem 26.18. *Let* $f : \overline{A} \to \mathbb{R}$ *be radial; that is, there exists* g *such that* $f(x) = g(r)$, $r = |x|$, $\forall x \in \overline{A}$; $\omega \in S^{N-1}$. *Assume that* $g \in AC^n([R_1, R_2])$, $n = \lceil \nu \rceil$, $\nu \geq 0$, *and*

$$g^{(k)}(R_1) = 0, \ k = 1, \ldots, n-1, \ \text{and} \ D_{*R_1}^{\nu} g \in L_{\infty}([R_1, R_2]).$$

Then

$$\left| f(R_1\omega) - \frac{\int_A f(y)\, dy}{Vol(A)} \right| =$$

$$\left| g(R_1) - \left(\frac{N}{R_2^N - R_1^N} \right) \int_{R_1}^{R_2} g(s)\, s^{N-1} ds \right| \leq$$

$$\left(\frac{N! \, \|D_{*R_1}^{\nu} g\|_{\infty,[R_1,R_2]}}{R_2^N - R_1^N} \right)$$

$$\left(\sum_{k=0}^{N-1} \frac{(-1)^k}{(N-k-1)!} R_2^{N-k-1} \frac{(R_2 - R_1)^{k+\nu+1}}{\Gamma(k+\nu+2)} \right). \tag{26.62}$$

Inequality (26.62) is sharp; namely it is attained by

$$g(s) := (s - R_1)^{\nu}, \ \nu > 0, \ s \in [R_1, R_2]. \tag{26.63}$$

We need

Definition 26.19. (See [58].) *Let* $F : \overline{A} \to \mathbb{R}$, $\nu \geq 0$, $n := \lceil \nu \rceil$ *such that* $F(\cdot\omega) \in AC^n([R_1, R_2])$, *for all* $\omega \in S^{N-1}$. *We call the Caputo radial fractional derivative the following function*

$$\frac{\partial_{*R_1}^{\nu} F(x)}{\partial r^{\nu}} := \frac{1}{\Gamma(n-\nu)} \int_{R_1}^{r} (r-t)^{n-\nu-1} \frac{\partial^n F(t\omega)}{\partial r^n}\, dt, \tag{26.64}$$

where $x \in \overline{A}$; *that is,* $x = r\omega$, $r \in [R_1, R_2]$, $\omega \in S^{N-1}$.
Clearly

$$\frac{\partial_{*R_1}^{0} F(x)}{\partial r^{0}} = F(x),$$

$$\frac{\partial_{*R_1}^\nu F(x)}{\partial r^\nu} = \frac{\partial^\nu F(x)}{\partial r^\nu}, \text{ if } \nu \in \mathbb{N}. \tag{26.65}$$

The above function (26.64) exists almost everywhere for $x \in \overline{A}$ (see [58]).

We continue Remark 26.17; we make

Remark 26.20. We treat here the general, not necessarily radial, case of f. We apply (26.62) to $f(r\omega)$, ω fixed, $r \in [R_1, R_2]$, under the following assumptions: $f(\cdot\omega) \in AC^n([R_1, R_2])$, for all $\omega \in S^{N-1}$, $\nu \geq 0$, $n := \lceil \nu \rceil$, where $f : \overline{A} \to \mathbb{R}$ is Lebesgue integrable;

$$\frac{\partial^k f}{\partial r^k}, \ k = 1, \ldots, n-1 \text{ vanish on } \partial B(0, R_1);$$

and

$$\frac{\partial_{*R_1}^\nu f}{\partial r^\nu} \in B(\overline{A}), \text{ along with}$$

$$D_{*R_1}^\nu f(\cdot\omega) \in L_\infty([R_1, R_2]), \ \forall \omega \in S^{N-1}.$$

So we have

$$\left| f(R_1\omega) - \left(\frac{N}{R_2^N - R_1^N}\right) \int_{R_1}^{R_2} f(s\omega) s^{N-1} ds \right| \leq$$

$$\left(\frac{N! \left\|\frac{\partial_{*R_1}^\nu f}{\partial r^\nu}\right\|_{\infty, \overline{A}}}{R_2^N - R_1^N}\right) \left(\sum_{k=0}^{N-1} \frac{(-1)^k}{(N-k-1)!} \frac{R_2^{N-k-1}(R_2-R_1)^{k+\nu+1}}{\Gamma(k+\nu+2)}\right)$$

$$=: \lambda_1. \tag{26.66}$$

Consequently it holds

$$\left| \frac{\int_{S^{N-1}} f(R_1\omega) d\omega}{\omega_N} - \frac{N}{(R_2^N - R_1^N)\omega_N} \int_{S^{N-1}} \right.$$

$$\left. \left(\int_{R_1}^{R_2} f(s\omega) s^{N-1} ds\right) d\omega \right| \leq \lambda_1. \tag{26.67}$$

That is,

$$\left| \frac{\Gamma(N/2)}{2\pi^{N/2}} \int_{S^{N-1}} f(R_1\omega) d\omega - \frac{\int_A f(x) dx}{Vol(A)} \right| \leq \lambda_1. \tag{26.68}$$

Therefore it holds for $x \in \overline{A}$ that

$$\left| f(x) - \frac{\int_A f(x) dx}{Vol(A)} \right| \leq \left| f(x) - \frac{\Gamma(N/2)}{2\pi^{N/2}} \int_{S^{N-1}} f(R_1\omega) d\omega \right| + \lambda_1. \tag{26.69}$$

We have established the next result.

Theorem 26.21. *Let* $f : \overline{A} \to \mathbb{R}$ *be Lebesgue integrable with* $f(\cdot\omega) \in AC^n([R_1, R_2])$, $\nu \geq 0$, $n := \lceil \nu \rceil$, $\forall \omega \in S^{N-1}$; $\partial^k f / \partial r^k$, $k = 1, \ldots, n-1$ *vanish on* $\partial B(0, R_1)$; $D^{\nu}_{*R_1} f(\cdot\omega) \in L_{\infty}([R_1, R_2])$, $\forall \omega \in S^{N-1}$; *and* $\partial^{\nu}_{*R_1} f / \partial r^{\nu} \in B(\overline{A})$ (*bounded functions on* $\overline{A}$). *Then for* $x \in \overline{A}$ *we have*

$$\left| f(x) - \frac{\int_A f(x)\, dx}{Vol(A)} \right| \leq \left| f(x) - \frac{\Gamma(N/2)}{2\pi^{N/2}} \int_{S^{N-1}} f(R_1\omega)\, d\omega \right|$$

$$+ \left(\frac{N! \left\| \frac{\partial^{\nu}_{*R_1} f}{\partial r^{\nu}} \right\|_{\infty, \overline{A}}}{R_2^N - R_1^N} \right) \left(\sum_{k=0}^{N-1} \frac{(-1)^k}{(N-k-1)!} \frac{R_2^{N-k-1} (R_2 - R_1)^{k+\nu+1}}{\Gamma(k+\nu+2)} \right).$$

$$(26.70)$$

We make

Remark 26.22. Let $f : \overline{B(0, R)} \to \mathbb{R}$ be a Lebesgue integrable function, that is not necessarily a radial function. Assume $f(\cdot\omega) \in AC^1([0, R])$, $\forall \omega \in S^{N-1}$; $0 \leq \nu < 1$, and $D^{\nu}_{*0} f(\cdot\omega) \in L_{\infty}([0, R])$, $\forall \omega \in S^{N-1}$.

Clearly here by (26.3) we obtain

$$f(s\omega) - f(0) = \frac{1}{\Gamma(\nu)} \int_0^s (s-t)^{\nu-1} D^{\nu}_{*0} f(t\omega)\, dt, \qquad (26.71)$$

$\forall \omega \in S^{N-1}$, $\forall s \in [0, R]$.

We further assume that

$$\left\| D^{\nu}_{*0} f(t\omega) \right\|_{\infty, (t \in [0, R])} \leq K, \quad \forall \omega \in S^{N-1}, \qquad (26.72)$$

where $K > 0$.

So we apply (26.47) for $f(\cdot\omega)$, $\forall \omega \in S^{N-1}$.

We obtain

$$\left| f(0) - \frac{N}{R^N} \int_0^R f(s\omega)\, s^{N-1} ds \right| \leq$$

$$\frac{\left(\left\| D^{\nu}_{*0} f(t\omega) \right\|_{\infty, (t \in [0, R])} \right) N R^{\nu}}{\Gamma(\nu+1)(\nu+N)} \overset{(26.72)}{\leq}$$

$$\frac{K N R^{\nu}}{\Gamma(\nu+1)(\nu+N)}. \qquad (26.73)$$

Consequently we find

$$\left| f(0) - \frac{N}{\omega_N R^N} \int_{S^{N-1}} \left(\int_0^R f(s\omega)\, s^{N-1} ds \right) d\omega \right|$$

$$\leq \frac{K N R^{\nu}}{\Gamma(\nu+1)(\nu+N)}. \qquad (26.74)$$

That proves

$$\left| f\left(0\right) - \frac{\int_{B(0,R)} f\left(x\right) dx}{Vol\left(B\left(0,R\right)\right)} \right| \leq \frac{KNR^{\nu}}{\Gamma\left(\nu+1\right)\left(\nu+N\right)}. \tag{26.75}$$

We have established our last result.

Theorem 26.23. *Let* $f : \overline{B\left(0,R\right)} \to \mathbb{R}$ *be a Lebesgue integrable function, not necessarily radial. Assume* $f\left(\cdot\omega\right) \in AC^{1}\left(\left[0,R\right]\right)$, $R > 0$, $\forall\omega \in S^{N-1}$; $0 \leq \nu < 1$, *and* $D_{*0}^{\nu}f\left(\cdot\omega\right) \in L_{\infty}\left(\left[0,R\right]\right)$, $\forall\omega \in S^{N-1}$.
Suppose also that

$$\left\| D_{*0}^{\nu}f\left(t\omega\right) \right\|_{\infty,\left(t\in[0,R]\right)} \leq K, \tag{26.76}$$

$\forall\omega \in S^{N-1}$, *where* $K > 0$.
Then

$$\left| f\left(0\right) - \frac{\int_{B(0,R)} f\left(x\right) dx}{Vol\left(B\left(0,R\right)\right)} \right| \leq \frac{KNR^{\nu}}{\Gamma\left(\nu+1\right)\left(\nu+N\right)}. \tag{26.77}$$

27
Appendix

27.1 Conversion Formulae for Different Kinds of Fractional Derivatives

Without loss of generality we work on $[0,1]$ and the anchor point is zero.

Let $\nu > 0$, $n \notin \mathbb{N}$, $n = [\nu]$. We denote by D_{R-L}^{ν} the Riemann–Liouville fractional derivative of order $\nu > 0$, by D_c^{ν} the Canavati fractional derivative of order $\nu > 0$, and by $^c\partial^{\nu}/\partial x^{\nu}$ the Caputo fractional derivative of order $\nu > 0$.

Consider
$$
C_0^{n+1}\left([0,1]\right) = \left\{ f \in C^{n+1}\left([0,1]\right) : \right.
$$
$$
\left. f\left(0\right) = f'\left(0\right) = \cdots f^{(n)}\left(0\right) = 0 \right\}.
$$

By Lemma 6, p. 64 of [101], we obtain that if $f \in C_0^{n+1}\left([0,1]\right)$, then $D_c^{\nu} f\left(x\right)$ exists and is continuous for any $x \in [0,1]$. We also have that $C^{n+1}\left([0,1]\right) \subset AC^{n+1}\left([0,1]\right)$. By Lemma 3.7, p. 41 of [134], for $f \in C^{n+1}\left([0,1]\right)$ we have that $^c\partial^{\nu} f\left(x\right)/\partial x^{\nu}$ exists and is continuous for any $x \in [0,1]$, and $^c\partial^{\nu} f\left(0\right)/\partial x^{\nu} = 0$.

By our Corollary 16.9 for the last case, we get equivalently that $D_{R-L}^{\nu} f\left(x\right)$ exists for any $x \in [0,1]$. So for $f \in C_0^{n+1}\left([0,1]\right)$ all three derivatives exist.

When $f \in C_0^{n+1}\left([0,1]\right)$, by Lemma 3.3, p. 39 of [134] we get that

$$
D_{R-L}^{\nu} f\left(x\right) = \frac{^c\partial^{\nu} f\left(x\right)}{\partial x^{\nu}},
$$

for any $x \in [0,1]$.

So we conclude that studying conversion formulae among the above three types of fractional derivatives makes perfect sense.

G.A. Anastassiou, *Fractional Differentiation Inequalities*,   635
DOI 10.1007/978-0-387-98128-4_27, © Springer Science+Business Media, LLC 2009

We mention again that

$$D_{R-L}^{\nu} f(x) = \frac{1}{\Gamma(n+1-\nu)} \left(\frac{d}{dx}\right)^{n+1} \int_0^x (x-t)^{n-\nu} f(t)\, dt, \qquad (27.1)$$

$$D_c^{\nu} f(x) = \frac{1}{\Gamma(n+1-\nu)} \frac{d}{dx} \int_0^x (x-t)^{n-\nu} f^{(n)}(t)\, dt, \qquad (27.2)$$

and

$$\frac{{}^c\partial^{\nu} f(x)}{\partial x^{\nu}} = \frac{1}{\Gamma(n+1-\nu)} \int_0^x (x-t)^{n-\nu} f^{(n+1)}(t)\, dt. \qquad (27.3)$$

Let us assume first that $f \in C_0^{n+1}([0,1])$. We set

$$y(x) = \int_0^x \frac{(x-t)^{n-1}}{(n-1)!} f(t)\, dt = \int_0^x \int_0^{x_1} \cdots \int_0^{x_{n-1}} f(x_n)\, dx_n \ldots dx_1, \qquad (27.4)$$

for any $x \in [0,1]$. Thus $y^{(n)}(x) = f(x)$, for all $x \in [0,1]$, and $y^{(i)}(0) = 0$, $i = 0, 1, \ldots, n$. Clearly $y \in C_0^{n+1}([0,1])$, so that $D_c^{\nu} y$ exists and is continuous.
We observe that

$$D_{R-L}^{\nu} f(x) = \frac{1}{\Gamma(n+1-\nu)} \left(\frac{d}{dx}\right)^n \frac{d}{dx} \int_0^x (x-t)^{n-\nu} f(t)\, dt$$

$$= \frac{1}{\Gamma(n+1-\nu)} \left(\frac{d}{dx}\right)^n \frac{d}{dx} \int_0^x (x-t)^{n-\nu} y^{(n)}(t)\, dt$$

$$= \left(\frac{d}{dx}\right)^n D_c^{\nu} y(x). \qquad (27.5)$$

So we have proved

Proposition 27.1. For $f \in C_0^{n+1}([0,1])$ it holds that

$$D_{R-L}^{\nu} f(x) = \left(\frac{d}{dx}\right)^n D_c^{\nu} y(x), \qquad (27.6)$$

for any $x \in [0,1]$, where y is as in (27.4).

Next assume that $f \in C_0^{n+2}([0,1])$. We observe that the following function exists and is continuous,

$$D_c^{\nu} f'(x) = \frac{1}{\Gamma(n+1-\nu)} \frac{d}{dx} \int_0^x (x-t)^{n-\nu} f^{(n+1)}(t)\, dt$$

$$= \frac{d}{dx} \left(\frac{{}^c\partial^{\nu} f(x)}{\partial x^{\nu}}\right). \qquad (27.7)$$

So we have

Proposition 27.2. *For* $f \in C_0^{n+2}([0,1])$ *it holds*

$$D_c^\nu f'(x) = \left(\frac{{}^c\partial^\nu f(x)}{\partial x^\nu} \right)', \tag{27.8}$$

for any $x \in [0,1]$, *and* ${}^c\partial^\nu f(0)/\partial x^\nu = 0$.

From the last we get

$$\frac{{}^c\partial^\nu f(x)}{\partial x^\nu} = \int_0^x D_c^\nu f'(t)\, dt. \tag{27.9}$$

Call $F = f'$; that is, $f(x) = \int_0^x F(t)\, dt$. Thus by (27.9) we obtain

$$\frac{{}^c\partial^\nu \left(\int_0^x F(t)\, dt \right)}{\partial x^\nu}(x) = \int_0^x (D_c^\nu F)(t)\, dt, \tag{27.10}$$

for any $x \in [0,1]$.

Again by (27.8) we derive

$$D_c^\nu F(x) = \left(\frac{{}^c\partial^\nu \left(\int_0^x F(t)\, dt \right)}{\partial x^\nu}(x) \right)', \tag{27.11}$$

for any $x \in [0,1]$; here $F \in C_0^{n+1}([0,1])$.

Next we assume that $f \in C^{n+1}([0,1])$.

We put

$$g(x) = \int_0^x \frac{(x-t)^n}{n!} f(t)\, dt; \tag{27.12}$$

that is, $g^{(n+1)}(x) = f(x)$, for any $x \in [0,1]$, and $g \in C^{n+1}([0,1])$. Therefore there exists ${}^c\partial^\nu g/\partial x^\nu$ and it is continuous on $[0,1]$.

We observe that

$$D_{R-L}^\nu f(x) = \frac{1}{\Gamma(n+1-\nu)} \left(\frac{d}{dx} \right)^{n+1} \int_0^x (x-t)^{n-\nu} g^{(n+1)}(t)\, dt$$

$$= \left(\frac{d}{dx} \right)^{n+1} \left(\frac{{}^c\partial^\nu g(x)}{\partial x^\nu} \right). \tag{27.13}$$

We have established

Proposition 27.3. *Let* $f \in C^{n+1}([0,1])$. *Then*

$$D_{R-L}^\nu f(x) = \left(\frac{d}{dx} \right)^{n+1} \left(\frac{{}^c\partial^\nu g(x)}{\partial x^\nu} \right), \tag{27.14}$$

for any $x \in [0,1]$, *where* g *is as in* (27.12).

Conclusion 27.4. Any two kinds of the above main fractional derivatives of the literature are connected indirectly via the above established formulae (27.6), (27.8), and (27.14); see also our Lemma 16.10, which is Lemma 3.2, p. 39 of [134]. Consequently treating all, but each one separately, regarding establishing fractional differentiation inequalities in this monograph, it makes perfect sense and is well justified by their many applications, for example, in differential equations, and their importance in mathematics.

27.2   Some Basic Fractional Derivatives

Here we mention the formulae of some basic functions' fractional derivatives. The anchor points are again zero.

Regarding the Canavati fractional derivative from [101] we get

$$D_c^\nu x^\mu = \frac{\Gamma(\mu+1)}{\Gamma(\mu-\nu+1)} x^{\mu-\nu}, \ 0 < \nu \le \mu. \tag{27.15}$$

Let $\mu = m$ an integer and $\nu = 1/2$; we have

$$D_c^{1/2} x^m = \frac{m!}{\Gamma(m+(1/2))} x^{m-(1/2)}$$

$$= \frac{m!}{\left(m - \left(\frac{1}{2}\right)\right)\left(m - \left(\frac{3}{2}\right)\right)\ldots\left(\frac{1}{2}\right)\sqrt{\pi}} x^{m-(1/2)}. \tag{27.16}$$

Regarding the Riemann–Liouville fractional derivative, from [333, pp. 309–310], with function variable $t > 0$, order $\alpha \in \mathbb{R}$, we get

$$D_{R-L}^\alpha (H(t)) = \frac{t^{-\alpha}}{\Gamma(1-\alpha)}, \tag{27.17}$$

where H is the Heavyside function,

$$D_{R-L}^\alpha (H(t-a)) = \begin{cases} \frac{(t-a)^{-\alpha}}{\Gamma(1-\alpha)}, & t > a \\ 0, & 0 \le t \le a, \end{cases} \tag{27.18}$$

and

$$D_{R-L}^\alpha (\delta(t)) = \frac{t^{-\alpha-1}}{\Gamma(-\alpha)}, \tag{27.19}$$

where δ is the delta function.

We continue with

$$D_{R-L}^\alpha t^\nu = \frac{\Gamma(\nu+1)}{\Gamma(\nu+1-\alpha)} t^{\nu+\alpha}, \ \nu > -1, \tag{27.20}$$

$$D_{R-L}^\alpha e^{\lambda t} = t^{-\alpha} E_{1,1-\alpha}(\lambda t), \tag{27.21}$$

where $E_{a,b}$ stands for the two-parameter Mittag–Leffler function; see p. 17 of [333],

$$D_{R-L}^\alpha \cosh\left(\sqrt{\lambda}t\right) = t^{-\alpha} E_{2,1-\alpha}\left(\lambda t^2\right), \tag{27.22}$$

$$D_{R-L}^\alpha \ln t = \frac{t^{-\alpha}}{\Gamma(1-\alpha)}(\ln t + \Psi(1) - \Psi(1-\alpha)), \tag{27.23}$$

where Ψ is the digamma function, and

$$D_{R-L}^\alpha t^{\beta-1} \ln t = \frac{\Gamma(\beta) t^{\beta-\alpha-1}}{\Gamma(\beta-\alpha)}(\ln t + \Psi(\beta) - \Psi(\beta-\alpha)), \ \mathrm{Re}\,\beta > 0. \tag{27.24}$$

Regarding the Caputo fractional derivative from [134, pp. 155–156], with order $n > 0$, $n \notin \mathbb{N}$, $m := \lceil n \rceil$, $i = \sqrt{-1}$, $\mathbb{N}_0 = \mathbb{N} \cup \{0\}$, we get

$$D_{*0}^n x^j = \begin{cases} 0, & \text{if } j \in \mathbb{N}_0 \text{ and } j < m, \\ \frac{\Gamma(j+1)}{\Gamma(j+1-n)} x^{j-n}, & \text{if } j \in \mathbb{N}_0 \text{ and } j \geq m, \\ & \text{or } j \notin \mathbb{N} \text{ and } j > m-1. \end{cases} \tag{27.25}$$

Next let $c > 0$, $j \in \mathbb{R}$; then

$$D_{*0}^n \left((x+c)^j \right) = \frac{\Gamma(j+1)}{\Gamma(j+1-m)} \frac{c^{j-m-1} x^{m-n}}{\Gamma(m-n+1)}$$

$$\phantom{D_{*0}^n}\; {}_2F_1 \left(1, m-j; m-n+1; -\frac{x}{c} \right), \tag{27.26}$$

where ${}_2F_1$ is the hypergeometric series.

For $j \in \mathbb{R}$ we have

$$D_{*0}^n \left(e^{jx} \right) = j^m x^{m-n} E_{m-n+1} (jx), \tag{27.27}$$

where E_a stands for the Mittag–Leffler function; see [333, p. 16], and

$$D_{*0}^n (\sin jx) = \begin{cases} \frac{j^m i(-1)^{m/2} x^{m-n}}{2\Gamma(m-n+1)} \left[-{}_1F_1 (1; m-n+1; ijx) \right. \\ \left. +{}_1F_1 (1; m-n+1; -ijx) \right]; & m \text{ even}, \\ \frac{j^m (-1)^{(m-1)/2} x^{m-n}}{2\Gamma(m-n+1)} \left[{}_1F_1 (1; m-n+1; ijx) \right. \\ \left. +{}_1F_1 (1; m-n+1; -ijx) \right]; & m \text{ odd}, \end{cases} \tag{27.28}$$

where ${}_1F_1$ is the confluent hypergeometric function, and so on.

As we see above, computing the fractional derivatives exactly leads to highly complicated formulae, hard to handle. Therefore in applications they numerically approximate the fractional derivatives, and numerically solve the fractional differential equations, and so on.

Excellent sources for the above are [134, 140] and [333].

References

[1] M. Abramowitz and I. A. Stegun, Handbook of Mathematical Functions, Dover, New York, 1964.

[2] G. Acosta and R. G. Durán, An Optimal Poincaré inequality in L' for convex domains, Proc. AMS 132 (2003), No. 1, 195–202.

[3] R. P. Agarwal, Sharp Opial-type inequalities involving $r-$derivatives and their applications, Tôhoku Math. J. 47 (1995), 567–593.

[4] R. P. Agarwal and P. Y. H. Pang, Opial Inequalities with Applications in Differential and Difference Equations, Kluwer Academic Publishers, Dordrecht, Boston, London, 1995.

[5] O. P. Agrawal, Response of a diffusion-wave system subjected to deterministic and stochastic fields, ZAMM, Z. Angew, Math. Mech. 83 (2003) No. 4, 265–274.

[6] O. P. Agrawal, Fractional variational calculus in terms of Riesz fractional derivatives, J. Phys. A: Math. Theor. 40 (2007) 6287–6303.

[7] H. Ahn, Y. Q. Chen and I. Podlubny, Robust stability test of a class of linear time-invariant interval fractional-order system using Lyapunov inequality, Appl. Math. Comput. 187 (2007), No. 1, 27–34.

[8] M. A. Al-Bassam, On fractional analysis and its applications. In: H. L. Manocha (ed.), Modern Analysis and its Applications, Prentice Hall of India Ltd., New Delhi 1986, 269–307.

641

[9] I. Ali, V. Kiryakova and S. L. Kalla, Solutions of fractional multi-order integral and differential equations using a Poisson-type transform, J. Math. Anal. Appl. 269 (2002), No. 1, 172–199.

[10] C. D. Aliprantis and B. Owen, Principles of Real Analysis, Third edition, Academic Press, Inc., San Diego, CA, 1998.

[11] B. N. Al-Saqabi and V. K. Tuan, Solution of a fractional differintegral equation, Integral Transform. Spec. Funct. 4 (1996), No. 4, 321–326.

[12] T. J. Anastasio, The fractional-order dynamics of brainstern vestibulo-oculomotor neurons, Biological Cybernetics, 72 (1994), 69–79.

[13] G. A. Anastassiou, Ostrowski type inequalities, Proc. Amer. Math. Soc. 123 (1995), No. 12, 3775–3781.

[14] G. A. Anastassiou, Multivariate Ostrowski type inequalities, Acta Math. Hungar. 76 (1997), No. 4, 267–278.

[15] G. A. Anastassiou, General fractional Opial type inequalities, Acta Appl. Math. 54 (1998), 303–317.

[16] G. A. Anastassiou, Opial type inequalities for linear differential operators, Math. Inequal. Appl. 1 (1998), No. 2, 193–200.

[17] G. A. Anastassiou, Opial type inequalities involving fractional derivatives of functions, Nonlinear Stud. 6 (1999), No. 2, 207–230.

[18] G. A. Anastassiou, Opial type inequalities involving functions and their ordinary and fractional derivatives, Comm. Appl. Anal. 4 (2000), No. 4, 547–560.

[19] G. A. Anastassiou, Quantitative Approximations, Chapman & Hall/CRC, Boca Raton, New York, 2001.

[20] G. A. Anastassiou, Probabilistic Ostrowski type inequalities, Stochastic Anal. Appl. 20 (2002), No. 6, 1177–1189.

[21] G. A. Anastassiou, Multivariate Montgomery identities and Ostrowski inequalities, Numer. Funct. Anal. Optim. 23 (2002), No. 3–4, 247–263.

[22] G. A. Anastassiou, Univariate Ostrowski inequalities, revisited, Monatsh. Math. 135 (2002), No. 3, 175–189.

[23] G. A. Anastassiou, Multidimensional Ostrowski inequalities, revisited, Acta Math. Hungar. 97 (2002), No. 4, 339–353.

[24] G. A. Anastassiou, Fractional Ostrowski type inequalities, Commun. Appl. Anal. 7 (2003), No. 2, 203–208.

[25] G. A. Anastassiou, Fuzzy Ostrowski type inequalities, Comput. Appl. Math. 22 (2003), No. 2, 279–292.

[26] G. A. Anastassiou, Opial type inequalities involving fractional derivatives of two functions and applications, Comput. Math. Appl. 48 (2004), 1701–1731.

[27] G. A. Anastassiou, Fractional Opial inequalities for several functions with applications, J. Comput. Anal. Appl. 7 (2005), No. 3, 233–259.

[28] G. A. Anastassiou, Fractional and other approximation of Csiszar's f−divergence, (Proc. FAAT 04-Ed. F. Altomare), in Rendiconti Del Circolo Matemitico Di Palermo, Serie II, Suppl. 76 (2005), 197–212.

[29] G. A. Anastassiou, Difference of general integral means, JIPAM. J. Inequal. Pure Appl. Math. 7 (2006), No. 5, Article 185, 13pp. (electronic).

[30] G. A. Anastassiou, Multivariate Euler type identity and Ostrowski type inequalities, Numerical analysis and approximation theory, Casa cartii de stiinta, Cluj-Napoca, 2006, 27–54.

[31] G. A. Anastassiou, Csiszar and Ostrowski type inequalities via Euler-type and Fink identities, Panamer. Math. J. 16 (2006), No. 2, 77–91.

[32] G. A. Anastassiou, Ostrowski type inequalities over balls and shells via a Taylor-Widder formula, JIPAM. J. Inequal. Pure Appl. Math. 8 (2007), No. 4, Article 106, 13pp.

[33] G. A. Anastassiou, Opial type inequalities for Widder derivatives, Panamer. Math. J. 17 (2007), No. 4, 59–69.

[34] G. A. Anastassiou, Optimal multivariate Ostrowski Euler type inequalities, Stud. Univ. Babes-Bolyai Math. 52 (2007), No. 1, 25–61.

[35] G. A. Anastassiou, Multivariate Fink type identitty and multivariate Ostrowski, comparison of means and Grüss type inequalities, Math. Comput. Modelling 46 (2007), No. 3–4, 351–374.

[36] G. A. Anastassiou, High order Ostrowski type inequalities, Appl. Math. Lett. 20 (2007), No. 6, 616–621.

[37] G. A. Anastassiou, Opial type inequalities for semigroups, Semigroup Forum 75 (2007), No. 3, 625–634.

[38] G. A. Anastassiou, Taylor-Widder representation formulae and Ostrowski, Grüss, integral means and Csiszar type inequalities, Comput. Math. Appl. 54 (2007), No. 1, 9–23.

[39] G. A. Anastassiou, Multivariate Euler type identitty and optimal multivariate Ostrowski type inequalities, Adv. Nonlinear Var. Inequal. 10 (2007), No. 2, 51–104.

[40] G. A. Anastassiou, Multivariate Fractional Taylor's formula, Commun. Appl. Anal. 11 (2007), No. 2, 189–199.

[41] G. A. Anastassiou, Hilbert-Pachpatte type general Integral inequalities, Appl. Anal. 86, (2007), No. 8, 945–961.

[42] G. A. Anastassiou, Hilbert-Pachpatte type general multivariate integral inequalities, Intern. J. Appl. Math. 20, (2007), No. 4, 549–573.

[43] G. A. Anastassiou, Multivariate fractional Ostrowski type inequalities, Comput. Math. Appl. 54 (2007), 434–447.

[44] G. A. Anastassiou, Fractional multivariate Opial type inequalities over spherical shells, Commun. Appl. Anal. 11 (2007), No. 2, 201–233.

[45] G. A. Anastassiou, Riemann-Liouville Fractional Multivariate Opial type inequalities over Spherical Shells, Bulletin of Allahabad Math. Soc., India 23 (2008), 65–140.

[46] G. A. Anastassiou, Riemann-Liouville fractional Opial inequalities for several functions with Applications, Commun. Appl. Anal. 12 (2008), No. 4, 377–398.

[47] G. A. Anastassiou, Reverse Riemann-Liouville fractional Opial inequalities for several functions, Complex Variables and Elliptic Equations, 53, (2008), No. 6, 523–544.

[48] G. A. Anastassiou, Opial type inequalities involving Riemann-Liouville fractional derivatives of two functions with applications, Math. Comput. Model. 48 (2008), 344–374.

[49] G. A. Anastassiou, Opial type inequalities for cosine and sine operator functions, Semigroup Forum 76 (2008), No. 1, 149–158.

[50] G. A. Anastassiou, Riemann-Liouville Fractional Multivariate Opial Inequalities on Spherical Shell, RGMIA Monographs, 2008, http://rgmia.vu.edu.au/

[51] G. A. Anastassiou, Converse fractional opial inequalities for several functions, CUBO 10, (2008), No. 1, 117–142.

[52] G. A. Anastassiou, Caputo fractional Ostrowski type inequalities, J. Appl. Func. Anal. 4 (2009), No. 2, 218–236.

[53] G. A. Anastassiou, Multivariate fractional Taylor's formula revisited, Sarajevo J. of Math. to appear 2009.

[54] G. A. Anastassiou, Hilbert-Pachpatte type fractional integral inequalities, Math. Comput. Model. 49 (2009) 1539–1550.

[55] G. A. Anastassiou, Riemann-Liouville and caputo fractional approximation of Csiszar's f−divergence, Sarajevo J. of Math. to appear 2009.

[56] G. A. Anastassiou, Fractional Sobolev type inequalities, Appl. Anal. 87 (2008), No. 5, 607–624.

[57] G. A. Anastassiou, Fractional Poincaré type inequalities, Indian J. Math. 50, (2008), No. 3, 533–571.

[58] G. A. Anastassiou, Caputo fractional multivariate Opial type inequalities on spherical shells, Accepted 2008, Proceedings of AMAT 2008, In "J. Concrete and Appl. Math.", "Internat. Conf. on Approx. Th. and Appl. Math.", G. Anastassiou (ed.), held at U. Memphis, Memphis, TN, USA, October 11-13, 2008.

[59] G. A. Anastassiou, G. R. Goldstein and J. A. Goldstein, Multidimensional Opial inequalities for functions vanishing at an interior point, Atti Accad. Naz. Lincei Cl. Sci. Fis. Mat. Natur. Rend. Lincei (9) Mat. Appl. 15 (2004), No. 1, 5–15.

[60] G. A. Anastassiou, G. R. Goldstein and J. A. Goldstein, Multidimensional weighted Opial inequalities, Appl. Anal. 85 (2006), No. 5, 579–591.

[61] G. A. Anastassiou and J. A. Goldstein, Fractional Opial type inequalities and fractional differential equations, Result. Math. 41 (2002), 197–212.

[62] G. A. Anastassiou and J. A. Goldstein, Ostrowski type inequalities over Euclidean domains, Atti Accad. Naz. Lincei Cl. Sci. Fis. Mat. Natur. Rend. Lincei (9) Mat. Appl. 18 (2007), No. 3, 305–310.

[63] G. A. Anastassiou and J. A. Goldstein, Higher order Ostrowski type inequalities over Euclidean domains, J. Math. Anal. Appl. 337 (2008), 962–968.

[64] G. A. Anastassiou, J. Koliha and J. Pecaric, Opial inequalities for fractional derivatives, Dynam Systems Appl. 10, (2001) 395–406.

[65] G. A. Anastassiou, J. Koliha and J. Pecaric, Opial type L_p-inequalities for fractional derivatives, Intern. J. Math. & Math. Sci. 31 (2002), No. 2, 85–95.

[66] G. A. Anastassiou and J. Pecaric, General weighted Opial inequalities for linear differential operators, J. Math. Anal. Appl. 239 (1999), No. 2, 402–418.

[67] M. Annaby and P. Butzer, Sampling in Paley-Wiener spaces associated with fractional integro-differential operators, J. Comput. Appl. Math. 171 (2004), No. 1–2, 39–57.

[68] T. Apostol, Mathematical Analysis, Addison-Wesley Publ. Co., London, 1969.

[69] R. Askey, Inequalities via fractional integration, Lect. Notes Math. 457 (1975), 106–115.

[70] M. Axtell and M. E. Bise, Fractional calculus applications in control systems, Proc. of the IEEE 1990 Nat. Aerospace and Electronics Conf., New York, 1990, 563–566.

[71] R. L. Bagley and R. A. Calico, Fractional order state equations for the control of viscoelastically damped structures, J. Guid. Contr. Dynam. 14 (1991), 304–311.

[72] R. L. Bagley and P. J. Torvik, Fractional calculus-a different approach to the analysis of viscoelastically damped structures, AIAA J. 21, (1983), No. 5, 741–748.

[73] R. L. Bagley and P. J. Torvik, A theoretical basis for the application of fractional calculus to viscoelasticity, J. Rheol 27 (1983), No. 3, 201–210.

[74] R. L. Bagley and P. J. Torvik, On the appeerence of the fractional derivative in the behavior of real materials, J. Appl. Mech. 51 (1984), 294–298.

[75] R. L. Bagley and P. J. Torvik, Fractional calculus in the transient analysis of viscoelastically damped structures, AIAA J. 23, (1985), No. 6, 918–925.

[76] D. Bainov and P. Simeonov, Integral Inequalities and Applications, Kluwer Academic Publishers, Dordrecht, Boston, London, 1992.

[77] E. Barkai, R. Metzler and J. Klafter, From continuous time random walks to the fractional Fokker-Planck equation, Phys. Rev. E 61 (2000), 132–138.

[78] N. S. Barnett, P. Cerone, S. S. Dragomir and A. Sofo, Approximating Csiszar's $f-$divergence by the use of Taylor's formula with integral remainder (paper # 10, pp. 16). In: S. S. Dragomir (ed.), Inequalities for Csiszar $f-$Divergence in Information Theory, Victoria University, Melbourne, Australia, 2000. On line: http://rgmia.vu.edu.au

[79] H. Bauer, Ma$\beta-$und Integrations-Theorie, de Gruyter, Berlin, 1990.

[80] P. R. Beesack, On an integral inequality of Z. Opial, Trans. Amer. Math. Soc. 104 (1962), 470–475.

[81] V. A. Belavin, R. Sh. Nigmatullin, A. I. Miroshnikov and N. K. Luiskaya, Fractional differentiation of oscillographic polarograms by means of an electrochemical two-terminal network (in Russian), Trudy Kazan Aviatsion Institutel 5 (1964), 144–152.

[82] F. Bella, P. F. Biagi, M. Caputo, G. Della Monica, A. Ermini, P. Manjgaladze, V. Sgrigna and D. Zilpimiani, Very slow-moving crustal strain disturbances, Tectonophysics 179 (1990), 131–139.

[83] D. A. Benson, S. W. Wheatcraft and M. M. Meerachaert, Application of a fractional advection dispersion equation, Water Resour. Res. 36 (2000), 1403–1412.

[84] D. A. Benson, S. W. Wheatcraft and M. M. Meerachaert, The fractional-order governing equation of Lévy motion, Water Resour. Res. 36 (2000), 1413–1424.

[85] H. Beyer and S. Kempfle, Definition of physically consistent damping laws with fractional derivatives, ZAMM. 75 (1995), 623–635.

[86] B. Bittner, Irregular sampling based on the fractional Fourier transform, Proceedings of the 1997 workshop on Sampling Theory and Applications, Samp TA-97, Aveiro, Portugal, 1997, 431–436.

[87] B. Bonilla, J. J. Trujillo and M. Rivero, Fractional order continuity and some properties about integrability and differentiability of real functions, J. Math. Anal. Appl. 231 (1999), No. 1, 205–212.

[88] B. Bonilla, M. Rivero and J. J. Trujillo, On systems of linear fractional differential equations with constant coefficients, Appl. Math. Comput. 187 (2007), No. 1, 68–78.

[89] B. Bonilla, M. Rivero, L. Rodriguez-Germá and J. J. Trujillo, Fractional differential equations as alternative models to nonlinear differential equations, Appl. Math. Comput. 187 (2007), No. 1, 79–88.

[90] P. L. Butzer, H. Dyckhoff, E. Görlich and R. L. Stens, Best trigonometric approximation, fractional order derivatives and Lipschitz classes, Canad. J. Math. 29 (1977), 781–793.

[91] P. L. Butzer, M. Hauss and M. Schmidt, Factorial functions and Stirling numbers of fractional order, Resultate Math. 16 (1989), 16–48.

[92] P. L. Butzer, A. A. Kilbas and J. J. Trujillo, Fractional calculus in the Mellin setting and Hadamard-type fractional integrals, J. Math. Anal. Appl. 269 (2002), 1–27.

[93] P. L. Butzer, A. A. Kilbas and J. J. Trujillo, Composition of Hadamard-type fractional integration operators and the semigroup property, J. Math. Anal. Appl. 269 (2002), No. 2, 387–400.

[94] P. L. Butzer, A. A. Kilbas and J. J. Trujillo, Mellin transform analysis and integration by parts for Hadamard-type fractional integrals, J. Math. Anal. Appl. 270 (2002), No. 1, 1–15.

[95] P. L. Butzer, A. A. Kilbas and J. J. Trujillo, Stirling functions of the second kind in the setting of difference and fractional calculus, Numer. Funct. Anal. Optim. 24 (2003), No. 7–8, 673–711.

[96] P. L. Butzer, A. A. Kilbas, L. Rodriguez-Germá and J. J. Trujillo, Stirling functions of first kind in the setting of fractional calculus and generalized differences, J. Difference Equ. Appl. 13 (2007), No. 8 9, 683–721.

[97] P. L. Butzer and R. L. Stens, Fractional Chebyshev operational calculus and best algebraic approximation, Approximation theory, II (Proc. Internat. Sympos., Univ. Texas, Austin, Tex., 1976), Academic Press, New York, 1976, 315–319.

[98] P. L. Butzer and R. L. Stens, Chebyshev transform methods in the solution of the fundamental theorem of best algebraic approximation in the fractional case. In: Fourier Analysis and Approximation Theory (Colloquia Math. Soc. János Bolyai, 19, Proc. Conf. Budapest, August 1976, Eds. G. Alexits-P. Turán) Amsterdam 1978, 191–212.

[99] P. L. Butzer and U. Westphal, An access to fractional differentiation via fractional difference quotients. In: Fractional Calculus and its Applications. Lecture Notes in Mathematics, Springer, Berlin 1975, 116–145.

[100] P. L. Butzer and U. Westphal, An introduction to fractional calculus, Ch. 1. In: R. Hilfer (ed.), Applications of Fractional Calculus in Physics, Singapore, World Scientific, 2000, 1–85.

[101] J. A. Canavati, The Riemann-Liouville integral, Nieuw Archief Voor Wiskunde 5 (1987), No. 1, 53–75.

[102] M. Caputo, Linear models of dissipation whose Q is almost frequency Independent-II. Geophys, J. R. Astr. Soc. 13 (1967), 529–539.

[103] M. Caputo, Elasticità e Dissipazione, Zanichelli, Bologna, 1969.

[104] M. Caputo, Wave-number-independent rheology in a sphere, Atti Accad. Naz. Lincei Rend. Cl. Sci. Fis. Math. Natur. (8) 81 (1987), No. 2, 175–207.

[105] M. Caputo, The rheology of an anelastic medium studied by means of the observation of the splitting of its eigenfrequencies, J. Acoust. Soc. Am. 86 (1989), No. 5, 1984–1987.

[106] M. Caputo, The splitting of the free oscillations of the Earth caused by the rheology, Rend. Fis. Acc. Lincei, Ser. 9, 1 (1990), 119–125.

[107] M. Caputo, Lectures on Seismology and Rheological Tectonics, Univ. degli studi di Roma "La Sapienza", 1992–1993.

[108] M. Caputo, Free modes splitting and alterations of electrochemically polarizable media, Rend. Fis. Acc. Lincei, Ser. 9, 4 (1993), 89–98.

[109] M. Caputo, Mean fractional-order-derivatives differential equations and filters, Ann. Univ. Ferrara Sez. VII (N.S.) 41 (1995), 73–84.

[110] M. Caputo, 3-dimensional physically consistent diffusion in anisotropic media with memory, Atti. Accad. Naz. Lincei Cl. Sci. Fis. Mat. Natur. Rend. Lincei (9) Mat. Appl. 9 (1998), No. 2, 131–143.

[111] M. Caputo, Distributed order differential equations modelling dielectric induction and diffusion, Fract. Calc. Appl. Anal. 4 (2001), No. 4, 421–442.

[112] M. Caputo and F. Mainardi, A new dissipation model based on memory mechanism, Pure Appl. Geophys. 91 (1971), No. 8, 134–147.

[113] M. Caputo and F. Mainardi, A new dissipation model based on memory mechanism, Pure Appl. Geophys. 91 (1971a), 134–147.

[114] M. Caputo and F. Mainardi, Linear models of dissipation in anelastic solid, Rivista del Nuovo Cimento 1 (1971b), 161–198.

[115] G. E. Carlson, Investigation of fractional capacitor approximations by means of regular Newton processes, Kansas State University Bulletin, Vol. 48, No. 1, Special report No. 42, 1964.

[116] G. E. Carlson and C. A. Halijak, Simulation of the fractional derivative operator $\sqrt{s}$ and the fractional integral operator $1/\sqrt{s}$. Kansas State University Bulletin, Vol. 45, No. 7 (1961), 1–22.

[117] G. E. Carlson and C. A. Halijak, Approximation of fractional capacitors $(1/s)^{1/n}$ by a regular Newton process, IEEE Trans. Circuit Theory, Vol. CT-10, No. 2 (1964), 210–213.

[118] A. Carpinteri and F. Mainardi (eds.), Fractals and Fractional Calculus in Continuum Mechanics, Springer Verlag, Vienna-New York, 1997.

[119] A. Chaves, A fractional diffusion equation to describe Lévy flights, Phys. Lett. A 239 (1998), 13–16.

[120] A. V. Chechkin, R. Gorenflo, I. M. Sokolov and V. Yu. Gonchar, Distributed order time fractional diffusion equation, Fract. Calc. Appl. Anal. 6 (2003), No. 3, 259–279.

[121] A. V. Chechkin, R. Gorenflo and I. M. Sokolov, Fractional diffusion in inhomogeneous media, J. Phys. A 38 (2005), No. 42, L679–L684.

[122] M.-P. Chen, H. Irmak and H. M. Srivastava, A certain subclass of analytic functions involving operators of fractional calculus, Comput. Math. Appl. 35 (1998), No. 5, 83–91.

[123] M.-P. Chen, H. M. Srivastava and C.-S. Yu, Some operators of fractional calculus and their applications involving a novel class of analytic functions, Appl. Math. Comput. 91 (1998), No. 2–3, 285–296.

[124] Y. Q. Chen and K. L. Moore, Discretization schemes for fractional-order differentiators and integrators, IEEE Trans. On Circuits and Systems-I. Fundamental Theory and Applications 49 (2002), No. 3, 363–367.

[125] Y. Chen, B. M. Vinagre and I. Podlubny, Continued fraction expansion approaches to discretizing fractional order derivatives-an expository review, Nonlinear Dynam. 38 (2004), No. 1–4, 155–170.

[126] S.-K. Chua and R. L. Wheeden, A note on sharp 1–dimensional Poincaré inequalities, Proc. AMS 134 (2006), No. 8, 2309–2316.

[127] I. Csiszar, Eine Informationstheoretische Ungleichung und ihre Anwendung auf den Beweis der Ergodizität von Markoffschen Ketten, Magyar Trud. Akad. Mat. Kutato Int. Közl. 8 (1963), 85–108.

[128] I. Csiszar, Information-type measures of difference of probability distributions and indirect observations, Studia Math. Hungarica 2 (1967), 299–318.

[129] J. B. Diaz and T. J. Osler, Differences of fractional order, Math. Comput. 28 (1974), 185–202.

[130] K. Diethelm, An algorithm for the numerical solution of differential equations of fractional order, Elec. Trans. Numer. Anal. 5 (1997), 1–6, ISSN 1068–9613.

[131] K. Diethelm, Fractional error constants for quadrature formulas, Approximation theory IX, Vol. I. (Nashville, TN, 1998), 113–118, Innov. Appl. Math., Vanderbilt Univ. Press, Nashville, TN, 1998.

650   References

[132] K. Diethelm, Estimation of Quadrature errors in terms of Caputo-type fractional derivatives, Fract. Calc. Appl. Anal. 2 (1999), No. 3, 313–327.

[133] K. Diethelm, Predictor-Corrector Strategies for Single- and Multi-Term Fractional Differential Equations. In: E. A. Lipitakis (ed.), Proceedings of the 5th Hellenic-European Conference on Computer Mathematics and its Applications, LEA Press, Athens, 2002, 117–122.

[134] K. Diethelm, Fractional Differential Equations, On line: http://www.tu-bs.de/~diethelm/lehre/f-dgl02/fde-skript.ps.gz,2003.

[135] K. Diethelm, Smoothness properties of solutions of Caputo-type fractional differential equations, Fract. Calc. Appl. Anal. 10 (2007), No. 2, 151–160.

[136] K. Diethelm, J. M. Ford, N. J. Ford and M. Weilbeer, Pitfalls in fast numerical solvers for fractional differential equations, J. Comput. Appl. Math. 186 (2006), No. 2, 482–503.

[137] K. Diethelm and N. J. Ford, Analysis of fractional differential equations, J. Math. Anal. Appl. 265 (2002), No. 2, 229–248.

[138] K. Diethelm and N. J. Ford, Multi-order fractional differential equations and their numerical solution, App. Math. Comput. 154 (2004), No. 3, 621–640.

[139] K. Diethelm, N. J. Ford and A. D. Freed, A predictor-corrector approach for the numerical solution of fractional differential equations, Fractional order calculus and its applications. Nonlinear Dynam. 29 (2002), No. 1–4, 3–22.

[140] K. Diethelm, N. J. Ford, A. D. Freed and Y. Luchko, Algorithms for the fractional calculus: A selection of numerical methods, Comput. Meth. Appl. Mech. Engrg. 194 (2005), No. 6–8, 743–773.

[141] K. Diethelm and A. D. Freed, On the solution of nonlinear fractional differential equations used in the modeling of viscoplasticity. In: F. Keil, W. Mackens, H. Voβ and J. Werther (eds.), Scientific Computing in Chemical Engineering II: Computational Fluid Dynamics, Reaction Engineering, and Molecular Properties, Springer, Heidelberg (1999a), 217–224.

[142] K. Diethelm and Y. Luchko, Numerical solution of linear multi-term initial value problems of fractional order, J. Comput. Anal. Appl. 6 (2004), No. 3, 243–263.

[143] K. Diethelm and G. Walz, Numerical solution of fractional order differential equations by extrapolation, Numer. Algorithms 16 (1997), 231–253.

[144] S. S. Dragomir, On some Gronwall type lemmas, Studia Univ. Babes-Bolyai Math. 33 (1988), 29–36.

[145] S. S. Dragomir (ed.), Inequalities for Csiszar f-Divergence in Information Theory, Victoria University, Melbourne, Australia, 2000. On line: http://rgmia.vu.edu.au

[146] S. S. Dragomir and C. E. M. Pearce, Selected Topics on Hermite-Hadamard Inequalities and Applications, Monograph, Victoria University, Melbourne, Adelaide, Australia, 2000. On line: http://rgmia.vu.edu.au

[147] S. S. Dragomir and Y.-H. Kim, Hilbert-Pachpatte type integral inequalities and their improvement, J. Inequal. Pure Appl. Math. 4 (2003), No. 1, Article 16, http://jipam.vu.edu.au/

[148] D. P. Drianov, Equivalence between fractional average modulus of smoothness and fractional K-functional, C. R. Acad. Bulg. Sci. 38 (1985), 1609–1612.

[149] M. M. Dzhrbashyan and A. B. Nersesyan, Fractional derivatives and the Cauchy problem for differential equations of fractional order, Izv. Akademii Nauk Arm. SSR 3 (1968), No. 1, 3–29.

[150] E. C. Eckstein, J. A. Goldstein and M. Leggas, The mathematics of suspensions: Kac walks and asymptotic analyticity, Fourth Mississippi State Conference on Differential Equations and Computational Simulations, Electronic J. Differential Equations, Conf. 03, 1999, 39–50, 2000.

[151] E. C. Eckstein, B. Ma, M. Leggas and J. A. Goldstein, Linking theory and measurements of tracer particle position in suspension flows, 1–8, Proceedings of ASME FEDSM'00 ASME 2000 Fluids Engineering Division Summer Meeting, June 11–15, 2000, Boston, MA.

[152] J. T. Edwards, N. J. Ford and A. C. Simpson, The numerical solution of linear multi-term fractional differential equations: Systems of equations, J. Comput. Appl. Math. 148 (2002), 401–418.

[153] A. M. A. El-Sayed, Fractional derivative and fractional differential equations, Bull Fac. Sci., Alexandria Univ. 28 (1998), 18–22.

[154] A. M. A. El-Sayed, Fractional order evolution equations, J. Fract. Calc. 7 (1995), 89–100.

[155] A. M. A. El-Sayed, Fractional order diffusion-wave equation, Int. J. Theor. Phys. 35 (1996), 311–322.

[156] N. Engheta, On fractional calculus and fractional multipoles in electromagnetism, IEEE Trans. on Antennas and Propagations, 44, (1996), No. 4, 554–566.

[157] A. Erdélyi, On fractional integration and its applications to the theory of Hankel transforms, Quart. J. Math. (Oxford), 11 (1940), 293–303.

[158] L. Euler, Memoire dans le tome V des Comment, Saint Petersberg Années, 55, 1730.

[159] L. C. Evans, Partial Differential Equations, Graduate studies in Mathematics Vol. 19, American Math. Soc., Providence, R. I., 1998.

[160] A. M. Fink, On Opial's inequality for $f^{(n)}$, Proc. Amer. Math. Soc. 115 (1992), 177–181.

[161] N. J. Ford and A. C. Simpson, The numerical solution of fractional differential equations: Speed versus accuracy, Numer. Algorithms 26 (2001), 333–346.

[162] J. B. J. Fourier, Théorie analytique de la chaleur, Ocuvres de Fourier, 1, Didot, Paria, 1822, 508.

[163] Ch. Friedrich, Rheological material functions for associating comb-shaped or H-shaped polymers: A fractional calculus approach, Psilosophical Magazine Lett. 66, (1992), No. 6, 287–292.

[164] L. Gaul, P. Klein and S. Kempfle, Damping description involving fractional operators, Mech. Syst. Signal Process. 5 (1991), 81–88.

[165] M. Giona, S. Gerbelli and H. E. Roman, Fractional diffusion equation and relaxation in complex viscoelastic materials, Physica A 191 (1992), 449–453.

[166] W. G. Glöckle and T. F. Nonnenmacher, Fractional integral operators and Fox functions in the theory of viscoelasticity, Macro-molecules 24 (1991), 6426–6436.

[167] W. G. Glöckle and T. F. Nonnenmacher, A fractional calculus approach to self-similar protein dynamics, Biophysical J. 68 (1995), 46–53.

[168] R. Gorenflo, Fractional calculus: Some numerical methods, CISM Lecture Notes, Udine, Italy, 1996.

[169] R. Gorenflo, Fractional calculus: Some numerical methods. In: A. Carpinteri, F. Mainardi (eds.), Fractals and Fractional Calculus in Continuum Mechanics, Springer, Wien, 1997, 277–290.

[170] R. Gorenflo and E. A. Abdel-Rehim, Discrete models of time-fractional diffusion in a potential well, Fract. Calc. Appl. Anal. 8 (2005), No. 2, 173–200.

[171] R. Gorenflo and E. A. Abdel-Rehim, Convergence of the Grünwald-Letnikov scheme for time-fractional diffusion, J. Comput. Appl. Math. 205 (2007), No. 2, 871–881.

[172] R. Gorenflo, Y. Luchko and F. Mainardi, Wright functions as scale-invariant solutions of the diffusion-wave equation. Higher transcendental functions and their applications, J. Comput. Appl. Math. 118 (2000), No. 1–2, 175–191.

[173] R. Gorenflo, Y. F. Luchko and S. R. Umarov, On some boundary value problems for pseudo-differential equations with boundary operators of fractional order, Fract. Calc. Appl. Anal. 3 (2000), No. 4, 453–468.

[174] R. Gorenflo and F. Mainardi, Fractional calculus: Integral and differential equations of fractional order, Fractals and Fractional Calculus in Continuum Mechanics (Udine, 1996), 223–276, CISM Courses and Lectures, 378, Springer, Vienna, 1997.

[175] R. Gorenflo and F. Mainardi, Fractional calculus and stable probability distributions, Fourth Meeting on Current Ideas in Mechanics and Related Fields (Kraków, 1997). Arch. Mech. (Arch. Mech. Stos.) 50 (1998), No. 3, 377–388.

[176] R. Gorenflo and F. Mainardi, Fractional relaxation of distributed order, Complexus mundi, 33–42, World Sci. Publ., Hackensack, NJ, 2006.

[177] R. Gorenflo, F. Mainardi, D. Moretti, G. Pagnini and P. Paradisi, Fractional diffusion: Probability distributions and random walk models, Non extensive thermodynamics and physical applications (Villasimius, 2001), Phys. A 305 (2002), No. 1–2, 106–112.

[178] R. Gorenflo, F. Mainardi, D. Moretti and P. Paradisi, Time fractional diffusion: A discrete random walk approach, Fractional order calculus and its applications, Nonlinear Dynam. 29 (2002), No. 1–4, 129–143.

[179] R. Gorenflo, F. Mainardi, E. Scalas and M. Raberto, Fractional calculus and continuous-time finance III. The diffusion limit, Mathematical finance (Konstanz, 2000), 171–180, Trends Math., Birkhäuser, Basel, 2001.

[180] R. Gorenflo, F. Mainardi and A. Vivoli, Continuous-time random walk and parametric subordination in fractional diffusion, Chaos Solitons Fractals 34 (2007), No. 1, 87–103.

[181] R. Gorenflo, A. Vivoli and F. Mainardi, Discrete and continuous random walk models for space-time fractional diffusion, Nonlinear Dynam. 38 (2004), No. 1–4, 101–116.

[182] R. Gorenflo, A. Vivoli and F. Mainardi, Discrete and continuous random walk models for space-time fractional diffusion, J. Math. Sci. (NY) 132 (2006), No. 5, 614–628.

[183] I. S. Gradstein and I. M. Ryzhik, Tables of Integrals, Series and Products, Academic Press., New York, 1980.

[184] A. Graham, G. W. Scott Blair and R. F. J. Withers, A methodological problem in rheology, Br. J. Philos. Sci. 11 (1961), 265–278.

[185] A. P. Grinprimenko and A. A. Kilbas, On compositions of generalized fractional integrals and evaluation of definite integrals with Gauss hypergeometric functions, J. Math. Res. Exposition 11 (1991), No. 3, 443–446.

[186] S. B. Hadid and Yu. F. Luchko, An operational method for solving fractional differential equations of an arbitrary real order, Panamer. Math. J. 6 (1996), No. 1, 57–73.

[187] G. D. Handley, J. J. Koliha and J. Pecaric, Hilbert-Pachpatte type integral inequalities for fractional derivatives, Fract. Calc. Appl. Anal. 4 (2001), No. 1, 37–46.

[188] G. D. Handley, J. J. Koliha and J. Pecaric, New Hilbert-Pachpatte type integral inequalities, J. Math. Anal. Appl. 257 (2001), 238–250.

[189] G. D. Handley, J. J. Koliha and J. Pecaric, Hilbert-Pachpatte type multidimensional integral inequalities, J. Inequal. Pure Appl. Math. 5 (2004), No. 2, Article 34, http://jipam.vu.edu.au/

[190] G. H. Hardy and J. E. Littlewood, Some properties of fractional integral, I. Math. Z., 27 (1928), 565.

[191] G. H. Hardy, J. E. Littlewood and G. Polya, Inequalities, Cambridge University Press, Cambridge, UK, 1934.

[192] H. Hayakawa, Fractional dynamics in phase ordering processes, Fractals 1 (1993), 947–953.

[193] N. Hayek, J. Trujillo, M. Rivero, B. Bonilla and J. C. Moreno, An extension of Picard-Lindelöff theorem to fractional differential equations, Appl. Anal. 70 (1999), No. 3–4, 347–361.

[194] R. Hilfer, Foundations of fractional dynamics, Fractals 3 (1995), 549–556.

[195] R. Hilfer, On fractional diffusion and its relation with continuous time random walks, In: R. Kutner, A. Pekalski and K. Sznajd-Weron (eds.), Anomalous Diffusion: From Basis to Applications, Springer Verlag, Berlin 1999, 77–82.

[196] R. Hilfer, Fractional calculus and regular variation in thermodynamics, Applications of Fractional Calculus in Physics, World Sci. Publ., River Edge, NJ, 2000, 429–463.

[197] R. Hilfer (ed.), Applications of Fractional Calculus in Physics, World Scientific, Singapore, 2000.

[198] R. Hilfer, Fractional time evolution, Applications of Fractional Calculus in Physics, World Sci. Publ., River Edge, NJ, 2000, 87–130.

[199] R. Hilfer, On fractional diffusion and continuous time random walks, Phys. A 329 (2003), No. 1–2, 35–40.

[200] R. Hilfer, Remarks on fractional time, Time, Quantum and Information, Springer, Berlin, 2004, 235–241.

[201] R. Hilfer and L. Anton, Fractional master equations and fractal time random walks, Phys. Rev. E 51 (1995), R848–R851.

[202] H. J. Holmgren, Om differentialkalkylen med indices of hvad nature sam helst, Kgl. Sv. Vetenskapsakademia Handl, 11, 1864, 1.

[203] T. Y. Hu and K. S. Lau, Fractal dimensions and singularities of the Weierstrass type functions, Trans. Amer. Math. Soc. 335 (1993), No. 2, 649–665.

[204] R. N. Kalia, Fractional calculus and bilateral expansions of incomplete beta and elliptic functions, Intern. J. Math. Ed. Sci. Tech. 22 (1991), No. 2, 203–206.

[205] R. N. Kalia, Fractional calculus: Its brief history and recent advances, Recent Advances in Fractional Calculus, Global Res. Notes Ser. Math., Global, Sauk Rapids, MN, 1993, 1–30.

[206] R. N. Kalia (ed.), Recent Advances in Fractional Calculus, Global Publ. Co., Sauk Rapids, MN, 1993.

[207] R. N. Kalia and S. Kalia, Integrals derivable from Feynman integrals, fractional calculus applications to the H-function, and certain Gaussian model in statistical mechanics, Math Sci. Res. Hot-Line 2 (1998), No. 7, 19–24.

[208] R. N. Kalia and S. Keith, Fractional calculus and expansions of incomplete gamma functions, Appl. Math. Lett. 3 (1990), No. 1, 19–21.

[209] A. A. Kilbas, Asymptotic expansions of fractional integrals and of the solutions of the Euler-Poisson-Darboux equations (Russian), Differ. Uravn. 24 (1988), No. 10, 1764–1778, 1837–1838; Translation in Differ. Equ. 24 (1988), No. 10, 1174–1185.

[210] A. A. Kilbas, Asymptotic properties of fractional integrals and their applications, Boundary Value Problems, Special Functions and Fractional Calculus (Russian), Belorus, Gos. Univ., Minsk, 1996, 141–158.

[211] A. A. Kilbas, B. Bonilla, Kh. Rodriges, Kh. Trukhillo and M. Rivero, Compositions of fractional integrals and derivatives with a Bessel-type function and the solution of differential and integral equations (Russian), Dokl. Nats. Akad. Nauk Belarusi 42 (1998), No. 2, 25–29, 123.

[212] A. A. Kilbas, V. Bonilla and Kh. Trukhillo, Nonlinear differential equations of fractional order in the space of integrable functions (Russian), Dokl. Akad. Nauk 374 (2000), No. 4, 445–449.

[213] A. A. Kilbas, B. Bonilla and J. J. Trujillo, Existence and uniqueness theorems for nonlinear fractional differential equations, Demonstratio Math. 33 (2000), No. 3, 583–602.

[214] A. A. Kilbas and A. A. Koroleva, Integral transform with the extended generalized Mittag-Leffler function, Math. Model. Anal. 11 (2006), No. 2, 173–186.

[215] A. A. Kilbas and S. A. Marzan, The Cauchy Problem for Differential Equations with Fractional Caputo Derivative, Doklady Mathematics 70 (2004), No. 3, 841–845, Translated from Doklady Akademii Nauk 399 (2004), No. 1, 7–11.

[216] A. A. Kilbas and S. A. Marzan, Nonlinear Differential Equations with the Caputo Fractional Derivative in the space of Continuously Differentiable Functions, Differential Equations 41 (2005), No. 1, 84–89, Translated from Differentsialnye Uravneniya 41 (2005), No. 1, 82–86.

[217] A. A. Kilbas, T. Pierantozzi, J. J. Trujillo and L. Vazquez, On the solution of fractional evolution equations, J. Phys. A: Math. Gen. 37 (2004), 3271–3283.

[218] A. A. Kilbas, M. Rivero and J. J. Trujillo, Existence and uniqueness theorems for differential equations of fractional order in weighted spaces of continuous functions, Fract. Calc. Appl. Anal. 6 (2003), No. 4, 363–399.

[219] A. A. Kilbas, L. Rodriguez and J. J. Trujillo, Asymptotic representations for hypergeometric-Bessel type function and fractional integrals, J. Comput. Appl. Math. 149 (2002), No. 2, 469–487.

[220] A. A. Kilbas and M. Saigo, On Mittag-Leffler type function, fractional calculus operators and solutions of integral equations, Integral Transform. Spec. Funct. 4 (1996), No. 4, 355–370.

[221] A. A. Kilbas, H. M. Srivastava and J. J. Trujillo, Fractional differential equations: An emergent field in applied and mathematical sciences, Factorization, Singular Operators and Related Problems (Funchal, 2002), 151–173, Kluwer Acad. Publ., Dordrecht, 2003.

[222] A. A. Kilbas, H. M. Srivastava and J. J. Trujillo, Theory and Applications of Fractional Differential Equations, Elsevier, Amsterdam, Netherlands, 2006.

[223] A. A. Kilbas and A. A. Tityura, A Marchaud-Hadamard-type fractional derivative and the inversion of the Hadamard-type fractional integrals (Russian), Dokl. Nats. Akad. Nauk Belarusi 50 (2006), No. 4, 5–10, 125.

[224] A. A. Kilbas and J. J. Trujillo, Computation of fractional integrals via functions of hypergeometric and Bessel type, Higher transcendental functions and their applications, J. Comput. Appl. Math. 118 (2000), No. 1–2, 223–239.

[225] A. A. Kilbas and J. J. Trujillo, Differential equations of fractional order: methods, results and problems, I. Appl. Anal. 78 (2001), No. 1–2, 153–192.

[226] A. A. Kilbas and J. J. Trujillo, Differential equations of fractional order: methods, results and problems, II. Appl. Anal. 81 (2002), No. 2, 435–493.

[227] A. A. Kilbas, J. J. Trujillo and A. A. Voroshilov, Cauchy-type problem for diffusion-wave equation with the Riemann-Liouville partial derivative, Fract. Calc. Appl. Anal. 8 (2005), No. 4, 403–430.

[228] Y. C. Kim and H. M. Srivastava, Fractional integral and other linear operators associated with the Gaussian hypergeometric function, Complex Variables Theory Appl. 34 (1997), No. 3, 293–312.

[229] V. S. Kiryakova, Generalized fractional integral and fractional derivative representations of hypergeometric functions $_pF_q$ for $p = q$ or $p = q + 1$, Constructive Theory of Functions (Varna, 1987), 260–269, Publ. House Bulgar. Acad. Sci., Sofia, 1988.

[230] V. Kiryakova, Generalized fractional calculus and applications, Pitman Research Notes in Math. Series, 301. Longman Scientific & Technical, Harlow, copublished in U.S.A with John Wiley & Sons, Inc., New York, 1994.

[231] V. S. Kiryakova, Multiple (multiindex) Mittag-Leffler functions and relations to generalized fractional calculus, Higher transcendental functions and their applications, J. Comput. Appl. Math. 118 (2000), No. 1–2, 241–259.

[232] V. S. Kiryakova and B. Al-Saqabi, Solutions of Erdelyi-Kober fractional integral, differential and differintegral equations of second kind, C. R. Acad. Bulgare Sci. 50 (1997), No. 1, 27–30.

[233] V. S. Kiryakova and G. P. Ivanov, Fractional calculus and Mittag-Leffler functions: Some applications in control theory, Applications of Mathematics in Engineering (Sozopol, 1998), 117–123, Heron Press, Sofia, 1999.

[234] H. Kober, On fractional integrals and derivatives, Quart. J. Math. (Oxford), 11 (1940), 193–211.

[235] R. C. Koeller, Applications of fractional calculus to the theory of viscoelasticity, J. Appl. Mech. 51 (1984), 299–307.

[236] R. C. Koeller, Polynomial operators, Stieltjes convolution, and fractional calculus in hereditary mechanics, Acta. Mech. 58 (1986), 251–264.

[237] V. Kokilashvili and S. Samko, On Sobolev theorem for Riesz-type potentials in Lebesgue spaces with variable exponent, Z. Anal. Anwendungen 22 (2003), No. 4, 899–910.

[238] K. M. Kolwankar and A. D. Gangal, Fractional differentiability of nowhere differentiable functions and dimensions, Chaos 6 (1996), No. 4, 505–513.

[239] D. Kreider, R. Kuller, D. Ostberg and F. Perkins, An Introduction to Linear Analysis, Addison-Wesley, Reading, MA, 1966.

[240] S. Kullback and R. Leibler, On information and sufficiency, Ann. Math. Statist. 22 (1951), 79–86.

[241] S. Kullback, Information Theory and Statistics, Wiley, New York, 1959.

[242] B. Kuttner, Some theorems on fractional derivatives, Proc. London Math. Soc., 3, (1953), 480.

[243] S. F. Lacroix, Traité du Calcul Differential et du Calcul Integral, Courcier, Paris, 3 (1819), 409–410.

[244] J. L. Lagrange, Sur une nouvelle espée de calcul relatif a la differentiation et a l' integration des quantités variables, Ouvres de Lagrange, Gauthier-Villars, Paris, 3 (1772), 441.

[245] J. L. Lavoie, T. J. Osler and R. Tremblay, Fractional derivatives and special functions, SIAM Rev. 18 (1976), 240–268.

[246] G. W. Leibnitz, Letter from Hanover, Germany, September 30, 1695, to G. A. L' Hospital, Leibnizen Mathematische Schriften 2, pp. 301–302. Olms Verlag, Hildesheim, Germany, 1962, First published in 1849.

[247] A. V. Letnikov, On the historical development of the theory of differentiation of an arbitrary order, Mat. Sb. 3 (1868), 85–112 (in Russian).

[248] Y. S. Liang and W. Y. Su, Connection between the order of fractional calculus and fractional dimensions of a type of fractal functions, Anal. Theory Appl. 23 (2007), No. 4, 354–362.

[249] Y. S. Liang and W. Y. Su, The relations between the fractal dimensions of a type of fractal functions and the order of their fractional calculus, Chaos, Solitons Fractals 34 (2007), 682–692.

[250] S.-D. Lin, J.-C. Shyu, K. Nishimoto and H. M. Srivastava, Explicit solutions of some general families of ordinary and partial differential equations associated with the Bessel equation by means of fractional calculus, J. Fract. Calc. 25 (2004), 33–45.

[251] S.-D. Lin and H. M. Srivastava, Some families of the Hurwitz-Lerch zeta functions and associated derivative and other integral representations, Appl. Math. Comput. 154 (2004), No. 3, 725–733.

[252] S.-D. Lin, H. M. Srivastava, S.-T. Tu and P.-Y. Yang, Some families of linear ordinary and partial differential equations solvable by means of fractional calculus, Int. J. Differ. Equ. Appl. 4 (2002), No. 4, 405–421.

[253] S.-D. Lin, S.-T. Tu and H. M. Srivastava, Explicit solutions of certain ordinary differential equations by means of fractional calculus, J. Fract. Calc. 20 (2001), 35–43.

[254] S.-D. Lin, S.-T. Tu, H. M. Srivastava and P.-Y. Wang, Some families of multiple infinite sums and associated fractional differintegral formulas for power and composite functions, J. Fract. Calc. 30 (2006), 45–58.

[255] J. Liouville, Memoire sur quelques quéstions de géometrie et de mécanique, et sur un noveau genre pour réspondre ces quéstions, Jour. Ecole Polytech. 13 (1832), 1–69.

[256] P. I. Lizorkin, Bounds for trigonometrical integrals and the Bernstein inequality for fractional derivatives (Russian), Izv. Akad. Nauk SSSR, Ser. Mat. 29 (1965), No. 1, 109–126.

[257] E. R. Love, Fractional derivatives of imaginary order, J. London Math. Soc. (2) 3 (1971), 241–259.

[258] E. R. Love, Two index laws for fractional integrals and derivatives, J. Australian Math. Soc. 14 (1972), 385–410.

[259] E. R. Love, Some inequalities for fractional integrals, Linear spaces and approximation (Proc. Conf., Math. Res. Inst., Oberwolfach, 1977), 177–184. Intern. Ser. Numer. Math., Vol. 40, Birkhäuser, Basel, 1978.

[260] E. R. Love, A third index law for fractional integrals and derivatives, Fractional calculus (Glasgow, 1984), 63–74, Res. Notes in Math., 138, Pitman, Boston, MA, 1985.

[261] E. R. Love, Inequalities like Opial's inequality, Rocznik naukowo-dydaktyczny WSP w krakowie, Pr. Mat. 97 (1985), 109–118.

[262] E. R. Love, Lebesgue points of fractional integrals, Real Anal. Exchange 12 (1986/87), No. 1, 327–336.

[263] E. R. Love and Y. C. Young, On fractional integration by parts, Proc. Lond. Math. Soc. 44 (1938), 1–35.

[264] Ch. Lubich, Discretized fractional calculus, SIAM J. Math. Anal. 17, (1986), No. 3, 704–719.

[265] Y. Luchko and R. Gorenflo, The initial value problem for some fractional differential equations with the Caputo derivatives, ftp://ftp.math.fu-berlin.de/pub/math/publ/pre/1998/Pr-A-98-08.ps

[266] Y. Luchko and R. Gorenflo, An operational method for solving fractional differential equations with the Caputo derivatives, Acta. Math. Vietnam. 24 (1999), No. 2, 207–233.

[267] Yu. F. Luchko and H. M. Srivastava, The exact solution of certain differential equations of fractional order by using operational calculus, Comput. Math. Appl. 29 (1995), No. 8, 73–85.

[268] Yu. Luchko and J. J. Trujillo, Caputo-type modification of the Erdélyi-Kober fractional derivative, Fract. Calc. Appl. Anal. 10 (2007), No. 3, 249–267.

[269] Yu. F. Luchko and S. B. Yakubovich, Convolutions of the generalized fractional integration operator, Complex Analysis and Generalized Functions (Varna, 1991), 199–211, Publ. House Bulgar. Acad. Sci., Sofia, 1993.

[270] J. Lützen, Joseph Liouville 1809–1882, Master of Pure and Applied Mathematics, Springer, New York, Berlin, Heidelberg, 1990.

[271] J. A. T. Machado, Analysis and design of fractional-order digital control systems, J. Syst. Anal. Model. Simul. 27 (1997), 107–122.

[272] F. Mainardi, On the initial value problem for the fractional diffusion-wave equation. In: S. Rionero and T. Ruggeri (eds.), Waves and Stability in Continuous Media, World Scientific, Singapore, 1994, 246–251.

[273] F. Mainardi, Fractional relaxation in anelastic solids, J. Alloys Compd., Vol. 211/212 (1994), 534–538.

[274] F. Mainardi, Fractional diffusive waves in viscoelastic solids. In: J. L. Wegner and F. R. Norwood (eds.), Nonlinear Waves in Solids, ASME/AMR, Fairfield NJ, 1995, 93–97.

[275] F. Mainardi, Fractional relaxation-oscillation and fractional diffusion-wave phenomena, Chaos, Solitons Fractals 7 (1996), 1461–1477.

[276] F. Mainardi, The fundamental solutions for the fractional diffusion-wave equation, Appl. Math. Lett. 9 (1996), No. 6, 23–28.

[277] F. Mainardi, Fractional calculus: Some basic problems in continuum and statistical mechanics, Fractals and Fractional Calculus in Continuum Mechanics (Udine, 1996), 291–348, CISM Courses and Lectures, 378, Springer, Vienna, 1997.

[278] F. Mainardi, Applications of fractional calculus in mechanics. In: P. Rusev, I. Dimovski and V. Kiryakova (eds.), Transform Methods and Special Functions, Varna '96, SCT Publishers, Singapore, 1997.

[279] F. Mainardi, Linear viscoelasticity, Acoustic interactions with submerged elastic structures, Part IV, 97–126, Ser. Stab. Vib. Control Syst. Ser. B, 5. World Sci. Publ., River Edge, NJ, 2002.

[280] F. Mainardi and R. Gorenflo, The Mittag-Leffler function in the Riemann-Liouville fractional calculus, Boundary value problems, special functions and fractional calculus (Russian) (Minsk, 1996), 215–225, Belorus. Gos. Univ., Minsk.

[281] F. Mainardi and R. Gorenflo, Fractional calculus: Special functions and applications, Advanced special functions and applications (Melfi, 1999), 165–188. Proc. Melfi Sch. Adv. Top. Math. Phys., 1, Aracne, Rome, 2000.

[282] F. Mainardi, M. Roberto, R. Gorenflo and E. Scalas, Fractional calculus and continuous-time finance II: The waiting-time distribution, Physica A 287 (2000), 468–481.

[283] F. Mainardi, R. Gorenflo and A. Vivoli, Renewal processes of Mittag-Leffler and Wright type, Fract. Calc. Appl. Anal. 8 (2005), No. 1, 7–38.

[284] F. Mainardi and G. Pagnini, The Wright functions as solutions of the time-fractional diffusion equation, Advanced special functions and related topics in differential equations (Melfi, 2001), Appl. Math. Comput. 141 (2003), No. 1, 51–62.

[285] F. Mainardi, G. Pagnini and R. Gorenflo, Some aspects of fractional diffusion equations of single and distributed order, Appl. Math. Comput. 187 (2007), No. 1, 295–305.

[286] R. J. H. Marks and M. W. Hall, Differintegral Interpolation from a Band-limited Signal's Samples, IEEE Transact Acoust Speech Signal Process 29 (1981), 872–877.

[287] C. Martinez, M. Sanz and M. D. Martinez, About fractional integrals in the space of locally integrable functions, J. Math. Anal. Appl. 167 (1992), 111–122.

[288] D. Matignon and G. Montseny (eds.), Fractional Differential Systems: Models, Methods, and Applications, ESAIM Proceedings, Vol. 5, SMAI, Paris (http://www.emath.fr/Maths/Proc/Vol.5/index.htm), 1998.

[289] A. C. McBride, Fractional Calculus and Integral Fransforms of Generalalized Functions, Pitman Research Notes in Mathematics, # 31, Pitman, London, 1979.

[290] A. C. McBride, Fractional Calculus, New York, Halsted Press, 1986.

[291] A. C. McBride and G. F. Roach, Fractional Calculus, Pitman, Research Notes in Mathematics, 138, 1985.

[292] M. M. Meerschaert, D. A. Benson and B. Baeumer, Multidimensional advection and fractional dispersion, Phys. Rev. E 59 (1999), 5026–5028.

[293] M. M. Meerschaert, D. A. Benson, H. P. Scheffler and B. Baeumer, Stochastic solution of space-time fractional diffusion equations, Phys. Rev. E 65 (2002), 1103–1106.

[294] J. Mikusinski, Operational Calculus, Pergamon Press, New York, 1959.

[295] K. Miller and B. Ross, An Introduction to the Fractional Calculus and Fractional Differential Equations, John Wiley and Sons, Inc., New York, 1993.

[296] K. S. Miller, Derivatives of noninteger order, Math. Mag. 68 (1995), 183–192.

[297] D. S. Mitrinović, J. E. Pečarić and A. M. Fink, Classical and New Inequalities in Analysis, Kluwer Acad. Publ., Dordrecht, 1993.

[298] G. M. Mittag-Leffler, Sur la nouvelle fonction $E_\alpha(x)$, C. R. Acad. Sci. Paris 137 (1903), 554–558.

[299] D. Mo, Y. Y. Lin, J. H. Tan, Z. X. Yu, G. Z. Zhou, K. C. Gong, G. P. Zhang and X.-F. He, Ellipsometric spectra and fractional derivative spectrum analysis of polyaniline films, Thin Solid Films 234 (1993), 468–470.

[300] G. Montseny, J. Audounet and D. Matignon, Fractional integro-differential boundary control of the Euler-Bernoulli beam, 36th IEEE Conference on Decision and Control, San Diego, California, December 1997, IEEE-CSS, SIAM, 4973–4978.

[301] R. R. Nigmatullin, Fractional integral and its physical interpretation, Soviet J. Theor. and Math. Phys. 90, (1992), No. 3, 354–367.

[302] K. Nishimoto, Fractional Calculus, New Haven, CT University of New Haven Press, 1989.

[303] K. Nishimoto (ed.), Fractional Calculus and its Applications, Nihon University, Koriyama, 1990.

[304] T. F. Nonnenmacher, Fractional integral and differential equations for a class of Lévy-type probability densities, J. Phys. A: Math. Gen. 23 (1990), L697–L700.

[305] T. F. Nonnenmacher and W. G. Glöckle, A fractional model for mechanical stress relaxation, Philosophical Magazine Lett. 64, (1991), No. 2, 89–93.

[306] T. F. Nonnenmacher and R. Metzler, On the Riemann-Liouville Fractional Calculus and some Recent Applications, Fractals 3 (1995), 557–566.

[307] Z. M. Odibat, Rectangular decomposition method for fractional diffusion-wave equations, Appl. Math. Comput. 179 (2006), 92–97.

[308] Z. M. Odibat, Approximations of fractional integrals and Caputo fractional derivatives, Appl. Math. Comput. 178 (2006), 527–533.

[309] Z. M. Odibat and N. T. Shawagfeh, Generalized Taylor's formula, Appl. Math. Comput. 186 (2007), 286–293.

[310] K. B. Oldham, The reformulation of an infinite sum via semi-integration, SIAM J. Math. Anal. 14 (1983), No. 5, 974–981.

[311] K. B. Oldham and J. Spanier, The Fractional Calculus, Academic Press, New York, 1959.

[312] K. B. Oldham and J. Spanier, The replacement of Fick's law by a formulation involving semidifferentiation, J. Electroanal. Chem. Interfacial Electrochem. 26 (1970), 331–341.

[313] K. B. Oldham and J. Spanier, The Fractional Calculus: Integrations and Differentiations of Arbitrary Order, New York, Academic Press, 1974.

[314] K. B. Oldham and J. Spanier, The Fractional Calculus: Theory and Applications of Differentiation and Integration to Arbitrary Order, Dover Publications, New York, 2006.

[315] Z. Opial, Sur une inégalité, Ann. Polon. Math. 8 (1960), 29–32.

[316] T. J. Osler, Leibniz rule for fractional derivatives, generalized and an application to infinite series, SIAM J. Appl. Math. 18 (1970), 658–674.

[317] T. J. Osler, A further extension of the Leibniz rule to fractional derivatives and its relation to Parseval's formula, SIAM J. Math. Anal. 3 (1972), 1–16.

[318] A. Ostrowski, Über die Absolutabweichung einer differentiebaren Funktion von ihrem Integralmittelwert, Comment. Math. Helv. 10 (1938), 226–227.

[319] B. G. Pachpatte, Existence and uniqueness of solutions of higher order hyperbolic partial differential equations, Chinese J. Math. 17 (1989), 181–189.

[320] B. G. Pachpatte, Inequalities similar to the integral analogue of Hilbert's inequalities, Tamkang J. Math. 30 (1999), No. 1, 139–146.

[321] B. G. Pachpatte, On Hilbert type inequality in several variables, An. Stiint. Univ. Al. I. Cuza Iasi. Mat. (N.S.), 46 (2000), 245–250.

[322] B. G. Pachpatte, On two new multidimensional integral inequalities of the Hilbert type, Tamkang J. Math. 31 (2000), 123–129.

[323] P. Y. H. Pang and R. P. Agarwal, On an Opial type inequality due to Fink, J. Math. Anal. Appl. 196 (1995), 748–753.

[324] I. Petras and L. Dorcak, The frequency methods for stability investigation of fractional control systems, In SACTA Journal 2 (1999), No. 1–2, 75–85, Durban, South Africa.

[325] I. Podlubny, Fractional derivatives: A new stage in process modelling and control, 4th International DAAAM Symposium, Brno, Czech Republic, September 16–18, 1993, 263–264.

[326] I. Podlubny, Fractional-Order Systems and Fractional-Order Controllers, Technical Report UEF-03-94, Institute for Experimental Physics, Slovak Acad. Sci, 1994.

[327] I. Podlubny, Numerical methods of the fractional calculus, Transactions of the Technical University of Kosice, 4 (1994), No. 3–4, 200–208.

[328] I. Podlubny, Numerical solution of initial value problems for ordinary fractional-order differential equations, Proceedings of the 14th World Congress on Computation and Applied Mathematics, W. F. Ames (ed.), July 11–15, 1994, Atlanta, Georgia, USA, Late Papers volume, pp. 107–111.

[329] I. Podlubny, Solution of linear fractional differential equations with constant coefficients. In: P. Rusev, I. Dimovski and V. Kiryakova (eds.), Transform Methods and Special Functions, SCT Publishers, Singapore, 1995, pp. 217–228.

[330] I. Podlubny, Numerical solution of ordinary fractional differential equations by the fractional difference method, Advances in Difference Equations (Veszprém, 1995), 507–515, Gordon and Breach, Amsterdam, 1997.

[331] I. Podlubny, Riesz potential and Riemann-Liouville fractional integrals and derivatives of Jacobi polynomials, Appl. Math. Lett. 10, (1997), No. 1, 103–108.

[332] I. Podlubny, Application of orthogonal polynomials to solution of fractional integral equations, Transform Methods & Special Functions, Varna '96, 350–359, Bulgarian Acad. Sci., Sofia, 1998.

[333] I. Podlubny, Fractional differential equations, An introduction to fractional derivatives, fractional differential equations, to methods of their solution and some of their applications, Mathematics in Science and Engineering, 198. Academic Press, Inc., San Diego, CA, 1999.

[334] I. Podlubny, Matrix approach to discrete fractional calculus, Fract. Calc. Appl. Anal. 3 (2000), No. 4, 359–386.

[335] I. Podlubny, Geometric and physical interpretation of fractional integration and fractional differentiation, Dedicated to the 60th anniversary of Prof. Francesco Mainardi, Fract. Calc. Appl. Anal. 5 (2002), No. 4, 367–386.

[336] I. Podlubny and I. Kostial, Fractional derivative based process models and their applications, 4th International DAAAM Symposium, Brno, Czech Republic, September 16–18, 1993, 265–266.

[337] I. Podlubny and J. Misanek, The use of fractional derivatives for solution of heat conduction problems, Proceedings of the 9th Conference on Process Control, Tatranske Matliare, May 1993, STU Bratislava, 270–273.

[338] I. Podlubny, L. Dorcak and J. Misanek, Application of Fractional-Order Derivatives to Calculation of Heat Load Intensity Change in Blast Furnace Walls, Transact. Tech. Univ. Kodice 5 (1995), 137–144.

[339] V. G. Ponomarenko, Modulus of smoothness of fractional order and the best approximation in L_p, $1 < p < \infty$ (Russian); In: Constructive Function Theory '81, (Proc. Conf. Varna, June 1981), Publ. House Bulg. Acad. Sci., Sofia 1983, 129–133.

[340] J. K. Prajapat, R. K. Raina and H. M. Srivastava, Some inclusion properties for certain subclasses of strongly starlike and strongly convex functions involving a family of fractional integral operators, Integral Transform. Spec. Funct. 18 (2007), No. 9–10, 639–651.

[341] A. I. Priesto, S. Salinas de Romero and H. M. Srivastava, Some fractional-calculus results involving the generalized Lommel-Wright and related functions, Appl. Math. Lett. 20 (2007), No. 1, 17–22.

[342] A. V. Pskhu, Solution of a Boundary Value Problem for a fractional Partial differential equation, Differential Equations 39 2003, No. 8, 1150–1158, Translated from Differentsial'nye Uravneniya 39 2003, No. 8, 1092–1099.

[343] H. Rafeiro and S. Samko, On a class of fractional type integral equations in variable exponent spaces, Fract. Calc. Appl. Anal. 10 (2007), No. 4, 399–421.

[344] R. K. Raina and H. M. Srivastava, A certain subclass of analytic functions associated with operators of fractional calculus, Comput. Math. Appl. 32 (1996), No. 7, 13–19.

[345] A. Rényi, On measures of entropy and information. In Proceedings of the 4th Berkeley Symposium on Mathematical Statistic and Probability, I, Berkekey, CA, 1960, 547–561.

[346] B. Riemann, Versuch einer allgemeinen Auffassung der Integration und Differentiation, Gesammelte Mathematische Werke und Wissenschaftlicher Nachlass. Teubner, Leipzig 1876 (Dover, New York, 1953), 331–344.

[347] M. Riesz, L'intégrale de Riemann-Liouville et le problème de Cauchy, Acta. Math. 81 (1949), 1–223.

[348] F. Riewe, Mechanics with fractional derivatives, Phys. Rev. E 55 (1997), No. 3, 3581–3592.

[349] B. Ross, Fractional Calculus and its Applications, Springer-Verlag, Berlin, New York, 1975.

[350] B. Ross, Fractional calculus: An historical apologia for the development of a calculus using differentiation and antidifferentiation of non integral orders, Mathematics Magazine 50 (1977), No. 3, 115–122.

[351] B. Ross, The development of fractional calculus 1695–1900, Historia Math. 4 (1977), 75–89.

[352] B. Ross, S. G. Samko, E. R. Love, Functions that have no first order derivative might have fractional derivatives of all orders less than one, Real Anal. Exchange 20 (1994/95), No. 1, 140–157.

[353] Yu. A. Rossikhin and M. V. Shitikova, Applications of fractional calculus to dynamic problems of linear and nonlinear hereditary mechanics of solids, Appl. Mech. Rev. 50 (1997), No. 1, 15–67.

[354] H. L. Royden, Real Analysis, Second edition, Macmillan, New York, 1968.

[355] B. Rubin, Fractional Integrals and Potentials, Pitman Monographs and Surveys in Pure Appl. Math. 82, Longman, Harlow, 1996.

[356] W. Rudin, Real and Complex Analysis, International Student Edition, Mc Graw Hill, London, New York, 1970.

[357] Th. Runst and W. Sickel, Sobolev spaces of fractional orders, Nemytskij Operators, and Nonlinear Partial Differential Equations, De Gruyter, Berlin, 1996.

[358] R. S. Rutman and R. Gorenflo, Simulation and inversion of fractional integration: A systems theory approach, Inverse Problems and Applications to Geophysics, Industry, Medicine and Technology (Ho Chi Minh City, 1995), 141–148, Publ. HoChiMinh City Math. Soc., 2, HoChiMinh City Math. Soc., Ho Chi Minh City, 1995.

[359] S. Samadi, M. O. Ahmad and M. N. S. Swamy, Exact fractional-order differentiators for polynomial signals, IEEE Signal Process. Lett. 11 (2004), No. 6, 529–532.

[360] S. G. Samko, The coincidence of Grü nwald-Letnikov differentiation with other forms of fractional differentiation, The periodic and non-periodic cases (Russian). In: Reports of the Extended Session of the Seminar of the I. N. Vekua Inst. Appl. Math., Tbilisi 1985. Tbiliss. Gos. Univ. 1.

[361] S. G. Samko, Fractional integration and differentiation of variable order, Anal. Math. 21 (1995), No. 3, 213–236.

[362] S. Samko, Hardy-Littlewood-Stein-Weiss inequality in the Lebesgue spaces with variable exponent, Fract. Calc. Appl. Anal. 6 (2003), No. 4, 421–440.

[363] S. G. Samko, Fractional Weyl-Riesz integrodifferentiation of periodic functions of two variables via the periodization of the Riesz kernel, Appl. Anal. 82 (2003), No. 3, 269–299.

[364] S. Samko, On inversion of fractional spherical potentials by spherical hypersingular operators, Singular integral operators, factorization and applications, 357–368. Oper. Theory Adv. Appl. 142, Birkhäuser, Basel, 2003.

[365] S. G. Samko, A. A. Kilbas and O. I. Marichev, Fractional Integrals and Derivatives and Some of their Applications, Nauka i technika, Minsk, 1987.

[366] S. G. Samko, A. A. Kilbas and O. I. Marichev, Fractional integrals and derivatives, Theory and applications, Edited and with a foreword by S. M. Nikolskii, Translated from the 1987 Russian origina,. Revised by the authors, Gordon and Breach Science Publishers, Yverdon, 1993.

[367] S. G. Samko and A. Ya. Yakubov, Zygmund estimate for moduli of continuity of fractional order of a conjugate function (Russian), Izv. Vyssh. Uchebn. Zaved. Mat. 77 (1985), No. 12, 49–53.

[368] S. G. Samko and A. Ya. Yakubov, A Zygmund estimate for hypersingular integrals in the case of moduli of continuity of fractional order (Russian), Izv. Severo-Kavkaz. Nauchn. Tsentra Vyssh. Shkoly Estestv. Nauk. 140 (1986), No. 3, 42–47.

[369] R. K. Saxena, On fractional integration operators, Math. Z. 96 (1967), 288–291.

[370] R. K. Saxena, S. L. Kalla and V. S. Kiryakova, Relations connecting multiindex Mittag-Leffler functions and Riemann-Liouville fractional calculus, Algebras Groups Geom. 20 (2003), No. 4, 363–385.

[371] T. Sekine, K. Tsurumi, S. Owa and H. M. Srivastava, Integral means inequalities for fractional derivatives of some general subclasses of analytic functions, Inequalities in univalent function theory and their applications (Japanese) (Kyoto, 2002) Surikaisekikenkyusho Kokyuroku No. 1276 (2002), 79–88.

[372] E. Scalas, R. Gorenflo and F. Mainardi, Fractional calculus and continuous-time finance, Phys. A 284 (2000), No. 1–4, 376–384.

[373] E. Scalas, R. Gorenflo, F. Mainardi and M. Raberto, Revisiting the derivation of the fractional diffusion equation, Scaling and disordered systems (Paris, 2000), Fractals 11 (2003), Suppl. 281–289.

[374] W. R. Schneider and W. Wyss, Fractional diffusion and wave equations, J. Math. Phys. 30 (1989), 134–144.

[375] V. B. Shakhmurov and A. I. Zayed, Fractional Wigner distribution and ambiguity functions, Fract. Calc. Appl. Anal. 6 (2003), No. 4, 473–490.

[376] W. Smit and H. de Vries, Rheological models containing fractional derivatives, Rheologica Acta 9 (1970), 525–534.

[377] W. Specht, Zur Theorie der elementaren Mittel, Math. Z. 74 (1960), 91–98.

[378] H. M. Srivastava, Fractional calculus and its applications, Cubo Mat. Educ. 5 (2003), No. 1, 33–48.

[379] H. M. Srivastava and M. K. Aouf, A certain fractional derivative operator and its applications to a new class of analytic and multivalent functions with negative coefficients II, J. Math. Anal. Appl. 192 (1995), No. 3, 673–688.

[380] H. M. Srivastava, S.-D. Lin, Y.-T. Chao, P.-Y. Wang, Explicit solutions of a certain class of differential equations by means of fractional calculus, Russ J. Math. Phys. 14 (2007), No. 3, 357–365.

[381] H. M. Srivastava, A. K. Mishra and M. K. Das, A class of parabolic starlike functions defined by means of a certain fractional derivative operator, Fract. Calc. Appl. Anal. 6 (2003), No. 3, 281–298.

[382] H. M. Srivastava and S. Owa (eds.), Univalent Functions, Fractional Calculus, and their Applications, Halsted (Horwood), Chichester, Wiley, New York, 1989.

[383] D. Stroock, A Concise Introduction to the Theory of Integration, Third Edition, Birkäuser, Boston, Basel, Berlin, 1999.

[384] R. Taberski, Contribution to fractional calculus and exponential approximation, Funct. Approx. Comment. Math. 15 (1986), 81–106.

[385] F. B. Tatom, The relationship between fractional calculus and fractals, Factals 3 (1995), No. 1, 217–229.

[386] J. J. Trujillo, M. Rivero and B. Bonilla, On a new generalized Taylor's formula, Transform methods & spectral functions, Varna '96, 508–516. Bulgarian Acad. Sci., Sofia, 1998.

[387] J. J. Trujillo, M. Rivero and B. Bonilla, On a Riemann-Liouville generalized Taylor's formula, J. Math. Anal. Appl. 231 (1999), 255–265.

[388] C. C. Tseng, Design of fractional order digital FIR differentiators, IEEE Signal Process. Lett. 8 (2001), 77–79.

[389] C. C. Tseng, S. C. Pei and S. C. Hsia, Computation of fractional derivatives using Fourier transform and digital FIR differentiator, Signal Process. 80 (2000), 151–159.

[390] S.-T. Tu, T.-C. Wu and H. M. Srivastava, Comommutativity of the Leibniz rules in fractional calculus, Comput. Math. Appl. 40 (2000), No. 2–3, 303–312.

[391] V. K. Tuan and R. Gorenflo, The Grunwald-Letnikov difference operator and regularization of the Weyl fractional differentiation, Z. Anal. Anwendungen 13 (1994), 537–545.

[392] V. K. Tuan and R. Gorenflo, On the regularization of fractional differentiation of arbitrary positive order, Numer. Funct. Anal. Optim. 15 (1994), No. 5–6, 695–711.

[393] V. K. Tuan and R. Gorenflo, Extrapolation to the limit for the numerical fractional differentiation, Z. angew. Math. Mech. 75 (1995), 646–648.

[394] V. K. Tuan, R. K. Raina and M. Saigo, Multidimensional fractional calculus operators involving the Gauss hypergeometric function, Int. J. Math. Sci. 5 (1996), No. 2, 141–160.

[395] V. K. Tuan and M. Saigo, Multidimensional modified fractional calculus operators, Math. Nachr. 161 (1993), 253–270.

[396] M. Unser and T. Blu, Fractional splines and wavelets, SIAM Review 42 (2000), No. 1, 43–67.

[397] B. M. Vinagre and Y. Q. Chen, Lecture notes on fractional calculus applications in automatic control and robotics, The 41st IEEE CDC2002 Tutorial Workshop No. 2, B. M. Vinagre and Y. Q. Chen (eds.), retrieved from http://mechatronics.ece.usu.edu/foc/cdc02_tw2_ln.pdf, Las Vegas, Nevada, 2002, 1–310.

[398] B. M. Vinagre, L. Podlubny, A. Hernandez and V. Feliu, Some approximations of fractional order operators used in control theory and applications, Fract. Calc. Appl. Anal. 3 (2000), No. 3, 231–248.

[399] A. A. Voroshilov and A. A. Kilbas, The Cauchy problem for diffusion-wave equation with the Caputo partial derivative (Russian), Differ. Uravn. 42 (2006), No. 5, 599–609, 717. Translation in Differ. Equ. 42 (2006) No. 5, 638–649.

[400] P.-Y. Wang, S.-D. Lin and H. M. Srivastava, Remarks on a simple fractional-calculus approach to the solutions of the Bessel differential equation of the general order and some of its applications, Comput. Math. Appl. 51 (2006), No. 1, 105–114.

[401] B. J. West, P. Grigolini, R. Metzler and T. F. Nonnenmacher, Fractional diffusion and Lévy stable processes, Physical Review E, 55 (1997), 99–106.

[402] U. Westphal, An approach to fractional powers of operators via fractional differences, Proc. London Math. Soc. (3) 29 (1974), 557–576.

[403] H. Weyl, Bemerkungen zum Begriff des Differentialquotienten gebrochener Ordnung. Vierteljschr. Naturforsch. Gesellsch., Zürich, 62, 1917, 296.

[404] E. T. Whittaker and G. N. Watson, A Course in Modern Analysis, Cambridge University Press, Cambridge, UK, 1927.

[405] D. V. Widder, A generalization of Taylor's series, Trans. Am. Math. Soc. 30 (1928), No. 1, 126–154.

[406] D. Willett, The existence-uniqueness theorem for an nth order linear ordinary differential equation, Amer. Math. Monthly 75 (1968), 174–178.

[407] W. Wyss, The fractional diffusion equation, J. Math. Phys. 27, (1986), No. 11, 2782–2785.

[408] K. Yao, W. Y. Su and S. P. Zhou, On the Fractional Calculus of a Type of Weierstrass Function, Chinese Annals of Mathematics 25:A (2004), 711–716.

[409] K. Yao, W. Y. Su and S. P. Zhou, On the connection between the order of fractional calculus and the dimensions of a fractal function, Chaos, Solitons Fractals. 23 (2005), 621–629.

[410] M. Zähle and H. Ziezold, Fractional derivatives of Weierstrass-type functions, J. Computat. Appl. Math., 76 (1996), 265–275.

[411] A. I. Zayed, Fractional fourier transform of generalized functions, Integral Transform. Spec. Funct. 7 (1998), No. 3–4, 299–312.

[412] A. I. Zayed, A class of fractional integral transforms: a generalizaton of the fractional Fourier transform, IEEE Trans. Signal Process. 50 (2002), No. 3, 619–627.

List of Symbols

Index